I0147903

THE BOOK OF FROGS

THE BOOK OF FROGS

A LIFE-SIZE GUIDE TO SIX HUNDRED SPECIES FROM AROUND THE WORLD

MARK O'SHEA AND TIM HALLIDAY

Second Edition

THE UNIVERSITY OF CHICAGO PRESS
Chicago and London

MARK O'SHEA MBE is professor of herpetology at the University of Wolverhampton. He has made ten expeditions to Papua New Guinea since 1986 and from 2009 to 2014 co-led the first herpetofaunal survey of Timor-Leste. O'Shea hosted four seasons of *O'Shea's Big Adventure* for Animal Planet/Discovery Channel and has authored ten books. He has conducted fieldwork on numerous expeditions since 1983 and is a fellow of the Explorers' Club of New York, Royal Geographical Society, Linnean Society of London, and Royal Society of Biology. He has co-authored descriptions of twelve new snake species and one gecko. In 2018 an Asian pipesnake was named *Cylindrophis osheai* in his honor. He received an MBE in the 2020 Queen's Birthday Honours. He lives in Shropshire, England, twenty miles from the birthplace of Charles Darwin.

TIM HALLIDAY was one of the world's foremost authorities on amphibians. Tim specialized in herpetology for three decades, and was involved in a highly successful captive breeding project to conserve the endangered Mallorcan Midwife Toad. He wrote a number of general market titles, including *Vanishing Birds* (1978), *Sexual Strategy* (1980), and the *Dorling Kindersley Handbook of Reptiles and Amphibians* (2002). He edited, with Kraig Adler, two editions of the highly regarded *Encyclopedia of Amphibians and Reptiles* (2002). He also wrote over 80 academic papers on the sexual behavior, reproductive biology, and conservation of amphibians. He passed away in 2019.

The University of Chicago Press, Chicago 60637
The University of Chicago Press, Ltd., London

Published 2025
Printed in Pontian, Johor, Malaysia
PC/Aug/2025

34 33 32 31 30 29 28 27 26 25   1 2 3 4 5

ISBN-13: 978-0-226-84426-8 (cloth)
ISBN-13: 978-0-226-84427-5 (e-book)
DOI: https://doi.org/10.7208/chicago/9780226844275.001.0001

Library of Congress Control Number: 2025933071

Authorized Representative for EU General Product Safety Regulation (GPSR) queries:
Easy Access System Europe – Mustamäe tee 50, 10621 Tallinn, Estonia, gpsr.requests@easproject.com
Any other queries: press.uchicago.edu/press/contact.html

Conceived, designed, and produced by The Bright Press,
An imprint of The Quarto Group,
1 Triptych Place, London, SE1 9SH, UK
T (0)20 7700 6700
www.quarto.com

Publishers: Susan Kelly and James Evans
Art Directors: Michael Whitehead and James Lawrence
Editorial Director: Isheeta Mustafi
Commissioning Editor: Kate Shanahan
Editor: Ellie Stores
Senior Project Editor: Caroline Earle
Project Editor: David Price-Goodfellow
Designer: Ginny Zeal
Illustrator: Sandra Pond
Picture Researchers: Hedda Lloyd-Roennevig, Alison Stevens, & Susannah Jayes

JACKET IMAGES
Dreamstime/Farinoza: *Ranoidea splendida*; GettyImages/alekseystemmer: *Pelobatrachus nasutus*; GettyImages/imv: *Alytes obstetricans*; GettyImages/kazakovmaksim: *Xenopus laevis*; GettyImages/Ken Griffiths: *Amazophrynella minuta*; GettyImages/NurPhoto: *Anaxyrus americanus*; Shutterstock: *Pelophylax perezi*; Shutterstock: *Rana palustris*; Shutterstock/Alex Stemmer: *Trachycephalus resinifictrix*; Shutterstock/Benzine: *Kaloula pulchra*; Shutterstock/Craig Cordier: *Kassina senegalensis*; Shutterstock/dwi putra stock: *Pyxicephalus edulis*; Shutterstock/Eric Isselee: *Bufo bufo*; Shutterstock/Eric Isselee: *Dendrobates leucomelas*; Shutterstock/Eric Isselee: *Mantella aurantiaca*; Shutterstock/Eric Isselee: *Theloderma corticale*; Shutterstock/Lauren Suryanata: *Chacophrys pierottii*; Shutterstock/Rudmer Zwerver: *Rana temporaria*; Shutterstock/Vitalii Hulai: *Bombina bombina*; Shutterstock/xpixel: *Rana dalmatina*

LITHOCASE IMAGES
GettyImages/alekseystemmer: *Pelobatrachus nasutus*
GettyImages/Ken Griffiths: *Amazophrynella minuta*
Shutterstock/Eric Isselee: *Dendrobates leucomelas*

CONTENTS

INTRODUCTION

ABOVE **The Red-eyed Tree Frog** (*Agalychnis callidryas*) lives in the forests of Central America. One of the world's most colorful, and most photographed, frogs, it is adapted for climbing—the large, adhesive disks on the tips of its fingers and toes enable it to climb up vertical surfaces.

Though they include some of the world's noisiest, most colorful, and most poisonous animals, frogs are typically shy and retiring creatures, rarely encountered by humans. Most are active only at night, avoiding the dry air and warm temperatures of the daytime. Capable of surviving for very long periods without food, many remain hidden for most of their lives. It is not surprising, therefore, that most people are totally unaware of the enormous number and diversity of frog species that inhabit the Earth.

For example, frogs are typically perceived as being able to hop large distances, but this does scant justice to the variety of ways in which they move through their environment. Some are beautifully adapted for swimming, others for digging into the ground, and others still for climbing vertical surfaces, while a few can glide through the air using their outstretched feet as parachutes.

A RAISED PROFILE

The profile of frogs has changed enormously in the last 30–40 years, for three reasons. First, in the 1970s and 1980s, more people started to study frogs. They discovered that many frogs show unexpectedly complex behavior. While it had long been known that male frogs call to attract mates, research showed that frog calls are also involved in interactions between males, enabling males to assess the strength of their rivals. For females, mating calls are more than a means to locate males; there is increasing evidence that they enable them to assess the genetic quality of prospective mates. While it had long been thought that the typical frog left its eggs and hence tadpoles to fend for themselves, new discoveries revealed many kinds of complex parental care, performed by one or both parents.

Second, frogs are in danger. The first global gathering of scientists interested in frogs took place in 1989. While there was much interest in new discoveries about frogs, conversation was dominated by anecdotal accounts, from all over the world, of once-common frogs becoming rare and, in some instances, disappearing. It soon became evident that frogs, and other amphibians, are declining everywhere, even within protected areas. It is now clear that amphibian declines are one manifestation of a major global extinction event, often referred to as the Sixth Mass Extinction.

Finally, there are many more frogs in the world than previously thought. Largely stimulated by the quest to find the causes of amphibian declines, there has been a major effort to establish just how many amphibians there are. There are many more species than imagined even 20 years ago. In 1986, 4,015 species of amphibians were known; in early 2025 there are 7,784.

In this second edition, three new families are included: Ericabatrachidae from Africa, and Caligophrynidae and Neblinaphrynidae from the north Amazon. Some species from the first edition have been synonymized within other species (or relocated to different genera) and genera to different families, so it has been necessary to heavily modify five, and completely replace twelve, species accounts, and update most of the remaining. Accounts of 600 species are presented. Life-size photographs of each are included, and the descriptions cover many aspects of each species' natural history and behavior, give their conservation status, and describe reasons for any decline. Species and genera are now arranged within subfamilies, within families. The AMNH Amphibian Species of the World amphibiansoftheworld.amnh.org website has been used for taxonomic and nomenclatural decisions.

ABOVE **The Puerto Rican Coqui Frog** (*Eleutherodactylus coqui*) gets its name from the male's two-note call. The low-pitched "co" note is perceived primarily by other males, the high-pitched "qui" primarily by females.

ABOVE **The Amazon Milk Frog** (*Trachycephalus resinifictrix*) lays its eggs in water-filled tree-holes, where the tadpoles feed partly on unfertilized eggs.

WHAT IS A FROG?

ABOVE **The European Common Toad** (*Bufo bufo*) spends most of its life on land, entering water only for a few days in spring to breed.

Frogs and toads comprise the order Anura, by far the largest of the three orders that make up the vertebrate class Amphibia, animals that typically live part of their lives on land, part in water. The name Anura means "without a tail" and refers to the most obvious difference between frogs and toads and the other amphibian orders, the Caudata (salamanders and newts) and the Gymnophiona (caecilians), both of which have long, flexible tails. Anurans make up 88 percent of all amphibian species.

Anurans share with other amphibians a number of features that have a profound influence on their natural history and distribution. Like all animals, except birds and mammals, they are ectothermic, meaning that they derive body heat from their environment, directly or indirectly from the sun. As a result, they are absent from the coldest parts of the Earth. For example, only one species, the Wood Frog (*Boreorana sylvatica;* page 548), is found north of the Arctic Circle, in North America. Being ectothermic also means that their reproductive activity is limited to warmer parts of the year, when their external environment provides the energy they require to sustain vigorous activities, such as calling and fighting. Amphibians differ from reptiles in having a thin, soft skin that is very permeable to water. This means that they are mostly restricted to habitats where conditions are always moist or wet, and that they typically breed during times of the year when rainfall is high.

LIFE CYCLE

Like other amphibians, frogs exhibit a complex life cycle, that involves four distinct phases: egg, larva, juvenile, and adult. The larvae of frogs, known as tadpoles, are wholly aquatic creatures that obtain their oxygen from the water they live in via a pair of external gills, and they swim by beating their tails. The transition from a tadpole to an adult, a process called metamorphosis, is perhaps the most familiar and remarkable characteristic of frogs. It involves major anatomical changes, including the loss of the larval tail, the development of four limbs, and a switch from breathing oxygen dissolved in water through gills, to breathing atmospheric oxygen via lungs and the skin.

In the majority of frogs, the eggs are laid in water, where they hatch into swimming tadpoles, making these species dependent on water for breeding. Some frogs, however, have become emancipated from water by laying their eggs on land, contained within a waterproof capsule, or within their own bodies. Frogs are associated only with fresh water. While some can survive in brackish water, there are no anurans that have colonized the sea, not even the so-called Marine Toad (aka the Cane Toad, *Rhinella marina*; page 201).

Anurans have a stout body and a very short backbone, containing only eight or fewer vertebrae. The hind limbs are typically longer than the forelimbs, largely due to the tarsal (ankle) bones being greatly lengthened. At rest, the hind limbs are folded under the body, but they can be rapidly straightened, enabling many frogs to lunge at their prey or escape from an enemy by jumping. They are often very muscular, making frogs' legs a very attractive food source for humans in some parts of the world. The head is wide, the neck is inflexible, and the mouth can typically be opened very wide, enabling many anurans to consume surprisingly large prey. Virtually all frogs are carnivores, feeding on

BELOW **A European Pool Frog** (*Pelophylax lessonae*, left), showing typical anatomical features of frogs. The European Tree Frog (*Hyla arborea*, right) has adhesive disks at the end of its fingers and toes that enable it to climb up smooth, vertical surfaces.

living prey, particularly insects. Important for detecting moving prey, their eyes are usually very large. In fact, they may be so large that they bulge into the mouth—some frogs assist the passage of food from mouth to gut by "rolling their eyes."

Frogs typically have four fingers on their forelimbs and five toes on their feet. The shape and length of the fingers and toes vary a great deal from one species to another and largely reflects their mode of locomotion. For example, very aquatic frogs have webbing between their toes, and often also between their fingers, enabling them to swim more effectively. Tree frogs typically have disk-like tips to their fingers and toes that act as adhesive pads, enabling them to climb up smooth, vertical surfaces. Many burrowing frogs have horny tubercles on their hind feet that act as spades for moving soil.

In many anurans, the eardrum, called the tympanum, is visible just behind the eye. Hearing is of great importance to frogs because they communicate with one another by calling. These calls are produced by passing air back and forth between the lungs and the mouth, and in many species the sound is amplified by one or two inflatable vocal sacs that open into the mouth. The primary function of the lungs is to breathe in air, but the majority of frogs actually obtain most of the oxygen they need through their moist skin.

The skin of frogs is thin and contains numerous glands that secrete mucus to keep the surface moist. There are also glands that secrete toxic compounds. These may be very mild in their effects but, in some frogs, they are lethal to any animal that ingests them. The skin of frogs contains numerous cells that consist of a variety of colors. In poisonous frogs, these

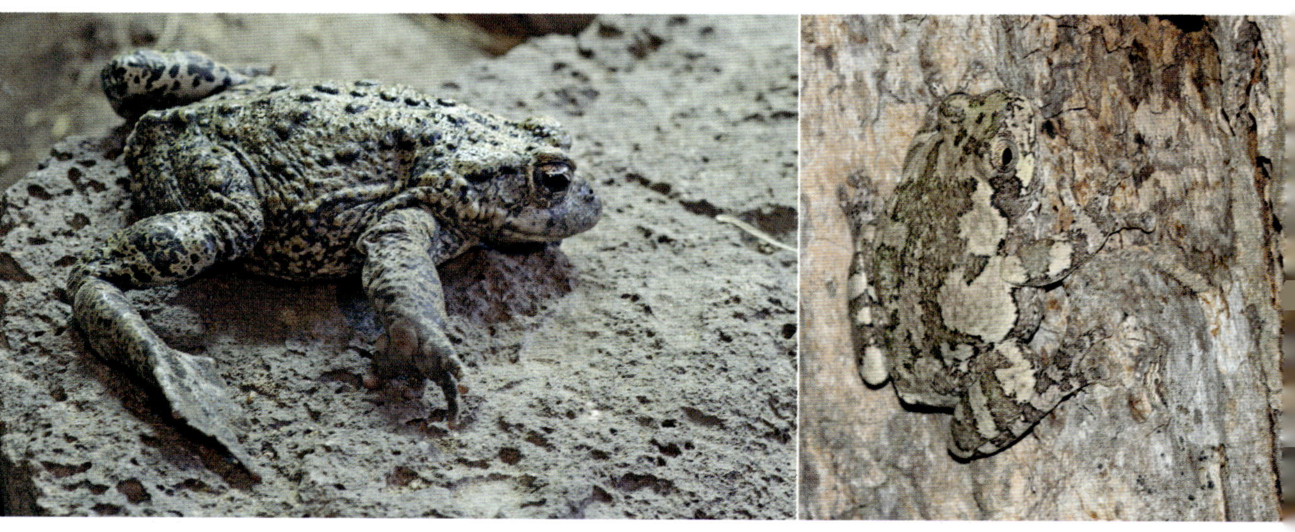

BELOW, THIS AND OPPOSITE PAGE **The coloration of many frogs** matches their typical habitat, making them inconspicuous to potential predators. Left to right: The Western Toad (*Anaxyrus boreas*) is sometimes found in semi-desert areas; the Eastern Gray Tree Frog (*Dryophytes versicolor*) changes its skin color to match its background; the South American Common Toad (*Rhinella margaritifera*) lives in leaf litter on the forest floor; the Tonkin Bug-eyed Frog (*Theloderma corticale*) rests by day on moss-covered logs.

colors are typically very conspicuous, providing a warning to potential enemies. In most frogs, however, the skin colors and patterns provide very effective camouflage. Many frogs can change the color of their skin, usually rather slowly, though some species are capable of changing color within a few minutes.

A distinction is often made between "frogs" and "toads," but this can be misleading because the two terms are used differently in different parts of the world. In Europe and North America, anurans with smooth skins and aquatic habits are known as frogs, and those with warty skins and more terrestrial habits as toads. However, anurans belonging to the African genus *Xenopus* (see pages 53–56) are often called clawed toads, despite being wholly aquatic and having very smooth skins. The distinction between frogs and toads is biologically meaningless; in this book, all anurans are referred to as frogs even though many are known colloquially as toads.

ABOVE **The bright colors of dendrobatid frogs**, such as the Strawberry Poison-dart Frog (*Oophaga pumilio*, right) and the Dyeing Poison Frog (*Dendrobates tinctorius*, left), protect them from predators by warning that their skin contains powerful toxins. As a result, they are unusual as they can be freely active during the day.

COMPLEX LIFE CYCLE

ABOVE **In the Common Midwife Toad** (*Alytes obstetricans*) the male provides parental care of the eggs. He carries them in strings twisted around his legs from mating until they are ready to hatch into tadpoles, at which time he deposits them in water.

In common with other amphibians, frogs have a life history that consists of four distinct stages: egg, larva or tadpole, juvenile, and adult. The length of time spent in each of these stages varies enormously, both within and between species.

EGGS

Amphibians produce eggs similar to those of fish. They have no shell but are bounded by a permeable membrane that encloses a spherical blob of protective jelly. The eggs thus require wet conditions; otherwise they will soon dry out and die. Each egg is supplied with a quantity of yolk on which, after hatching, the tadpole feeds until it can feed itself. The amount of yolk allocated varies greatly between species depending on their mode of reproduction. Frogs that lay their eggs in large waterbodies containing lots of food tend to produce many small eggs, each containing only a little yolk. Conversely, those laying their eggs in tiny pools in the leaf axils of plants, where there is very little for tadpoles to eat, lay a few, large eggs, each containing sufficient yolk to see the tadpole through most or all of its development. Likewise, frogs that show parental care of their eggs typically lay far fewer, larger eggs than frogs that release their eggs into water to fend for themselves.

Water is a hazardous environment for frogs' eggs and tadpoles, containing many fishes, insect larvae, and other predators that readily eat them. Some frogs have evolved modes of egg-laying that emancipate them from open water and thus reduce this predation risk. Numerous tree frogs attach their eggs to leaves high up in trees, for example. Several frog species lay their eggs in foam nests that not only protect them from predators but also provide a reliably moist environment in which the tadpoles can develop. Some species lay their eggs on land and some carry them about on their backs. Most remarkable of all are those frogs that carry their eggs within some part of their body, such as their skin, their gut, their vocal sacs, or their oviducts.

In many tropical terrestrial frogs, the eggs are large, are laid on land, and hatch to produce tiny froglets. Called "direct development," this life history appears to have eliminated the tadpole stage, but this is not the case. The egg hatches into a tadpole that completes its development within the confines of a capsule, feeding only on yolk provided by the mother. Direct-developing frogs have thus evolved a life cycle that makes them independent of ponds or streams, though they are still restricted to damp habitats.

The clutch size of frogs ranges from several thousand in larger, aquatic frogs to single figures in small, terrestrial species with parental care. Within a species, the clutch size of an individual female is typically related to her size, larger females producing more eggs and, in some species, eggs containing more yolk.

The eggs of frogs that are laid in open water tend to be black because they have a layer of the pigment melanin around them. This protects the genetic material within the egg, DNA, from being broken up by ultraviolet radiation in sunlight. Frogs' eggs that are laid underground or in shady places tend to be white or cream in color.

TADPOLES

Tadpoles are rather simple creatures that swim, eat continuously, and grow as fast as possible. Most tadpoles are very palatable to many predators, and the sooner they can reach metamorphosis and escape to the relative safety of land, the better. Rapid development is also vital for tadpoles living in ephemeral ponds, which often lack predators but may soon dry out. The tadpoles of some desert-living frogs can reach metamorphosis just eight days after hatching from the egg. Typically, tadpoles live in water and obtain oxygen by means of gills. They are usually herbivorous and their rate of growth is dependent on temperature as well as food

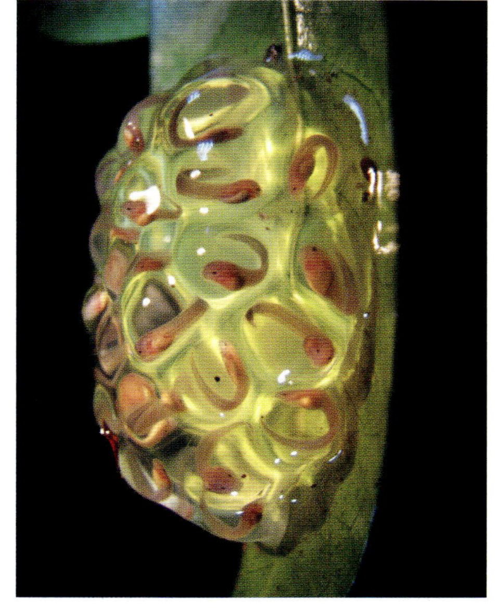

BELOW **The eggs of frogs** consist of a jelly-filled capsule that protects the developing embryo; shown here are the eggs of a glass frog, attached to a plant.

ABOVE **The tadpoles of the Copan Brook Frog** (*Duellmanohyla soralia*) of Central America have funnel-shaped mouths for filtering tiny food items from the water. They are unusual in having iridescent green spots.

supply. A herbivorous diet requires a long gut—the distinctive globular body shape of tadpoles is primarily due to their long, coiled gut. Many tadpoles feed by scraping algae off plants, stones, and other objects, and to help them in this their mouths are surrounded by rows of tiny teeth made of keratin (the material that forms fingernails in humans). The tadpoles of some species live in open water, often in shoals. They usually lack teeth and feed by swimming around open-mouthed, filtering tiny food items from the water. The mouths of the tadpoles of stream-breeding frogs are often surrounded by a sucker, which enables them to attach themselves to rocks to avoid being swept away by the current. In many species, the tadpole diet shifts from plant material to animal tissue as metamorphosis approaches. They become scavengers, feeding on the corpses of dead animals, including one another. In a few species, larger tadpoles become predators, hunting and feeding on the tadpoles of their own and other species.

Tadpoles appear to be simple, unsophisticated animals, but research has shown that they are capable of altering their behavior in response to changes in their environment. For example, in many species, tadpoles gather together in dense schools when they sense the presence of a predator. They do this is by detecting waterborne chemicals that are released when a tadpole is attacked. Most remarkably, the tadpoles of some species preferentially cluster together with their own immediate kin. This kin recognition is probably mediated by means of odor.

At metamorphosis, the larval anatomy is totally transformed into the adult form, with the acquisition of lungs and legs, and the loss of gills and the tail. There is also a radical change of diet, with froglets developing the ability to detect and catch small insects and other moving prey. This involves development and enlarging of the eyes.

JUVENILES

After metamorphosis, young frogs live on land, sometimes for several years, before returning to water to breed. This juvenile phase of a frog's life cycle is the least well known, froglets being small and very secretive in their habits. In some frogs, the juvenile phase lasts for only a few months, but in others it may last for several years. In the Pacific Tailed Frog (*Ascaphus truei*; page 40), for example, adulthood is achieved at four years of age.

ADULTS

Adult frogs continue to grow throughout life, albeit at a slower and slower rate, producing much variation in adult body size. This is why, in this book, the body length of frogs is usually given as a range rather than a single value. In many frogs, females are larger than males; the reverse is true in only a few frog species. In some species, such as the European Common Toad (*Bufo bufo*; page 157), the size difference arises because males reach sexual maturity one year earlier, and at a smaller size than females. This difference in the age of maturity also produces a biased sex ratio, with adult males greatly outnumbering adult females.

PARENTAL CARE

In many frogs and toads, the female makes a modest contribution to the survival of her eggs and the resultant tadpoles. The eggs are deposited in water during mating and are then left to fend for themselves. Eggs and

LEFT **Metamorphosing Caatinga Horned Frogs** (*Ceratophrys joazeirensis*) from Brazil. They have all the anatomical features of adults, such as arms and legs, but have yet to lose their tails fully.

larvae typically suffer very high mortality through predation, desiccation, freezing, or disease, and the chances that at least some will survive are increased by producing eggs in very large numbers. However, several frogs and toads have evolved an alternative strategy, in which fewer eggs are produced, and one or both parents provide some form of care that enhances each egg's chances of survival.

In almost all frogs, the eggs are fertilized externally, the male shedding sperm onto them as they are ejected from the female's body. In about 12 species, however, fertilization is internal, the male inserting sperm into the female's oviduct. In such species, the young can be retained in the female's body for part or all of their development, providing a high degree of protection. In a recently discovered fanged frog from Sulawesi (*Limnonectes larvaepartus*), the young are born as tadpoles; in the Kihansi Spray Toad (*Nectophrynoides asperginis*; page 183), they are born as tiny froglets.

Several tropical frogs protect their eggs by laying them in a foam nest, made from a secretion, usually produced by the female and whipped into a foam by both male and female beating their legs. Some frogs, such as the African leaf-folding frogs (*Afrixalus*; see pages 438–440), protect their eggs by folding a leaf around them. In the midwife toads (*Alytes*; see pages 47–48) of Europe, the eggs are laid in strings, fertilized externally, and then gathered up by the male. He carries the eggs, twisted around his legs, until they are ready to hatch, keeping them moist by occasionally entering a pool of water.

BELOW **A pair of Red-eyed Tree Frogs** (*Agalychnis callidryas*) in amplexus. As in the majority of frogs, the male is smaller than the female. In this species, the eggs are laid attached to leaves hanging over a pool. When they hatch, the tadpoles fall into the water below.

In most species with direct development, one or both parents remain close to the eggs, defending them against predators and keeping them clear of fungal infection.

In a number of poison frogs, such as the Black and Yellow Poison Frog (*Dendrobates leucomelas*; page 375), the male provides parental care, defending the eggs until they hatch, and then carrying them on his back to tiny pools in the axils of

ABOVE **This male Sky Blue Poison Dart Frog** (*Hyloxalus azureiventris*) from Peru guards the eggs and then carries the resulting tadpoles to streams, where they complete their development.

plants, called phytotelmata, where they complete their development. In poison frogs of the genus *Ranitomeya* (see pages 385–388), males carry the tadpoles to phytotelmata and females then feed them with unfertilized eggs. In at least one species, the Imitating Poison Frog (*R. imitator*; page 386), male and female form a stable, monogamous relationship that lasts for the whole time it takes them to produce and rear their young.

There are frogs that hold their developing young within their bodies, but not in their reproductive tract. The male of the Hip-pocket Frog (*Assa darlingtoni*; page 105) has a pouch of skin on each side of its body in which the tadpoles develop. In the now extinct Southern Gastric Brooding Frog (*Rheobatrachus silus*; page 117), the tadpoles developed in their mother's stomach. In the endangered Darwin's Frog (*Rhinoderma darwinii*; page 257) the tadpoles develop in the male's vocal sac. Most remarkably, the young of the Surinam Toad (*Pipa pipa*; page 58) complete their entire development, from egg to froglet, in pockets in the skin on their mother's back.

Some frogs carry their offspring on their backs after they have metamorphosed into froglets. Males of the Horned Land Frog (*Sphenophryne cornuta*; page 482), from Papua New Guinea, carry tiny froglets for several days, moving around after dark. Each night, a few jump off and disappear into the forest in what appears to be an adaptation that ensures the young are widely dispersed.

AN ACOUSTIC WORLD

Frog choruses are among the most spectacular events occurring in the natural world. In many tropical and semitropical habitats, male frogs gather in huge numbers within a restricted area and, usually at night, generate a sound that can be heard up to a mile (1.5 km) away. Some frog choruses contain only one species, but, especially in tropical habitats, they can contain up to 20 different species, all calling together. Choruses are typically formed around a pond or larger body of water that is the site where female frogs will mate and lay their eggs.

Because calling is associated with mating, the calls of male frogs are often referred to as "mating calls." It is more correct, however, to call

BELOW **European Edible Frogs** (*Pelophylax* kl. *esculentus*) resting by day on a log. Males call mostly during the night, while floating in the water.

them "advertisement calls," because their role in behavioral interactions between males is just as important as their role in mating. Males respond to one another's calls, sometimes by fighting, but more often by simply changing their position so as to ensure that they are not too close to a rival. The spacing that occurs makes it easier for females to discriminate among the calls of individual males as they move through a chorus.

ABOVE **Male North American Bullfrogs** (*Aquarana catesbeiana*) fighting. The male bullfrog is territorial, defending an area from which he calls to attract females. Fighting is rare, most disputes being resolved by an exchange of calls.

SPECIES RECOGNITION

In a multi-species chorus, the most important task for a female is to identify a male of her own species and to avoid mating with a male of a different one. Species recognition in many frogs is ensured by the nature of the male's call and by the sensitivity of the female's ears. Male calls are very stereotyped and species-specific, containing a limited range of sound frequencies that distinguish them from other species. Unlike human ears, which are sensitive to a wide range of sound frequencies, those of frogs are sensitive only to those frequencies present in the advertisement calls of their own species. Females are thus essentially deaf to the calls of males belonging to species other than their own.

The Puerto Rican Coqui (*Eleutherodactylus coqui*; page 402) gets its name from the male's advertisement call, which consists of a low-frequency "co" note, followed by a high-frequency "qui." The ears of males contain many auditory cells that respond to only the "co" note, whereas those of females are particularly sensitive to the "qui" note. This unusual example thus illustrates both the dual function of the male's call, and the way in which the ears of frogs are selectively tuned to the calls of their own species.

CALL VARIATION

There is much more to vocal interactions between frogs than species recognition, however. Detailed analysis of recordings of frog calls reveals that, within a species, they are quite variable. Individual males may call more often, for longer, or at a slightly different sound frequency than other males of the same species, for example. In frog species that establish choruses that are maintained for several weeks or months, this enables males to recognize one another as individuals. The North American

ABOVE **A male Painted Reed Frog** (*Hyperolius marmoratus*) calling, his vocal sac fully extended. In this South African species, males expend a lot of energy to produce a remarkably loud call. Those males that are able to call for several nights have higher mating success than those that can call for only one night.

Bullfrog (*Aquarana catesbeiana*; page 545) has an extended breeding season, males spacing themselves out around the edge of a pond. They are much more aggressive to an intruder with an unfamiliar call than they are to their immediate neighbors, whose calls they have learned to recognize, a phenomenon called "dear enemy recognition."

For females, variation in male calls provides the opportunity for mate choice, enabling them to mate preferentially with males whose calls have certain characteristics. In a few species, females prefer larger males, whose voices are slightly deeper than those of small males. More commonly, females prefer those males that put more energy into their calling, calling more often or for longer, for example. For many frogs, calling is the most energetic activity in which they engage throughout their lives, and the ability to call vigorously for long periods may reflect their general health and vigor. Female choice for particular call properties may thus enable them to mate preferentially with genetically superior males.

The energetic nature of calling means that males may rapidly run out of the nutrient reserves they use while calling, limiting the number of nights for which they can call. Exhausted males that have become silent may not be sexually inactive, however. In some species, such males adopt a "satellite" strategy, sitting silently close to a calling male and attempting to intercept and clasp any female that approaches him.

In many frogs the loudness of the male's call is increased by the possession of vocal sacs that become inflated as air in the lungs is passed over the vocal cords into the mouth. Some frogs have a single vocal sac under the chin, while others have paired sacs, one on each side of the mouth. The vocal sacs are often white or bright yellow in color, and there is evidence from some species that females can see inflated vocal sacs, even on very dark nights, enhancing their ability to find a mate.

DISTRESS AND RELEASE CALLS

Many frogs can make other sounds that are distinct from the advertisement call. For example, the European Common Frog (*Rana temporaria*; page 571) produces a loud, piercing scream when attacked by a cat. This so startles the cat that it drops the frog, allowing it to escape. In many species, males are indiscriminate during mating activity, attempting to clasp any frog that moves, including other males of their own species and females of other species. Individuals clasped by an inappropriate partner often give a "release call" that causes the clasping male to release them. In the European Common Toad (*Bufo bufo*; page 157), the release call plays a role in fights between males over females. Larger males have deeper voices than small males and, as males grapple with one another, small males tend to give up if they hear the deeper release call of a larger rival.

BELOW **A male European Edible Frog** (*Pelophylax* kl. *esculentus*) calling while floating in the water. His paired vocal sacs serve to amplify his call.

FROGS, TOADS & PEOPLE

ABOVE **There are many versions** of the fairytale "The Little Princess and the Golden Ball," in which a princess kisses a frog and turns him into a handsome prince. This illustration by Mabel Dearmer comes from *The Frog Princess*, published in 1897.

The phenomenon of metamorphosis, by which aquatic tadpoles are transformed into terrestrial frogs, may be the origin of several transformation myths in human folklore. Witches are said to be able to turn people into frogs, and the kiss of a princess to turn an ugly frog into a handsome prince. In ancient Egypt, the annual appearance of millions of frogs following the flooding of the Nile was associated with the frog-goddess Heqet. Because flooding restored fertility to formerly barren land, she was a symbol of fertility.

FROGS AS FOOD AND PETS

The long, muscular legs of some frog species have long been regarded as a gastronomic delicacy by many people, particularly in some parts of Europe, North America, and Asia. For example, the California Red-legged Frog (*Amerana draytonii*) was hunted for food during the 1849 gold rush and by the 1870s its numbers had declined dramatically around the San Francisco area. Moreover, at the turn of the century, Californian frog legs were being exported to France. Depletion of natural populations led to the development of frog farms in Europe and elsewhere, where the large North American Bullfrog (*Aquarana catesbeiana*; page 545) was reared.

ABOVE **The American Bullfrog** (*Aquarana catesbeiana*) is bred in frog farms in many parts of the world as a source of food. Unfortunately, the frogs frequently escape and form invasive populations that are a threat to native species.

The export of frogs from one part of the world to another has created a new set of problems. Bullfrogs that escaped from frog farms founded invasive populations that have displaced native frog species and spread diseases such as chytridiomycosis. Intensive efforts are now being made to eradicate the North American Bullfrog from France, Britain, and other parts of Europe.

In some parts of the world, notably West Africa, frogs are more than a delicacy; they can be a vital source of protein. As a result, the populations of some large-bodied species, such as the Goliath Frog (*Conraua goliath*; page 527), have been seriously depleted. Weighing up to 6½ lb (3 kg), the Goliath Frog is the world's largest frog. As well as being eaten in large numbers by local people, the species has been collected for zoos and the pet trade, and is now listed as Endangered.

There is an extensive international trade in frogs, kept as pets in terraria. Most popular are the poison frogs from Central and South America (pages 363–388), and the mantellas from Madagascar (pages 624–639). Both groups are small and brightly colored, and some species thrive and even breed in a terrarium environment. Collecting for the pet trade has had a serious negative impact on the wild populations of some frogs and has been a major cause of their decline. While some frogs thrive and breed readily in captivity, many do not and collection for the pet trade is a constant threat to wild populations.

FROGS IN SCIENCE

Frogs are important in medicine and biological research. The most intensively studied frog in the world is the African Clawed Frog (*Xenopus laevis*; page 54), an aquatic species that thrives in captivity. Through the 1940s and 1950s, it was widely used for human pregnancy testing. The urine of a pregnant woman, injected into a female *Xenopus*, causes it to produce eggs. This procedure has been replaced by other methods, but the African Clawed Frog is widely used as a model organism for studying how animals develop from egg to adult and was the first vertebrate to be cloned.

More recently, the discovery that many frogs secrete antibacterial compounds from their skin has excited considerable interest. Since existing antibiotics lose their effectiveness, there is an urgent need to find new compounds that can be used to control infections and it may be that frogs can supply at least part of the solution to this serious problem. Poison-dart frogs of the genus *Epipedobates* (see pages 367–368) produce a unique substance called epibatidine, which is a much more effective painkiller than morphine and, unlike that drug, appears not to be addictive. The skin secretions of poison-dart frogs have long been used in hunting by Central and South American Indians, and those of several toad species are used in many parts of the world as hallucinogens. A number of frog species are the source of cures for a variety of conditions in traditional Chinese medicine. This has led to the overcollection of several Asian frog species.

BELOW **The skin secretions** of the Phantasmal Poison Frog (*Epipedobates tricolor*) contain a powerful painkiller known as epibatidine.

An attempt to use a species of toad as a biological control agent has proved to be a major ecological disaster. Native to Central and South America, the large Marine Toad (*Rhinella marina*; page 201) is known in Australia as the Cane Toad because it was introduced in order to control insect pests on sugarcane farms in Queensland in 1935. This was never likely to be effective, as Cane Toads feed at night and the pests are active by day. Moreover, Cane Toads are not able

LEFT **The African Clawed Frog** (*Xenopus laevis*) is bred in large numbers and used for medical research.

to climb up sugarcane plants to reach the insect pests. They have flourished in Australia, feasting on native fauna and building up much denser populations than in their native habitat. They spread rapidly, increasing their range by around 19 miles (30 km) each year, and they now occur throughout western and northern Australia. They have had a serious adverse effect on biodiversity in Australia, not only causing population declines among native frogs, but also among many reptiles, which are poisoned by the Cane Toad's skin secretions.

LEFT **A Cane Toad** (*Rhinella marina*) in a field of sugarcane in Queensland, Australia, where it was introduced in an attempt to control insect pests.

POPULATION DECLINES

Since the 1980s, biologists who study frogs and other amphibians throughout the world have become increasingly alarmed by rapid declines in the populations of their focus animals. The precise scale of the problem became apparent in 2022 with the release of the Second Global Amphibian Assessment, which concluded that 33.5 percent of the world's amphibian species were threatened with extinction, representing nearly 2,700 species (see Conservation Status table below). Since the 1970s around 200 amphibian species are believed to have become extinct and nearly half of all species have populations that are declining.

CONSERVATION STATUS OF AMPHIBIAN SPECIES

Conservation Status	Number of species	Percentage of species
DATA DEFICIENT (DD)	909	11.3
LEAST CONCERN (LC)	3,739	46.7
NEAR THREATENED (NT)	451	5.6
VULNERABLE (VU)	811	10.1
ENDANGERED (EN)	1,264	15.8
CRITICALLY ENDANGERED (CR)	612	7.6
CR PROBABLY EXTINCT (CRPE)	186	2.3
EXTINCT IN THE WILD (EW)	2	0.02
EXTINCT (EX)	37	0.5
TOTAL	8,011	

Results of the Second Global Amphibian Assessment, released in 2022, showing the number of amphibian species in seven of the nine IUCN Red List categories. The three categories Vulnerable, Endangered, and Critically Endangered are grouped together as Threatened (T); this group consists of 2,687 species, representing 33.5 percent of all known amphibian species.

Source: Re:wild, Synchronicity Earth, IUCN SSC Amphibian Specialist Group. 2023. State of the World's Amphibians: The Second Global Amphibian Assessment. Texas, USA: Re:wild.

ABOVE **The icon of the amphibian decline phenomenon** is the Golden Toad (*Incilius periglenes*) of Costa Rica, which disappeared around 35 years ago. The causes of its decline remain unclear but are thought to be climate change and the disease chytridiomycosis.

Amphibians are by no means a special case. All kinds of animals and plants are declining, in what many biologists believe to be a mass extinction event, the sixth to have occurred in the history of planet Earth. Whereas the previous mass extinctions were caused by natural events—the fifth, for example, was caused by an impact with a giant asteroid 66 million years ago—the current sharp increase in extinctions is caused by the activities of human beings. In 2024, the Worldwide Fund for Nature (WWF) published the latest update of its *Living Planet Report*, an analysis of the status of wildlife populations based on the continuous monitoring of many species. It concluded that the global population of vertebrates (fishes, amphibians, reptiles, birds, and mammals) had declined by 73 percent between 1970 and 2020. Amphibian population declines are thus one component in a general decline in biodiversity on Earth.

Rates of decline and extinction vary among ecosystems. The global population of freshwater animals of all kinds has declined by 83 percent, a greater decline than in any other ecosystem. Amphibians, being totally dependent on fresh water, are thus indicators of the global decline in the extent and quality of the Earth's freshwater resources. They typically do not occur in larger lakes or rivers, many of which are afforded some degree of conservation protection because of their value as a source of fish. Rather,

ABOVE **The most important cause** of amphibian population declines and of extinctions around the world is habitat destruction. Here, rainforest in Ecuador is being destroyed to make way for agriculture.

they breed in small ponds, marshes, swamps, and mountain streams, which are generally poorly protected.

Certain features of amphibians make them especially vulnerable to the various factors that adversely affect their environment. Foremost among these is their dependence on water, especially for breeding. Another important feature is the lack of an outer protective layer, especially in the egg and larval stages, making them very vulnerable to a wide variety of pollutants. As will become apparent throughout this book, many amphibians have a tiny geographic range, making them particularly vulnerable to habitat destruction and to ecological changes resulting from climate change.

CAUSES OF AMPHIBIAN POPULATION DECLINES

The most common cause of amphibian population declines is habitat destruction. Throughout the world, natural habitats are being destroyed to make way for agriculture and for human habitation. Particularly important in relation to amphibians are the destruction of forests and the draining of wetlands. Many amphibians have been adversely affected by chemical contamination of freshwater habitats, particularly by agrochemicals such as pesticides, herbicides, and fertilizers. In some parts of the world, amphibians have declined as the result of the introduction of alien species, particularly fishes that eat amphibian larvae. Such introductions have sometimes been deliberate, as in the introduction of trout for sport and of mosquitofish (*Gambusia*) to control insect pests. Some amphibians have declined as a result of overexploitation by humans, for food, as pets, or as a source of traditional medicines.

Human activities have caused alterations in the Earth's atmosphere, including erosion of the ozone layer, which protects living organisms from harmful UV-B radiation from the sun. In some parts of the world, elevated UV-B levels damage the DNA in amphibian eggs, leading to abnormal development and death in the early life stages. Climate change is having a significant effect on many amphibians, causing them to breed earlier and altering the length of time for which their breeding sites retain enough water for larvae to complete their development. Climate change is expected to have a serious effect on many amphibians in the future, because of the way it changes rainfall, vegetation, and other environmental factors.

In addition to the anthropogenic factors discussed, all of which are implicated in the decline and extinction of animals other than amphibians, amphibians have been severely impacted by a number of diseases that are specific to them. The most important of these is chytridiomycosis, a fungal disease that is remarkable in two respects. First, it is affecting amphibians worldwide and, second, it is capable of infecting a very large number of species. Most diseases of wildlife infect one or few species.

None of the environmental changes listed above is involved in all amphibian declines, and few declines can be attributed to only one factor. Importantly, in many instances the various factors act synergistically with one another. For example, in the Cascade Mountains of northwestern USA, a series of prolonged droughts have caused water levels to fall in the ponds in which the Western Toad (*Anaxyrus boreas*; page 136) breeds. This has increased the intensity of UV-B radiation reaching the eggs, in turn leading to an increased rate of genetic mutation and making developing embryos more susceptible to infection by the fungus *Saprolegnia*.

Many of the environmental factors that have adversely affected amphibians are subtle in their effects, causing sublethal changes that can be detected only by careful research. Among these are a variety of chemical contaminants, sometimes referred to as "gender benders," which affect the reproductive system. For example, the widely used herbicide atrazine has been shown to have a feminizing effect on males of a variety of fish and frog species. This will have deleterious consequences for the recruitment of young to the population. Pollution of waterbodies by nitrogenous fertilizers rarely reaches levels that actually kill tadpoles, but they can impair their growth, leading eventually to small froglets that are less likely to survive. In Britain, climate change is causing milder winters, during which the European Common Toad (*Bufo bufo*; page 157) is more active than it is in cold winters. As a result, the females use up more of the fat reserves that are required for egg production in the spring and so lay fewer eggs.

BELOW **The Western Toad** (*Anaxyrus boreas*) has declined over much of its extensive range, probably due to multiple factors, including habitat degradation, elevated ultraviolet radiation and the disease chytridiomycosis.

AMPHIBIAN DISEASES

Like all animals, including humans, amphibians are susceptible to a variety of infectious diseases, caused by a variety of microbes, called pathogens. From time to time, a particular disease may show a dramatic increase in its incidence, its virulence, its range, or in the range of species it infects. Examples of such "emerging infectious diseases" that have affected humans recently are AIDS and Ebola. Emerging infectious diseases have had a devastating effect on many amphibian species in recent years, on local, regional, and global scales.

Ranaviruses, first discovered in the 1960s in the USA, are thought to have evolved from fish viruses and infect freshwater reptiles as well as amphibians. They cause skin lesions and swollen limbs, and can lead to mass mortality events. The viruses are associated with major die-offs

RIGHT **European Common Frogs** (*Rana temporaria*) in a breeding pond. When large numbers of frogs come together to breed it becomes very much easier for pathogens, such as viruses, bacteria, and fungi, to spread among them.

LEFT **A dead European Common Frog** (*Rana temporaria*) in a garden in London, UK. The red color on the underside of the thighs suggests that its death is due to red-legged disease.

among European Common Frogs (*Rana temporaria*; page 571) in the southeast of the United Kingdom in the late 1980s. These have been most common in artificial garden ponds in urban areas, where frog populations can become locally very dense and where they often share their breeding ponds with introduced fish, such as goldfish.

Red-legged disease is caused by the bacterium *Aeromonas hydrophila*. It gets its name from its most obvious symptom—hemorrhages on the insides of the thighs. It is implicated in mass mortality events among Mountain Yellow-legged Frogs (*Amerana muscosa*; page 543) in California in 1979, and among Western Toads (*Anaxyrus boreas*; page 136) in Colorado in the 1970s and 1980s.

CHYTRIDIOMYCOSIS

Much the most serious emerging infectious disease to affect amphibians is chytridiomycosis. It is the only wildlife disease known to be able to cause serious population declines and extinctions in species living in pristine habitats, and thus not subject to other environmental stressors. Discovered in 1993 in Queensland, Australia, chytridiomycosis is caused by the fungus *Batrachochytrium dendrobatidis* (*Bd*). *Bd* infects the outer layers of amphibian skin, which contain keratin, a fibrous protein. In tadpoles, keratin is found only in the hard mouthparts, and *Bd* has no adverse effects on them. At metamorphosis, however, *Bd* begins to invade the keratinized skin of adult frogs, interrupting the complex physiological processes that take place in the skin, and eventually causing death by cardiac arrest.

There are many aspects of chytridiomycosis that are not yet understood. It is not known, for example, whether the disease can be

RIGHT **A frog** that has died from the fungal disease chytridiomycosis.

spread by animals other than amphibians. Nor is it clear why some species are much more susceptible than others. Chytridiomycosis is known to have caused extensive population declines, including extinctions, in at least 200 amphibian species, but there are several species that are unaffected by it. These include the North American Bullfrog (*Aquarana catesbeiana*; page 545) and the African Clawed Frog (*Xenopus laevis*; page 54). Both these species can carry *Bd* and have been distributed around the world by humans. It is thus very likely that humans are partly responsible for the global spread of chytridiomycosis.

It is clear that chytridiomycosis can spread very quickly. Biologists have been following the distribution of the disease, notably through Central America, where it has spread southward, and in eastern Australia, where it has moved northward. In some places, it has moved as much as 60 miles (100 km) in a year. While chytridiomycosis appears to be a new disease in some parts of the world, it has been present in some places for a very long time. A recent study of museum specimens collected in Illinois, USA, between 1888 and 1989, has revealed that it has been present there for at least 120 years. In Africa, *Bd* has been found in museum specimens dating back to 1933 in Cameroon and 1938 in South Africa.

There are two major hypotheses about the origins of chytridiomycosis. One is that *Bd* is a newly evolved pathogen that is spreading across the world. The other is that it is a disease that has been long established in some parts

LEFT **A mass mortality event** caused by chytridiomycosis among Mountain Yellow-legged Frogs (*Amerana muscosa*), high in the Sierra Nevada, California. Frogs that live at higher altitudes are particularly susceptible to this lethal disease.

of the world but has become a global disease as a result of being spread by human activities. It has also been suggested that one or more environmental stressors, such as chemical pollution, elevated UV-B, or climate change, may have made amphibians more susceptible to chytridiomycosis.

To date, at least 500 amphibian species are known to be infected by chytridiomycosis. *Bd* has different effects on different species of frogs. In some species it is lethal, but in others there are no apparent effects, though they are able to pass the fungus on to susceptible species. Among the most susceptible species are frogs that breed in mountain streams, where conditions are cool. *Bd* seems to thrive at lower temperatures and, in a number of Australian frogs, it has caused serious declines at high altitudes but not, in the same species, in lowland areas.

Considerable research efforts are being made to develop a cure for chytridiomycosis. In the laboratory, changes in water temperature, antifungal agents, and various other treatments have proved effective, but none is appropriate for curing the disease in wild populations. The best chance of preventing the spread of chytridiomycosis is to try to restrict the movement of infected amphibians from one area to another. Field biologists across the world now routinely take great care to clean their boots and equipment before moving from one location to another. Preventing the movement of amphibians in the global networks that comprise food and pet trade networks is urgently needed, but poses an enormous practical and legal challenge.

DISTRIBUTION & CLASSIFICATION

Frogs and toads are found on all continents of the world except Antarctica. The greatest number of species is found in the South American tropics. New frog species are continually being described, Colombia, for example, is home to 798 frog species, 384 of which are endemic, meaning that they are found nowhere else in the world. By contrast, the United Kingdom is home to 11 species, only seven of which are native and none endemic. Frogs and toads are more widely distributed than the salamanders and newts, which occur mostly in the northern hemisphere and are most diverse in North America, and the caecilians, which occur only in tropical regions of South and Central America, Africa, and southern Asia.

The classification of frogs has recently undergone a radical change. Prior to 2006, they were divided into 28 families. Recent analysis of DNA sequences to compare one species with another has revolutionized anuran classification, and now 59 families are recognized. The largest family, the Strabomantidae, did not exist under the previous system. It currently includes 815 species, all found in South America. The next largest family, the Bufonidae, did exist under the previous system. Containing 659 species, it has a global distribution with the exception of Australia, where it is represented by the introduced Cane Toad (*Rhinella marina*; page 201).

Frogs and toads evolved in a world where plate tectonics caused the early continents to break up and drift apart. As a result, some older families, like the Bufonidae, are found on many continents, while others, of more recent origin, are restricted to just one. The island of Madagascar, for example, is home to 425 species, 285 of which belong to the endemic family Mantellidae.

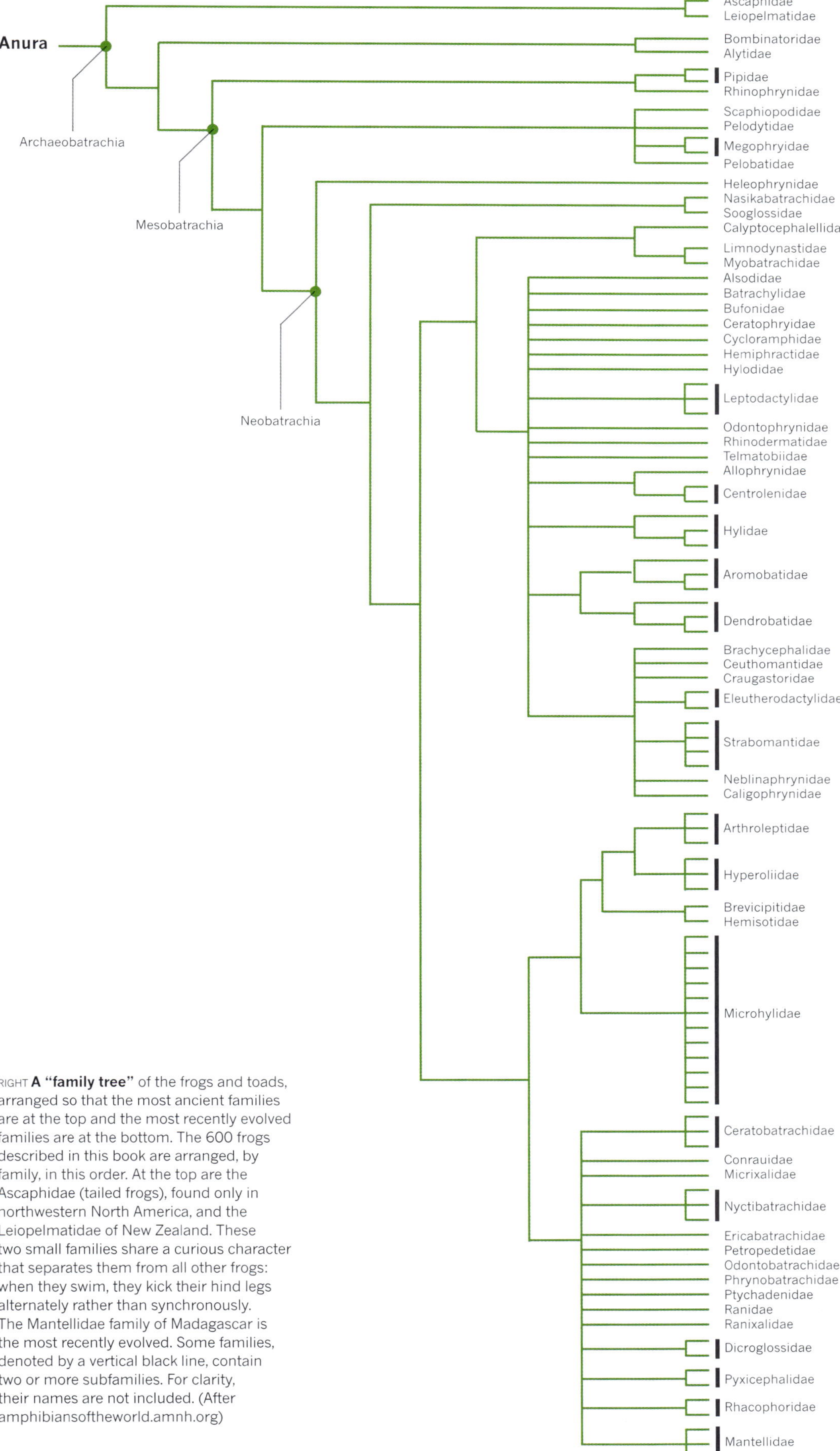

RIGHT **A "family tree"** of the frogs and toads, arranged so that the most ancient families are at the top and the most recently evolved families are at the bottom. The 600 frogs described in this book are arranged, by family, in this order. At the top are the Ascaphidae (tailed frogs), found only in northwestern North America, and the Leiopelmatidae of New Zealand. These two small families share a curious character that separates them from all other frogs: when they swim, they kick their hind legs alternately rather than synchronously. The Mantellidae family of Madagascar is the most recently evolved. Some families, denoted by a vertical black line, contain two or more subfamilies. For clarity, their names are not included. (After amphibiansoftheworld.amnh.org)

WHY THE NUMBER OF FROG SPECIES IS INCREASING

Although nearly two million species of animals and plants have been described, our ignorance of most species living on Earth remains considerable: It has been estimated that only about ten percent of all species have been named and documented. It is unlikely that many new species of mammals or birds are still to be found, but biologists have been surprised by the recent flood of accounts of newly discovered amphibians. In 2002, 5,339 species of amphibian were known to science; by the end of 2024, the figure had risen to 8,808 species. By early 2025 the AMNH Amphibian Species of the World database recognized 8,839 amphibian species. A number of recently described frog species are included in this book.

BELOW **One of the world's largest frogs**, the Titicaca Water Frog (*Telmatobius culeus*) is found only in Lake Titicaca, on the border of Bolivia and Peru. Exploited as a source of food, it is now Critically Endangered.

ABOVE **The Lesser Tree Frog** (*Dendropsophus minutus*) has a huge range, covering much of South America. Analysis of variation in its DNA across its range suggests that it may actually be several species.

Traditionally, new species have been found by documenting the animals and plants of previously unexplored parts of the world. Some of the new frog species are the result of such exploration, particularly of little studied parts of the world such as Papua New Guinea and inaccessible mountainous regions such as the Andes. Most amphibians are small, are very secretive in their habits, and are active for only short periods of the year. They are thus easily overlooked, particularly by expeditions that visit a country only for a short period. Not all new species have been found in exotic places, however. The Atlantic Coast Leopard Frog (*Lithobates kauffeldi*; page 559), which lives within the New York metropolitan area, was newly described in 2014.

DNA ANALYSIS

More commonly (as in the case of the Atlantic Coast Leopard Frog, for example), discoveries of new species are examples of "cryptic species" that have been revealed by modern DNA analyses, these uncovering patterns of variation at the genetic level that are not apparent from visual inspection of specimens. Such variation usually has a strong geographic component, with frogs from different localities showing marked genetic differences. As a result, biologists have split some widespread species into several new ones. Recently, for example, the African Clawed Frog (*Xenopus laevis*; page 54) has been shown to consist of four distinct species, differentiated by geography and genetics, but not by their appearance. The Lesser Tree Frog (*Dendropsophus minutus*) is distributed widely across South America. A recent study of DNA variation in this species has revealed no fewer than 43 distinct genetic lines, many of which could be separate species.

THE FROGS

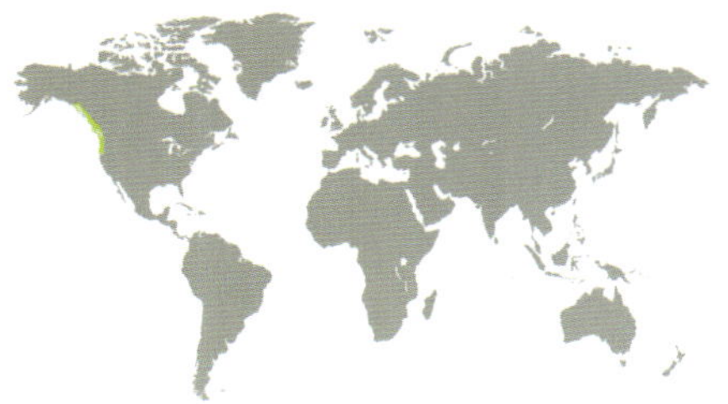

FAMILY	Ascaphidae
OTHER NAMES	Coastal Tailed Frog, Western Tailed Frog
DISTRIBUTION	Northwestern USA, Canada
ADULT HABITAT	Humid forest up to 6,500 ft (2,000 m)
LARVAL HABITAT	Mountain streams
CONSERVATION STATUS	IUCN Least Concern. Sensitive to logging and road construction

ADULT LENGTH
1–2 in
(25–51 mm);
males are usually
smaller than females

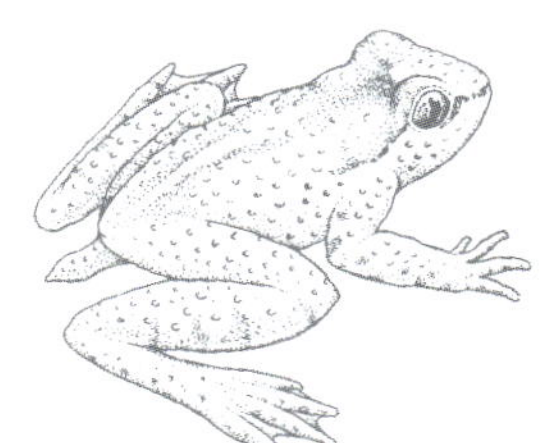

ASCAPHUS TRUEI

PACIFIC TAILED FROG

(STEJNEGER, 1899)

Actual size

The Pacific Tailed Frog gets its name from the male's "tail," which the female, shown here, lacks. In the breeding season, the male has greatly enlarged forearms. The eyes are large and have gold irises. The back is cream, gray, red, or black, with a variable pattern of dark streaks and blotches.

This very unusual frog has internal fertilization, the eggs starting their development inside the female's body. During mating, which takes place in water, the male uses a tail-like extension of his cloaca to insert sperm into the female. The female then lays 28–96 eggs in strings under rocks in fast-flowing mountain streams. The tadpoles are adapted to life in running water, having sucker-like mouths that enable them to attach themselves to rocks. Though essentially terrestrial, adults have webbed hind feet, enabling them to swim strongly when they do enter water. Unlike most frogs, male tailed frogs do not call.

SIMILAR SPECIES

The Rocky Mountain Tailed Frog (*Ascaphus montanus*) was recently separated from *A. truei* on the basis of genetic differences. It is found in several inland mountain ranges, particularly in Montana; and also in Washington, Idaho, and Oregon. It reportedly avoids warm water and is thought to breed only in alternate years.

FAMILY	Leiopelmatidae
OTHER NAMES	None
DISTRIBUTION	Two restricted areas in North Island, New Zealand
ADULT HABITAT	Moist forest, at altitudes of 650–3,300 ft (200–1,000 m)
LARVAL HABITAT	Within eggs, in shallow nests in the ground
CONSERVATION STATUS	IUCN Critically Endangered. Was relatively common in 1996

LEIOPELMA ARCHEYI

ARCHEY'S FROG

(TURBOTT, 1942)

ADULT LENGTH
Male
up to 1¼ in (31 mm)

Female
up to 1$^{7}/_{16}$ in (37 mm)

This primitive frog, which lacks a protrusible tongue, catches its prey by lunging at it open-mouthed. Mating occurs in moist, shallow depressions under logs, where the large eggs are laid in strings. Males defend the eggs and discharge antimicrobial secretions onto them. The tadpoles develop within the eggs until metamorphosis, when the froglets, still with tails, climb onto their father's back. He carries them around for several weeks until metamorphosis is complete. When attacked, adults assume a stiff-legged defensive posture. An intensive conservation program seeks to protect Archey's Frog by breeding it in captivity and releasing young animals into the wild.

Actual size

Archey's Frog is variable in color, being green or brown with dark patches. It has smooth skin, a broad head, and round pupils. There are numerous defensive granular glands in the skin, notably in six longitudinal rows on the back. Fossil evidence suggests that Archey's Frog and its relatives have changed little for 200 million years, making them "living fossils."

SIMILAR SPECIES

There are four species in the genus *Leiopelma*; all are restricted to New Zealand and have declined dramatically in recent years, largely because of the disease chytridiomycosis. They are nocturnally active and lack the vocal apparatus necessary to produce a mating call. Hochstetter's Frog (*L. hochstetteri*) is the most widespread species and it is classed as Least Concern. It lays its eggs in water. See also Hamilton's Frog (page 42).

FAMILY	Leiopelmatidae
OTHER NAMES	Stephens Island Frog
DISTRIBUTION	Stephens Island (Takapourewa), off South Island, New Zealand
ADULT HABITAT	Deep crevices among boulders in forest
LARVAL HABITAT	Within eggs
CONSERVATION STATUS	IUCN Vulnerable. Population is thought to number fewer than 300

ADULT LENGTH
Male
up to 1 11/16 in (43 mm)
Female
up to 2 in (52 mm)

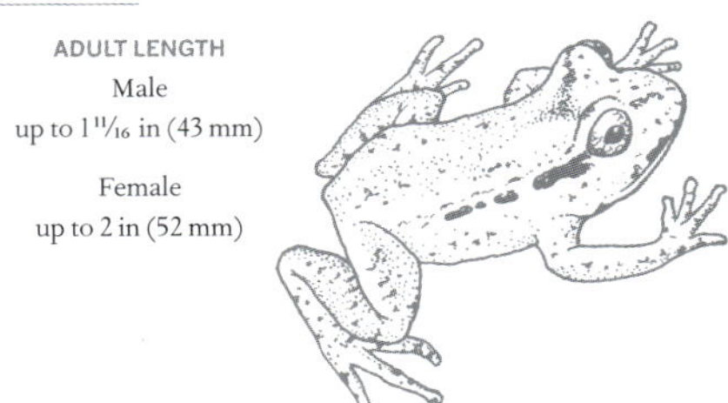

LEIOPELMA HAMILTONI

HAMILTON'S FROG

(MCCULLOCH, 1919)

Actual size

Hamilton's Frog is cryptically colored, relying on camouflage to avoid detection by predators. It is usually brown (occasionally green) with black patterning, and there is a dark stripe through the eye. There are granular glands, producing defensive secretions all over its body as well as forming six rows on its back.

Formerly widespread on both of New Zealand's main islands, Hamilton's Frog has been largely wiped out, initially through predation by rats and other introduced predators, and more recently by the disease chytridiomycosis. It is now confined to the tiny Stephens Island (Takapourewa) in the Cook Strait, where it survives in an area of just 6,500 sq ft (600 sq m). Males lack a vocal apparatus and do not call. During mating, the female produces 7–19 large eggs in a string, which are guarded by the male. The tadpoles develop within the eggs, which hatch to produce tiny froglets that are carried by the male.

SIMILAR SPECIES

The Maud Island Frog (*Leiopelma pakeka*) has only recently been recognized as a separate species. It occurs naturally on Maud Island (Te Hoire) in the Marlborough Sounds, and it too is Vulnerable. It has been successfully bred in captivity, opening the possibility to be introduced to other predator-free islands. In common with the other *Leiopelma* species and the tailed frogs (*Ascaphus*), adults swim by kicking their hind legs alternately rather than synchronously, as most frogs do.

FAMILY	Bombinatoridae
OTHER NAMES	Busuanga Flat-headed Frog, Busuanga Disk-tongued Frog
DISTRIBUTION	Busuanga, Culion, Balabac, and Palawan islands, Philippines
ADULT AND LARVAL HABITAT	Pristine, clear, fast-flowing lowland rainforest streams up to 2,600 ft (800 m) altitude
CONSERVATION STATUS	IUCN Near Threatened. Threats include water pollution, from agricultural runoff, and habitat fragmentation

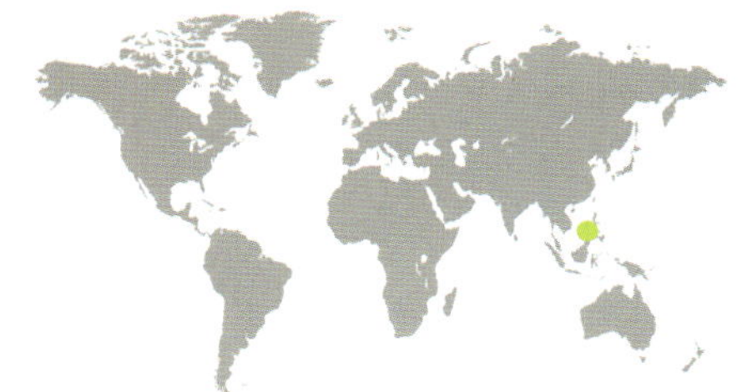

ADULT LENGTH
2¾–3 in (68–76 mm)

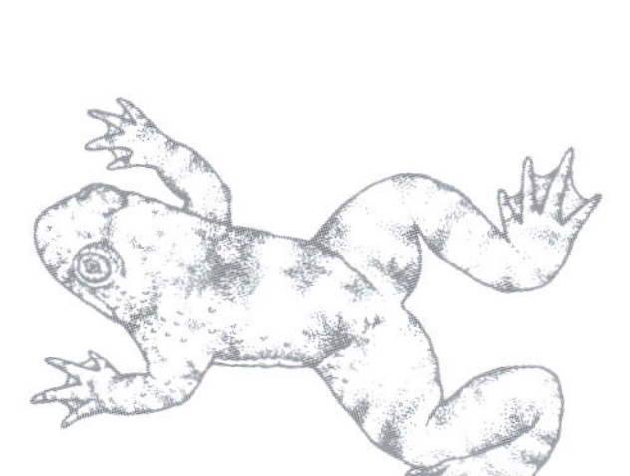

BARBOURULA BUSUANGENSIS

PHILIPPINE FLAT-HEADED FROG

TAYLOR & NOBLE, 1924

The Philippine Flat-headed Frog is a large, highly aquatic species that has a flattened head and body for hugging the rocks in fast-flowing lowland rainforest streams. It has dorsally positioned eyes and nostrils but lacks external tympana, suggesting a fully aquatic existence without the requirement to hear sounds, i.e. frog calls. It floats at the surface but dives quickly and hides under rocks when threatened. Its closest relative, the Bornean Flat-headed Frog (*Barbourula kalimantanensis*), is the world's only known lungless frog, breathing through its skin. It is not known if the Philippine species does the same. Females contain very large eggs, and because no tadpoles had been found, it was thought the Philippine Flat-headed Frog may have direct development, meaning eggs hatch directly into small froglets, but in 2024 the mystery was solved with the discovery of black-spotted, translucent-white tadpoles with large oral suckers to attach themselves to rocks.

The Philippine Flat-headed Frog has a very flattened body, with eyes and nostrils positioned on top of its head and stout limbs with fully webbed fingers and toes. The skin is covered in tiny tubercles that may be sensory. It is drably patterned to blend in with the algae-covered rocks and pebbles of its underwater world.

SIMILAR SPECIES

The two species of flat-headed frogs were thought to be related to the midwife toads (*Alytes*; pages 47–48) and the painted frogs (*Discoglossus*; pages 49–50), but they are now considered more closely related to the fire-bellied and bell toads (*Bombina*; pages 44–46).

Actual size

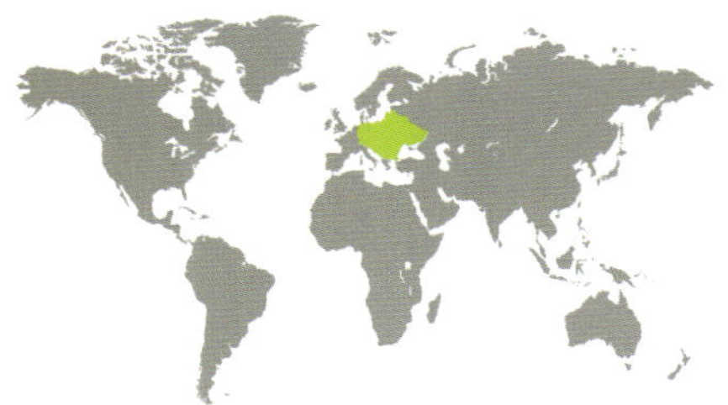

FAMILY	Bombinatoridae
OTHER NAMES	None
DISTRIBUTION	Central and eastern Europe
ADULT HABITAT	In or near wetlands, ponds, and small lakes. In places it is tolerant of polluted water
LARVAL HABITAT	Ponds and small lakes
CONSERVATION STATUS	IUCN Least Concern. Declining in parts of its range due to the draining of wetlands

ADULT LENGTH
up to 2⅜ in
(60 mm)

BOMBINA BOMBINA

EUROPEAN FIRE-BELLIED TOAD

(LINNAEUS, 1761)

The European Fire-bellied Toad has a squat body, warty skin, and upward-pointing eyes. It is dark gray (occasionally green) with black markings above, and red or orange with black marking below. This coloration represents a dual defense system: The dorsal side provides camouflage but, if attacked, the toad exposes warning coloration on its belly in a posture called the unken reflex.

Fire-bellied toads breed in ponds and small lakes, the males calling while floating in open water. The male clasps the female just above her hind legs and has rough black nuptial pads on his forearms that ensure a secure grip. The female lays between 80 and 300 eggs, and the tadpoles take 2–2½ months to reach metamorphosis. Adults mature at 2–4 years old and can live for at least 12 years. Glands in the skin produce a secretion that protects them against fungal infection and deters potential predators; it causes sneezing fits in humans.

SIMILAR SPECIES

The Yellow-bellied Toad (*Bombina variegata*; page 46) is similar in size and appearance to *B. bombina*, except that its belly is yellow rather than red. It occurs across much of western and southern Europe, and there are many places where the two species occur together, though *B. bombina* generally prefers larger ponds than *B. variegata*. There is a narrow zone of hybridization between the two species running from northern Germany to Romania. Genus *Bombina* contains seven species.

Actual size

FAMILY	Bombinatoridae
OTHER NAMES	Oriental Bell Toad, Tuti Toad
DISTRIBUTION	Northeastern China, Korea, southern Japan, eastern Russia
ADULT HABITAT	A wide range of habitats near stagnant or slow-moving water
LARVAL HABITAT	Stagnant or slow-moving water
CONSERVATION STATUS	IUCN Least Concern. Common in some agricultural landscapes

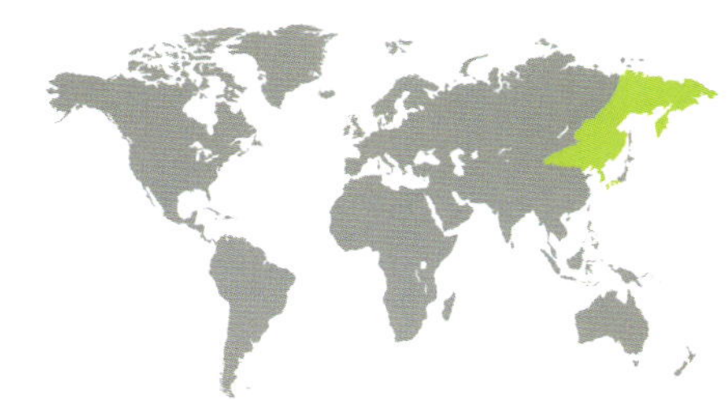

BOMBINA ORIENTALIS

ORIENTAL FIRE-BELLIED TOAD

(BOULENGER, 1890)

ADULT LENGTH
Male up to 2⅛ in (55 mm)
Female up to 2⅝ in (65 mm)

This small toad-like frog breeds readily in captivity. In nature, it has a prolonged breeding season from May to August. Males develop nuptial pads on their first and second fingers; they use these to clasp females firmly around the pelvis. Females lay several batches of 3–45 eggs, laying between about 40–260 over the course of a season. Adults can live for at least 20 years. If attacked, the toad produces a noxious milky secretion from the skin on its hind legs and assumes an unken posture that exposes its vividly colored belly.

SIMILAR SPECIES

Three Asian *Bombina* species are classified as Vulnerable; all are threatened by loss of their habitat as a result of deforestation. The Large-spined Bell Toad (*Bombina fortinuptialis*) has a restricted range, the Lichuan Bell Toad (*B. lichuanensis*) is known from only two restricted sites, and both are Vulnerable. The Hubei Firebelly Toad (*B. microdeladigitora*) is unusual in calling from, and laying its eggs in, tree-holes.

The Oriental Fire-bellied Toad gets its name from its vividly colored ventral surface, which is red to yellow with black spots. The dorsal side is cryptically colored green, gray, or brown with black spots. The upper surfaces of the body are covered in numerous prominent warts. Its eyes, which have triangular pupils, point upward and are all that is visible above water when the toad is floating.

Actual size

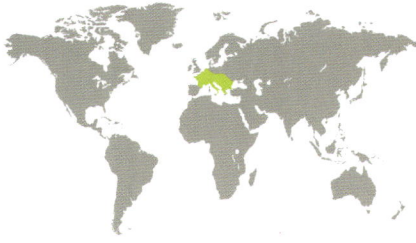

FAMILY	Bombinatoridae
OTHER NAMES	Yellowbelly Toad
DISTRIBUTION	Central, southern, and eastern Europe
ADULT HABITAT	Woodland near wetlands, ponds, and small lakes. In places it is tolerant of polluted water
LARVAL HABITAT	Ponds and small lakes, occasionally streams
CONSERVATION STATUS	IUCN Least Concern. Declining in some parts of its range due to draining of wetlands

ADULT LENGTH
1¾–2⅛ in
(45–55 mm)

BOMBINA VARIEGATA

YELLOW-BELLIED TOAD

(LINNAEUS, 1758)

Actual size

The Yellow-bellied Toad has a slightly flattened body, warty skin, stubby fingers and toes, and upwardly pointing eyes, the latter with triangular pupils. It is dark olive-gray on top, making it well camouflaged against many backgrounds. The underside is bright yellow with black spots; this pattern is revealed when the toad adopts an unken defensive posture.

This small frog is known to live for more than 20 years in nature and has lived for 27 years in captivity. An opportunistic breeder, responding rapidly to heavy rainfall, it can breed three times in a single year. It breeds in all kinds of waterbodies, but prefers larger ponds that do not dry out before the larvae reach metamorphosis. Males form choruses in ponds and lakes, producing a "poop… poop… poop" call. They have black nuptial pads on their hands and forearms that enable them to clasp females firmly. Adults spend the winter in burrows or under rocks and logs.

SIMILAR SPECIES

The Apennine Yellow-bellied Toad (*Bombina pachypus*) has been synonymized with *B. variegata*. In many parts of Europe its range overlaps with the slightly larger European Fire-bellied Toad (*B. bombina*; page 44) and there is a narrow zone of hybridization between the two species running from northern Germany to Romania.

FAMILY	Alytidae
OTHER NAMES	Ferreret
DISTRIBUTION	Remote gorges in the Serra de Tramuntana, Mallorca, Spain
ADULT HABITAT	Rocky crevices
LARVAL HABITAT	Pools in mountain streams
CONSERVATION STATUS	IUCN Endangered. This status is due to its small population, restricted habitat, and sensitivity to disease

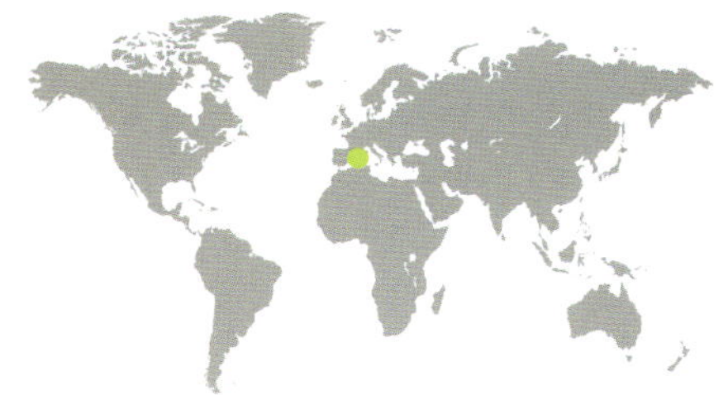

ADULT LENGTH
Male
1–1⅜ in (25–35 mm)
Female
1–1½ in (26–38 mm)

ALYTES MULETENSIS

MALLORCAN MIDWIFE TOAD

(SANCHIZ & ADROVER, 1979)

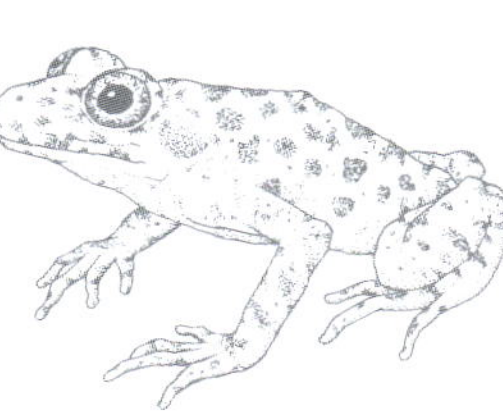

Known as a fossil long before it was discovered alive, this small frog was found as recently as 1977. It once lived throughout Mallorca and is now confined to a few localities in the west of the island. Its natural habitat is protected and a captive breeding program produces young animals for release into the wild. It declined in Roman times when non-native animals were introduced to Mallorca. The Viperine Snake (*Natrix maura*) is a predator and the Iberian Water Frog (*Pelophylax perezi*; page 566) is a competitor, its tadpoles feeding on the same kind of food. Both these species thrive at low altitudes, but have not been able to colonize Mallorca's higher mountains.

Actual size

The Mallorcan Midwife Toad is a small frog with a relatively large head. It has large protruding eyes, and long legs, fingers, and toes. Its skin is yellow, pale brown, or pale green with dark green or black spots, and there are a few warts on its back. The tadpoles grow to be very large and there is little further growth after metamorphosis.

SIMILAR SPECIES

Four species of midwife toads live on the European mainland, including the Common Midwife Toad (*Alytes obstetricans*; see page 48). They are larger than the Mallorcan Midwife Toad and lay more, but smaller, eggs. All six species are susceptible to the disease chytridiomycosis.

FAMILY	Alytidae
OTHER NAMES	Olive Midwife Toad
DISTRIBUTION	Western Europe. Introduced to the UK
ADULT HABITAT	Woodland, gardens, drystone walls, quarries, rock slides, up to 6,500 ft (2,000 m) altitude. Thrives in agricultural and urban habitats
LARVAL HABITAT	Ponds, slow-moving streams and rivers
CONSERVATION STATUS	IUCN Least Concern. Becoming less common across most of its range due to habitat loss, predation by introduced fish, and the disease chytridiomycosis, which has caused mass mortality, e.g., in Spain

ADULT LENGTH
up to 2⅛ in
(55 mm)

ALYTES OBSTETRICANS

COMMON MIDWIFE TOAD

(LAURENTI, 1768)

Actual size

The Midwife Toad has numerous warts on its back. These produce secretions that make the toad unpalatable to potential predators and that also protect the eggs from fungal infection; if a male loses any of his eggs, they soon become infected and die. While carrying eggs, a male's movements are severely restricted; unable to pursue food, he loses weight.

This small frog is famous for the fact that the male carries out prolonged parental care of the eggs. In spring, males call from a burrow, producing a high-pitched "poo, poo, poo." During an elaborate mating procedure, the female produces her eggs in a string; the male fertilizes them and then wraps them around his hind legs. He carries them for 3–6 weeks, seeking damp places and occasionally entering water to moisten them. A female can produce up to four egg clutches each year, and a male can carry up to three clutches, usually from different females, at the same time.

SIMILAR SPECIES

There are five other species of Midwife Toad, all slightly smaller than *Alytes obstetricans*: the Iberian Midwife Toad (*A. cisternasii*) occurs in southwestern Spain and southern Portugal; the Endangered Betic Midwife Toad (*A. dickhilleni*) in southeastern Spain; the Catalonian Midwife Toad (*A. almogavarii*) in the Pyrenees; and the Moroccan Midwife Toad (*A. maurus*), an Endangered species, in Morocco. The Mallorcan Midwife Toad (*A. muletensis*; page 47) is confined to a few high-altitude sites in Mallorca and is classed as Endangered.

FAMILY	Alytidae
OTHER NAMES	None
DISTRIBUTION	Portugal, western Spain
ADULT HABITAT	Ponds, swamps, mountain streams. Also thrives in urban ponds
LARVAL HABITAT	Ponds, swamps, mountain streams
CONSERVATION STATUS	IUCN Least Concern

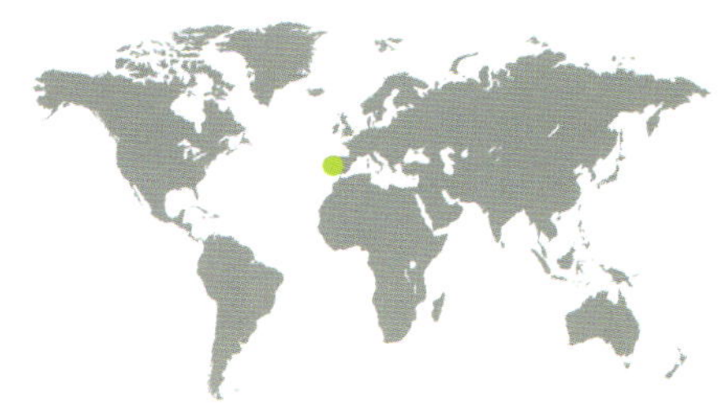

DISCOGLOSSUS GALGANOI

IBERIAN PAINTED FROG

CAPULA, NASCETTI, LANZA, BULLINI & CRESPO, 1985

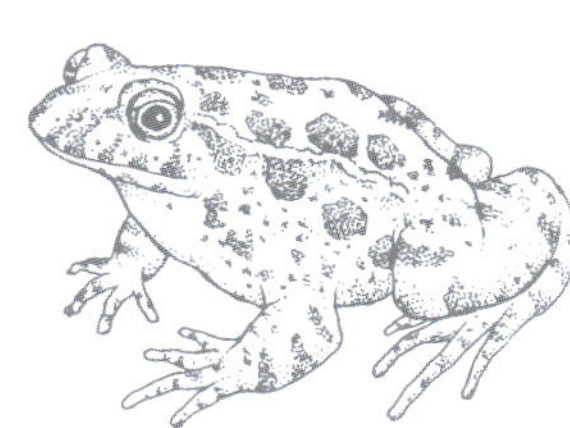

ADULT LENGTH
Male
up to 3⅛ in (80 mm)
Female
up to 2¹⁵⁄₁₆ in (75 mm)

This frog has an unusual reproductive strategy. The breeding season is very long, lasting from October to late summer. Stimulated by higher temperatures and humid conditions, males call at night to attract females. A female can lay up to six clutches, each containing 300–1,500 eggs, over the season. Most unusually, the female lays only 20–50 eggs each time she mates, so that her extremely large number of eggs—around 5,000 over the season—are fertilized by many different males. As a result, her progeny are much more genetically varied than most frogs.

SIMILAR SPECIES

The Corsican Painted Frog (*D. montalentii*) is classified as Near Threatened, its numbers having declined due to predation by alien fish introduced to Corsica. The Tyrrhenian Painted Frog (*D. sardus*) occurs mainly on Sardinia and Corsica; some of its populations have been severely affected by chytridiomycosis. The East Iberian Painted Frog (*D. jeanneae*) has been synonymized with *D. galganoi*.

The Iberian Painted Frog has three skin pattern variants: plain-colored, spotted, and striped. The background color is pale brown or green, and the spots or stripes are darker brown or black. The body is plump, the head is flat, and the hind legs are long and muscular. The eyes are large and prominent, and have pupils shaped like an upside-down droplet.

Actual size

FAMILY	Alytidae
OTHER NAMES	None
DISTRIBUTION	Northern Africa, Sicily, Malta, Gozo. Introduced to parts of southern mainland Europe
ADULT HABITAT	A wide variety of waterbodies, including those manmade
LARVAL HABITAT	Ponds, swamps, mountain streams
CONSERVATION STATUS	IUCN Least Concern. Has declined in areas of intensive agriculture and where wetlands have been drained

ADULT LENGTH
Male
up to 2¹³⁄₁₆ in (70 mm)

Female
up to 2¾ in (68 mm)

DISCOGLOSSUS PICTUS

PAINTED FROG

OTTH, 1837

This frog is unusual in that its geographic range is expanding. Originally native to Algeria, Tunisia, and the Mediterranean islands of Sicily, Malta, and Gozo, it has been introduced by humans to southern France and the Girona province in northeastern Spain. Here its range is expanding at the rate of about 3.8 sq miles (10 sq km) every 6–7 years. It occurs in a wide range of habitats, from coastal areas to mountains, and thrives in many manmade waterbodies such as irrigation ditches. There is evidence, in Spain, that it is having an adverse effect on native species.

SIMILAR SPECIES

This frog is very similar to the Iberian Painted Frog (*Discoglossus galganoi*; page 49) in having a prolonged breeding season during which females produce a large number of eggs fertilized by many different males. A fifth species, the Moroccan Painted Frog (*D. scovazzi*), occurs in North Africa. All the painted frogs are unusual in that males are slightly larger than females.

Actual size

The Painted Frog is, as its names implies, often brightly colored. Skin patterns are very variable: It can be plain, spotted, or striped. Spots and stripes are often dark with pale edges. It has a plump body, a flat head, a pointed snout, and prominent eyes. There is a pronounced fold of skin along each side of the body.

FAMILY	Alytidae
OTHER NAMES	Palestinian Painted Frog
DISTRIBUTION	Hula wetlands, northern Israel
ADULT HABITAT	Wetlands
LARVAL HABITAT	Not known, but probably wetlands
CONSERVATION STATUS	IUCN Critically Endangered. Formerly considered to be Extinct

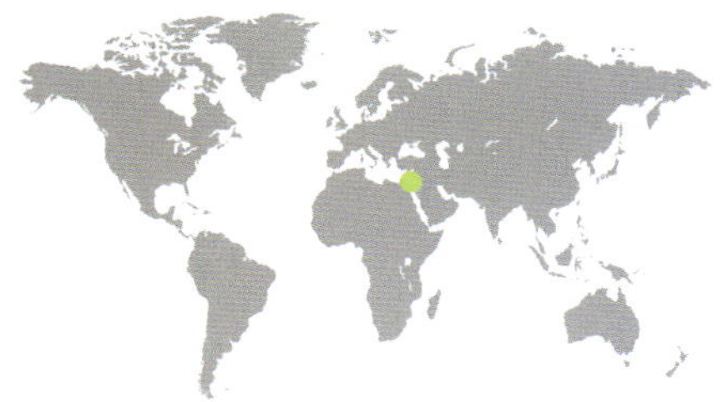

ADULT LENGTH
around 1$^{9}/_{16}$ in
(40 mm);
based on one specimen

LATONIA NIGRIVENTER

HULA PAINTED FROG

(MENDELSSOHN & STEINITZ, 1943)

Described as a "living fossil," this extremely rare frog was believed to have become extinct when the Hula marshes were drained to eliminate malaria and create agricultural land in the 1950s. In 1964, a tiny remnant of the original wetlands, centered on a single pond, was declared a nature reserve. Here, the Hula Painted Frog, last seen in 1955 and declared Extinct in 1996, was rediscovered, as a single specimen, in 2011. Nothing is known of its life history, ecology, or behavior. It is possible that the species may also occur in southern Syria.

SIMILAR SPECIES

Formerly known as *Discoglossus nigriventer*, this frog is very similar to other *Discoglossus* species (see pages 49 and 50). Compared to the Painted Frog (*D. pictus*), it has longer forelimbs and a less prominent snout, and its eyes are farther apart. Fossil remains of frogs that lived 15,000 ago have been classified as belonging to the genus *Latonia*.

Actual size

The Hula Painted Frog gets its name from its black belly, which is covered in white spots. Its back is pale or reddish brown with black or dark olive-gray markings. It has a globular body, a pointed snout, and upward-pointing eyes. Analysis of fossil remains suggests that the genera *Discoglossus* and *Latonia* separated about 32 million years ago.

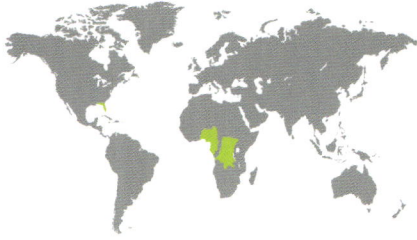

FAMILY	Pipidae: Dactylethrinae
OTHER NAMES	Congo Dwarf Clawed Frog
DISTRIBUTION	Cameroon, Democratic Republic of Congo, Gabon, Nigeria. Introduced to Florida, USA
ADULT AND LARVAL HABITAT	Still water in forest
CONSERVATION STATUS	IUCN Least Concern. Declining in places due to deforestation and habitat degradation

ADULT LENGTH
Male
up to 1$\frac{3}{8}$ in (35 mm)
Female
up to 1$\frac{9}{16}$ in (40 mm)

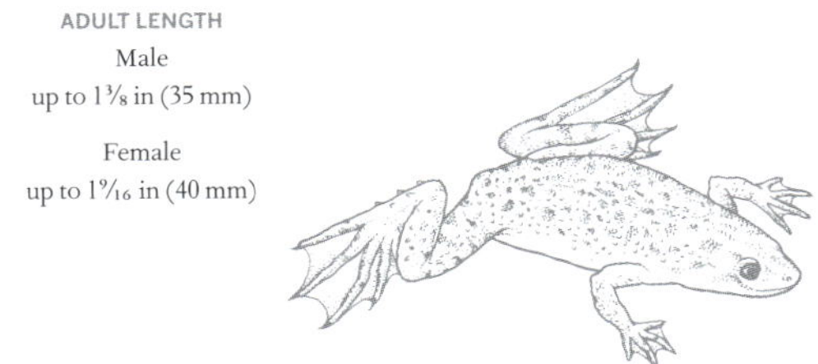

HYMENOCHIRUS BOETTGERI

ZAIRE DWARF CLAWED FROG

(TORNIER, 1897)

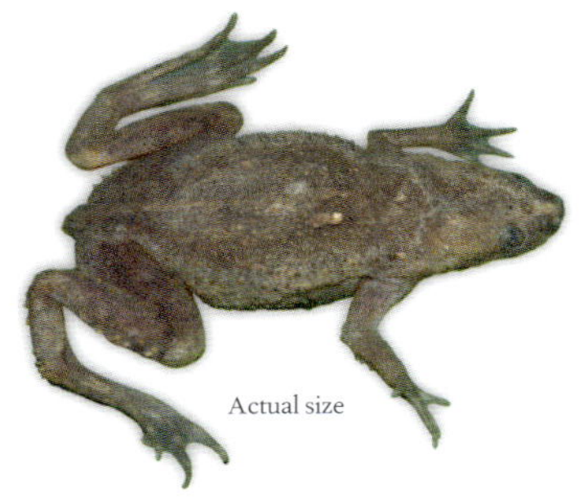

The Zaire Dwarf Clawed Frog is brown or gray in color, with small, dark spots. It has a flattened body, a pointed head, long legs, and claws on its hind feet. There is webbing between the fingers and toes. Its eyes are positioned toward the top of its head and are directed upward.

This fully aquatic frog mates at night, in a process that takes several hours, with only a few eggs being laid and fertilized at a time. Males attract females both by producing clicking sounds and by secreting a sexual attractant from glands in their skin. Egg-laying is carried out in an upside-down position, the eggs being released at the water's surface. Adults lack a tongue and ingest their prey by suction, like many fish. The tadpoles are also suction feeders and are unusual in being predators rather than herbivores.

SIMILAR SPECIES

There are three similar species, all found in central Africa. The Western Dwarf Clawed Frog (*Hymenochirus curtipes*) and Eastern Dwarf Clawed Frog (*H. boulengeri*) are native to the Democratic Republic of the Congo, and the Gabon Dwarf Clawed Frog (*H. feae*) to Gabon. The latter has fully webbed hands and feet. Very little is known about the biology or conservation status of these three species. Merlin's Dwarf Clawed Frog (*Pseudhymenochirus merlini*), from West Africa (Sierra Leone and Guinea-Bissau), is the sister-species to *Hymenochirus*.

FAMILY	Pipidae: Dactylethrinae
OTHER NAMES	Cape Clawed Toad, Gill's Clawed Frog
DISTRIBUTION	Coastal lowlands, Cape Peninsula to Cape Agulhas, Western Cape, South Africa
ADULT AND LARVAL HABITAT	Acidic black-water lakes associated with fynbos vegetation
CONSERVATION STATUS	IUCN Endangered. Threatened by habitat loss and hybridization

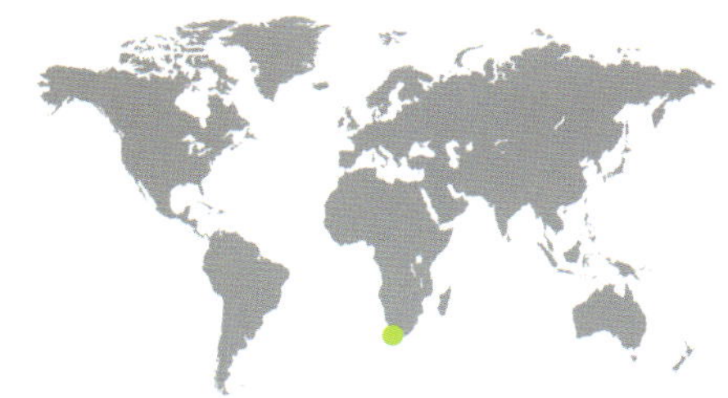

XENOPUS GILLI

CAPE PLATANNA

ROSE & HEWITT, 1927

ADULT LENGTH
Male
$\frac{9}{16}$–$\frac{3}{4}$ in (15–20 mm)
Female
$1\frac{15}{16}$–$2\frac{3}{8}$ in (50–60 mm)

This extremely rare frog was once recorded at 75 localities in the coastal lowlands at the southern tip of Africa, but by 2004 it occurred in only four. Its rapid decline was primarily due to the destruction of its very specific habitat, acidic black-water pools associated with the region's unique fynbos vegetation. As the habitat became less suitable for the Cape Platanna, it became more suitable for the more widespread African Clawed Frog (*Xenopus laevis*; see page 54). The two species tend to engage in hybrid matings, which lead to the production of sterile males. The Cape Platanna is now protected within Table Mountain National Park.

Actual size

The Cape Platanna has long, powerful hind limbs, with four fully webbed toes. Its hands are not webbed and its fingers have sharp claws that it uses to tear up its food. It is yellowish brown in color, with two or four bands of dark brown spots along the back; in some specimens these spots are fused to form stripes.

SIMILAR SPECIES

The platannas or clawed frogs, of which there are 29 species across sub-Saharan Africa, are wholly aquatic. They leave the water only after heavy rain to make short migrations to other ponds. *Xenopus gilli* has a more pointed head and more forward-pointing eyes than other members of the genus, and one less toe on its hind limbs than other platannas.

FAMILY	Pipidae: Dactylethrinae
OTHER NAMES	Common Platanna
DISTRIBUTION	Much of sub-Saharan Africa. Introduced to the UK, USA, and Chile
ADULT AND LARVAL HABITAT	Any still and slow-moving water in savanna
CONSERVATION STATUS	IUCN Least Concern

ADULT LENGTH
Male
$1\frac{13}{16}$–$3\frac{13}{16}$ in (46–98 mm)
Female
$2\frac{3}{16}$–$5\frac{7}{8}$ in (57–147 mm)

XENOPUS LAEVIS

AFRICAN CLAWED FROG

(DAUDIN, 1802)

The African Clawed Frog has a flattened body and head. It has long, muscular hind limbs with long, webbed toes; the three inner toes have black claws. The forelimbs are small and have only four long fingers. The eyes are small, with circular pupils, and point upward. The skin is dark gray to greenish brown on the back, and paler on the underside.

Probably the most intensively studied frog in the world, this wholly aquatic species thrives in captivity. Through the 1940s and 1950s, it was widely used for human pregnancy testing. It is also a model organism for studying how animals develop from egg to adult and was the first vertebrate to be cloned. It has been released in many parts of the world, and thriving populations exist in several countries, including the UK, USA, and Chile. In parts of Africa it is a source of food.

SIMILAR SPECIES

Currently, 29 species are recognized in the genus *Xenopus*, all with a streamlined body shape adapted for swimming. Recent research suggests, however, that *X. laevis* may consist of four genetically distinct, but similar-looking species. "Stitch marks" in the skin contain lateral-line organs, similar to those of fish, which detect vibrations in water. Males lack vocal sacs but call underwater, making a clicking sound. In the breeding season males develop nuptial pads on their hands, helping them grip the female during mating.

Actual size

FAMILY	Pipidae: Dactylethrinae
OTHER NAMES	None
DISTRIBUTION	Lake Oku, northern Cameroon
ADULT AND LARVAL HABITAT	Shallow lake
CONSERVATION STATUS	IUCN Critically Endangered. Restricted to a single lake

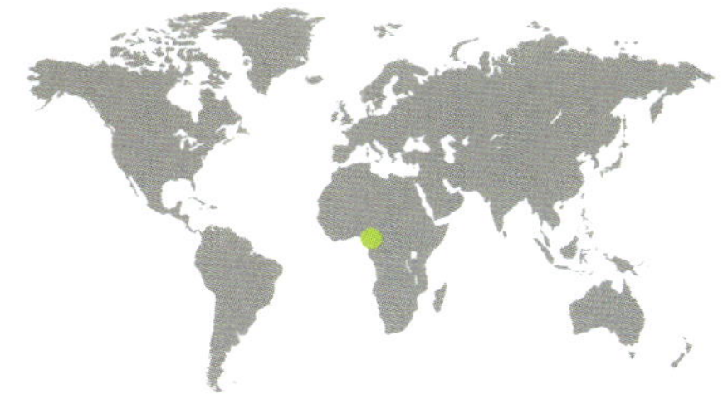

ADULT LENGTH
Male
$1\frac{1}{8}$–$1\frac{1}{4}$ in (28–31 mm)

Female
$1\frac{1}{4}$–$1\frac{7}{16}$ in (32–36 mm)

XENOPUS LONGIPES

LAKE OKU CLAWED FROG

LOUMONT & KOBEL, 1991

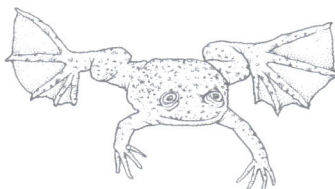

Mystery surrounds a mass mortality event that affected this unusual frog in August 2006, when large numbers of dead and moribund frogs were found with empty stomachs and covered in skin lesions. Tests indicated that the well-known amphibian diseases chytridiomycosis and ranavirus were not responsible, and it is possible that the cause was chemical pollution, either of agricultural origin or resulting from volcanic activity. Lake Oku is a volcanic crater and is so isolated that it supports no native fish. Conservation efforts include preventing the introduction of alien fish that would feed on this unusual frog and its tadpoles.

SIMILAR SPECIES

Smaller than any of the other 29 *Xenopus* species distributed across sub-Saharan Africa, *X. longipes* is also a less powerful swimmer. This is possibly a reflection of the lack of natural predators in Lake Oku. Other *Xenopus* species are preyed on by fish and birds, such as herons.

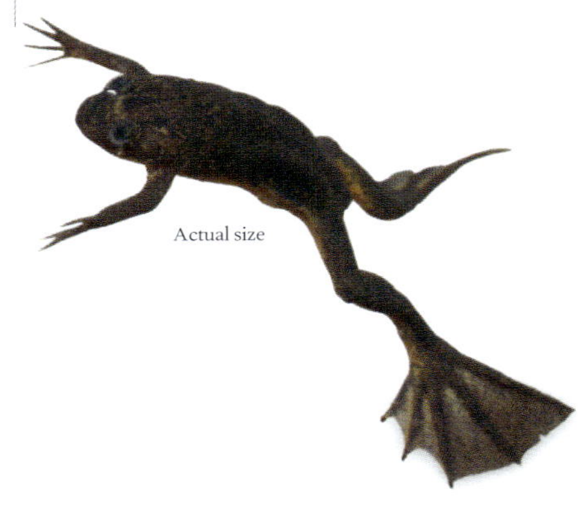

The Lake Oku Clawed Frog has long, slender limbs, long toes, and less webbing between its toes than other *Xenopus* species. Its pear-shaped body is brown above and orange below, and is speckled with tiny black dots. These dots are tiny pointed spicules in the skin, and are more numerous in males. Its large eyes point upward, enabling the frog to see above the water surface while floating.

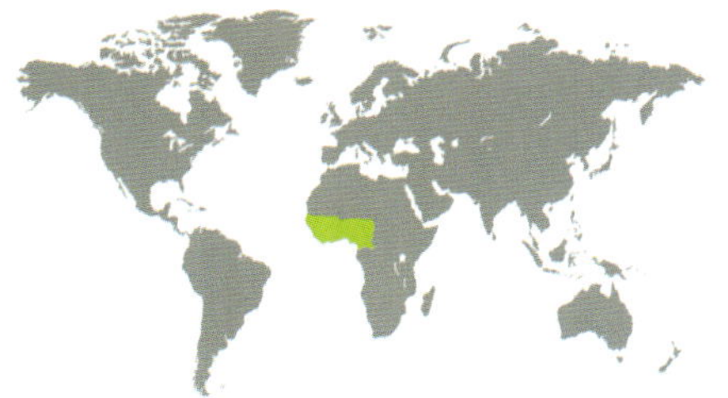

FAMILY	Pipidae: Dactylethrinae
OTHER NAMES	Western Clawed Frog
DISTRIBUTION	Western Africa, from Senegal to Cameroon
ADULT HABITAT	Pools and slow-moving streams in tropical forest
LARVAL HABITAT	Large forest pools
CONSERVATION STATUS	IUCN Least Concern. Sensitive to deforestation

ADULT LENGTH
Male
$1\frac{1}{4}$–$1\frac{9}{16}$ in (32–39 mm)
Female
$1\frac{7}{8}$–$2\frac{1}{8}$ in (48–55 mm)

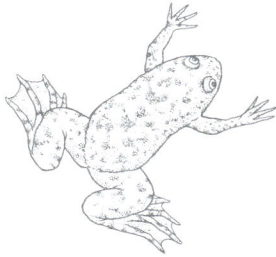

XENOPUS TROPICALIS

TROPICAL CLAWED FROG

(GRAY, 1864)

Although this small frog is widely used in research laboratories around the world, rather little is known about its natural history. While essentially aquatic, it can apparently move considerable distances over land after heavy rain. Its mating call, produced underwater, is a low-frequency trill. According to some reports, the eggs float at the surface; others suggest that they are attached individually to water plants. The tadpoles have been observed swimming around ponds in very tight shoals. This may be a defense against predators or a means of stirring the mud to flush out their prey.

SIMILAR SPECIES

Xenopus tropicalis is increasingly used as a laboratory animal in preference to the African Clawed Frog (*X. laevis*; page 54), because it is smaller, has a shorter generation time (less than five months), and lays more eggs. Its closest relative is the Cameroon Clawed Frog (*X. epitropicalis*), which occurs in central Africa, including Cameroon, Gabon, Congo, and the Democratic Republic of the Congo.

The Tropical Clawed Frog is pale to dark brown on its dorsal side, with numerous small dark spots; ventrally, it is white or yellow, mottled with black. It has very small eyes, beneath each of which is a small tentacle. On its hind feet, it has black claws on toes 3–5 (breeding males only), and a fourth claw on the metatarsal joint.

Actual size

FAMILY	Pipidae: Pipinae
OTHER NAMES	None
DISTRIBUTION	Eastern and northeastern Brazil
ADULT HABITAT	Wetlands, marshes, ponds, lakes
LARVAL HABITAT	Skin pockets on their mother's back
CONSERVATION STATUS	IUCN Least Concern. Sensitive to habitat loss and pollution where land is cleared for agriculture

PIPA CARVALHOI

CARVALHO'S SURINAM TOAD

(MIRANDA-RIBEIRO, 1937)

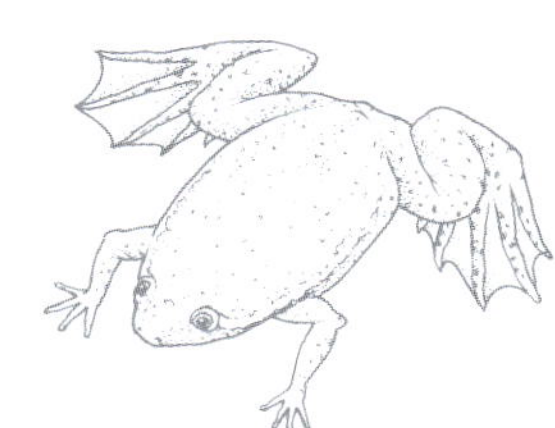

ADULT LENGTH
Male
1¼–2³⁄₁₆ in (32–57 mm)
Female
1⅝–2¾ in (41–68 mm)

This fully aquatic frog can cross land after heavy rain. Males defend positions under water against other males, calling to attract females. They have an elaborate mating procedure like that of the Surinam Toad (*Pipa pipa*; see page 58), during which fertilized eggs are pressed into the skin on the female's back. There they hatch and develop, emerging as tadpoles, not as froglets as happens in the Surinam Toad. Because the reproductive process is therefore much shorter, they can reproduce several times in a year. The tadpoles develop to a large size; they feed by sucking muddy water into their mouths and filtering out tiny animals.

Carvalho's Surinam Toad has a wide triangular head and large protruding eyes. The body is covered in conical tubercles, especially on the lower part of the back. The skin color is variable, ranging from either pale or dark brown to gray. The lateral-line organs along its flanks are white.

SIMILAR SPECIES

This species differs from the other members of the genus in having pronounced fang-like teeth. The body is less compressed than in other species. It is most similar to the Sabana Surinam Toad (*Pipa parva*) and Myer's Surinam Toad (*P. myersi*), which also produces tadpoles rather than froglets. Myer's Surinam toad, from Colombia and Panama, is endangered by habitat destruction. The skin of all *Pipa* species is equipped with lateral-line organs, which detect tiny movements in the water, such as those made by moving prey.

Actual size

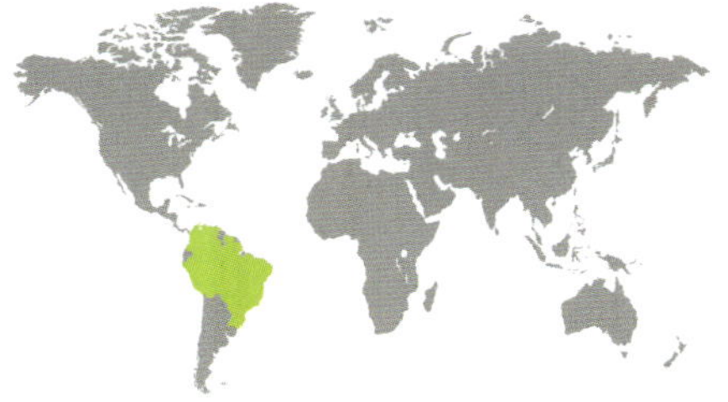

FAMILY	Pipidae: Pipinae
OTHER NAMES	Star-fingered Toad
DISTRIBUTION	Northern South America (Peru, Colombia, Venezuela, Suriname, Guinea, Brazil, Bolivia)
ADULT HABITAT	Sluggish rivers and canals with muddy bottoms
LARVAL HABITAT	Skin pockets on their mother's back
CONSERVATION STATUS	IUCN Least Concern. Sensitive to habitat loss

ADULT LENGTH
Male
up to 6 $^{1}/_{16}$ in (154 mm)
Female
up to 6¾ in (171 mm)

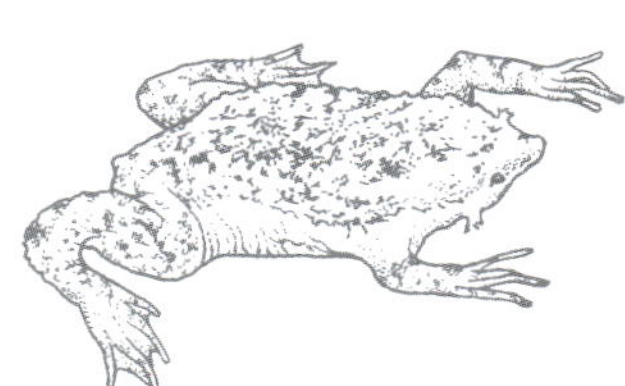

PIPA PIPA

SURINAM TOAD

(LINNAEUS, 1758)

Actual size

This strange-looking frog has a remarkable mode of reproduction. The male clasps the female from above and the pair perform a repeated series of movements in which they turn upside down. During this, the eggs are extruded, fertilized, and trapped in the space between the male's belly and the female's back. From there, they are pressed into the skin of the female's back, where they are absorbed. In pockets in their mother's skin, the eggs hatch into tadpoles that then metamorphose into tiny frogs. After 3–4 months, the froglets, now ¾ in (2 cm) long, wriggle their way out into the water.

SIMILAR SPECIES

All seven *Pipa* species are totally aquatic. They find food in the mud by feeling with their fingers, which have star-shaped tactile organs on their tips. When they find food, they lunge at it and, lacking a tongue, use suction to draw it into their mouth. Arrabal's Surinam Toad (*P. arrabali*), the Albina Surinam Toad (*P. aspera*), and Utinga Surinam Toad (*P. snethlageae*) also produce froglets.

The Surinam Toad looks rather like a dead leaf, with its brown coloration, flattened body, and overall shape. It has a large triangular head, small eyes, and nostrils at the ends of narrow tubes. There are many spine-like tubercles on the body. The front limbs are short and weak; the hind limbs are long and very muscular, and have webbed feet.

FAMILY	Rhinophrynidae
OTHER NAMES	Cone-nosed Frog
DISTRIBUTION	Central America, Mexico, extreme southeastern Texas, USA
ADULT HABITAT	Savanna and seasonally dry forest in lowland coastal areas
LARVAL HABITAT	Ephemeral waterbodies
CONSERVATION STATUS	IUCN Least Concern. Occurs in many protected areas; protected by law in Mexico

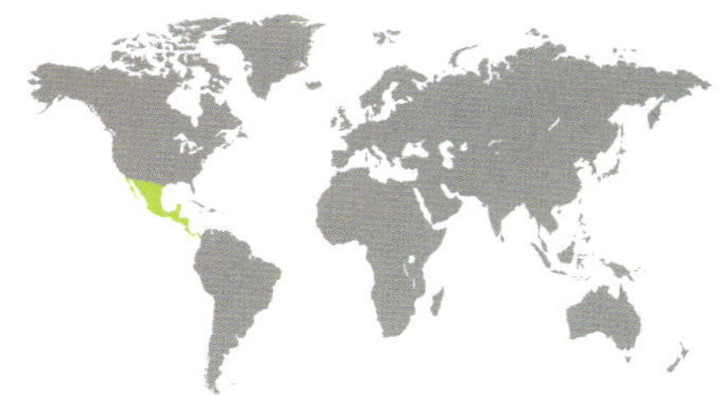

RHINOPHRYNUS DORSALIS

MEXICAN BURROWING TOAD

(DUMÉRIL & BIBRON, 1841)

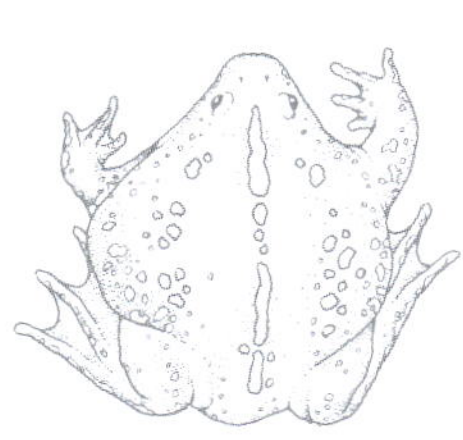

ADULT LENGTH
Male
up to 2 15/16 in (75 mm)
Female
up to 3½ in (89 mm)

This unmistakable frog lives largely underground, digging itself backward into the soil with hard tubercles on its hind feet. It emerges only after rain to feed on ants, termites, and other insects. Breeding occurs at the start of the rainy season, with males calling loudly in choruses than can be heard 2–2.5 miles (3–4 km) away. Mating occurs in ephemeral pools, and the female produces several thousand eggs. The tadpoles often gather together in compact shoals. The frog's pointed snout is used to probe into ant and termite nests, and its tongue, attached at the back of the mouth, can be flicked out to catch prey.

SIMILAR SPECIES

As the sole member of its family, this species has no close relatives. Similar frogs are known from fossils from the Oligocene (23–34 million years ago) in Saskatchewan, Canada, and the Paleocene (56–66 million years ago) in Wyoming, USA. *Rhinophrynus* is thought to have separated from all other frogs 190 million years ago. It shares some characteristics with other unrelated burrowing frogs, such as its rotund body shape and the tubercles on its hind feet.

The Mexican Burrowing Toad has a globular, flaccid body covered with loose skin, and is black or brown with yellow or orange markings. Its head and eyes are tiny, and its limbs are short and robust. The hind feet are webbed and bear a hard tubercle. The end of its snout is usually callused as a result of probing into the ground for food.

Actual size

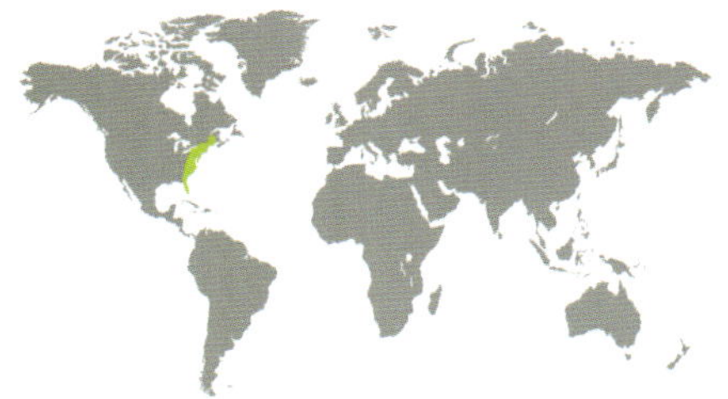

FAMILY	Scaphiopodidae
OTHER NAMES	None
DISTRIBUTION	Eastern USA
ADULT HABITAT	Deciduous and coastal pine forests with sandy and light soils
LARVAL HABITAT	Ephemeral pools
CONSERVATION STATUS	IUCN Least Concern. Sensitive to urbanization

ADULT LENGTH
1¾–2⅞ in
(44–72 mm)

SCAPHIOPUS HOLBROOKII

EASTERN SPADEFOOT

(HARLAN, 1835)

Emerging from their burrows after heavy rain, Eastern Spadefoots can travel more than 3,000 ft (910 m) to the ephemeral ponds where they breed. Males have a large vocal sac and make a loud nasal, croaking call. Females produce around 4,000 eggs, which are attached to submerged vegetation. These soon hatch into tadpoles that develop quickly, ensuring that they reach metamorphosis before their pond dries up. The keratinized "spade" on each hind foot enables the frogs to burrow backward into the ground. When attacked, they produce a noxious secretion from their skin; this deters predators and causes sneezing and weeping in people.

The Eastern Spadefoot gets its name from the elongated, sickle-shaped black tubercles on its hind feet, which it uses for excavating soil. It has large protruding eyes with vertical, slit-like pupils. The skin is pale gray or yellow with dark markings. In many individuals two light bands make an hourglass pattern on the back.

SIMILAR SPECIES

Couch's Spadefoot (*Scaphiopus couchii*) is greenish in color and is found across western USA and Mexico, occurring even in areas with very low annual rainfall, including deserts. The very rapid development of its eggs and tadpoles enables it to breed in ephemeral pools after occasional rain. Hurter's Spadefoot (*S. hurterii*) is yellow or brown and is found in central-southern USA.

Actual size

FAMILY	Scaphiopodidae
OTHER NAMES	None
DISTRIBUTION	Central USA, Canada, Mexico
ADULT HABITAT	Prairie grassland, arid plains, deserts
LARVAL HABITAT	Ephemeral pools
CONSERVATION STATUS	IUCN Least Concern

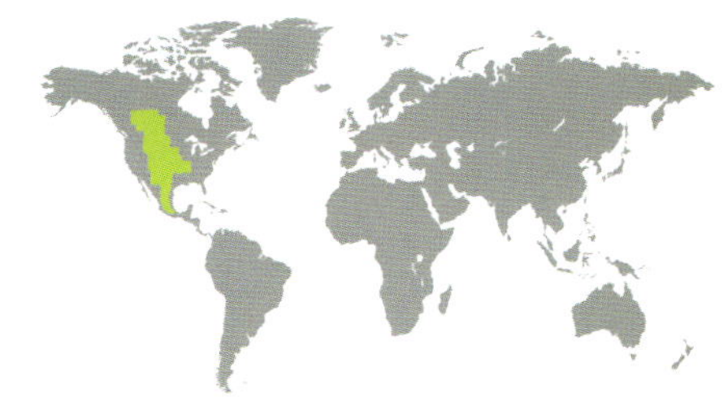

ADULT LENGTH
Male
up to 1½ in (38 mm)

Female
up to 1 9/16 in (40 mm)

SPEA BOMBIFRONS

PLAINS SPADEFOOT

(COPE, 1863)

This small frog lives mostly underground, emerging only at night to feed or to breed. It can dig its own burrow but often lives in the burrows of other animals. After heavy rain, Plains Spadefoots make their way to ephemeral pools, sometimes as far as half a mile (1 km) away. Males call together in a chorus, producing a snore-like bleat, and have rough nuptial pads on their hands with which they clasp females. Each female lays around 2,000 eggs in clumps, each containing 10–250 eggs. If the water is warm, the eggs hatch within 24 hours.

SIMILAR SPECIES

The Great Basin Spadefoot (*Spea intermontana*) is found across several of the western states of the USA. Its tadpoles occur in two forms depending on local conditions. One is herbivorous, while the other is carnivorous and has a horny beak for tearing up its prey. The same is true of the New Mexico Spadefoot (*S. multiplicata*), which is found in the southwestern states and in Mexico.

Actual size

The Plains Spadefoot has a rotund toad-like body. The skin on its back is gray to brown or sometimes greenish in color, and with darker spots and blotches. The underside is white. There is a bony lump, called a boss, between the prominent eyes. The tubercles on the hind feet, used for digging into the ground, are black.

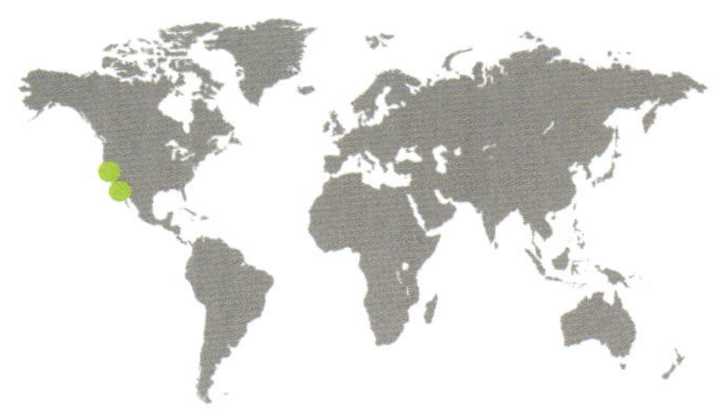

FAMILY	Scaphiopodidae
OTHER NAMES	None
DISTRIBUTION	Central Valley and coastal mountains of California; Baja California, Mexico
ADULT HABITAT	A wide variety of habitats with sandy soil
LARVAL HABITAT	Ephemeral ponds and streams
CONSERVATION STATUS	IUCN Near Threatened. Has declined by about 80 percent since the 1950s

ADULT LENGTH
1 7/16–2 5/8 in
(37–65 mm)

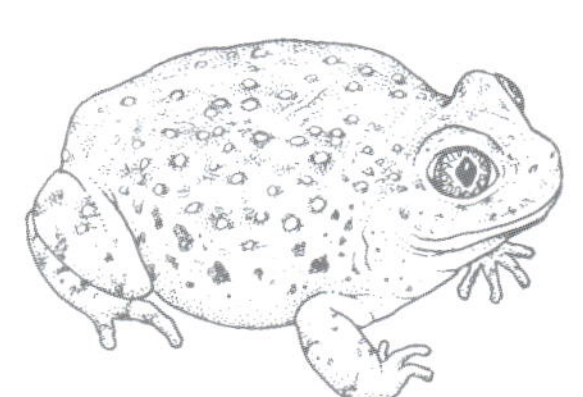

SPEA HAMMONDII

AMERICAN WESTERN SPADEFOOT

(BAIRD, 1859)

This burrowing frog has declined dramatically over the last 60 years, primarily as a result of loss of suitable habitat. Urbanization and the spread of intensive agriculture, typically involving irrigation, have destroyed the temporary ponds and streams that it needs for reproduction. It has also been adversely affected by introduced Bullfrogs (*Aquarana catesbeiana*; page 545) and mosquitofish (*Gambusia* spp.); Bullfrog larvae compete with the spadefoot tadpoles and mosquitofish eat them. The tadpoles are unusual in that they are often seen hanging vertically from the water's surface, breathing or feeding. In some localities the tadpoles feed on other tadpoles, of the same and other species.

The American Western Spadefoot has a plump body, short legs, a rounded snout, and large protruding eyes with vertical pupils. Its skin is loose and there are many wart-like tubercles on its back. The toes of the hind feet are fully webbed. The skin is light green to gray with scattered darker splotches; the tubercles are orange or red.

SIMILAR SPECIES

Formerly, *Spea hammondii* and the New Mexico Spadefoot (*S. multiplicata*) were considered to be a single wide-ranging species, but they have been separated on the basis of differences in several characters. For example, *S. hammondii* has more elongated spades on its feet and has pale gold rather than copper irises.

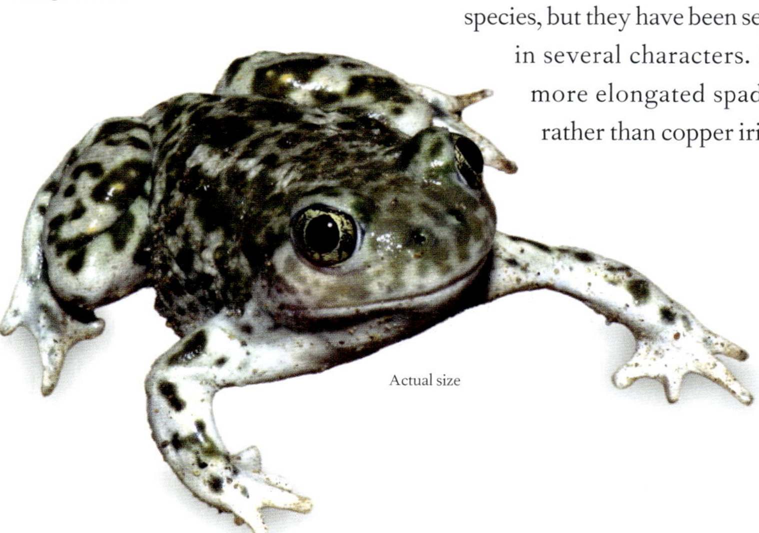

Actual size

FAMILY	Pelodytidae
OTHER NAMES	None
DISTRIBUTION	Belgium, France, Italy, Luxembourg, Portugal, Spain
ADULT HABITAT	Stony and sandy areas in forests, open country, agricultural areas
LARVAL HABITAT	Ponds
CONSERVATION STATUS	IUCN Least Concern. Declining over much of its range due to anthropogenic habitat change

PELODYTES PUNCTATUS

PARSLEY FROG

(DAUDIN, 1802)

ADULT LENGTH
Male
up to 1⅜ in (35 mm)

Female
up to 1¾ in (45 mm)

During breeding, which takes place at night, male Parsley Frogs call from underwater and females respond with their own, quieter call. The male grasps the female in amplexus around her pelvis for as long as five hours before a clutch of 40–200 eggs is laid and attached to water plants. Individuals can breed as often as three times in a year. The tadpoles can grow to a large size, achieving a length of 2⅝ in (65 mm). This species is negatively affected by human activities such as the draining of wetlands and the canalization of rivers. It is also threatened, especially in Spain and Portugal, by introduced Louisiana Crayfish (*Procambarus clarkii*).

Actual size

The Parsley Frog gets its name from the green patches on its skin, which is otherwise pale gray or yellowish. It is very agile and has long hind legs, a flat head, and large bulging eyes with vertical pupils. It can climb by pressing its belly against a vertical surface, using it as a kind of sucker. The eggs smell of fish and adults often smell of garlic.

SIMILAR SPECIES

There are four other parsley frog species. The Iberian Parsley Frog (*Pelodytes ibericus*) occurs in southern Spain and Portugal while the Lusitanian Parsley Frog (*P. atlanticus*) is endemic to Portugal. The Caucasian Parsley Frog (*P. caucasicus*) is found in Azerbaijan, Georgia, Russia, and Turkey, and is classified as Near Threatened. It is subject to the destruction and chemical pollution of its habitat and to predation by introduced North American Racoons (*Procyon lotor*).

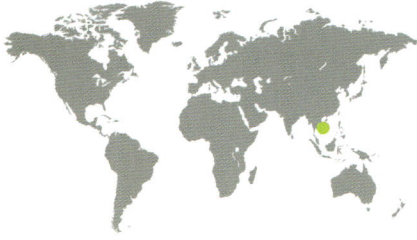

FAMILY	Megophryidae: Leptobrachiinae
OTHER NAMES	None
DISTRIBUTION	Song Thanh Nature Reserve, Vietnam
ADULT HABITAT	Forest in mountains at altitudes of 4,250–5,000 ft (1,300–1,500 m)
LARVAL HABITAT	Small mountain streams
CONSERVATION STATUS	IUCN Endangered. Known from only one location. Affected by habitat loss through deforestation

ADULT LENGTH
Male
¾–⅞ in (20–21 mm);
based on five specimens
Female
⅞ in (22 mm);
based on one specimen

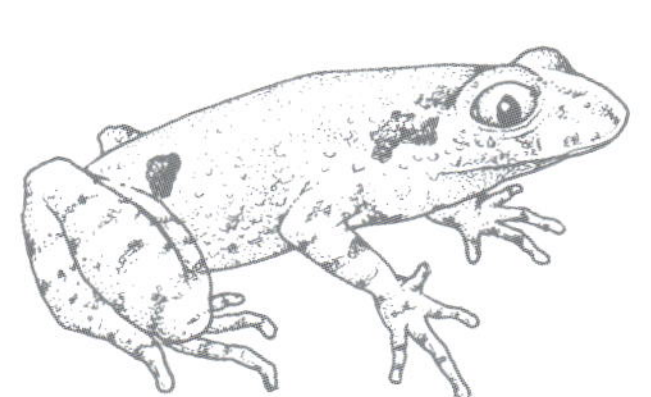

LEPTOBRACHELLA APPLEBYI

APPLEBY'S ASIAN TOAD

(ROWLEY & CAO, 2009)

Actual size

Appleby's Asian Toad has a rounded snout, a clearly visible tympanum, and eyes with gold or copper irises and vertical pupils. The fingers have slightly expanded tips. The skin is gray or brown with black blotches on the flanks and a black stripe through the eye. The belly is decorated with small bluish spots.

This small species, described in 2009 in the genus *Leptolalax*, exemplifies an important aspect of our knowledge of frogs. In the last 20 years, more than 2,000 species, previously unknown to science, have been described. This increase in knowledge is partly the result of exploration of little-known parts of the world, and partly the application of new techniques to differentiate one species from another. Appleby's Asian Toad is very similar in appearance to several close relatives but can be differentiated by its advertisement call. It produces a rasping call containing four to five notes. Only six individuals have been collected, all from just one location in Vietnam.

SIMILAR SPECIES

The genus *Leptolalax*, known as "slender litter frogs," consisted of four species in 1983, 20 in 2009, when *L. applebyi* was described, and 59 in 2018 when it was synonymized with *Leptobrachella*, which now contains 105 species. Seventeen species occur in Borneo alone.

FAMILY	Megophryidae: Leptobrachiinae
OTHER NAMES	Gracile Litter Frog, Matang Asian Toad
DISTRIBUTION	Borneo, Peninsular Malaysia
ADULT HABITAT	Tropical rainforest at altitudes of 500–3,600 ft (150–1,100 m)
LARVAL HABITAT	Streams with clear water and sandy bottoms
CONSERVATION STATUS	IUCN Least Concern, although it has lost much of its habitat through deforestation and siltation of streams

LEPTOBRACHELLA GRACILIS

SARAWAK SLENDER LITTER FROG

(GÜNTHER, 1872)

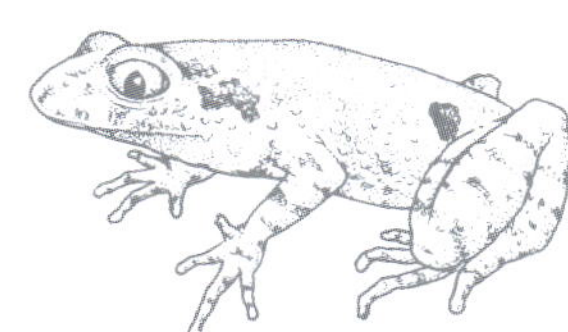

ADULT LENGTH
Male
1¼–1$^{9}/_{16}$ in
(31–40 mm)
Female
1$^{9}/_{16}$–1$^{15}/_{16}$ in
(40–50 mm)

As its common name implies, this small frog lives mostly in leaf litter, although it may also be found perching on low branches in rainforest shrubs. It feeds primarily on small insects. It breeds in fast-flowing, clear streams, males producing a long series of loud notes to attract females. It is thought that the female lays her eggs under stones in streams. The tadpoles have long, slender bodies and long, muscular tails with low fins, enabling them to wriggle their way between stones.

SIMILAR SPECIES

The Sabah Slender Litter Frog (*Leptobrachella fritinniens*), described in 2013, is found in northern Borneo and appears to tolerate alteration of its forest habitat. It is also known as the Twittering Litter Frog because of its call, a long series of short notes. The White-bellied Slender Litter Frog (*L. hamidi*) is found in lowland forest in Borneo. Its habitat is subject to logging and it is listed as Vulnerable to Extinct. See also Appleby's Asian Toad (*L. applebyi*; page 64) and the Sung Toad (*L. sungi*; page 67).

The Sarawak Slender Litter Frog has a slender body and legs, and large eyes. The skin is rough on the back and flanks. Its back is brown or gray with dark markings, the flanks are decorated with pale warts and dark spots, the upper arms are cream, and the upper half of the iris is orange.

Actual size

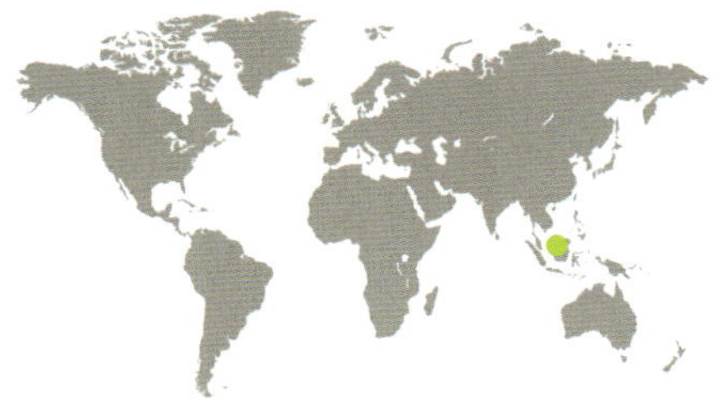

FAMILY	Megophryidae: Leptobrachiinae
OTHER NAMES	None
DISTRIBUTION	Southeastern Sarawak, Malaysia; Pulau Serasan, Indonesia
ADULT HABITAT	Hilly lowland forest up to 1,150 ft (350 m) altitude
LARVAL HABITAT	Small streams
CONSERVATION STATUS	IUCN Near Threatened. Threatened by habitat loss

ADULT LENGTH
Male
11/16–3/4 in (17–19 mm)
Female
up to 3/4 in (20 mm)

LEPTOBRACHELLA SERASANAE

STRIPED DWARF LITTER FROG

DRING, 1984

Actual size

This tiny frog lives in leaf litter on the forest floor, feeding on small insects and other invertebrates. It is usually found close to the streams where it breeds. Males call from the edge of a stream, producing a high-pitched metallic trill. Mating and egg-laying have not been reported. Being very small, it is easily overlooked and it is possible that it is found over a wider range than is currently reported. Nonetheless, it is considered to be Vulnerable because of extensive deforestation in Borneo and Indonesia.

SIMILAR SPECIES

Genus *Leptobrachella* contains 105 species of tiny litter frogs found from China to Borneo. Four species are Critically Endangered including the Liapso Litter Frog, *L. palmata*, from Sabah, which is known from only one location. *Leptobrachella natunae*, reported only from Great Natuna Island, has not been seen since its discovery 100 years ago.

The Striped Dwarf Litter Frog has a slender body, a narrow head, long legs, and pointed fingertips. It is dark brown in color with darker markings. A black stripe running from the eye to the groin distinguishes it from other frogs in the genus.

FAMILY	Megophryidae: Leptobrachiinae
OTHER NAMES	None
DISTRIBUTION	Northern Vietnam, southern China
ADULT HABITAT	Subtropical lowland forest up to 3,600 ft (1,100 m) altitude
LARVAL HABITAT	Streams
CONSERVATION STATUS	IUCN Least Concern, although its habitat is threatened in places by urban development and tourism

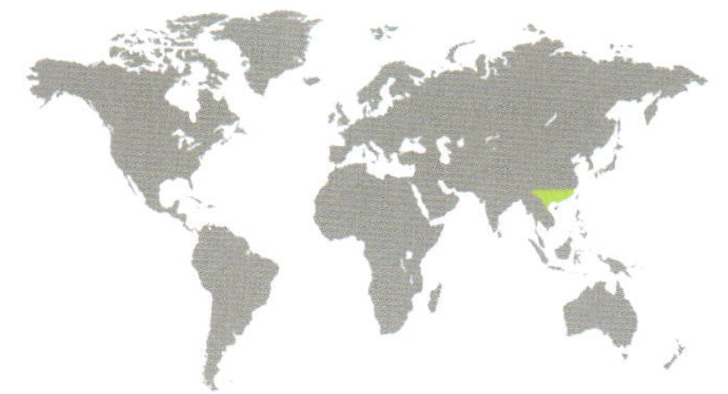

LEPTOBRACHELLA SUNGI

SUNG TOAD

(LATHROP, MURPHY, ORLOV & HO, 1998)

ADULT LENGTH
Male
1⅞–2$^{1}/_{16}$ in (48–53 mm)
Female
2$^{3}/_{16}$–2¼ in (57–59 mm)

This cryptically colored frog is most commonly found close to streams within forests. It is one of the largest of the 40 or so species that make up this scarcely known genus, called Asian toads or slender litter frogs. They live among leaf litter on the forest floor, feeding primarily on insects. It is believed that the Sung Toad lays its eggs under stones in streams. Its tadpoles have slender bodies and long tails, enabling them to wriggle their way between stones.

SIMILAR SPECIES

The Thao Asian Toad (*Leptobrachella pelodytoides*) is a smaller species that is found in Myanmar, Laos, and southern China. The Painted Slender Litter Frog (*L. picta*) occurs in Sabah on Borneo and is listed as Least Concern even though it lost much of its forest habitat through deforestation. *Leptobrachella botsfordi*, described in 2013, occurs at high altitude in Vietnam and is Critically Endangered. See also Appleby's Asian Toad (*L. applebyi*; page 64) and the Sarawak Slender Litter Frog (*L. gracilis*; page 65).

The Sung Toad is a rather plump frog, with large eyes and small warts on its back. The body and legs are brown or gray, with darker spots on the body and stripes on the legs. There is a fold of skin running from the eye, above the rather indistinct tympanum to the corner of the mouth. The irises of the eyes are iridescent green.

Actual size

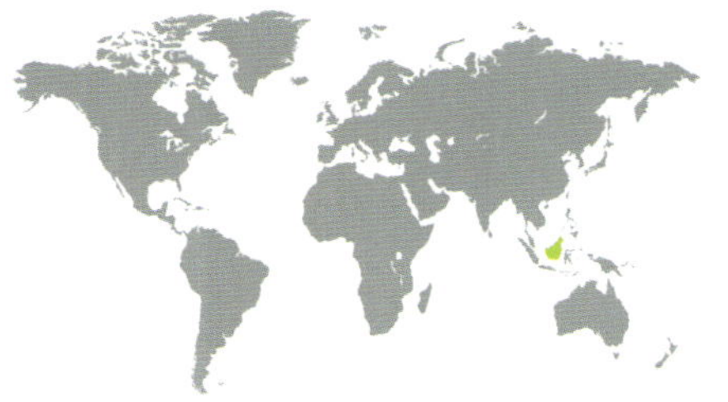

FAMILY	Megophryidae: Leptobrachiinae
OTHER NAMES	None
DISTRIBUTION	Borneo
ADULT HABITAT	Forest below 3,300 ft (1,000 m) altitude
LARVAL HABITAT	Pools
CONSERVATION STATUS	IUCN Least Concern. Has a large range, but this is declining due to habitat loss

ADULT LENGTH
Male
1 11⁄16–2 15⁄16 in (43–75 mm)
Female
2 3⁄8–3 11⁄16 in (60–95 mm)

LEPTOBRACHIUM ABBOTTI

LOWLAND LITTER FROG

(COCHRAN, 1926)

This large frog lives in leaf litter on the forest floor, where it often assumes a very erect sitting posture. It feeds on large insects. Its hind limbs are short and lack the power required for leaping, so it moves in small hops. The male calls from a pool, producing a single resonating squawk. Males appear to call alone and do not form a chorus. Mating and egg-laying have not been described, but the tadpoles are known to grow as long as 2 15⁄16–3 1⁄2 in (75–90 mm).

SIMILAR SPECIES

The Mountain Litter Frog (*Leptobrachium montanum*) is smaller than *L. abbotti* and is its montane counterpart in Borneo, being found at altitudes greater than 3,000 ft (900 m). The Spotted Litter Frog (*L. hendricksoni*) has a wide range, being found in Thailand, Malaysia, Indonesia, and Borneo. The Black-eyed Litter Frog (*L. nigrops*) is found in coastal areas of Malaysia. Genus *Leptobrachium* contains 38 species.

The Lowland Litter Frog has a broad head, a wide mouth, and bulging black eyes. The legs are short and slender, and seem too small for a frog of this size. The skin is smooth and dark brown to black in color. The belly is mottled black and white.

Actual size

FAMILY	Megophryidae: Leptobrachiinae
OTHER NAMES	None
DISTRIBUTION	Northeastern India
ADULT HABITAT	Forest at altitudes of 6,400 m (1,950 m)
LARVAL HABITAT	Unknown
CONSERVATION STATUS	IUCN Least Concern. Seems to have a very restricted range

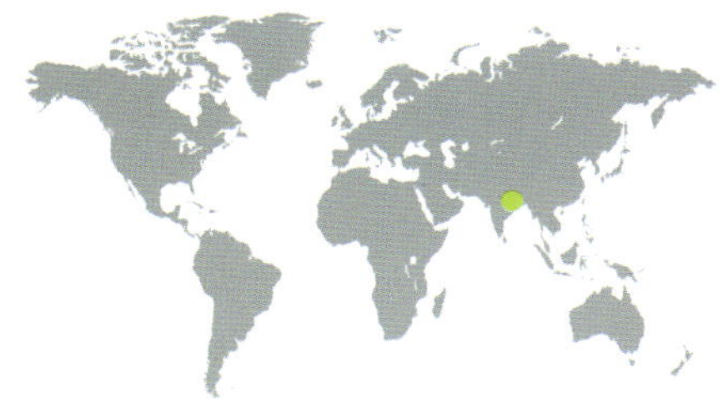

ADULT LENGTH
around 1⅞ in (47 mm); based on one male specimen

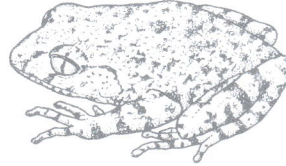

LEPTOBRACHIUM BOMPU

INDIAN BLUE-EYED LITTER FROG

SONDHI & OHLER, 2011

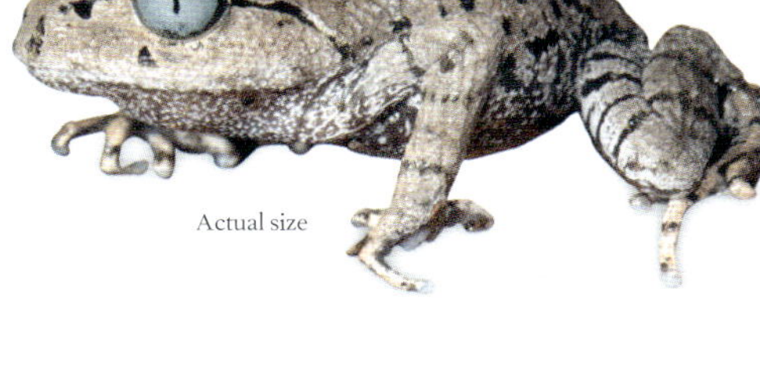

Actual size

Most frogs have brightly colored eyes, but in a few, as in this recently described species, the irises are bright blue. Iris color is not known to have any biological significance, but it can be very useful in enabling biologists to separate one species from another. This species is known only from the Eaglenest Wildlife Sanctuary in the Arunachal Pradesh region of northeastern India. It is not an agile or athletic species but crawls across the ground in a rather laborious manner. After heavy rain, males call along streams, producing a loud "kek… kek… kek" sound.

The Indian Blue-eyed Litter Frog has a wide head with a rounded snout; long, slender limbs; and skin that is covered in numerous tiny wrinkles. The upper surfaces are gray or brown with black blotches on the back and black stripes on the legs. The large eyes have vertical slit-like pupils and pale blue irises.

SIMILAR SPECIES

The only other blue-eyed species in the genus *Leptobrachium* is *L. waysepuntiense* from Sumatra, described in 2010. Smith's Litter Frog (*L. smithi*) is a common species in India, Thailand, Myanmar, and other parts of Asia. The upper halves of its irises are yellow, orange, or red. See also the Lowland Litter Frog (*L. abbotti*; page 68) and the Emei Moustache Frog (*L. boringii*; page 70).

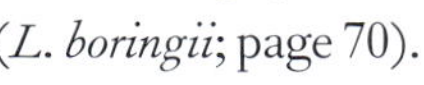

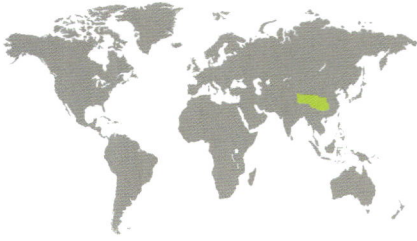

FAMILY	Megophryidae: Leptobrachiinae
OTHER NAMES	Taosze Spiny Toad
DISTRIBUTION	Three separate areas in central China
ADULT HABITAT	Temperate forest, grassland, and arable land, at altitudes of 2,000–5,500 ft (600–1,700 m)
LARVAL HABITAT	Slow-flowing streams
CONSERVATION STATUS	IUCN Endangered. Subject to overcollecting for the pet trade, and threatened by degradation and loss of its habitat

ADULT LENGTH
Male
2⅝–3⅝ in (65–93 mm)
Female
2 ⅛–3$^{3}/_{16}$ in (55–82 mm)

LEPTOBRACHIUM BORINGII

EMEI MOUSTACHE FROG

(LIU, 1945)

Breeding males of this species, arguably one of the world's most bizarre frogs, develop an array of 10–16 sharp spines projecting outward from the upper jaw. These fall off at the end of the breeding season. Males, which are larger than females, build and defend aquatic nests in streams. They call underwater to attract females and to defend their nests. Fights between males lead to change of possession of nests and can result in injuries. Larger males are more successful in defending nests and mate more often than small males.

The Emei Moustache Frog has a wide head with a rounded snout. Breeding males develop sharp black spines around the upper jaw. The eyes are large, with irises that are pale blue above the pupil and black below. The body is gray or brown with darker markings.

SIMILAR SPECIES

The genus *Leptobrachium* contains 38 species, known collectively as large-eyed litter frogs. They are found across mainland Asia and on the many islands of Southeast Asia. *Leptobrachium bompu* (page 69), from India, is notable for its pale blue eyes. *Leptobrachium leishanense* is found in China and is categorized as Endangered; it is harvested for food by local people. *Leptobrachium kantonishikawai* from Malaysian Borneo is Critically Endangered.

Actual size

FAMILY	Megophryidae: Leptobrachiinae
OTHER NAMES	None
DISTRIBUTION	Sabah, eastern Malaysia
ADULT HABITAT	Forest at altitudes of 5,000–7,200 ft (1,500–2,200 m)
LARVAL HABITAT	Streams
CONSERVATION STATUS	IUCN Near Threatened. Its range has been greatly reduced by deforestation and it is now confined to one protected area

LEPTOBRACHIUM GUNUNGENSE

KINABALU LARGE-EYED LITTER FROG

MALKMUS, 1996

ADULT LENGTH
Male
1 3/16–2½ in (46–63 mm)
Female
1 15/16–2⅝ in (50–65 mm)

This terrestrial frog is found during most of the year among leaf litter, in rock crevices, and in holes in the ground. Its legs are short and it is capable of only short hops. It breeds in quiet pools in forest streams, where males call to attract females, producing a series of loud, quacking calls. The tadpoles have powerful, muscular tails and can grow to be as much as 2 13/16 in (70 mm) in length. They hide under rocks by day, emerging at night to feed on dead vegetation with their beak-like mouths.

SIMILAR SPECIES

Leptobrachium gunungense is very similar in appearance and habits to another Bornean frog, the Montane Litter Frog (*L. montanum*), which lives at lower altitudes and has a different call. See also the Lowland Litter Frog (*L. abbotti*; page 68), *L. bompu* (page 69), and the Emei Moustache Frog (*L. boringii*; page 70).

The Kinabalu Large-eyed Litter Frog has a large head, a rounded snout, large, protruding eyes, and short legs. The upper surfaces are dark gray or brown, and the underside is pale. There are dark transverse stripes on the legs, and the pupils of the large black eyes are surrounded by white sclera.

Actual size

FAMILY	Megophryidae: Leptobrachiinae
OTHER NAMES	None
DISTRIBUTION	Southern Sichuan Province, China
ADULT HABITAT	Montane wet forest at altitudes of 9,380–9,850 ft (2,860–3,000 m)
LARVAL HABITAT	Small mountain streams
CONSERVATION STATUS	IUCN Critically Endangered. Known from only one stream, with a population thought to number fewer than 100 individuals

ADULT LENGTH
Male
average 2 in (52 mm)
Female
average 2⅜ in (60 mm)

OREOLALAX LIANGBEIENSIS

LIANGBEI TOOTHED TOAD

LIU & FEI, 1979

The Liangbei Toothed Toad gets its name from the teeth in its upper jaw. It has many spiny warts on its back. The fingers have rounded tips and the toes have fringed edges. The male has spines on its upper arms and chest. The back is brownish yellow with circular black spots, and there are yellow stripes on the legs and arms.

The male of this species has an impressive array of spines on its back, fingers, and chest. It is likely that these are used in either fighting or mating, but neither has been described. The frog is terrestrial for most of the year, living in ravines and the surrounding forest, but moves to small mountain streams in May to breed. Females lay clutches of around 350 eggs, stuck in ringlike clusters to the underside of rocks. Hiding between stones, the tadpoles develop in slower water. The species has a very restricted range, as its future is under threat from deforestation.

SIMILAR SPECIES

The genus *Oreolalax* currently includes 19 species, all of which are found in southwestern China and Vietnam. Several of them are threatened with extinction to varying degrees. Six species, including the Omei Toothed Toad (*O. omeimontis*), are listed as Endangered, three as Vulnerable, and two as Near Threatened.

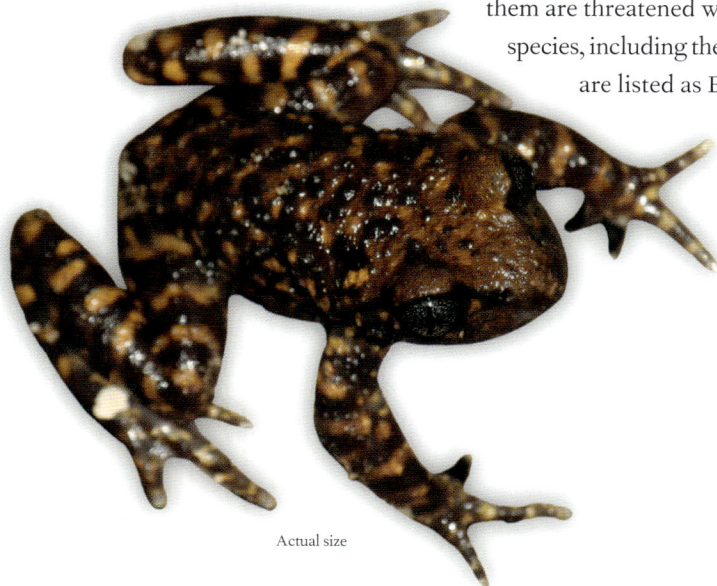

Actual size

FAMILY	Megophryidae: Leptobrachiinae
OTHER NAMES	Boulenger's Lazy Toad, Himalayan Stream Toad
DISTRIBUTION	Southern China; Nepal
ADULT HABITAT	Alpine meadows and forest, at altitudes of 10,800–16,700 ft (3,300–5,100 m)
LARVAL HABITAT	Streams
CONSERVATION STATUS	IUCN Least Concern. Has a large range and occurs within some protected areas

ADULT LENGTH
Male average
2¹⁄₁₆ in (53 mm)
Female average
2½ in (62 mm)

SCUTIGER BOULENGERI

XIZANG ALPINE TOAD

(BEDRIAGA, 1898)

This common, slow-moving frog is found in alpine habitats in the Himalayas, living close to the headwaters of streams and the edges of glacial lakes. It lays its eggs in mountain streams. The water in such streams is very cold and it can take as long as five years for the tadpoles to reach metamorphosis. They may grow up to 2½ in (62 mm) long and, during the winter months, can be seen living under the ice.

Actual size

The Xizang Alpine Toad is a large, rather fat frog with short hind legs. There are numerous round warts on its back. The back is gray or olive-green with dark spots and the underside is yellowish white. There is a pale triangular patch on the snout between the eyes.

SIMILAR SPECIES

The Nyingchi Alpine Toad (*Scutiger nyingchiensis*) is another alpine species, occurring at 9,800–16,400 ft (3,000–5,000 m) in China, India, Nepal, and Pakistan. It is listed as of Least Concern, but is threatened by the diversion of natural water sources for irrigation. The Muli Cat-eyed Toad (*S. muliensis*) and the Hopachai Lazy Toad (*S. glandulatus*) differ from other *Scutiger* species in that males are larger than females. Genus *Scutiger* contains 30 species.

FAMILY	Megophryidae: Megophryinae
OTHER NAMES	*Megophrys brachykolos*, Peak Spadefoot Toad
DISTRIBUTION	Hong Kong, southern China
ADULT HABITAT	Forest
LARVAL HABITAT	Streams
CONSERVATION STATUS	IUCN Endangered. It has a very restricted range and is threatened by urbanization and chemical pollution

ADULT LENGTH
Male
1 5/16–1 9/16 in (34–40 mm)

Female
1 9/16–1 7/8 in (40–48 mm)

BOULENOPHRYS BRACHYKOLOS

SHORT-LEGGED HORNED TOAD

(INGER & ROMER, 1961)

Actual size

This small, little-known frog breeds in streams in Hong Kong's hilly country parks. Males produce a series of short notes from hidden positions in holes or under rocks. As with other frogs in this genus, the tadpoles have upward-pointing funnels around their mouths, which they use to feed from the water's surface. The species is thought to have been adversely affected by the insecticides used in Hong Kong to control mosquitoes. It is suspected that it may occur outside its current known range, in southern China and Vietnam.

SIMILAR SPECIES

Boettger's Spadefoot Toad (*Boulenophrys boettgeri*) lives in tropical forest in northeastern India and central China, and breeds in mountain streams up to 8,200 ft (2,500 m) altitude. *Boulenophrys elongata* is a small frog found in streams on Mt. Lianhua in eastern China. Described in 2024, it brings the number of species in the genus *Boulenophrys* in China to 63, from a total of 68 species in the genus.

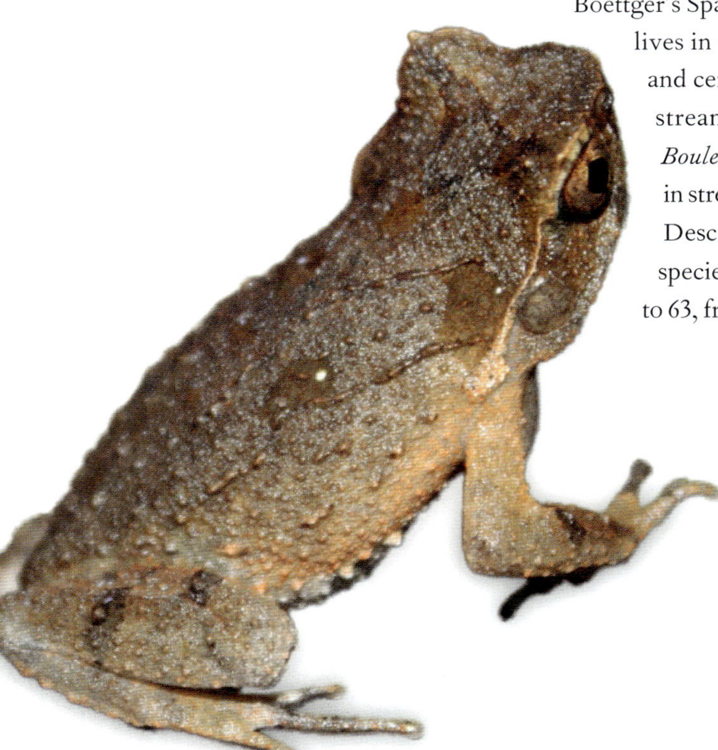

The Short-legged Horned Toad, as its name implies, has short hind legs and fleshy projections above its eyes. The snout is pointed. The upper surfaces are brown or gray with darker blotches and small white spots, and the legs have dark cross-stripes. A thin white line runs from the eye to the armpit. The underside is pale.

FAMILY	Megophryidae: Megophryinae
OTHER NAMES	None
DISTRIBUTION	South-central Vietnam
ADULT HABITAT	Tropical forest
LARVAL HABITAT	Forest streams
CONSERVATION STATUS	IUCN Least Concern. Has a small range that is subject to habitat loss

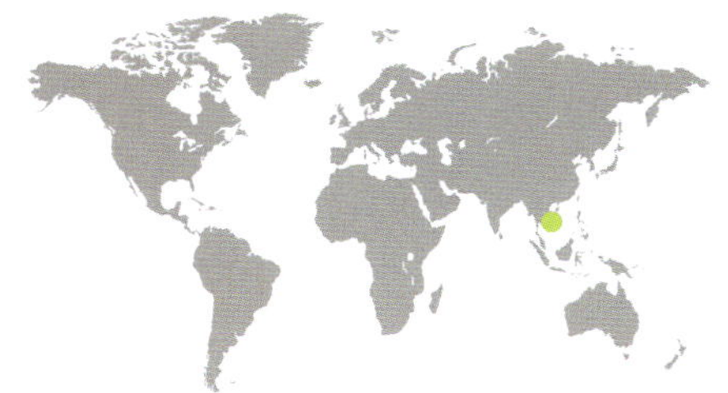

BRACHYTARSOPHRYS INTERMEDIA

ANNAM BROAD-HEADED TOAD

(SMITH, 1921)

ADULT LENGTH
Male
up to 4¹¹⁄₁₆ in (118 mm)
Female
up to 5½ in (139 mm)

This large frog is known only from the Tay Nguyen Plateau of Vietnam, where it was common in the early twentieth century. Its numbers have decreased as a result of destruction of its forest habitat; it was also harvested for food. An ambush predator, its color and shape make it well camouflaged among leaves on the forest floor. Ridges on its back and head make it look as if it is covered in armor. It breeds in mountain streams, where its eggs are attached in clumps to rocks. Males have been heard calling from crevices under stones.

The Annam Broad-headed Toad has a very wide head, adorned with prominent projections above the eyes. Its wide mouth enables it to take very large prey. Ridges on the back and head give it an armored appearance. It is light to reddish brown in color, with dark bands on its legs.

SIMILAR SPECIES

There are eight other species in the genus *Brachytarsophrys*, none of them well known. They include the Karin Hills Frog (*B. carinense*), found in Myanmar and Thailand; the Chuanan Short-legged Toad (*B. chuannanensis*), found in Sichuan Province, China; the Kakhien Hill Frog (*B. feae*) from southern China, Myanmar, Thailand, and Vietnam; and *B. platyparietus* which occurs in southern and southwestern China. There are reports that the last three species are harvested for food by local people.

Actual size

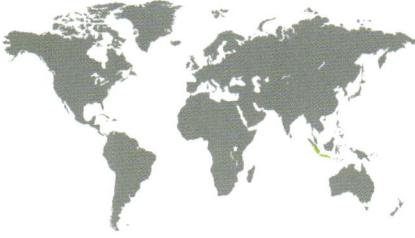

FAMILY	Megophryidae: Megophryinae
OTHER NAMES	Montane Horned Toad, Asian Spadefoot Toad
DISTRIBUTION	Java and western Sumatra, Indonesia
ADULT HABITAT	Rainforest and woodland leaf litter up to 4,000 ft (1,200 m) altitude
LARVAL HABITAT	Forest streams
CONSERVATION STATUS	IUCN Least Concern. May be locally threatened by habitat alteration or intensive grazing

ADULT LENGTH
Male
3 1/8–3 9/16 in (80–92 mm)

Female
3 1/2–4 5/16 in (90–111 mm)

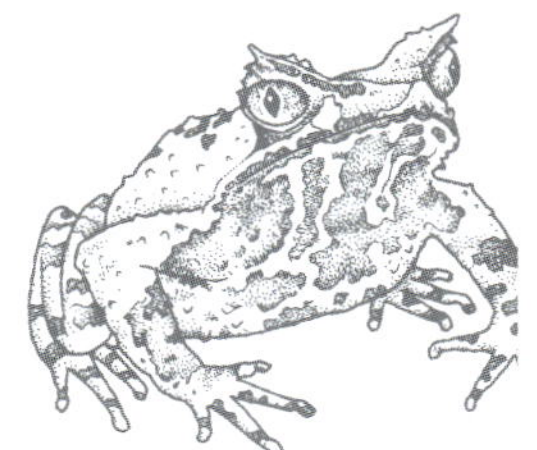

MEGOPHRYS MONTANA

JAVAN HORNED TOAD

KUHL & VAN HASSELT, 1822

The Javan Horned Toad is a large, nocturnal, terrestrial rainforest leaf-litter inhabitant that relies heavily on camouflage to avoid detection. It has large, angular, leaflike projections ("horns") over its large eyes and a projecting fleshy appendage on its snout. Combined with its cryptic patterning, comprising a variety of russet shades, the frog blends into the dead leaves, unless it moves. It can jump but is not especially good at it due to its short, slender limbs and heavy body. Its eyes are dark and usually surrounded by dark pigmentation that may serve to break up their outline. The larger toad species, like the Javan Horned Toad, can be voracious predators of large invertebrates, including potentially dangerous centipedes and scorpions. The Javan Horned Toad is thought to breed along shallow streams, with males making loud "kang" sounds, especially on nights with a full moon. The brown tadpoles feed upside down on the surface meniscus.

The Javan Horned Toad is a large, camouflaged rainforest frog with a stout body patterned with red, browns, and grays and distinctive projections over its eyes and on its snout. Its limbs are slender and short, and its fingers and toes are unwebbed. Many of these frogs are difficult to identify without knowing the geographical origin.

SIMILAR SPECIES

Megophrys was once a much larger genus, but many species have been allocated to other megophrynine genera, including *Boulenophrys* (page 74), *Brachytarsophrys* (page 75), *Pelobatrachus* (pages 77–79), *Xenophrys* (page 80), and five other genera, so it now contains only five species, the other four being Sumatran endemics.

Actual size

FAMILY	Megophryidae: Megophryinae
OTHER NAMES	*Megophrys baluensis*, Balu Spadefoot Toad
DISTRIBUTION	Sabah, northern Borneo
ADULT HABITAT	Montane forest
LARVAL HABITAT	Streams
CONSERVATION STATUS	IUCN Near Threatened. It has a very restricted range but occurs within two protected areas

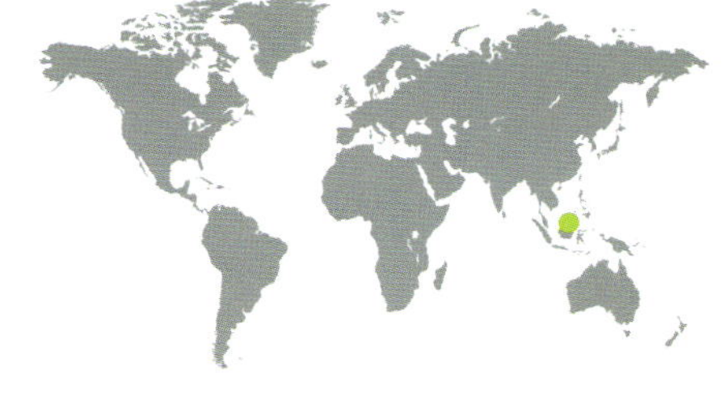

PELOBATRACHUS BALUENSIS

KINABALU HORNED FROG

(BOULENGER, 1899)

ADULT LENGTH

Male
1⅝–1¾ in (41–45 mm)

Female
2⅛–2 13/16 in (55–70 mm)

This rare frog is very well camouflaged among the leaf litter on the forest floor, where adults and juveniles are found. It breeds in the calmer stretches of rocky mountain streams. The tadpoles of this and other species in the genus are unusual in that the lips around their mouth are greatly enlarged to form a funnel. As the tadpoles swim up to the water surface, the funnel opens out to form a device that channels water and small food particles, such as pollen and bacteria, into the mouth. The particles are filtered from the water inside the gill chamber.

SIMILAR SPECIES

Genus *Pelobatrachus* contains seven species, all of which were previously in the genera *Megophrys* and/or *Xenophrys*. *Pelobatrachus* also includes both *P. edwardinae* (page 78) and *P. nasutus* (page 79). Another species, *P. kalimantanensis*, is named for Kalimantan, the Indonesia part of Borneo, but it also inhabits the Crocker Range in Sarawak, a state of Malaysian Borneo.

The Kinabalu Horned Frog is a stocky frog with a wide head, a blunt snout, and short, slender hind limbs. A triangular "horn," adorned with several small spines, projects from each eyelid. There are several ridges and bumps on the back. The upper surfaces are brown or reddish, with darker blotches on the back and stripes on the legs.

Actual size

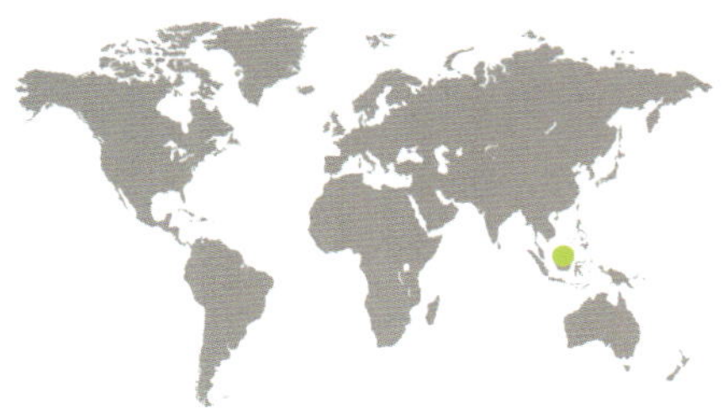

FAMILY	Megophryidae: Megophryinae
OTHER NAMES	Edwardina's Horned Frog
DISTRIBUTION	Sarawak and western Sabah, Malaysia; may be more widespread elsewhere in Borneo
ADULT HABITAT	Hilly lowland rainforest
LARVAL HABITAT	Clear rocky streams
CONSERVATION STATUS	IUCN Least Concern. Threatened by habitat loss due to logging

ADULT LENGTH
Male
$1\frac{9}{16}$–$1\frac{5}{8}$ in (39–42 mm)

Female
$2\frac{13}{16}$–$3\frac{3}{16}$ in (70–82 mm)

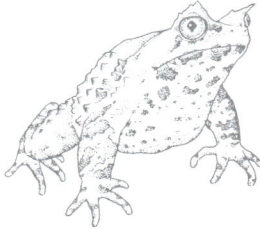

PELOBATRACHUS EDWARDINAE

ROUGH HORNED FROG

(INGER, 1989)

An inhabitant of the forest floor, this secretive frog has a heavy body and rather slender legs. Consequently, it is not capable of jumping far and relies more on camouflage than agility to escape from predators. It hides under dead leaves and logs by day, and emerges to feed at night. It has a very wide head and, consequently, a large mouth that enables it to eat large prey, including snails, insects, centipedes, and scorpions. It has not been observed breeding, but is believed to do so in clear rocky streams.

The Rough Horned Frog has many irregularly shaped projections protruding from the skin on its head and back. These break up its outline and enhance its camouflage. It is clay-brown in color, with black bars on the side of the head and on its legs and arms. There are numerous tubercles on its body and limbs; some of these are rounded, while others have sharp points.

SIMILAR SPECIES

Pelobatrachus edwardinae used to be included within the genus *Megophrys*, whose members it closely resembles. Known as horned frogs, *Megophrys* and *Pelobatrachus* species have fleshy projections from their eyebrows and are found in the Malay Peninsula and the Malay Archipelago.

Actual size

FAMILY	Megophryidae: Megophryinae
OTHER NAMES	Bornean Horned Frog, Long-nosed Horned Frog
DISTRIBUTION	Borneo, Sumatra, Indonesia, Peninsular Malaysia, Thailand
ADULT HABITAT	Tropical rainforest, from sea level up to 5,250 ft (1,600 m) altitude
LARVAL HABITAT	Streams
CONSERVATION STATUS	IUCN Least Concern. Has a large range, including several protected areas

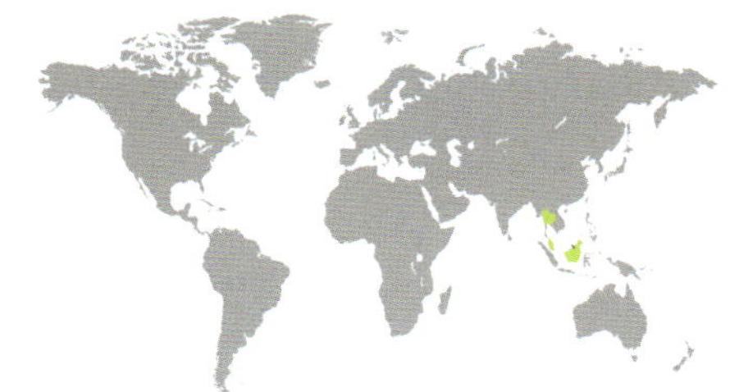

ADULT LENGTH
Male
2 13/16–4 1/8 in (70–105 mm)

Female
3 1/2–4 15/16 in (90–125 mm)

PELOBATRACHUS NASUTUS

BORNEO HORNED FROG

(SCHLEGEL, 1858)

This large frog has a voracious appetite, feeding on a wide variety of animals, including spiders, small rodents, lizards, crabs, scorpions, and other frogs. It does not forage for its prey but waits for it to come close, relying on camouflage to hide among leaves on the forest floor. Breeding occurs in streams, males producing a loud "honk" call in response to an approaching storm that heralds the time to breed. The tadpoles have upward-pointing mouths and feed on plant debris floating on the water's surface. Common in many parts of its range, this species is popular in the international pet trade and has been bred successfully in captivity.

The Borneo Horned Frog gets its name from the hornlike skin projection in the middle of its snout. It also has longer "horns" above the large eyes. There are two pairs of folds in the skin on the back, resembling the veins of dead leaves. The legs are short and slender. It is brown or gray in color, with darker markings and a black patch around the eye.

SIMILAR SPECIES

The Javan Horned Toad (*Megophrys montana*; page 76) was previously considered to be the same species as *P. nasutus* when it too was included in the genus *Megophrys*. *Megophrys montana* is found in Java, and perhaps in Sumatra, and differs from *P. nasutus* by being smaller, lacking the "horn" on the tip of its snout, and the folds of skin on its back.

Actual size

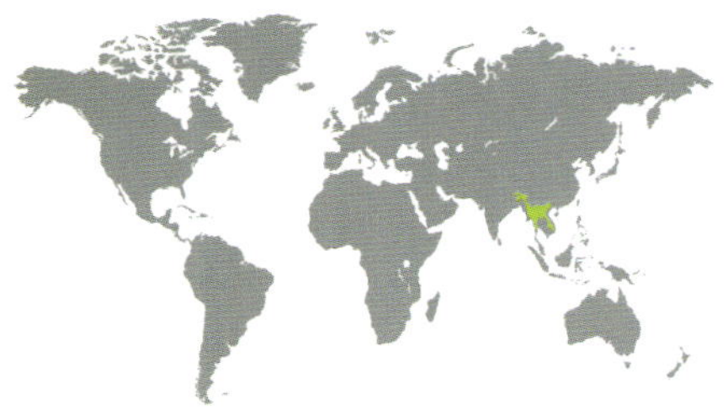

FAMILY	Megophryidae: Megophryinae
OTHER NAMES	*Megophrys major*, Anderson's Spadefoot Toad, Doi Suthep Spine-eyed Frog, Great Stream Horned Toad
DISTRIBUTION	Northeastern India, Bhutan, Thailand, southern China, Laos, Vietnam
ADULT HABITAT	Evergreen forest at altitudes of 820–8,200 ft (250–2,500 m)
LARVAL HABITAT	Streams
CONSERVATION STATUS	IUCN Least Concern. Declining in parts of its range but occurs within some protected areas

ADULT LENGTH
Male
2¾–2 15/16 in (68–75 mm)
Female
3⅛–3 9/16 in (80–92 mm)

XENOPHRYS MAJOR

WHITE-LIPPED HORNED TOAD

(BOULENGER, 1908)

This large frog gets its name from the pointed fleshy projections above each eye. It is found on the forest floor, close to the streams where it breeds. It favors streams in which the water is clear and swift flowing. As with other frogs in this genus, the tadpoles have upward-pointing funnels around their mouths with which they feed from the water's surface. It is reported that, in Bhutan, the species is eaten by local people and it is possible that it is eaten elsewhere in its large range.

The White-lipped Horned Toad has a fleshy "horn" projecting from each eyelid, a number of folds in the skin on its back, and large warts on its flanks. The upper surfaces are gray, green, or reddish brown, and there is usually a dark triangular patch on the head between the eyes. There is a white stripe along the upper lip and a narrow white stripe above the eyes.

SIMILAR SPECIES

Xenophrys once contained 46 species but many species are now in different genera, for example, the Tianzishan Horned Toad (*Boulenophrys tuberogranulatus*), a very warty Chinese species, the Malacca Spadefoot Toad (*Grillitschia longipes*), which is from Thailand and Peninsular Malaysia and is listed as Near Threatened, the Kinabalu Horned Frog (*Pelobatrachus baluensis*; page 77) and the Short-legged Horned Toad (*Boulenophrys brachykolos*; page 74). *Xenophrys* is still a large genus, with 30 species.

Actual size

FAMILY	Pelobatidae
OTHER NAMES	None
DISTRIBUTION	France, Portugal, Spain
ADULT HABITAT	Open landscapes such as dunes, fields, and meadows
LARVAL HABITAT	Still waterbodies
CONSERVATION STATUS	IUCN Vulnerable. Declining in parts of its range due to habitat loss, contamination, and introduced fish and crayfish

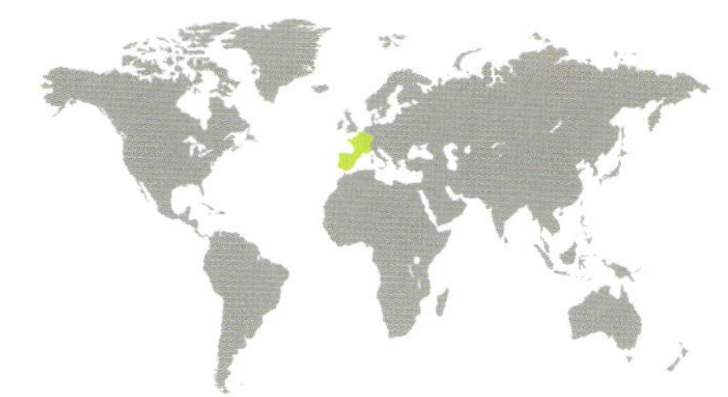

PELOBATES CULTRIPES

WESTERN SPADEFOOT

(CUVIER, 1829)

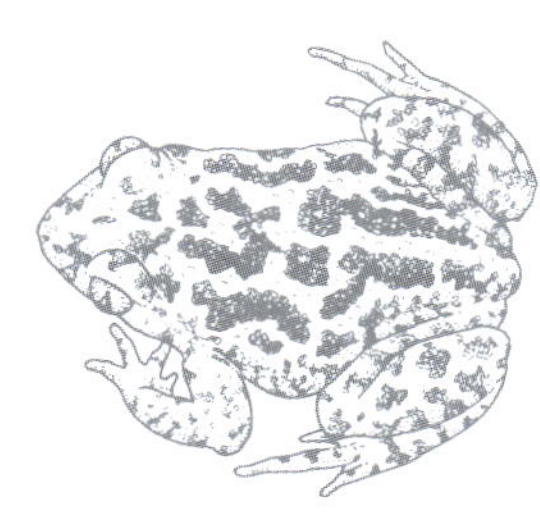

ADULT LENGTH
Male
up to 3½ in (90 mm)
Female
up to 4¾ in (120 mm)

Western Spadefoots emerge at night from their burrows to breed sometime between October and May, depending on latitude and altitude. Males arrive at breeding sites before females and call underwater, producing a "co… co… co" sound. They generally outnumber females and fights between them are common. Amplexus lasts for 72 hours on average and results in up to 7,000 eggs being laid in a long band. Under favorable conditions, tadpoles can grow very large, delaying metamorphosis until they are 4¾ in (120 mm) long. Often, however, there is mass mortality of tadpoles when their pond dries up before they reach metamorphosis.

SIMILAR SPECIES

The Moroccan Spadefoot (*Pelobates varaldii*) occurs in north-western Morocco; genetic analysis suggests that it became separated from *P. cultripes* in the Pliocene, 2.6–5.3 million years ago. The species is categorized as Endangered, its numbers having declined considerably as a result of habitat loss due to land drainage, pollution by agrochemicals, and introduced predators, particularly mosquitofish (*Gambusia* spp.).

Actual size

The Western Spadefoot gets its name from the horny black tubercles on its hind feet. It uses these to dig down vertically into soil. It has a robust body, a large head, and short hind legs with webbed toes. Its smooth skin is variable in color, being gray, brown, yellow, or green with dark brown or black markings.

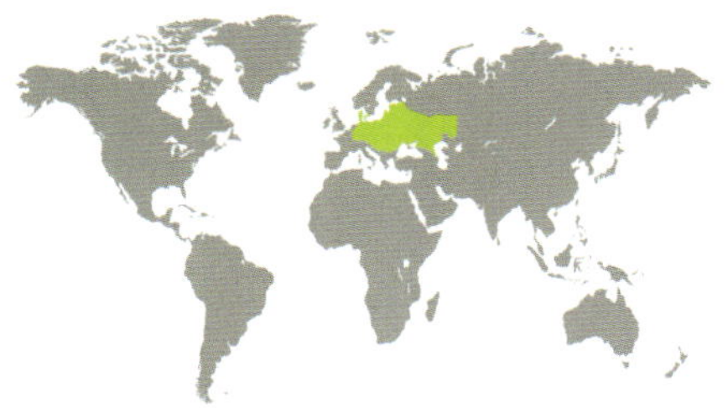

FAMILY	Pelobatidae
OTHER NAMES	None
DISTRIBUTION	Europe and Asia, from Germany to Kazakhstan
ADULT HABITAT	Open areas in forests
LARVAL HABITAT	Permanent waterbodies
CONSERVATION STATUS	IUCN Least Concern. Very sensitive to changes in soil structure and water quality

ADULT LENGTH
Male
up to 2⁵⁄₈ in (65 mm)

Female
up to 3⅛ in (80 mm)

PELOBATES FUSCUS

COMMON SPADEFOOT

(LAURENTI, 1768)

Common Spadefoots are noted for their breeding site fidelity, individuals returning year after year to the same pond. They breed most successfully in large ponds with steep banks that contain few predators. Their breeding behavior is similar to that of the Western Spadefoot (*Pelobates cultripes*; page 81), though they are less fecund, laying clutches of 480–3,000 eggs, with older females laying more eggs than younger ones. When attacked by a predator or molested by a human, they hold their ground, puff up their body, stand up on all four limbs, and emit a squeal.

The Common Spadefoot has horny tubercles on its hind feet with which it digs down vertically into soil. It has a robust body, a large head, and short hind legs with webbed toes. There is a distinct dome on top of the head. Its smooth skin varies in color, being gray, brown, yellow, or green with darker markings. Some individuals have numerous small red spots.

SIMILAR SPECIES

Pelobates contains six species. Eastern Spadefoot (*Pelobates syriacus*) occurs in sandy and stony areas from the Balkan Peninsula to the Middle East, from Serbia and Greece, through Turkey, to northern Iran. It differs from *P. fuscus* in being larger in size, lacking the dome on the head, and in having pale-colored "spades" on its hind feet. The Balkan Spadefoot (*P. balcanicus*) also inhabits the Balkans and Greece, while Pallas' Spadefoot (*P. vespertinus*) occurs in Ukraine and Russia.

Actual size

FAMILY	Heleophrynidae
OTHER NAMES	Natal Cascade Frog
DISTRIBUTION	Maluti and Drakensberg Mountains, South Africa, Lesotho, and Swaziland
ADULT HABITAT	Montane forest and grassland at altitudes of 1,900–8,775 ft (580–2,675 m)
LARVAL HABITAT	Fast-flowing forest streams
CONSERVATION STATUS	IUCN Least Concern. Sensitive to deforestation, stream degradation, water extraction, and introduced trout

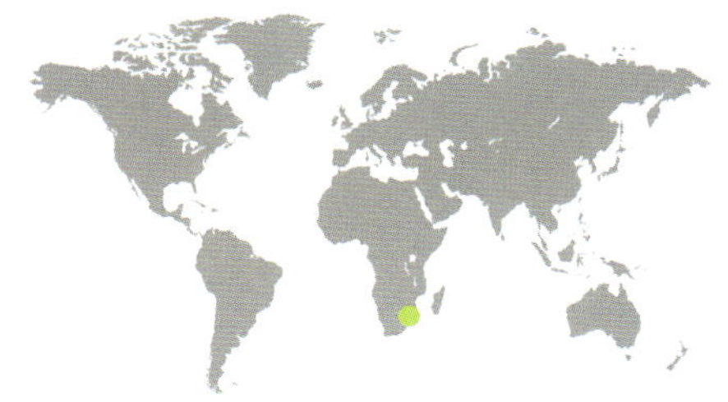

HADROMOPHRYNE NATALENSIS

NATAL GHOST FROG

(HEWITT, 1913)

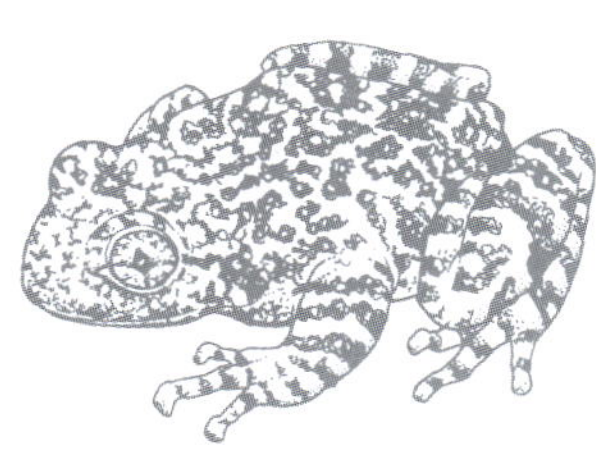

ADULT LENGTH
Male
up to 1¾ in (45 mm)

Female
up to 2⅝ in (65 mm)

Adult Natal Ghost Frogs are secretive and nocturnal, and hence seen less often than their tadpoles. They breed in fast-flowing mountain streams and, because the water is cold, it takes the tadpoles two years to reach metamorphosis, by which time they can be as long as 3⁵⁄₁₆ in (85 mm). The tadpoles browse on algae and have a sucker-like mouth with which they attach themselves to rocks. Mating occurs between March and May. Males develop spines on the fingers and chest, and call from near streams, producing a melodious "ting." Clutches of 50–200 eggs are laid under stones. The skin produces a toxic secretion similar to that of a wasp sting.

SIMILAR SPECIES

Hadromophryne natalensis is the only species in its genus and was formerly included within the genus *Heleophryne* (see pages 84–85). It is likely that future research will reveal that it is in fact more than one species. It is distributed across a number of mountain ranges, separated by unsuitable habitat, and there is negligible gene flow between populations.

The Natal Ghost Frog has large protuberant eyes with vertical pupils. The head and body are flattened, and the toes are partially webbed. Its fingers and toes have triangular tips. The back has green or yellow reticulated markings on a dark brown, purple-brown, or black background. Its underside is dirty white with light brown markings on the throat.

Actual size

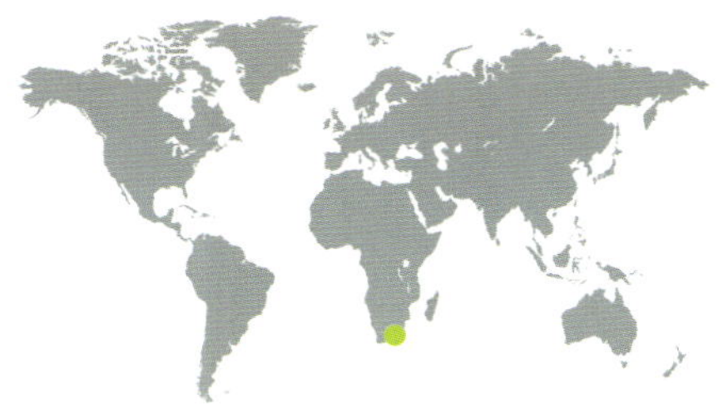

FAMILY	Heleophrynidae
OTHER NAMES	None
DISTRIBUTION	Eastern Langeberg Mountains, Western Cape, South Africa
ADULT HABITAT	Forest patches surrounded by fynbos at altitudes of 700–1,650 ft (215–500 m)
LARVAL HABITAT	Fast-flowing mountain streams
CONSERVATION STATUS	IUCN Least Concern. Its range is largely covered by protected areas

ADULT LENGTH
Male
up to 1⅜ in (35 mm)
Female
up to 1¹³⁄₁₆ in (46 mm)

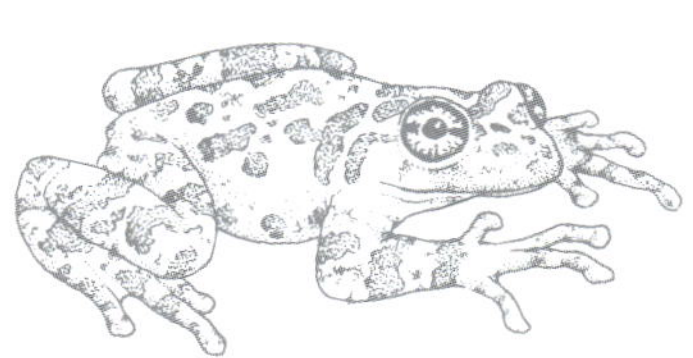

HELEOPHRYNE ORIENTALIS

EASTERN GHOST FROG

FITZSIMONS, 1946

In early summer, when the flow in the waterways of South Africa's Langeberg Mountains is decreasing, male Eastern Ghost Frogs make their way to the edge of streams and call by night and by day, producing a high-pitched, ringing "ik." The calling males have loose dorsal skin, swollen forearms, and spines on their chest and fingers. The eggs are laid, not in the water as with other ghost frogs, but under moss-covered rocks beside streams, in clutches of 110–190. After hatching, the tadpoles enter streams, where a large sucker-like disk around the mouth enables them to attach themselves to rocks.

SIMILAR SPECIES

There are six species in the genus *Heleophryne*, each found on a different mountain range in South Africa and separated from each other by habitat unsuitable for ghost frogs. The Cederberg Ghost Frog (*H. depressa*) is confined to the Cederberg Range and Hewitt's Ghost Frog (*H. hewitti*) to the Elandsberg Range. The latter species is categorized as Endangered, having been adversely affected by commercial deforestation.

The Eastern Ghost Frog has a flattened body, webbed toes, and large protruding eyes. There are conspicuous horizontal and vertical stripes through the otherwise golden eye, giving the impression of a cross. Seen from above, the frog is beige or olive-green in color, with darker markings. The fingers and toes end in triangular toe pads.

Actual size

FAMILY	Heleophrynidae
OTHER NAMES	Rose's Ghost Frog
DISTRIBUTION	Eastern side of Table Mountain, Western Cape, South Africa
ADULT HABITAT	Forest and fynbos at altitudes of 790–3,475 ft (240–1,060 m)
LARVAL HABITAT	Mountain streams
CONSERVATION STATUS	IUCN Critically Endangered. Declining rapidly, despite its range lying within a national park

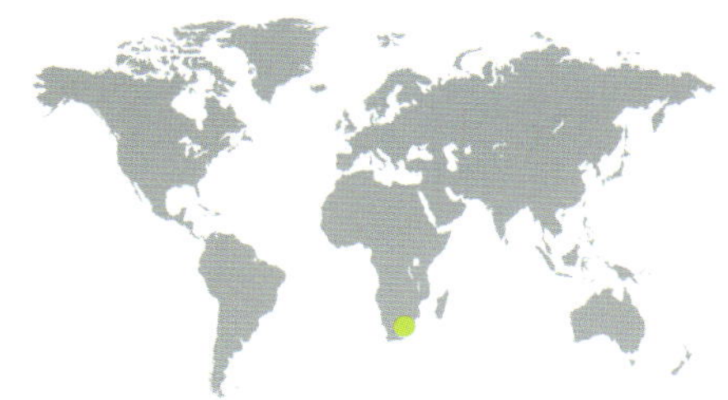

ADULT LENGTH
Male
up to 1$\frac{15}{16}$ in (50 mm)
Female
up to 2$\frac{3}{8}$ in (60 mm)

HELEOPHRYNE ROSEI

TABLE MOUNTAIN GHOST FROG

HEWITT, 1925

One of world's rarest frogs, the Table Mountain Ghost Frog is confined to an area of only 2.7–3 sq miles (7–8 sq km) close to the city of Cape Town. It is threatened by invasive non-native vegetation, frequent bush fires, the damming of streams to create reservoirs, and tourism; it has also been infected by chytridiomycosis. Its tadpoles have large sucker-like mouths and climb out of the water and up rocks at night. Because its tadpoles require 12 months to reach metamorphosis, it is essential that the streams in which they live flow year-round. This becomes unlikely if streams are blocked by vegetation or if too much water is extracted to supply the local human population.

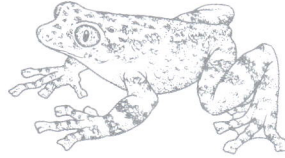

Actual size

The Table Mountain Ghost Frog has a squat, flattened body that enables it to hide in very narrow crevices. It has large, prominent eyes, triangular tips to its fingers and toes, and webbed hind feet. It is pale green in color, with purple or reddish-brown blotches.

SIMILAR SPECIES

The Southern Ghost Frog (*Heleophryne regis*) is found in the coastal mountains of Eastern Cape and Western Cape provinces. The Cape Ghost Frog (*H. purcelli*) occurs in the Cederberg Range in Western Cape. Both species, unlike the Table Mountain Ghost Frog, have a dark horizontal stripe across the iris of the eye.

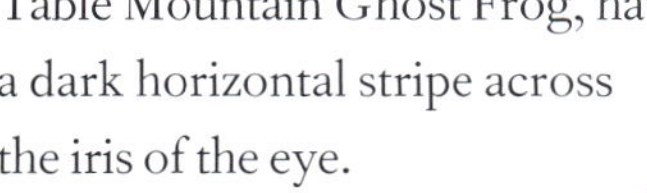

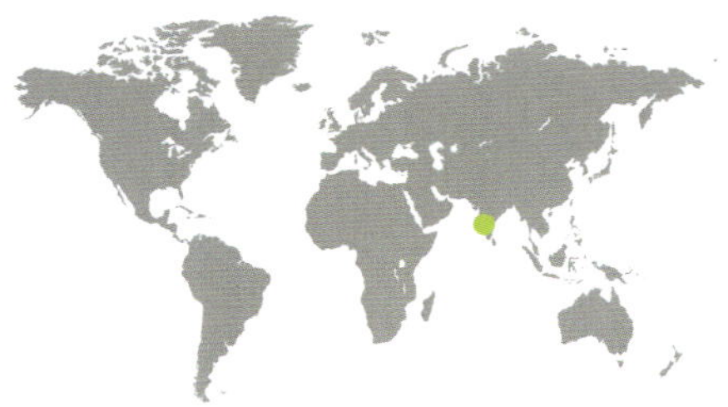

FAMILY	Nasikabatrachidae
OTHER NAMES	Sahyadri Pig-nosed Frog, Paschi Pig-nosed Frog
DISTRIBUTION	Two restricted localities in the Western Ghats, India
ADULT HABITAT	Forest
LARVAL HABITAT	Ponds, streams
CONSERVATION STATUS	IUCN Endangered. Has a very restricted range that is subject to deforestation

ADULT LENGTH
Male
up to 2⅜ in (60 mm)
Female
up to 3½ in (90 mm)

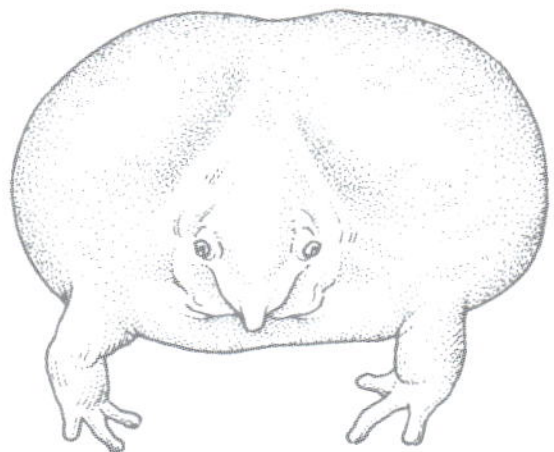

NASIKABATRACHUS SAHYADRENSIS

PURPLE FROG

(BIJU & BOSSUYT, 2003)

It is not surprising that very little is known about this unusual frog. For all but about two weeks each year it lives buried deep underground, feeding predominantly on termites, and it was not discovered until 2003. After the first rains of the pre-monsoon season, it emerges to breed in flooded pools and streams. Males attract females by calling from shallow burrows near water. Being much smaller than females, males cannot clasp females as most frogs do, but instead hold onto their spine, possibly with the help of sticky skin secretions. The female produces eggs in large numbers.

The Purple Frog has a bloated appearance, with a globular body and a small head. It has small eyes and its pointed snout ends in a white protuberance. The smooth, shiny skin is purple above and gray below. The limbs are short, ending in partially webbed feet with rounded toes. Both hind feet carry white wart-like protuberances that are used when it digs backward into the soil.

SIMILAR SPECIES

A second species, the Critically Endangered Bhupathi's Pig-nosed Frog (*N. bhupathi*), inhabits India's Western Ghats. Their ancestors separated from their closest relatives, the Seychelles frogs (Sooglossidae; pages 87–88), 130 million years ago. It bears little resemblance to them, but is somewhat similar in shape to other, unrelated frogs that live buried underground, such as the Mexican Burrowing Toad (*Rhinophrynus dorsalis*; page 59) and the Spotted Burrowing Frog (*Hemisus guttatus*; page 466) of Africa.

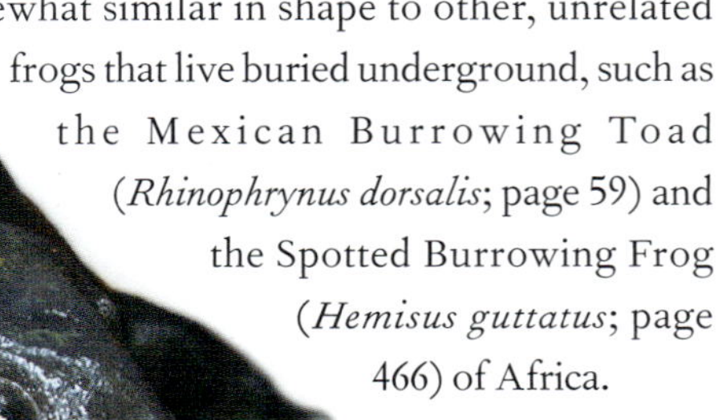

Actual size

FAMILY	Sooglossidae
OTHER NAMES	None
DISTRIBUTION	Mahé and Silhouette islands, Seychelles
ADULT HABITAT	Rainforest
LARVAL HABITAT	Within eggs, laid on the ground
CONSERVATION STATUS	IUCN Endangered. Threatened by climate change and habitat degradation

SECHELLOPHRYNE GARDINERI

GARDINER'S SEYCHELLES FROG

(BOULENGER, 1911)

ADULT LENGTH
Male
up to 5/16 in (8 mm)
Female
up to 1/2 in (12 mm)

Climate change is thought to be the most serious threat to this tiny frog. Over a period of 16 years, at lower altitudes in its range where rainfall patterns have changed, its population has declined by 67 percent. At higher altitudes (over 3,251 ft, or 991 m), populations are currently stable, but if climate change continues it is predicted that suitable habitat will disappear. This is a ground-dwelling frog, living among leaf litter and low plants. It is nocturnally active and breeds year-round, laying clutches of 8–16 eggs on the ground. This species shows direct development, the tadpole stage being completed within the eggs, which hatch to produce tiny frogs.

Actual size

Gardiner's Seychelles Frog is possibly the world's smallest frog; newly metamorphosed frogs are only 1/16 in (1.6 mm) long. It has a pointed snout and large, protruding eyes. Its forelimbs are small, but its hind limbs are relatively large and powerful. It is variable in color, ranging from light tan to reddish brown, and there is a dark stripe along each flank.

SIMILAR SPECIES

There is one other species in this genus, the Critically Endangered Seychelles Palm Frog (*Sechellophryne pipilodryas*). It is confined to an area of only 5.8 sq miles (15 sq km) on Silhouette Island, where it is associated with a particular palm. It is declining at lower altitudes, due to the presence of invasive alien plants. Nothing is known about its mode of reproduction.

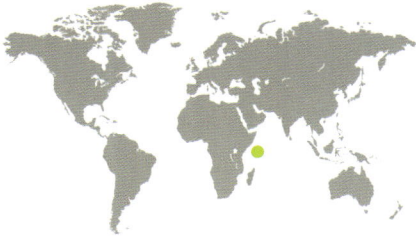

FAMILY	Sooglossidae
OTHER NAMES	None
DISTRIBUTION	Higher altitudes on Mahé, Praslin, and Silhouette islands, Seychelles
ADULT HABITAT	Leaf litter in wet forest
LARVAL HABITAT	On the back of a parent
CONSERVATION STATUS	IUCN Endangered. Threatened by climate change and by habitat degradation

ADULT LENGTH
Male
up to 9/16 in (15 mm)
Female
up to 3/4 in (20 mm)

SOOGLOSSUS SECHELLENSIS

SEYCHELLES FROG

(BOETTGER, 1896)

Actual size

The Seychelles Frog is one of world's smallest frogs. It is golden brown in color, with black bands and spots on the head, back, and legs. There is a dark triangular patch on the top of the head and the underside is white. It is rather secretive, hiding in leaf litter or rock crevices, emerging only after rain.

Although common where it occurs, the Seychelles Frog is classed as Endangered because the extent of suitable habitat in which it can live is very limited and is likely to decrease as a result of climate change. The male has a complex advertisement call, with one primary and four secondary notes. A clutch of 6–15 eggs is laid in nests on the ground and is guarded by a parent. When they hatch, the tadpoles climb onto the parent's back, where they complete their development until they are tiny froglets. It has yet to be determined whether parental care is provided by the father, the mother, or both.

SIMILAR SPECIES

Thomasset's Frog (*Sooglossus thomasseti*) is larger than *S. sechellensis* and lives in mossy forest and boulder fields on Mahé and Silhouette islands. It is considered Critically Endangered because of a trend for reduced rainfall early in the year. The closest relative of the Seychelles frogs is the Purple Frog (*Nasikabatrachus sahyadrensis*; page 86) of India. The species became separated when the Seychelles split away from India around 65 million years ago.

FAMILY	Calyptocephalellidae
OTHER NAMES	Chilean Giant Frog, Wide Mouth Toad
DISTRIBUTION	Central Chile, lowlands up to 1,600 m (500 m) altitude
ADULT HABITAT	Lakes, rivers, ponds
LARVAL HABITAT	Large ponds
CONSERVATION STATUS	IUCN Vulnerable. Declining due to habitat degradation and being traded for food

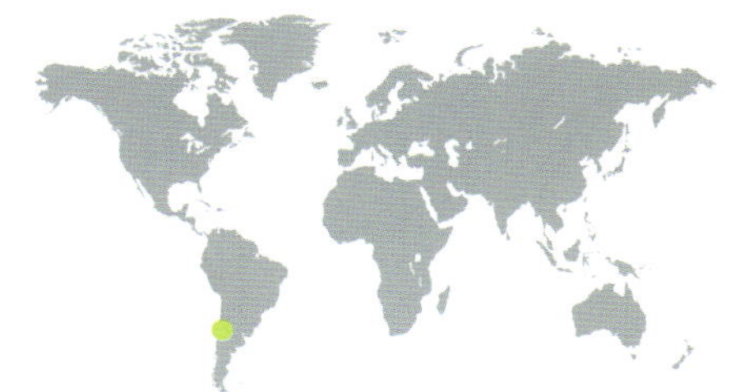

CALYPTOCEPHALELLA GAYI

HELMETED WATER TOAD

(DUMÉRIL & BIBRON, 1841)

ADULT LENGTH
Male
up to 4¾ in (120 mm)

Female
up to 12⅝ in (320 mm)

This large aquatic frog is aggressive when attacked or molested, inflating its body, opening its mouth, and lunging toward its attacker. It has a very varied diet, feeding on insect larvae, fish, other frogs, small birds, and mammals. In the breeding season, during September and October, males call loudly and clutches of 1,000–10,000 eggs are laid in shallow water. The tadpoles take two years to reach metamorphosis. Hunted as a food source for people, the species has declined. It is also threatened by habitat loss, pollution, and introduced trout, which eat the tadpoles.

The Helmeted Water Toad has a robust body and a large head with a rounded snout. The eyes are small and have vertical pupils. There are elongated bumps in the skin on the back and the hind feet are partially webbed. It is yellow or green in color when young, turning to dark gray as it ages.

SIMILAR SPECIES

The only species placed in its genus, *Calyptocephalella gayi* belongs to a family of only four species, the other three being the smaller false toads (genus *Telmatobufo*; page 90). The false toads are rarely seen and, like the Helmeted Water Toad, occur only in Chile.

Actual size

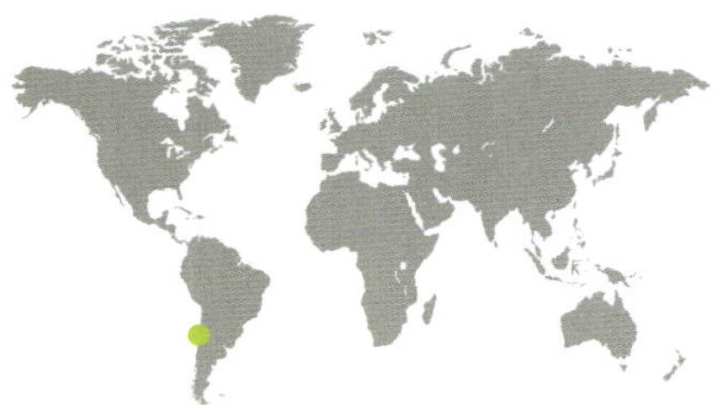

FAMILY	Calyptocephalellidae
OTHER NAMES	Bullock's Mountains False Toad
DISTRIBUTION	Coastal mountains, Arauco Province, Chile
ADULT HABITAT	Montane forest
LARVAL HABITAT	Mountain streams
CONSERVATION STATUS	IUCN Endangered. Threatened by deforestation and degradation of its larval habitat

ADULT LENGTH
2½–3¼ in
(62–83 mm)

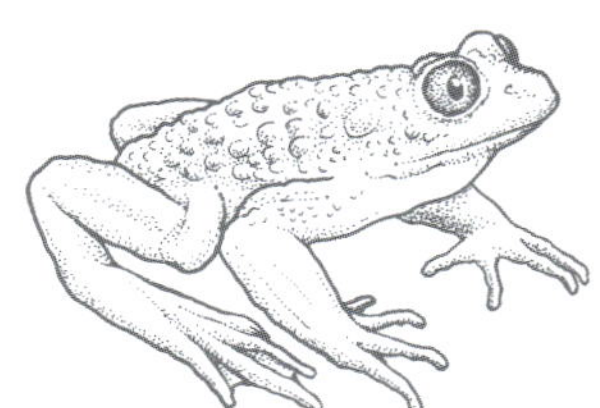

TELMATOBUFO BULLOCKI

BULLOCK'S FALSE TOAD

SCHMIDT, 1952

This stocky, toad-like frog is one of the world's rarest frogs. Discovered in 1952, recent intensive searches have found it in only a handful of locations. In the breeding season, the male develops a cluster of sharp spines on his thumbs, presumably to enable him to grasp the female securely in fast-flowing water. The tadpoles are adapted for living in mountain streams. They have powerful tails that enable them to swim upstream, and large sucker-like mouths with which they attach themselves firmly to rocks. Deforestation is causing the streams in which this species breeds to become silted up and thus unsuitable for the tadpoles.

Bullock's False Toad has a large head, a rounded snout, and a robust body covered in many prominent warts. The parotoid glands behind the eyes are large and oval in shape. The frog has long, slender legs and webbing between the toes. It is grayish brown on its back and there is a distinctive yellow stripe across the top of its head.

SIMILAR SPECIES

The genus *Telmatobufo* is endemic to Chile. The Pelado Mountains False Toad (*T. australis*) is found only in the Pelado Mountains in the south and is listed as Vulnerable. The Chile Mountains False Toad (*T. venustus*) and Los Queules False Toad (*T. ignotus*) have limited ranges and are Endangered. They are threatened by deforestation and by introduced trout, which eat their tadpoles, and by pollution and habitat alteration respectively.

Actual size

FAMILY	Limnodynastidae
OTHER NAMES	None
DISTRIBUTION	Coastal Queensland and New South Wales, Australia
ADULT HABITAT	Open country and forests, up to 1,300 ft (400 m) altitude
LARVAL HABITAT	Streams, ponds
CONSERVATION STATUS	IUCN Least Concern. Sensitive to habitat loss and, at higher altitudes, to chytridiomycosis

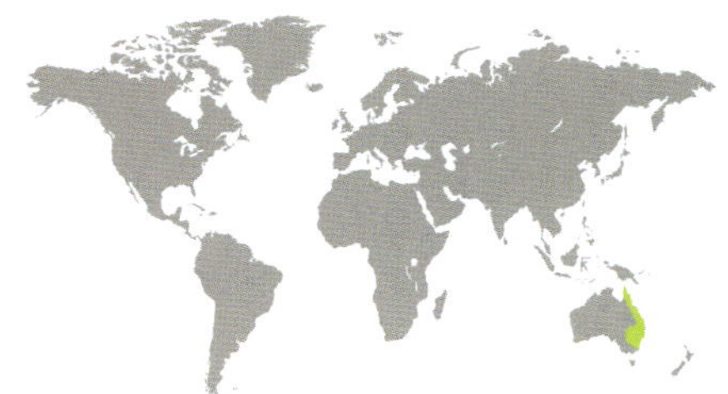

ADELOTUS BREVIS

TUSKED FROG

(GÜNTHER, 1863)

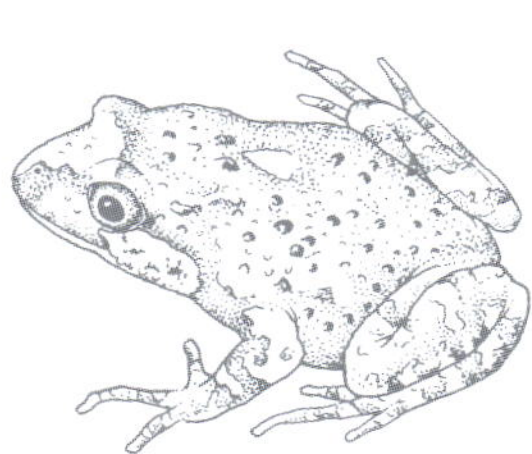

ADULT LENGTH

Male

$1\frac{5}{16}$–$1\frac{15}{16}$ in (34–50 mm)

Female

$1\frac{1}{8}$–$1\frac{9}{16}$ in (29–40 mm)

This frog is very unusual in two respects. First, males are larger than females and, second, it has two tusks in its lower jaw. These are modified teeth and are present in both sexes, but are larger in males. Males use their tusks when fighting one another for the possession of calling sites close to ponds and streams. The call is a soft "chuluk." More than 600 eggs are laid in a foam nest that is usually hidden beneath foliage. Detailed studies of the mating dynamics of Tusked Frogs have revealed that larger males enjoy higher reproductive success than smaller ones.

SIMILAR SPECIES

Adelotus brevis is the only species in its genus. Its large head and tusks make it readily distinguishable from other frogs.

Actual size

The Tusked Frog gets its name from the paired tusks projecting from its lower jaw. The head of the male is often so large that it is equivalent in size to the rest of its body. The back is covered in small warts. The upper surfaces are brown or slate-gray and there is a paler brown patch on the nose. The lower surfaces are mottled black and white, and there is a red patch on the groin and hind legs.

FAMILY	Limnodynastidae
OTHER NAMES	Eastern Owl Frog
DISTRIBUTION	Great Dividing Range, southeastern Australia
ADULT HABITAT	Coastal woodland and forest with sandy soil
LARVAL HABITAT	Ephemeral ponds
CONSERVATION STATUS	IUCN Endangered. Declining due to loss of its forest habitat

ADULT LENGTH
Male
2⅜–3¹⁄₁₆ in (60–78 mm)
Female
2⅞–3¹³⁄₁₆ in (73–97 mm)

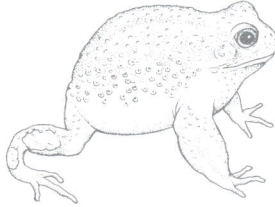

HELEIOPORUS AUSTRALIACUS

GIANT BURROWING FROG

(SHAW & NODDER, 1795)

A large frog that spends much of its life in an underground burrow, the Giant Burrowing Frog emerges to breed in warm, wet weather between August and Marsh. Males call from partially flooded crayfish burrows near creeks, producing an owl-like hooting call. They have sharp spines on their fingers, arms, and chest that enable them to maintain a firm grip on the slippery females during mating. Eggs are laid in foamy masses of 770–1,240 eggs in ephemeral pools. The tadpoles are plump and grow to a large size. This species is threatened by destruction of its forest habitat.

SIMILAR SPECIES

Of the six species in the genus *Heleioporus*, this is the only one found in eastern Australia. The Western Marsh Frog (*H. barycragus*) is similar in appearance and habits but is found only in the Darling Range in Western Australia. Because of its large size, it is sometimes mistaken for the Cane Toad (*Rhinella marina*; see page 201).

The Giant Burrowing Frog has a rotund body and stout, muscular arms and legs. It has large protruding eyes and its skin is covered in warts that are tipped with white spines. It is dark brown to blue-back above and paler below. There are yellow spots on the flanks and a yellow stripe along the upper lip.

Actual size

FAMILY	Limnodynastidae
OTHER NAMES	None
DISTRIBUTION	Coastal region, southwestern Australia
ADULT HABITAT	Bushland, sandy swamps
LARVAL HABITAT	Foam nest in burrow, then ephemeral pools
CONSERVATION STATUS	IUCN Least Concern. Locally common and with a stable population

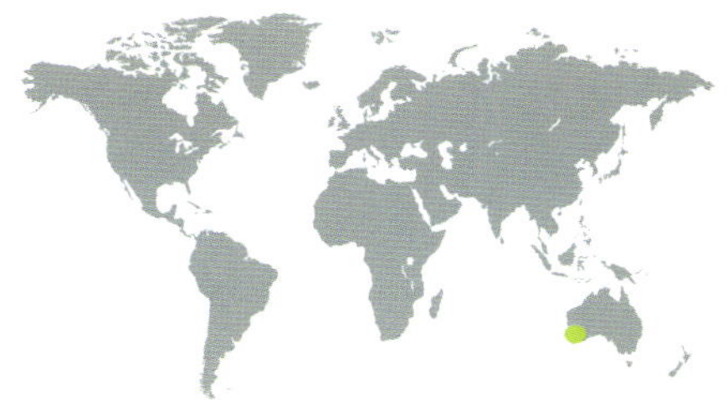

ADULT LENGTH
Male
1¾–2¹¹⁄₁₆ in (45–66 mm)
Female
1¹³⁄₁₆–2½ in (46–63 mm)

HELEIOPORUS EYREI

MOANING FROG

(GRAY, 1845)

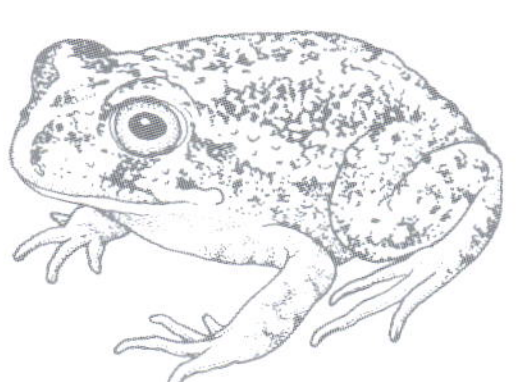

In the breeding season during April and May, males move to sandy swamps where they dig horizontal burrows. They call from within their burrows, producing a long, low moan, and repeating it frequently. Females approach them and mating occurs in the burrow. Between 80 and 500 eggs are laid in a foamy mass at the deepest point of the burrow. Here they hatch and develop until heavy rain washes the tadpoles out of the burrow and into ephemeral pools. During the hot, dry summer, adults dig themselves into the ground to estivate until rain returns.

SIMILAR SPECIES

Heleioporus eyrei is very similar in appearance to two other species, which are also found in southwestern Australia but can be distinguished by their calls. The slightly larger Plains Frog (*H. inornatus*) makes a "woop-woop-woop" sound, while the slightly smaller Sand Frog (*H. psammophilus*) makes a "put-put-put" sound like an outboard engine. *Heleioporus eyrei* also lacks the nuptial spines on its forelimbs that are characteristic of the genus. The sixth species in the genus is the White-spotted Frog (*H. albopunctatus*) from southwestern Australia.

The Moaning Frog has a rotund body, a large head, and large bulbous eyes. The limbs are rather weak in comparison to those of many burrowing frogs. Its back is brown with marbled patterns in yellow, white, or gray. The underside is white. There is a pale ridge under the eye and numerous white spots on the flanks.

Actual size

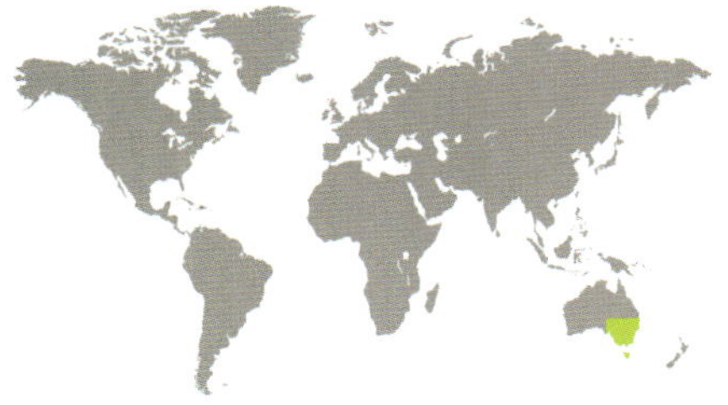

FAMILY	Limnodynastidae
OTHER NAMES	Bullfrog, Eastern Banjo Frog
DISTRIBUTION	New South Wales, Victoria, South Australia, and Tasmania, Australia
ADULT HABITAT	A wide range of habitats, including gardens, dams, and swamps
LARVAL HABITAT	Dams, small lakes, marshes, slow-moving streams
CONSERVATION STATUS	IUCN Least Concern. Wide-ranging and common in many places

ADULT LENGTH
Male
2–$2\frac{13}{16}$ in (52–70 mm)
Female
2–3¼ in (52–83 mm)

LIMNODYNASTES DUMERILII

POBBLEBONK

PETERS, 1863

A large frog, the Pobblebonk lives in a burrow, emerging to feed or breed after rain. It breeds during winter, in dams, small lakes, marshes, and slow-moving streams. Males may travel up to 0.6 miles (1 km) to reach a breeding site. They conceal themselves in floating vegetation and make a loud, resonant "bonk" call, repeated every few seconds. The female lays 3,900–4,000 eggs in a floating foam nest. Across the species' large range, five subspecies are recognized, differentiated by coloration, body size, and mating call.

The Pobblebonk is a large, stout-bodied frog with a broad, rounded head and short, muscular limbs. The eyes are large and protruding. The skin on the back is rough, warty, and pale brown in color with dark brown markings. In some populations there is a pale stripe down the middle of the back. The flanks have a bronze or purple sheen with black mottling.

SIMILAR SPECIES

Of the 13 species of *Limnodynastes*, those most similar to *L. dumerilii* are the Northern Bullfrog (*L. terraereginae*), distributed from New South Wales to the northern tip of Queensland, and the Western Bullfrog (*L. dorsalis*), from the southwestern corner of Australia. It does not have scarlet coloration in its groin like those species.

Actual size

FAMILY	Limnodynastidae
OTHER NAMES	Carpenter Frog
DISTRIBUTION	Northeastern Australia
ADULT HABITAT	Rocky habitats, e.g., hills, gorges, screes, caves
LARVAL HABITAT	Permanent or temporary streams
CONSERVATION STATUS	IUCN Least Concern. Has an extensive range; there is no evidence of population decline

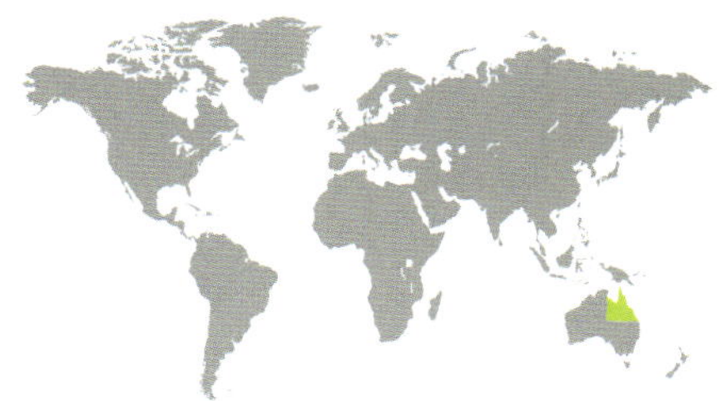

ADULT LENGTH
Male
$1^{11}/_{16}$–$2^{1}/_{2}$ in (43–62 mm)
Female
$1^{7}/_{8}$–$2^{3}/_{8}$ in (47–61 mm)

LIMNODYNASTES LIGNARIUS

WOODWORKER FROG

(TYLER, MARTIN & DAVIES, 1979)

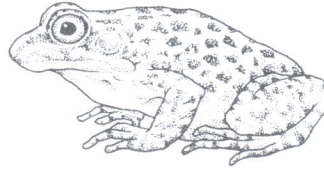

A mating call sounding like the tap of a hammer on a piece of wood gives this frog its common name. It is found in very dry habitats, often hiding deep in caves. It breeds in spring or summer, making its way to temporary or permanent streams. Males develop nuptial pads on their forearms consisting of several black-tipped spines. They call from beneath rocks in or near streams, and the females lay 350–400 eggs in a foam nest in a rocky pool. The tadpoles take around nine weeks to reach metamorphosis.

Actual size

SIMILAR SPECIES

This species was previously placed in its own genus, *Megistolotis*; this has recently been merged with *Limnodynastes*. It has no close relatives and is readily distinguished from other frogs by its very large, conspicuous tympanum.

The Woodworker Frog has a bulky body and short, muscular limbs. There is no webbing on its hands or feet. The dorsal surface is dull slate to dark olive, occasionally mottled with darker or paler markings. The underside is white or purplish brown. The very large tympanum is conspicuous behind the eye.

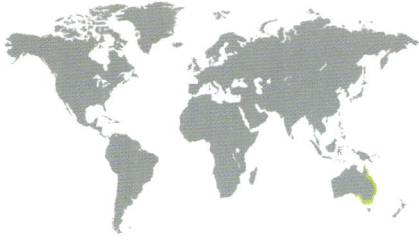

FAMILY	Limnodynastidae
OTHER NAMES	Brown Frog, Peron's Marsh Frog
DISTRIBUTION	Coastal areas of eastern Australia, from Queensland to South Australia, and extreme north of Tasmania, Australia
ADULT HABITAT	Wet areas in forests, woodland, open country, and gardens
LARVAL HABITAT	Still or slow-moving water
CONSERVATION STATUS	IUCN Least Concern. Common over an extensive range. May be spreading in Queensland

ADULT LENGTH
Male
$1\frac{7}{8}$–$2\frac{3}{4}$ in (48–69 mm)

Female
$1\frac{13}{16}$–$2\frac{7}{8}$ in (46–73 mm)

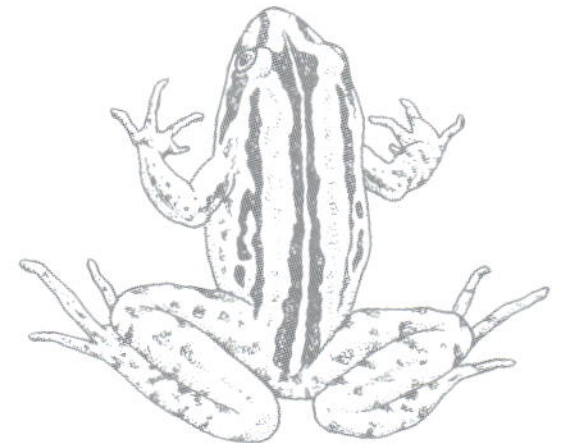

LIMNODYNASTES PERONII

STRIPED MARSH FROG

(DUMÉRIL & BIBRON, 1841)

Known for its voracious feeding habits, the Striped Marsh Frog eats a wide variety of animals, including other, smaller frogs. Breeding occurs between August and March. It is secretive by day, but males will call from dense vegetation. At night, they call while floating in water, producing a soft yet explosive "whuck." During the breeding season the male develops very swollen forearms and a sharp nuptial spine consisting of a bone protruding from its first finger. The female lays 700–1,000 eggs in a foam nest that is tangled in aquatic vegetation.

SIMILAR SPECIES

The Salmon-striped Frog (*Limnodynastes salmini*) is similar in appearance to *L. peronii*, but has three brown to pink stripes on its back. Breeding males do not develop the nuptial spine. The Spotted Grass Frog (*L. tasmaniensis*) is much smaller than *L. peronii* and has a dorsal skin pattern consisting of large spots rather than stripes.

The Striped Marsh Frog has long, powerful legs with long toes and a pointed snout. There is no webbing between either fingers or toes. The dorsal side is strikingly patterned with longitudinal dark and light brown stripes. There is often a pale stripe along the midline. The underside is white. The male's throat is yellowish with brown spots.

Actual size

FAMILY	Limnodynastidae
OTHER NAMES	Painted Frog, Painted Spadefoot Toad
DISTRIBUTION	Southeastern South Australia, western Victoria, Australia
ADULT HABITAT	Open grassland, woodland, mallee scrub
LARVAL HABITAT	Ponds and other still waterbodies
CONSERVATION STATUS	IUCN Least Concern. Has an extensive range and no known threats

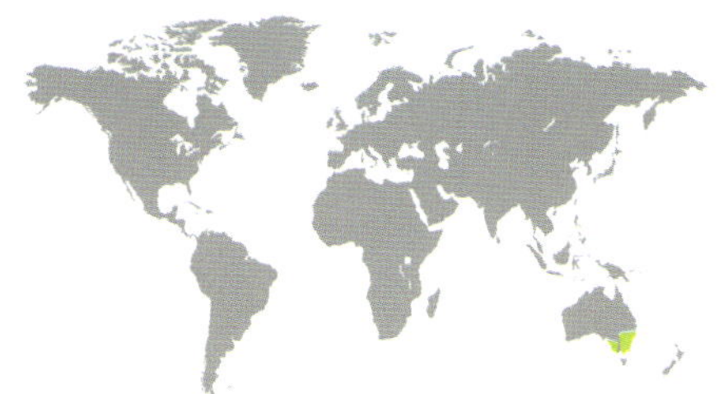

NEOBATRACHUS PICTUS

PAINTED BURROWING FROG

PETERS, 1863

ADULT LENGTH
Male
1 13/16–2 1/4 in (46–58 mm)
Female
1 7/8–2 1/8 in (48–55 mm)

A powerful burrower, this colorful frog lives in arid areas and spends much of its time underground. It emerges to breed after heavy rain, and migrates to ponds and other still waterbodies. Males call while floating in the water, producing a long, musical trill. Breeding males develop tiny, sharp projections on their back in the breeding season. It is thought that this deters other males from clasping them by mistake. The female lays around 1,000 eggs in a string, which is entwined around submerged vegetation. When threatened, the frog may inflate its body and rear up on straightened legs.

Actual size

SIMILAR SPECIES

There are ten species in this genus, all similar in appearance and in their habits. Seven are found in Western Australia, including the Shoemaker Frog (*Neobatrachus sutor*; page 98). Sudell's Frog (*N. sudelli*) has the widest range, across most of southern Australia. None is of conservation concern.

The Painted Burrowing Frog is large and stockily built. It has prominent eyes with vertical pupils but the ears are not visible. The limbs are short and robust, and there is no webbing on the fingers or toes. The back is olive-green in color with diffuse darker markings; the underside is white.

FAMILY	Limnodynastidae
OTHER NAMES	None
DISTRIBUTION	Central Western Australia
ADULT HABITAT	Clay or loamy soils in arid shrubland and deserts
LARVAL HABITAT	Ephemeral ponds on clay soil
CONSERVATION STATUS	IUCN Least Concern. Has a large range and no known threats

ADULT LENGTH
Male
$1\frac{3}{8}$–$1\frac{5}{8}$ in (35–42 mm)
Female
$1\frac{5}{16}$–2 in (34–51 mm)

NEOBATRACHUS SUTOR

SHOEMAKER FROG

MAIN, 1957

Actual size

The Shoemaker Frog has a plump, round body, short legs, and large protruding eyes. The toes are fully webbed and there is a horny ridge on the hind feet that it uses to shift soil when digging. The back is yellow to gold with brown blotches, and the belly is white.

This small burrowing frog gets its common name from its mating call, which resembles the "tap-tap-tap" of a cobbler's hammer. Most of its life is spent deep underground, protecting itself in a cocoon that covers its entire body except its nostrils. In this state, called estivation, its metabolic rate slows down; it can last for several months like this, even years if there is no rain. Shoemaker Frogs emerge after rain to feed on termites and, in the summer, to breed. They enter ephemeral pools and males call while floating in the water. Females lay strings of 200–1,000 eggs that take 40 days to complete their development.

SIMILAR SPECIES

The Shoemaker Frog is one of the smallest *Neobatrachus* burrowing frog species in Western Australia. Species in this genus are very similar in appearance and habits but can be differentiated by their calls. Humming Frog (*N. pelobatoides*) males produce a low-pitched hum, while Goldfield's Bullfrog (*N. wilsmorei*) males make a "plonk-plonk-plonk" sound. The latter species is also distinguished by having longitudinal yellow stripes on its back.

FAMILY	Limnodynastidae
OTHER NAMES	Catholic Frog, Crucifix Toad, Crucifix Frog, Holy Cross Frog
DISTRIBUTION	Inland Queensland and New South Wales, Australia
ADULT HABITAT	Arid savanna, woodland, mallee scrub
LARVAL HABITAT	Ephemeral ponds
CONSERVATION STATUS	IUCN Least Concern. Has a large range and no known threats

NOTADEN BENNETTI

HOLY CROSS TOAD

GÜNTHER, 1873

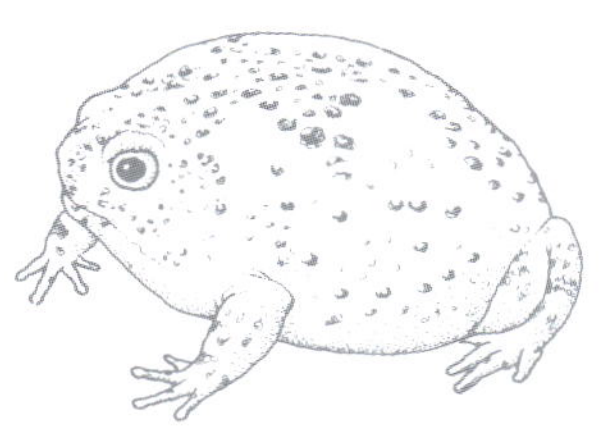

ADULT LENGTH
Male
1⅝–2½ in (42–63 mm)
Female
1¹³⁄₁₆–2¾ in (46–68 mm)

Nearly spherical in shape, this burrowing frog is a vigorous hopper when above ground. Found in arid areas, it mostly lives underground, emerging after rain to forage for ants and termites. It has an unusual form of defense: When attacked, it produces a very sticky, protein-rich skin secretion that deters its enemies. This secretion is also important during mating, enabling the male, whose arms are too short to grasp the female properly, to attach himself to her back. During breeding, the male produces an owl-like "woop." Eggs are laid in ephemeral pools, and development of eggs and tadpoles is completed in six weeks.

Actual size

The Holy Cross Toad has been described as resembling a colored ping-pong ball. The body is round and the legs are very short. The hind feet carry horny tubercles that serve as spades during digging. The back is bright yellow with a multicolored cross that includes black, white, and red markings. The flanks are bluish and the belly is white.

SIMILAR SPECIES

There are four species in the genus *Notaden*, all rather similar but found in different, non-overlapping regions of Australia. The Northern Spadefoot Toad (*N. melanoscaphus*), from the extreme north, is black in color and runs like a mouse. Weigel's Spadefoot Toad (*N. weigeli*) occupies a tiny range in Western Australia and is known only from female specimens. The Desert Spadefoot Toad (*N. nichollsi*) has the widest range across western and central Australia.

FAMILY	Limnodynastidae
OTHER NAMES	None
DISTRIBUTION	Confined to the Baw Baw Plateau near Melbourne, Victoria, Australia
ADULT HABITAT	Wet heath, streamside thickets
LARVAL HABITAT	Foam nest in pools
CONSERVATION STATUS	IUCN Critically Endangered. Has a very restricted range and has declined dramatically since the mid-1980s

ADULT LENGTH
Male
$1\frac{11}{16}$–$1\frac{13}{16}$ in (43–46 mm)

Female
$1\frac{7}{8}$–$2\frac{1}{8}$ in (47–55 mm)

PHILORIA FROSTI

BAW BAW FROG

SPENCER, 1901

Actual size

The Baw Baw Frog has a flattened body and head, and short limbs. The fingers and toes are not webbed. There are several warts on the back and prominent parotoid glands behind the head. The skin is dull brown to slate-gray in color, dappled with yellow markings.

Found only in very wet places, the Baw Baw Frog breeds between October and December. Males call to attract females with a series of moans and grunts. When a female approaches him, he clasps her around her pelvis. She produces 50–180 eggs and then uses her first and second fingers, which are equipped with special flanges, to whip the mucus around the eggs into a stiff foam; this is then concealed under a log or rock. The tadpoles remain in or close to the foam nest for 5–8 weeks, during which time they live off the yolk provided by their mother.

SIMILAR SPECIES

There are five other species in the genus *Philoria*, all found in rather restricted areas in southeastern Queensland and northwestern New South Wales. All are secretive frogs, living in water-saturated habitats, and all lay their eggs in foam nests in which the tadpoles feed only on yolk. Four of these species are classified as Endangered, and one Vulnerable. The status of the recently described Mt. Ballow Frog (*P. knowlesi*) is unknown.

FAMILY	Limnodynastidae
OTHER NAMES	Fletcher's Frog, Sandpaper Frog
DISTRIBUTION	Coastal Queensland and New South Wales, Australia
ADULT HABITAT	Forested areas with high rainfall
LARVAL HABITAT	Ephemeral waterbodies
CONSERVATION STATUS	IUCN Least Concern. Has become rare in some parts of its range as a result of deforestation

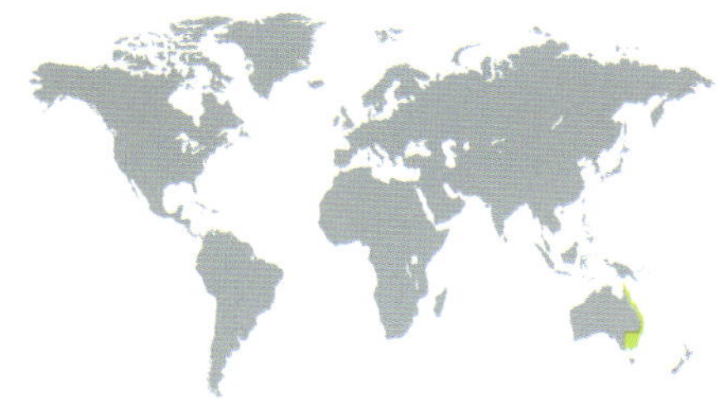

PLATYPLECTRUM FLETCHERI

BLACK-SOLED FROG

(BOULENGER, 1890)

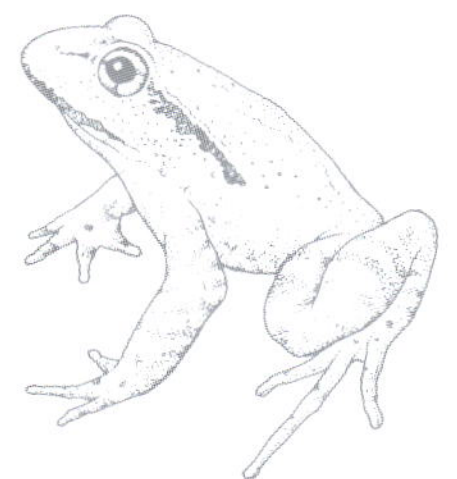

ADULT LENGTH
Male
1⅝–1⅞ in (42–48 mm)
Female
1¾–2⅛ in (45–54 mm)

This well-camouflaged frog spends much its time hiding under leaves and other ground cover, and is also often found in tree-holes. It breeds in summer after heavy rain, the males calling from the ground in or near water, making a "gar-r-r-up" sound. Mating occurs in ephemeral waterbodies, each female laying about 300 eggs in a foam nest. The tadpoles are notable for being cannibalistic, feeding on tadpoles of their own and of other species, but this is probably a response to a lack of their normal plant food as a result of overcrowding.

Actual size

The Black-soled Frog has a flattened head and body, long and muscular legs, and long fingers and toes that are not webbed. In the breeding season, the skin becomes rough and feels like sandpaper. The dorsal surface is yellow, pale brown, or dark brown. There is a narrow dark stripe through the eye and there are prominent stripes on the legs.

SIMILAR SPECIES

There are five other species in the genus *Platyplectrum*, three from New Guinea and two from Australia. New Guinea species were previously in genus *Lechriodus* and called cannibal frogs because both tadpoles and adults eat their counterparts, of the same and of other species. The males of the Arfak Cannibal Frog (*P. platyceps*) have formidable spines on their forelimbs that help them to clasp females securely. Males of *P. aganopsis* have smaller, softer spines; the species' scientific name means "gentle husband."

FAMILY	Limnodynastidae
OTHER NAMES	Ornate Frog
DISTRIBUTION	Northern and eastern Australia
ADULT HABITAT	Woodland and forest, including both arid and wet areas
LARVAL HABITAT	Ephemeral pools
CONSERVATION STATUS	IUCN Least Concern. Has an extensive range and no known threats

ADULT LENGTH
Male
1⅛–1⁷⁄₁₆ in (29–37 mm)
Female
1⅜–1⅝ in (35–42 mm)

PLATYPLECTRUM ORNATUM

ORNATE BURROWING FROG

(GRAY, 1842)

Actual size

The Ornate Burrowing Frog is a small, stubby frog with protruding eyes and partial webbing between the toes. It is very variable in color, being gray, brown, or yellow, often with darker markings. There is usually a pale butterfly-shaped patch behind the eyes. The dorsal surface is covered in red-tipped warts, and the ventral side is smooth and white.

The tadpoles of this burrowing frog often have to develop in small pools that become very hot. Recent research has shown that they are unusually tolerant of both high temperatures and high levels of ultraviolet radiation. Across much of its range, the Ornate Burrowing Frog lives in arid conditions and remains underground for long periods, emerging after rain to feed and to breed. Males produce a nasal "unk" mating call while floating in a pond. Females then lay up to 1,600 eggs in a foam nest that collapses after a few hours to form a floating layer.

SIMILAR SPECIES

Spencer's Burrowing Frog (*Platyplectrum spenceri*) is found across a large area of central Australia, from Western Australia to Queensland. Its habitat and natural history are similar to that of *P. ornatum*, though the male's mating call is very different, sounding like "ho-ho-ho." There are no members of this genus that are considered in danger.

FAMILY	Myobatrachidae
OTHER NAMES	Creek Frog
DISTRIBUTION	Southwestern Western Australia
ADULT HABITAT	Dense vegetation on clay soils
LARVAL HABITAT	Within the egg, in a burrow
CONSERVATION STATUS	IUCN Critically Endangered. Has a very restricted distribution within a greatly altered and fragmented habitat

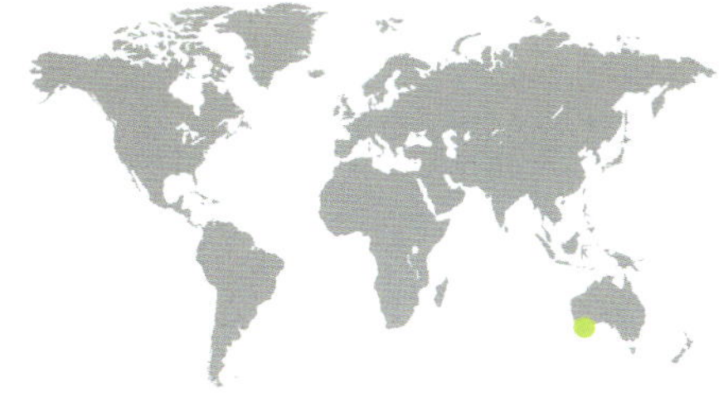

ANSTISIA ALBA

WHITE-BELLIED FROG

(WARDELL-JOHNSON & ROBERTS, 1989)

ADULT LENGTH
Male
3/4–15/16 in (20–24 mm)
Female
1 11/16 in (17 mm)

The total population of this small frog was estimated to be about 250 in 1998. The species occupies a very small range of 50 sq miles (130 sq km) but within this, it occupies a mere 1 sq mile (2.5 sq km). It exists in isolated populations, separated by inhospitable habitat, and there is virtually no movement of individuals between populations. Its habitat is threatened by vegetation change, fertilizer use, livestock, and bush fires, but several populations are now within protected areas. It has direct development, the larval stage being completed within the eggs, which are laid in shallow tunnels under dense vegetation.

Actual size

The White-bellied Frog gets its name from its white underside, which distinguishes it from other *Geocrinia* species. It is a small, squat frog with a warty back and a smooth belly. The back is pale gray or light brown, and speckled with darker markings.

SIMILAR SPECIES

Anstisia contains four species previously included in *Geocrinia* (see page 108). They are all confined to tiny areas in extreme southwestern Australia. The West Australian Orange-bellied Frog (*A. vitellina*) is considered Vulnerable, while two other species, Walpole's Frog (*A. lutea*) and the Karri Frog (*A. rosea*), are Low Concern.

FAMILY	Myobatrachidae
OTHER NAMES	Northern Sandhill Frog
DISTRIBUTION	Coastal areas near Shark Bay, Western Australia
ADULT HABITAT	Sand dunes
LARVAL HABITAT	Within eggs, in deep burrows
CONSERVATION STATUS	IUCN Least Concern. Has a very small range but occurs at high population densities

ADULT LENGTH
Male
1–1 3/16 in (26–30 mm)

Female
1 1/8–1 5/16 in (28–33 mm)

ARENOPHRYNE ROTUNDA

SANDHILL FROG

TYLER, 1976

Actual size

The Sandhill Frog has a broad head, a wide, slightly flattened body, and baggy skin, especially around the hind legs, as if its skin is too large for its body. Associated with its burrowing habits are powerful forelimbs with spade-like hands, and a protective pad on the nose. Its skin is white, cream, or pale green with blotches of black and brick-red.

A burrowing frog that, unusually, digs headfirst, the Sandhill Frog is totally emancipated from standing water, able to derive all the moisture it needs from damp sand beneath the ground surface. It lives underground during the day, emerging at night to forage for ants and other insects, which it catches by flicking out a long, slender tongue. It crawls rather than hops, and can travel up to 100 ft (30 m) from its daytime burrow in search of food. Males and females together dig a deep burrow in which up to 11 large eggs are deposited. Development is direct, the eggs hatching after two months to release tiny froglets.

SIMILAR SPECIES

The Southern Sandhill Frog (*Arenophryne xiphorhyncha*), described in 2008, is found in the Kalbarri National Park, to the south of Shark Bay. It is darker than *A. rotunda* and has a more pointed snout; genetic differences indicate that the two species diverged 5–7 million years ago. Both species show similarities to the Turtle Frog (*Myobatrachus gouldi*; page 112), which is also a headfirst burrower.

FAMILY	Myobatrachidae
OTHER NAMES	Marsupial Frog, Pouched Frog
DISTRIBUTION	Southern Queensland and northern New South Wales, Australia
ADULT HABITAT	Leaf litter in damp montane forest
LARVAL HABITAT	Pouches on their father's body
CONSERVATION STATUS	IUCN Vulnerable. Has declined where forests have been cleared, but its range includes protected areas

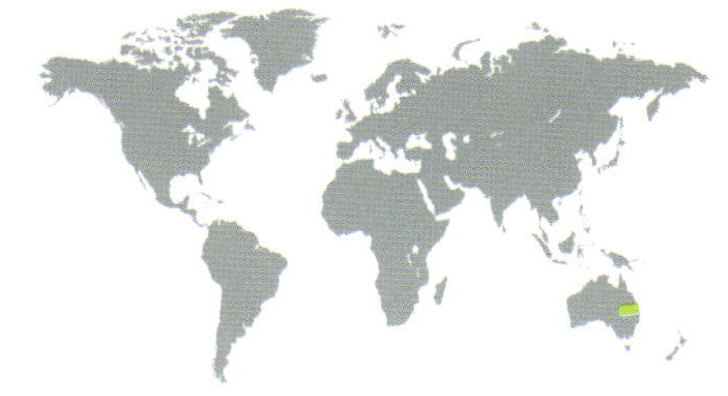

ASSA DARLINGTONI

HIP-POCKET FROG

(LOVERIDGE, 1933)

ADULT LENGTH
Male
9/16–3/4 in (15–19 mm)
Female
11/16–1 in (18–25 mm)

During the breeding season, males of this unique frog call from leaf litter. Mating, which has never been described, leads to the production of about ten eggs in a mass of jelly on the ground. Both parents guard the eggs until they hatch, at which point the male approaches the tadpoles and allows them to wriggle onto his back, and then into pouches on each side of his body close to his hind legs. After two months, the young emerge from these pouches as tiny froglets. Adult frogs live on the ground and move about by crawling rather than hopping.

Actual size

The Hip-pocket Frog has a broad body and slender arms, fingers, and toes. It is gray or reddish brown on the back, with dark V-shaped markings. There is a prominent fold of skin running from the eye to the groin. The skin is smooth on the back, and rough and warty on the flanks.

SIMILAR SPECIES

This species had been the only member of its genus since 1972, before which it was placed in the genus *Crinia* (see pages 106–107), but because of its unique mode of reproduction, it has been placed in its own genus. In 2021 a second species, the Mt. Wollumbin Hip-pocket Frog (*A. wollumbin*), was described. Both are small frogs, with smooth skin and long, unwebbed fingers and toes. The Hip-pocket Frog can be confused with the endangered Loveridge's Frog (*Philoria loveridgei*). The latter has thicker arms and folded skin on its back.

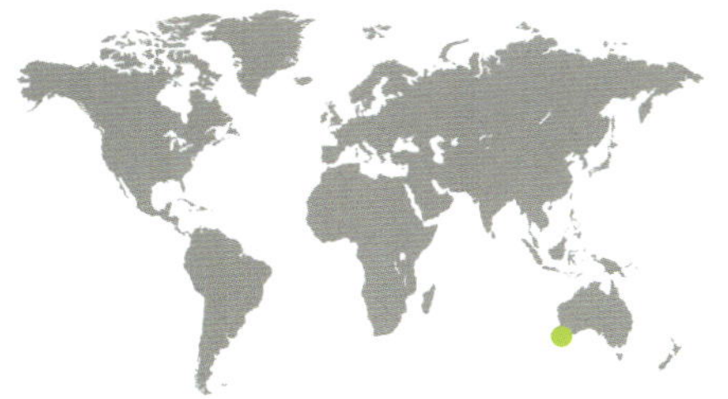

FAMILY	Myobatrachidae
OTHER NAMES	Red-thighed Froglet, Tschudi's Froglet
DISTRIBUTION	Southwestern corner of Western Australia
ADULT HABITAT	Coastal plains and forest where there is winter rain
LARVAL HABITAT	Ponds and other small waterbodies
CONSERVATION STATUS	IUCN Least Concern. Has a large range and no apparent threats

ADULT LENGTH
Male
$\frac{15}{16}$–$1\frac{1}{4}$ in (24–32 mm)

Female
$1\frac{3}{16}$–$1\frac{7}{16}$ in (30–36 mm)

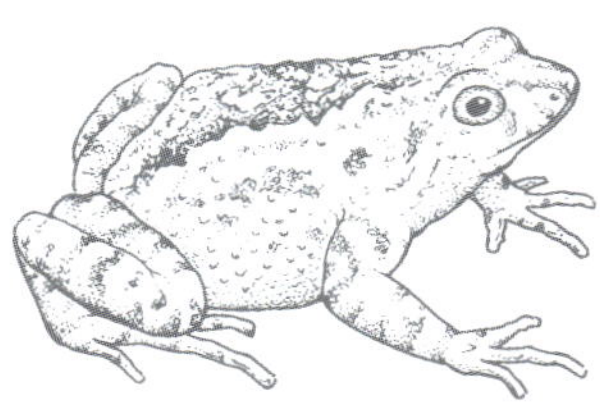

CRINIA GEORGIANA

QUACKING FROG

TSCHUDI, 1838

Actual size

The Quacking Frog has a squat, flattened body, a large head, smooth skin, and short limbs. It is gray, brown, or black with lighter or darker longitudinal stripes, and there is a red patch in the groin and over much of the thigh. The upper eyelids are red or golden.

As its name implies, this small frog has a mating call like the quack of a duck. Breeding occurs on cold nights between July and October, with males forming choruses around ponds. Individuals produce a variable number of quacks, and tend to match their calling to that of their near neighbors so as to be equally attractive to females. As in most frogs, the male clasps the female from behind but, unusually, in nearly half of pairings a second male will clasp her from her ventral side. Paternity studies have revealed that this results in her eggs being fathered by more than one male.

SIMILAR SPECIES

There are 17 species in the genus *Crinia*. All have long, unwebbed fingers and toes and lay their eggs in small clumps in water. These frogs are highly variable in terms of skin color and texture. Each species has a distinctive call. For example, the Bilingual Froglet (*C. bilingua*), found in northwestern Australia, has a two-part call, which consists of a short note followed by a prolonged trill.

FAMILY	Myobatrachidae
OTHER NAMES	Beeping Froglet, Plains Froglet
DISTRIBUTION	Southeastern Australia
ADULT HABITAT	Woodland, floodplains, open and disturbed areas
LARVAL HABITAT	Ponds, dams, swamps, etc.
CONSERVATION STATUS	IUCN Least Concern. Has a large range and no apparent threats

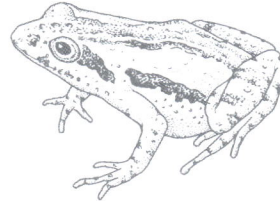

ADULT LENGTH
Male
11/16–7/8 in (18–22 mm)
Female
7/8–15/16 in (21–23 mm)

CRINIA PARINSIGNIFERA

EASTERN SIGN-BEARING FROGLET

MAIN, 1957

Unusually, this small, cryptically colored frog calls year-round, often during the day. It hides by day under logs and plant debris, sometimes in large numbers. It breeds in midwinter, males calling from emergent vegetation at the edge of a pond. The call is an often repeated "eeeeek" that has been likened to the sound of a wet finger being drawn across an inflated balloon. The eggs are small and are found singly or in loose clumps at the bottom of the pond. The tadpoles reach metamorphosis in 11–12 weeks.

Actual size

The Eastern Sign-bearing Froglet is very variable in color but is usually brown on the back and gray on the belly. Like some other froglets in this genus, individuals may have smooth, wrinkled, or warty skin. There is no webbing on the fingers or the toes.

SIMILAR SPECIES

The Common Froglet (*Crinia signifera*) is highly variable in both coloration and skin texture. Its skin may be smooth, warty, or wrinkled. It is readily distinguished from *C. parinsignifera* by its call, which is a "crick...crick...crick" sound. The Endangered Sloane's Froglet (*C. sloanei*) is less variable and is usually mustard-yellow in color.

FAMILY	Myobatrachidae
OTHER NAMES	Victorian Smooth Froglet
DISTRIBUTION	Southeastern Victoria, Australia
ADULT HABITAT	Moist areas in forests, woodlands, shrubland, and grassland
LARVAL HABITAT	Dams, ditches, ponds
CONSERVATION STATUS	IUCN Least Concern. Has a large range and no apparent threats

ADULT LENGTH
Male
$^{15}/_{16}$–1$^{1}/_{8}$ in (24–28 mm)

Female
$^{7}/_{8}$–1$^{5}/_{16}$ in (21–33 mm)

GEOCRINIA VICTORIANA

EASTERN SMOOTH FROG

(BOULENGER, 1888)

Actual size

This small frog is secretive in its habits but has a distinctive and extremely interesting call consisting of two parts: one to three "wa-a-a-rk" notes, followed by up to 50 short chirps. Experiments using recorded calls have shown that the introductory notes act as territorial calls between males, and that the chirps serve to attract females. Eggs are laid in clutches of 90–160 in moist leaf litter or at the bases of grass tussocks. The tadpoles can survive within the eggs for up to four months, until heavy rain allows them to emerge from the eggs and enter standing waterbodies.

SIMILAR SPECIES

The Smooth Frog (*Geocrinia laevis*) is physically indistinguishable from *G. victoriana* but does not occur in the same areas and has a very different call, consisting of a number of "cra-a-a-ack" notes, of which the first is the longest. It is found in eastern Victoria and parts of Tasmania.

The Eastern Smooth Frog has a rounded snout, short limbs, and a smooth belly, and lacks webbing on its fingers or toes. The back is brown or gray and there is often a pink patch in the armpit, the groin, or both. The backs of the thighs are pink, marbled with black. The throat may be yellowish.

FAMILY	Myobatrachidae
OTHER NAMES	Forest Toadlet
DISTRIBUTION	Southwestern corner of Western Australia
ADULT HABITAT	Karri (*Eucalyptus diversicolor*) forest
LARVAL HABITAT	Within the egg
CONSERVATION STATUS	IUCN Least Concern. Has a large range and shows no evidence of decline

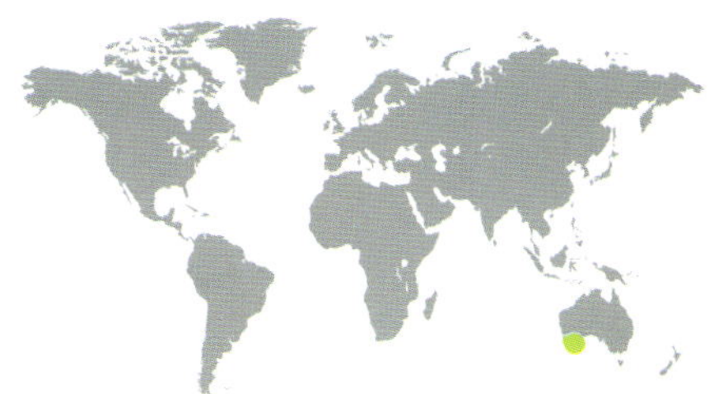

METACRINIA NICHOLLSI

NICHOLL'S TOADLET

(HARRISON, 1927)

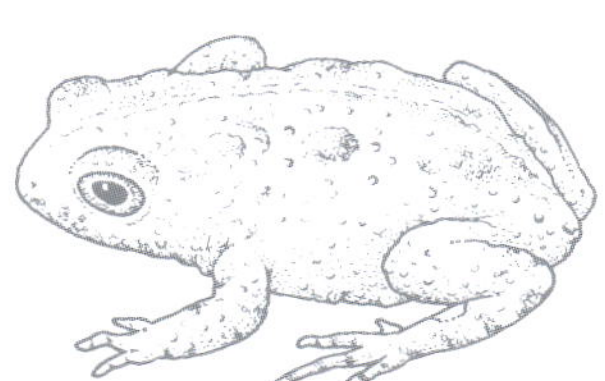

ADULT LENGTH

Male

¾–⅞ in (19–21 mm)

Female

⅞–15/16 in (22–24 mm)

This tiny frog is entirely freed from standing water, though it does require damp conditions. It is found in forests in leaf litter, under stones, or under logs. It breeds in summer, following rain, with the male producing a short "ark" mating call. The female lays 25–30 eggs in a moist spot in leaf litter or under a log. This species shows direct development, the entire tadpole phase taking place within the egg, which hatches to produce a tiny froglet. This process takes around two months.

Actual size

SIMILAR SPECIES

The only species in the genus *Metacrinia*, Nicholl's Toadlet is rather similar to toadlets of the genus *Pseudophryne* (see pages 114–116), which have a similar call but do not show direct development. Only one of these toadlets is found in the same area, Günther's Toadlet (*P. guentheri*; page 116), but it is larger than *M. nichollsi* and lacks ventral yellow or orange markings.

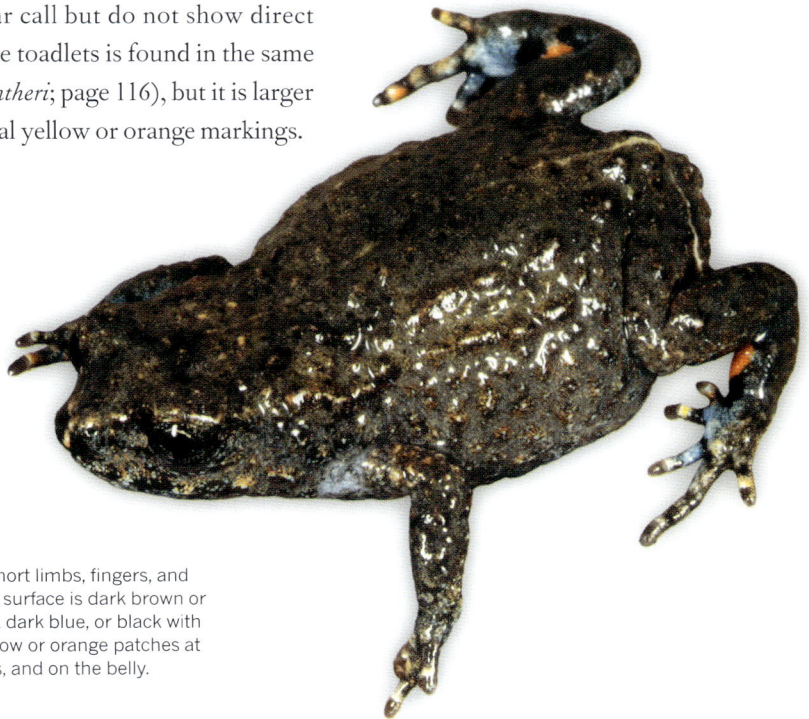

Nicholl's Toadlet is a squat frog with short limbs, fingers, and toes, and a very warty back. The dorsal surface is dark brown or black with pink flecks. The belly is gray, dark blue, or black with a marbled white pattern. There are yellow or orange patches at the base of each limb, under the thighs, and on the belly.

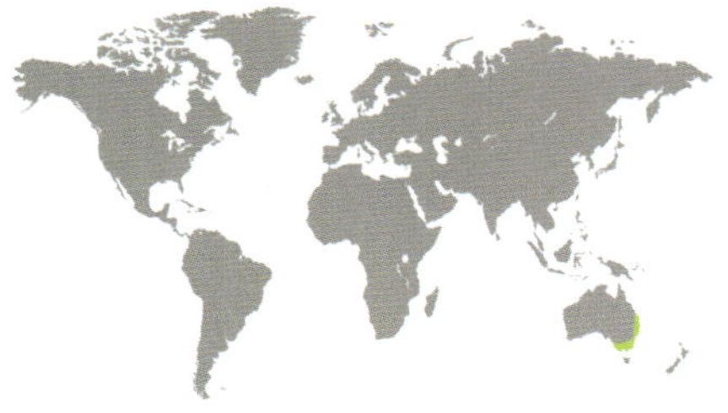

FAMILY	Myobatrachidae
OTHER NAMES	Silver-eyed Barred Frog, Southern Barred Frog
DISTRIBUTION	Coastal mountains of New South Wales and Victoria, Australia
ADULT HABITAT	Rainforest at altitudes of 65–4,600 ft (20–1,400 m)
LARVAL HABITAT	Permanent forest streams
CONSERVATION STATUS	IUCN Vulnerable. Declined dramatically in the mid-1980s and is now known from only a few localities

ADULT LENGTH
Male
$2\frac{3}{8}$–$2\frac{1}{2}$ in (60–63 mm)

Female
$2\frac{15}{16}$–$3\frac{1}{8}$ in (74–80 mm)

MIXOPHYES BALBUS

STUTTERING FROG

STRAUGHAN, 1968

This large frog breeds between September and April. Males produce a soft, grating trill from hidden spots beside streams. During mating, a nest is dug in shallow, flowing water and around 500 eggs are laid in it, pasted firmly onto rocks. The tadpoles take a year to reach metamorphosis. The causes of the dramatic decline of the Stuttering Frog are not known for certain but are thought to include degradation of streams as a result of upstream human activities such as logging, the disease chytridiomycosis, and introduced fish, which eat tadpoles. There is an ongoing captive breeding program for this species.

The Stuttering Frog is a plump frog with long, muscular legs. There is partial webbing between the toes but no webbing between the fingers. The dorsal surface and legs are yellow-gray, brown, or olive-green with darker blotches, and there are dark bars on the legs and arms. There is a dark stripe running from the tip of the nose to behind the eye.

SIMILAR SPECIES

There are marked genetic differences between southern and northern populations, and the animals from the south, in Victoria, are probably a distinct species. The captive breeding program is keeping the two populations separate to maintain genetic distinctiveness. The related Great Barred Frog (*Mixophyes fasciolatus*) is a larger frog, found in the coastal mountains of New South Wales and Queensland, and is not of conservation concern.

Actual size

FAMILY	Myobatrachidae
OTHER NAMES	None
DISTRIBUTION	Conondale Range, southeastern Queensland and Richmond Range, northeastern New South Wales, Australia
ADULT HABITAT	Wet montane forest
LARVAL HABITAT	Mountain streams
CONSERVATION STATUS	IUCN Endangered. Has a restricted range. May be recovering from a population decline during the 1970s and 1980s

MIXOPHYES FLEAYI

FLEAY'S BARRED FROG

CORBEN & INGRAM, 1987

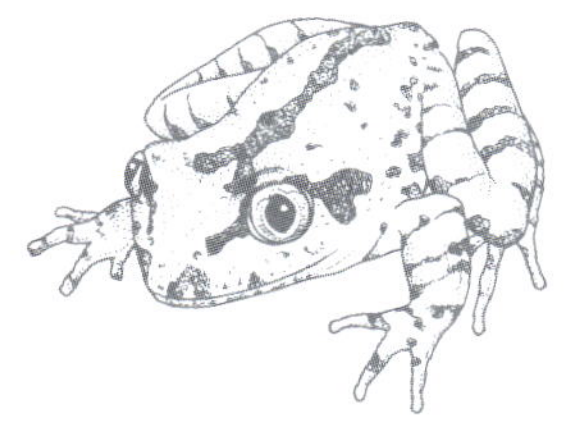

ADULT LENGTH
Male
2½–2^{13}⁄$_{16}$ in (63–70 mm)
Female
3^{1}⁄$_{16}$–3½ in (79–89 mm)

This handsome frog provides clear evidence that at least some frog populations can recover from the devastating effects of the disease chytridiomycosis. Found in only two small mountain ranges, it suffered severe population declines, beginning in the 1970s, along with many other stream-living frogs in eastern Australia. However, a seven-year study at the beginning of this century has revealed three- to ten-fold increases in numbers in several populations. The breeding biology of this species is very similar to that of the Stuttering Frog (*Mixophyes balbus*; page 110).

SIMILAR SPECIES

The Giant or Southern Barred Frog (*Mixophyes iteratus*) is larger than *M. fleayi* and has a more extensive range in coastal Queensland and New South Wales. It is classed as Vulnerable, but its decline seems to have been due more to the fragmentation and degradation of its forest habitat than to chytridiomycosis. There are nine species in the genus *Mixophyes*.

Fleay's Barred Frog has a blunt snout, and large protruding eyes with vertical pupils. The toes are partially webbed and the fingers are not webbed. The back is light brown with darker mottled markings and the underside is white or yellow. Between the eyes is a darker Y-shaped marking that extends down the back.

Actual size

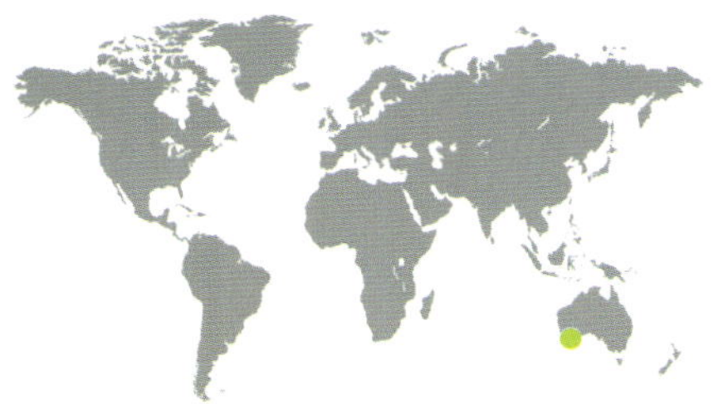

FAMILY	Myobatrachidae
OTHER NAMES	None
DISTRIBUTION	Southwestern corner of Western Australia
ADULT HABITAT	Sandy soils in arid woodland and shrubland
LARVAL HABITAT	Within the egg, deep underground
CONSERVATION STATUS	IUCN Least Concern. Has no known threats

ADULT LENGTH
Male
1$\frac{5}{16}$–1$\frac{5}{8}$ in (34–42 mm)
Female
1$\frac{3}{4}$–1$\frac{15}{16}$ in (44–50 mm)

MYOBATRACHUS GOULDII

TURTLE FROG

(GRAY, 1841)

Actual size

The Turtle Frog is so called because it looks rather like a turtle without its shell. It has a small head and very small eyes. The body is spherical and the limbs are short. The forelimbs are powerful, reflecting their importance for digging into the ground. The skin is pink, light brown, or dark brown in color.

This weird-looking frog lives almost all its life deep underground. It is unusual in that it digs its way headfirst into the ground. It may be found under logs and is often associated with the nests of termites, on which it feeds. It breeds in summer after heavy rain, when the males produce abrupt deep croaks with only their heads above ground. Mating occurs within the male's burrow and results in around 40 very large eggs buried as much as 4 ft (1.2 m) underground. Development takes up to two months and is direct, occurring within the egg, the young emerging as tiny frogs.

SIMILAR SPECIES

The only member of its genus, *Myobatrachus gouldii* is quite unlike any other frog. It is most similar to the two species of sandhill frogs (genus *Arenophryne*; page 104), which also dig into the ground headfirst and have direct development.

FAMILY	Myobatrachidae
OTHER NAMES	Red-groined Froglet
DISTRIBUTION	Coastal areas of New South Wales and eastern Victoria, Australia
ADULT HABITAT	Near water in forest and heathland
LARVAL HABITAT	Permanent water
CONSERVATION STATUS	IUCN Least Concern. Has a large range. There are no apparent threats but it may be adversely affected by future human development

PARACRINIA HASWELLI

HASWELL'S FROGLET

(FLETCHER, 1894)

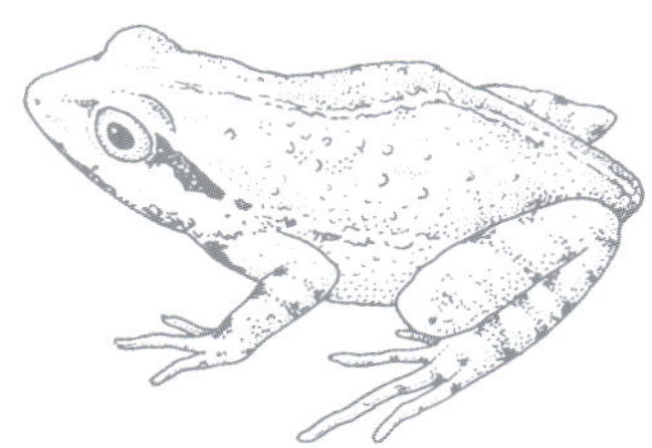

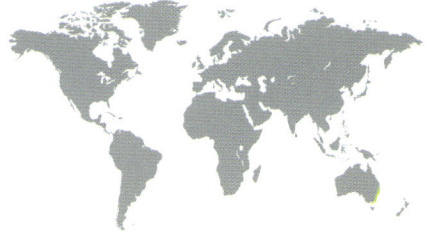

ADULT LENGTH
Male
1¼–1⁵⁄₁₆ in (31–33 mm)
Female
1¼–1⅜ in (32–35 mm)

This little-known frog is found around streams, dams, and swamps in coastal areas, often hiding under stones. It breeds in permanent waterbodies in spring and summer, males calling from grasses and sedges that emerge from the water. The call is said to resemble the quacking of distant ducks. Eggs are laid in loose clusters of 8–80. It is not a common species but its population appears to be stable. It may, however, be threatened in the future by tourism development of its coastal habitat.

SIMILAR SPECIES

Paracrina haswelli is the only species in the genus, which was previously grouped with froglets of the genus *Crinia* (see pages 106–108). It differs from the other Australian froglets in having rather longer legs.

Actual size

Haswell's Froglet has a slender body and head, long legs, and long fingers and toes. The back is dull brown in color, often with a thin, pale stripe down its middle. There is also a conspicuous black stripe running down each side of the head through the eye, and a brick-red patch in the groin and on the back of the thighs.

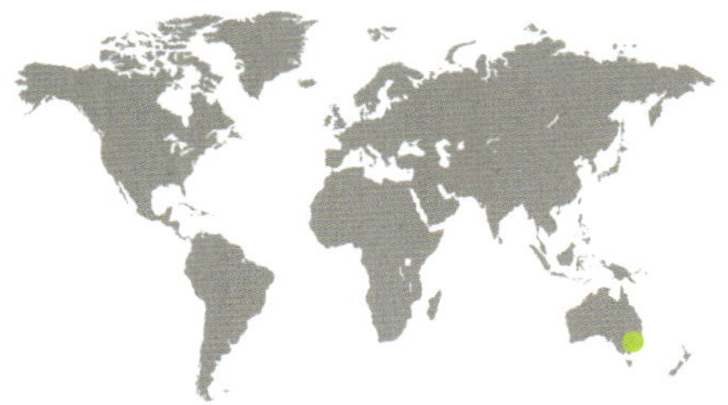

FAMILY	Myobatrachidae
OTHER NAMES	None
DISTRIBUTION	Hawkesbury Sandstone region around Sydney, New South Wales, Australia
ADULT HABITAT	Woodland, under logs or stones in damp, marshy areas
LARVAL HABITAT	Ephemeral creeks
CONSERVATION STATUS	IUCN Near Threatened. Has a very restricted range and its habitat is subject to intense pressure from urbanization

ADULT LENGTH
Male
⅞–1⅛ in (22–28 mm)
Female
1–1⅛ in (25–29 mm)

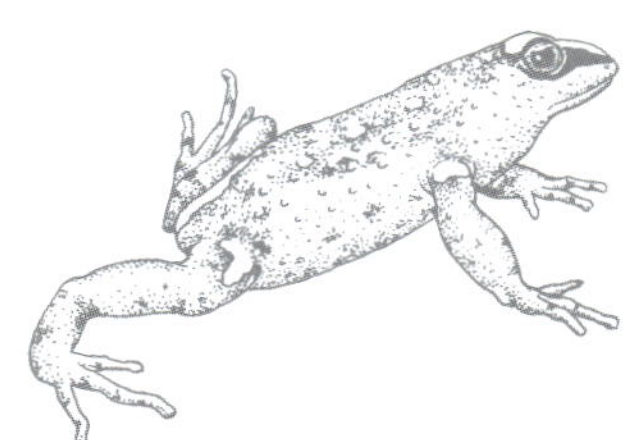

PSEUDOPHRYNE AUSTRALIS

RED-CROWNED TOADLET

(GRAY, 1835)

Actual size

The Red-crowned Toadlet gets its name from the T-shaped, bright red or orange patch on its head, between the ears, and extending to the snout. There is also a red stripe on the rump. Otherwise the back is dark brown, with red, black, and white spots. There is a white spot on the upper arm and the belly has a marbled pattern of black and white markings.

This small, colorful frog is unusual in that it may breed at any time of year, other than winter, with females able to produce several clutches of eggs in a year. It is somewhat gregarious, with 20–30 animals found close together in one area. The large eggs are laid in clutches of about 20 in a crude nest on the ground, the male staying close to them until rain enables the tadpoles to escape into ephemeral pools. Urbanization close to the species' habitat has seriously reduced its abundance and, in a 1998 study, only 56 breeding sites were found.

SIMILAR SPECIES

There are 14 species in this genus, ten of them occurring in eastern and southeastern Australia, three in the west, and one in south-central Australia. Several of the southeastern species have overlapping ranges and hybrids among them are quite common. This may be related to the fact that they have rather similar calls and poorly developed ears, suggesting that mating calls are not as effective in species recognition as they are in many frogs.

FAMILY	Myobatrachidae
OTHER NAMES	None
DISTRIBUTION	Australian Alps, New South Wales, Australia
ADULT HABITAT	*Sphagnum* bogs at altitudes of 4,070–5,610 ft (1,240–1,710 m)
LARVAL HABITAT	Pools in *Sphagnum* bogs
CONSERVATION STATUS	IUCN Critically Endangered. Has a very restricted range and is threatened by climate change

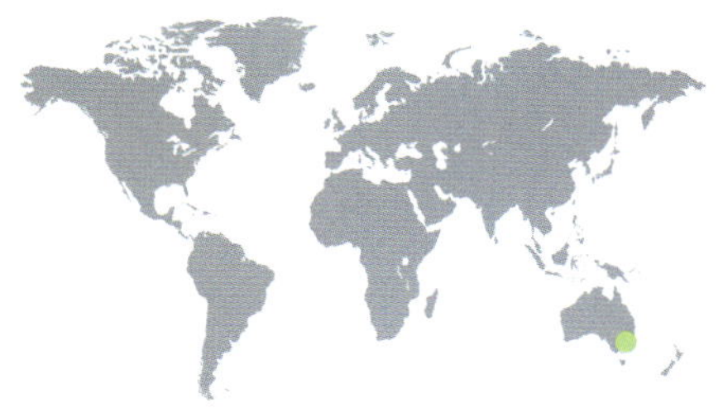

PSEUDOPHRYNE CORROBOREE

SOUTHERN CORROBOREE FROG

MOORE, 1953

ADULT LENGTH
Male
1–1⅛ in (26–28 mm)
Female
1–1$^{3}/_{16}$ in (26–30 mm)

An icon of the dramatic decline that has occurred among many Australian frogs, the Southern Corroboree Frog is the subject of an intensive conservation program in which animals are bred in captivity and released into the wild. It is a "sit and wait" predator, feeding mostly on ants and termites. In the breeding season, between January and March, males dig burrows from which they call to attract females. A clutch of 10–38 eggs is deposited in the burrow and the male protects them for 2–4 weeks. Individuals reach sexual maturity at three years.

Actual size

The Southern Corroboree Frog is distinctively colored, with a bright yellow back marked with longitudinal black stripes. The belly is black and white. The bright coloration provides a warning to potential predators that glands in its skin produce toxic secretions. Unlike most frogs, which derive skin toxins from their diet, corroboree frogs synthesize their own.

SIMILAR SPECIES

The Northern Corroboree Frog (*Pseudophryne pengilleyi*), formerly regarded as a separate population of the same species, has narrower stripes and a yellow or greenish background color. Also classified as Critically Endangered, it lost 42 percent of its breeding sites during a severe drought between 1997 and 2009. The disease chytridiomycosis has contributed to its decline in some parts of its range, and in places invasive alien plants have made the habitat unsuitable.

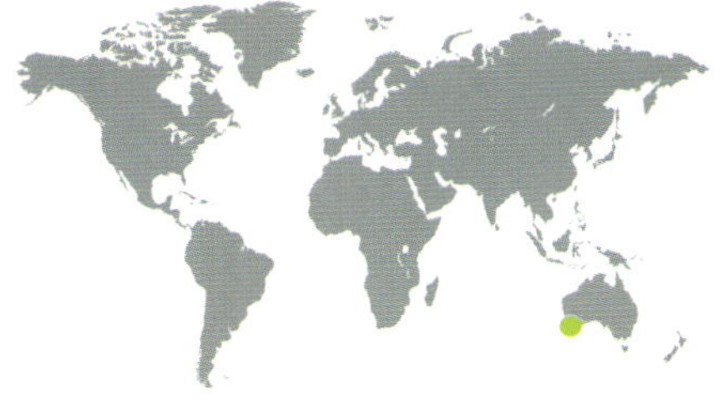

FAMILY	Myobatrachidae
OTHER NAMES	Crawling Frog
DISTRIBUTION	Southwestern corner of Western Australia
ADULT HABITAT	Wet and dry forest, grassland, farmland
LARVAL HABITAT	Ephemeral pools
CONSERVATION STATUS	IUCN Least Concern. May be threatened in future by reduced rainfall brought about by climate change

ADULT LENGTH
Male
1–1 3/16 in (26–30 mm)
Female
1 1/8–1 5/16 in (29–33 mm)

PSEUDOPHRYNE GUENTHERI

GÜNTHER'S TOADLET

BOULENGER, 1882

Actual size

Günther's Toadlet has very warty skin, which forms folds where it converges between the shoulders. There are two large horny tubercles on each foot. The skin is pale gray or brown with darker spots and blotches on the back, and there is a pale T-shaped patch on the head. The belly has a marbled black and white pattern.

This small frog breeds in the autumn, following rain. The male's call is a short squawk. The eggs are large and are deposited in damp soil under logs or rocks, or in tunnels. It has frequently been observed that one of the parents remains close to the eggs. The tadpoles emerge from the eggs at an advanced stage of development, when heavy rain washes them into ephemeral pools. Western Australia is a region experiencing long-term climate change, and associated reduced rainfall may pose a long-term threat to this species.

SIMILAR SPECIES

Two other toadlet species are found in Western Australia. Douglas's Toadlet (*Pseudophryne douglasi*) occurs in the north of the state and is the only member of the genus to breed in open water rather than on land. The Orange-crowned Toadlet (*P. occidentalis*) is a tiny frog that is adapted to very arid habitats.

FAMILY	Myobatrachidae
OTHER NAMES	Southern Platypus Frog
DISTRIBUTION	Blackall and Conondale mountain ranges, southeastern Queensland, Australia
ADULT HABITAT	Forest streams at altitudes of 1,300–2,600 ft (400–800 m)
LARVAL HABITAT	Within the mother's stomach
CONSERVATION STATUS	IUCN Extinct. Last seen in the wild in 1981

RHEOBATRACHUS SILUS

SOUTHERN GASTRIC BROODING FROG

LIEM, 1973

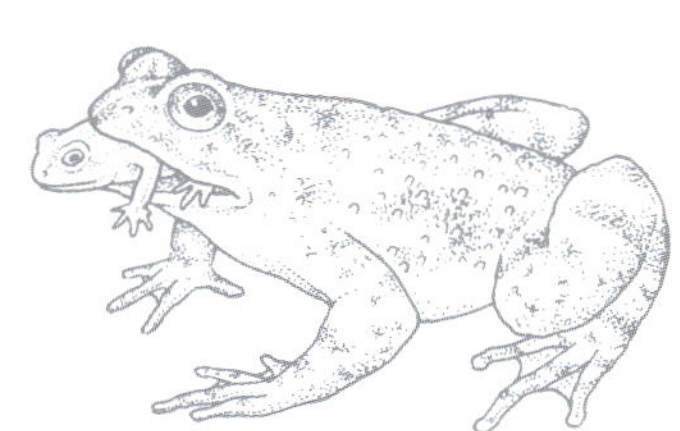

ADULT LENGTH
Male
1$\frac{5}{16}$–1$\frac{5}{8}$ in (33–41 mm)

Female
1$\frac{3}{4}$–2$\frac{1}{8}$ in (44–54 mm)

The extinct gastric brooding frogs were unique in that the tadpoles developed in their mother's stomach, a process that lasted 6–7 weeks, at the end of which tiny froglets emerged from the mother's mouth. The mating process was never observed but it is assumed that the female swallowed the eggs just after they were fertilized by the male. Females are known to have produced about 40 eggs, but brooded only 21–26 of them. They did not feed during brooding, and secretions from the developing young prevented the production of acid in the maternal stomach.

SIMILAR SPECIES

The Northern Gastric Brooding Frog (*Rheobatrachus vitellinus*) was discovered in 1984 but had disappeared in the wild by 1985. It occupied a very small range in rainforest within the Eungella National Park, Queensland. While both species were subject to deterioration of their stream habitat by feral pigs and invasive plants, it is thought that they were exterminated by the disease chytridiomycosis.

The Southern Gastric Brooding Frog lived in or very close to water. A strong swimmer, it had fully webbed hind feet. The fingers were long and not webbed. The eyes were large and projected upward, enabling it to see above the water's surface. It was dull gray to slate-gray on the back, with darker and lighter patches, and paler on the underside.

Actual size

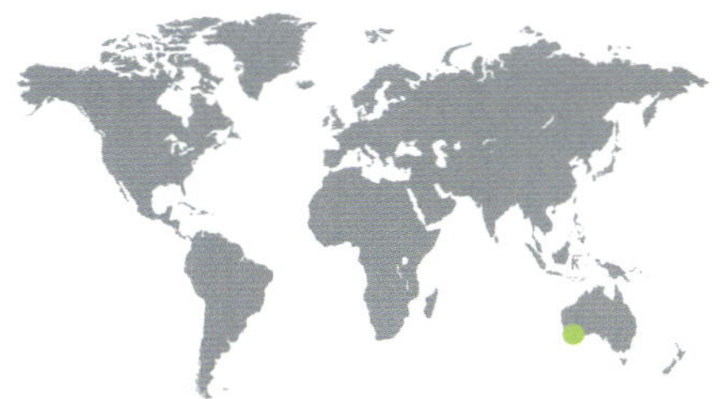

FAMILY	Myobatrachidae
OTHER NAMES	None
DISTRIBUTION	An area of less than 7.7 sq miles (20 sq km) in southwestern Australia
ADULT AND LARVAL HABITAT	Peat swamps in high-rainfall areas
CONSERVATION STATUS	IUCN Endangered. Has a very restricted range that is vulnerable to bush fires

ADULT LENGTH
Male
1⅛–1⅜ in (29–35 mm)
Female
1¼–1⁷⁄₁₆ in (31–36 mm)

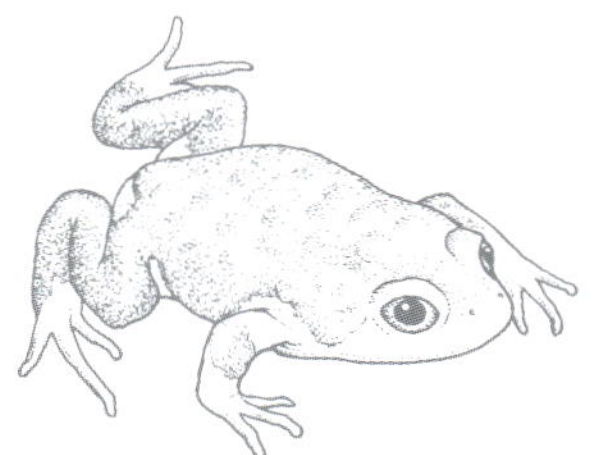

SPICOSPINA FLAMMOCAERULEA

SUNSET FROG

ROBERTS, HORWITZ, WARDELL-JOHNSON, MAXSON & MAHONY, 1997

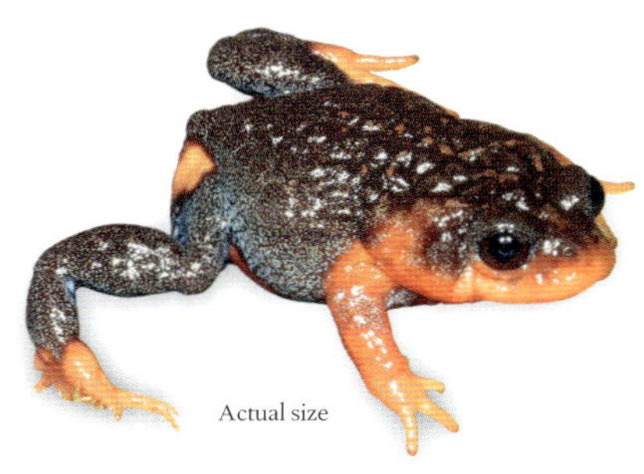

Actual size

This tiny ground-living frog was discovered as recently as 1994, when 27 populations were found in a restricted area near the coast of Western Australia. It is a habitat specialist, found only in peat swamps. Males call from October to December from pools, seepages, and streams, and females produce fewer than 200 eggs, laid singly on algae. The very small range of this frog makes it particularly susceptible to adverse environmental events—one population was badly affected by a bush fire in 1994. It is the focus of a captive breeding and reintroduction program.

SIMILAR SPECIES

Spicospina flammocaerulea is the only member of its genus. Its closest relatives are the toadlets of the genus *Uperoleia* (see pages 121–123), which have similar eggs and tadpoles. The Sunset Frog's present habitat is thought to be a relic of a much more extensive area of peat swamp that existed 5–6 million years ago.

The Sunset Frog gets its name from its remarkable coloration: purple, black, or dark gray on the back, and bright orange with blue spots on the belly. Its hands and feet are yellow or red and have no webbing. It has large parotoid glands behind its bulbous eyes, and numerous prominent glands on its back.

FAMILY	Myobatrachidae
OTHER NAMES	Mount Glorious Day Frog, Southern Day Frog
DISTRIBUTION	Blackall, Conondale, and D'Aguilar mountain ranges, eastern Queensland, Australia
ADULT HABITAT	In or near forest streams at altitudes of 1,600–2,600 ft (500–800 m)
LARVAL HABITAT	Forest streams
CONSERVATION STATUS	IUCN Extinct. Not seen since 1979

ADULT LENGTH
Male
⅞–1¹⁄₁₆ in (22–27 mm)
Female
⅞–1¼ in (22–31 mm)

TAUDACTYLUS DIURNUS

MOUNT GLORIOUS TORRENT FROG

STRAUGHAN & LEE, 1966

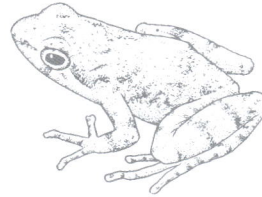

This frog lived in and close to fast-flowing streams, a habitat that is always damp, allowing it to be active by day. Breeding occurred in warm, wet weather between October and May. Eggs were laid in clumps of 24–36, attached to submerged rocks or branches. Despite the fact that most of its range lay within a national park, the Mount Glorious Torrent Frog population declined dramatically over a period of four years in the late 1970s. Its decline may have been due to disturbance of its stream habitat by feral pigs and invasion by alien plants, but was more likely caused by the disease chytridiomycosis.

Actual size

The Mount Glorious Torrent Frog had smooth skin and lacked webbing on its hands and feet. It was gray or brown with darker mottling above, and cream or white with dark spots below. It had adhesive disks at the ends of its fingers and toes. Males had a very soft, chuckling call and lacked vocal sacs.

SIMILAR SPECIES

Taudactylus diurnus was one of six members of the genus, from coastal Queensland, and most of them associated with fast-flowing streams. The Sharp-nosed Day Frog (*T. acutirostris*) is also now believed Extinct, the Tinkling Frog (*T. rheophilus*) and Kroombit Tinker Frog (*T. pleione*) are Critically Endangered, while the Eungella Torrent Frog (*T. eungellensis*) is Endangered. It is thought that their decline is probably also a result of the disease chytridiomycosis.

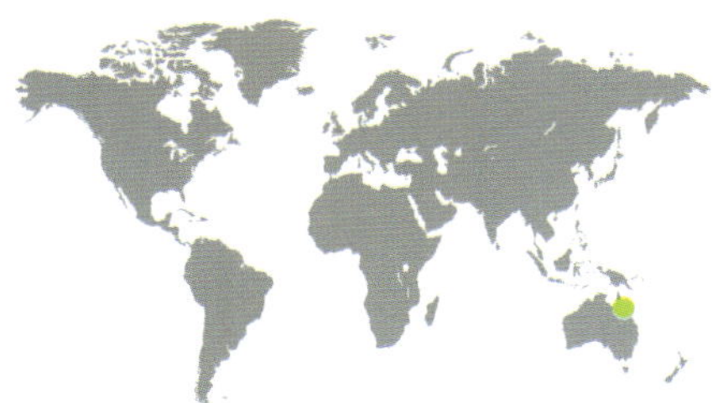

FAMILY	Myobatrachidae
OTHER NAMES	Northern Timber Frog
DISTRIBUTION	Coastal Queensland, around Cairns, Australia
ADULT HABITAT	Fast-flowing streams in upland rainforest
LARVAL HABITAT	Thought to be fast-flowing mountain streams
CONSERVATION STATUS	IUCN Critically Endangered. Confined to five mountain tops

ADULT LENGTH
Male
15/16–1 1/16 in (24–27 mm)
Female
15/16–1 1/4 in (24–31 mm)

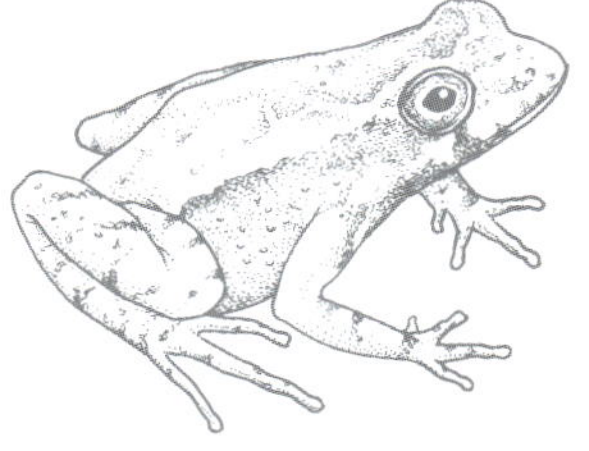

TAUDACTYLUS RHEOPHILUS

TINKLING FROG

LIEM & HOSMER, 1973

Actual size

The Tinkling Frog has disks on the ends of its fingers and toes, but no webbing on its hands or feet. It has a gray to reddish or dark brown back with irregular dark markings. There is a dark bar on the head between the eyes. A narrow, pale stripe runs from the eye to the groin, with a broader dark stripe beneath it. There are also dark stripes on the legs.

The aptly named Tinkling Frog lives under rocks and logs close to fast-flowing mountain streams. Males call round the clock, but mostly by day, producing a metallic "tink" that is repeated four or five times in quick succession. Eggs and tadpoles have not been observed, but gravid females have been found carrying 35–50 eggs. One of the world's rarest frogs, this species declined so much in the 1980s and 1990s that it was thought to have become extinct. In 1996, however, calling males were heard and, subsequently, a few juveniles have been found.

SIMILAR SPECIES

Two of the six species in the genus *Taudactylus*, the Mount Glorious Torrent Frog (*T. diurnus*; see page 119) and Sharp-nosed Day Frog (*T. acutirostris*), are Extinct, and the others are threatened with extinction. Together with other frogs living at higher altitudes in eastern Australia, they share certain characteristics that make them very susceptible to the lethal disease chytridiomycosis. These include low fecundity, specialized habitat requirements, and reproduction in flowing streams.

FAMILY	Myobatrachidae
OTHER NAMES	Eastern Gungan
DISTRIBUTION	Eastern Australia, from Queensland, through New South Wales to Victoria
ADULT HABITAT	Forest, woodland, grassland
LARVAL HABITAT	Ponds, lagoons, lakes, dams
CONSERVATION STATUS	IUCN Least Concern. Has a large range, but may be threatened in the future by urban development

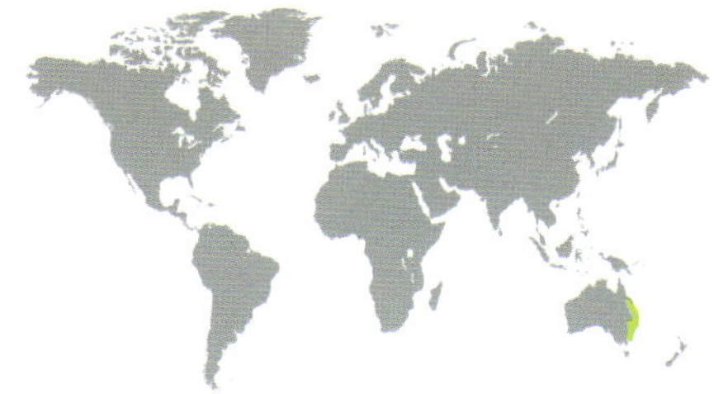

UPEROLEIA LAEVIGATA

SMOOTH TOADLET

KEFERSTEIN, 1867

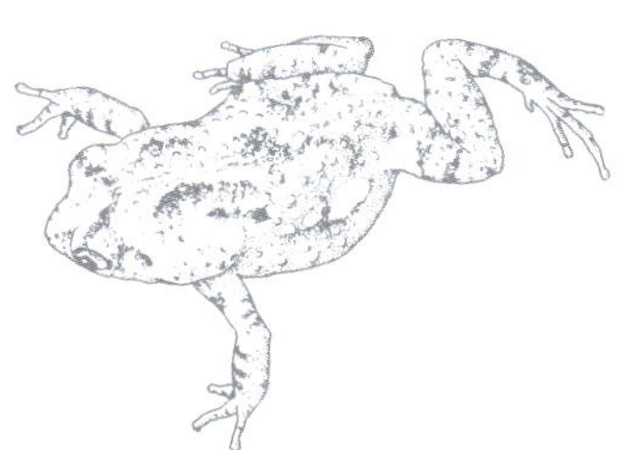

ADULT LENGTH
Male
¾–1⅛ in (20–28 mm)
Female
⅞–1¼ in (22–32 mm)

Males of this diminutive toad-like frog gather around ponds in spring and summer to establish small territories. They have three calls: an advertisement call that attracts females; an encounter call in response to intrusion by rival males; and a courtship call in response to visiting females. Males fight in defense of their territories, with heavier males more likely to win. Fights are prolonged between males of similar weight, and lighter males tend to avoid fights with heavier males. Gravid females spend up to three nights at a pond, visiting several males before choosing one as a mate. The eggs are attached individually to submerged vegetation.

Actual size

The Smooth Toadlet has prominent parotoid glands behind its eyes, and a pale triangular patch on its head. The back is pale gray or brown with darker markings. There is a pale yellow patch in the armpits, and a yellow or orange patch in the groin. The latter is exposed when it adopts a head-down, back-raised defensive posture.

SIMILAR SPECIES

The range of *Uperoleia laevigata* overlaps that of two related species, the Dusky Toadlet (*U. fusca*) and the Near Threatened Tyler's Toadlet (*U. tyleri*). The Dusky Toadlet can be distinguished by its darker belly, and Tyler's Toadlet by its larger parotoid glands. When attacked or molested, all *Uperoleia* species produce noxious secretions from their skin.

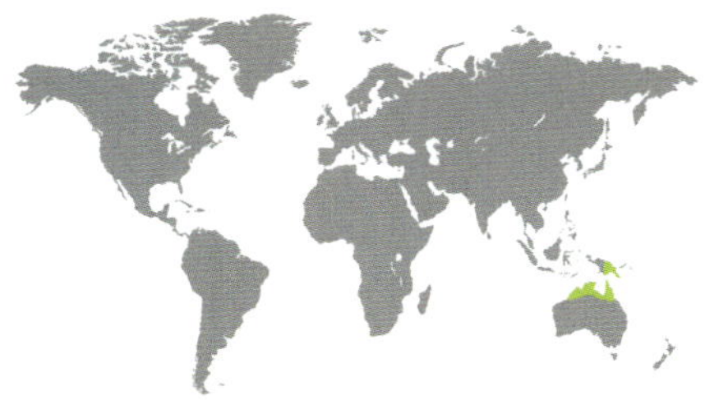

FAMILY	Myobatrachidae
OTHER NAMES	Stonemason Gungan
DISTRIBUTION	Northern Australia; Papua New Guinea
ADULT HABITAT	Grassland and woodland subject to flooding
LARVAL HABITAT	Ephemeral waterbodies
CONSERVATION STATUS	IUCN Least Concern. Has a large range and no apparent threats

ADULT LENGTH
Male
$^{11}/_{16}$–$1^{1}/_{16}$ in (17–27 mm)
Female
$^{15}/_{16}$–$1^{1}/_{8}$ in (24–29 mm)

UPEROLEIA LITHOMODA

STONEMASON TOADLET

TYLER, DAVIES & MARTIN, 1981

This small frog gets its common name from the male's call, which sounds like two stones being struck together. It breeds in winter following rain, with the males calling from cover near temporary pools. The eggs are laid in clumps, which fall to the bottom of the pond. During periods of drought, the frog goes underground. Other aspects of its natural history are not well known.

SIMILAR SPECIES

Fifteen other *Uperoleia* species occur in the northern half of Australia, all with smaller ranges than *U. lithomoda*. Similar in appearance and habits, they are most readily differentiated by their calls. The Small Toadlet (*U. minima*) is known only from male specimens, which do not exceed ⅞ in (21 mm) in length. Overall the genus contains 28 species, and it occurs in every state except Tasmania, with 13 species in Western Australia, ten in Northern Territory, and eight in Queensland. Most are not threatened, but Tyler's Toadlet (*U. tyleri*) is Near Threatened, Martin's Toadlet (*U. martini*) is Vulnerable, and the Howard River Toadlet (*U. daviesae*) is Endangered.

The Stonemason Toadlet has rough, wrinkled skin. It is dull gray or brown in color, with a narrow cream line down the middle of the head and yellow, cream, or gold spots along the flanks. There is an orange patch in the groin. There is no webbing on the hands or feet. The hind feet have horny tubercles that are used when digging a burrow.

Actual size

FAMILY	Myobatrachidae
OTHER NAMES	Chubby Gungan, Red-groined Toadlet
DISTRIBUTION	Southern Queensland and New South Wales, Australia
ADULT HABITAT	Forest, woodland, grassland
LARVAL HABITAT	Dams, ponds
CONSERVATION STATUS	IUCN Least Concern. Has a large range and no apparent threats

ADULT LENGTH
Male
11/16–1 1/4 in (18–32 mm)
Female
11/16–1 3/16 in (18–30 mm)

UPEROLEIA RUGOSA

WRINKLED TOADLET

(ANDERSSON, 1916)

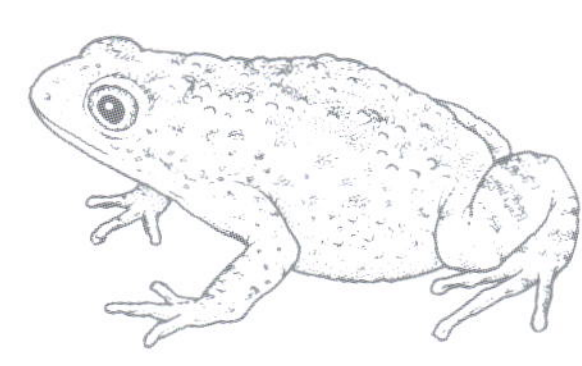

This small frog breeds at any time of year, other than winter, males forming choruses in flooded grassland and billabongs. The male's call is a high-pitched click. During mating, each egg is individually fertilized by the male and then individually attached to submerged vegetation by the female. Wrinkled Toadlets seem to thrive in disturbed land and males have been observed calling from water-filled hollows in the ground left by livestock. Individuals vary in size across the species' range, being smaller near the coast.

SIMILAR SPECIES

The genus *Uperoleia* contains 28 species, all rather similar in appearance, spread across the whole of Australia. The natural history of most species is not well known. The range of *U. rugosa* overlaps that of the Smooth Toadlet (*U. laevigata*; page 121), which is similar in size but has smooth skin.

Actual size

The Wrinkled Toadlet is a small, squat frog with short limbs and tubercles on its hind feet, which it uses for digging a burrow. It is gray to brown above, with darker markings and yellow-tipped tubercles on its back. The parotoid glands are yellowish. There is a reddish-orange patch in the groin, extending onto the thighs.

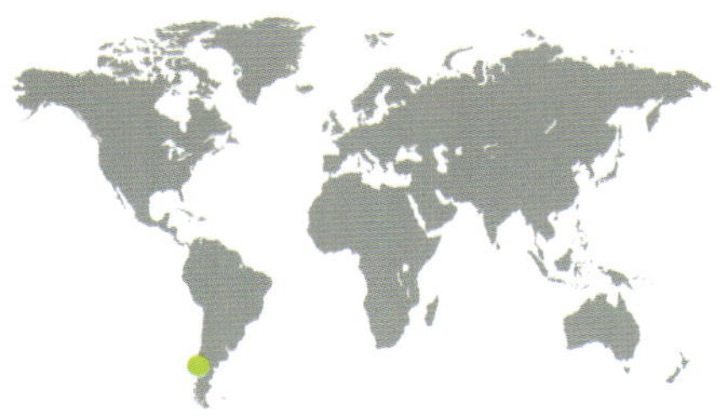

FAMILY	Alsodidae
OTHER NAMES	None
DISTRIBUTION	Nahuelbuta Mountains, southern Chile
ADULT HABITAT	Forest
LARVAL HABITAT	Pools in streams
CONSERVATION STATUS	IUCN Endangered. Has a very restricted range. Threatened by the replacement of natural forest with plantations

ADULT LENGTH
Male
$1\frac{7}{16}$–2 in (36–52 mm)
Female
$1\frac{5}{8}$–$2\frac{1}{8}$ in (42–54 mm)

ALSODES VANZOLINII

VANZOLINI'S SPINY-CHEST FROG

(DONOSO-BARROS, 1974)

Actual size

Vanzolini's Spiny-chest Frog has a large head, large eyes, and long legs. There is a distinctive pale triangle on the snout and broad black stripes running through the eyes. The back is pale brown with dark brown patches, and the legs are striped in pale and dark brown.

Found only on the western slopes of the Nahuelbuta Range in Chile, the population of this extremely rare frog declined by 80 percent between 2000 and 2010, by which time it appeared to occupy an area of less than 4 sq miles (10 sq km). However, in 2010 three new populations were reported nearby. It remains threatened because its natural forest habitat is being replaced by plantations of pine and eucalyptus. It inhabits ravines in remnants of native forest and breeds in streams. During the breeding season, males develop two clusters of sharp black spines on their chests. It is assumed that these help the male hold the female during mating.

SIMILAR SPECIES

There are 19 species in the genus *Alsodes*, all found in the Andes in Chile and Argentina; most are close to extinction. Pehuenche Spiny-chest Frog (*A. pehuenche*) and Cantillana Spiny-chest Frog (*A. cantillanensis*) are Critically Endangered. Both the Mountain Spiny-chest Frog (*A. montanus*) and La Parva Spiny-chest Frog (*A. tumultuosus*) are Vulnerable, being found at single locations that are being developed as ski resorts. The Island Spiny-chest Frog (*A. monticola*) is known from only one specimen, collected by Charles Darwin.

FAMILY	Alsodidae
OTHER NAMES	Contulmo Ground Frog
DISTRIBUTION	Chile
ADULT HABITAT	Temperate forest
LARVAL HABITAT	Small pools
CONSERVATION STATUS	IUCN Endangered. Its range has been reduced to a single location by deforestation

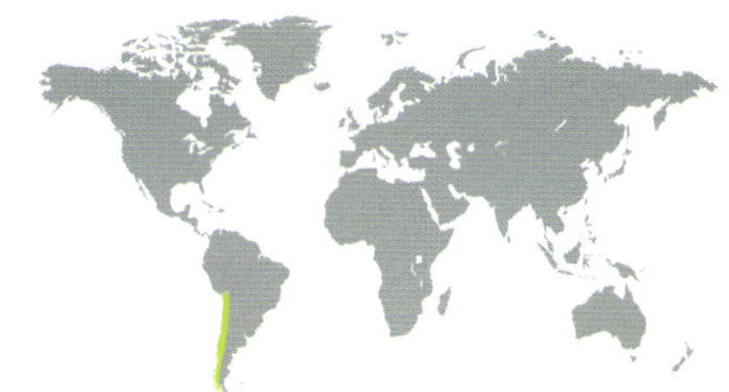

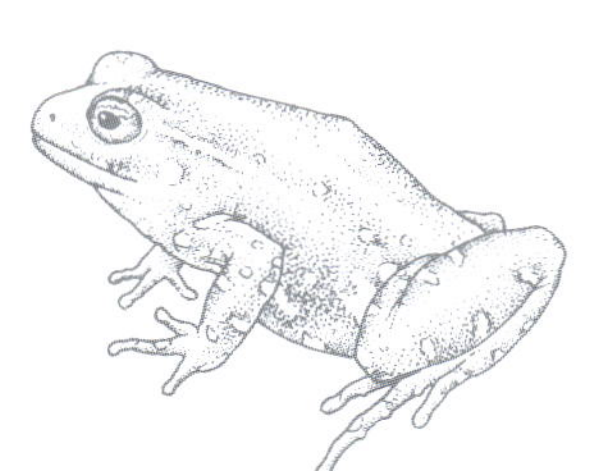

ADULT LENGTH
1$^{5}/_{16}$–1$^{11}/_{16}$ in
(34–43 mm)

EUPSOPHUS CONTULMOENSIS

CONTULMO TOAD

ORTIZ, IBARRA-VIDAL & FORMAS, 1989

This small frog is usually found under logs and rocks on the forest floor. Males have a simple mating call, consisting of a single short note. During mating, the male clasps the female around her pelvis. The eggs are laid in very small water-filled holes on hillsides. Other species in this genus have tadpoles that do not feed, but rely on large amounts of yolk to sustain them until metamorphosis; the same may be true of this species. Much of the forest habitat of the Contulmo Toad has been destroyed and the one location where it survives is subject to heavy pressure from tourism.

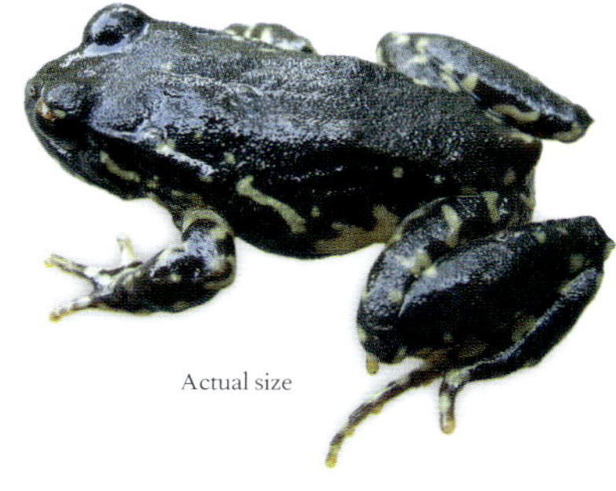

Actual size

SIMILAR SPECIES

There are ten species in the genus *Eupsophus*, known as ground frogs and found in Chile and Argentina. Emilio's Ground Frog (*E. emiliopugini*) is common and found in both countries; males of the species call from inside burrows. The Mocha Island Ground Frog (*E. insularis*) occurs only in Chile and is Critically Endangered due to habitat alteration.

The Contulmo Toad has a wide head, a round snout, and long, thin toes. The skin is smooth. The upper surfaces are dark brown, black, or purple. The belly, fingers, and toes are all yellow, and there are yellow markings on the sides of the body and the legs.

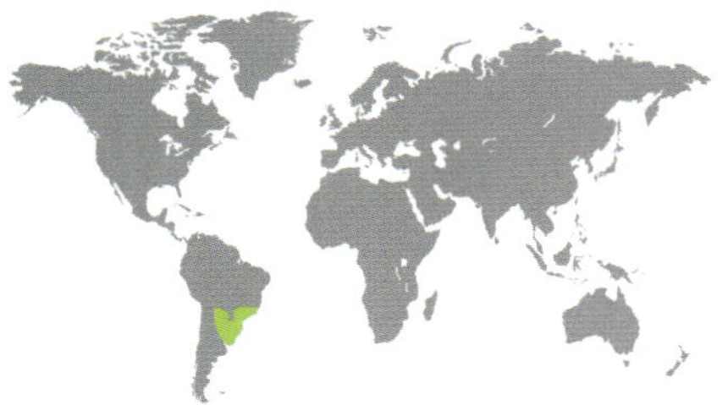

FAMILY	Alsodidae
OTHER NAMES	Dwarf Toad, Lesser Malacca Toad, Stream Toad
DISTRIBUTION	Southern Brazil, Uruguay, northern Argentina
ADULT HABITAT	Along rivers in open country and forest
LARVAL HABITAT	Small pools
CONSERVATION STATUS	IUCN Least Concern. Declining in places as a result of habitat destruction, but occurs within some protected areas

ADULT LENGTH
Male
1⁷⁄₁₆–2⅛ in (37–55 mm)

Female
1⁹⁄₁₆–2½ in (39–63 mm)

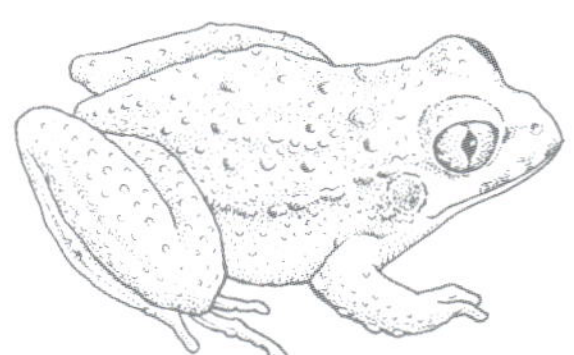

LIMNOMEDUSA MACROGLOSSA

RAPIDS FROG

(DUMÉRIL & BIBRON, 1841)

Actual size

The Rapids Frog has a compact body, a short, rounded snout, and large, prominent eyes. There are warts on its back and its feet are partially webbed. The upper surfaces are pale brown or pale gray, sometimes with dark brown patches, and there are transverse stripes on the legs.

This frog has specialized breeding habits based on rocky streams. Mating takes place between August and February. Males call just after dusk from underneath a rock and partially submerged in the water. When pairs are formed, the female carries the male around as she seeks a suitable pool for her eggs. The eggs are laid in small pools close to streams, and at times when water levels in streams rise to encompass the pools, the tadpoles make their way to slow-moving sections.

SIMILAR SPECIES

Limnomedusa macroglossa is the only species in its genus. It is most closely related to two genera of frogs found in Patagonia, the spiny-chest frogs (genus *Alsodes*; page 124) and the ground frogs (genus *Eupsophus*; page 125).

FAMILY	Batrachylidae
OTHER NAMES	None
DISTRIBUTION	Laguna Blanca National Park, eastern Argentina
ADULT HABITAT	Lakes and ponds up to 3,940 ft (1,200 m) altitude
LARVAL HABITAT	Ponds
CONSERVATION STATUS	IUCN Endangered. Has declined dramatically, due to competition from introduced fish and infection from viral and fungal diseases

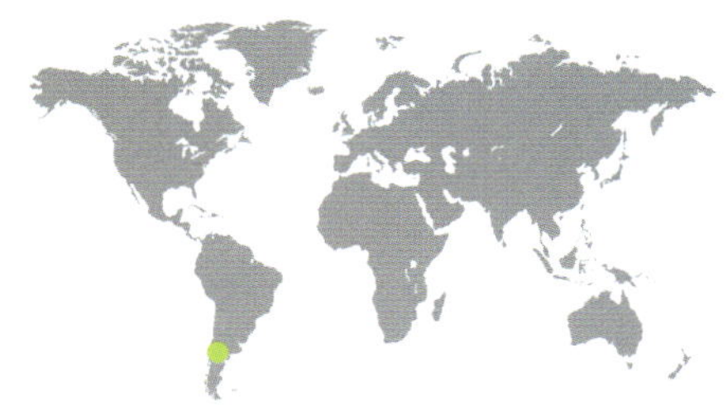

ADULT LENGTH
Male
1³⁄₁₆–1⅝ in (30–42 mm)
Female
1³⁄₁₆–1⅞ in (30–48 mm)

ATELOGNATHUS PATAGONICUS

PATAGONIA FROG

(GALLARDO, 1962)

This increasingly rare frog exists in two forms and is able to switch from one to the other depending on environmental conditions. The aquatic form has smooth, loose skin and feeds on aquatic prey, while the rough-skinned terrestrial form has a diet of insects and lives around the shore. Individuals of any age can switch from one form to another, apparently depending on which environment provides the better conditions. Confined to a restricted area, the species has declined dramatically since the mid-1980s. Introduced trout have destroyed its food supply in larger lakes and it is heavily infected by ranavirus, a common disease of amphibians, and by chytridiomycosis.

Actual size

The Patagonia Frog has a small head with a pointed snout. In its aquatic form (shown here), it has smooth, loose skin that forms folds on the body and thighs. The hind limbs are slender and the toes are fully webbed. The upper surfaces are gray or olive-brown with small dark spots, and the underside is orange.

SIMILAR SPECIES

There are five species in the genus *Atelognathus*, all called Patagonian frogs, and all found in southern Argentina and southern Chile. Most have restricted ranges at high altitudes and little is known about any of them. The Zapala Frog (*A. praebasalticus*), which occurs at high altitudes in Argentina, has declined as a result of the introduction of alien predatory fish such as perch and trout. It is listed as Endangered.

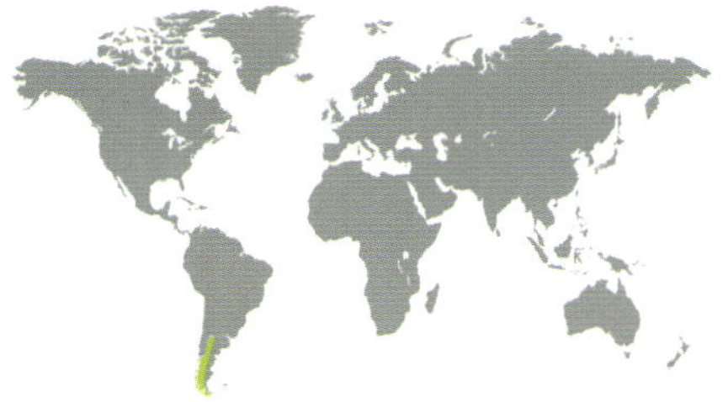

FAMILY	Batrachylidae
OTHER NAMES	None
DISTRIBUTION	Southern Chile, western Argentina
ADULT HABITAT	Forest, swamps
LARVAL HABITAT	Ephemeral or permanent shallow pools
CONSERVATION STATUS	IUCN Least Concern. Has declined in the north of its range. Appears to have adapted to altered habitats

ADULT LENGTH
15/16–1⅝ in
(23–42 mm);
females are larger
than males

BATRACHYLA TAENIATA

BANDED WOOD FROG

(GIRARD, 1855)

Actual size

The Banded Wood Frog has a slender body, a pointed snout, large eyes, and long, thin legs. Its fingers and toes are long and thin. The upper surfaces are coffee-colored, reddish brown, or yellowish, with a white stripe along each flank. A conspicuous black stripe runs through the eye.

An unusual feature of the breeding biology of this frog is that it lays its eggs on land. It does so under logs or in leaf litter, before heavy rain floods the area, allowing the eggs to complete their development. Egg development stops at a particular stage and does not resume until the eggs are fully immersed in water, a process called intracapsular resistance. This species has declined in the north of its range, where forests have been felled to make way for agriculture and human settlements.

SIMILAR SPECIES

There are five species in the genus *Batrachyla*, all of which are found in Chile and Argentina. The Marbled Wood Frog (*B. antartandica*) and the Gray Wood Frog (*B. leptopus*) live and breed in the same localities as *B. taeniata*. The three species have distinctive mating calls, ensuring that they do not interbreed. *Batrachyla fitzroya* is confined to an island within a lake in Argentina, and is Vulnerable. Nibaldo's Wood Frog (*B. nibaldoi*) is endemic to the eastern Andes of Chile.

FAMILY	Batrachylidae
OTHER NAMES	None
DISTRIBUTION	Southern Chile, Argentina
ADULT HABITAT	Forest and wetlands
LARVAL HABITAT	Ephemeral ponds
CONSERVATION STATUS	Data Deficient. This species is rarely seen and its population status is unknown

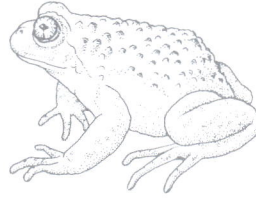

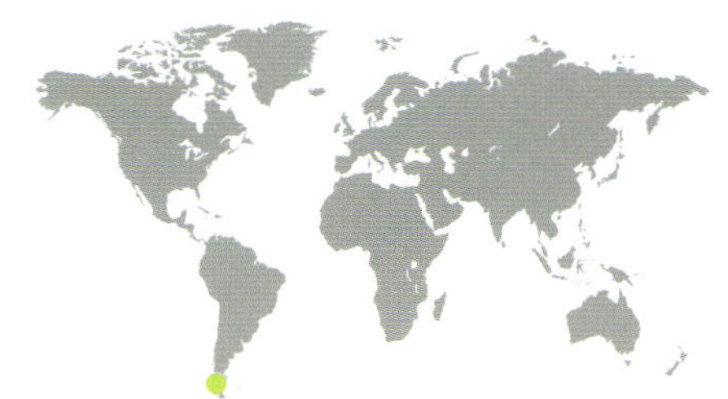

ADULT LENGTH
1⁵⁄₁₆–1¹³⁄₁₆ in
(33–46 mm)

CHALTENOBATRACHUS GRANDISONAE

PUERTO EDEN FROG

(LYNCH, 1975)

This little-known frog is named after the town of Puerto Eden on Wellington Island, off the coast of Chile, where it was first discovered. It has since been found, in 1997, at two locations in Argentina. Patagonia, the region in which it lives, has a generally cold and wet climate with very long winters. The frog breeds in October, laying clutches of up to 30 eggs, in clusters attached to branches or stones under the water. In some places, the tadpoles reach metamorphosis in December, before their ponds dry out. In others, where ponds are permanent, they may take more than a year to reach metamorphosis.

SIMILAR SPECIES

Chaltenobatrachus grandisonae is the only species in its genus. It was previously included in the genus *Atelognathus* (see page 127) but has recently been found to be genetically distinct.

Actual size

The Puerto Eden Frog is a heavily built frog with a rounded snout and large, protruding eyes with bright orange irises. The upper surfaces are bright green; the undersides of the body and legs are white. The back is covered in large brown or red warts.

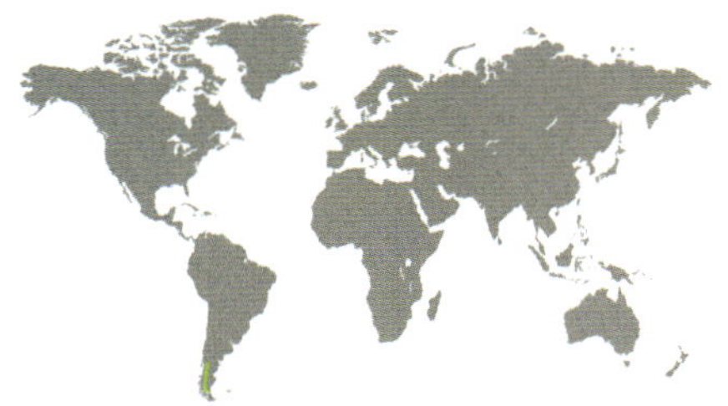

FAMILY	Batrachylidae
OTHER NAMES	None
DISTRIBUTION	Eastern slopes of the Andes in Chile and Argentina
ADULT HABITAT	Forest
LARVAL HABITAT	Pools, ponds
CONSERVATION STATUS	IUCN Least Concern. Has a restricted range and could be vulnerable to deforestation

ADULT LENGTH
Male
2¹/₁₆–2³/₁₆ in (53–56 mm)

Female
2³/₈–2¹¹/₁₆ in (60–66 mm)

HYLORINA SYLVATICA

EMERALD FOREST FROG

BELL, 1843

Found in deciduous forest on the eastern slopes of the southern Andes, this frog is remarkable for changing color with the time of day. At night it is dark green, but by day it becomes emerald-green with iridescent copper markings. Outside the breeding season, it is hard to find as it remains in the thicker parts of the undergrowth. In summer it moves to open areas around the lakes and lagoons where it breeds. Males call while floating on the water's surface and eggs are laid close to the water's edge.

SIMILAR SPECIES

Hylorina sylvatica is the only species in its genus. It belongs to a small family of frogs, found only in Chile and Argentina, which includes the genera *Atelognathus* (see page 127) and *Batrachyla* (see page 128). It also includes the Puerto Eden Frog (*Chaltenobatrachus grandisonae*; page 129), from Wellington Island off the Chile coast, which is known only from a single specimen, described more than 25 years ago.

The Emerald Forest Frog has a large head with a rounded snout, very prominent eyes with vertical pupils, and long, unwebbed fingers and toes. By day, the skin is bright emerald-green with iridescent gold or copper markings. In particular, two broad bands run from the nostrils back along the length of the body.

Actual size

FAMILY	Bufonidae
OTHER NAMES	None
DISTRIBUTION	Southern Sri Lanka
ADULT HABITAT	Rainforest up to 5,500 ft (1,700 m) altitude
LARVAL HABITAT	Permanent pools in mountain streams
CONSERVATION STATUS	IUCN Endangered. Threatened by loss and pollution of its forest habitat

ADENOMUS KELAARTII

KELAART'S DWARF TOAD

(GÜNTHER, 1858)

ADULT LENGTH
Male
1–1 5/16 in (25–33 mm)
Female
1 7/16–1 15/16 in (36–50 mm)

This small toad is usually found in leaf litter near the mountain streams in which it breeds. Males call at night from boulders in streams and females lay up to 1,000 eggs in a string. The species may also be arboreal—it has been found at the top of a palm tree, 50 ft (15 m) above the ground. The forest in which it lives is being cut down, largely to make way for cardamom plantations, and the streams in which it breeds are susceptible to pollution by pesticides, fertilizers, and other agricultural chemicals.

Actual size

Kelaart's Dwarf Toad is small and slender, with a pointed snout. Its body is covered with warts that may be smooth or spiny. The back is light to deep brown, marbled with dark brown markings and sometimes with red and blue spots. The underside is white to pale yellow.

SIMILAR SPECIES

There is a second species of Sri Lankan stream-breeding forest-dwelling *Adenomus*. The Kandyan Dwarf Toad (*A. kandianus*) had not been seen since 1876 and was classified as Extinct until it was rediscovered in 2009. Threatened by urbanization, it is classified as Endangered. Das's Toad (*A. dasi*), which was known only from a single location, has been synonymized with *A. kandianus*.

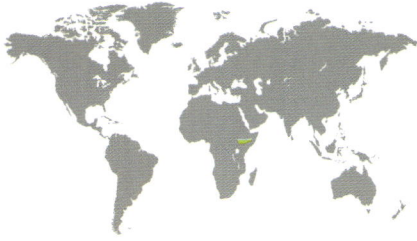

FAMILY	Bufonidae
OTHER NAMES	Ethiopian Mountain Toad
DISTRIBUTION	Bale Mountains, southern Ethiopia
ADULT HABITAT	Forest and moorland at altitudes of 10,500–13,100 ft (3,200–4,000 m)
LARVAL HABITAT	Within the egg
CONSERVATION STATUS	IUCN Endangered. Has a restricted range that is threatened by habitat destruction

ADULT LENGTH
Male
up to 7/8 in (22 mm)
Female
up to 1 1/4 in (32 mm)

ALTIPHRYNOIDES MALCOLMI

MALCOLM'S ETHIOPIAN TOAD

(GRANDISON, 1978)

Actual size

Malcolm's Ethiopian Toad is a plump toad with rows of large warts running down its back. The back is variable in color, being gray, greenish, brown, or black. There may also be green or reddish stripes with black dots, and some individuals have a pale stripe running down the middle of the back.

Found only in the Bale Mountains of Ethiopia, this small toad has a very unusual mode of reproduction. The female may be clasped by several males. Fertilization is internal and it appears that it is achieved by the male clasping the female from below, rather than from above. The female retains the eggs briefly in her oviducts before laying them, surrounded by mucus, in wet grass close to a pond. The male stays with the eggs and attracts more females to the clutch, until a communal egg mass of up to 20 clutches is formed. Larval development is completed within the eggs.

SIMILAR SPECIES

There is one other species in this genus, Osgood's Ethiopian Toad (*Altiphrynoides osgoodi*; sometimes placed alone in the genus *Spinophrynoides*), which breeds in temporary pools. This is found in the montane forests of south-central Ethiopia, where, like *A. malcolmi*, it is threatened by deforestation. It is categorized as Critically Endangered. The disease chytridiomycosis is found among Ethiopian frogs and toads, and may have contributed to the general decline of these and other species.

FAMILY	Bufonidae
OTHER NAMES	None
DISTRIBUTION	Amazon Basin, South America
ADULT HABITAT	Lowland tropical forest
LARVAL HABITAT	Small ephemeral pools
CONSERVATION STATUS	IUCN Least Concern. Has a very large range and there is no evidence of general decline

AMAZOPHRYNELLA MINUTA

TINY TREE TOAD

(MELIN, 1941)

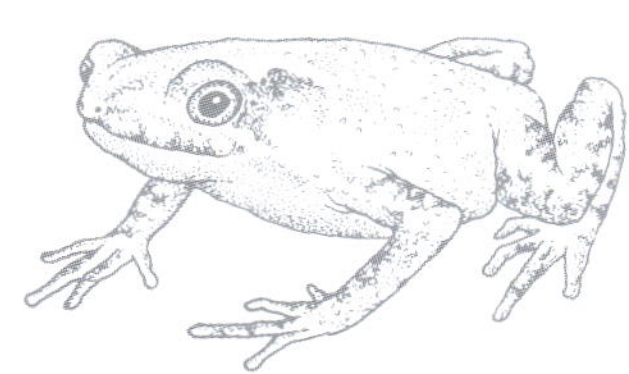

ADULT LENGTH
Male
9/16–11/16 in (14–17 mm)
Female
11/16–15/16 in (18–23 mm)

Its small size and cryptic coloration make this tiny frog hard to see among leaf litter on the forest floor. Males can be considerably smaller than females. The species is active by day and feeds on ants and other insects. The male's call is a frequently repeated, high-pitched trill. Breeding takes place in the wet season, between November and May, and females lay 70–250 eggs on tree roots, tree trunks, or fallen leaves above rain-filled hollows. The tadpoles are most often found in small pools beside larger pools.

Actual size

The Tiny Tree Toad has a pointed snout and long, slender legs. The body is slender, though becomes swollen in gravid females. The back is dappled in various shades of brown and some individuals have small green spots. The belly is yellow toward the head, white behind, with a black marbled pattern.

SIMILAR SPECIES

There are 12 other species in the genus *Amazophrynella*, which is centered on Brazil and Peru, but also occurs from Suriname to Bolivia. *Amazophrynella bokermanni* is known from only one location but is assumed to be more widely distributed; it has a similar reproductive biology to *A. minuta*. All the other species have been described since 2012 and some are poorly known.

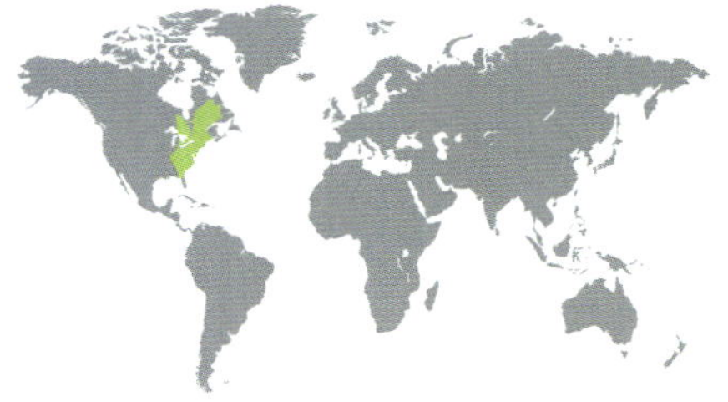

FAMILY	Bufonidae
OTHER NAMES	None
DISTRIBUTION	Eastern North America
ADULT HABITAT	Woodland, gardens
LARVAL HABITAT	Ephemeral ponds and other small waterbodies
CONSERVATION STATUS	IUCN Least Concern. Has a very large range

ADULT LENGTH
Male
2⅛–3⁵⁄₁₆ in (54–85 mm)

Female
2³⁄₁₆–4⁵⁄₁₆ in (56–111 mm); both sexes grow to a larger size in northern parts of the species' range

ANAXYRUS AMERICANUS

AMERICAN TOAD

(HOLBROOK, 1836)

This common toad breeds between April and July, males gathering around ephemeral waterbodies and producing a melodic trill lasting 5–30 seconds. Whereas males breed first at two years of age, females delay breeding until they are three years old. As a result, there are usually many more males than females in breeding groups and fights among males are common. Larger males tend to win and succeed in mating. Females lay 4,000–8,000 eggs, deposited in two strings. The black tadpoles often move around in compact schools; detailed research has found that these tend to consist of close kin.

The American Toad has a rounded snout and dry, warty skin. There is a reddish-brown crest above the eyes. The back is brown, brick-red, gray, or olive-green with lighter and darker patches. The skin also has darker spots, the warts are variously colored, and there is often a pale stripe down the middle of the back.

SIMILAR SPECIES

Anaxyrus americanus is easily confused with Fowler's Toad (*A. fowleri*), which occurs in the same parts of the USA but not in Canada. The two species commonly hybridize. Fowler's Toads tend to breed a little later, usually have spots with pale edges, and have a call that is described as a wailing scream. Genus *Anaxyrus* contains 25 species.

Actual size

FAMILY	Bufonidae
OTHER NAMES	Baxter's Toad
DISTRIBUTION	Wyoming, USA
ADULT HABITAT	Prairie
LARVAL HABITAT	Lakes in floodplains
CONSERVATION STATUS	IUCN Extinct in the wild. Persists as one population, sustained by captive breeding

ANAXYRUS BAXTERI

WYOMING TOAD

(PORTER, 1964)

ADULT LENGTH

Male

1⅞–2⅜ in (48–60 mm)

Female

1⅞–2¾ in (48–68 mm)

A relatively common species in the 1950s, this burrowing toad declined dramatically in the 1960s and 1970s, and by the mid-1980s it was thought to be extinct. In 1987, however, a few surviving individuals were found and, from these, a captive-bred population has been established, from which juvenile toads are periodically released into a single protected site at Mortenson Lake, Wyoming. In nature, the species is threatened by drought, mammalian predators, and, most seriously, the disease chytridiomycosis. The male's call is a harsh trill, lasting 3–5 seconds. Females lay 1,000–6,000 eggs in two strings.

SIMILAR SPECIES

Previously considered to be a subspecies of the Canadian Toad (*Anaxyrus hemiophrys*; page 139), *A. baxteri* is a relict, persisting at high altitude, and becoming isolated as North America became warmer at the end of the Pleistocene glaciation. It is smaller than *A. hemiophrys*.

The Wyoming Toad has very warty skin and horny tubercles on its hind feet for digging. The hard cranial crests fuse to form a grooved boss between the eyes. The back is brown, gray, or greenish with dark blotches, and there is usually a pale stripe down the line of the spine.

Actual size

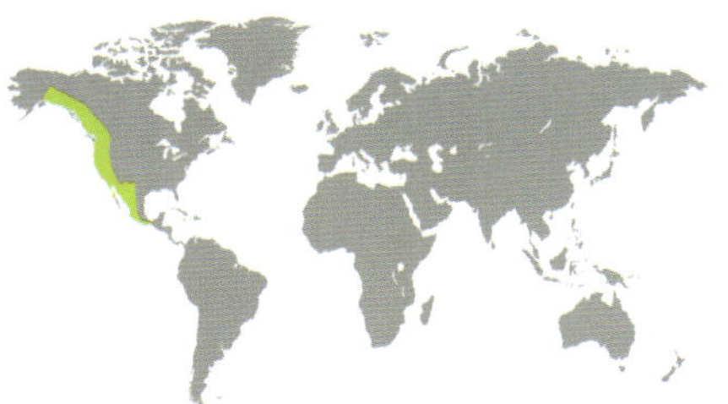

FAMILY	Bufonidae
OTHER NAMES	Boreal Toad, California Toad (subspecies *halophilus*)
DISTRIBUTION	Western North America, from Alaska to Mexico
ADULT HABITAT	Mountain meadows and woodlands, desert springs and streams
LARVAL HABITAT	Ephemeral and permanent waterbodies
CONSERVATION STATUS	IUCN Near Threatened. Has declined dramatically over much of its large range

ADULT LENGTH
Male
2 3/16–4 1/4 in (56–108 mm)
Female
2 3/8–4 15/16 in (60–125 mm)

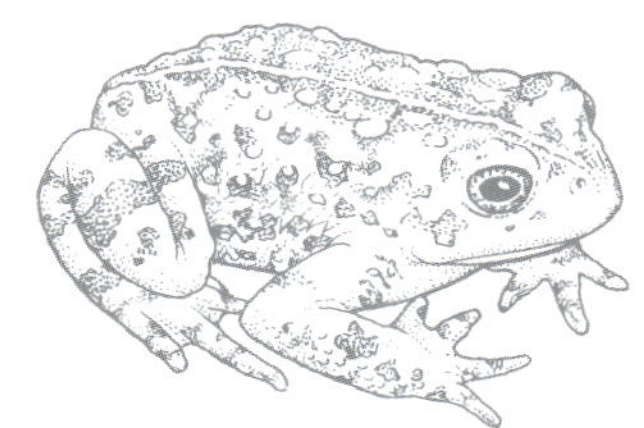

ANAXYRUS BOREAS

WESTERN TOAD

(BAIRD & GIRARD, 1852)

The Western Toad may be black, gray, brown, reddish, yellow, or tan in color. The back and sides are covered with brown or reddish warts, bordered in black. There is usually a narrow white or cream stripe running down the middle of the back. The prominent parotoid glands are oval in shape.

This robust toad generally walks rather than hops and shows considerable variation, both in color and in the timing of breeding, across its huge range. Males do not have a mating call that attracts females from a distance, and lack vocal sacs. During mating, they produce a soft, birdlike chirp that probably serves as a short-range threat signal between males, which often greatly outnumber females. The Western Toad has declined dramatically in many parts of its range. At higher altitudes, this is thought to be due to the adverse effects on its eggs of increased ultraviolet radiation. In many places, it is susceptible to the disease chytridiomycosis.

SIMILAR SPECIES

A currently recognized subspecies, the California Toad (*Anaxyrus boreas halophilus*) may be raised to species status; it is generally paler than toads from other parts of the range. Closely related species, all with very small ranges, are the Amargosa Toad (*A. nelsoni*) from Nevada; the Black Toad (*A. exsul*) from Deep Springs, California; and the Yosemite Toad (*A. canorus*) from the Sierra Nevada. The Amargos Toad is Critically Endangered, while the other two species are Vulnerable.

Actual size

FAMILY	Bufonidae
OTHER NAMES	California Toad, Mexican Arroyo Toad
DISTRIBUTION	Southern California, northern Mexico
ADULT HABITAT	Sand, gravel, and stony ground in seasonal river courses
LARVAL HABITAT	Ephemeral pools
CONSERVATION STATUS	IUCN Endangered. Threatened by habitat destruction, drought, and introduced predators (North American Bullfrog – *Aquarana catesbeiana* – and fish)

ANAXYRUS CALIFORNICUS

ARROYO TOAD

(CAMP, 1915)

ADULT LENGTH
Male
2–2 11/16 in (51–67 mm)
Female
2 11/16–3 1/16 in (66–78 mm)

Arroyos are steep-sided watercourses that are dry and sandy, except after heavy rain. The Arroyo Toad lives in this apparently unpromising habitat by spending the day buried in the sand and emerging to hunt at night. After heavy rain, males gather around quieter parts of a river and call, producing a long, musical trill. In this unpredictable habitat, the eggs face dual threats: drying out; and being washed away by a flash flood. The Arroyo Toad has vanished from around 70 percent of its former range and its overall population has halved in the last ten years.

SIMILAR SPECIES

Anaxyrus californicus was previously regarded as a subspecies of the Arizona Toad (*A. microscaphus*), a common species living around streams and arroyos in semiarid areas across a large part of southwestern USA and Mexico. Where the damming of rivers and irrigation schemes have produced more permanent waterbodies, *A. californicus* frequently hybridizes with Woodhouse's Toad (*A. woodhousii*; page 144).

The Arroyo Toad is a stocky toad with a blunt snout, short limbs, and oval parotoid glands. Its skin is covered in small, dark warts. It is variable in color, being gray, olive, or brown, and there is usually a pale Y-shaped patch on the top of the head, between the eyes, and a pale stripe running down the lower back.

Actual size

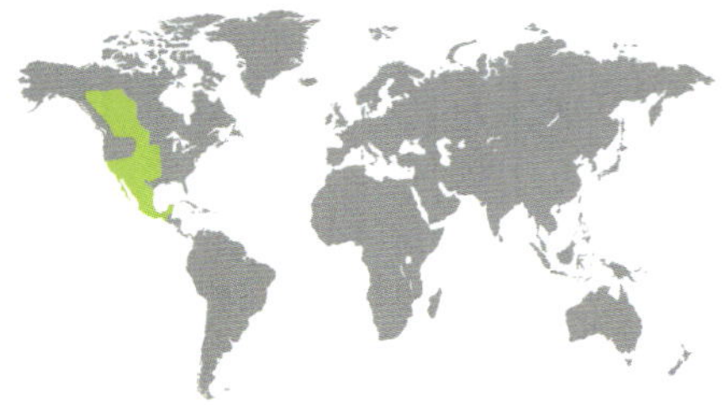

FAMILY	Bufonidae
OTHER NAMES	None
DISTRIBUTION	Canada, central and southwestern USA, Mexico
ADULT HABITAT	Open, dry country, including prairie, agricultural land, and desert (in the southwest)
LARVAL HABITAT	Ponds, ditches, ephemeral waterbodies
CONSERVATION STATUS	IUCN Least Concern. Has a very large range and does not seem to be declining

ADULT LENGTH
Male
1⅞–4 in (47–103 mm)
Female
1$^{15}/_{16}$–4½ in (49–115 mm)

ANAXYRUS COGNATUS

GREAT PLAINS TOAD

(SAY, 1823)

The Great Plains Toad is a robustly built toad with prominent parotoid glands and cranial crests that meet to form a V-shaped boss on its head. The pale skin is decorated with paired green or dark gray markings. There are numerous warts on the body, head, and legs. There is a sharp-edged tubercle on each hind foot.

This boldly patterned toad spends most of its days in an underground burrow it digs using spade-like tubercles on its hind feet. The first spring rain triggers mass movements of toads to breeding sites, where males produce a call that lasts a minute and which has been likened to the sound of a pneumatic drill. In some places, large choruses are formed, but within these not all males call. Many adopt a silent "satellite" behavior, sitting close to a calling male and attempting to intercept females that are attracted to his call. Smaller females lay 1,300 eggs, but larger ones can lay up to 45,000 eggs.

SIMILAR SPECIES

The Texas Toad (*Anaxyrus speciosus*) is smaller than *A. cognatus* and much plainer in appearance, being rather uniformly gray or olive-green in color. It is found in Texas, New Mexico, and Mexico; it, too, spends the day in an underground burrow. It has a raucous, high-pitched call and, like *A. cognatus*, has a large vocal sac that is sausage-shaped when it is fully inflated.

Actual size

FAMILY	Bufonidae
OTHER NAMES	Dakota Toad
DISTRIBUTION	Central Canada, northern Midwest region of USA
ADULT HABITAT	Prairie, near lakes and other permanent waterbodies
LARVAL HABITAT	Lakes, ponds, ditches
CONSERVATION STATUS	IUCN Least Concern. Declining in some parts of its range due to urbanization and draining of wetlands

ANAXYRUS HEMIOPHRYS

CANADIAN TOAD

(COPE, 1886)

ADULT LENGTH

Male

2 3/16–2 3/4 in (56–68 mm)

Female

2 3/16–3 1/8 in (56–80 mm)

More aquatic than most toads, Canadian Toads are active by day, retiring to burrows at night. In winter, they hibernate underground, often in Mima mounds, large mounds of looser soil created by the activities of gophers. Across much of their range they breed in May, in permanent waterbodies. The male advertisement call is a melodic trill. The female lays up to 7,000 eggs in a single string. If attacked or molested, Canadian Toads either swim quickly into deep water or, when on land, produce a distasteful secretion from their parotoid glands.

SIMILAR SPECIES

The Canadian Toad is similar to both the American Toad (*A. americanus*; page 134, with which it hybridizes in southeastern Manitoba, Canada, and Woodhouse's Toad (*A. woodhousii*; page 144), with which its geographic range overlaps.

The Canadian Toad has a dome-like boss on its head, running from the tip of the snout to the eyes. The parotoid glands are narrow. The back is olive-green or gray in color, with many dark patches that each enclose red or brown warts. The belly is spotted. There are horny tubercles on the hind feet used for digging.

Actual size

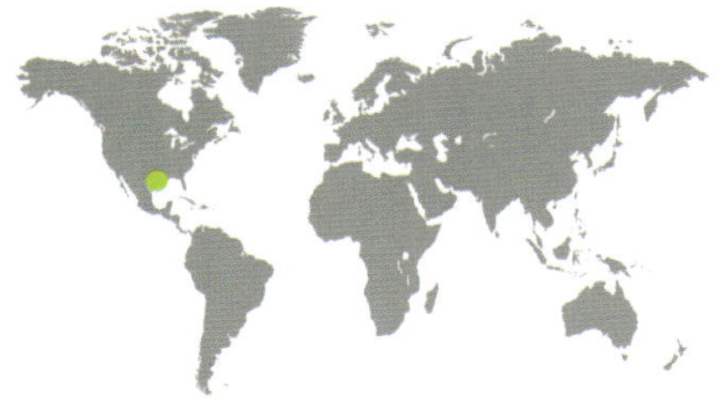

FAMILY	Bufonidae
OTHER NAMES	None
DISTRIBUTION	Southeastern Texas, USA
ADULT HABITAT	Areas with sandy soils, in and around pine forests
LARVAL HABITAT	Ephemeral and permanent pools
CONSERVATION STATUS	IUCN Endangered. Declining as a result of habitat deterioration

ADULT LENGTH
Male
$1\frac{15}{16}$–$2\frac{11}{16}$ in (49–66 mm)
Female
$2\frac{3}{16}$–$3\frac{1}{8}$ in (57–80 mm)

ANAXYRUS HOUSTONENSIS

HOUSTON TOAD

(SANDERS, 1953)

Spending much of its life underground, the Houston Toad requires soft, sandy soil that it can easily borrow into. It breeds in early spring, spending the day in its burrow and moving on warm, humid nights to ponds up to 130 ft (40 m) away. The male produces a musical trill lasting around 15 seconds. The species became extinct in the Houston area in the 1960s, through a combination of prolonged drought and loss of its habitat to urbanization. It is now most common in Bastrop State Park, but even here its numbers are declining. Attempts to breed it in captivity have been unsuccessful.

SIMILAR SPECIES

Anaxyrus houstonensis closely resembles the American Toad (*A. americanus*; page 134) but is rather smaller. It is thought to be a relict of the latter species, whose range moved northward with the melting of the ice sheet 10,000 years ago. The two species have a similar call, though that of *A. houstonensis* has a higher pitch.

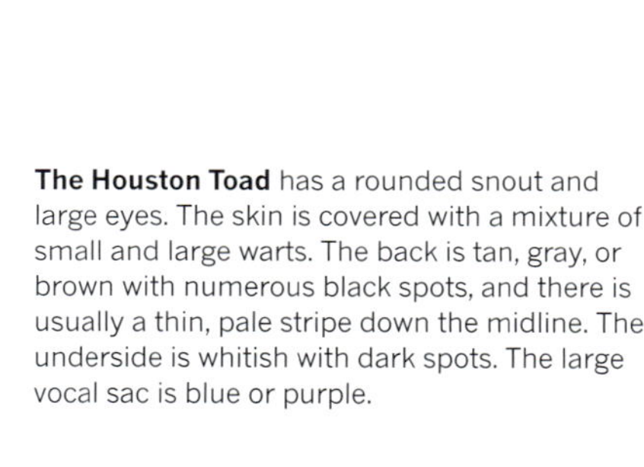

Actual size

The Houston Toad has a rounded snout and large eyes. The skin is covered with a mixture of small and large warts. The back is tan, gray, or brown with numerous black spots, and there is usually a thin, pale stripe down the midline. The underside is whitish with dark spots. The large vocal sac is blue or purple.

FAMILY	Bufonidae
OTHER NAMES	None
DISTRIBUTION	Southeastern USA, Mexico
ADULT HABITAT	Rocky areas in arid and semiarid forests, grassland, and deserts
LARVAL HABITAT	Creeks, pools, and springs
CONSERVATION STATUS	IUCN Least Concern. There is no evidence of population decline and it occurs within several protected areas

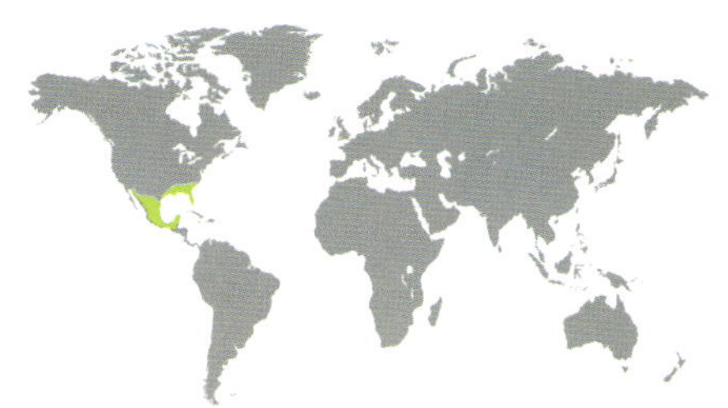

ADULT LENGTH
Male
1½–2½ in (38–63 mm)

Female
2–3 in (52–76 mm)

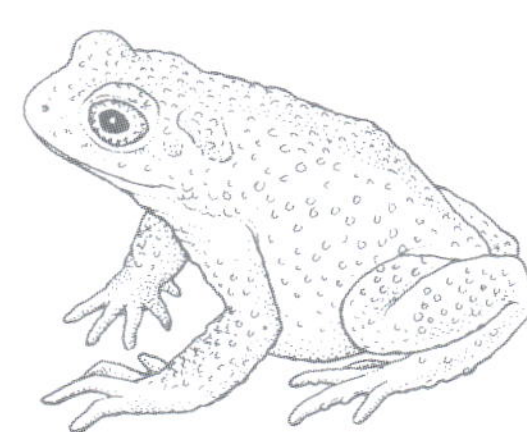

ANAXYRUS PUNCTATUS

RED-SPOTTED TOAD

(BAIRD & GIRARD, 1852)

Adapted for life in harsh environments, including deserts, Red-spotted Toads are remarkably tolerant of extreme variations in temperature. They are very adept at finding hiding places among rocky outcrops that enable them to be warmer or cooler than the ambient temperature. They breed between March and September, migrating to breeding sites after heavy rain. Males form small choruses at creeks, pools, and springs, producing a high-pitched trill rather like that of tree crickets. Uniquely among North American toads, the females lay their eggs—up to 5,000 of them—singly, rather than in strings.

SIMILAR SPECIES

Anaxyrus punctatus may be found together with the Arizona Toad (*A. microscaphus*) in southeastern USA, sometimes using the same breeding sites. In Arizona and Mexico, it occurs alongside, and sometimes hybridizes with, the Sonoran Green Toad (*A. retiformis*), a somewhat smaller species noted for its bright green and yellow coloration.

The Red-spotted Toad gets its name from the numerous bright red or orange warts on its back, sides, and legs. The skin is otherwise gray, tan, olive-green or brown. The belly is white. It has a flattened head, and the parotoid glands are circular in shape and about the same diameter as the eyes.

Actual size

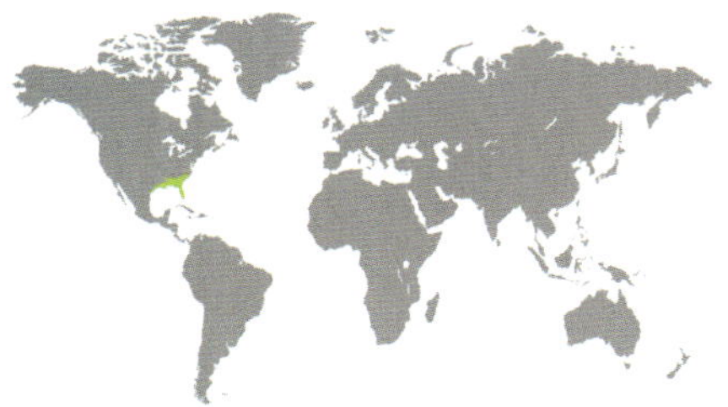

FAMILY	Bufonidae
OTHER NAMES	None
DISTRIBUTION	Southeastern USA
ADULT HABITAT	Oak and pine woodlands
LARVAL HABITAT	Ephemeral ponds and flooded areas
CONSERVATION STATUS	IUCN Least Concern. There is no evidence of population decline

ADULT LENGTH
Male
¾–1 in (20–26 mm)
Female
$^{15}/_{16}$–1$^{3}/_{16}$ in (24–30 mm)

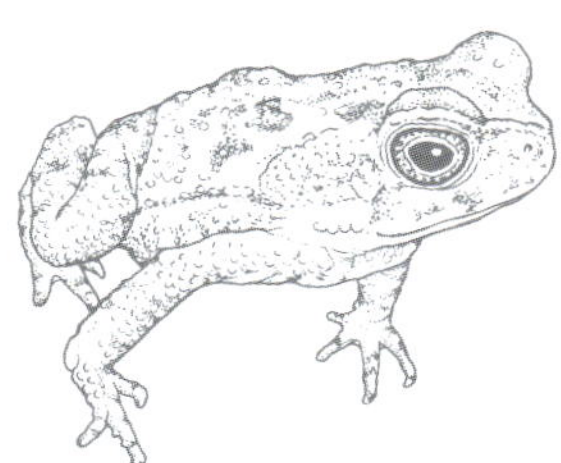

ANAXYRUS QUERCICUS

OAK TOAD

(HOLBROOK, 1840)

Actual size

The Oak Toad has a blunt snout and prominent parotoid glands behind the eyes. Its back is pale gray to black, with dark patches arranged more or less symmetrically either side of a yellow stripe running down the middle. Numerous orange or red warts are scattered on the back and legs.

Described as a tiny toad of many colors, the Oak Toad is North America's smallest toad. Its bright coloration provides effective camouflage in leaf litter, and it is unusual among toads in being active by day as well as by night. It breeds between April and October, after heavy, warm rain. Males call from near flooded areas, producing a high-pitched and very loud "cheep… cheep… cheep." A chorus of several males is deafening to humans. The male's vocal sac is sausage-shaped and curves upward in front of his snout when fully inflated.

SIMILAR SPECIES

While many other American toads (genus *Anaxyrus*) lay their eggs in long strings, in *A. quercicus* egg strings are broken up into short bands of five or six eggs. The Little Mexican Toad (*A. kelloggi*) is another small toad found in woodland along the Pacific coastal plain of western Mexico. It also lays its eggs in short strands, sometimes singly.

FAMILY	Bufonidae
OTHER NAMES	Carolina Toad, Southeastern Toad
DISTRIBUTION	Southeastern USA
ADULT HABITAT	Oak and pine woodland. Especially common where soil is sandy
LARVAL HABITAT	Ephemeral and semipermanent ponds, pools, and ditches
CONSERVATION STATUS	IUCN Least Concern. There is no evidence of general decline

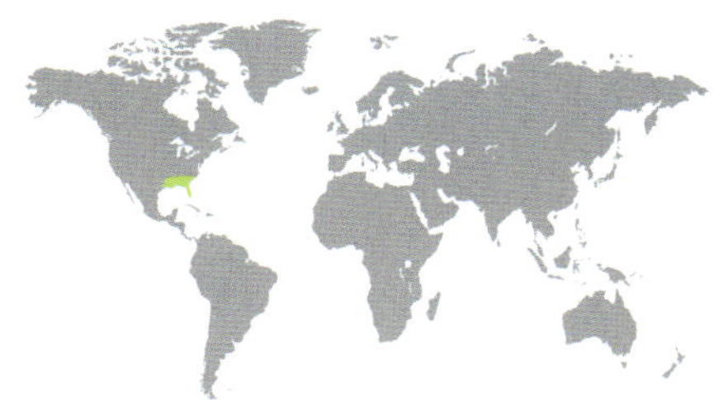

ANAXYRUS TERRESTRIS

SOUTHERN TOAD

(BONNATERRE, 1789)

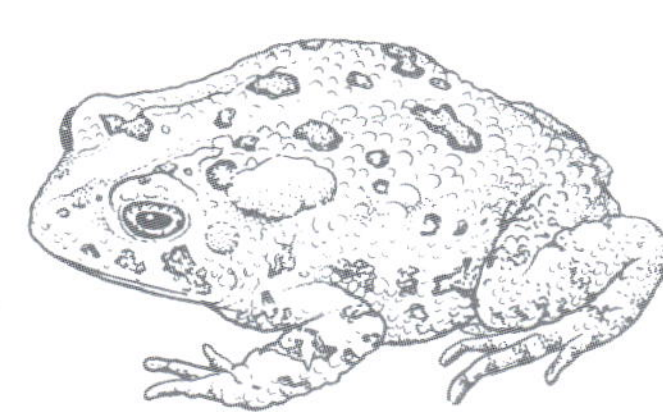

ADULT LENGTH
Male
1⅝–3¼ in (42–82 mm)
Female
1¾–3⁹⁄₁₆ in (44–92 mm)

This common, medium-sized toad is active at night, spending the day underground, usually in a burrow that it has dug itself. It thrives in urban areas and is often seen at night catching insects under streetlights. It breeds early in the year, between February and May. The male's advertisement call is a shrill trill, lasting 4–8 seconds. Males may call from the pond edge and wait for females to come to them, or may move around the breeding site, attempting to clasp any frog or toad that they encounter. Females lay 2,500–4,000 eggs in long strings.

SIMILAR SPECIES

The Southern Toad is rather similar in appearance to its northern relative, the American Toad (*Anaxyrus americanus*; page 134). They sometimes hybridize, and hybrid males have calls intermediate between the two species. To the west, the Coastal Plain Toad (*Incilius nebulifer*) is found from Mississippi to Costa Rica. It has a rather flattened appearance and the male has a particularly large vocal sac.

The Southern Toad has pronounced knobs on the top of its head between and behind the eyes. The upper surfaces are covered in warts and are brown, red, or gray with black and red spots. Individuals tend to turn almost black when the weather is cold and wet.

Actual size

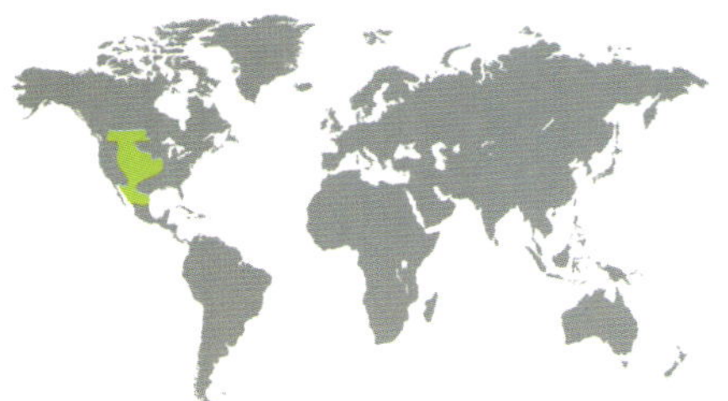

FAMILY	Bufonidae
OTHER NAMES	Rocky Mountain Toad
DISTRIBUTION	Central USA, southern Canada, northern Mexico
ADULT HABITAT	Near water in open habitats, e.g., grassland, farmland, desert scrub, woodland, city gardens
LARVAL HABITAT	Still waterbodies
CONSERVATION STATUS	IUCN Least Concern. Population is stable across its very large range

ADULT LENGTH
Male
2¾–3¹³⁄₁₆ in (69–98 mm)
Female
3¼–4¼ in (84–109 mm)

ANAXYRUS WOODHOUSII

WOODHOUSE'S TOAD

(GIRARD, 1854)

Common across the prairie states of the USA, this large toad breeds from February to August, usually after rain. Larger males defend calling positions at the water's edge, producing a nasal, buzzing call lasting 2–4 seconds. Males occasionally fight over call sites and the largest males generally win. As in many frogs and toads, male mating success increases the more nights they attend a chorus. Smaller males often do not call or engage in fights, but instead adopt a "satellite" strategy, seeking to intercept females approaching calling males. Larger females can lay as many as 28,000 eggs in long strings.

Woodhouse's Toad has a blunt snout and a warty skin, the warts often tipped with red spots. It is very variable in color, being brown, yellowish, gray, or green with dark spots and patches. There is usually a pale stripe running down the middle of the back, and the underside is pale.

SIMILAR SPECIES

Once regarded as a subspecies of *Anaxyrus woodhousii*, Fowler's Toad (*A. fowleri*) is found in eastern USA and, where the two species meet, they frequently hybridize. Males of the two species have very similar calls. In other parts of its range, *A. fowleri* hybridizes with the American Toad (*A. americanus*; page 134).

Actual size

FAMILY	Bufonidae
OTHER NAMES	Kinabalu Slender Toad
DISTRIBUTION	Northeastern Borneo
ADULT HABITAT	Montane forest at altitudes of 2,460–5,250 ft (750–1,600 m)
LARVAL HABITAT	Mountain streams
CONSERVATION STATUS	Low Concern. Much of its habitat has been destroyed by deforestation, but it occurs in at least two protected areas

ANSONIA HANITSCHI

KADAMAIAN STREAM TOAD

INGER, 1960

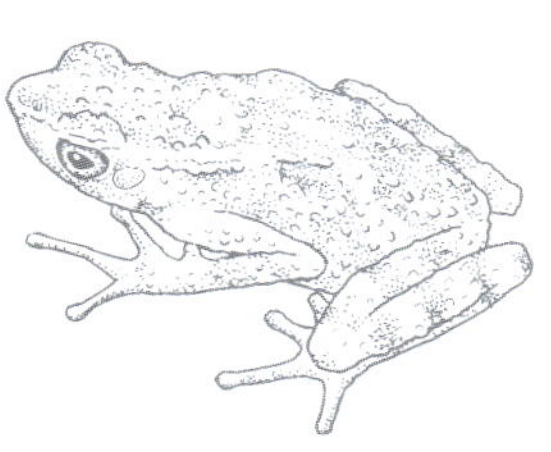

ADULT LENGTH
Male
¾–1⅛ in (20–28 mm)
Female
1⅛–1⅜ in (28–35 mm)

This small toad lives on the forest floor and is most commonly found close to mountain streams. Males gather along stream edges, sometimes in large numbers, and produce a high-pitched chirp or a short trill. The tadpoles are small and are adapted for life in fast-flowing water: They have a streamlined shape and a mouth that is modified to form a sucker, with which they attach themselves to rocks. Logging in the forests of Borneo not only destroys the habitat of this and other frogs, but also causes the streams in which they breed to silt up.

The Kadamaian Stream Toad has a slender, rather flattened body, thin arms and legs, a pointed snout and numerous small warts on its head, back, and limbs. The feet are partially webbed. The upper surfaces are greenish gray or reddish brown, with paler and darker spots on the back, and dark stripes on the legs.

SIMILAR SPECIES

The genus *Ansonia* currently contains 38 species, some of them very recently described. Called slender toads or stream toads, they are found in southern India and Southeast Asia. They are toad-like in having dry, rough skin, but do not have the large parotoid glands behind the eyes that are typical of most toads. They have slender, elongated bodies and most feed on ants. The Cave-dwelling Stream Toad (*A. khaochangensis*), from Thailand, and Mesilau Stream Toad (*A. guibei*), from Borneo, are Critically Endangered.

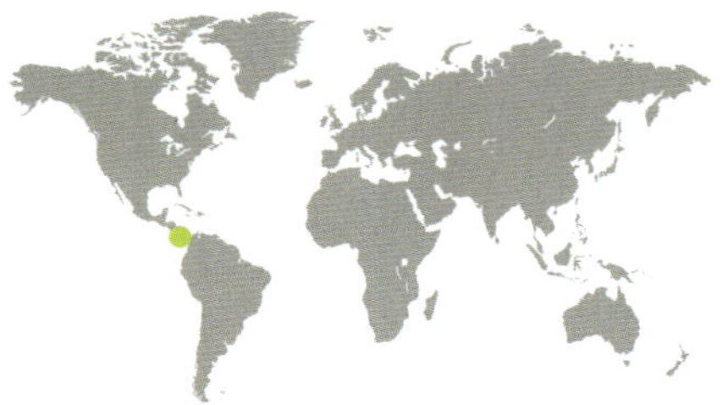

FAMILY	Bufonidae
OTHER NAMES	Lewis' Stubfoot Toad
DISTRIBUTION	Talamanca mountain range, Costa Rica and Panama
ADULT HABITAT	Cloud forest at altitudes of 4,600–8,200 ft (1,400–2,500 m)
LARVAL HABITAT	Streams
CONSERVATION STATUS	IUCN Extinct. Has not been seen in Costa Rica since 1996 and may be extinct

ADULT LENGTH
Male
$1\frac{1}{8}$–$1\frac{5}{16}$ in (28–34 mm)
Female
$1\frac{7}{16}$–$1\frac{15}{16}$ in (36–49 mm)

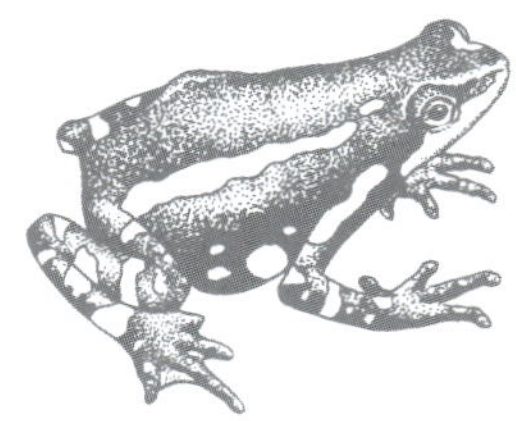

ATELOPUS CHIRIQUIENSIS

CHIRIQUI HARLEQUIN TOAD

SHREVE, 1936

Actual size

The Chiriqui Harlequin Toad has a narrow body, long, thin arms and legs, and a glandular protuberance on the tip of its snout. It is very variable in color overall, and males and females are strikingly different from one another in this respect. Males may be yellow, yellowish green, green, brown, or red. Females are equally variable, tend to be darker than males, and may have brightly colored spots or stripes.

This colorful frog inhabits high-altitude cloud forest, where the cool, damp conditions seem to be ideal for the fungus that causes the disease chytridiomycosis. This disease has swept through Central America in recent years, causing catastrophic population declines and several species extinctions. Its bright colors signal that it is toxic—its skin contains the nerve poison tetrodotoxin. Safe from predators, it is active by day, males defending territories close to streams by calling and, if this fails to deter intruders, by wrestling. When a pair forms, they dive into the water for 15–30 minutes, during which time the female lays her eggs.

SIMILAR SPECIES

To date, 100 species of *Atelopus*, known as harlequin frogs or stubfoot toads, have been described. Many of them are threatened with extinction to various degrees. Another inhabitant of the Talamanca range in Costa Rica, the Pass Stubfoot Toad (*A. senex*), declined dramatically in 1987 and 1988, and it is now categorized as Extinct. The Maracay Harlequin Frog (*A. vogli*) from Venezuela is also Extinct, while 66 of the remaining 97 species are Critically Endangered.

FAMILY	Bufonidae
OTHER NAMES	Veragua Stubfoot Toad
DISTRIBUTION	Northern Venezuela
ADULT HABITAT	Montane forest up to 7,200 ft (2,200 m) altitude
LARVAL HABITAT	Swift-flowing streams
CONSERVATION STATUS	IUCN Critically Endangered. Has declined across its entire range as a result of infection by chytridiomycosis

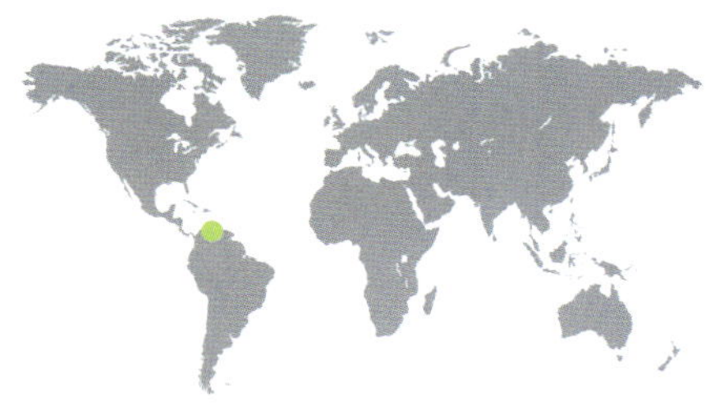

ADULT LENGTH
Male
$1\frac{1}{8}$–$1\frac{3}{8}$ in (28–35 mm)
Female
$1\frac{9}{16}$–$1\frac{15}{16}$ in (39–50 mm)

ATELOPUS CRUCIGER

RANCHO GRANDE HARLEQUIN FROG

(LICHTENSTEIN & MARTENS, 1856)

This beautifully patterned little frog was thought to be extinct, having not been seen since 1986, but then in 2004 a single small population was discovered. Living in the mountains along the northern coast of Venezuela, it breeds in fast-flowing streams. Males take up positions along streams, waiting for females to emerge from the forest. When they do arrive, most already have a male riding on their back. Although their ears are rather poorly developed, males have a varied vocal repertoire of buzzes, whistles, and chirps. The tadpoles are adapted for life in torrents, having long tails and abdominal suckers.

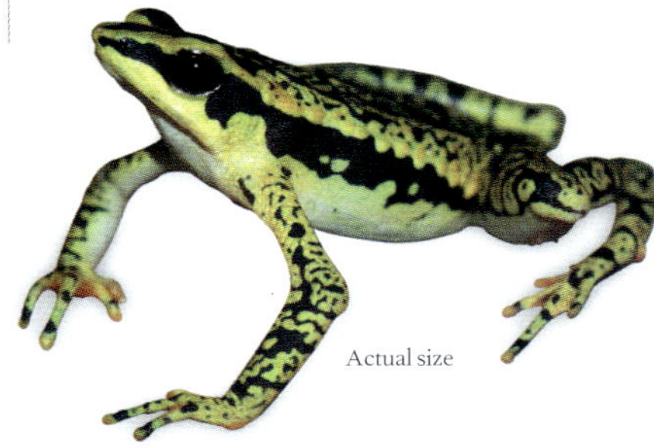

Actual size

The Rancho Grande Harlequin Frog has a slender body, long, slender limbs, and a pointed snout. The upper surfaces vary between yellow and green, and are decorated with a complex pattern of black spots and lines. Most individuals have a broad black stripe along the flanks and there is often an X-shaped marking on the back of the head.

SIMILAR SPECIES

In addition to *Atelopus cruciger*, at least three other Venezuelan members of this genus are listed as Critically Endangered. These are the La Carbonera Stubfoot Toad (*A. carbonerensis*), the Mucubaji Stubfoot Toad (*A. mucubajiensis*), and the Scarlet Harlequin Frog (*A. sorianoi*). All are found in forest at high altitudes, and all have declined as a result of infection by chytridiomycosis.

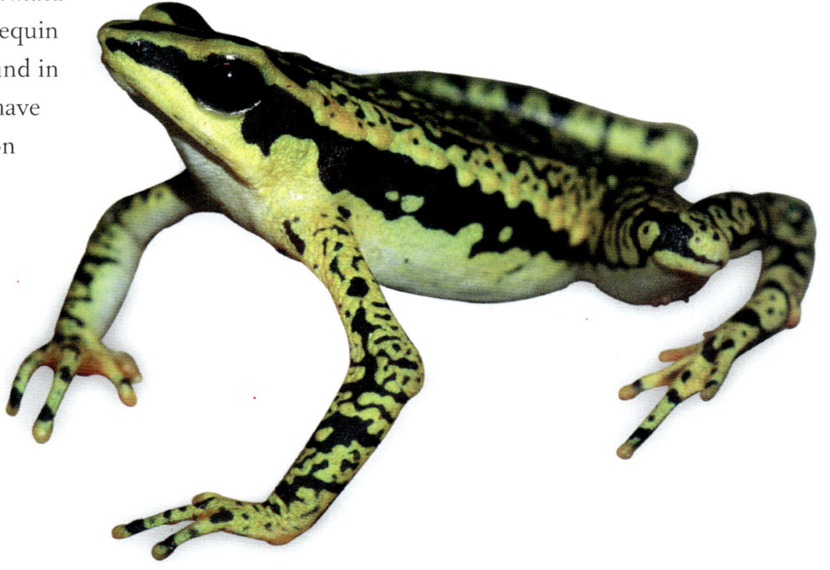

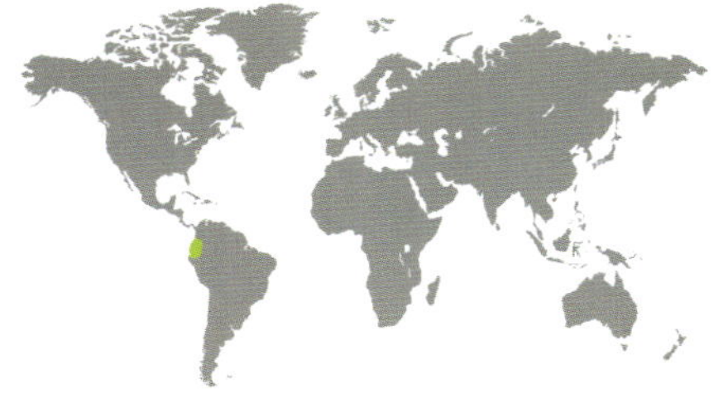

FAMILY	Bufonidae
OTHER NAMES	Black Andean Toad, Quito Stubfoot Toad
DISTRIBUTION	Andes of Ecuador at altitudes of 9,200–13,800 ft (2,800–4,200 m)
ADULT HABITAT	Forest
LARVAL HABITAT	Mountain streams
CONSERVATION STATUS	IUCN Critically Endangered, may be Extinct

ADULT LENGTH
Male
$1\frac{5}{16}$–$1\frac{5}{8}$ in (34–42 mm)

Female
$1\frac{3}{8}$–$1\frac{7}{8}$ in (35–48 mm)

ATELOPUS IGNESCENS

JAMBATO TOAD

(CORNALIA, 1849)

Actual size

The Jambato Toad had a narrow body and head, and a pointed snout. Its body was covered in small warts. The upper surfaces were a uniform, glossy black and its belly was orange.

This toad epitomizes what has happened in the last 30 years to the harlequin frogs, or stubfoot toads, of Central and northern South America. Formerly abundant, it lived in association with fast-flowing streams in high-altitude forests and was last seen in 1988. While it was affected in some parts of its range by habitat destruction, the primary cause of its decline was the disease chytridiomycosis, probably exacerbated by climate change. It may be significant that 1987, the last year it was seen, was exceptionally warm and dry in the Ecuadorian Andes. Around the world, chytridiomycosis particularly affects stream-living frogs at higher altitudes.

SIMILAR SPECIES

Like *Atelopus ignescens*, the Carrikeri Harlequin Frog (*A. carrikeri*), also known as the Guajara Stubfoot Toad, is black in color. Listed as Endangered, it is found in Colombia and has exceptionally long tadpoles. Another Endangered species from Ecuador is *A. exiguus*. This is smaller than most *Atelopus* species and several individuals have only four toes, rather than five, on their hind feet.

FAMILY	Bufonidae
OTHER NAMES	None
DISTRIBUTION	Andes of Peru at altitudes of 2,000–3,000 ft (600–900 m)
ADULT HABITAT	Lowland tropical forest
LARVAL HABITAT	Streams
CONSERVATION STATUS	IUCN Vulnerable. Has declined dramatically as a result of habitat loss and chytridiomycosis

ATELOPUS PULCHER

RÍO HUALLAGA STUBFOOT TOAD

(BOULENGER, 1882)

ADULT LENGTH
Male
1–1⅛ in (25–29 mm)

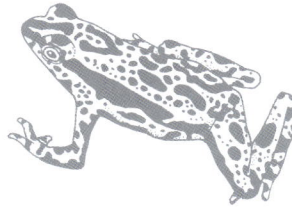

Female
1¼–1⅜ in (32–35 mm)

This very rare frog shows marked sexual dimorphism, the male being much smaller than the female. This is fortunate from the female's point of view, as *Atelopus* species typically maintain amplexus, the male riding on her back, for several days. Like its relatives, this species lives on or close to the ground and is active by day. In the ten years up to 2006, its numbers declined by at least 80 percent. It was found to be infected by the disease chytridiomycosis in 2003, and it has also been adversely impacted by the clearing of forests to create agricultural land. Río Huallaga Stubfoot Toad is the subject of a captive breeding program.

Actual size

The Río Huallaga Stubfoot Toad has a slender body, a pointed snout, and long hind limbs. Its upper surfaces are brown or bluish black, with a complex pattern of green longitudinal stripes and spots. The underside is bright red in females and cream in males, and the undersurfaces of the hands and feet are red.

SIMILAR SPECIES

Río Huallaga Stubfoot Toad is very similar to the Pebas Stubfoot Toad (*A. spumarius*; page 150) and to *A. loettersi*. The latter, described as recently as 2011, is nearly identical in external appearance, but it lacks a middle ear and is genetically distinct. The species is found in southeastern Peru, and its conservation status is Near Threatened.

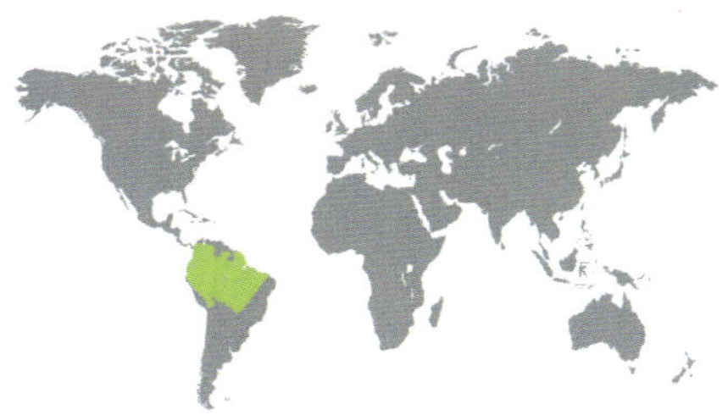

FAMILY	Bufonidae
OTHER NAMES	None
DISTRIBUTION	Several, widely separated areas in South America: Brazil, Colombia, Ecuador, French Guiana, Peru, Suriname
ADULT HABITAT	Tropical forest up to 2,000 ft (600 m) altitude
LARVAL HABITAT	Forest streams
CONSERVATION STATUS	IUCN Least Concern, but potentially threatened by deforestation in places and may be susceptible to chytridiomycosis

ADULT LENGTH
Male
1–1⅛ in (26–29 mm)
Female
1¼–1 9/16 in (31–39 mm)

ATELOPUS SPUMARIUS

PEBAS STUBFOOT TOAD

COPE, 1871

Actual size

The Pebas Stubfoot Toad has a flattened body, a narrow head, and a pointed snout. The back is black or brown, with a complex, netlike pattern in green. The belly is white at the anterior end of the body and red toward the back. The undersides of the legs, hands, and feet are red.

This small, ground-dwelling toad is active by day, when it can be found in leaf litter and on fallen logs near streams. It is capable of reproducing at any time of the year, laying its eggs in strings in streams or streamside pools. When attacked, it assumes a defensive posture in which it lies on its belly, lifting up its hands and feet to expose their bright red undersides. It produces a skin secretion containing a poison called tetrodotoxin, found in other amphibians and some fish. The species has not been reported in Ecuador since 1994 and has declined in Peru, but remains common in eastern parts of its range.

SIMILAR SPECIES

It has been suggested that *Atelopus spumarius* may, in fact, be more than one species. At least two subspecies have been elevated as species: *A. hoogmoedi*, from Surinam, which is yellow and black; and *A. barbotini*, from French Guiana, which is black with delicate red or pink markings.

FAMILY	Bufonidae
OTHER NAMES	Bolivian Stubfoot Toad, Three-colored Stubfoot Toad
DISTRIBUTION	Southeastern Peru, Bolivia
ADULT HABITAT	Humid montane forest at altitudes of 2,000–8,200 ft (600–2,500 m)
LARVAL HABITAT	Streams
CONSERVATION STATUS	IUCN Critically Endangered. Its habitat is declining due to the spread of smallholder farming

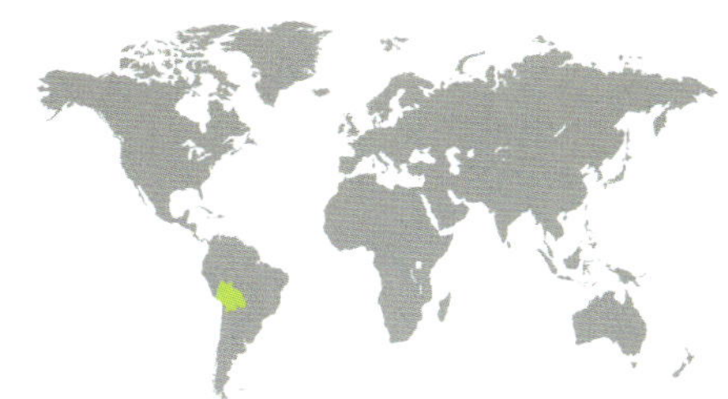

ATELOPUS TRICOLOR

THREE-COLORED HARLEQUIN TOAD

BOULENGER, 1902

ADULT LENGTH
Male
¾–1 1/16 in (19–27 mm)
Female
average 1 7/16 in (36 mm)

Found in the eastern Andes, this small frog appears not to have declined as much as other *Atelopus* species. It is generally seen perching above the ground in low vegetation. It breeds in streams, males gathering to call in groups of four to ten close to the water. Over some of its range it is threatened by habitat loss due to forest clearance for smallholder farms growing coffee, coca, and chili peppers, and by habitat deterioration as a result of increased stream sedimentation. The disease chytridiomycosis has yet to be found in this species, but may be a major threat in the future.

Actual size

The Three-colored Harlequin Toad has a slim body and head, a pointed snout, and long hind limbs. There are numerous small warts on its back, which is black with stripes and spots of yellow or green. The hands and feet are tinged with red, especially on their undersides.

SIMILAR SPECIES

Atelopus tricolor is thought to be closely related to the Carabaya Stubfoot Toad (*A. erythropus*), which can be found in cloud forest in the Andes of Peru and is listed as being Critically Endangered. The Cayenne Studfoot Toad (*A. flavescens*) is a species found in the lowlands up to altitudes of 1,000 ft (300 m) in French Guiana; it is currently listed as Low Concern.

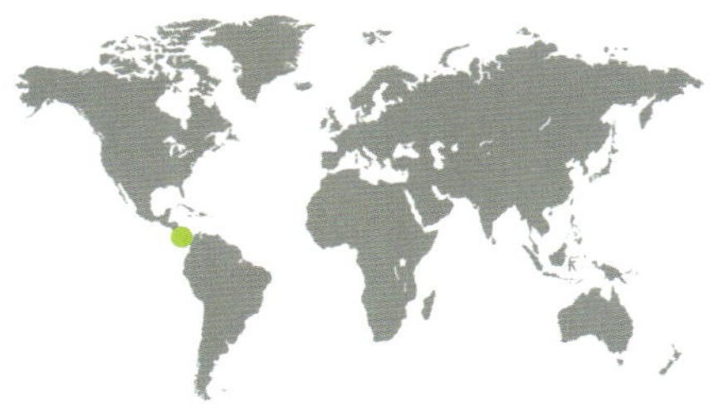

FAMILY	Bufonidae
OTHER NAMES	Clown Frog
DISTRIBUTION	Costa Rica, Panama
ADULT HABITAT	Tropical forest up to 6,500 ft (2,000 m) altitude
LARVAL HABITAT	Forest streams
CONSERVATION STATUS	IUCN Critically Endangered. Has undergone recent population decline; now known from only one location

ADULT LENGTH
Male
1–$1\frac{5}{8}$ in (25–41 mm)
Female
$1\frac{5}{16}$–$2\frac{3}{8}$ in (33–60 mm)

ATELOPUS VARIUS

HARLEQUIN FROG

(LICHTENSTEIN & MARTENS, 1856)

This variably colored frog is active by day, sitting on rocks or logs close to streams. Its bright coloration warns potential predators that its skin contains toxic compounds. This does not, however, protect it from a parasitic fly that lays an egg in its thigh. The egg hatches into a larva that burrows into the frog, killing it within a few days. This may have contributed to its dramatic decline, but more important factors are climate change and the disease chytridiomycosis. Harlequin Frogs were also collected in large numbers for the international pet trade. The species appeared to have become extinct by 1996, but a single population was found in Costa Rica in 2003.

SIMILAR SPECIES

Atelopus varius is one of several members of the genus that have all but vanished in Central America in recent years. The principal cause is the disease chytridiomycosis, which since 1974 has been making its way westward through Central America. The Chiriqui Harlequin Toad (*A. chiriquiensis*; page 146) and the Pass Stubfoot Toad (*A. senex*) both occupy smaller ranges than *A. varius* and are now listed as Extinct.

Actual size

The Harlequin Frog has long, slender limbs, a flattened body, a narrow head, prominent eyes, and a pointed snout. The back is yellow, orange, or lime-green with black markings, and the belly is white, yellow, orange, or red.

FAMILY	Bufonidae
OTHER NAMES	Golden Arrow Poison Frog, Zetek's Poison Frog
DISTRIBUTION	Central Panama
ADULT HABITAT	Tropical mountain forest
LARVAL HABITAT	Forest streams
CONSERVATION STATUS	IUCN Critically Endangered. Threatened by chytridiomycosis, habitat degradation, and collecting for the pet trade. Has been successfully bred in captivity

ADULT LENGTH
Male
1⅜–1⅞ in (35–48 mm)
Female
1¾–2½ in (45–63 mm)

ATELOPUS ZETEKI

PANAMANIAN GOLDEN FROG

DUNN, 1933

The national animal of Panama, this brightly colored frog may be extinct in the wild. It declined by 80 percent between 2000 and 2010, primarily due to the disease chytridiomycosis, spreading through Panama from west to east. Active by day, its vivid coloration signals that its skin secretes a lethal nerve poison, zetekitoxin. It breeds in fast-flowing streams, where its mating call is often inaudible. Instead, males attract females by stereotyped hand-waving and foot-raising movements. Males and females remain in amplexus for a few days, and females lay 200–600 eggs attached to rocks in streams.

SIMILAR SPECIES

There are currently 100 species in the genus *Atelopus*, found across Central and South America. The closest relative of *A. zeteki* is the equally brightly colored Harlequin Frog (*A. varius*; page 152), also found in Panama.

The Panamanian Golden Frog is golden yellow in color, decorated with a variable number of irregular black spots. Its head is longer than it is wide and the snout is pointed. The limbs are long. Newly metamorphosed frogs are vivid green with dark markings, making them cryptically colored against moss.

Actual size

FAMILY	Bufonidae
OTHER NAMES	Brongersma's Toad
DISTRIBUTION	Morocco, Western Sahara, northwestern Algeria
ADULT HABITAT	Semiarid areas up to 5,250 ft (1,600 m) altitude
LARVAL HABITAT	Ephemeral pools and ponds
CONSERVATION STATUS	IUCN Least Concern. Population is declining as a result of habitat deterioration

ADULT LENGTH
Male
up to 2 in (51mm)
Female
up to 1⅞ in (48 mm)

BARBAROPHRYNE BRONGERSMAI

TIZNIT TOAD

(HOOGMOED, 1972)

Actual size

This small toad lives in semiarid, hilly areas with rather sparse vegetation, up to altitudes of 5,250 ft (1,600 m). It is also found in plowed fields, and usually hides beneath stones during the day. The ephemeral ponds and pools it breeds in are generally located in rocky areas, and it has also been found in modified waterbodies such as water channels. It can be surprisingly abundant in suitable and unaltered habitat, but its population is in decline because its habitat is becoming increasingly arid. This represents a regional trend lasting more than 2,000 years: There is clear evidence that the region was wetter and more verdant in Roman times.

SIMILAR SPECIES

Barbarophryne brongersmai is the only species in its genus. Among toads, it is similar to, but considerably smaller than, the Natterjack Toad (*Epidalea calamita*; page 167) and the Green Toad (*Bufotes viridis*; page 162).

The Tiznit Toad is unusual among toads in that males and females are apparently similar in size, though data on body size in this species are scarce. It has small but prominent parotoid glands behind its eyes and red- or orange-tipped warts on its back. The upper surfaces are gray, straw-colored, or red, with green patches containing black dots.

FAMILY	Bufonidae
OTHER NAMES	Previously *Xanthophryne tigerina*, Yellow Tiger Toad
DISTRIBUTION	Western Ghats, India
ADULT HABITAT	Forest, scrubland, and grassland
LARVAL HABITAT	Temporary puddles
CONSERVATION STATUS	IUCN Endangered. Has a very restricted range that is threatened by deforestation.

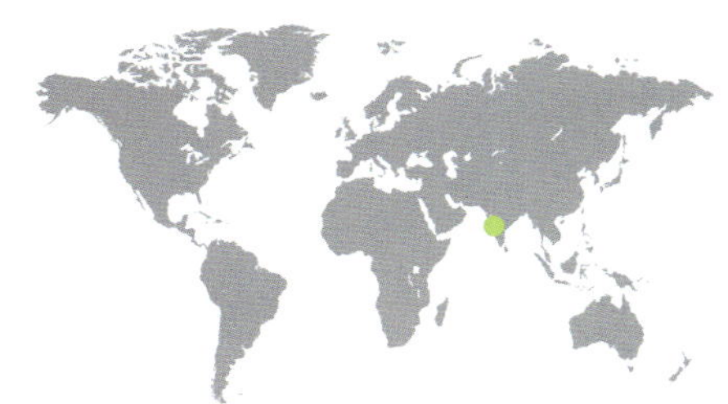

ADULT LENGTH
Male
$1^{5}/_{16}$ in (33 mm)

Female
$1^{3}/_{8}$ in (35 mm)

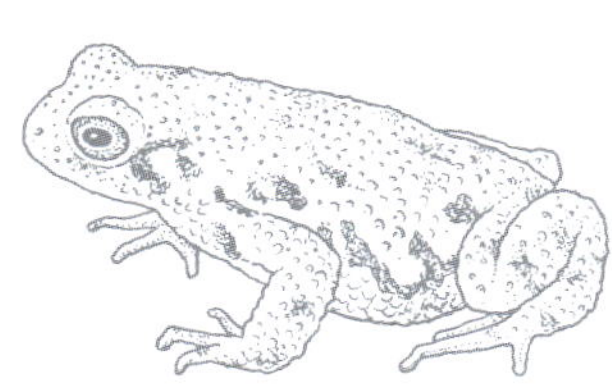

BEDUKA AMBOLI

AMBOLI LATERITIC TOAD

DUBOIS, OHLER & PYRON, 2021

This tiny, recently described toad is found at only one location, Amboli in the Western Ghats of India, and occupies an area of less than 4 sq miles (10 sq km). Its numbers have declined considerably since 2001, as has the amount of suitable habitat available to it. It hides by day in holes and crevices in rocks, emerging to forage at night. It breeds during the monsoon season, laying a clutch of 30–35 eggs in puddles. Mating has not been reported.

Actual size

The Amboli Lateritic Toad is a small toad with flat parotoid glands behind its eyes, and numerous spine-tipped warts on its back. There is no webbing on the hands or feet. It gets its common and scientific names from a number of irregular yellow stripes running from the back to the flanks.

SIMILAR SPECIES

Beduka koynayensis, known as the Humbali Village Toad, the Chrome Yellow Toad, or the Koyna Toad, is the only other species in this genus. It is found at only two locations in the Western Ghats, where its habitat is being reduced by deforestation, and is categorized at Endangered. It is largely yellow in color.

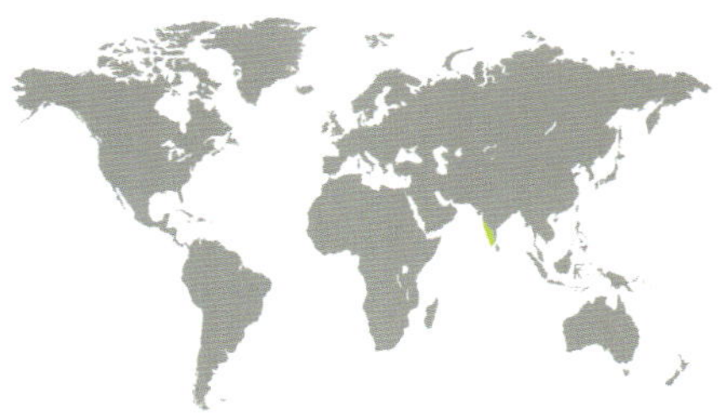

FAMILY	Bufonidae
OTHER NAMES	Black Torrent Toad, Ornate Stream Toad
DISTRIBUTION	Southwestern India
ADULT HABITAT	Tropical forest at altitudes of 2,000–3,300 ft (600–1,000 m)
LARVAL HABITAT	Fast-flowing streams
CONSERVATION STATUS	IUCN Endangered. Has a very small range and its habitat is at risk from deforestation

ADULT LENGTH
Male
1$\frac{1}{16}$–1$\frac{3}{16}$ in (27–30 mm)

Female
around 1$\frac{3}{8}$ in (35 mm);
based on one specimen

BLAIRA ORNATA

MALABAR TORRENT TOAD

(GÜNTHER, 1876)

Actual size

The Malabar Torrent Toad has a slender body, a short, pointed snout, warty skin, and webbing between its toes. The upper surfaces are brown or black, with yellow spots and patches of very variable size and extent. There is a reddish flush on the flanks, and the belly is bright red with yellow spots.

The Western Ghats are a range of mountains running down the western side of India. They are considered a biodiversity hotspot since they are home to many species that occur nowhere else. Of around 180 amphibian species found there, 80 percent are unique to the region, especially its tropical forest. This small, colorful toad is found among moss-covered boulders along fast-flowing forest streams and is known from only one small area, where the forest is being cleared to make way for coffee plantations. Its tadpoles have a powerful sucker around the mouth, enabling them to cling to rocks in flowing water.

SIMILAR SPECIES

The only other species in the genus *Blaira* is the Kerala Stream Toad, also known as the Red Stream Toad (*B. rubigina*). Listed as Near Threatened, it occurs in two locations in the Western Ghats, one of them protected, the Silent Valley National Park. It frequents torrential mountain streams at altitudes of 3,300–6,600 ft (1,000–2,000 m).

FAMILY	Bufonidae
OTHER NAMES	Common Toad
DISTRIBUTION	Europe, Asia, northwestern Africa
ADULT HABITAT	Woods, gardens, and fields
LARVAL HABITAT	Larger ponds and small lakes
CONSERVATION STATUS	IUCN Least Concern. Has a very extensive range. No general threats are reported, but it is declining in some parts of its range, such as the UK

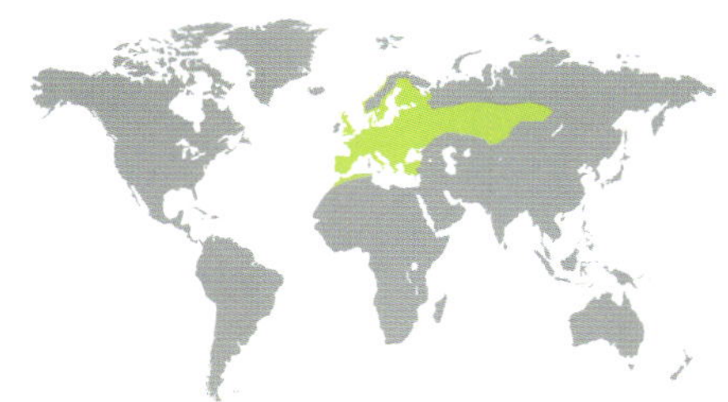

ADULT LENGTH
Male
1 13/16–5 1/8 in (46–130 mm)
Female
2–5 13/16 in (52–150 mm)

BUFO BUFO

EUROPEAN COMMON TOAD

(LINNAEUS, 1758)

For most of the year, Common Toads lead solitary lives, hiding by day under logs, rocks, or rotting vegetation, and emerging at night to feed on insects and other invertebrates. In spring they migrate to ponds, often in large numbers, to form mating groups characterized by apparently aimless wrestling and calling. Males greatly outnumber females and fight one another to be in amplexus with a female when she spawns. The males' quiet calls enable them to assess one another's size, smaller males making higher-pitched calls. The female lays 3,000–6,000 eggs in two long strings that are wrapped around vegetation.

SIMILAR SPECIES

There are a total of 26 species in the genus *Bufo*, distributed across Europe, Asia, and North Africa. The Giant Toad (*B. spinosus*), formerly considered a subspecies of *B. bufo*, is found in Spain, Portugal, southwestern France, and North Africa. It is larger than *B. bufo* and gets its name from the sharp spines that tip the warts on its back. See also the Caucasian Toad (*B. verrucosissimus*; page 160). Once a huge genus, *Bufo* now contains many fewer species.

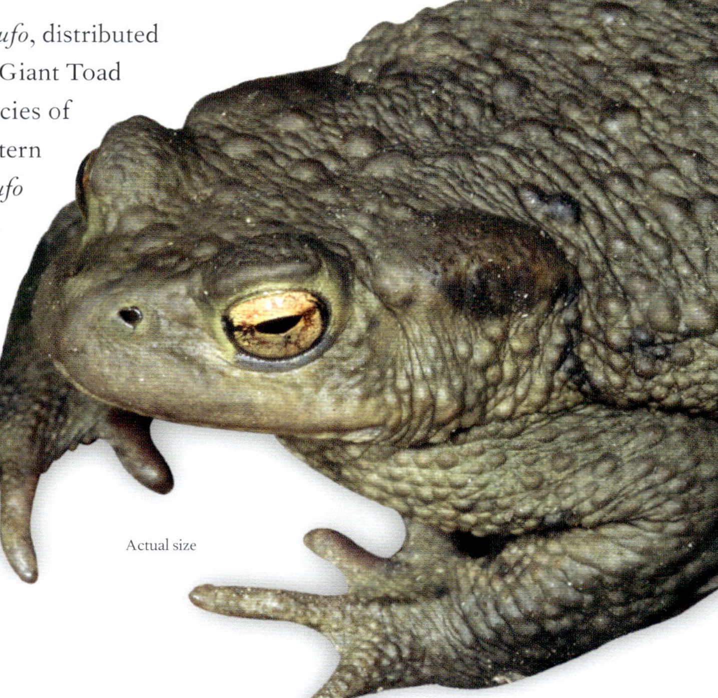

Actual size

The European Common Toad has a robust build, warty skin, prominent eyes, and large parotoid glands on its head, behind the eyes. It is green, gray, or brown, with darker markings. In southern parts of its range (e.g., in Italy), individuals grow to a very large size and their warts are tipped by spines.

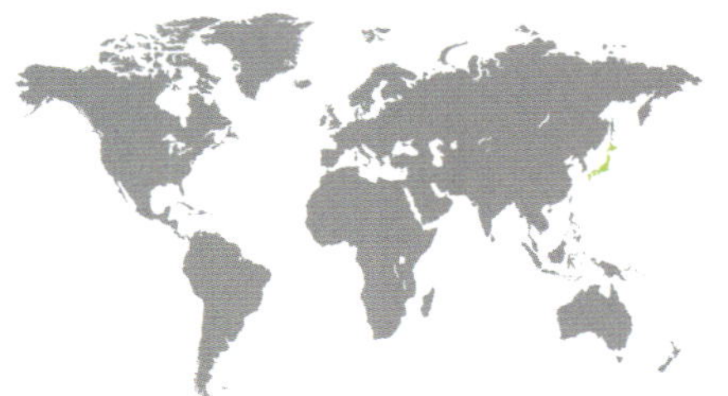

FAMILY	Bufonidae
OTHER NAMES	Includes *Bufo japonicus*, Japanese Common Toad
DISTRIBUTION	Japan
ADULT HABITAT	Woods, from sea level to high altitudes
LARVAL HABITAT	Ponds, swamps, and puddles
CONSERVATION STATUS	IUCN Least Concern. No general threats have been reported, but declining in the western part of its range

ADULT LENGTH
Male
1[13]/16–6[7]/16 in (46–161 mm)
Female
2[1]/16–6[15]/16 in (53–176 mm)

BUFO FORMOSUS

EASTERN JAPANESE COMMON TOAD

(BOULENGER, 1883)

Breeding in this large toad occurs between February and March and, as in the European Common Toad (*Bufo bufo*; page 157), involves intense "scramble competition" among males, who may outnumber females by as many as ten to one. The biased sex ratio of this and other toads arises primarily because females reach sexual maturity one or two years later than males. Eastern Japanese Common Toads show strong breeding site fidelity, returning to the same pond year after year, even when other breeding ponds are nearby. The secretion of the species' parotoid glands has been used in traditional medicine to dress cuts and burns.

SIMILAR SPECIES

The Japanese Stream Toad (*Bufo torrenticola*) is unusual among members of the genus in that it breeds in running water. It lives in mountainous regions, has a restricted range, and mates and lays its eggs in mountain streams. Its tadpoles have large oral disks that enable them to hold onto rocks while feeding on algae. The species is also unusual in that adults are sometimes found in trees.

Actual size

The Eastern Japanese Common Toad has a robust body, short, stout limbs, and numerous warts on its back. It is variable in color, being green, yellow-brown, brown, or red on its back. There is a broad black band along the flanks, with a narrower, very pale band above it. The eyes have gold irises and horizontal pupils.

FAMILY	Bufonidae
OTHER NAMES	Chinese Toad, Chusan Island Toad, Zhoushan Toad
DISTRIBUTION	Eastern China, Korean Peninsula, eastern Russia, Japan
ADULT HABITAT	Woodland, grassland, meadows
LARVAL HABITAT	Ponds, lakes, streams, rivers
CONSERVATION STATUS	IUCN Least Concern. Has a very large range, a very varied habitat, and is not declining

ADULT LENGTH
Male
$2\frac{3}{16}$–$3\frac{5}{16}$ in (56–86 mm)

Female
$2\frac{1}{2}$–$3\frac{15}{16}$ in (62–102 mm)

BUFO GARGARIZANS

ASIATIC TOAD

CANTOR, 1842

This large toad occurs at very high population densities in parts of eastern China and Russia, but is rare elsewhere. It is found in a wide variety of habitats, and feeds on insects, mollusks, millipedes, and arachnids. The eggs are laid in two long strings containing 1,200–7,400 eggs, larger females producing more eggs. In winter it retreats, either to hiding places on land, where it may gather in large groups, or to deep water, where it can survive beneath ice. The toxins in its skin have long been used in Asian medicine and are also of interest to the pharmaceutical industry for their antimicrobial and anticancer properties.

SIMILAR SPECIES

Bufo gargarizans is very similar to the European Common Toad (*B. bufo*; page 157), but is slightly larger, has spines on its back, and has a distinctive dark band along the side of its head and body. The Taiwan Common Toad (*B. bankorensis*) is larger still and occurs only on Taiwan. The Miyako Toad (*B. gargarizans miyakonis*) is a subspecies of the Asiatic Toad and is found on some of the smaller islands of Japan.

Actual size

The Asiatic Toad is a large, plump toad with large parotoid glands behind the eyes and numerous large warts on its back. The hind feet are webbed. It is dark brown on its back, and pale brown on its flanks and belly. A dark band begins below the parotoid gland and, to varying degrees in different individuals, runs along the flank to the groin.

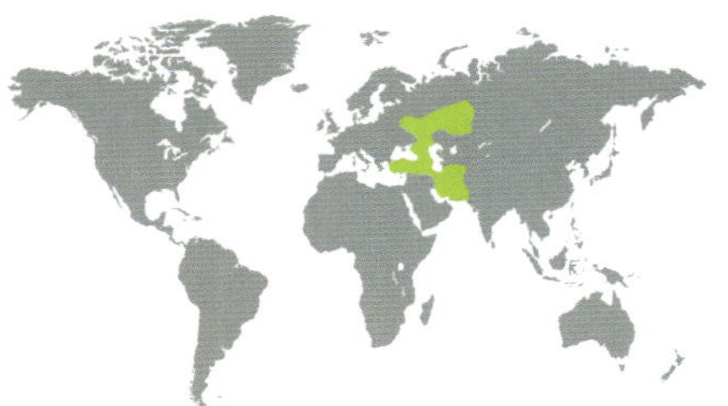

FAMILY	Bufonidae
OTHER NAMES	Caucasian Common Toad, Colchic Toad
DISTRIBUTION	Georgia; Russia, Turkey, Azerbaijan, Iran
ADULT HABITAT	Mountain forests
LARVAL HABITAT	Pools, ponds, lakes, and streams
CONSERVATION STATUS	IUCN Least Concern, but declining in parts of its range due to habitat destruction

ADULT LENGTH
Male
2¹³⁄₁₆–3⁵⁄₁₆ in (70–85 mm)

Female
3⁷⁄₈–7⁹⁄₁₆ in (100–190 mm)

BUFO VERRUCOSISSIMUS

CAUCASIAN TOAD

(PALLAS, 1814)

The Caucasian Toad has a robust build, very warty skin and very prominent parotoid glands. Its upper surfaces are green, gray, brown, or reddish with darker markings, and the belly is gray or yellowish. The eyes have golden irises and horizontal pupils.

This large toad shows a more marked size difference between males and females than any other *Bufo* species. Breeding occurs in April and May at lower altitudes, and as late as August at high altitudes. In many parts of its range, pools suitable for breeding are few and far between, and males and females, already in amplexus, wander about looking for a place to mate. If a larger, more permanent waterbody forms, males gather around it and wait for females to arrive. Depending on their size, females lay 870–10,500 eggs in two long strings.

SIMILAR SPECIES

Bufo verrucosissimus is very similar to, but considerably larger than, the European Common Toad (*B. bufo*; page 157). Eichwald's Toad (*B. eichwaldi*) is a rare species found in Azerbaijan and Iran. It is declining rapidly as a result of deforestation but its conservation status has not yet been assessed.

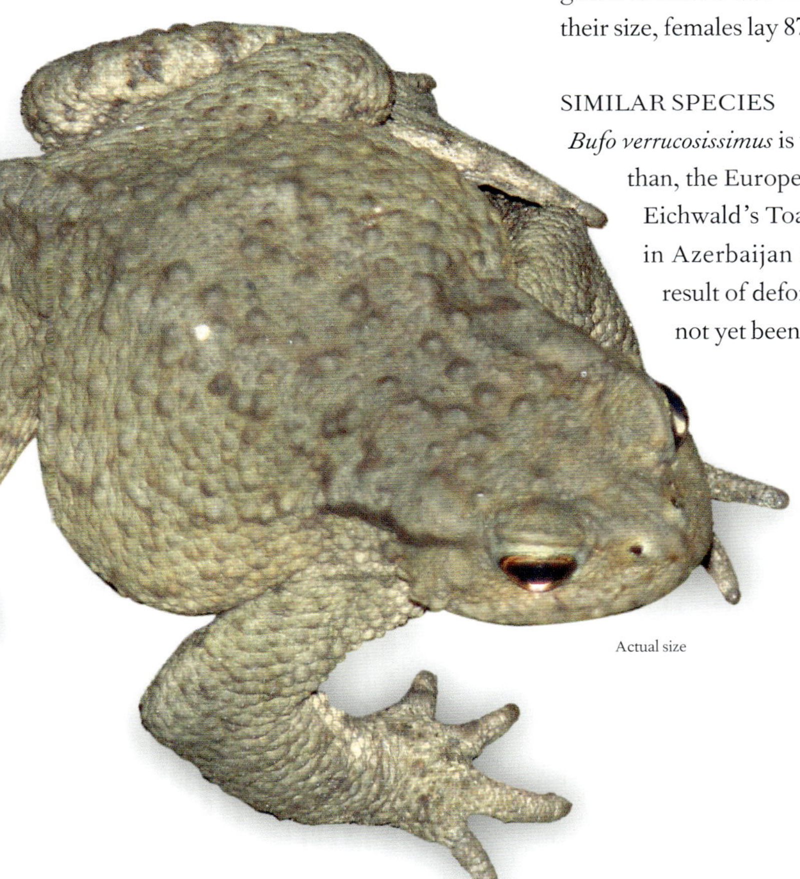

Actual size

FAMILY	Bufonidae
OTHER NAMES	Khasi Hill Rock Toad, Rock Toad
DISTRIBUTION	India, Bangladesh
ADULT HABITAT	Wet forest
LARVAL HABITAT	Small water-filled holes
CONSERVATION STATUS	IUCN Critically Endangered. Threatened by habitat loss, due to deforestation and quarrying

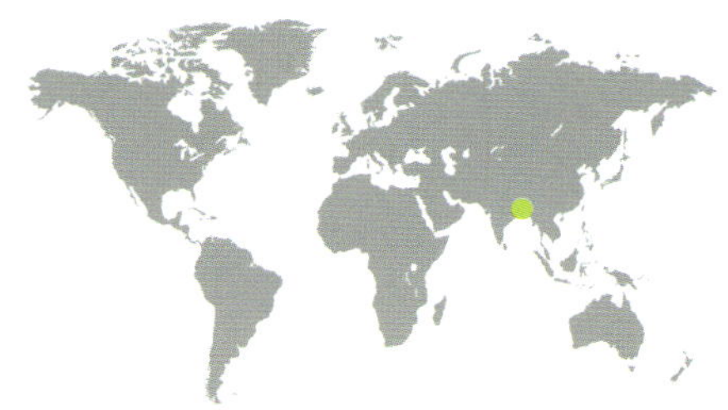

ADULT LENGTH
1⁷⁄₁₆–1⁹⁄₁₆ in (37–39 mm)

BUFOIDES MEGHALAYANUS

MAWBLANG TOAD

(YAZDANI & CHANDA, 1971)

This scarcely known little toad is apparently confined to two rocky areas, one in northeastern India and the other in Bangladesh. It is found either on the ground, sometimes in rock crevices, or in palm-like screwpine trees (*Pandanus* spp.). Mating and egg-laying occur either in water-filled leaf axils or in small pools among rocks and boulders. Its total range is thought to be little more than 115 sq miles (300 sq km) and it is under threat from the clear-cutting of forests and rock-blasting to quarry for stone.

Actual size

The Mawblang Toad has a slim body, a small head, slender limbs, and very warty skin. The toes are webbed and the fingers have slightly swollen tips. Its upper surfaces are dark brown or black, and there is often a scattering of yellow spots on its flanks.

SIMILAR SPECIES

The Mizoram Rock Toad (*B. bhupathyi*) and the Garo Hills Toad (*B. kempi*) are both found in India. *Bufoides* is most closely related to the Asian genus *Ansonia* (see page 145), called slender toads or stream toads. These are small, slim toads, of which there are 28 species, many of them found in Borneo.

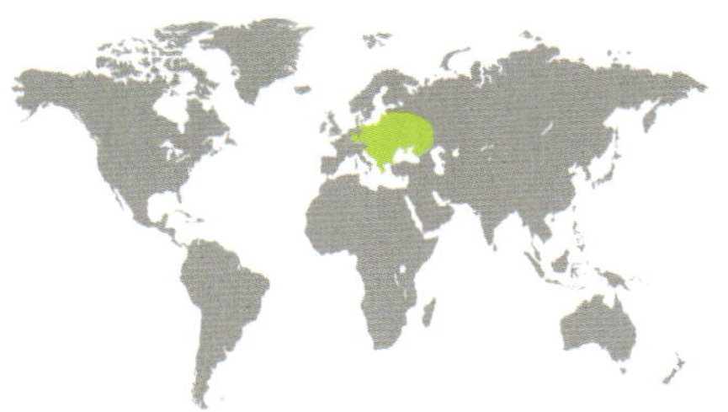

FAMILY	Bufonidae
OTHER NAMES	*Bufo viridis*, *Pseudepidalea viridis*
DISTRIBUTION	From Germany in the west to eastern Russia, and south to Greece and Crete
ADULT HABITAT	Wide range of habitats with dry, sandy soil
LARVAL HABITAT	Small and large bodies of still water
CONSERVATION STATUS	IUCN Least Concern. Very common in many parts of its very extensive range

ADULT LENGTH
Male
1⅞–3½ in (48–90 mm)

Female
2¹³⁄₁₆–4¾ in (70–120 mm)

BUFOTES VIRIDIS

GREEN TOAD

(LAURENTI, 1768)

More colorful than most toads, the Green Toad is noted for its adaptability. It has a huge range, is found in a wide variety of habitats, breeds in all kinds of standing water, and, in many places, lives in agricultural and urban habitats. It is more tolerant of high temperatures and drought conditions than most amphibians. In spring, males form choruses at night, individuals defending their call sites and producing a high-pitched trill. Females lay 2,000–30,000 eggs in long strings. If molested or attacked, Green Toads produce an evil-smelling white secretion from their skin.

SIMILAR SPECIES

There is considerable uncertainty about the relationships among the 12 species of *Bufotes* and other toad genera found across eastern Europe and Asia, as far as China. *Bufotes variabilis*, from Turkey and Kazakhstan, is now treated as a synonym of *B. viridis*; *B. siculus*, from Sicily, is a synonym of the African Green Toad (*B. boulengeri*), and *B. zamdaensis*, from China, is a synonym of the Swat Green Toad (*B. pseudoraddei*). All are known locally as "green toads."

Actual size

The Green Toad has a rounded snout with prominent parotoid glands and large, protruding eyes. The skin is white to pale yellow or gray, patterned with green markings and, in many individuals, flecks of red. The belly is white. Males have more muscular forearms than females and are generally less colorful.

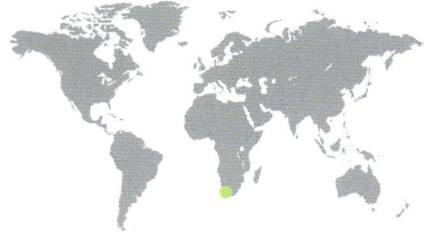

FAMILY	Bufonidae
OTHER NAMES	Cape Mountain Toad, Rose's Mountain Toadlet
DISTRIBUTION	Western Cape Province, South Africa
ADULT HABITAT	Mountains with fynbos vegetation
LARVAL HABITAT	Small pools
CONSERVATION STATUS	IUCN Critically Endangered. Habitat loss has reduced it to a few isolated populations

CAPENSIBUFO ROSEI

ROSE'S MOUNTAIN TOAD

(HEWITT, 1926)

ADULT LENGTH
Male
average 7/8 in (22 mm),
up to 1 1/8 in (28 mm)

Female
average 1 1/4 in (31 mm),
up to 1 9/16 in (39 mm)

This small toad is unusual among frogs, and unique in southern Africa, in that it does not call. It also lacks a tympanum, so probably cannot hear. It has short, rather weak hind legs and moves about by walking rather than hopping. It breeds after rain, in August and September, and, in the past, sometimes gathered around pools in very large numbers. Females lay around 100 eggs in amber-colored strings. This species has disappeared from many places where it was previously common, including the western section of its range, on Table Mountain in the Cape Peninsula.

Actual size

Rose's Mountain Toad has an elongated body and very short hind legs. The skin is smooth, and there are blister-like ridges and warts on its back and sides. The upper surfaces are gray or brown, decorated with light-colored bands and large dark spots. There are red or orange spots on the parotoid glands and along the sides of the body.

SIMILAR SPECIES

There are now five species in the genus *Capensibufo*. The Tradouw Mountain Toad (*C. tradouwi*), also known as Tradouw's Toad, is found on several mountain ranges in Western Cape Province. In contrast to Rose's Mountain Toad, it does have a tympanum and males do call, producing a creaking sound. It breeds in winter. The other three species are poorly known.

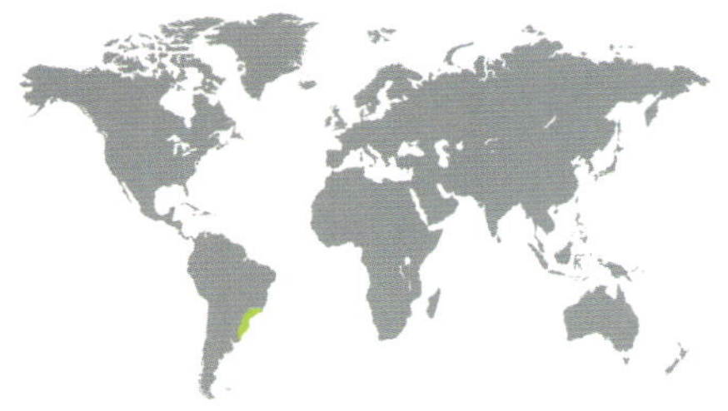

FAMILY	Bufonidae
OTHER NAMES	None
DISTRIBUTION	Coastal mountains in southeastern Brazil, up to 3,000 ft (910 m) altitude
ADULT HABITAT	Forest
LARVAL HABITAT	Water-filled leaf axils of bromeliads
CONSERVATION STATUS	IUCN Least Concern. The population is stable

ADULT LENGTH
Male
½–⅝ in (12–16 mm)
Female
11/16–⅞ in (17–21 mm)

DENDROPHRYNISCUS BREVIPOLLICATUS

COASTAL TREE TOAD

JIMÉNEZ DE LA ESPADA, 1870

Actual size

The Coastal Tree Toad is a tiny frog with a pointed snout and many small warts on its back, flanks, and legs. The upper surfaces are tan, brown, or bronze with darker markings. There are dark cross-bands on the legs. The underside is cream.

This diminutive frog is found on the ground or in foliage near the ground. It lays its eggs in phytotelmata, tiny pools of water that form between the leaves of ground-living or tree-living bromeliads. As the food supply in phytotelmata is typically insufficient for the developing tadpoles, they lack a mouth and are equipped with sufficient yolk to enable them to reach metamorphosis. Adults of this species represent an example of miniaturization, an evolutionary trend that has occurred in several amphibian groups, whereby descendent species are much smaller than their ancestors.

SIMILAR SPECIES

There are 17 species in the genus *Dendrophryniscus*, some of them found in Amazonia and others in the Atlantic Forest along the east coast of Brazil. *Dendrophryniscus carvalhoi* is a bromeliad-breeding species known from only two locations in the Atlantic Forest. Davor's Tree Toad (*D. davori*) is Critically Endangered, while Lauro's Tree Toad (*D. lauroi*) is Endangered. Both are localized Brazilian species.

FAMILY	Bufonidae
OTHER NAMES	None
DISTRIBUTION	Equatorial Guinea, Cameroon, Nigeria
ADULT HABITAT	Montane forest up to at least 3,300 ft (1,000 m) altitude
LARVAL HABITAT	Unknown
CONSERVATION STATUS	IUCN Vulnerable. Has a small, fragmented range that is subject to deforestation

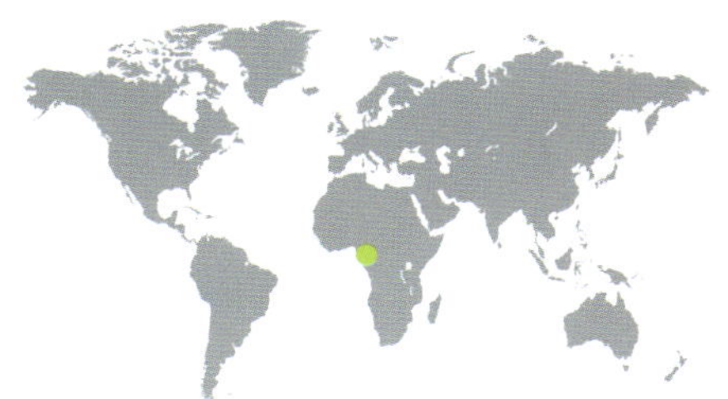

ADULT LENGTH
around ¾ in (20 mm); based on one specimen

DIDYNAMIPUS SJOSTEDTI

FOUR-DIGIT TOAD

ANDERSSON, 1903

So barely known is this tiny frog that only one individual has been measured. It gets its name from the fact that it has just four toes, instead of the usual five. In addition, two of its fingers and two of its toes are reduced to rudiments. Reduction in the size and number of digits is a feature of the evolutionary phenomenon called miniaturization (see page 171). It is known from only six widely separated locations in West Africa, and is reported to form aggregations of up to 40 animals, of mixed ages and sex, in forest clearings. Its reproductive behavior has not been observed, and it has been variously suggested that its tadpoles develop in water, within the egg, or inside their mother.

Actual size

The Four-digit Toad has an elongated body, slender legs, and a pointed snout. All but two fingers and toes are reduced to rudiments or are lost altogether. The upper surfaces are brown with darker and lighter marbled patterns, and the underside is white.

SIMILAR SPECIES

Didynamipus sjostedti is the only species in its genus and is thought to be closely related to the Western Nimba Toad (*Nimbaphrynoides occidentalis*; page 185), a small West African frog that gives birth to fully developed young.

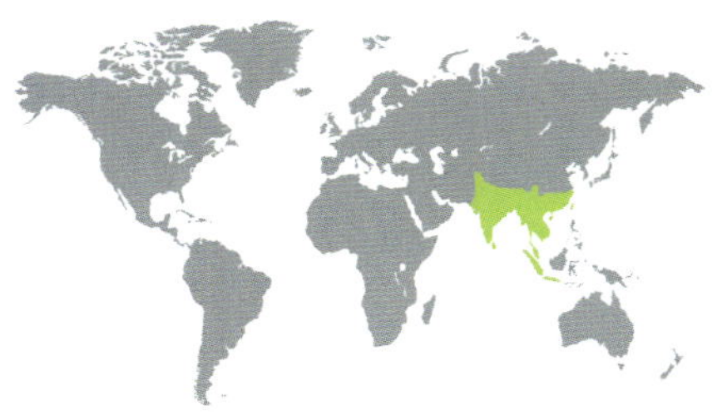

FAMILY	Bufonidae
OTHER NAMES	Asian Common Toad, Common Sunda Toad (Borneo), Black-spined Toad
DISTRIBUTION	Southeast Asia, from Pakistan in the west to Taiwan in the east, and south to Java
ADULT HABITAT	Disturbed lowland habitats, including farmland and urban areas
LARVAL HABITAT	Ephemeral pools
CONSERVATION STATUS	IUCN Least Concern. It has adapted to altered habitats and is becoming more common in parts of its range

ADULT LENGTH
Male
$2\frac{3}{16}$–$3\frac{1}{4}$ in (57–83 mm)

Female
$2\frac{5}{8}$–$3\frac{5}{16}$ in (65–85 mm)
in Borneo; in Pakistan
up to 6 in (150 mm)

DUTTAPHRYNUS MELANOSTICTUS

SOUTHEAST ASIAN TOAD

(SCHNEIDER, 1799)

One of very few amphibians that are becoming more common, this large toad has been described as timid and lethargic. It has rather short hind legs and moves about in short hops. It hides by day, and comes out at night to feed. In urban areas it is commonly seen waiting under streetlights for insects to fall to the ground. It breeds in temporary pools, the males attracting females with a low, rattling trill. Females lay their eggs in strings wound around submerged vegetation. In some places it grows to a very large size; in Pakistan, females can be as long as 6 in (150 mm).

The Southeast Asian Toad has a relatively small head, short hind limbs, large oval parotoid glands, and clearly defined tympana. The back, flanks, and limbs are covered in warts tipped with small spines. The back is gray or reddish brown with darker markings, and dark markings around the eye give it the appearance of wearing spectacles.

SIMILAR SPECIES

There are a total of 23 species in the genus *Duttaphrynus*, found in Africa and Asia. The Himalayan Toad (*D. himalayanus*) is a rare species, found across the Himalayas at altitudes of 6,500–11,500 ft (2,000–3,500 m). Noellert's Toad (*D. noellerti*) is a little-known species found in Sri Lanka that is listed as Endangered.

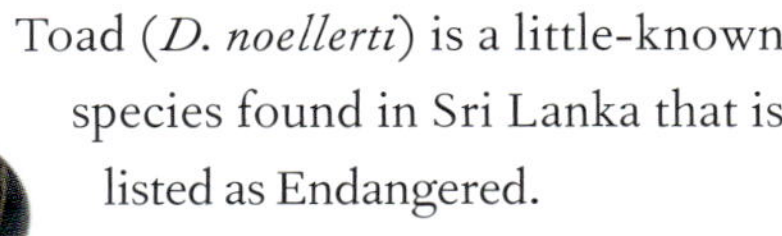

Actual size

FAMILY	Bufonidae
OTHER NAMES	None
DISTRIBUTION	Western and central Europe, UK, Ireland
ADULT HABITAT	Open landscapes with light, sandy soils, including heathland and sand dunes
LARVAL HABITAT	Shallow, ephemeral pools
CONSERVATION STATUS	IUCN Least Concern. Common in much of Europe, but Endangered and protected in the UK and Ireland

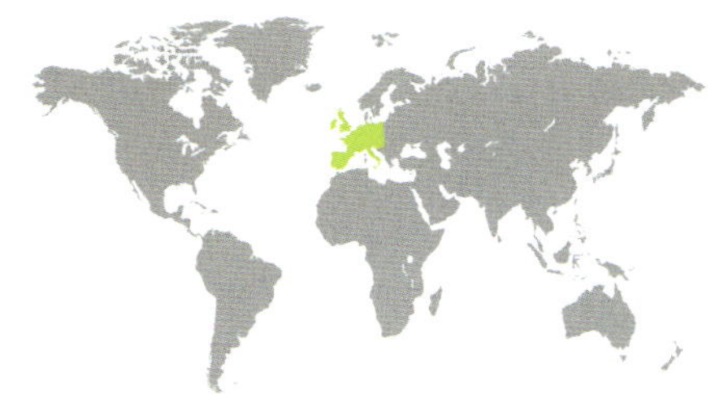

EPIDALEA CALAMITA

NATTERJACK TOAD

(LAURENTI, 1768)

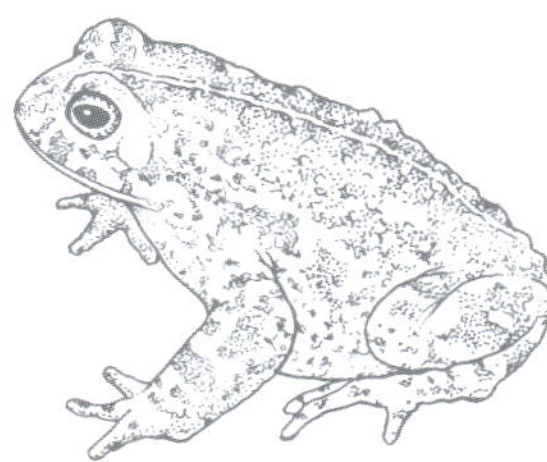

ADULT LENGTH
Male
1¼–2¹¹⁄₁₆ in (45–70 mm)

Female
1¹⁵⁄₁₆–3⅛ in (50–80 mm); both sexes can be larger in Spain

With shorter legs than most toads, the Natterjack moves in a scuttling motion, like a mouse. It can breed anytime between April and July, and some individuals breed more than once a year. Males produce a loud, rolling croak, lasting one or two seconds, that is audible several miles away. This attracts females to shallow, ephemeral pools, where they lay 1,500–7,500 eggs in long strings. Mass mortality among eggs or tadpoles is quite common if a breeding pond dries up. Outside the breeding season, adults move large distances, and experiments have shown that they use the Earth's magnetic field to navigate.

SIMILAR SPECIES

The only species in its genus, *Epidalea calamita* is most similar to, and, in parts of its range, lives alongside, the European Common Toad (*Bufo bufo*; page 157) and the Green Toad (*Bufotes viridis*; page 162). It differs from both in having shorter legs and a yellow stripe down the middle of its back.

The Natterjack Toad has prominent parotoid glands just behind the eyes. The back is very variable in color, but is often gray or olive-green with numerous dark spots, and there is a narrow yellow stripe running down its middle. There are numerous yellow or red warts on the back and sides, and the belly is pale.

Actual size

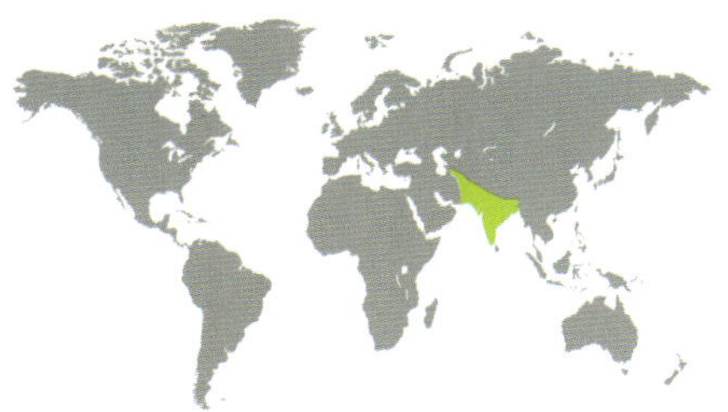

FAMILY	Bufonidae
OTHER NAMES	Assam Toad, Indus Valley Toad
DISTRIBUTION	Asia, from Iran in the west to Bangladesh and Bhutan in the east
ADULT HABITAT	Varied lowland and upland habitats, including grassland, woodland, farmland, and urban areas
LARVAL HABITAT	Ephemeral and permanent pools
CONSERVATION STATUS	IUCN Least Concern. It has adapted to agricultural habitats

ADULT LENGTH
average 3½ in (90 mm)

FIROUZOPHRYNUS STOMATICUS

MARBLED TOAD

(LÜTKEN, 1864)

Especially common in the Indus Valley in Pakistan, this toad is the first amphibian to appear in flooded areas after early summer rain. Males gather around pools and attract females with a low, guttural call. The eggs are laid in strings wrapped around submerged vegetation. The tadpoles form dense schools when they are small, as a defense against predators. Many tadpoles die because their pond dries up before they reach metamorphosis. As in many toads around the world, large numbers of adults living in urban areas are killed on roads as they make their way to breeding sites or hunt for insects under streetlights.

SIMILAR SPECIES

The Olive Toad (*Firouzophrynus olivaceus*) is a common toad that is found in Iran, Pakistan, and India. It is very similar in its natural history to *F. stomaticus* but has a much less extensive range. All five species in this genus, which were formerly in *Duttaphrynus*, are Least Concern.

The Marbled Toad has a blunt snout and kidney-shaped parotoid glands. Its skin is covered in many tightly packed warts, which look like tiny marbles. The upper surfaces are gray, olive-green, or almost black, with a mottled pattern of darker patches. The upper lip is cream.

Actual size

FAMILY	Bufonidae
OTHER NAMES	None
DISTRIBUTION	Bahia, eastern Brazil
ADULT HABITAT	Forest
LARVAL HABITAT	Unknown
CONSERVATION STATUS	IUCN Least Concern

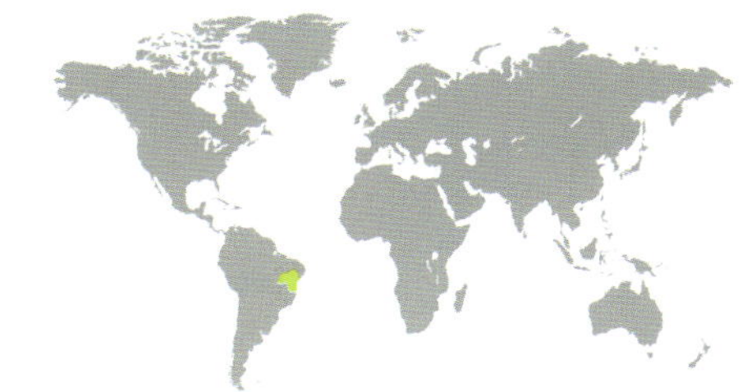

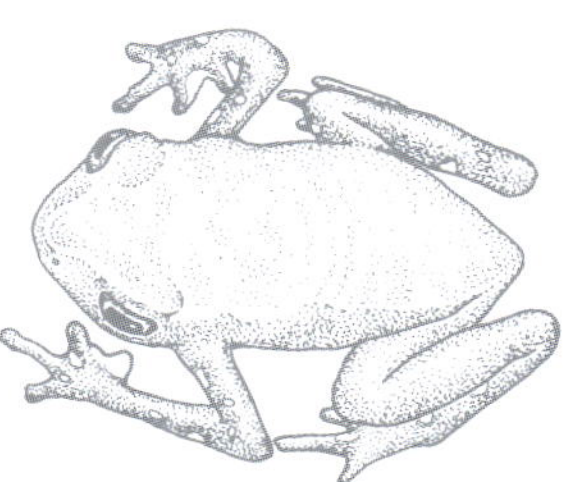

ADULT LENGTH
Male
average 1⅛ in (28 mm)

Female
unknown

FROSTIUS ERYTHROPHTHALMUS

RED-EYED FROST'S TOAD

PIMENTA & CARAMASHI, 2007

This small frog, previously considered to be a member of the genus *Atelopus*, or harlequin frogs, is found within a small area of the Atlantic Forest in Brazil. There it lives in the leaf litter and among bromeliads. Males call while perched on tree trunks or in low bushes. Its reproductive behavior has not been observed but it is thought that it lays its eggs in bromeliads and that its tadpoles are endotrophic, meaning that they lack mouths and are dependent on yolk provided by their mother.

Actual size

The Red-eyed Frost's Toad has a slender body, thin legs, a pointed snout, and warty skin. It is dark gray or black all over, with yellow spots on its belly. The eyes have bright red irises, although they may sometimes be yellow.

SIMILAR SPECIES

The only other species in the genus *Frostius* is Pernambuco Frost's Toad (*F. pernambucenis*), which is found in northeastern Brazil. It is distinguished from *F. erythropthalmus* in having bright yellow rather than red irises and by having a different call. Interactions between males involve calling and limb-waving. It has an extensive range along the Brazilian coastal mountains and is listed as of Least Concern.

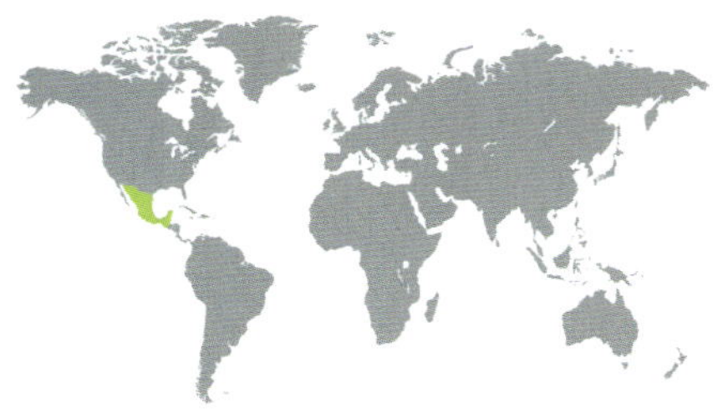

FAMILY	Bufonidae
OTHER NAMES	None
DISTRIBUTION	Mexico, Guatemala
ADULT HABITAT	Cloud forest
LARVAL HABITAT	Streams and small rivers
CONSERVATION STATUS	IUCN Endangered. Under severe threat because its small range is subject to habitat destruction

ADULT LENGTH
Male
2⅛–2¾ in (54–68 mm)
Female
2¹⁄₁₆–3⅛ in (53–80 mm)

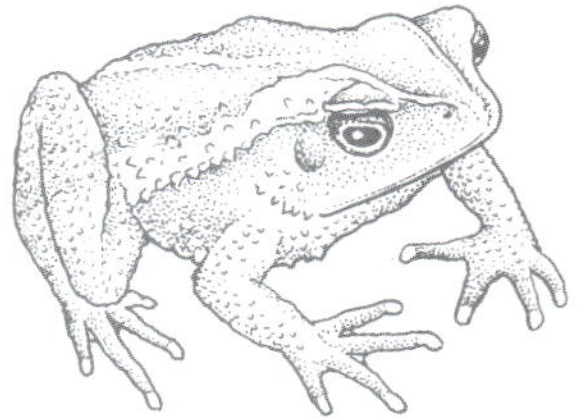

INCILIUS AURARIUS

CUCHUMATAN GOLDEN TOAD

MENDELSON, MULCAHY, SNELL, ACEVEDO & CAMPBELL, 2012

This recently described toad is related to Costa Rica's extinct Golden Toad (*Incilius periglenes*; page 174) and, in common with it, lives at high altitude in cloud forest. Unlike most toads, which lay their eggs in ponds and pools, it lays its eggs in streams. It does so in the dry season, when water levels are low—its tadpoles are not adapted to life in fast-flowing water, for example by having suckers, and would probably not survive in swollen streams. The species' future is uncertain, not least because of the difficulties associated with assessing its status in the field.

SIMILAR SPECIES

There are 39 species in the genus *Incilius*, known as Central American toads. Some of these species are called forest toads, including Campbell's Forest Toad (*I. campbelli*), found in Mexico, Guatemala, Honduras, and Belize. It is listed as Least Concern and differs from *I. aurarius* in that males are brown rather than gold. The Large-crested Toad (*I. macrocristatus*) is found in Mexico and Guatemala and is listed as Near Threatened.

The Cuchumatan Golden Toad is unusual in that males and females are very different in color. Whereas females are brown with dark markings (as shown here), males are pale gold. The toad has bony crests on its relatively large head and oval parotoid glands. There is a line of large warts running along each side of the body.

Actual size

FAMILY	Bufonidae
OTHER NAMES	None
DISTRIBUTION	Costa Rica, Panama
ADULT HABITAT	Rainforest at altitudes of 2,500–6,900 ft (760–2,100 m)
LARVAL HABITAT	Shallow pools
CONSERVATION STATUS	IUCN Critically Endangered. Has recently declined dramatically across its small range and may be extinct

INCILIUS FASTIDIOSUS

PICO BLANCO TOAD

(COPE, 1875)

ADULT LENGTH
Male
$1\frac{11}{16}$–2 in (43–52 mm)
Female
$1\frac{9}{16}$–$2\frac{3}{8}$ in (40–60 mm)

This small toad lives most of its life underground, emerging only to breed following heavy rain. The male has no mating call but does have a short release call that he gives when clasped by another male. Males clasp females around the pelvis and, because males greatly outnumber females, a single female may be clasped by between two and ten males. Occasionally, a female will drown because of the weight of males hanging on to her. The eggs are laid in strings with constrictions in the jelly between each one, so that they resemble rosary beads.

Actual size

The Pico Blanco Toad has prominent crests on its head and short limbs, especially the hind legs. The back is covered in numerous warts and has a line of large warts along each side. The upper surfaces are brown or black and the warts are red, rust-colored, or pink. There is usually a pale stripe running down the middle of the back.

SIMILAR SPECIES

Incilius fastidiosus is one of a number of similar species found in Central America that appear to be victims of the disease chytridiomycosis. The Almirante Trail Toad (*I. peripetetes*), a little-known toad from Panama, is also categorized as Critically Endangered, as is Holdridge's Toad (*I. holdridgei*), known only from the mountains of Costa Rica, and which may be extinct. See also the Golden Toad (*I. periglenes*; page 174).

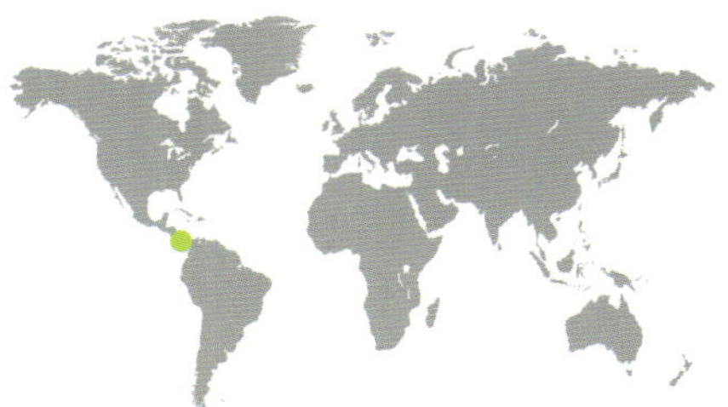

FAMILY	Bufonidae
OTHER NAMES	Dark Green Toad
DISTRIBUTION	Costa Rica; Panama
ADULT HABITAT	Wet forest from lowlands up to 3,500 ft (1,080 m) altitude
LARVAL HABITAT	Streams
CONSERVATION STATUS	IUCN Least Concern. A widespread species, showing no evidence of decline

ADULT LENGTH
Male
1$^{11}/_{16}$–2$^{15}/_{16}$ in (43–74 mm)
Female
2$^{5}/_{8}$–4$^{3}/_{16}$ in (65–107 mm)

INCILIUS MELANOCHLORUS

WET FOREST TOAD

(COPE, 1877)

Most toads are sexually dimorphic, with females larger than males, a disparity that is particularly marked in this Central American toad. It breeds in the dry season, when water levels in streams are low, reducing the risk that its eggs or tadpoles will be swept downstream. The male's call is a short trill, repeated frequently. The toad is active at night and may be found hiding in leaf litter during the day. As a stream-breeding species, there is concern that it may become infected with the fungal disease chytridiomycosis.

SIMILAR SPECIES

The Yellow Toad (*Incilius luetkenii*) is a large, lowland-living toad, found from Mexico to Costa Rica, that is notable for being uniformly yellow or green in color. The Southern Round-gland Toad (*I. coccifer*) is a smaller toad, found across much of Central America. The Green Climbing Toad (*I. coniferus*) breeds in pools and is unusual in that it climbs up into vegetation; it is found from Nicaragua to Ecuador.

Actual size

The Wet Forest Toad has a relatively large head, large eyes, and short hind legs. There is a line of large, pointed warts along each flank and similar warts on the hind legs. It is pale brown or gray in color, with large, irregular dark gray or black blotches. There is usually a thin, pale stripe down the middle of its back.

FAMILY	Bufonidae
OTHER NAMES	None
DISTRIBUTION	USA: Mississippi and Texas; eastern Mexico
ADULT HABITAT	Wide range of habitats, including forest, agricultural land, and suburban areas
LARVAL HABITAT	Still water of any kind
CONSERVATION STATUS	IUCN Least Concern. Has a large range and its population is stable

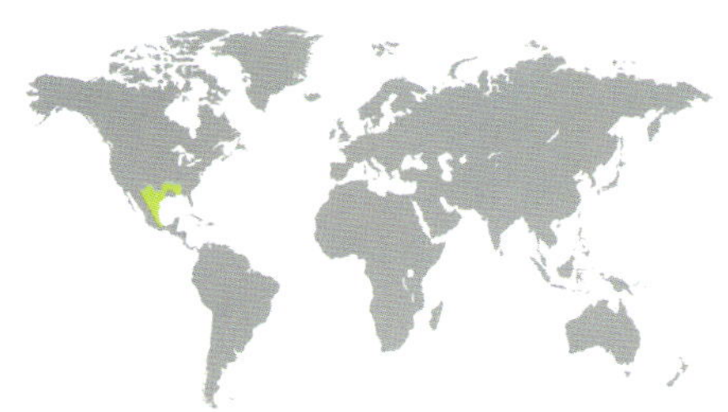

ADULT LENGTH
Male
2 1/16–3 13/16 in (53–98 mm)
Female
2 1/8–4 15/16 in (54–125 mm)

INCILIUS NEBULIFER

COASTAL PLAIN TOAD

(GIRARD, 1854)

This large toad lives mostly in coastal areas from Mississippi to southern Mexico, but also occurs far inland in Texas. Unusually for a toad, it climbs trees and can be found as high as 15 ft (4.5 m) above the ground. It is typically found near water, and it breeds from March to August, after heavy rain, in flooded meadows and fields. The male has a particularly large vocal sac and his call is a rattling trill, lasting 4–6 seconds. The female can lay up to 20,000 eggs, produced in long strings. When alarmed, this toad flattens itself against the ground.

SIMILAR SPECIES

The Gulf Coast Toad (*Incilius valliceps*) is so similar in appearance to *I. nebulifer* that, until recently, they were considered to be the same, wide-ranging species. It occurs in Central America, from southern Mexico to northern Costa Rica. Nocturnally active, it lives in forest and a variety of disturbed habits.

Actual size

The Coastal Plain Toad has a squat, flattened body and there is a depressed area between prominent bony crests on the top of its head. The skin is covered in warts. Generally dark brown in color, it has an obvious pale stripe down the middle of its back and broad pale stripes down each flank.

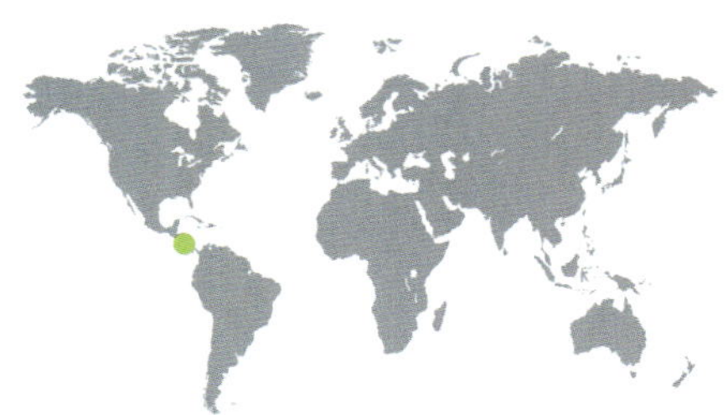

FAMILY	Bufonidae
OTHER NAMES	None
DISTRIBUTION	Monteverde Cloud Forest Reserve, Costa Rica
ADULT HABITAT	Cloud forest
LARVAL HABITAT	Shallow pools
CONSERVATION STATUS	IUCN Extinct. Had a very restricted range; thought to have been extirpated by climate change or chytridiomycosis

ADULT LENGTH
Male
1⅝–1⅞ in (41–48 mm)

Female
1⅞–2⅛ in (47–54 mm)

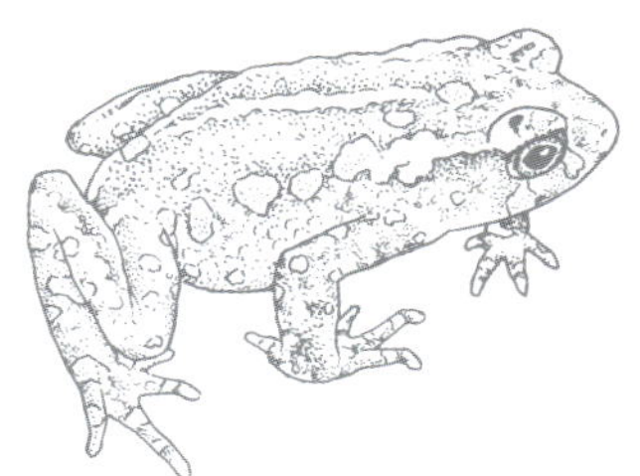

INCILIUS PERIGLENES

GOLDEN TOAD

(SAVAGE, 1967)

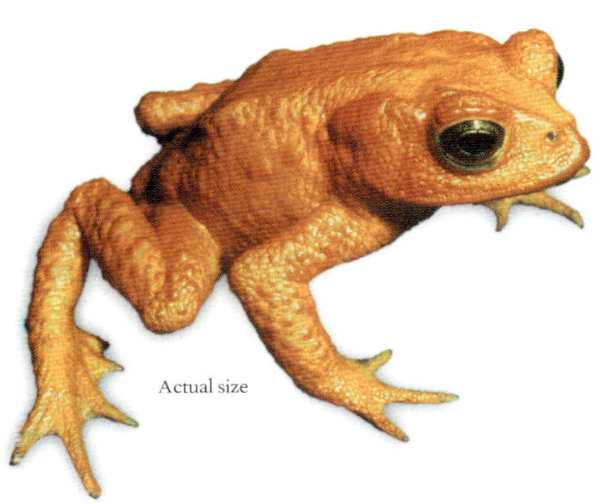

Actual size

This remarkable toad provides a striking example of a catastrophic population decline. Living in an area of less than 4 sq miles (10 sq km), 1,500 individuals were counted in 1987, 9 in 1988, and 1 in 1989; since then, none has been seen. Outside a short breeding season, it lived underground. It was unusual among toads in being both brightly colored and sexually dimorphic in terms of color. Its decline in the late 1980s coincided with a succession of unusually dry rainy seasons and was probably due to chytridiomycosis, which had not been discovered at that time.

SIMILAR SPECIES

A number of members of the genus *Incilius* are found in the mountains of Central America. The Cerro Utyum Toad (*I. epioticus*) is a rare species found in Costa Rica and Panama. The little-known *I. guanacaste* is found in Costa Rica, but is Endangered, and the recently described *I. karenlipsae*, from Panama, is Data Deficient and may be extinct. See also the Cuchumatan Golden Toad (*I. aurarius*; page 170).

The Golden Toad was a small toad whose skin was covered with small warts topped by tiny black spines. Males were uniformly bright orange, while females were greenish yellow to black with red spots edged with yellow. The eyes had black irises and horizontal pupils, and there was no tympanum. The toes were partially webbed. The slightly smaller male had a more pointed snout and longer legs than the female.

FAMILY	Bufonidae
OTHER NAMES	Forest Toad, Malayan Dwarf Toad
DISTRIBUTION	Sumatra, Indonesia; Borneo
ADULT HABITAT	Lowland forest
LARVAL HABITAT	Pools and slow-moving streams
CONSERVATION STATUS	IUCN Least Concern. A widespread species, its numbers are declining due to deforestation

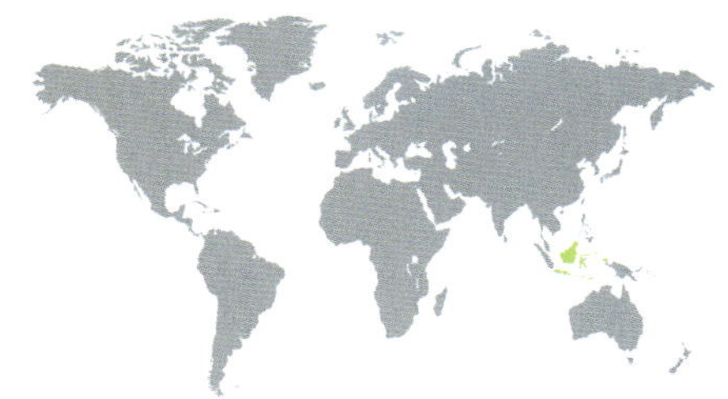

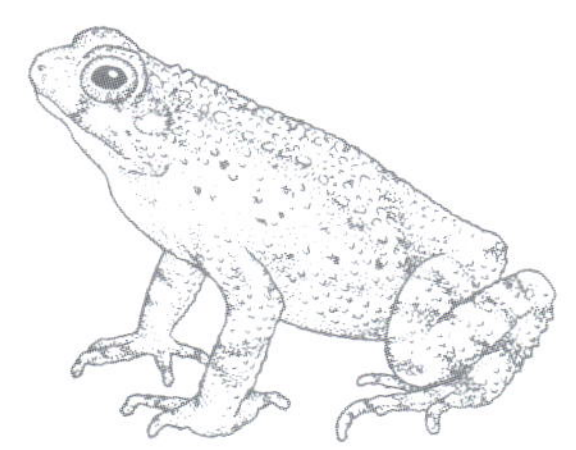

ADULT LENGTH
Male
1⅛–1$^{11}/_{16}$ in (28–43 mm)
Female
1$^{7}/_{16}$–2⅛ in (36–55 mm)

INGEROPHRYNUS DIVERGENS

CRESTED TOAD

(PETERS, 1871)

This small, stocky toad lives in leaf litter on the forest floor, its coloration and skin patterns providing camouflage among the dead leaves. It is active by night, hopping about in search of its food, which consists mostly of ants and termites. Breeding takes place at rain-filled pools or slow-moving streams, males producing a rasping, rising trill that they sing in choruses. Females lay their eggs in strings. Like most toads, the species' skin contains secretions that are distasteful to potential predators.

Actual size

SIMILAR SPECIES

To date, 12 species have been described in the genus *Ingerophrynus*, found in Southeast Asia and called Hainan toads. The Indonesian Toad (*I. biporcatus*) is a larger toad than *I. divergens*; it also lives in forest and has prominent crests on its head. Most *Ingerophrynus* are Least Concern, but see also the Lesser Toad (*I. parvus*; page 176).

The Crested Toad has a wide head with a prominent pair of bony crests and short, slender legs. There are many spiny warts on its back and flanks. The upper parts are pale, reddish, or dark brown with black spots and, sometimes, black chevrons on the back. The underside is yellow or pale brown.

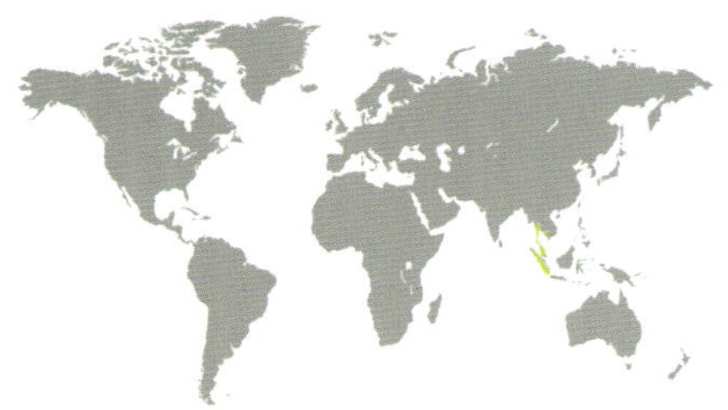

FAMILY	Bufonidae
OTHER NAMES	Dwarf Toad, Lesser Malacca Toad, Stream Toad
DISTRIBUTION	Cambodia, Indonesia, Myanmar, Thailand, Peninsular Malaysia
ADULT HABITAT	Forest
LARVAL HABITAT	Pools or slow-moving streams
CONSERVATION STATUS	IUCN Least Concern. It is declining in places but occurs within some protected areas

ADULT LENGTH
Male
1³⁄₁₆–1⅜ in (30–35 mm)
Female
1⁹⁄₁₆–1¹⁵⁄₁₆ in (40–50 mm)

INGEROPHRYNUS PARVUS

LESSER TOAD

(BOULENGER, 1887)

Actual size

The Lesser Toad has a short snout, slender forelimbs, and bony ridges on the top of its head. Its skin is wrinkled and covered in warts, and its toes are partially webbed. The upper surfaces are gray, brown, reddish, or black, and there is often a pair of black spots in the middle of the back. The underside is pale.

This small toad is a good jumper and can climb up into vegetation. Rain at any time of year stimulates males to gather next to pools and streams, sometimes in large numbers, and produce a monotonous, rasping call. Females, however, come out to breed only when rain is especially heavy. The species occurs in rubber plantations and gardens, suggesting that it is not wholly dependent on its natural forest habitat. The fungus that causes chytridiomycosis was found in this species in 2011, but there is no evidence that it is causing mass mortality. The toad's skin secretions are very toxic.

SIMILAR SPECIES

A new species in this genus, The Gollum Toad (*I. gollum*), was found in Peninsular Malaysia as recently as 2007, suggesting that there may be other species waiting to be discovered. The Kumquat Frog (*I. kumquat*), described in 2001, lives in peat swamps in Peninsular Malaysia, where its habitat is being drained. Both species are listed as Endangered. See also the Crested Toad (*I. divergens*; page 175).

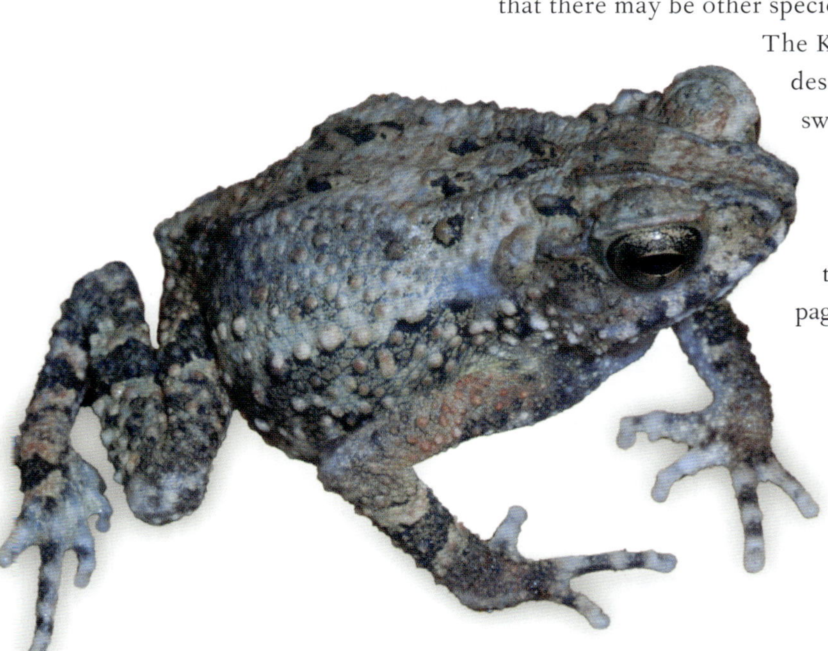

FAMILY	Bufonidae
OTHER NAMES	Fire Toad, Indonesia Tree Toad
DISTRIBUTION	Western Java, Indonesia
ADULT HABITAT	Forest at altitudes of 3,300–6,600 ft (1,000–2,000 m)
LARVAL HABITAT	Slow-moving streams
CONSERVATION STATUS	IUCN Critically Endangered. It has a small range and has suffered a major population decline

LEPTOPHRYNE CRUENTATA

BLEEDING TOAD

(TSCHUDI, 1838)

ADULT LENGTH
Male
¾–1³⁄₁₆ in (20–30 mm)

Female
1–1⁹⁄₁₆ in (25–40 mm)

This small, colorful toad was common within a restricted area in the mountains of western Java before the active volcano Mt. Galunggung erupted in 1982. In 1987, it was extremely rare, and in 2003 only one individual was found, suggesting that volcanic activity had caused its decline. A survey for the disease chytridiomycosis in 2007 found a number of individuals, and evidence of a low level of infection; three other frog species were more heavily infected. It is thus not clear if volcanic activity or disease caused the decline of the Bleeding Toad. It is found sitting on rocks in the spray zone of mountain streams.

SIMILAR SPECIES

The Ciremai Bleeding Toad (*L. javanica*), from Indonesia, is also Critically Endangered, while the Hourglass Toad (*Leptophryne borbonica*), found in lowland forests in Indonesia, Malaysia, and Thailand, is Least Concern. The latter gets its name from a darker hourglass-shaped or X-shaped patch on its back. It lives among leaf litter on the forest floor and its skin produces highly toxic secretions.

Actual size

The Bleeding Toad gets its name from the red and yellow spots on its otherwise black back and legs. There are two color morphs: one has a black hourglass marking bordered with red and yellow; and the other has black and yellow spots scattered all over its back. The underside is red or yellow. A small, slender toad, it has very warty skin and webbed toes.

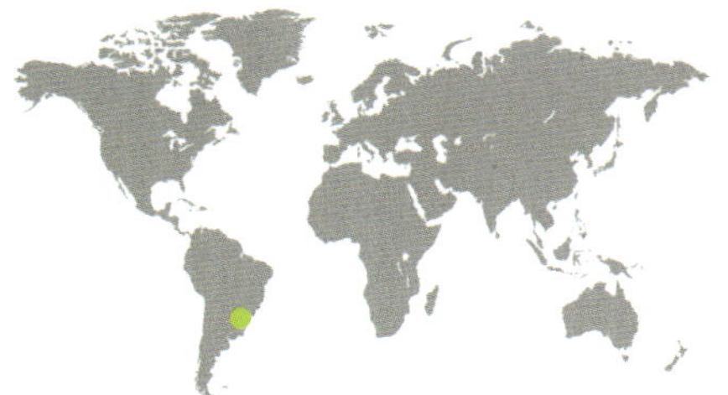

FAMILY	Bufonidae
OTHER NAMES	None
DISTRIBUTION	Forqueta River, southern Brazil
ADULT HABITAT	Forest
LARVAL HABITAT	Pools at the margins of a river
CONSERVATION STATUS	IUCN Critically Endangered. Has a tiny range, and is threatened by habitat destruction and collecting

ADULT LENGTH
$1\frac{1}{8}$–$1\frac{5}{8}$ in (29–41 mm)

MELANOPHRYNISCUS ADMIRABILIS

RED-BELLY TOAD

DI-BERNARDO, MANEYRO & GRILLO, 2006

Actual size

This tiny, colorful toad is found only along a 2,300 ft (700 m) stretch of the Forqueta River in southern Brazil. A terrestrial species, it is active by day, living in rocky areas beside the river. It breeds after rain in September or October, pairs laying more than one clutch of about 20 eggs in different pools at the river's edge. A proposed hydroelectric scheme threatens to destroy the species' habitat and thus bring about its extinction. It is also subject to collection because of its attractive and unusual appearance. When molested, it reveals its red underside, a warning that its skin contains toxic secretions.

SIMILAR SPECIES

There are 31 species in the genus *Melanophryniscus*, all found in South America. Many of them are restricted to extremely small ranges, and most are smaller than *M. admirabilis*. *Melanophryniscus pachyrhynus* was known from only two specimens, collected in 1905, but was rediscovered in 2008 in southern Brazil and Uruguay. Its conservation status has been assessed as Low Concern. See also the Montevideo Red-belly Toad (*M. montevidensis*; page 179).

The Red-belly Toad has many large, bulbous glands on its head, body, and legs. On its upper surfaces these are yellow or pale green against a darker green background, while on the belly they are yellow or green against a black background. The abdomen is bright red, as are the undersides of the hands and feet.

FAMILY	Bufonidae
OTHER NAMES	Darwin's Toad
DISTRIBUTION	Uruguay, extreme south of Brazil
ADULT HABITAT	Coastal areas with dunes, sandy soils, and marshland
LARVAL HABITAT	Ephemeral pools
CONSERVATION STATUS	IUCN Near Threatened. Common in places but declining in others due to habitat loss

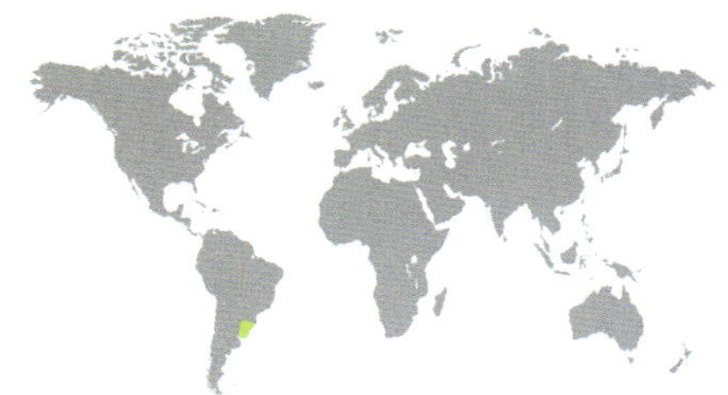

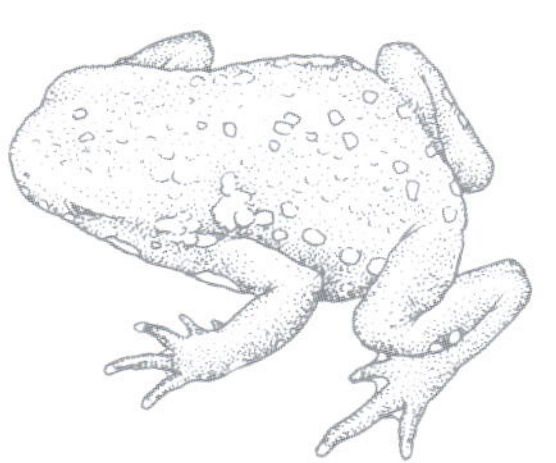

ADULT LENGTH
Male
¾–15⁄16 in (19–24 mm)
Female
⅞–1⅛ in (22–28 mm)

MELANOPHRYNISCUS MONTEVIDENSIS

MONTEVIDEO RED-BELLY TOAD

(PHILIPPI, 1902)

This very small toad is active by day and is an "explosive breeder," meaning that mating activity lasts just a few days. It breeds on warm days following rain, males producing a call consisting of a whistle followed by a train of pulses. The eggs are laid in temporary pools that do not contain fish or other predators. When attacked, the toad adopts a defensive posture in which it shows off the bright red color on its belly and feet. Its coastal habitat is being lost to urbanization and tree plantations, and is predicted to become drier and hence less favorable owing to climate change.

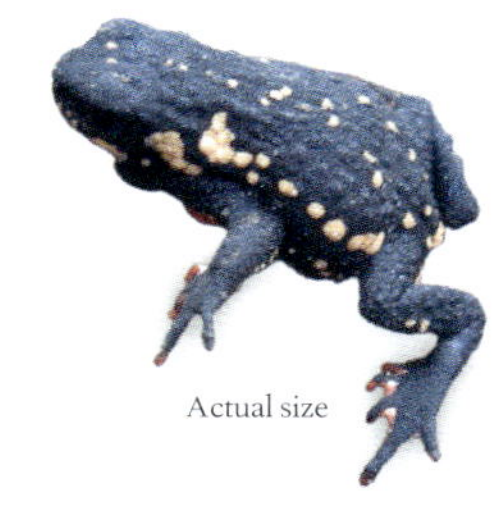

Actual size

The Montevideo Red-belly Toad is a very small toad with warty skin. Its upper surfaces are black, with scattered yellow spots on the back and a line of larger yellow spots along the flanks. The underside is black with red or yellow spots or blotches. The abdomen is bright red, as are the palms of the hands and soles of the feet.

SIMILAR SPECIES

Melanophryniscus montevidensis is very similar in size and coloration to the Uruguay Red-belly Toad (*M. atroluteus*), a common species found in interior grassland areas of Uruguay and southern Brazil. *Melanophryniscus setiba*, described as recently as 2012, is found in the restinga habitat of woodland and shrubs that occurs in many places along the coast of Brazil. See also the Red-belly Toad (*M. admirabilis*; page 178).

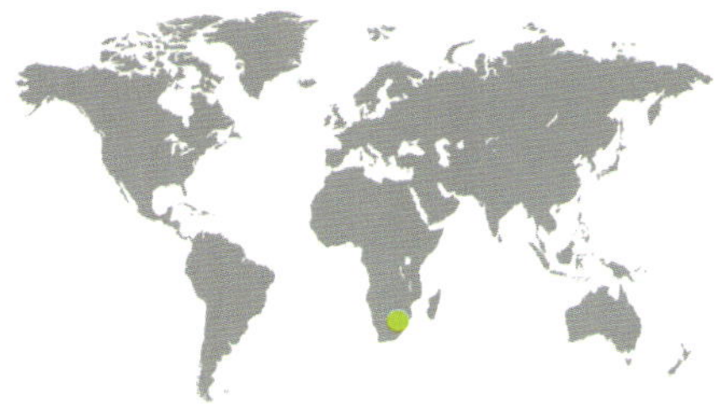

FAMILY	Bufonidae
OTHER NAMES	Boulenger's Earless Toad, Mashonaland Toad
DISTRIBUTION	Zimbabwe–Mozambique border
ADULT HABITAT	Forest
LARVAL HABITAT	Pools among the buttress roots of trees
CONSERVATION STATUS	Low Concern, but it has a restricted range, threatened by habitat destruction

ADULT LENGTH
Male
1–$1\frac{3}{8}$ in (25–35 mm)
Female
$1\frac{1}{4}$–$1\frac{13}{16}$ in (32–46 mm)

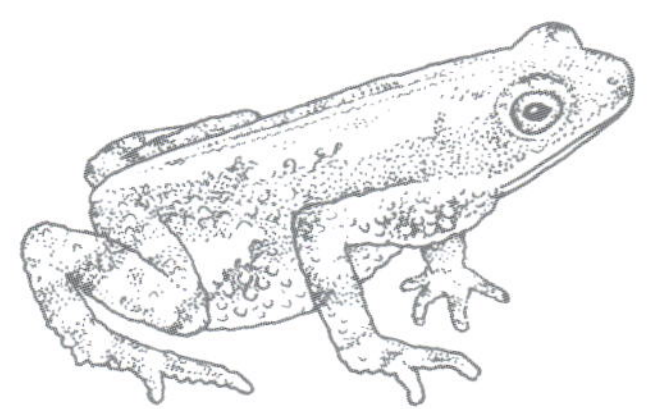

MERTENSOPHRYNE ANOTIS

CHIRINDA TOAD

(BOULENGER, 1907)

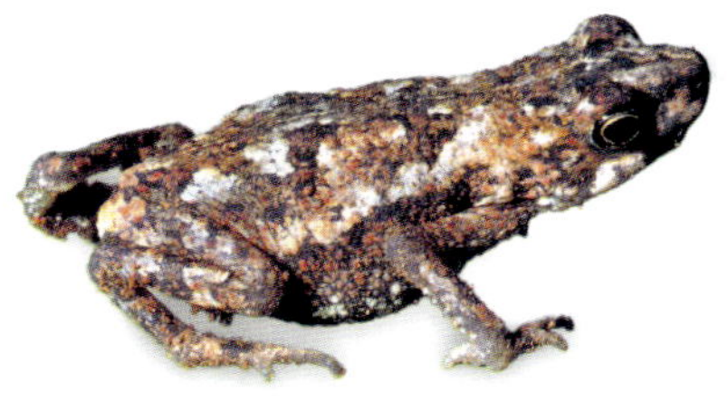

Actual size

The Chirinda Toad has a flat back and head, giving it a box-like shape. The snout is pointed, there is no tympanum, and the parotoid glands are broad and flattened. It is pale brown in color, with darker brown spots and stripes on the legs. The back and shoulders are pale brown.

This tiny toad lives among dead leaves on the forest floor, relying on camouflage to avoid detection. Despite lacking ears, males call in summer to attract females, producing a plaintive chirp. Mating has not been reported. Eggs are laid in strings of about 100, in tiny pools of water formed between the buttress roots of trees. The tadpoles have a crown of spongy tissue on the top of their heads. It is thought that this enables them to breathe from the water's surface while browsing just beneath it, an adaptation to the lack of dissolved oxygen in tree pools.

SIMILAR SPECIES

Fourteen species have been described from Central and East Africa, particularly in the mountains of Tanzania, where the Usambara Toad (*M. usambarae*) is Critically Endangered, and the Uzungwe Toad (*M. uzunguensis*) is Vulnerable. The Woodland Toad (*M. micranotis*) is a relatively common arboreal species from Kenya and Tanzania and is only $\frac{15}{16}$ in (24 mm) long. It is believed to have internal fertilization and it has been reported that, during mating, the male and female press their ventral surfaces together.

FAMILY	Bufonidae
OTHER NAMES	Emerald Forest Toad, Patagonian Toad
DISTRIBUTION	Southern Chile, Argentina
ADULT HABITAT	Humid forest, bogs and tundra, up to 6,600 ft (2,000 m) altitude
LARVAL HABITAT	Ephemeral pools
CONSERVATION STATUS	IUCN Least Concern. Common and has a stable population

ADULT LENGTH
average 1¾ in (45 mm); female is larger than male

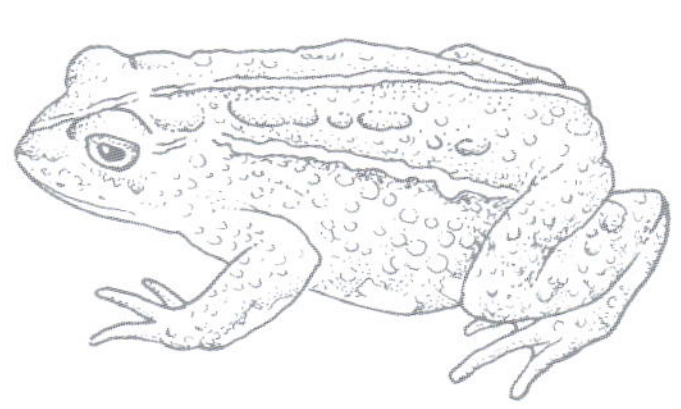

NANNOPHRYNE VARIEGATA

EDEN HARBOUR TOAD

GÜNTHER, 1870

Together with the Marbled Wood Frog (*Batrachyla antartandica*), this is the world's most southerly distributed amphibian. It lives in woodland containing southern beech (*Nothofagus*) trees and in more open habitat. Although it is abundant in some places, very little is known about its natural history. It lays its eggs in strings, these hatching into the small black tadpoles typical of true toads. The young are unusual in being strikingly colored: They are black with pale green or white longitudinal stripes.

Actual size

The Eden Harbour Toad is a small toad with a short snout and many large warts on its flanks and legs. It is unusual in having two pairs of parotoid glands, together with several prominent glands on its back and legs. The upper surfaces are dark green or chestnut-brown, decorated with three or five longitudinal yellow stripes. The underside is white with black spots.

SIMILAR SPECIES

The other three species in the genus *Nannophryne* all occur in the Andes and are poorly known. The Paramo Toad (*N. cophotis*) and the Abra Malaga Toad (*N. corynetes*) have restricted ranges in Peru. *Nannophryne apolobambica* is known only from a small area of cloud forest in Bolivia. All three are Endangered or Critically Endangered.

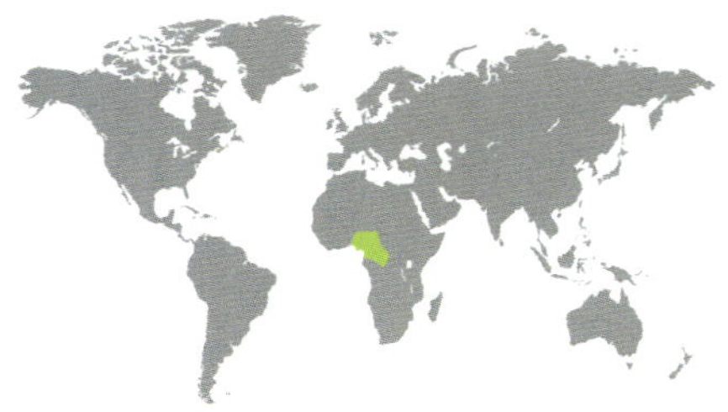

FAMILY	Bufonidae
OTHER NAMES	None
DISTRIBUTION	Western and central Africa, from Nigeria in the west to eastern Democratic Republic of the Congo in the east
ADULT HABITAT	Lowland forest
LARVAL HABITAT	In small pools in tree-holes
CONSERVATION STATUS	IUCN Least Concern. Has a very large range but little is known about its population status

ADULT LENGTH
1–1½ in (25–38 mm)

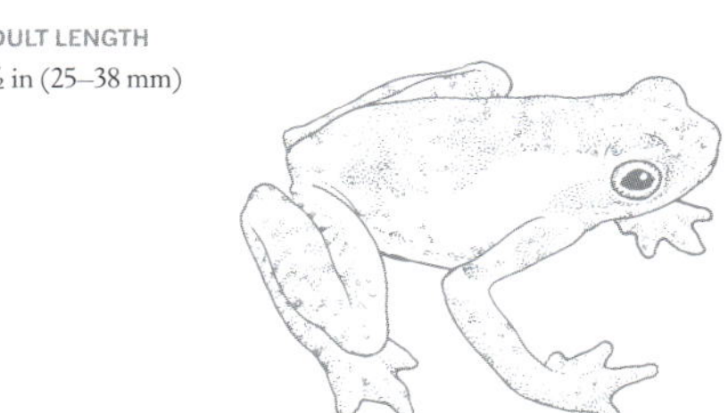

NECTOPHRYNE AFRA

AFRICAN TREE TOAD

BUCHHOLZ & PETERS, 1875

Actual size

The African Tree Toad has long, thin hind legs and very extensive webbing on its hands and feet. The back is greenish brown, with three transverse bands of green or yellow. There is a broad green or yellow stripe running along the flanks. The legs are brown with green blotches. The underside is largely yellow.

This small toad is said to be slow and clumsy in its movements. It lives on the ground by day and climbs up into the trees at night. Its breeding behavior is centered on small tree-holes that contain shallow pools of water. The male calls from such a tree-hole and the female lays her eggs there. The male guards the eggs until they hatch and subsequently attends the resulting tadpoles for up to two weeks. To ensure that the tadpoles have an adequate supply of oxygen, he swims in the pool, beating his legs vigorously to aerate the water.

SIMILAR SPECIES

The only other species in the genus *Nectophryne* is Bates' Tree Toad (*N. batesii*). Its range is very similar to that of *N. afra*, but even less is known about its biology or population status.

FAMILY	Bufonidae
OTHER NAMES	None
DISTRIBUTION	Udzungwa Mountains, Eastern Arc Mountains, Tanzania
ADULT HABITAT	Spray zone of a river gorge
LARVAL HABITAT	Within the mother
CONSERVATION STATUS	IUCN Extinct in the wild. Survives only in captive populations

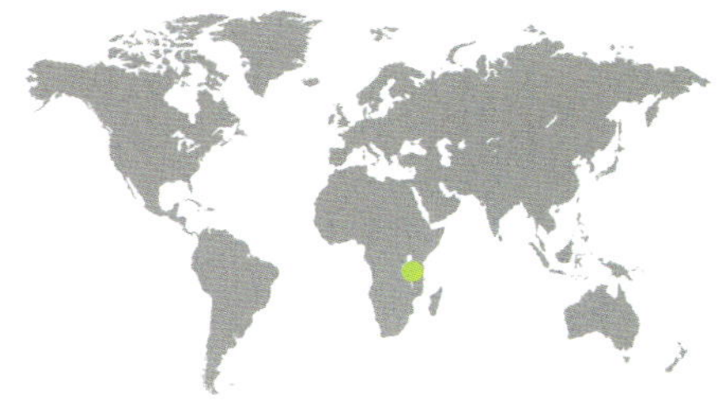

ADULT LENGTH

Male

9/16–11/16 in (15–17 mm)

Female

11/16–3/4 in (18–20 mm)

NECTOPHRYNOIDES ASPERGINIS

KIHANSI SPRAY TOAD

POYNTON, HOWELL, CLARKE & LOVETT, 1999

Described in 1999, this tiny toad had become extinct in nature by 2005. When discovered, it was abundant within an area of only 5 acres (2 ha), consisting of a gorge with vegetation kept moist by the spray from waterfalls. In 1999, the gorge was dammed, eliminating the spray, drying up the vegetation, and causing the toad population to collapse. Attempts to replicate the spray artificially have been unsuccessful and the species exists only in captive populations, where it has been bred successfully. Fertilization is internal and the tadpoles develop and metamorphose within the mother, emerging as toadlets a mere 3/16 in (5 mm) long.

Actual size

The Kihansi Spray Toad is a tiny toad with long, slender limbs and large, dark eyes. The back is yellow or golden, with yellow or brown speckles, and there are broad brown stripes running along each side from the eye to the groin. The skin on the underside is transparent, such that the internal organs, and developing tadpoles, are visible.

SIMILAR SPECIES

To date, 13 species have been described in the genus *Nectophrynoides*, all of which are found in the forests and wetlands of the Eastern Arc Mountains of Tanzania. All are viviparous, the eggs being retained within the mother until their development is complete. Except Tornier's Forest Toad (*N. tornieri*; page 184), and the Morogoro Tree Toad (*N. viviparus*), all species are threatened with extinction.

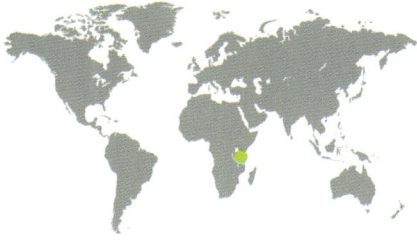

FAMILY	Bufonidae
OTHER NAMES	Usumbara Viviparous Toad
DISTRIBUTION	Eastern Arc Mountains, Tanzania
ADULT HABITAT	Forest and forest edges up to 5,900 ft (1,800 m) altitude
LARVAL HABITAT	Within the mother
CONSERVATION STATUS	IUCN Least Concern. Found in several mountain ranges, but its habitat is declining due to deforestation

ADULT LENGTH
Male
up to 1⅛ in (28 mm)
Female
up to 1$\frac{5}{16}$ in (34 mm)

NECTOPHRYNOIDES TORNIERI

TORNIER'S FOREST TOAD

(ROUX, 1906)

Actual size

Tornier's Forest Toad is a tiny toad with a wide head, a slender body, and long, slender limbs. The eyes are large and protruding, and the tympana are clearly defined. The fingers and toes have blunt tips. The upper surfaces are pale brown, and the underside is white or gray in males and translucent in females.

This small toad is most commonly seen perching in low vegetation. The male's call consists of a series of high-pitched clicks. This species and its close relatives are unusual in being viviparous, meaning that fertilization is internal and the eggs and tadpoles develop and metamorphose within the mother, emerging as tiny toadlets. The toad is found in several forested mountain ranges in Tanzania's Eastern Arc, a biodiversity hotspot that is threatened by habitat destruction. The fact that it is also found in banana plantations suggests that it may be more adaptable than many frogs.

SIMILAR SPECIES

Of the 13 species that have been described so far in the genus *Nectophrynoides*, *N. tornieri* and the Morogoro Tree Toad (*N. viviparus*) are the only species that are not considered under threat. The Uluguru Forest Toad (*N. cryptus*) has not been seen since 1927 and is listed as Endangered. The Dwarf Forest Toad (*N. minutus*), also Endangered, is less than ⅞ in (22 mm) in length. See also the Kihansi Spray Toad (*N. asperginis*; page 183).

FAMILY	Bufonidae
OTHER NAMES	None
DISTRIBUTION	Mt. Nimba, on the border of Guinea, Côte d'Ivoire, and Liberia, West Africa
ADULT HABITAT	Montane grassland at altitudes greater than 4,000 ft (1,200 m)
LARVAL HABITAT	Within the mother
CONSERVATION STATUS	IUCN Critically Endangered. Has a very limited range, threatened by habitat destruction through mining

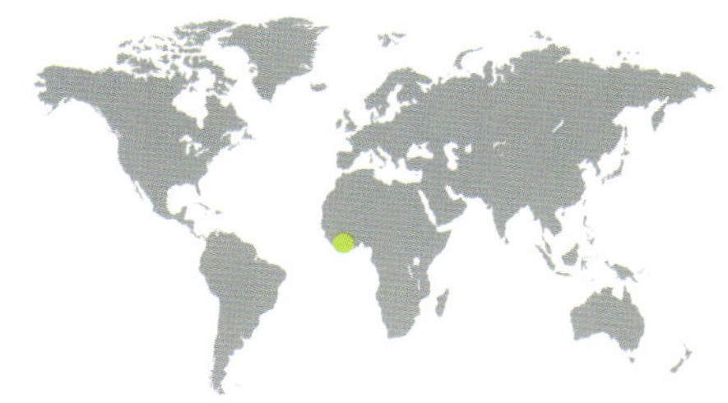

NIMBAPHRYNOIDES OCCIDENTALIS

WESTERN NIMBA TOAD

(ANGEL, 1943)

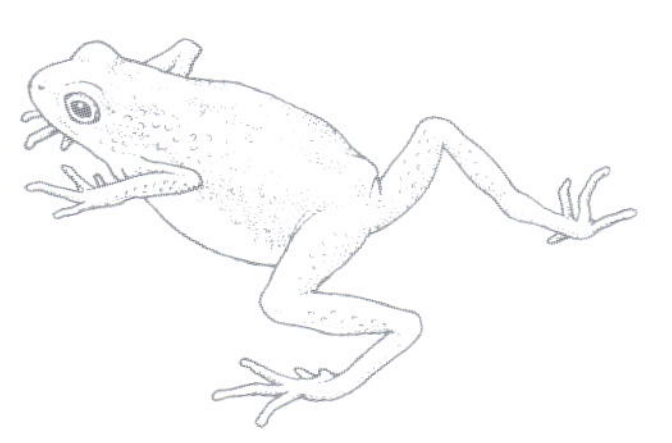

ADULT LENGTH

Male
average $^{11}/_{16}$ in (18 mm)

Female
average $^{7}/_{8}$ in (22 mm)

Despite being a UNESCO World Heritage Site, Mt. Nimba is being extensively mined for iron ore and bauxite. It is home to several unusual species, including this tiny toad, which is the only truly viviparous frog yet discovered. After mating, during the wet season between July and September, females carrying 4–35 fertilized eggs retreat underground. The tadpoles develop in their mother's oviducts, first using up their egg yolk, then feeding on "uterine milk" secreted by glands in the oviduct walls. Nine months after mating, females emerge from underground to give birth to tiny toadlets, $^{5}/_{16}$ in (7.5 mm) long.

Actual size

The Western Nimba Toad has a pointed snout and long, unwebbed fingers and toes. Pregnant females have obviously swollen bellies. The color on the back is brown or black with paler patches, and the underside is white. Some individuals have a broad black stripe along each flank. The legs are pale brown with darker brown stripes.

SIMILAR SPECIES

Nimbaphrynoides occidentalis is the only species in its genus. A slightly larger frog occurs nearby, around Mt. Alpha in Liberia, and was classified as a separate species, *N. liberiensis*, but is now regarded as a subspecies of *N. occidentalis*. It is also viviparous and is slightly larger than the frogs on Mt. Nimba.

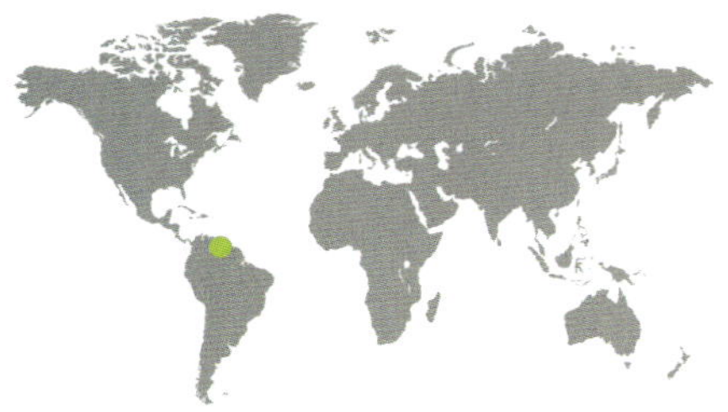

FAMILY	Bufonidae
OTHER NAMES	Roraima Black Frog
DISTRIBUTION	Mt. Roraima, on the border of Brazil, Venezuela, and Guyana
ADULT HABITAT	Bare rock faces and swamps at altitudes of 7,500–9,200 ft (2,300–2,800 m)
LARVAL HABITAT	Within the egg
CONSERVATION STATUS	IUCN Endangered. It has a very restricted range and is often disturbed by tourists

ADULT LENGTH
Male
5/8–15/16 in (16–24 mm)
Female
3/4–1 3/16 in (20–30 mm)

OREOPHRYNELLA QUELCHII

RORAIMA BUSH TOAD

(BOULENGER, 1895)

Actual size

The Roraima Bush Toad is a small toad with a wide head and, being a species that walks rather than hops, has thin arms and legs. It has stubby fingers and toes, large eyes, and many warts on its back. The upper surfaces are black and glossy, and the belly is yellow, mottled with black.

This remarkable little toad belongs to a genus whose members display a unique escape behavior. If attacked, for example, by a frog-eating tarantula, it tucks its legs and head into its body, rolls itself into a ball, and tumbles, like a small pebble, down a slope. The toad has a slow, deliberate gait and lives at the top of the tepuis, or flat-topped mountains, that are characteristic of eastern Venezuela and western Guyana. It lays 9–13 eggs in cavities under moss, and some accounts report that several pairs may breed communally. The tadpoles develop directly within the eggs, which hatch to produce tiny toadlets.

SIMILAR SPECIES

There are eight species in the genus *Oreophrynella*, all found on tepuis in Venezuela, Guyana, and Brazil. Because they live high up on mountains, their ranges are small and separated from those of related species, and most are rare and of conservation concern. For example, the Venezuelan Pebble Toad (*O. nigra*) has a very small range in Venezuela and is categorized as Vulnerable. See also Seegobin's Tepui Toad (*O. seegobini*; page 187).

FAMILY	Bufonidae
OTHER NAMES	None
DISTRIBUTION	Guyana
ADULT HABITAT	Dense, low vegetation and peat bogs
LARVAL HABITAT	Within the egg
CONSERVATION STATUS	IUCN Vulnerable with a very restricted range and is thought to be very sensitive to climate change

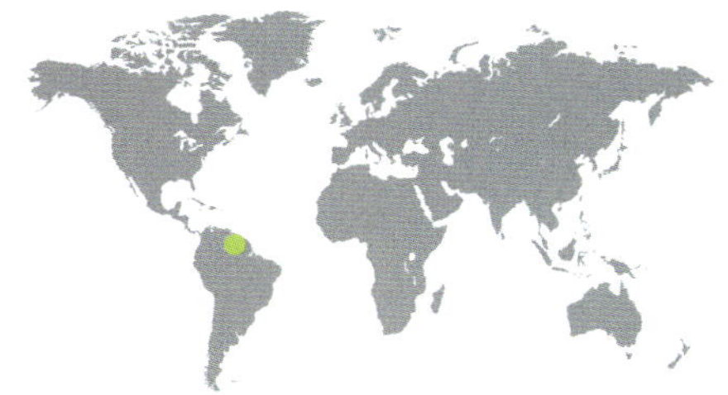

ADULT LENGTH
Male
up to ⅞ in (21 mm)
Female
unknown

OREOPHRYNELLA SEEGOBINI

SEEGOBIN'S TEPUI TOAD

KOK, 2009

Maringma-tepui is a flat-topped mountain, 6,850 ft (2,088 m) high, in Guyana. It is one of many tepuis that are scattered across eastern Venezuela and western Guyana. The summit supports a unique vegetation of dwarf forest, bushes, and peat bogs in an area of only 420 acres (170 ha). This small, very cryptic toad is active by day and lives on the ground. All that is known of its reproductive biology is that males produce a soft peeping sound, but it is assumed that, like other tepui toads, females lay eggs on land and that there is direct development, the eggs hatching to produce tiny toadlets.

Actual size

Seegobin's Tepui Toad is a tiny toad with a wide head and bony crests just behind the eyes. The fingers and toes are stubby and webbed, and there are numerous small and large warts on the back and limbs. The upper surfaces are black or brown, and the underside is orange-brown.

SIMILAR SPECIES

Oreophrynella seegobini is the most recently described of the eight species in the genus *Oreophrynella*, known as tepui toads or bush toads. All are listed as Vulnerable or Data Deficient. Mt. Roraima is the largest of the tepuis, with an area of 12 sq miles (31 sq km), and is situated where the borders of Venezuela, Guyana, and Brazil meet. It is home to the Roraima Bush Toad (*O. quelchii*; page 186).

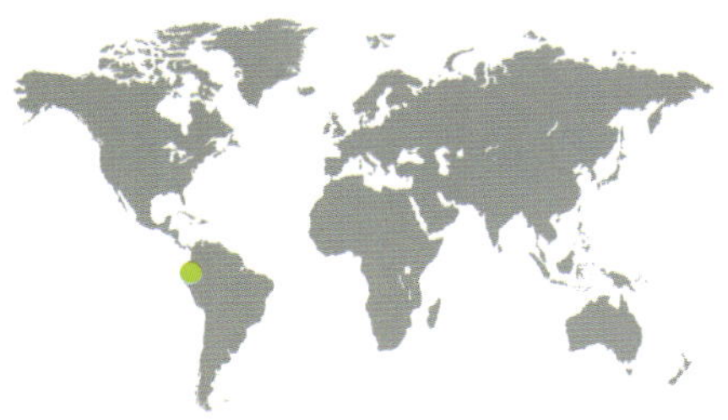

FAMILY	Bufonidae
OTHER NAMES	None
DISTRIBUTION	Ecuador
ADULT HABITAT	Cloud forest at altitudes of around 5,000 ft (1,500 m)
LARVAL HABITAT	Unknown
CONSERVATION STATUS	IUCN Endangered, and known from only one location, within a protected area

ADULT LENGTH
Male
$^{11}/_{16}$–1 in (17–26 mm)

Female
around $1^{5}/_{16}$ in (33 mm)

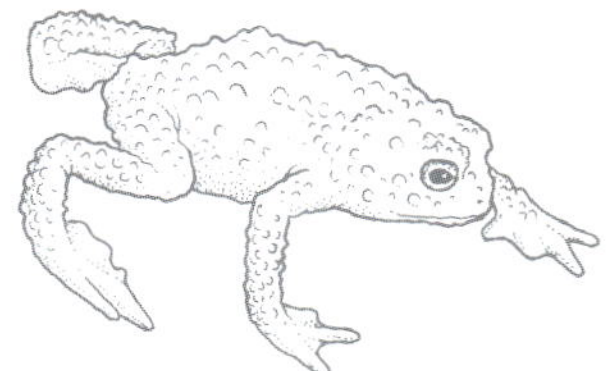

OSORNOPHRYNE SIMPSONI

SIMPSON'S PLUMB TOAD

PÁEZ-MOSCOSO, GUAYASAMIN & YÁNEZ-MUÑOZ, 2011

Actual size

Simpson's Plumb Toad has a short, pointed snout; long arms with webbed fingers; and long, slender legs with webbed feet. There are many warts on the head and back, and large conical pustules on the flanks. The upper surfaces of this toad are dark to light brown with pale patches, the underside is brown, and the hands and feet are orange.

So little is known about this small toad that only one female has been described in detail. The species has been found at night sitting on bromeliads and ferns. It belongs to a genus of small, plump toads, all found at high altitude in the Andes of Ecuador and Colombia, and most with very small ranges. Females are considerably larger than males and there is evidence that they lay large eggs on the ground, which probably hatch directly into tiny toads. Members of the genus are threatened by loss of their habitat to logging and illegal agriculture, and by the use of crop sprays.

SIMILAR SPECIES

There are currently 11 species in the genus *Osornophryne*. The Guacamayo Plump Toad (*O. guacamayo*) has a protruding proboscis on its snout. It is found in Colombia and northeastern Ecuador and is listed as Vulnerable. *Osornophryne sumacoensis* occurs only in cloud forest around a crater lake on Volcán Sumaco in Ecuador and is also listed as Vulnerable. A volcanic eruption is a potential future threat to its existence.

FAMILY	Bufonidae
OTHER NAMES	St. Andrew's Cross Toadlet, Short-legged Dwarf Toad
DISTRIBUTION	Borneo
ADULT HABITAT	Wet lowland forest up to 3,300 ft (1,000 m) altitude
LARVAL HABITAT	Unknown
CONSERVATION STATUS	Low Concern, but under threat from deforestation

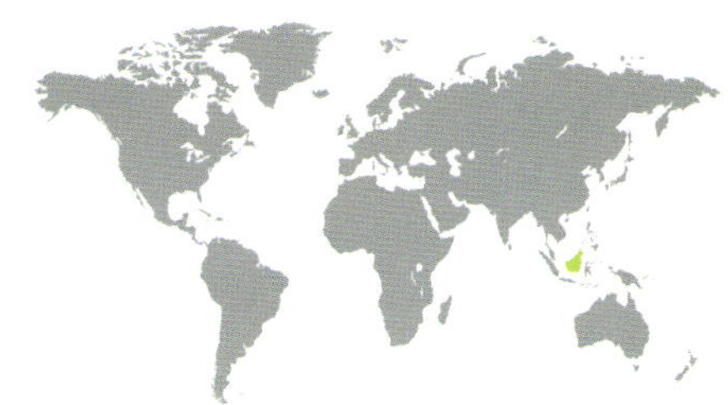

ADULT LENGTH
Male
$\frac{9}{16}$–$\frac{11}{16}$ in (14–17 mm)
Female
$\frac{5}{8}$–$\frac{11}{16}$ in (16–18 mm)

PELOPHRYNE SIGNATA

LOWLAND DWARF TOAD

(BOULENGER, 1894)

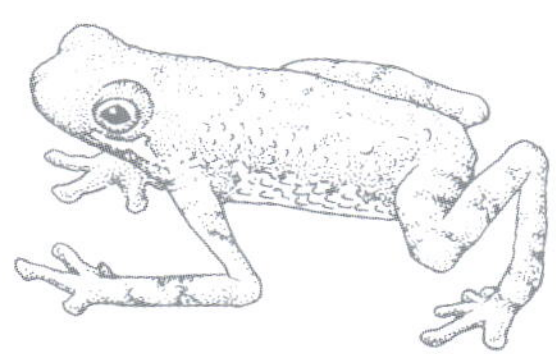

This tiny toad is so small that it is easily confused with the juveniles of larger toad species. It is usually seen on the ground but occasionally climbs up into vegetation. Its breeding behavior has not been reported but, like other dwarf toads, it probably lays small numbers of large white eggs in very small pools. Males produce a high-pitched call, mostly at dusk. The population status of the species is uncertain, but its habitat is being reduced by deforestation.

Actual size

The Lowland Dwarf Toad has a blunt snout, long, thin legs, and disks on the tips of its fingers. The back is covered with warts and is dark brown in color, with small black spots. Some individuals have a dark X-shaped mark on the back. There is a broad cream band running down the flanks from eye to groin.

SIMILAR SPECIES

There are 13 species in the Asian genus *Pelophryne*, known as dwarf toads. The Southeast Asian Toadlet (*P. brevipes*) is found in many parts of Southeast Asia. It lays its eggs in the water-filled axils of leaves; its tadpoles do not eat, but complete their development nourished only by yolk. The Kinabalu Dwarf Toad (*P. misera*) is black in color and is found at high elevations in Borneo. The Linanit Dwarf Toad (*P. linanitensis*), also from Borneo, is Critically Endangered.

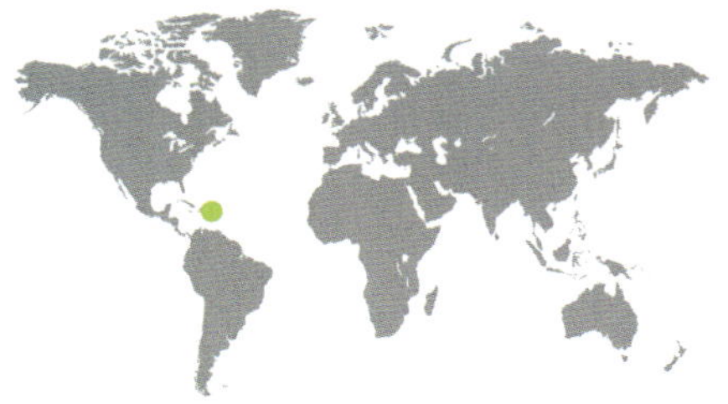

FAMILY	Bufonidae
OTHER NAMES	Lowland Caribbean Toad
DISTRIBUTION	Puerto Rico; formerly also British Virgin Islands and US Virgin Islands
ADULT HABITAT	Semiarid rocky outcrops in forest
LARVAL HABITAT	Permanent and ephemeral pools
CONSERVATION STATUS	IUCN Endangered. Threatened by habitat destruction; currently sustained by captive breeding

ADULT LENGTH
Male
$2\tfrac{9}{16}$–$3\tfrac{5}{16}$ in (64–85 mm)

Female
$2\tfrac{9}{16}$–$4\tfrac{3}{4}$ in (64–120 mm)

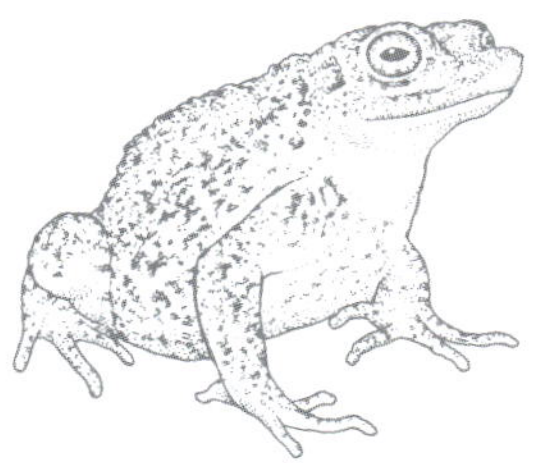

PELTOPHRYNE LEMUR

PUERTO RICAN CRESTED TOAD

COPE, 1868

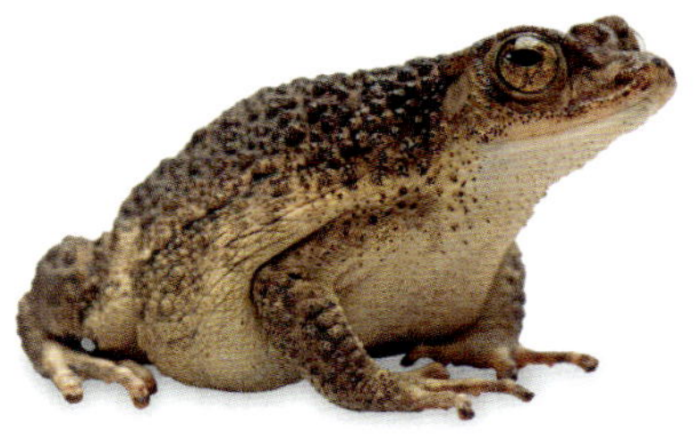

The Puerto Rican Crested Toad gets its name from a prominent bony crest above its eyes. It has an upturned snout and very warty skin. The back is yellowish brown, the underside is yellow, and the irises of the eyes are gold, marbled in black. Females have rougher skin and larger crests than males, and males tend to be yellower than females.

The range of this odd-looking toad has been reduced to a single location on the south coast of Puerto Rico. Even here, one of its last breeding ponds was filled in to create a parking lot. Previously, it was also found on the island's north coast and in the British Virgin Islands and US Virgin Islands. It has lost most of its natural habitat through urbanization. Its population is now stable, and may be expanding, as a result of the construction of new breeding ponds and the release of animals bred in captivity. It breeds in permanent or temporary pools, females laying up to 15,000 eggs in strands.

SIMILAR SPECIES

All 14 species in the genus *Peltophryne* are found on Cuba and other Caribbean islands. The Western Giant Toad (*P. fustiger*) is a large yellow and brown toad that is quite common on Cuba. In contrast, the Hispaniolan Crestless Toad (*P. fluviatica*), of the Dominican Republic, has not been seen since its discovery in 1972. Categorized as Critically Endangered, it is probably extinct.

Actual size

FAMILY	Bufonidae
OTHER NAMES	None
DISTRIBUTION	Cuba
ADULT HABITAT	Upland pinewoods and broadleaf forest
LARVAL HABITAT	Forest streams
CONSERVATION STATUS	IUCN Critically Endangered. Threatened by habitat destruction and chytridiomycosis

ADULT LENGTH
average 1 in (26 mm)

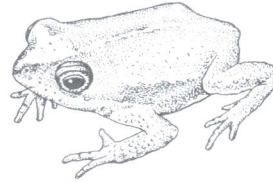

PELTOPHRYNE LONGINASUS

CUBAN LONG-NOSED TOAD

(STEJNEGAR, 1905)

This small toad is found in three widely separated locations in western, central, and eastern Cuba. No animals have been recorded for the eastern population since the 1900s, and the remaining populations are threatened by deforestation. Recently, the disease chytridiomycosis has been detected in this species. The toad is active on the ground during the day and is believed to move up into trees at night. The male calls while floating in a stream, where the eggs are laid.

Actual size

The Cuban Long-nosed Toad gets its name from its pointed snout. A small toad, its skin is smooth and there are no bony ridges on the head. The back is bronze to purple in color, the upper lip is white, and there is a dark stripe running down each side from the nose, through the eye, to the groin.

SIMILAR SPECIES

The Cuban Spotted Toad (*Peltophryne taladai*) has a mating call described as sounding like a machine gun. It protects itself from its enemies by inflating its body and is categorized as Least Concern. Much rarer is the Endangered *P. florentinoi*, known only from a single coastal swamp. It may be threatened by sea-level rises caused by climate change, which could flood its habitat with seawater.

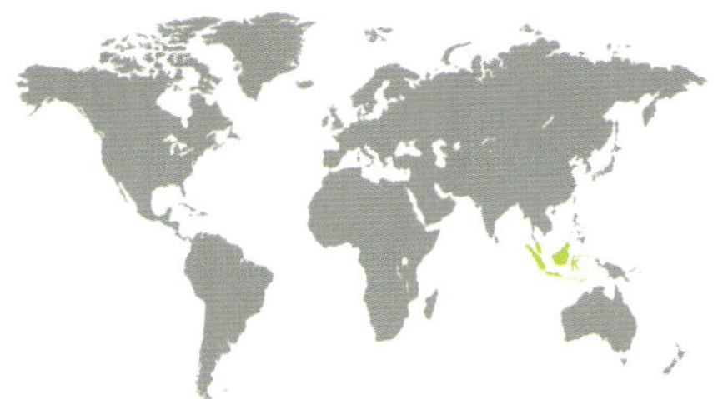

FAMILY	Bufonidae
OTHER NAMES	Asian Giant Toad, Rough Toad
DISTRIBUTION	Borneo; Java and Sumatra, Indonesia; Peninsular Malaysia
ADULT HABITAT	Lowland rainforest
LARVAL HABITAT	Small to medium-sized forest streams
CONSERVATION STATUS	IUCN Least Concern. Has a large range but is threatened in places by deforestation

ADULT LENGTH
Male
$2^{13}/_{16}$–$3^{7}/_{8}$ in (70–100 mm)
Female
$3^{11}/_{16}$–$5^{1}/_{2}$ in (95–140 mm)

PHRYNOIDIS ASPERA

RIVER TOAD

(GRAVENHORST, 1829)

Known for being rather sedentary in its habits, this toad lives close to streams and individuals are often found in the same spot from day to day. It breeds at all times of year, males producing a rasping chirp from the edge of a stream. Calling tends to peak at the full moon, males keeping apart from one another and not forming a chorus. The female lays an average of 12,800 eggs in streams, and the tadpoles tend to gather where the current is least strong. River Toads are eaten by local people in Borneo and Peninsular Malaysia.

SIMILAR SPECIES

The only other species in this genus, the Giant River Toad (*Phrynoidis juxtaspera*), is even larger than *P. aspera*, females reaching a body length of $8^{1}/_{4}$ in (215 mm). This toad is less sedentary in its movements and is found wandering through the lowland forests of Borneo and Sumatra. It is renowned for its disgusting smell.

Actual size

The River Toad is a large toad with a broad head, lacking bony crests. The parotoid glands are circular or oval. The skin is rough and covered in warts, and is uniformly dark brown, black, gray, or green in color. The irises of the eyes are golden.

FAMILY	Bufonidae
OTHER NAMES	None
DISTRIBUTION	Northern Namibia
ADULT HABITAT	Semiarid grassland and bushveld
LARVAL HABITAT	Ephemeral pools
CONSERVATION STATUS	Data Deficient. This species lives in a remote area and is rarely seen

POYNTONOPHRYNUS DAMARANUS

DAMARALAND PYGMY TOAD

(MERTENS, 1954)

ADULT LENGTH
Males
average 1¼ in (32 mm)
Female
average 1⁷⁄₁₆ in (36 mm)

Situated between the Namib Desert to the west and the Kalahari Desert to the east, Damaraland is noted for the abundance and diversity of its reptiles but is an unlikely habitat for a frog. It receives only 13¾ in (350 mm) of rain per year, most of it falling in January and February, when this small toad enters temporary ponds to breed. Little is known of the species' breeding behavior other than that it lays its eggs in strings. With local air temperatures reaching 77–95°F (25–35°C) during the day, it is not surprising that the frog is active only at night.

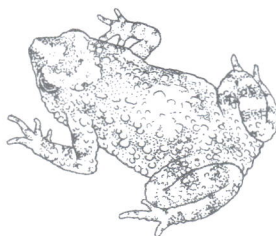

Actual size

SIMILAR SPECIES

There are 15 species in the African genus *Poyntonophrynus*, known as pygmy toads. Small and flattened in shape, they hide by day under rocks, in burrows, or in rock crevices. The Dombe Pygmy Toad (*P. dombensis*) is found in Angola and northwestern Namibia, while the Southern Pygmy Toad (*P. vertebralis*) lives on the high veldt of South Africa and breeds in rocky pools. See also the Northern Pygmy Toad (*P. fenoulheti*; page 194).

The Damaraland Pygmy Toad has a flattened body, a broad head, and rather short legs. The back and thighs are densely covered in warts and there are small spines on the head. The upper surfaces are olive-brown with darker blotches edged with black, and there is a dark stripe between the eyes. The underside is white.

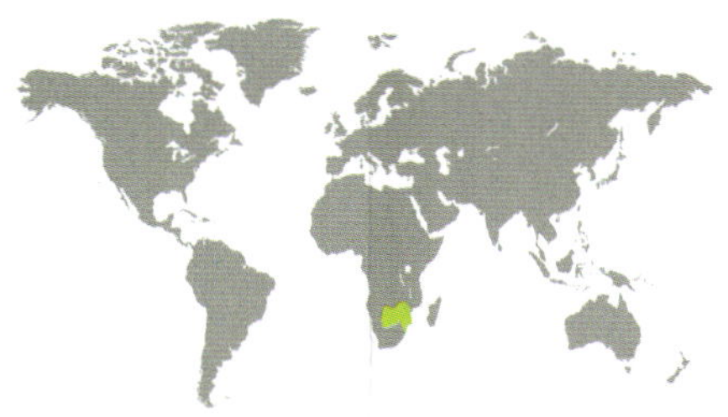

FAMILY	Bufonidae
OTHER NAMES	Dwarf Toad, Fenoulhet's Toad, Newington Toad, Transvaal Pygmy Toad
DISTRIBUTION	Zimbabwe, Botswana, Mozambique, northeastern South Africa, Swaziland
ADULT HABITAT	Savanna, grassland, and bushveld up to 5,600 m (1,700 m) altitude
LARVAL HABITAT	Ephemeral pools
CONSERVATION STATUS	IUCN Least Concern. There are no apparent threats to the species or its habitat

ADULT LENGTH
Male
$1\frac{1}{8}$–$1\frac{5}{16}$ in (28–33 mm)
Female
$1\frac{3}{16}$–$1\frac{11}{16}$ in (30–43 mm)

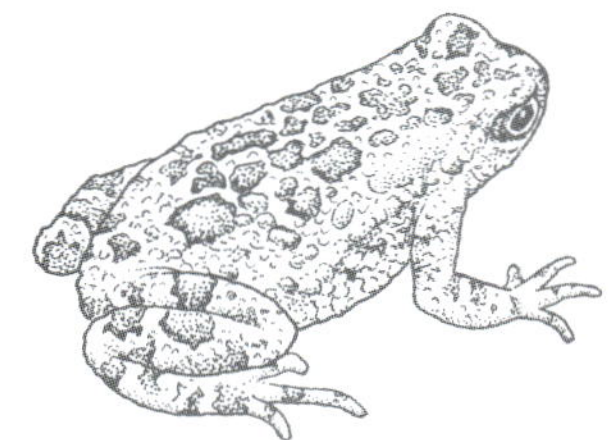

POYNTONOPHRYNUS FENOULHETI

NORTHERN PYGMY TOAD

(HEWITT & METHUEN, 1912)

Actual size

This small, squat toad is associated with rocky outcrops, where it is found by day in crevices, often in groups of five or six. It often shares its hiding places with scorpions and lizards. It breeds after heavy rain, between October and February, at which time males develop bright yellow or orange throats. They call antiphonally from the edges of rain-filled rock pools, producing a high-pitched creak. The female produces as many as 2,000 eggs, laid in strings. They hatch within 24 hours and the tadpoles take only 19 days to reach metamorphosis.

SIMILAR SPECIES

Poyntonophrynus fenoulheti is similar in appearance to the Southern Pygmy Toad (*P. vertebralis*), but the latter's call is a cricket-like chirp. The Beira Pygmy Toad (*P. beiranus*) occurs in two separate areas, in Malawi and Mozambique. Its call is a high-pitched buzz. The Kavango Pygmy Toad (*P. kavangensis*) is found in sandy areas in Angola, Botswana, Zimbabwe, and Namibia. See also the Damaraland Pygmy Toad (*P. damaranus*; page 193).

The Northern Pygmy Toad has a flattened body, a broad head, and rather short legs. The back is covered in warts tipped with small spines. The parotoid glands behind the eyes are flattened. The upper surfaces are pale gray with brown blotches and red spots. The underside is white, and the male's throat is yellow.

FAMILY	Bufonidae
OTHER NAMES	Aquatic Swamp Toad
DISTRIBUTION	Borneo; Sumatra, Indonesia; Peninsular Malaysia
ADULT AND LARVAL HABITAT	Coastal and peat swamps
CONSERVATION STATUS	IUCN Least Concern. Declining in places due to draining of wetlands

ADULT LENGTH
Male
3–3⅝ in (77–94 mm)

Female
3 9/16–6 3/16 in (92–155 mm)

PSEUDOBUFO SUBASPER
FALSE TOAD
TSCHUDI, 1838

Very little is known about this large toad, primarily because it is cryptically colored. It has been described as "lethargic" in its behavior, it lives in a habitat in which it is very easy to hide, and when disturbed, it dives into water. Males have been reported to call from mats of vegetation hanging over water, but the call has not been described. Nothing is known about the tadpoles. Distributed over many parts of Southeast Asia, the species is at risk from pollution and drainage of its habitat.

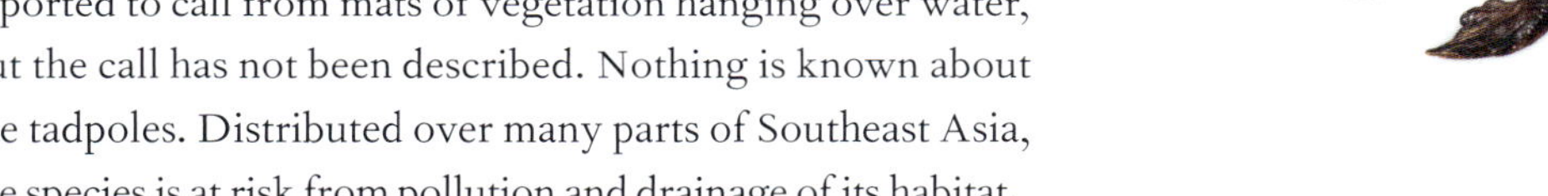

SIMILAR SPECIES
Pseudobufo subasper is the only species in its genus and, in terms of body shape, is unlike other toads. The name "false toad" is also used for members of the genus *Telmatobufo* (e.g., Pelado Mountains False Toad, *T. australis*).

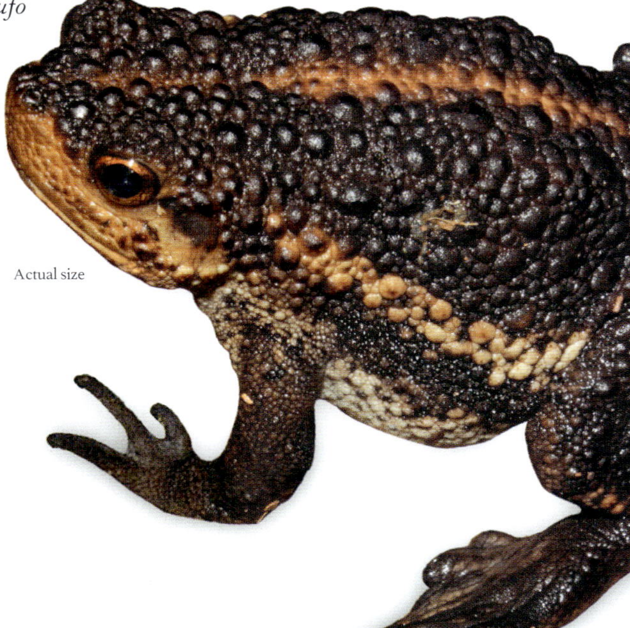

Actual size

The False Toad has a very rotund body, covered in warts, and a small head. The nostrils are positioned on top of the snout. The hind feet are fully webbed and the fingers are long, slender, and blunt. The upper surfaces are brown or black with yellowish stripes along the flanks and down the middle of the back. The underside is yellowish.

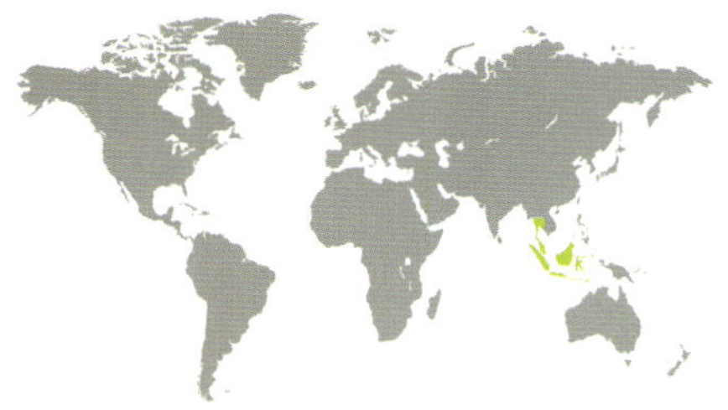

FAMILY	Bufonidae
OTHER NAMES	Previous name *Pedostibes hosii*, Brown Tree Toad
DISTRIBUTION	Borneo; Sumatra, Indonesia; Peninsular Malaysia; southern Thailand
ADULT HABITAT	Forest up to 2,300 ft (700 m) altitude
LARVAL HABITAT	Forest streams
CONSERVATION STATUS	IUCN Least Concern. Has a large range, but is threatened in places by deforestation

ADULT LENGTH
Male
1 15/16–3 1/8 in (50–80 mm)
Female
3 1/2–4 1/8 in (89–105 mm)

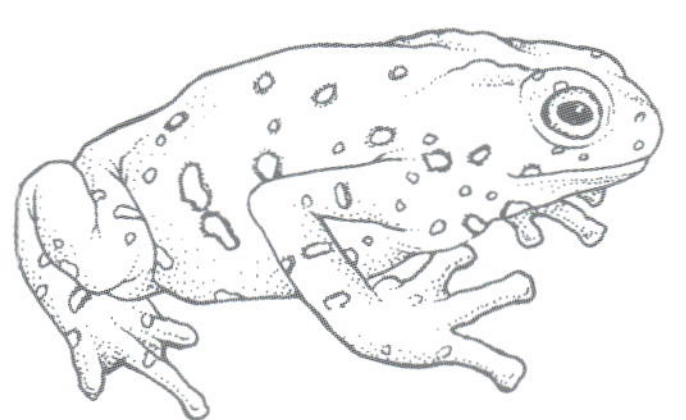

RENTAPIA HOSII

BOULENGER'S ASIAN TREE TOAD

(BOULENGER, 1892)

Unusually for a toad, this species is arboreal, and can be found in trees as high as 20 ft (6 m) above the ground. Its fingers and toes have adhesive pads at their tips, similar to those of tree frogs. It feeds on ants and other insects. It descends to the ground to breed, gathering along forest streams and males produce a grating squawk. The eggs are laid in strings and the tadpoles are found among leaf litter in pools at the edge of streams. There is no distinct breeding season, with mating aggregations forming at any time of year.

SIMILAR SPECIES

There were five species in the tree toad genus *Pedostibes* until the Southeast Asian species were moved to the genus *Rentapia*, and one of them, the Marbled Green Toad (*P. rugosus*), was synonymized with Everett's Asian Tree Toad (*R. everetti*). That left just the Malabar Tree Toad (*P. tuberculosus*), which is endemic to the riparian forests of the Western Ghats of India, in genus *Pedostibes*.

Boulenger's Asian Tree Toad has prominent bony ridges on its head running from the eyes to the parotoid glands. The head is otherwise smooth, and there are few warts on the back. Both toes and fingers have disk-shaped tips and the toes are webbed. All males and some females are light brown to chocolate-brown in color, but some females are black or purple with conspicuous yellow spots.

Actual size

FAMILY	Bufonidae
OTHER NAMES	Previous name *Andinophryne colomai*
DISTRIBUTION	Northern Ecuador
ADULT HABITAT	Forest at altitudes of 3,870–4,600 ft (1,180–1,400 m)
LARVAL HABITAT	Unknown
CONSERVATION STATUS	A very rare species that has not been seen since 1984

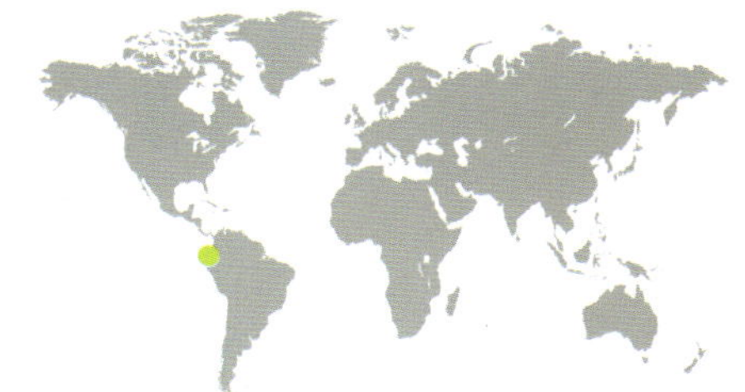

RHAEBO COLOMAI

CARCHI ANDES TOAD

(HOOGMOED, 1985)

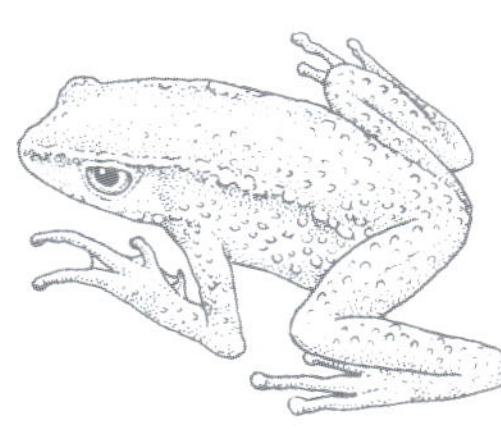

ADULT LENGTH
Male
1^{11}/$_{16}$ in (43 mm); based on one specimen
Female
around 1^{5}/$_{16}$ in (34 mm); based on two specimens

Only a few specimens of this very rare toad have been found, all of them close to mountain streams in the Andes in northwestern Ecuador. The gut contents of one specimen indicate that it feeds on large ants. It was last seen in 1984, leading it to be included in a list of more than 200 "lost species," but intensive searches have failed to find it since, and it may be extinct. Much of its forest habitat has been destroyed by logging, land clearance for agriculture, and the extensive use of herbicides.

Actual size

SIMILAR SPECIES

Rhaebo colomai is one of 12 species in the genus *Rhaebo*, but it was previously in the genus *Andinophryne* with two other rare species, including the Tandayapa Andes Toad (*R. olallai*), a Critically Endangered species only known from one locality in Ecuador and one in Colombia. The latter species had not been seen since 1970, when its rediscovery was reported in 2014. The young of this species have a beautifully mottled skin pattern that becomes uniform brown as they get older.

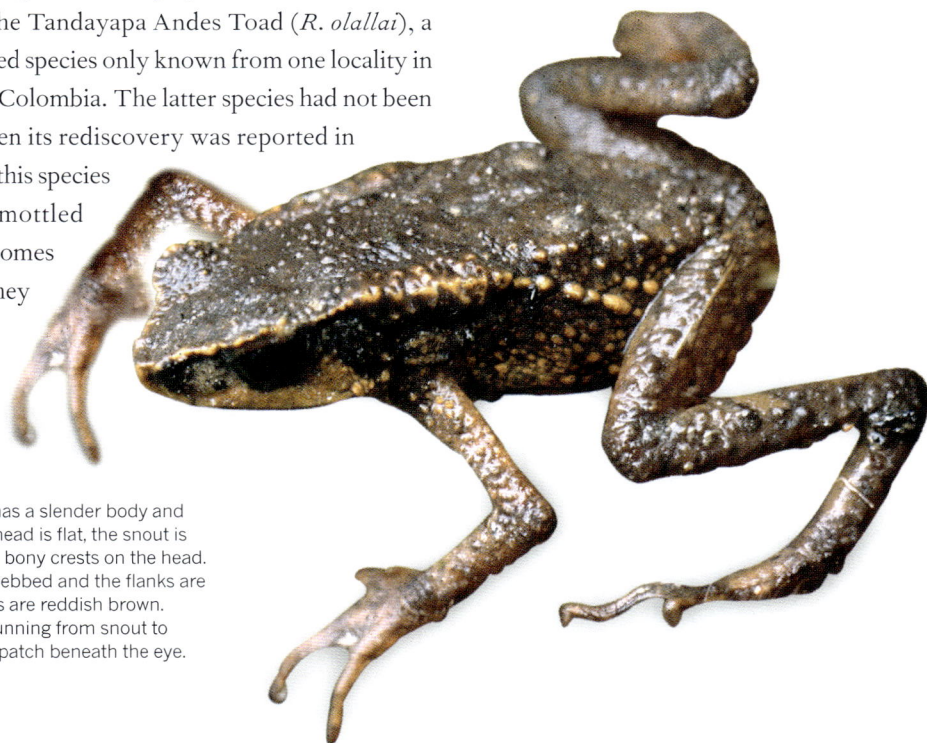

The Carchi Andes Toad has a slender body and long, slender limbs. The head is flat, the snout is protruding, and there are bony crests on the head. The hands and feet are webbed and the flanks are warty. The upper surfaces are reddish brown. There is a cream stripe running from snout to groin, and a large cream patch beneath the eye.

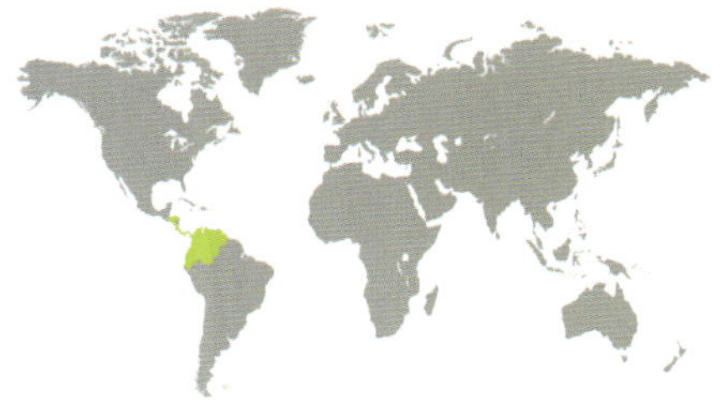

FAMILY	Bufonidae
OTHER NAMES	None
DISTRIBUTION	Central and South America, from Honduras in the north to Ecuador in the south and Venezuela in the east
ADULT HABITAT	Lowland wet forest
LARVAL HABITAT	Rocky pools
CONSERVATION STATUS	IUCN Least Concern. Declining in northern parts of its range due to chytridiomycosis

ADULT LENGTH
Male
1⅝–2½ in (42–62 mm)
Female
1¹⁵⁄₁₆–3⅛ in (50–80 mm)

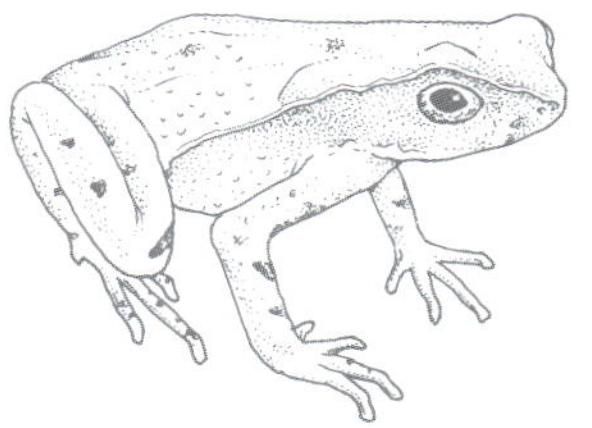

RHAEBO HAEMATITICUS

TRUANDO TOAD

COPE, 1862

The distinctive coloration of this ground-living toad is described as a "dead-leaf" pattern. The relatively pale back looks like the top of a leaf, with the dark band along the front its shadow. Between them, a narrow, pale stripe looks like a leaf edge. The toad breeds "explosively" after heavy rain between March and July. The eggs are laid in two long strings in rocky pools beside forest streams or rivers. Along with other streamside frogs, it was commonly seen in Panama until 1996 and 1997, when it became very rare. Several dead animals were found, killed by chytridiomycosis.

SIMILAR SPECIES

Ten species removed from the genus *Bufo* were placed in the genus *Rhaebo* with three species originally in the genus *Andinophryne* (see *Rhaebo colomai*; page 197). All occur in Central and South America. The Spotted Toad (*R. guttatus*) has a huge range covering most of the Amazon Basin and may be a complex of more than one species. The Cundinamarca Toad (*R. glaberrimus*) lives in the foothills of mountains in Colombia, Peru, and Venezuela, and is threatened in places by dam construction, but despite this it is listed as Least Concern.

The Truando Toad has long limbs and very large parotoid glands behind the eyes. There are warts on the back and on the parotoid glands. The back and limbs are tan to grayish purple, sometimes with black spots and orange blotches. The flanks are dark brown, below a narrow white stripe. The underside is beige or yellow.

Actual size

FAMILY	Bufonidae
OTHER NAMES	None
DISTRIBUTION	Southern Ecuador
ADULT HABITAT	Marshes at altitudes of 6,725–7,200 ft (2,050–2,200 m)
LARVAL HABITAT	Probably in small pools
CONSERVATION STATUS	IUCN Critically Endangered. Has a very small range and its habitat has been largely lost to agriculture and urbanization

RHINELLA AMABILIS

LOVELY TOAD

(PRAMUK & KADIVAR, 2003)

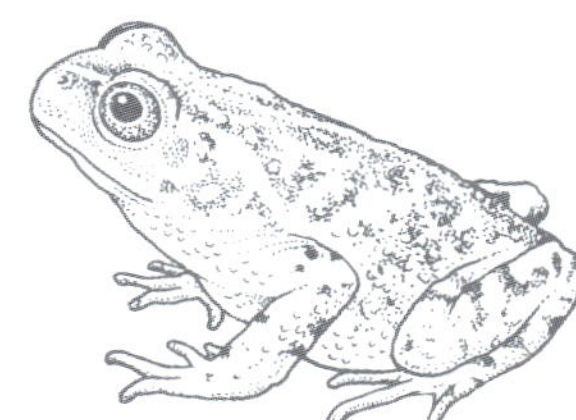

ADULT LENGTH
Male
average 3³⁄₁₆ in (82 mm)
Female
average 2⅞ in (73 mm)

Last collected in 1968, this little-known toad was not described until 2003, from museum specimens. It is confined to a high-altitude valley in the Andes, in Loja province, Ecuador. Surveys of its range in 1989 and 2001 failed to find any specimens and it is possible that it is extinct. Males differ from females in having glands and spines on their flanks and legs; the function of these is not known. It is assumed that the toad lays its eggs in ponds. The name *amabilis* as given in the original description of this species refers to the "particularly lovely demeanor of this and all toads."

SIMILAR SPECIES

The Lovely Toad is similar to the Warty Toad (*R. spinulosa*), a large toad that has a very extensive range, occurring in the Andes from Peru in the north to Argentina in the south. It is common around ponds, lakes, and streams, and also occurs on arable farmland.

The Lovely Toad is a large toad with a long and triangular snout, long hind legs, large eyes, and a large and very obvious tympanum. Its upper surfaces are yellowish green with numerous dark blotches. The underside is white.

Actual size

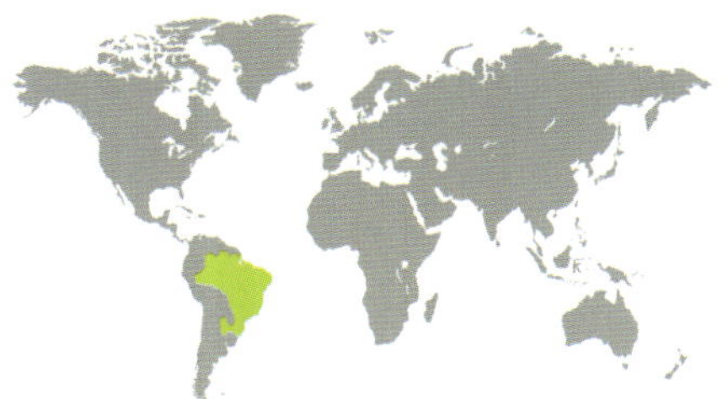

FAMILY	Bufonidae
OTHER NAMES	None
DISTRIBUTION	Brazil, eastern Paraguay, northeastern Argentina
ADULT HABITAT	Forest and open grassland up to 3,900 m (1,200 m) altitude
LARVAL HABITAT	In lakes, ponds, and puddles
CONSERVATION STATUS	IUCN Least Concern. Has a large range and its population is stable

ADULT LENGTH
Male
$3\frac{7}{8}$–$6\frac{5}{8}$ in (100–166 mm)
Female
$5\frac{5}{16}$–$7\frac{9}{16}$ in (135–190 mm)

RHINELLA ICTERICA

YELLOW CURURU TOAD

(SPIX, 1824)

This large toad is a "sit and wait" predator, feeding mostly on beetles and ants. Breeding occurs between August and January, males calling by day and by night from the water. The call is described as a "melodious tremolo." The female lays several thousand eggs in long strings. In November and December, the tiny toadlets all leave their pond together, migrating in huge numbers away from the water. Males and females are strikingly different in appearance: Males are uniformly yellow, while females have several large, dark blotches on their back, symmetrical on either side of the midline.

SIMILAR SPECIES

Rhinella icterica is similar to the Cane Toad (*R. marina*; page 201). The Argentine Toad (*R. arenarum*) has a large range, occurring in Brazil, Uruguay, and Bolivia, as well as Argentina. By contrast, the Alto Marañón Toad (*R. vellardi*) is found only in a restricted area in northern Peru. Its conservation status is now listed as Endangered.

Actual size

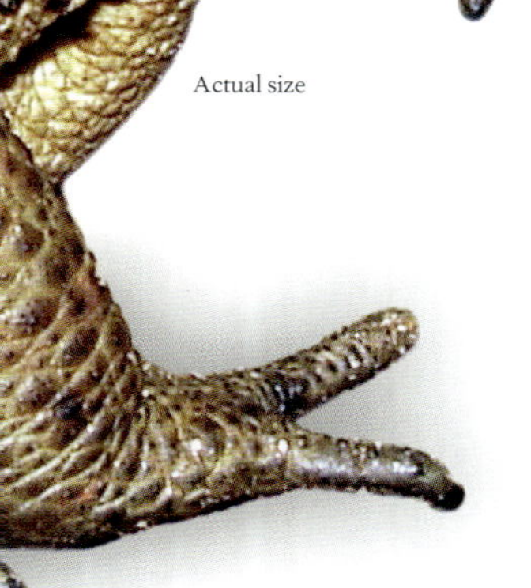

The Yellow Cururu Toad is a large toad with very large parotoid glands behind the eyes and many blunt, thorny warts on its back. The upper surfaces are yellow to pale brown, sometimes with a greenish tinge. The underside is white with a marbled brown pattern. Females have symmetrical black blotches on their back and black stripes on their legs.

FAMILY	Bufonidae
OTHER NAMES	Formerly *Bufo marinus*, *Chaunus marinus*; Giant Toad, Marine Toad
DISTRIBUTION	From the southern tip of Texas to southern Brazil. Introduced to Australia, and some Caribbean and Pacific islands, including Hawaii
ADULT HABITAT	Near rivers and wetlands, including brackish water and mangrove swamps. Also found in towns and gardens
LARVAL HABITAT	Small and large bodies of still water
CONSERVATION STATUS	IUCN Least Concern

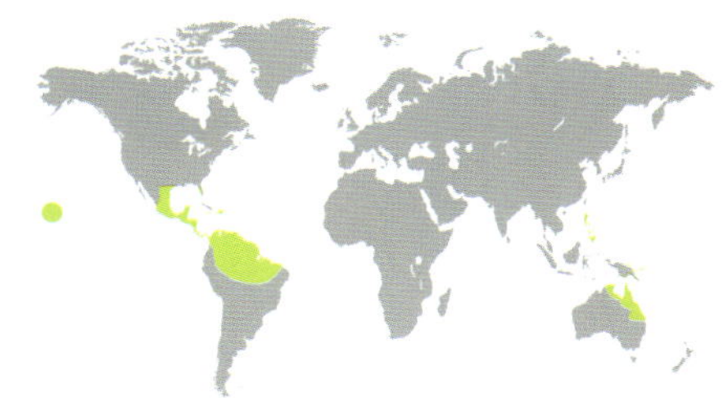

RHINELLA MARINA

CANE TOAD

(LINNAEUS, 1758)

ADULT LENGTH
Male
up to 6 in (150 mm)
Female
up to 8¾ in (225 mm)

One of the world's largest frogs, the formidable Cane Toad has a voracious appetite. Called the Marine Toad in its native habitat, it is unusual in tolerating brackish water around estuaries and in mangrove swamps. It is attracted to human settlements and is often seen waiting under streetlights for insects to fall. Larger females can lay more than 20,000 eggs in a single clutch. Males attract females with a slow, low-pitched trill, sounding like a distant tractor. They have few enemies; as eggs, tadpoles, and adults Cane Toads are distasteful or toxic to potential predators. This has had a negative impact in Australia, where native reptiles and mammals, as well as domestic cats and dogs, die when they eat them.

SIMILAR SPECIES

There are 97 species in the genus *Rhinella*; all are found in South and Central America. The South American Common Toad (*Rhinella margaritifera*) is common over a wide range from Panama to Brazil. Some *Rhinella* species are threatened by deforestation, such as the Critically Endangered Mesopotamian Beaked Toad (*R. rostrata*), known from a single locality in Colombia.

Actual size

The Cane Toad is a large, robust toad with very warty skin, prominent eyes and very large, swollen-looking parotoid glands behind the eyes. It is usually yellowish brown with darker markings and is especially yellow on the flanks and throat. The fingers and toes have hardened tips that are often dark brown in color.

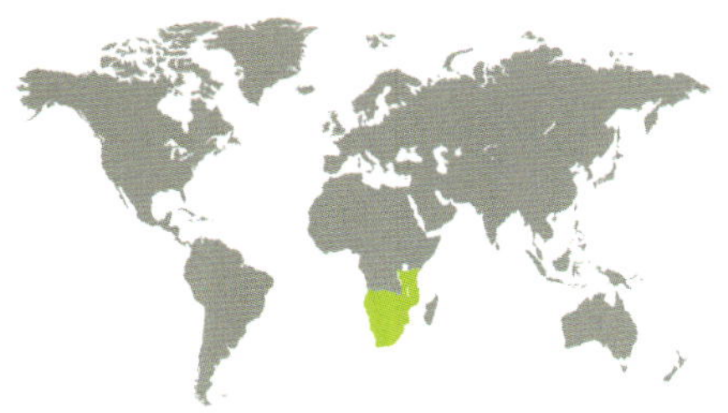

FAMILY	Bufonidae
OTHER NAMES	African Split-skin Toad
DISTRIBUTION	East and southern Africa, from Tanzania to South Africa
ADULT HABITAT	Savanna and scrubland
LARVAL HABITAT	Deep waterbodies
CONSERVATION STATUS	IUCN Least Concern. Has a very large range and can thrive around human settlements

ADULT LENGTH
Male
up to $3\frac{7}{16}$ in (88 mm)
Female
up to $3\frac{9}{16}$ in (92 mm)

SCHISMADERMA CARENS

RED TOAD

(SMITH, 1848)

Red Toads breed in spring after heavy rain, gathering among emergent vegetation in deep, muddy ponds. The male calls while floating, inflating a large white vocal sac and producing a long, loud "whoop." Large choruses can form at night and on overcast days, males struggling with one another to clasp females. The eggs are laid in double strings and estimates of clutch size range from 2,500 to 20,000 eggs. The black tadpoles form very dense schools that move slowly up and down in the water. Red Toads often shelter in houses and are able to climb up into the eaves.

SIMILAR SPECIES

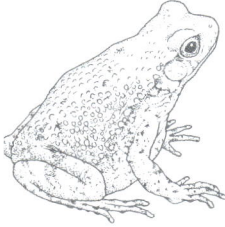

Schismaderma carens was the only species in its genus, until 2021, when the Angolan Red Toad (*S. branchi*) was described. *Schismaderma* is estimated to have separated from other toads about 55 million years ago.

The Red Toad has a distinctive ridge, rich in glands, running from the large tympanum to the groin. This separates the reddish-brown (sometimes pink) color of the back from the whitish flanks and belly. There is a pair of dark spots on the lower back and a less obvious pair of spots on the shoulders. There are no parotoid glands and, compared with most toads, very few warts.

Actual size

FAMILY	Bufonidae
OTHER NAMES	Includes *Amietophrynus rangeri*
DISTRIBUTION	South Africa, Lesotho, Swaziland
ADULT HABITAT	Grassland and fynbos, typically at altitudes greater than 3,300 ft (1,000 m)
LARVAL HABITAT	Permanent ponds, pools, dams. Occasionally slow-moving rivers
CONSERVATION STATUS	IUCN Least Concern. Has a large range, but is declining in northern and eastern parts

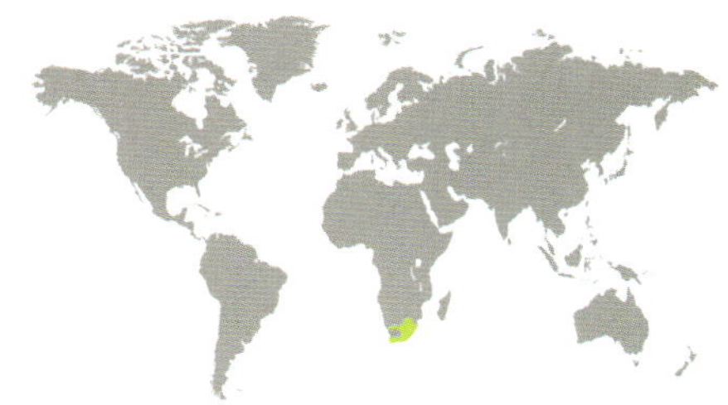

SCLEROPHRYS CAPENSIS

RAUCOUS TOAD

TSCHUDI, 1858

ADULT LENGTH
Male
2⅜–3½ in (60–90 mm)
Female
3½–4⁹⁄₁₆ in (90–117 mm)

Named for its incessant duck-like quacking, this large toad breeds over a period of up to four months in spring and summer. Males call from floating vegetation and females move among them before choosing one, nudging him to initiate amplexus. While calling, males lose weight and, from time to time, leave the chorus for some days to feed. The males that mate with the most females are those that attend the chorus and call for the most nights. Females lay up to 10,000 eggs in two long strings. Raucous Toads seem to thrive alongside people and are quite common in farm dams and garden ponds.

SIMILAR SPECIES

Like the Guttural Toad (*Sclerophrys gutturalis*; page 206), *S. capensis* does well in altered habitats—but apparently not as well as the former species, which appears to be displacing it in northern and eastern parts of South Africa. The two species also frequently hybridize. *Sclerophrys capensis* differs from the Guttural Toad, as well as the Eastern Olive Toad (*S. garmani*; page 205), as it has no red markings on its thighs.

The Raucous Toad has a blunt snout, long legs, large parotoid glands, and very warty skin. The skin on the back is olive-gray to brown, with paired, irregularly shaped, dark brown patches. There is a dark bar on the head between the eyes and, in many individuals, a thin, pale line down the middle of the back.

Actual size

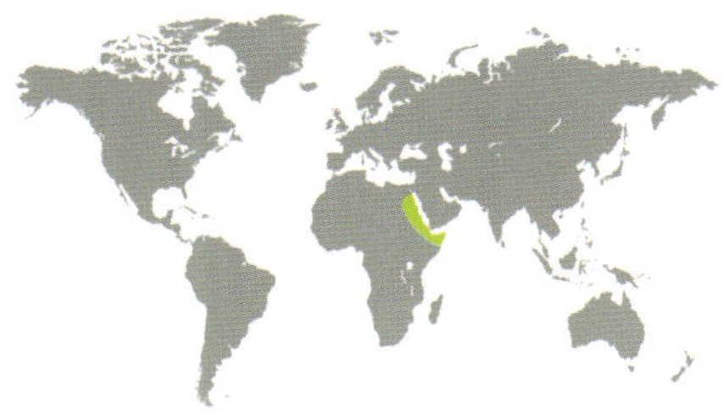

FAMILY	Bufonidae
OTHER NAMES	*Amietophrynus dodsoni*
DISTRIBUTION	Northeastern Africa, from southeastern Egypt to Djibouti and Somalia
ADULT HABITAT	Dry riverbeds and caves, from sea level to 5,900 ft (1,800 m) altitude
LARVAL HABITAT	Ephemeral pools
CONSERVATION STATUS	IUCN Least Concern. Common in many parts of its range

ADULT LENGTH
Male
around $2\frac{1}{16}$ in (53 mm); based on one specimen

Female
up to $2\frac{9}{16}$ in (64 mm)

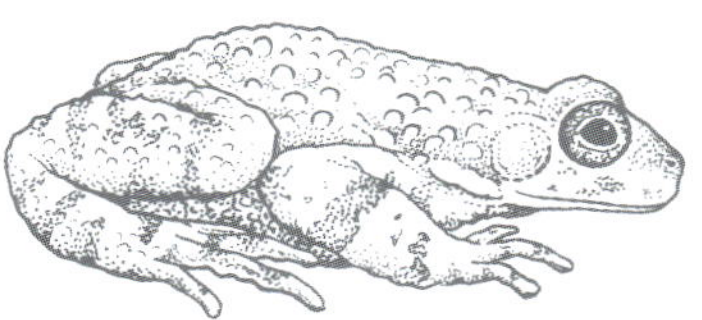

SCLEROPHRYS DODSONI

DODSON'S TOAD

(BOULENGER, 1895)

Widespread in Djibouti and Somalia, this small toad lives in an environment that is very dry for most of the year. As a result, it remains dormant for long periods, hiding deep in rock crevices and in caves. It emerges after rain, to breed in temporary pools. The male's mating call is a single note, likened to the bark of a dog. Females lay up to around 470 eggs. The tadpoles develop very quickly as the pools in which they live start to dry out, and metamorphosis occurs about six weeks after egg-laying.

SIMILAR SPECIES

Sclerophrys dodsoni is a member of an Afro-Arabian clade previously included in the genus *Duttaphrynus*, which is now confined to Asia (see page 166). African and Arabian species are placed in the genus *Sclerophrys*, which now numbers 46 species, while *Duttaphrynus* contains 23 species. Most *Sclerophrys* are Least Concern, but the Tai Toad (*S. taiensis*), from Ivory Coast, is Endangered.

Dodson's Toad has a short, rounded snout, short hind legs, a conspicuous tympanum, and prominent oval parotoid glands on its head. There are numerous warts on its back, which is olive, gray, or yellowish in color with small black and orange spots. There are darker transverse stripes on the legs.

Actual size

FAMILY	Bufonidae
OTHER NAMES	Garman's Toad
DISTRIBUTION	Eastern Africa, from Ethiopia and Somalia to Botswana and northern South Africa
ADULT HABITAT	Well-wooded, low-lying areas in savanna
LARVAL HABITAT	Ephemeral pools, farm dams, garden ponds
CONSERVATION STATUS	IUCN Least Concern. Has a large range and no apparent threats

SCLEROPHRYS GARMANI

EASTERN OLIVE TOAD

(MEEK, 1897)

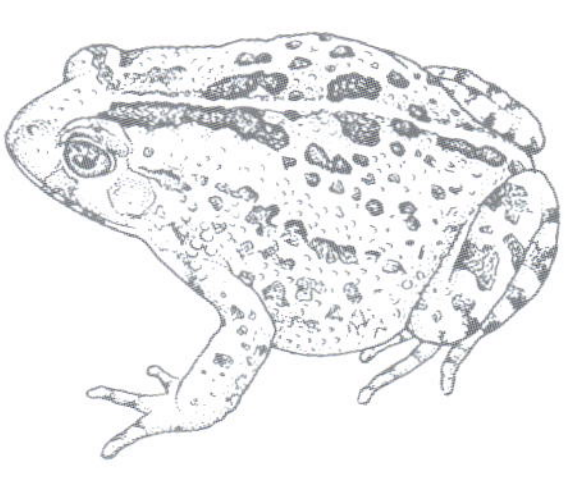

ADULT LENGTH
Male
2½–4⅛ in (63–105 mm)
Female
2⅝–4½ in (65–115 mm)

This large, strikingly marked toad breeds after the first heavy rain in spring or summer. Males call from exposed or partially concealed positions close to water, producing a loud "kwaak." Individual males return to their chosen call site, even after being displaced by 0.6 miles (approx. 1 km). The female produces 12,000–20,000 black eggs in two long strings and these hatch within 24 hours. The tadpoles can be either light or dark brown in color, matching the color of the mud in their pond. During the breeding season, some adults are eaten by young crocodiles.

SIMILAR SPECIES

The Western Olive Toad (*Sclerophrys poweri*) is barely distinguishable from *S. garmani*, both in its appearance and its call. It is found in Namibia, Botswana, western Zimbabwe, and western South Africa, and occurs in some very dry habitats, including the Kalahari Desert.

The Eastern Olive Toad is a robust toad with a blunt snout, large eyes, and well-developed parotoid glands behind the eyes. The back is covered with warts, each with a black tip. It is olive-green or light brown in color, with a pattern of paired square brown patches, these often being flushed red.

Actual size

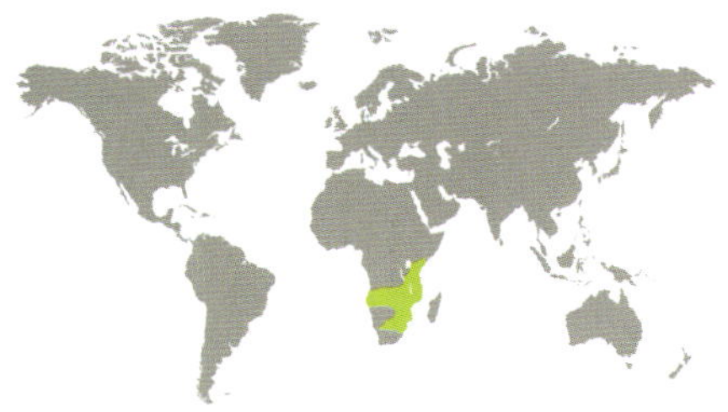

FAMILY	Bufonidae
OTHER NAMES	African Common Toad
DISTRIBUTION	Eastern and southern Africa, from Angola in the west, to Kenya in the north, and South Africa in the south
ADULT HABITAT	Moist areas in forests, woodland, shrubland, and grassland
LARVAL HABITAT	Small permanent waterbodies
CONSERVATION STATUS	IUCN Least Concern. Has a large, expanding range. Tolerant of habitat change

ADULT LENGTH
Male
2$\frac{9}{16}$–3½ in (64–90 mm)
Female
2½–4¾ in (62–120 mm)

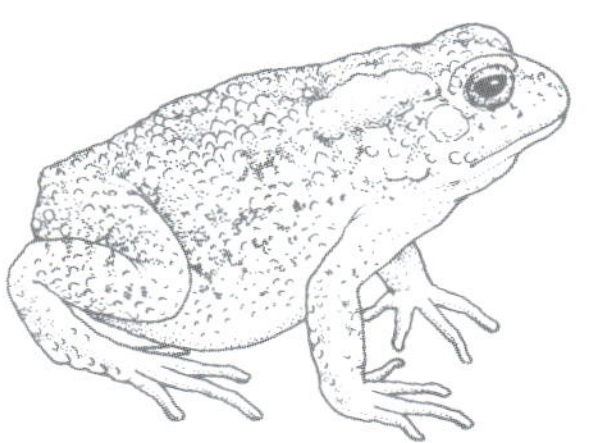

SCLEROPHRYS GUTTURALIS

GUTTURAL TOAD

(POWER, 1927)

This large toad gets its name from its mating call, a loud, extended snore. At night males form a chorus around a pond and neighboring males tend to call antiphonally with one another. As a female approaches her chosen male, she may be intercepted and grasped by one or more other males, leading to the formation of a "mating ball" consisting of one female and several males. Females lay up to 25,000 eggs in two strings. Guttural Toads thrive in many human-modified habitats, such as farms, where they are welcome as predators on insect pests, slugs, and snails.

The Guttural Toad has a blunt snout and large eyes, behind which are prominent parotoid glands. The back is yellow or brown in color with dark brown patches and a pale stripe down its midline. There are red patches on the thighs. Two pairs of dark patches on the head form a pale cross between the eyes.

SIMILAR SPECIES

As Guttural Toads expand their range southward in South Africa, they increasingly overlap with the Raucous Toad, with which they commonly hybridize. Further north, they overlap with the Flat-backed Toad (*A. maculatus*; page 207), which has similar markings on its head.

Actual size

FAMILY	Bufonidae
OTHER NAMES	Hallowell's Toad, Lesser Cross-marked Toad
DISTRIBUTION	West, East, and South Africa
ADULT HABITAT	Near rivers in humid savanna
LARVAL HABITAT	Pools left in riverbeds when water levels are low
CONSERVATION STATUS	IUCN Least Concern. Has a very large range

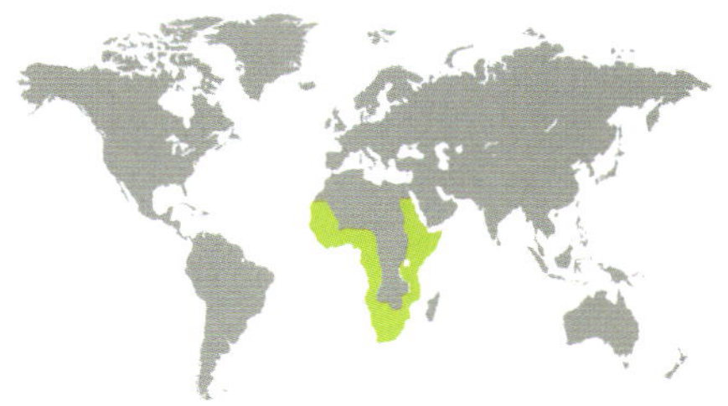

SCLEROPHRYS MACULATA

FLAT-BACKED TOAD

(HALLOWELL, 1854)

ADULT LENGTH
Male
$1\frac{1}{2}$–$2\frac{9}{16}$ in (38–64 mm)
Female
$1\frac{5}{8}$–$3\frac{5}{16}$ in (41–85 mm)

This medium-sized toad has a geographic range that covers much of sub-Saharan Africa. In Burkina Faso and Nigeria it is collected and eaten as it is regarded as a significant source of protein for some human populations. When breeding, males form choruses around the edge of pools in rivers, producing a "quork… quork" sound, alternating their calls with those of their nearest neighbor. The female produces 2,000–8,000 eggs in two strings. The tadpoles develop in pools left when water levels have fallen. In response to attack by a predator, they form dense aggregations, sometimes of several thousand individuals.

SIMILAR SPECIES

Sclerophrys maculata lacks the red markings on the thighs seen in the Guttural Toad (*S. gutturalis*; see page 206) and the Eastern Olive Toad (*S. garmani*; see page 205). It is also smaller and usually paler in color than those species. In West Africa it occurs with the Subdesert Toad (*S. xeros*), which is found in drier habitats.

The Flat-backed Toad is so called because its parotoid glands, which are prominent in many toad species, are flattened. The skin is covered in warts, which in males especially may have dark, sharp tips. The back is pale or light brown with paired darker patches and a pale, thin line running down the midline. There is a pale cross on top of the head.

Actual size

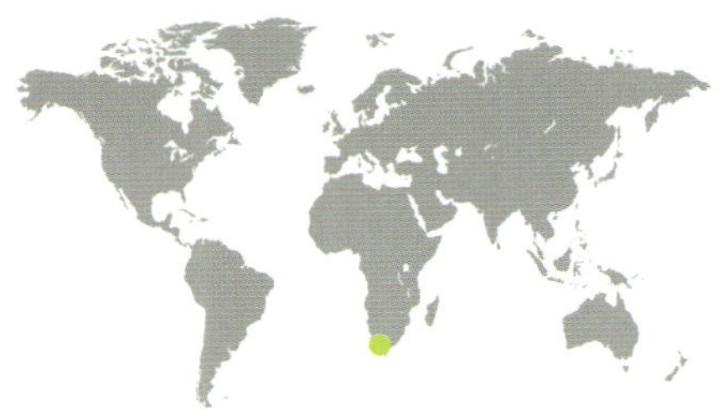

FAMILY	Bufonidae
OTHER NAMES	Panther Toad
DISTRIBUTION	Southwestern tip of Africa
ADULT HABITAT	In and close to wetlands, vleis, dams, and ponds
LARVAL HABITAT	Ephemeral ponds
CONSERVATION STATUS	IUCN Endangered

ADULT LENGTH
Male
up to 3⅞ in (100 mm)
Female
up to 5½ in (140 mm)

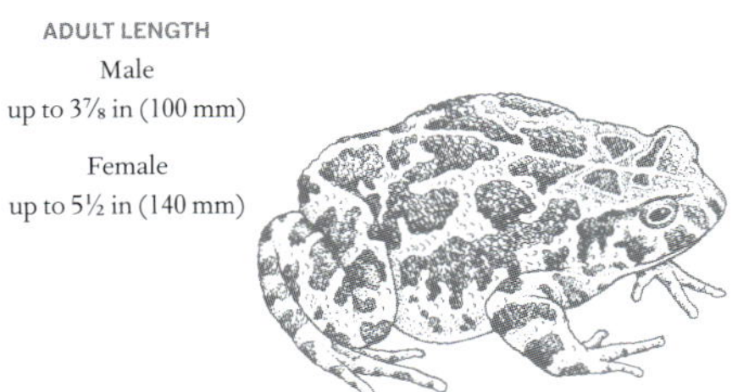

SCLEROPHRYS PANTHERINA

WESTERN LEOPARD TOAD

(SMITH, 1828)

Described as an "explosive breeder," the Western Leopard Toad mates and lays eggs over a period of only four or five days in spring. In the evening and during the night, males call to attract females from vegetation around a pond or from open water. Their call is a prolonged, slow snoring sound. Females approach males and initiate amplexus, during which they lay up to 25,000 eggs. Unlike other toads, males do not attempt to displace rival males from the backs of females, even though they commonly outnumber females during the breeding period. This species is also unusual in that it eats snails.

SIMILAR SPECIES

Sclerophrys pantherina is one of several toad species found in southern Africa. It is similar in appearance to the Eastern Leopard Toad (*S. pardalis*; page 209) although their ranges do not overlap. Guttural Toads (*S. gutturalis*; page 206), whose natural range is far to the north, have been introduced to Western Cape province. They threaten *S. pantherina* by competing for breeding habitat, and hybrid matings between the two species may occur.

Actual size

The Western Leopard Toad gets its name from the striking symmetrical spots on its back. It has a rounded snout, large eyes, large parotoid glands behind the eyes, and very warty skin. The upper surfaces are yellow, decorated with large, red-brown patches, edged in black. There is a thin, pale stripe down the middle of the back.

FAMILY	Bufonidae
OTHER NAMES	Snoring Toad
DISTRIBUTION	Eastern Cape, South Africa
ADULT HABITAT	Grassy bushveld, parks, gardens
LARVAL HABITAT	Large permanent waterbodies
CONSERVATION STATUS	IUCN Least Concern. However, there is concern that it is declining, due to human impacts on its habitat

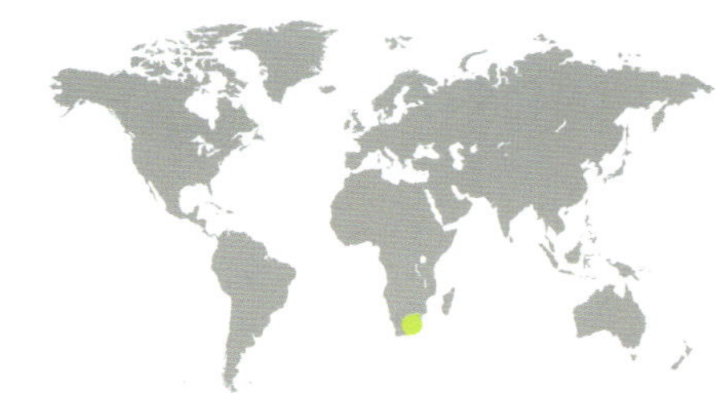

SCLEROPHRYS PARDALIS

EASTERN LEOPARD TOAD

(HEWITT, 1935)

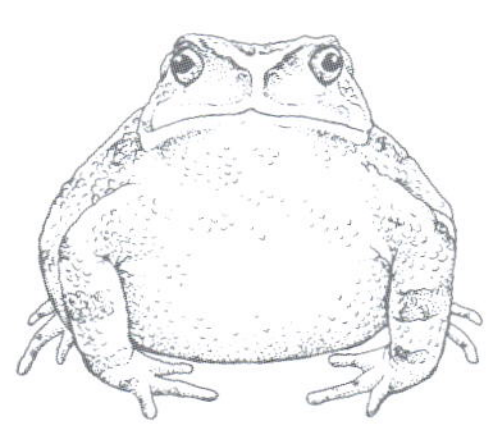

ADULT LENGTH
Male
3⅛–3⅞ in (80–100 mm)
Female
3⅞–5⅞ in (100–147 mm)

This large, colorful toad has a restricted range along the coast in the Eastern Cape province of South Africa. There is concern that, across much of its range, agricultural expansion and urbanization are causing population declines. In some places, very large numbers are killed on roads as they migrate to their breeding sites. It does occur, however, in a number of protected areas. It breeds in large ponds and lakes in late winter and spring. Males call while floating in the water, holding vegetation with one hand, producing a sound like a drawn-out snore.

The Eastern Leopard Toad is a robust toad with large eyes and prominent parotoid glands. Its back is mainly brown, but has symmetrical large reddish-brown patches edged with black and yellow. There is a narrow yellow stripe down the midline of the back and a dark bar on the head behind the eyes. The underside is pale and devoid of markings.

SIMILAR SPECIES

Sclerophrys pardalis is very similar in appearance and behavior to the Western Leopard Toad (*S. pantherina*; page 208). Until 1998, they were regarded as a single species, even though they are separated by a considerable distance. *Sclerophrys pardalis* differs from both the Guttural Toad (*S. gutturalis*; page 206) and the Eastern Olive Toad (*S. garmani*; page 205) in having no red markings on its thighs.

Actual size

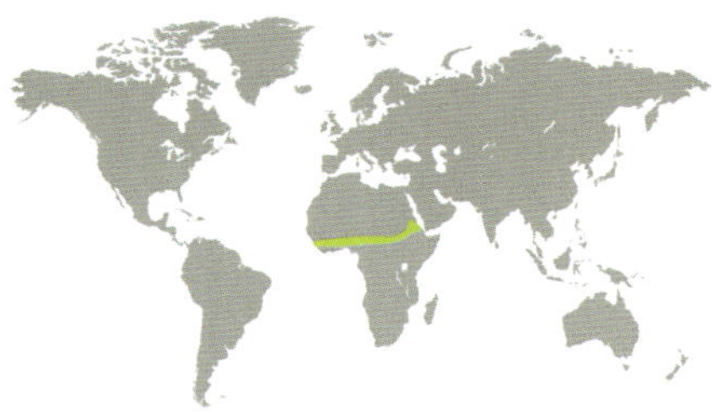

FAMILY	Bufonidae
OTHER NAMES	*Amietophrynus pentoni*, Shaata Gardens Toad
DISTRIBUTION	North Africa, south of the Sahara, from Gambia in the west to Eritrea in the east
ADULT HABITAT	Dry savanna and semidesert
LARVAL HABITAT	Ponds and pools
CONSERVATION STATUS	IUCN Least Concern. No evidence of general decline across its large range

ADULT LENGTH
Male
$2\frac{1}{8}$–$2\frac{15}{16}$ in (54–74 mm)
Female
$2\frac{1}{4}$–$3\frac{11}{16}$ in (58–95 mm)

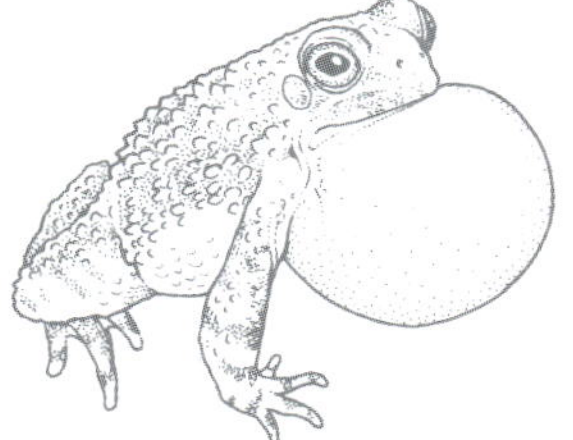

SCLEROPHRYS PENTONI

PENTON'S TOAD

(ANDERSON, 1893)

This distinctively shaped toad occupies a range that spans the whole width of Africa, bounded to the north by desert and to the south by tropical forest. A species of very arid habitats, it has been recorded within the Sahara Desert. It lives mostly underground but has been seen emerging to feed opportunistically on swarming termites. It breeds after rain, gathering in choruses of up to 20 males around small pools. Such choruses may last for only a single night. The pools in which the eggs are laid tend to dry up very quickly and the species shows remarkably rapid development, reaching metamorphosis in 10–13 days.

Penton's Toad has a stout body, a very blunt head, and massive thighs. It has large eyes, just behind which are large, flattened parotoid glands. The toad's upper surfaces are pale or dark brown and are covered in warts, often reddish, and the underside is white.

SIMILAR SPECIES

The Subsaharan Toad (*Sclerophrys xeros*) occupies much the same range as *S. pentoni* and, like that species, is adapted to arid conditions and breeds in temporary waterbodies. The large Moroccan Toad (*S. mauritanicus*) is found in Morocco and Algeria. Its habitat is being reduced by water extraction to meet the needs of an expanding human population.

Actual size

FAMILY	Bufonidae
OTHER NAMES	African Bouncing Toad, African Toad, Egyptian Toad, Reuss's Toad, Square-marked Toad
DISTRIBUTION	Sub-Saharan Africa, from Senegal in the west to Kenya in the east, south to Angola, and north to Egypt
ADULT HABITAT	Savanna grassland
LARVAL HABITAT	Slow-moving rivers
CONSERVATION STATUS	IUCN Least Concern. Has a huge range, including a wide diversity of habitats, and its population is stable

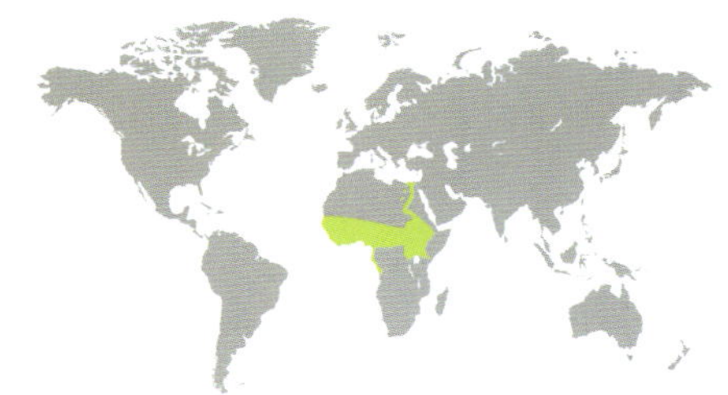

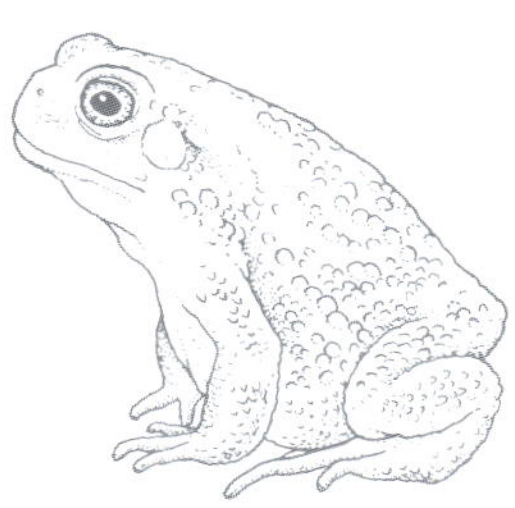

ADULT LENGTH
Male
$2\frac{1}{2}$–$3\frac{9}{16}$ in (62–91 mm)
Female
$2\frac{13}{16}$–$5\frac{1}{8}$ in (70–130 mm)

SCLEROPHRYS REGULARIS

COMMON AFRICAN TOAD

(REUSS, 1833)

This common toad feeds primarily on ants and termites, and can lay up to 13,000 eggs, deposited in two strings. It is expanding its range in Egypt, where it is taking advantage of irrigation schemes. Because of the toxins in their skin, especially in the parotoid glands on their heads, toads do not make good eating for humans, but this species is harvested for food in parts of its range, notably in Nigeria and Burkina Faso. The toads are beheaded, skinned, disemboweled, washed, and dried, a complex procedure that eliminates all the toxic parts of the body.

SIMILAR SPECIES

Given its enormous and diverse range, it is likely that *S. regularis* includes more than one species. In Egypt, it occurs alongside the Nile Valley Toad (*S. kassasii*), a more aquatic species that can form very large populations in flooded rice fields. The Lake Victoria Toad (*S. vittata*) is a little-known species that occurs in grassland to the west of Lake Victoria in Uganda.

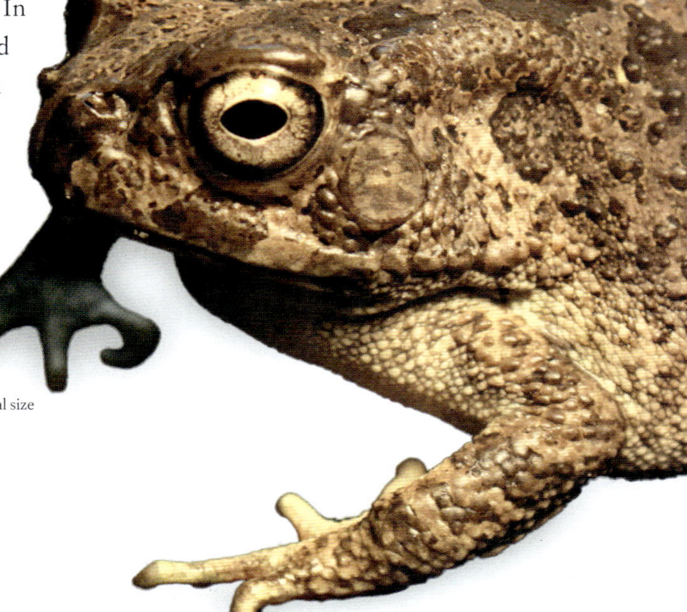

Actual size

The Common African Toad has a plump body, a rounded snout, and very warty skin. There are large, prominent parotoid glands behind the eyes. It is very variable in color, but is most commonly olive-green or brown, darker on the back than on the belly. There are several darker blotches on the back, these often being rectangular in shape.

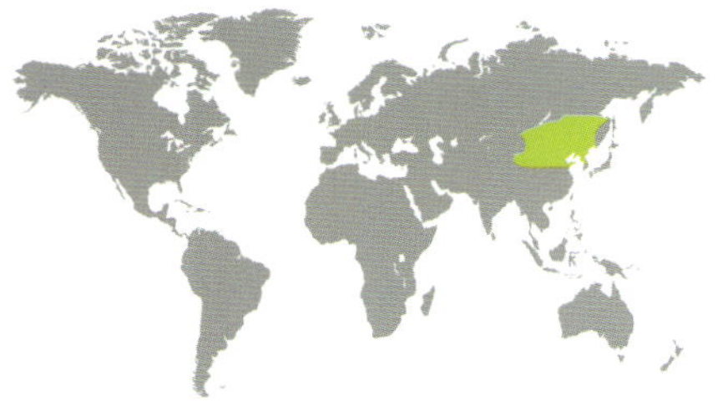

FAMILY	Bufonidae
OTHER NAMES	Piebald Toad, Siberian Toad, Tengger Desert Toad
DISTRIBUTION	Mongolia, China, Russia, Korean Peninsula
ADULT HABITAT	Grassland and semidesert. Also urban and agricultural areas
LARVAL HABITAT	Ponds
CONSERVATION STATUS	IUCN Least Concern. Has an enormous range and its population is stable

ADULT LENGTH
Male
1⁹⁄₁₆–2¹⁵⁄₁₆ in (40–75 mm)
Female
1¹³⁄₁₆–3½ in (46–89 mm)

STRAUCHBUFO RADDEI

MONGOLIAN TOAD

(STRAUCH, 1876)

This relatively small toad is found across a huge area and in a wide variety of habitats, including the Gobi Desert. It prefers soft, sandy soils into which it can burrow deep down during the cold winter months. It breeds between March and July, males and females gathering—sometimes in large numbers—at ponds, which can be few and far between. The female lays 1,000–6,000 eggs in two long strings. Ponds sometimes dry out before the tadpoles reach metamorphosis, and tadpole mortality can be very high. Where ponds do not dry out, tadpoles may overwinter in the water, metamorphosing the following spring.

SIMILAR SPECIES

The only species in its genus, *Strauchbufo raddei* is very similar to toads in the genus *Bufo*, in which it was formerly included. Its closest relative is another former *Bufo* species, the Green Toad (*Bufotes viridis*; page 162). The Swat Green Toad (*B. pseudoraddei*) lives in a similar dry habitat, sometimes at high altitude, in Afghanistan and Pakistan. Like *S. raddei*, it thrives in many agricultural areas where artificial ponds have been created.

The Mongolian Toad has warty skin, a rounded snout, and large parotoid glands behind the eyes. The back and legs are pale olive-green or gray with large, dark blotches, these often enclosing red dots. There is a narrow stripe, clear of markings, down the middle of the back. The underside is pale gray with a few dark spots.

Actual size

FAMILY	Bufonidae
OTHER NAMES	None
DISTRIBUTION	Eastern Cape, South Africa
ADULT HABITAT	Moist grassland at altitudes of 4,600–5,900 ft (1,400–1,800 m)
LARVAL HABITAT	Shallow pools
CONSERVATION STATUS	IUCN Critically Endangered. Has a very restricted range that is subject to habitat loss

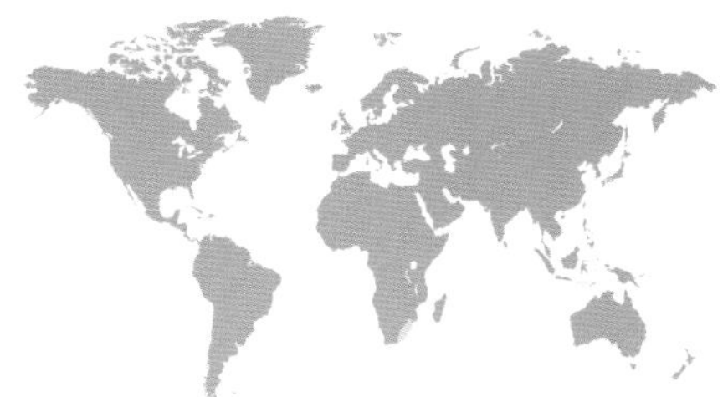

VANDIJKOPHRYNUS AMATOLICUS

AMATOLA TOAD

(HEWITT, 1925)

ADULT LENGTH
Male
average 7/8 in (23 mm)
Female
average 1 3/16 in (30 mm),
up to 1 7/16 in (37 mm)

This small toad is on the edge of extinction. It had not been seen since around 1998, until, in 2011, a single female and some eggs were found. Confined to a small area in the Winterberg and Amatola mountains, its high-altitude grassland habitat has been seriously degraded and reduced by overgrazing and by the planting of trees for timber. The toad breeds after heavy rain in October to December, males congregating around ponds to form a chorus of brief nasal squawks. The female lays several hundred eggs in a single string, wrapped around submerged vegetation.

Actual size

The Amatola Toad is a small toad whose back is covered in flattened warts. The parotoid glands are prominent. The back is dark gray, olive-green, or brown in color, and in many individuals there is a narrow, pale stripe down its middle. The underside is pale.

SIMILAR SPECIES

Vandijkophrynus amatolicus was formerly considered a subspecies of the Cape Sand Toad (*V. angusticeps*), a species found in Western Cape, in two very different types of habitat: Coastal lowlands and mountainous areas. It is associated with the fynbos vegetation that is unique to South Africa. It is considerably larger than *V. amatolicus*.

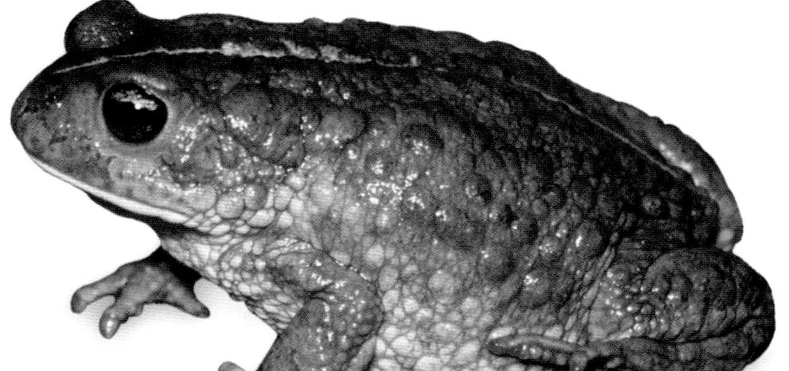

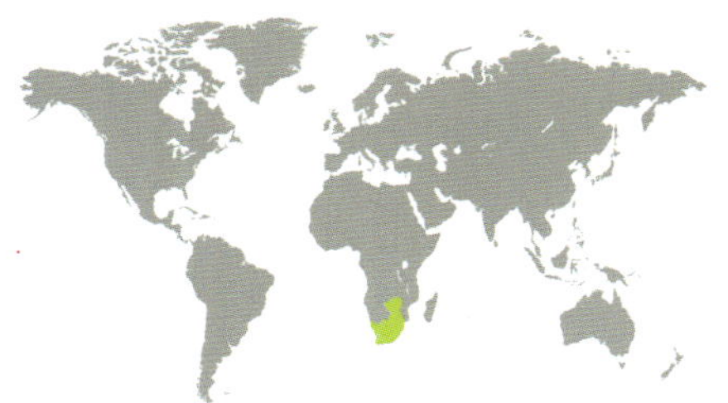

FAMILY	Bufonidae
OTHER NAMES	Drakensberg Toad, Gariep Toad, Mountain Toad
DISTRIBUTION	South Africa, Lesotho, Zimbabwe
ADULT HABITAT	Grassland, thorn bush, fynbos
LARVAL HABITAT	Ephemeral pools
CONSERVATION STATUS	IUCN Least Concern. A common and widespread species that occurs in several protected areas

ADULT LENGTH
Male
average 2⅜ in (61 mm)

Female
average 3 in (77 mm),
up to 3¹¹⁄₁₆ in (95 mm)

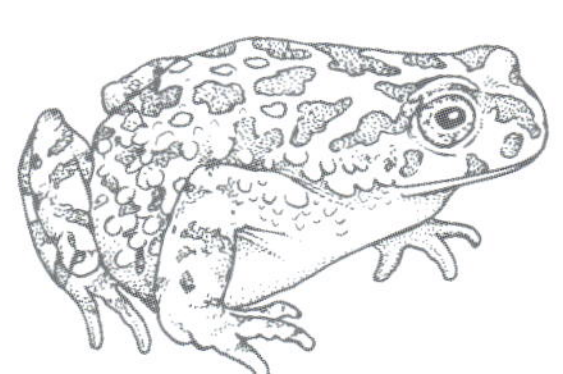

VANDIJKOPHRYNUS GARIEPENSIS

KAROO TOAD

(SMITH, 1848)

This large toad typically runs and walks rather than hops. Commonly found along riverbanks, it can survive in very dry habitats and at high altitudes where the winter is severe. It is sometimes found in termite mounds. It is an opportunistic breeder, using almost any small pool, including animal footprints. It usually breeds between September and February, males attracting females by means of a series of raucous squawks. The eggs are small and are laid in long strings, often wrapped around waterplants. The tadpoles often aggregate in dense schools and reach metamorphosis after 20 days.

The Karoo Toad has very large parotoid glands behind its eyes, especially individuals living in southern parts of the species' range. It has rough skin and its back is covered in smooth bumps. The upper surfaces are pale gray or olive, with large, irregular blotches of dark green, brown, or maroon. The underside is dirty white.

SIMILAR SPECIES

There are five species in the southern African genus *Vandijkophrynus* (formerly included in *Bufo*). The Paradise Toad (*V. robinsoni*) has smaller parotoid glands than *V. gariepensis*, and is found near springs and temporary water sources in the extreme west of South Africa. The Inyanga Toad (*V. inyangae*) is found in arid areas at high altitudes in eastern Zimbabwe, while the Cloud Frog (*V. nubicola*) occurs at elevation in Lesotho. See also the Amatola Toad (*V. amatolicus*; page 213).

Actual size

FAMILY	Bufonidae
OTHER NAMES	None
DISTRIBUTION	Cameroon
ADULT HABITAT	Forest at altitudes of 2,300–3,900 ft (700–1,200)
LARVAL HABITAT	Fast-flowing mountain streams
CONSERVATION STATUS	IUCN Endangered. Agriculture is encroaching on the species' natural habitat and it does not occur in any protected areas

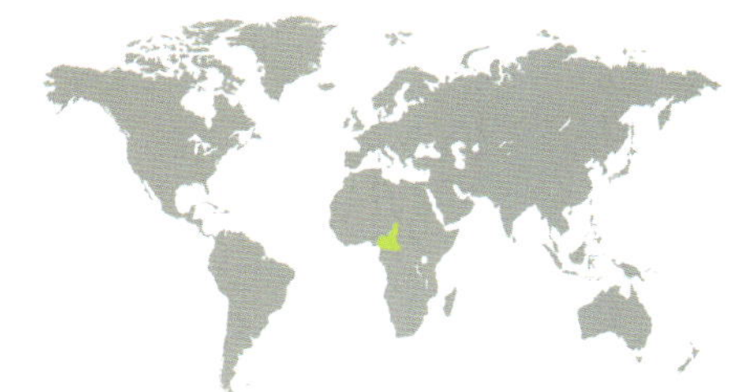

ADULT LENGTH
Male
1$\frac{3}{16}$–1$\frac{3}{4}$ in (30–44 mm)
Female
1$\frac{11}{16}$–1$\frac{15}{16}$ in (43–49 mm)

WERNERIA PREUSSI

BUEA SMALLTONGUE TOAD

(MATSCHIE, 1893)

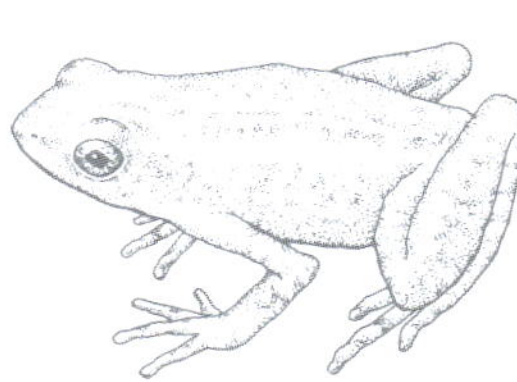

This small toad is one of six species of smalltongue or torrent toads that are found in and around rocky mountain streams on various forested mountains in West Africa. Areas of this habitat are few and far between, and are typically being destroyed to make way for agriculture. The toad hides under stones by day, emerging at night to feed on beetles. The eggs are laid in strings attached to rocks in streams. The tadpoles are adapted for life in fast-flowing water, with flattened bodies and large suckers around their mouths, with which they cling to rocks.

Actual size

SIMILAR SPECIES

Werneria preussi and the Hill Torrent Toad (*W. submontana*) are Endangered. The other four *Werneria* species are listed as Critically Endangered. Tandy's Smalltongue Toad (*W. tandyi*), the Bamboutos Smalltongue Toad (*W. bambutensis*), and Mertens' Smalltongue Toad (*W. mertensiana*) all inhabit high elevations in Cameroon. The Iboundy Torrent Toad (*W. iboundji*) is a species that is endemic to Gabon.

The Buea Smalltongue Toad has a slender body and long, slender legs. The male has more extensively webbed feet than the female. The upper and lower surfaces are dark brown. Females are uniformly colored, while males and young females have a brick-red, yellowish, or gray stripe running from eyelid to groin.

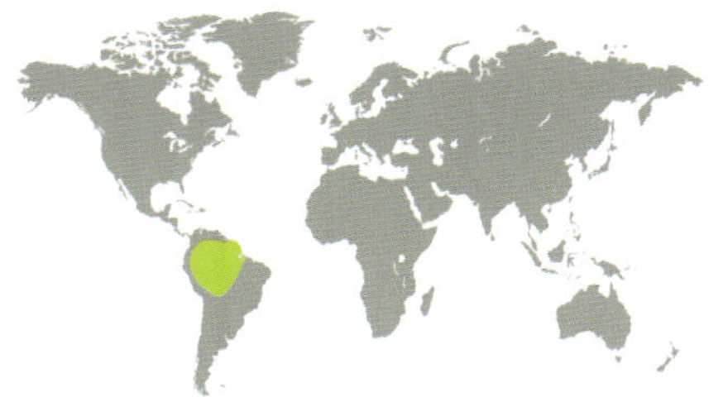

FAMILY	Ceratophryidae
OTHER NAMES	Amazonian Horned Frog
DISTRIBUTION	Amazon Basin
ADULT HABITAT	Leaf litter in tropical forest
LARVAL HABITAT	Pools and ponds
CONSERVATION STATUS	IUCN Least Concern. Collecting for the pet trade may affect some populations

ADULT LENGTH
Male
up to 2⅞ in (72 mm)
Female
up to 6 in (150 mm)

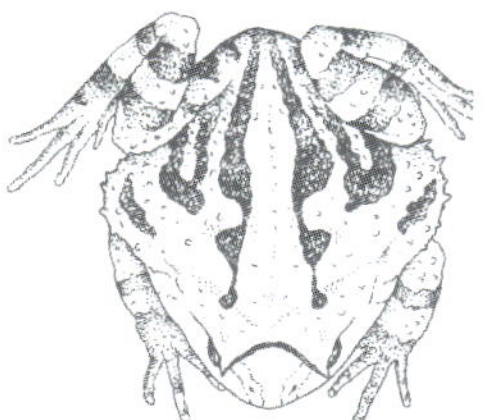

CERATOPHRYS CORNUTA

SURINAM HORNED FROG

(LINNAEUS, 1758)

The Surinam Horned Frog has hornlike projections above the eyes. Its highly variable color patterns provide very effective camouflage. It is unusual in having a tadpole that feeds not on plant food, but on other tadpoles, of its own and of other frog species. Like their parents, the tadpoles are voracious predators and have a beak and several rows of sharp "teeth."

Surinam Horned Frogs live among leaf litter throughout the Amazon Basin. They are "sit and wait" predators, remaining motionless until their prey comes close, when, with mouth wide open, they lunge forward. They have enormous heads and mouths, enabling them to eat not only ants and beetles, but also larger creatures, such as other frogs, lizards, and mice. They are active at night, burrowing backward into the soil with only their camouflaged heads above ground. Following rain, males call to attract females, making a loud "baaaa" sound. The female produces 300–600 eggs, which she carries to a small pool after fertilization.

SIMILAR SPECIES

There are eight species in the genus *Ceratophrys*, most with smaller ranges than *C. cornuta*. They are popular in the international pet trade where, because of their shape, they are often known as "Pac-Man frogs." *Ceratophrys cranwelli* is persecuted in Argentina, the southern part of its range, because of an erroneous belief that it is venomous. Its eggs are sold internationally for scientific research. See also Bell's Horned Frog (*C. ornata*; page 217).

Actual size

FAMILY	Ceratophryidae
OTHER NAMES	Argentine Horned Frog, Ornate Horned Frog
DISTRIBUTION	Argentina, southern Brazil, Uruguay
ADULT HABITAT	Leaf litter in tropical forest
LARVAL HABITAT	Pools and ponds
CONSERVATION STATUS	IUCN Near Threatened. Has declined due to habitat loss and collecting for the pet trade

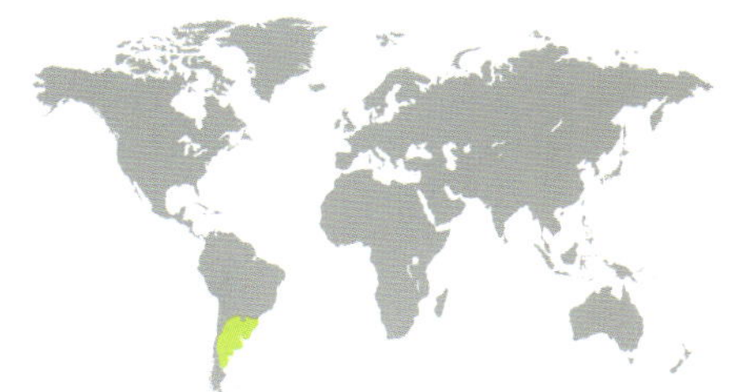

CERATOPHRYS ORNATA

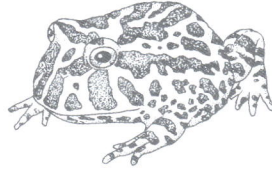

BELL'S HORNED FROG

(BELL, 1843)

ADULT LENGTH
Male
up to 4½ in (115 mm)
Female
up to 6⅝ in (165 mm)

This large frog lives in leaf litter in rainforest. A "sit and wait" predator, it feeds primarily on other frogs but can also eat birds, rodents, and snakes. It has toothlike processes in both jaws. During the fall and winter, when conditions are dry, it burrows into the ground and surrounds itself with a waterproof cocoon. It emerges in spring to breed in temporary ponds. Its tadpoles are carnivorous and are remarkable in being able to communicate with one another. If attacked by a predator, they produce a "distress call" that alerts other tadpoles to the danger.

Bell's Horned Frog has a wide, flattened body and is almost circular when seen from above. It has hornlike fleshy projections above the eyes. It has a very wide head and, consequently, an enormous mouth, enabling it to take large prey. It is variable in color, but is most commonly bright green with red, black, and brown stripes.

SIMILAR SPECIES

There are eight species in the genus *Ceratophrys*, distributed across much of South America. All have very similar natural history. The Brazilian Horned Frog (*C. aurita*) is found in Brazil, the Venezuelan Horned Frog (*C. calcarata*) in Colombia and Venezuela, and the Ecuadorian Horned Frog (*C. testudo*) in northern Brazil and Ecuador. The Pacific Horned Frog (*C. stolzmanni*), from Ecuador and Peru, in classed as Vulnerable. See also the Surinam Horned Frog (*C. cornuta*; page 216).

Actual size

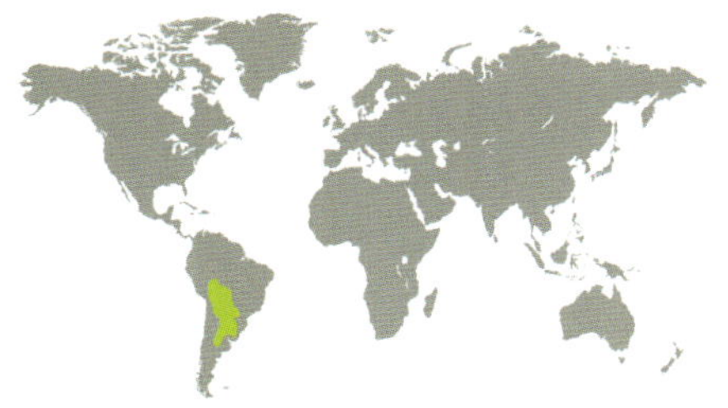

FAMILY	Ceratophryidae
OTHER NAMES	Chacoan Burrowing Frog, Lesser Chini Frog
DISTRIBUTION	Gran Chaco region of Argentina, Bolivia, and Paraguay
ADULT HABITAT	Shrubland and forest
LARVAL HABITAT	Ephemeral pools
CONSERVATION STATUS	IUCN Least Concern. Declining due to habitat destruction and collecting for the pet trade

ADULT LENGTH
average 2⅛ in (55 mm)

CHACOPHRYS PIEROTTII

CHACO HORNED FROG

(VELLARD, 1948)

Adults of this frog are very rarely seen because they spend much of their time underground, emerging after rain to feed and breed. They are described as "explosive breeders," appearing at ephemeral pools after heavy rain and completing mating and egg-laying over the course of just a few nights. The adults eat a variety of insects and other creatures, including smaller frogs. When threatened, they inflate their body with air and stand up on extended legs. The species' Chaco habitat is disappearing as it is converted to farmland, and the toad is also subjected to intense collecting during the breeding season for sale in the pet trade.

SIMILAR SPECIES

Chacophrys pierotti is the only species in its genus and is closely related to Budgett's Frog (*Lepidobatrachus laevis*; page 219). It has been suggested that it is not a true species, but a hybrid between two other inhabitants of the Gran Chaco, *Lepidobatrachus llanensis* and *Ceratophrys cranwelli*, but detailed genetic research has disproved this hypothesis.

The Chaco Horned Frog has a large head, relative to its body, short limbs, large, prominent eyes, and a blunt, rounded snout. It is very variable in color, with some individuals being predominantly green, some brown, and some a mixture of both. There is often a broad, pale band running down the middle of the back.

Actual size

FAMILY	Ceratophryidae
OTHER NAMES	Hippo Frog, Wide-mouth Frog
DISTRIBUTION	Gran Chaco region of Argentina, Paraguay, and Bolivia
ADULT HABITAT	Ephemeral pools or underground
LARVAL HABITAT	Ephemeral pools
CONSERVATION STATUS	IUCN Least Concern. Declining due to habitat loss, and susceptible to chytridiomycosis

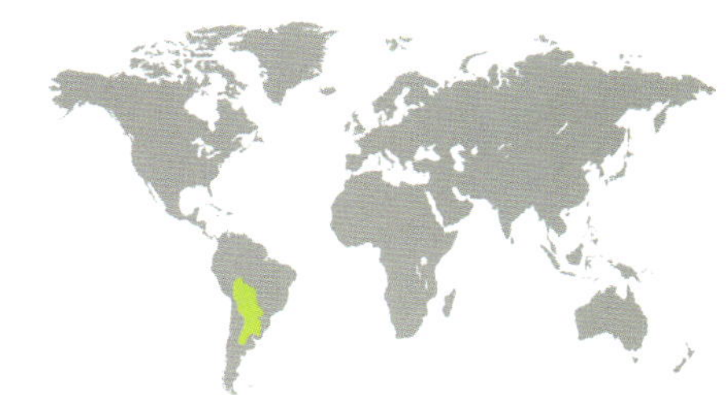

ADULT LENGTH
Male
up to 2⅜ in (60 mm)

Female
up to 3⅞ in (100 mm)

LEPIDOBATRACHUS LAEVIS

BUDGETT'S FROG

BUDGETT, 1899

The Gran Chaco is a large, semiarid lowland in which, after summer rains, ephemeral pools, or pozos, are formed. These are home to this very aggressive frog, which has large teeth in its upper jaw and two fangs in its lower jaw. A nocturnal feeder, it hides underwater, with only its eyes above the surface, waiting for its prey to come near. Its huge mouth enables it to eat large prey, including other frogs. During the dry winter months it lives underground, surrounded by a hard cocoon that prevents water loss from its body. When attacked, it inflates its body, raises itself on its legs, and shrieks loudly.

Budgett's Frog looks almost circular from above. It has a huge head, nostrils and eyes that point upward, and short arms and legs. The toes are fully webbed and there are hard black tubercles on the hind feet for digging. The skin is smooth, except for a V-shaped row of raised glands on the back. These contain sensory organs that detect vibrations in water.

SIMILAR SPECIES

There are two other species in this genus, *Lepidobatrachus asper* and *L. llanensis*, also found in the Gran Chaco. They have a life history similar to that of *L. laevis*. All three species have tadpoles that develop very rapidly. The tadpoles are carnivorous, and have adult-like jaws that enable them to swallow their prey whole. Their diet includes other tadpoles, including even those of their own species.

Actual size

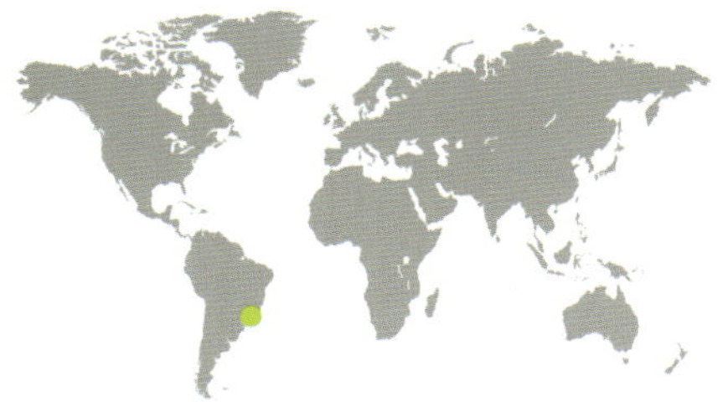

FAMILY	Cycloramphidae
OTHER NAMES	None
DISTRIBUTION	Serra do Mar mountains, southeastern Brazil, up to 3,300 ft (1,000 m) altitude
ADULT HABITAT	Forest
LARVAL HABITAT	In water covering rocks near waterfalls
CONSERVATION STATUS	IUCN Endangered. Though locally common in places, it is thought to be generally declining due to habitat loss

ADULT LENGTH
Male
1⅛–1½ in (29–38 mm)
Female
1¼–1¾ in (31–44 mm)

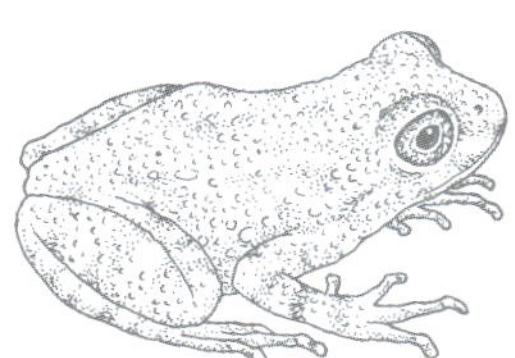

CYCLORAMPHUS IZECKSOHNI

IZECKSOHN'S BUTTON FROG

HEYER, 1983

Actual size

Izecksohn's Button Frog has a robust elliptical body, a rounded snout, and protuberant eyes. The skin is covered in small warts, some of them whitish in color, and there is webbing between the toes. The back is dark brown with some paler and darker blotches, and there are pale and dark stripes on the legs.

This small frog is associated with the fast-flowing forest streams in which it breeds. The tadpoles live and develop out of the water, on the surfaces of wet rocks in the splash zone close to waterfalls. They have very long, thin tails with low tail fins, and a large oral disk around the mouth with which they adhere to rocks. The forest habitats of this and many other frog species are threatened as they are cleared to make way for mining, agriculture, and human settlements.

SIMILAR SPECIES

There are 30 species in the genus *Cycloramphus*, all found in the Atlantic Forest along the southeastern coast of Brazil. *Cycloramphus faustoi* is found only on the small island of Isla de Alcatrazes. This is used by the Brazilian Navy for artillery training, which sometimes causes bush fires. As a result of its restricted range and the threat to its habitat, the species is listed as Vulnerable.

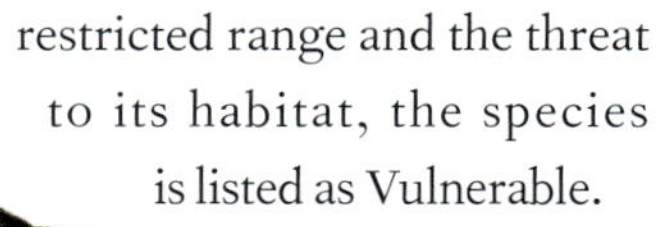

FAMILY	Cycloramphidae
OTHER NAMES	Rock River Frog
DISTRIBUTION	Southeastern Brazil
ADULT HABITAT	Rocky areas in forest
LARVAL HABITAT	Wet rock faces
CONSERVATION STATUS	IUCN Least Concern. Has disappeared from parts of its former range and may be at risk from chytridiomycosis

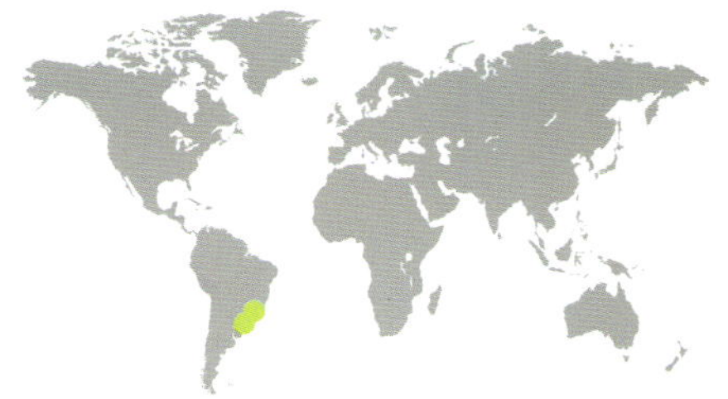

THOROPA MILIARIS

MILITARY RIVER FROG

(SPIX, 1824)

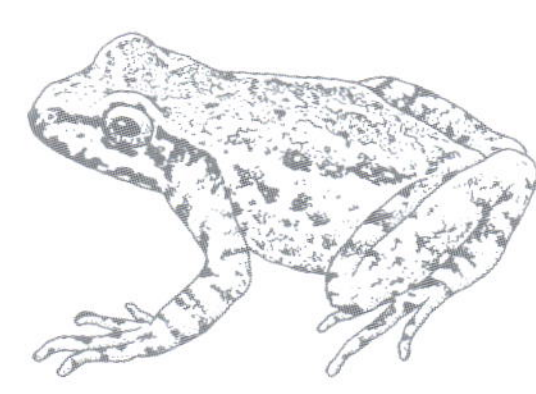

ADULT LENGTH
Male
2⅛–2¹³⁄₁₆ in (54–71 mm)
Female
2⅝–3⅛ in (65–81 mm)

This frog, and the other species in its genus, are unusual in having semiterrestrial tadpoles that live clinging to wet rock faces. They have flattened bodies and long, finless tails, and can cling to vertical rock faces, where they feed initially on algae. Larger, early developing tadpoles become cannibalistic, feeding on eggs of their own species that have not yet hatched. Breeding occurs mostly in December and January, and is centered on crevices where there are areas of rock that are permanently wet. These are scarce and can become very crowded with eggs and developing tadpoles.

SIMILAR SPECIES

The seven species in the genus *Thoropa*, called rock frogs, are all found in southeastern Brazil. In the Petropolis River Frog (*T. petropolitana*), males have been observed defending long-term calling sites in rock crevices, and also attending their eggs. This species and Lutz's River Frog (*T. lutzi*) have both declined dramatically in recent years and are listed as Critically Endangered. It is thought their decline may be due to the disease chytridiomycosis.

The Military River Frog has a rounded snout, large eyes, and very visible tympana. The fingers and toes are long and unwebbed, and do not end in adhesive disks. The back is tan or brown, the legs are brown with darker cross-stripes, and there is a pale stripe on the head between the eyes.

Actual size

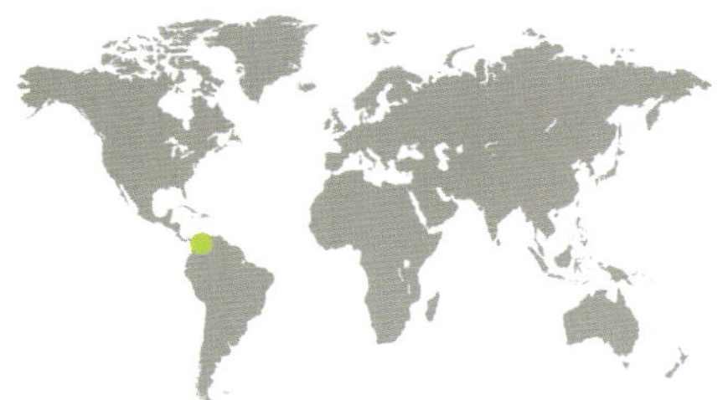

FAMILY	Hemiphractidae
OTHER NAMES	None
DISTRIBUTION	Northern Colombia
ADULT HABITAT	Mountain streams at altitudes of 4,040–8,860 ft (1,230–2,700 m)
LARVAL HABITAT	Within the egg, on the mother's back
CONSERVATION STATUS	IUCN Vulnerable. Has a very restricted range that is threatened by habitat destruction

ADULT LENGTH
average 1⅜ in (35 mm)

CRYPTOBATRACHUS BOULENGERI

BOULENGER'S BACKPACK FROG

RUTHVEN, 1916

Actual size

Boulenger's Backpack Frog is a small frog with a relatively large head and prominent eyes. It has adhesive disks on the ends of its fingers and toes. Its skin is pale brown with darker brown markings.

This small frog is so named because the female carries her developing young on her back. Mating has not been observed, but it results in fewer than 50 large eggs being attached firmly to the female. Unlike many other species in this family, known as "marsupial frogs," the eggs are not contained in a pouch but are exposed to the air. Development is direct, meaning that the tadpole phase is completed within the eggs, which hatch to release tiny froglets. The species frequents mountain streams at altitudes of 4,040–8,860 ft (1,230–2,700 m) on one mountain range in Colombia, the Sierra Nevada de Santa Marta, where its habitat is being reduced by deforestation.

SIMILAR SPECIES

There are five species described to date in this little-known genus. All occur in Colombia, with Pedro Ruiz's Backpack Frog (*C. pedroruizi*) also extending into Venezuela. All species are confined to tiny ranges at high altitude on various mountain chains. *Cryptobatrachus pedroruizi* and Ruthven's Backpack Frog (*C. ruhveni*) are classed as Endangered as a result of habitat loss. The Perijá Backpack Frog (*C. remotus*) is classed as Vulnerable.

FAMILY	Hemiphractidae
OTHER NAMES	Dwarf Marsupial Frog, Fitzgerald's Marsupial Frog
DISTRIBUTION	Trinidad and Tobago, northern Venezuela
ADULT HABITAT	Humid forest
LARVAL HABITAT	Small pools in the leaf axils of bromeliads
CONSERVATION STATUS	IUCN Least Concern, but threatened by habitat destruction

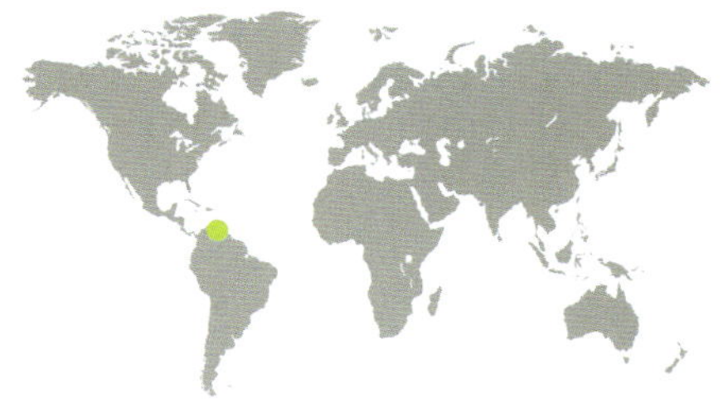

FLECTONOTUS FITZGERALDI

MOUNT TUCUCHE TREE FROG

(PARKER, 1933)

ADULT LENGTH
Male
⅝–¾ in (16–19 mm)
Female
¾–¹⁵⁄₁₆ in (19–24 mm)

The female of this tiny frog has a pouch of skin on her back, in which she keeps up to six very large eggs until they hatch. She then releases the resulting tadpoles into the water-filled leaf axils of bromeliads. The tadpoles do not feed, but complete their development, in 5–20 days, using yolk provided by their mother. Mating has not been described, but males can be heard year-round producing a cricket-like chirp just after sunset. It is possible that females can produce more than one brood in a year.

Actual size

The Mount Tucuche Tree Frog is a very small frog with a rounded snout and large eyes. It is an agile climber and has adhesive disks on the tips of its fingers and toes. The smooth skin on the back is pale brown or yellowish in color. As seen here, there is a pouch of skin that opens down the middle of the back.

SIMILAR SPECIES

The only other species in this genus, *Flectonotus pygmaeus*, is rather larger and can carry more eggs. It is found in the northern mountains of Venezuela and Colombia. Very little is known about this species, but it has been recorded that, during mating, the male inserts the eggs into the female's pouch. These two species are closely related to the genus *Fritziana* (see page 224).

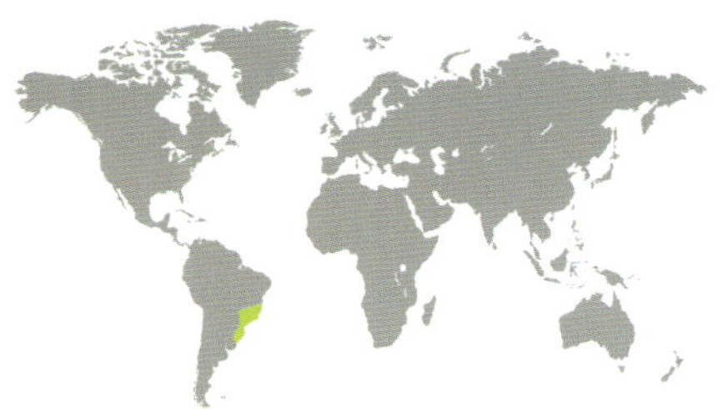

FAMILY	Hemiphractidae
OTHER NAMES	None
DISTRIBUTION	Southeastern Brazil
ADULT HABITAT	Humid forest
LARVAL HABITAT	Small pools in the leaf axils of bromeliads
CONSERVATION STATUS	IUCN Least Concern. A common frog that shows no evidence of decline

ADULT LENGTH
Male
$\frac{7}{8}$–$1\frac{1}{4}$ in (23–32 mm)
Female
$1\frac{1}{4}$–$1\frac{3}{8}$ in (32–35 mm)

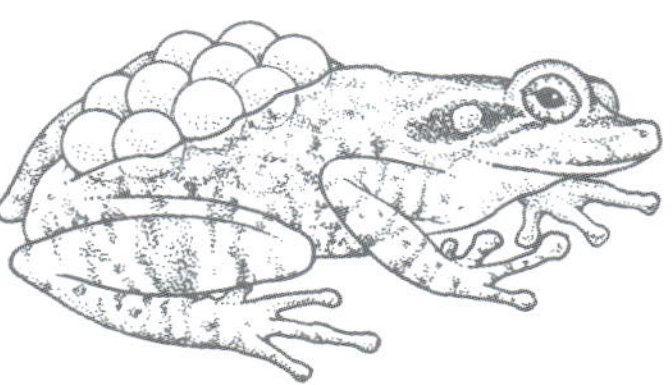

FRITZIANA GOELDII

GOULD'S BACKPACK FROG

(BOULENGER, 1895)

Actual size

Gould's Backpack Frog is a small, slender frog with a pointed snout and large eyes. It is an agile climber and has adhesive disks on the tips of its fingers and toes. The skin on the back is pale brown or gray in color, with darker markings. This female is carrying 12 eggs on her back.

Mating in this small frog is complex and prolonged. While a pair are in amplexus, the female secretes mucus from her cloaca. The male gathers this with his hind feet and whips it into a sticky pad on the female's back. He then moves fertilized eggs forward and presses them into this pad with his feet. This is repeated until the female is carrying a clutch of 9–22 eggs. The mucus pad hardens and cannot be removed without injuring the female. After 19 days the female deposits the eggs into a pool in a bromeliad. They hatch into tadpoles, feeding on yolk, and metamorphose after 21–25 days.

SIMILAR SPECIES

There are six other species in this genus, also found in southeastern Brazil. The female Split-back Frog (*F. fissilis*) develops skin folds on its back, that partially enclose the eggs, which hatch into tadpoles before being released by the female. Ule's Backpack Frog (*F. ulei*) and three recently described species, Izecksohn's Backpack Frog (*F. izecksohni*), the Southern Backpack Frog (*F. mitus*), and Almeida's Split-back Frog (*F. tonini*), are associated with bromeliads. Only Ohaus' Split-back Frog (*F. ohausi*) is associated with giant bamboos, laying its eggs in broken stems. Genus *Fitzania* is related to *Flectonotus* (see page 223).

FAMILY	Hemiphractidae
OTHER NAMES	None
DISTRIBUTION	Eastern slopes of the Andes in Colombia and Ecuador
ADULT HABITAT	Cloud forest with abundant epiphytes
LARVAL HABITAT	Within an egg in a pouch on the mother's back
CONSERVATION STATUS	IUCN Least Concern, but deforestation and spraying of illegal crops are placing it under threat

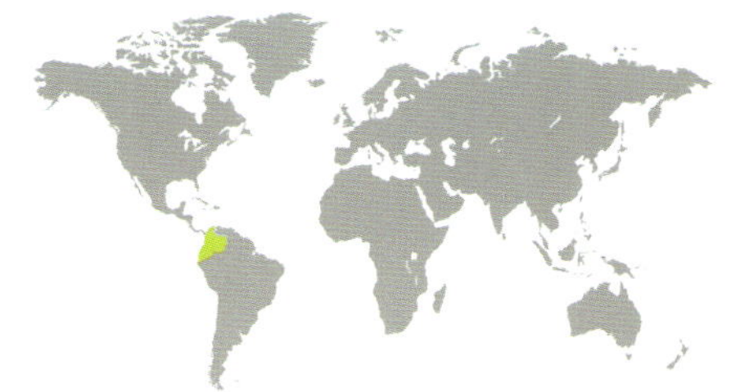

ADULT LENGTH
up to 3 in (77 mm)

GASTROTHECA ANDAQUIENSIS

ANDES MARSUPIAL FROG

RUIZ-CARRANZA & HERNÁNDEZ-CAMACHO, 1976

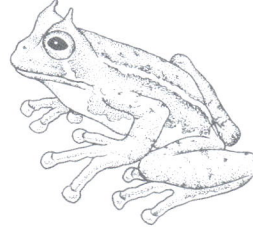

Marsupial frogs are so named because the young develop in a pouch on the female's back. The Andes Marsupial Frog is arboreal and is most commonly found in vegetation near water. During mating, around ten eggs are produced, fertilized, and, with the male's assistance, inserted into the female's pouch. In this species, there is direct development, meaning that there is no free-living tadpole stage, the young completing their development in the pouch and emerging as tiny frogs after hatching. It has been reported that this species is sexually dimorphic in color, males being brown and females green, but green males have also been recorded.

The Andes Marsupial Frog is a large frog with a rounded snout and hornlike pointed appendages above its eyes. There are also spines on its heels. It is variable in color, being brown or green, or a mixture of both. It has large adhesive disks on its fingers and toes that are bright green in color.

SIMILAR SPECIES

There are at least 77 species in this genus, only some of which exhibit direct development. *Gastrotheca ovifera* lives among bromeliads in cloud forest in the coastal mountains of Venezuela and Bolivia, and is listed as Vulnerable. Ten other species are Vulnerable, four Near Threatened, 17 Endangered, and five Critically Endangered, including the Horned Marsupial Frog (*G. cornuta*; page 226). Its eggs are ⅜ in (9.8 mm) in diameter, the largest known of any amphibian.

Actual size

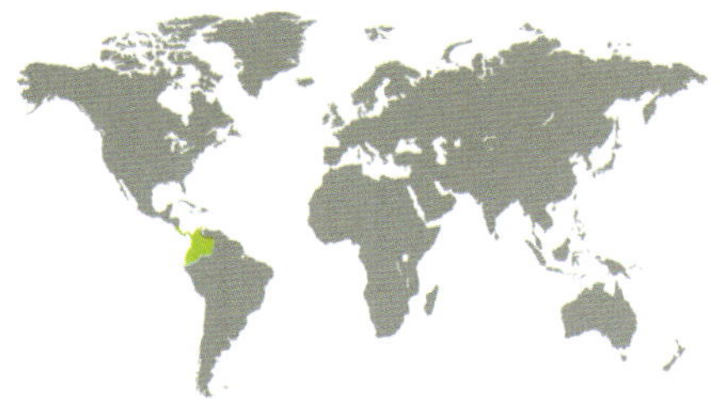

FAMILY	Hemiphractidae
OTHER NAMES	None
DISTRIBUTION	Costa Rica, Panama, Colombia, Ecuador
ADULT HABITAT	Wet lowland forest up to 3,300 ft (1,000 m) altitude
LARVAL HABITAT	Within a pouch on the mother's back
CONSERVATION STATUS	IUCN Critically Endangered. Much of its habitat has been destroyed by deforestation and it has been badly affected by chytridiomycosis. It has not been seen in Costa Rica since 1996

ADULT LENGTH
Male
2$\frac{11}{16}$–3$\frac{1}{8}$ in (66–81 mm)

Female
average 3 in (77 mm)

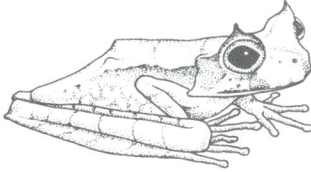

GASTROTHECA CORNUTA

HORNED MARSUPIAL FROG

(BOULENGER, 1898)

This rather large frog lives high up in the forest canopy near rivers and creeks, and is active at night. The male's call is like the popping of a champagne cork, repeated at intervals of 8–12 minutes. The eggs of this species are $\frac{3}{8}$ in (9.8 mm) in diameter, the largest known for any frog. They develop in a pouch on the female's back, each egg within a separate chamber. The tadpole's gills act like a mammalian placenta, attaching to the walls of their chamber, so that there is gas exchange between the tadpole's and the mother's blood. The young eventually emerge from their brood pouch as tiny frogs.

SIMILAR SPECIES

Marsupial frogs of the genus *Gastrotheca* are found in Central America and across South America as far south as Argentina. The Common Marsupial Frog (*G. marsupiata*) occurs in central Peru and southern Bolivia. The Gold-spotted Marsupial Frog (*G. aureomaculata*), which is found only in Colombia, is listed as Endangered. The Abra Acanacu Marsupial Frog (*G. excubitor*) is found in southern Peru and is categorized as Vulnerable.

The Horned Marsupial Frog has a broad head, a blunt snout, and a triangular flap of skin projecting from above each eye. There are large disks on the fingers and toes, and a short spike on each heel. The upper surfaces are dark brown by day and pale brown at night, and there are several dark ridges running transversely across the back.

Actual size

FAMILY	Hemiphractidae
OTHER NAMES	None
DISTRIBUTION	Southern Ecuador, northern Peru
ADULT HABITAT	Humid montane forest at altitudes of 6,200–10,400 ft (1,900–3,180 m)
LARVAL HABITAT	Mountain streams
CONSERVATION STATUS	IUCN Least Concern. Has a large range and there is no evidence of population decline

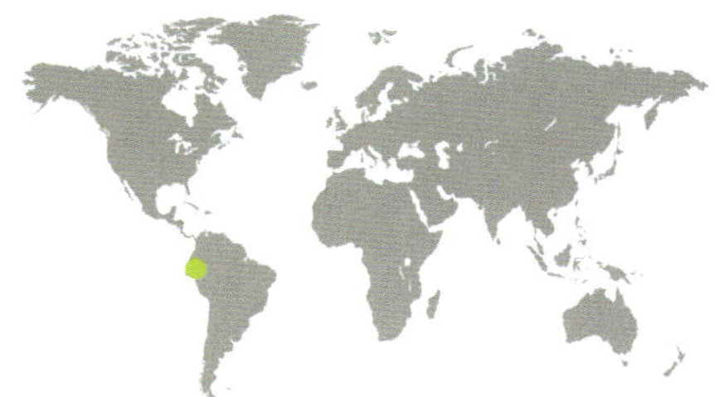

ADULT LENGTH
2¼–2⅞ in (58–73 mm)

GASTROTHECA MONTICOLA

MOUNTAIN MARSUPIAL FROG

BARBOUR & NOBLE, 1920

This species is one of a number of marsupial frogs in which development of the young in the mother's pouch continues only until the eggs hatch. At this point the mother releases the resulting tadpoles into a stream. Adults feed on insects and other invertebrates, and hide under logs during winter and dry periods. At least one other marsupial frog, the San Lucas Marsupial Frog (*G. pseustes*), is known to have been infected by chytridiomycosis, and the disease may become a threat to all marsupial frogs.

SIMILAR SPECIES

Several other *Gastrotheca* species carry their young until the tadpole stage. These include the Salta Marsupial Frog (*G. chrysosticta*), an Endangered species from Argentina, and the San Lucas Marsupial Frog (*G. pseustes*), a species from Ecuador that is listed as Near Threatened. The Riobamba Marsupial Frog (*G. riobambae*), from Colombia and Ecuador, was once collected for the international pet trade but the widespread destruction of its habitat has led to this species being listed as Vulnerable.

The Mountain Marsupial Frog has a broad head, large, protruding eyes, and adhesive disks on its fingers and toes. It is green or brown in color. As shown in the illustration above, a female carrying eggs in her dorsal pouch has a bloated appearance.

Actual size

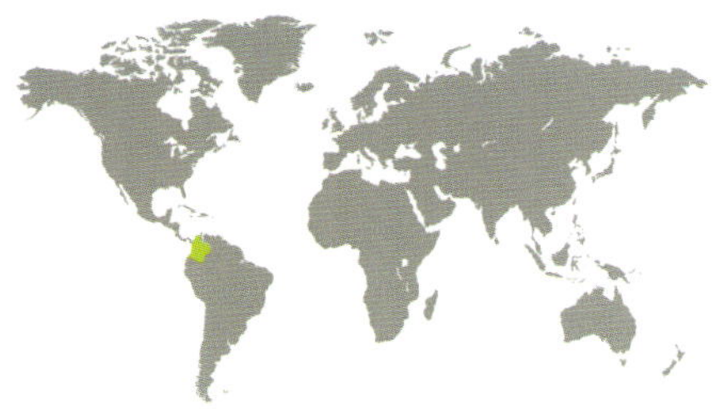

FAMILY	Hemiphractidae
OTHER NAMES	None
DISTRIBUTION	Colombia
ADULT HABITAT	Moist mountain forest at altitudes of 7,120–8,330 ft (2,170–2,540 m)
LARVAL HABITAT	Ponds or pools in mountain streams
CONSERVATION STATUS	IUCN Endangered. Its range has been reduced by habitat loss and other factors to just five known locations

ADULT LENGTH
Male
up to 2 in (50 mm)
Female
1$^{9}/_{16}$–2¾ in (40–68 mm)

GASTROTHECA TRACHYCEPS

CERRO MUNCHIQUE MARSUPIAL FROG

DUELLMAN, 1987

This species is one of only two among the marsupial frogs in which the skin on the head is fused to its skull. In other frogs in which there is such "co-ossification" on the head, it is associated with hiding by day in a hole, with the head closing the entrance. This behavior may reduce water loss during the day and also act as a defense against predators. This species is known to hide in bromeliads during the day. As in other marsupial frogs, the fertilized eggs are transferred during mating to a pouch on the female's back. The young are released, as tadpoles, into pools or mountain streams.

SIMILAR SPECIES

The other marsupial frog with ossified head skin is Niceforo's Marsupial Frog (*Gastrotheca nicefori*), a relatively abundant species found in Panama, Colombia, and Venezuela. It retains its young in its dorsal pouch until the froglet stage. So, too, does Günther's Marsupial Frog (*G. guentheri*), from the Andes in Colombia and Ecuador. This species, listed as Data Deficient, is the only frog known to have teeth in both lower and upper jaws.

The Cerro Munchique Marsupial Frog has a blunt snout and very long limbs, fingers, and toes. There are well-developed adhesive pads on the ends of its fingers and toes. The top of the head is rough to the touch. It is generally green or brown with dark stripes on its back, and cream on its underside.

Actual size

FAMILY	Hemiphractidae
OTHER NAMES	None
DISTRIBUTION	Upper Amazon Basin and lower Andes of Colombia, Ecuador, and Peru
ADULT HABITAT	Humid mountain forest up to 4,000 ft (1,200 m) altitude
LARVAL HABITAT	Within the egg, on the mother's back
CONSERVATION STATUS	IUCN Least Concern. A generally rare frog, vulnerable to habitat loss

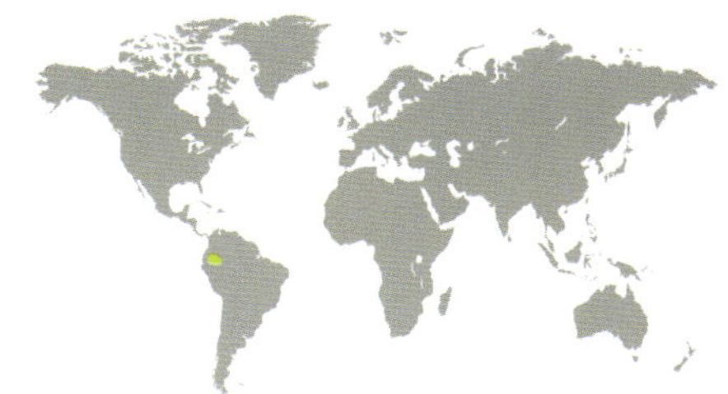

HEMIPHRACTUS PROBOSCIDEUS

SUMACO HORNED TREE FROG

(JIMÉNEZ DE LA ESPADA, 1871)

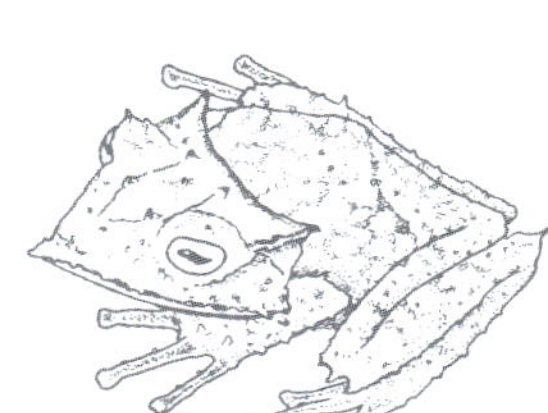

ADULT LENGTH
Male
$1^{11}/_{16}$–$1^{15}/_{16}$ in (43–50 mm)
Female
$2^{3}/_{16}$–$2^{11}/_{16}$ in (57–66 mm)

This strange-looking frog hunts at night for other frogs, lizards, and large insects. Its ability to deal with such large prey is enhanced by the possession of large, toothlike fangs in its lower jaw. It is generally found 3–8 ft (1–2.5 m) above the ground. When attacked, it opens its large mouth to reveal a bright yellow interior and tongue. The female carries up to 26 eggs on her back, attached with a gelatinous secretion. There is direct development, the larvae developing inside the eggs, which hatch to produce tiny froglets.

SIMILAR SPECIES

There are eight other species in the genus *Hemiphractus*. All are found in northwestern South America, and all have a similar appearance and similar natural history. They are also all rather rare, and are most commonly found where the frogs on which they prey are most abundant. The Banded Horned Tree Frog (*H. fasciatus*) is found in Colombia, Ecuador, and Panama and is categorized as Vulnerable, its numbers having declined due to habitat destruction.

The Sumaco Horned Tree Frog has a large triangular head, the points of which are formed by fleshy projections on the snout and eyelids. The body is flattened and there are sharp spurs on the heels. The back is brown or tan with green, brown, or gray stripes and spots. The underside is tan with orange spots. The tongue and interior of the mouth are bright yellow.

Actual size

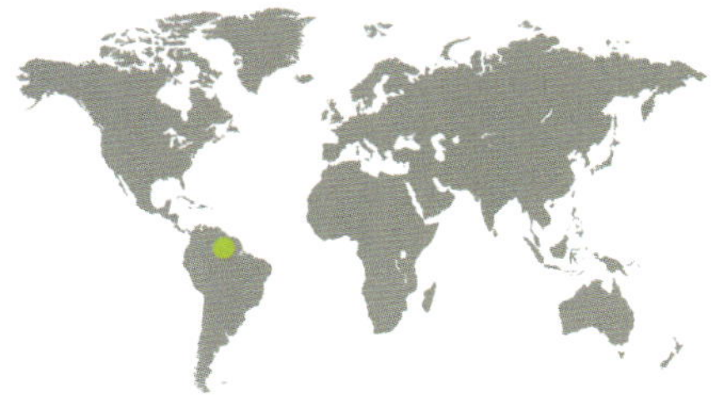

FAMILY	Hemiphractidae
OTHER NAMES	Evans' Stefania
DISTRIBUTION	West-central Guyana
ADULT HABITAT	Lowland and montane forest
LARVAL HABITAT	Within the egg, on the mother's back
CONSERVATION STATUS	Data Deficient

ADULT LENGTH
Male
up to 2 1/16 in (53 mm)
Female
up to 3 13/16 in (98 mm)

STEFANIA EVANSI

GROETE CREEK TREE FROG

(BOULENGER, 1904)

The Groete Creek Tree Frog is a large frog with a pointed snout, large eyes, and adhesive disks on its fingers and toes. Generally brown in color, it occurs in two color morphs, one plain and the other striped on its back. The legs have pale and dark brown stripes.

Frogs of the genus *Stefania*, known as "carrying frogs," are characterized by their mode of parental care. The female carries the eggs on her back, stuck to her in a layer of mucus, until they have developed into tiny frogs. Development is direct, meaning that there is no free-swimming tadpole stage. The tadpole stage is completed in the egg, which hatches to release a tiny frog 2–3 months after fertilization. The female can carry between 11 and 30 young, depending on her size. Mating takes place in the wet season but has not otherwise been described.

SIMILAR SPECIES

To date, 23 species of *Stefania* have been described. Most occur on the isolated, flat-topped mountains, or tepuis, that are a feature of northern South America. Six species are listed as Near Threatened, seven as Vulnerable, but the natural history of some species is so little known that they are listed as Data Deficient. Brewer's Carrying Frog (*S. breweri*) is known from only a single specimen, found at 4,100 ft (1,250 m) altitude on an isolated tepui called Cerro Autana. It is listed as Vulnerable, while the Mt. Roraima Carrying Frog (*S. roraimae*) is currently the only Endangered species in the genus.

Actual size

FAMILY	Hylodidae
OTHER NAMES	Brazilian Spiny-thumb Frog
DISTRIBUTION	Southeastern Brazil
ADULT HABITAT	Forest
LARVAL HABITAT	Streams
CONSERVATION STATUS	IUCN Least Concern. Has declined in some parts of its range

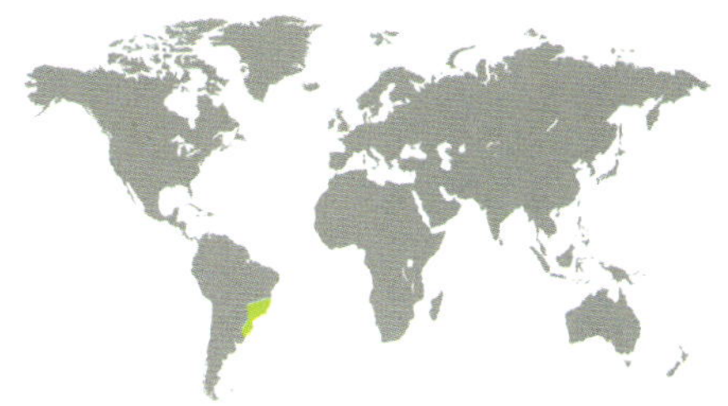

CROSSODACTYLUS GAUDICHAUDII

GAUDICHAUD'S FROG

DUMÉRIL & BIBRON, 1841

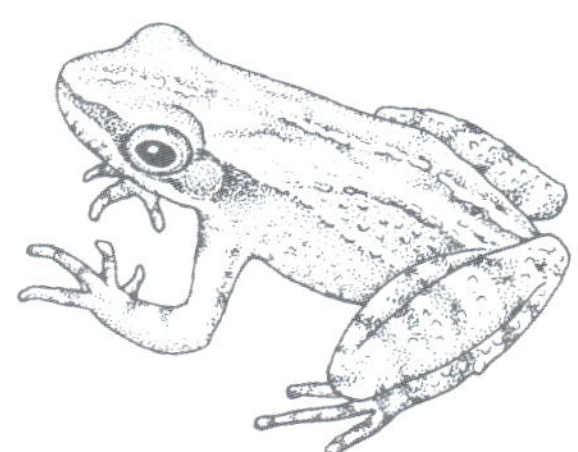

ADULT LENGTH
Male
⅞ in (22 mm);
based on one individual

Female
unknown

This small, diurnally active frog is found close to streams in the Atlantic Forest in southeastern Brazil. Males are territorial, calling from the edge of a stream and, from time to time, diving into the water to excavate a nest site underneath a rock. When a female responds positively to a male, they dive together to the nest site to mate and lay eggs. After a while, the male dives again to cover the nest entrance with stones. Tadpoles at various stages of development are found in streams throughout the year, suggesting that breeding can occur at any time.

Actual size

Gaudichaud's Frog has two spines on each thumb. The fingers and toes are long, are not webbed, and do not have adhesive disks. The back is brown with broken dark stripes down each side and there may be a pale stripe down its middle.

SIMILAR SPECIES

The 13 species in this genus are called spiny-thumb frogs, because of the spines on the thumbs of both sexes. It is not known what function these serve. Indeed, as recently as 2016 these frogs were scarcely known, with eight species listed as Data Deficient. Today, five are listed as Critically Endangered, including Boulenger's Spiny-thumb Frog (*C. boulengeri*) and the recently described Rio São Francisco Spiny-thumb Frog (*C. franciscanus*), two species are Endangered, one Vulnerable, and one Near Threatened. Populations are threatened by habitat destruction, pollution, and the introduction of North American Bullfrogs (*Aquarana catesbeiana*; page 545).

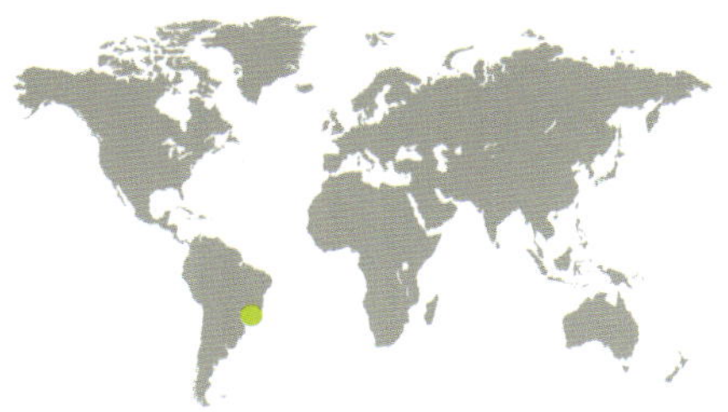

FAMILY	Hylodidae
OTHER NAMES	Warty Tree Toad
DISTRIBUTION	Southeastern Brazil
ADULT HABITAT	Forest up to 4,000 ft (1,200 m) altitude
LARVAL HABITAT	Forest streams
CONSERVATION STATUS	IUCN Least Concern. Its range includes several protected areas

ADULT LENGTH
Male
1⁹⁄₁₆–1⅝ in (39–42 mm)
Female
1¹¹⁄₁₆–1¹⁵⁄₁₆ in (43–50 mm)

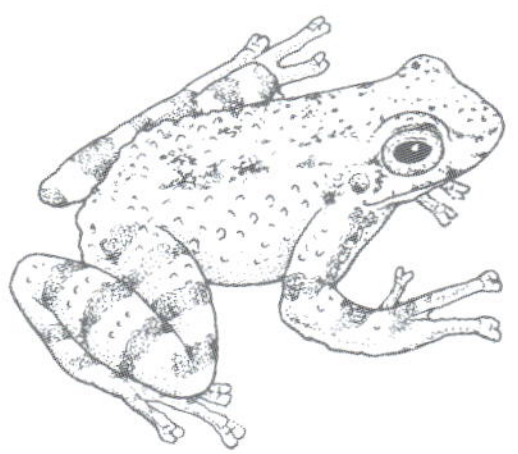

HYLODES ASPER

BRAZILIAN TORRENT FROG

(MÜLLER, 1924)

Actual size

The Brazilian Torrent Frog has a pointed snout, prominent eyes, and warty skin. The upper surfaces are khaki in color, with darker patches and some small pale spots, the flanks are orange, and the underside is pale. The fingers and toes have well-developed adhesive disks. The male has paired vocal sacs and silvery coloration on the edges of his toes.

This frog is best known for the male's use of foot-waving to attract females. It lives close to fast-flowing streams, a noisy environment where calling is less effective. Spray from streams keeps the air moist, allowing frogs to be active by day. Males call from elevated positions, from sunrise to sunset, and extend and wave one hind leg at a time. This exposes the silvery coloration on the hind feet. Foot-waving serves both to signal territorial ownership to male rivals and to attract females. From time to time, a male will dive into the stream to excavate an underwater chamber where the eggs will be laid.

SIMILAR SPECIES

There is a total of 26 species in this little-studied genus, all found in Brazil and Argentina. The fungus that causes the disease chytridiomycosis has been found in the Endangered São Paulo Tree Toad (*Hylodes magalhaesi*) from southeastern Brazil. Vanzolini's Tree Toad (*H. vanzolinii*) and three other species are Critically Endangered. This is a cause for concern because, in other parts of the world, this disease has particularly affected stream-living frogs.

FAMILY	Hylodidae
OTHER NAMES	None
DISTRIBUTION	Southeastern Brazil
ADULT HABITAT	Mountain streams in forest
LARVAL HABITAT	Calmer parts of mountain streams
CONSERVATION STATUS	IUCN Least Concern. Its range includes protected areas. Has tested positive for chytridiomycosis

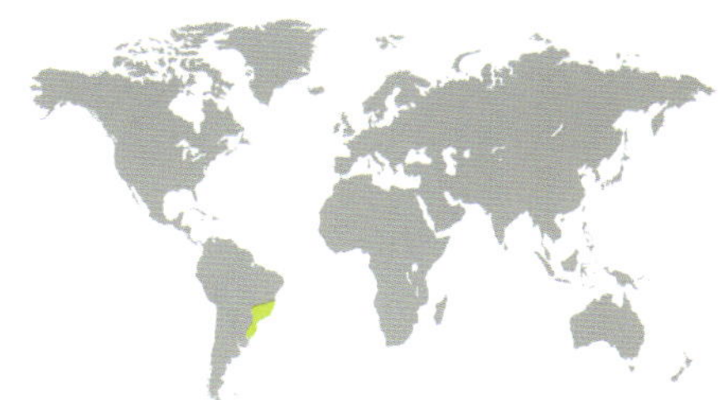

MEGAELOSIA GOELDII

RIO BIG-TOOTH FROG

(BAUMANN, 1912)

ADULT LENGTH
Male
average 3½ in (90 mm)
Female
average 3⅞ in (100 mm)

This large frog is very difficult to collect because it lives in and around mountain streams and dives into deep water at any sign of a threat. It is one of many frog species that are found only in the Atlantic Forest that runs along the southeastern coast of Brazil. It is most commonly seen sitting on rocks emerging from the water. Its mating call has not been recorded, but the fact that tadpoles are found in streams suggests that the eggs are laid in the water. It feeds on insects, other invertebrates, and smaller frogs.

SIMILAR SPECIES

In 2016 the genus *Megaelosia* contained seven species but in 2021 six of them were moved to the genus *Phantasmarana*, which now contains eight species. All are found in streams in the Atlantic Forest. Two of these, Massart's Big-toothed Frog (*P. massarti*) and the Agile Big-toothed Frog (*P. apuana*), are larger than *M. goeldii*. *Phantasmarana massarti* and the Boticário Big-toothed Frog (*P. boticariana*) are Endangered.

The Rio Big-tooth Frog has a large head with a rounded snout, and muscular arms and legs. There are small disks on the fingers and toes, which are not webbed. The back and limbs are dark brown in color, with black mottling on the back and black cross-bands on the legs.

Actual size

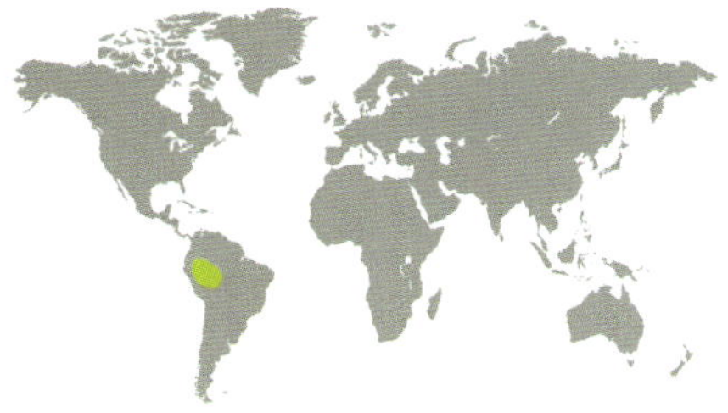

FAMILY	Leptodactylidae: Leiuperinae
OTHER NAMES	None
DISTRIBUTION	Lower slopes of the eastern Andes in Peru, Ecuador, Colombia, and Brazil
ADULT HABITAT	Leaf litter in lowland forest
LARVAL HABITAT	Ephemeral pools
CONSERVATION STATUS	IUCN Least Concern. Has a very large range that includes protected areas

ADULT LENGTH
15/16–1 7/16 in (24–37 mm)

EDALORHINA PEREZI

PEREZ'S SNOUTED FROG

JIMÉNEZ DE LA ESPADA, 1870

Actual size

Perez's Snouted Frog is a toad-like frog with a rounded snout that has a cone-shaped tubercle on its tip. There are small skin projections above the eyes, looking rather like eyelashes. There is a fold in the skin, running down each side from the eye to the groin. The frog is brown above, black on its flanks, and white on its belly.

This small frog is active by day and uses its resemblance to dead leaves as camouflage on the forest floor. Males call in isolation from one another, producing a series of low-pitched, short whistles. A male and female can remain in amplexus for as long as six days. Experiments have shown that they are selective when seeking a pool for their eggs, avoiding those that contain insect predators and those that already contain tadpoles. The female produces a secretion that the male whips into a foam with his hind legs. The pair make several foam nests, each containing 30–40 eggs.

SIMILAR SPECIES

The only other species in this genus is the Common Snouted Frog (*Edalorhina nasuta*), which has a long, fleshy snout. It is found in Peru and its natural history is little known.

FAMILY	Leptodactylidae: Leiuperinae
OTHER NAMES	None
DISTRIBUTION	Central and northern South America, Trinidad and Tobago
ADULT HABITAT	Lowland forest, near natural and manmade ponds
LARVAL HABITAT	Temporary pools, puddles, and potholes
CONSERVATION STATUS	IUCN Least Concern. It is very common, has a stable population, and is not subject to any obvious threats

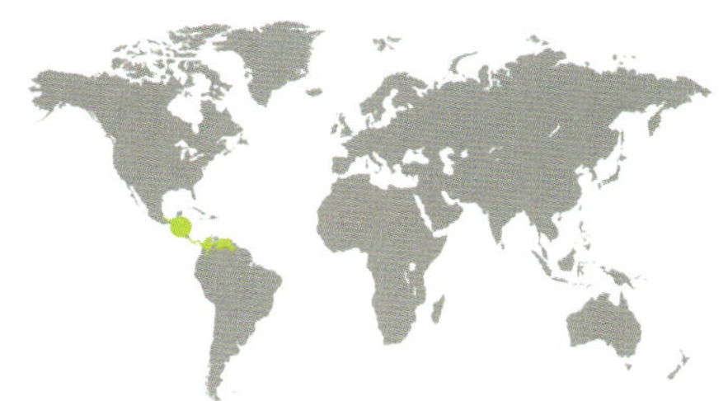

ENGYSTOMOPS PUSTULOSUS

TÚNGARA FROG

(COPE, 1864)

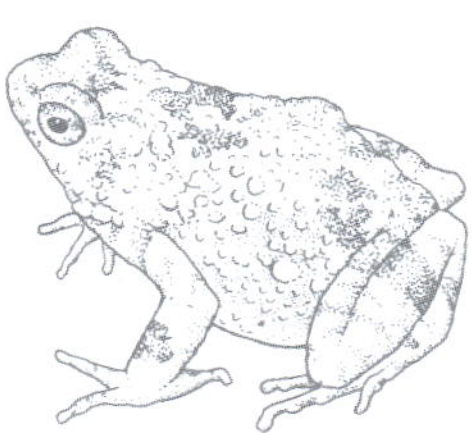

ADULT LENGTH
Male
up to 1⁵⁄₁₆ in (33 mm)
Female
up to 1⅜ in (35 mm)

Drab in appearance, the Túngara Frog has an exciting, hazardous sex life. After heavy rain, males enter any kind of pool, including flooded wheel-ruts, and start to call loudly. Their call consists of a "whine," followed by one to six "chucks." The number of "chucks" depends on the level of competition for females; females prefer calls with several "chucks" and so a male has to respond to calling neighbors by performing more "chucks." However, this strategy is risky; their calls also attract the predatory Fringe-lipped Bat (*Trichops cirrhosus*), which is more likely to eat males producing calls with more "chucks."

Actual size

The Túngara Frog is a small grayish-brown frog with skin covered in warts and pustules, giving it a toad-like appearance. The snout is pointed and the eyes are protruding. There may be lighter or darker blotches on the back and transverse stripes on the legs.

SIMILAR SPECIES

There are nine species in this genus, distributed across Central and South America. All lay their eggs in floating foam nests, consisting of a secretion produced by the female and whipped up by both her and the male beating their hind legs. The foam keeps the eggs and the developing tadpoles moist and cool, and also protects them from pathogens and predators.

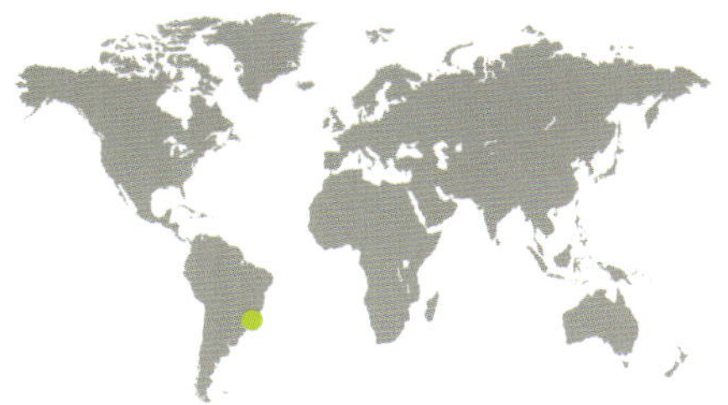

FAMILY	Leptodactylidae: Leiuperinae
OTHER NAMES	None
DISTRIBUTION	Atlantic Forest, southeastern Brazil
ADULT HABITAT	Formerly forest, now more open habitats
LARVAL HABITAT	Ponds, marshes
CONSERVATION STATUS	IUCN Least Concern. Population declining, as a result of habitat destruction and, possibly, chytridiomycosis

ADULT LENGTH
average 1 in (26 mm)

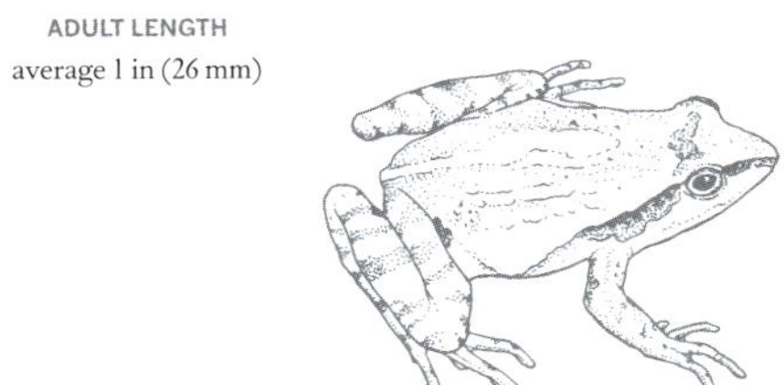

PHYSALAEMUS BARRIOI

BOCAINA DWARF FROG

BOKERMANN, 1967

Actual size

The Bocaina Dwarf Frog has an elliptical body and a triangular head. The fingers and toes are long and thin and are not webbed. The back is brown or dark green, and the legs are striped. There is a dark stripe through the eye and yellow, and red and black patches in the groin.

Known only from a single forested mountain range, and living at an altitude of 4,000 ft (1,200 m) above sea level, this small frog seems to have a precarious future. The forest where it was first described has largely been destroyed and it is now found around permanent marshes in more open areas. Males call at night between August and February, and the eggs are laid in foam nests anchored to vegetation at the water's edge. After hatching, the tadpoles leave the protection provided by the foam nest and disperse into the water.

SIMILAR SPECIES

There are 50 species in the genus *Physalaemus*, found across Central and South America. *Physalaemus barrioi* is closely related to the Graceful Dwarf Frog (*P. gracilis*), a common species from southern Brazil and Uruguay. It lives in forest borders and Cerrado grassland, breeding in ephemeral ponds. It has adapted well to human activities, and is found in disturbed and polluted habitats. The Santa Cruz Dwarf Frog (*P. soaresi*) and Angra dos Reis Dwarf Frog (*P. angrensis*), both from Brazil, are Critically Endangered.

FAMILY	Leptodactylidae: Leiuperinae
OTHER NAMES	None
DISTRIBUTION	Eastern Amazon Basin of Brazil, Guyana, French Guiana, Suriname, Venezuela
ADULT HABITAT	Lowland forest clearings and edges
LARVAL HABITAT	Ephemeral pools
CONSERVATION STATUS	IUCN Least Concern. Thrives in disturbed habitats, degraded forest, and near human settlements

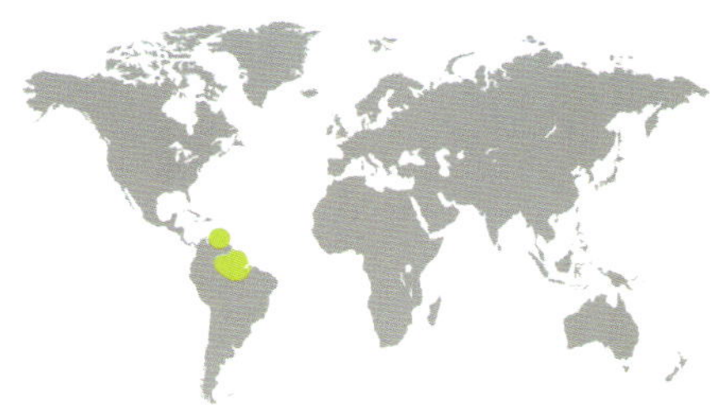

ADULT LENGTH
1–1⁵⁄₁₆ in (26–33 mm)

PHYSALAEMUS EPHIPPIFER

STEINDACHNER'S DWARF FROG

(STEINDACHNER, 1864)

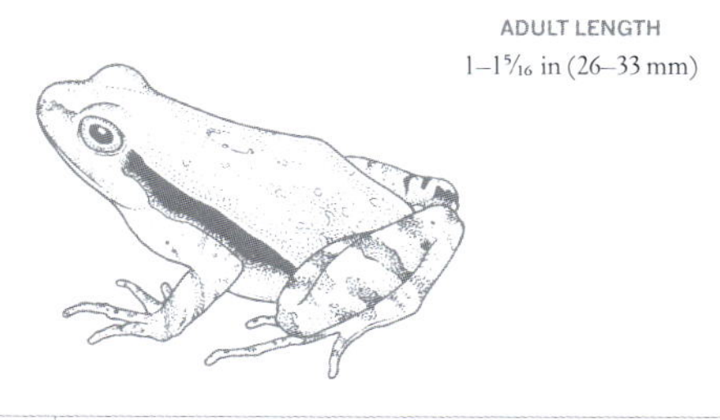

This small and very common frog breeds over a three-month period, beginning in February when the rainy season starts. Males form noisy choruses after dark around shallow pools. The call consists of a "whine," followed by a variable number of "chucks." A pair takes about 40 minutes to mate and creates a foam nest in which the eggs are laid. The nest is made from a secretion produced by the female's oviducts that is whipped up by the male's legs. The eggs hatch after three days and the nest disintegrates after 4–6 days, releasing the tadpoles into the water.

SIMILAR SPECIES

The call of *Physalaemus ephippifer* is very similar to that of the Túngara Frog (*Engystomops pustulosus*; page 235). All species in this genus make foam nests, which protect their eggs from a variety of threats. In *P. ephippifer*, it is thought that the foam prevents eggs being eaten by tadpoles of the same species. In the closely related Barker Frog (*P. cuvieri*), the main threat to eggs is a predatory fly.

Actual size

Steindachner's Dwarf Frog is a stockily built little frog with a pointed snout. The fingers and toes are not webbed. The back is brown with darker spots, the flanks are dark brown, and the underside is cream. There are dark brown transverse bars on the legs, and orange or red patches in the armpit and the groin.

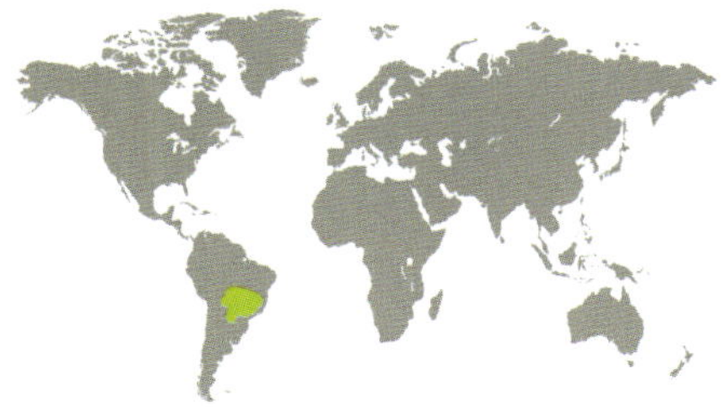

FAMILY	Leptodactylidae: Leiuperinae
OTHER NAMES	Previously *Eupemphis nattereri*
DISTRIBUTION	Bolivia, southern Brazil, Paraguay
ADULT HABITAT	South American grassland and savanna (Cerrado)
LARVAL HABITAT	Permanent and ephemeral pools
CONSERVATION STATUS	IUCN Least Concern. Has a large range but is declining in many places where its habitat is cleared for agriculture

ADULT LENGTH
Male
$1\frac{11}{16}$–$2\frac{1}{8}$ in (43–55 mm)

Female
$1\frac{11}{16}$–$2\frac{3}{16}$ in (43–56 mm)

PHYSALAEMUS NATTERERI

CUYABA DWARF FROG

(STEINDACHNER, 1863)

The Cuyaba Dwarf Frog is renowned for the two large black and white eyespots on its lower back. When threatened, it inflates its body, lowers its head, and lifts up its rear end, displaying these false eyes to startle its attacker (see illustration above). If this display fails to deter a predator, glands within the eyespots expel a noxious secretion into its mouth. Living underground for much of the time, it emerges to breed after heavy rain, between October and January. Eggs are laid in foam nests on steep slopes close to water, with nests of several pairs often touching each other.

SIMILAR SPECIES

Physalaemus nattereri was for a while placed in its own monotypic genus, *Eupemphix*, before being returned to *Physalaemus*. It lives in the same habitat as the Menwig Frog (*P. albonotatus*), a smaller frog that, unlike *P. nattereri*, thrives in agricultural land, even where it is polluted by pesticides.

The Cuyaba Dwarf Frog has a plump toad-like body, a blunt snout, and short, muscular limbs. There is no webbing on its fingers or toes. Its skin is decorated with a complex marbled pattern in darker and lighter shades of brown. The black eyespots on its lower back are normally hidden by the hind legs.

Actual size

FAMILY	Leptodactylidae: Leiuperinae
OTHER NAMES	None
DISTRIBUTION	The Andes in Bolivia, Peru, Chile, and Argentina
ADULT HABITAT	High-altitude grassland and scrubland
LARVAL HABITAT	Slow-moving streams and small ponds
CONSERVATION STATUS	IUCN Vulnerable

ADULT LENGTH
Male
up to $1\frac{1}{8}$ in (28 mm)

Female
up to $1\frac{1}{4}$ in (32 mm)

PLEURODEMA MARMORATUM

MARBLED FOUR-EYED FROG

(DUMÉRIL & BIBRON, 1840)

Males of this colorful frog have the distinction of having the largest testes, relative to body size, of any frog—they occupy a third of the body cavity. There is no explanation as to why the frogs should be blessed in this way. In Peru, as climate change causes the glaciers to melt, so this frog—an inhabitant of higher altitudes in the Andes—is expanding its range upward, breeding in ponds that previously were frozen. This species has tested positive for the fungus that causes chytridiomycosis, but there is no evidence that it is causing mortality.

Actual size

The Marbled Four-eyed Frog has a rounded snout, large eyes, rather short hind legs, and toes that are not webbed but do have fringes of skin along them. The back is tan or green, with small and large, dark brown or dark green markings and some patches of red.

SIMILAR SPECIES

Many of the 15 species in this genus, known as four-eyed frogs, have large spots or color patches on their lower back, which they expose when attacked by enemies such as the Cuyaba Dwarf Frog (*Physalaemus nattereri*; page 238). The Chilean Four-eyed Frog (*P. thaul*), for example, has two large black eyespots, which it exposes in a head-down, rear-end-raised posture. Despite its common name, *Pleurodema marmoratum* lacks this character.

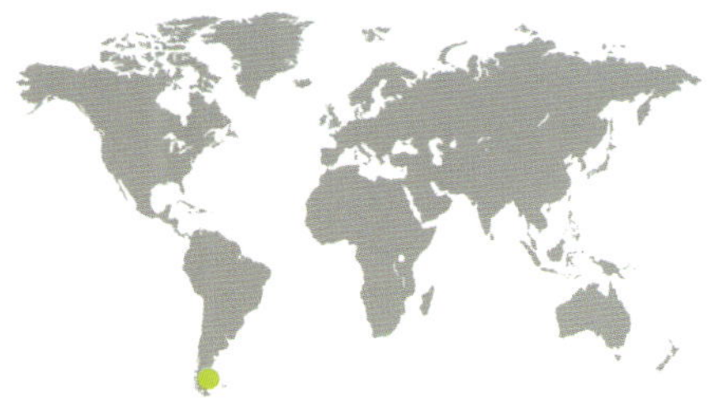

FAMILY	Leptodactylidae: Leiuperinae
OTHER NAMES	*Pleurodema somuncurense*, El Rincon Stream Frog
DISTRIBUTION	Southern Argentina
ADULT HABITAT	Geothermal wetlands
LARVAL HABITAT	Thermal springs and streams
CONSERVATION STATUS	IUCN Critically Endangered. Has a very restricted range that is badly fragmented by habitat degradation

ADULT LENGTH
Male
unknown

Female
average 1½ in (38 mm)

PLEURODEMA SOMUNCURENSIS

SOMUNCURA FROG

(CEI, 1969)

Actual size

The Somuncura Frog has a flattened body, smooth skin, and long hind legs and toes. The eyes are large and point upward. The upper surfaces are yellow-brown with irregular dark spots, and there is a narrow, pale yellow stripe down the middle of the back. The underside is yellow with a dark, reticulated pattern.

This wholly aquatic frog is found only on the Somuncura Plateau, an isolated area of volcanic origin in Argentina's Rio Negro province. It is typically found under stones in streams that are fed by thermal springs. Its mating behavior, eggs, and tadpoles have not been described. It has been very badly affected by introduced Rainbow Trout (*Oncorhynchus mykiss*), which eat its tadpoles. The area is heavily grazed by sheep and goats, and these have caused degradation and pollution of streams. The area falls within a protected area, but this is ineffectively managed.

SIMILAR SPECIES

At one time this wholly aquatic frog was contained in its own monotypic genus *Somuncuria* and it has also been included in *Telmatobius* which contains the fully aquatic Titicaca Water Frog (*T. culeus*; page 258) and the Lake Junin Frog (*T. macrostomus*; page 259), but those frogs belong to a different family, Telmatobidae.

FAMILY	Leptodactylidae: Leiuperinae
OTHER NAMES	None
DISTRIBUTION	Southern Brazil, southeastern Paraguay, Uruguay, northeastern Argentina
ADULT HABITAT	Grassland up to 3,300 ft (1,000 m) altitude
LARVAL HABITAT	Ephemeral pools and ditches
CONSERVATION STATUS	IUCN Least Concern. A common species that has a large range

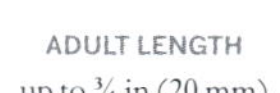

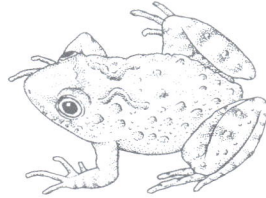

ADULT LENGTH
up to ¾ in (20 mm)

PSEUDOPALUDICOLA FALCIPES

HENSEL'S SWAMP FROG

(HENSEL, 1867)

Although this small frog is very common, it is scarcely known. While its natural habitat is open grassland, it has thrived where this has been replaced by agricultural land. Rice fields in particular have compensated for the loss of the wetland areas that it requires for breeding. It also breeds in drainage ditches. The male, which has a single large vocal sac, calls from the ground. The eggs are laid in a mass in shallow water, and are not enclosed within a foam nest like those of many closely related frogs.

Actual size

Hensel's Swamp Frog has a plump body, a rounded snout, and large eyes. The skin on the back is warty. The upper surfaces are brown or gray, with darker markings on the back and cross-stripes on the legs. Some individuals have a conspicuous yellow stripe down the middle of the back.

SIMILAR SPECIES

Twenty-six species have been described in the genus *Pseudopaludicola*, fifteen of them since 2010. Known as swamp frogs, they are distributed across much of South America. The Bolivian Swamp Frog (*Pseudopaludicola boliviana*) is a common terrestrial frog that has a huge range in central South America, from Colombia and Venezuela in the north, to Argentina in the south. The Leticia Swamp Frog (*Pseudopaludicola ceratophryes*) lives in the leaf litter of forests in Colombia and Peru.

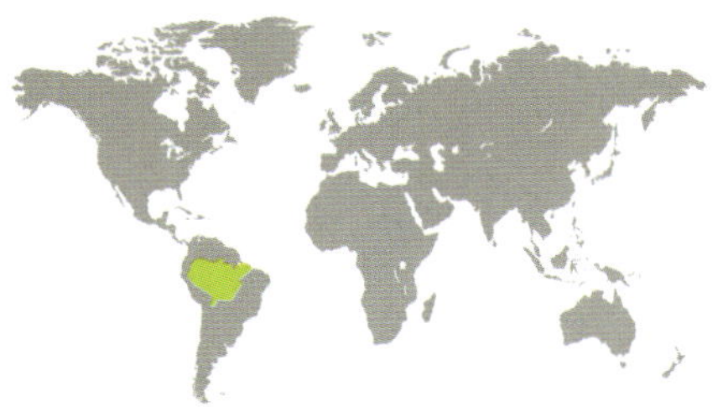

FAMILY	Leptodactylidae: Leptodactylinae
OTHER NAMES	None
DISTRIBUTION	Amazon Basin, South America
ADULT HABITAT	Tropical forest
LARVAL HABITAT	Within the egg, in a foam nest underground
CONSERVATION STATUS	IUCN Least Concern. Has a very large range and no evidence of general decline

ADULT LENGTH
Male
$^{11}/_{16}$–$^{15}/_{16}$ in (18–24 mm)
Female
$^{7}/_{8}$–$1^{1}/_{8}$ in (23–28 mm)

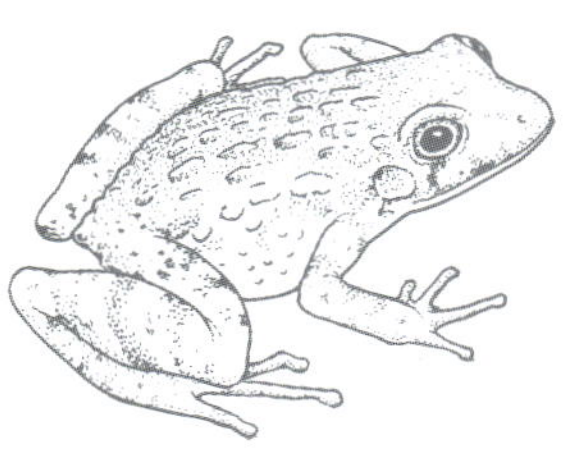

ADENOMERA ANDREAE

LOWLAND TROPICAL BULLFROG

(MÜLLER, 1923)

Actual size

The Lowland Tropical Bullfrog has a long, pointed snout and longitudinal rows of low tubercles on its back. The back is gray or tan with dark brown spots, and there are sometimes cream or pink stripes along each side. There is usually a triangular brown patch on the back of the head.

This small frog is active by day and night and lives on the forest floor. The male's call is a single, harsh note that has been likened to the cry of a kitten. Around 20 eggs are laid in a foam nest within a flask-shaped cavity excavated in the ground by the male. The eggs contain a large amount of yolk that sustains the tadpoles as they develop within them. Development is direct, the eggs hatching to release tiny frogs.

SIMILAR SPECIES

There is a total of 32 species in the genus *Adenomera*. These are little-known frogs that appear to have a similar life history. All are found in northern South America and none is of conservation concern. Most are more easily distinguished by their calls than by their appearance. In *A. simonstuarti*, from Bolivia and Peru, adult males have a ridge on the snout that they use for digging.

FAMILY	Leptodactylidae: Leptodactylinae
OTHER NAMES	Moaning River Frog
DISTRIBUTION	Amazon Basin of Brazil, Colombia, Peru, Bolivia, and French Guiana
ADULT HABITAT	Flooded lowland forest
LARVAL HABITAT	Unknown
CONSERVATION STATUS	IUCN Least Concern. Has a large range and there is no evidence of population decline

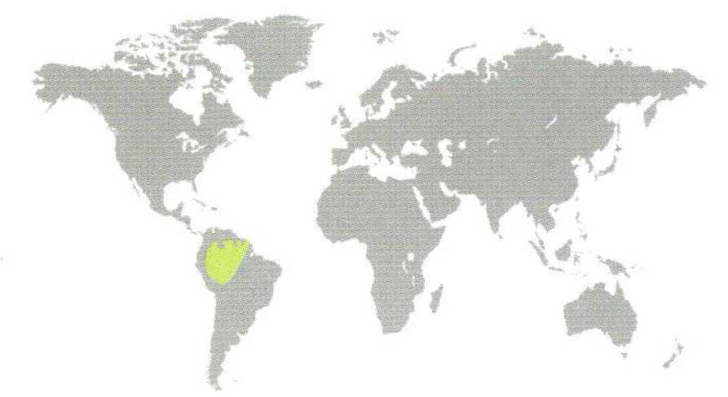

HYDROLAETARE SCHMIDTI

SCHMIDT'S FOREST FROG

(COCHRAN & GOIN, 1959)

ADULT LENGTH
Male
3 7/16–4½ in (88–115 mm)
Female
3½–4¾ in (90–120 mm)

Although this large frog has been reported from several, widely spaced locations in the Amazon Basin, it appears to be uncommon. It is nocturnally active and is found in marshes and swamps in the forest. It is an aquatic species, typically seen floating with just its eyes protruding above the water surface. It has been reported that males excavate a water-filled hole in the soil from which they call, but nothing is known about other aspects of its reproductive biology.

SIMILAR SPECIES

Hydrolaetare schmidti was once thought to be the only member of its genus, but two other species have since been described. The Feijo White-lipped Frog (*H. dantasi*), from Brazil, was described in 2003, and the Caparu Frog (*H. caparu*) was discovered in Bolivia in 2007. The call of *H. dantasi* is unusual in that the initial part of it consists of a loud thump, made by its vocal sac hitting the ground.

Schmidt's Forest Frog has a pointed snout and protruding eyes. Its toes are fully webbed and its back is covered in small warts. The back is brown, gray, and green in color, and there is a broad, dark band down its middle. The underside is cream or pale orange, with a dark brown marbled pattern.

Actual size

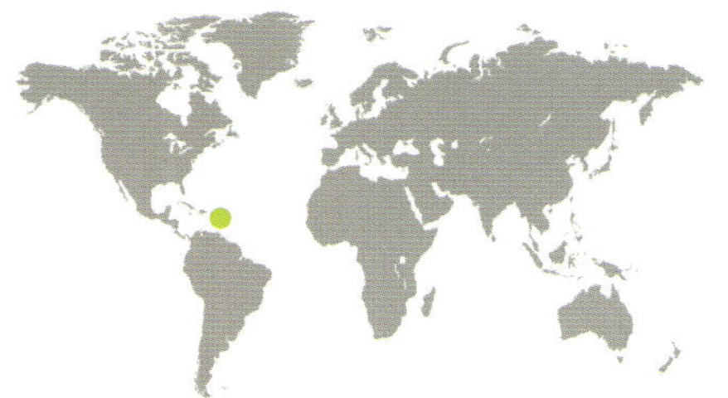

FAMILY	Leptodactylidae: Leptodactylinae
OTHER NAMES	Giant Ditch Frog
DISTRIBUTION	Dominica, Montserrat. Extinct on some other Caribbean islands
ADULT HABITAT	Forest
LARVAL HABITAT	Foam nest in an underground burrow
CONSERVATION STATUS	IUCN Critically Endangered. Has declined drastically since 1995 as a result of volcanic activity and chytridiomycosis. Is being successfully bred in captivity

ADULT LENGTH
Male
3¹¹⁄₁₆–7¾ in (95–195 mm)
Female
3⅞–8⅛ in (99–210 mm)

LEPTODACTYLUS FALLAX

MOUNTAIN CHICKEN

MÜLLER, 1926

This spectacular frog gets its odd name from long being considered a gastronomic delicacy in the West Indies. Today, it is categorized as Critically Endangered. Volcanic activity in Montserrat since 1995 has destroyed most of its habitat there, and the arrival of chytridiomycosis on Dominica has affected the island's population to the point where it may be extinct. The species has a highly unusual mode of reproduction. Male and female together defend a burrow and create a foam nest within which the eggs are laid. The eggs hatch into very long, eel-like tadpoles that the female feeds by depositing unfertilized eggs in the nest. After 45 days, 26–43 froglets emerge; these have been fed between 10,000 and 25,000 eggs.

SIMILAR SPECIES

To date, there are 83 species described in the genus *Leptodactylus*. The only other species found in the West Indies is the White-lipped Frog (*L. albilabris*), a smaller species that lives in Puerto Rico, the Virgin Islands, and Hispaniola. It has been reported that males call from the burrows of land crabs. Eggs are laid in a foam nest hidden under a rock, and the species is not threatened by extinction.

Actual size

The Mountain Chicken is one of the world's largest frogs, with females larger than males. Its long, powerful hind legs were much sought after as food. Its back is chestnut-brown, sometimes with darker spots and stripes, its flanks are orange, and its belly is yellow. There are dark stripes on the thighs.

FAMILY	Leptodactylidae: Leptodactylinae
OTHER NAMES	None
DISTRIBUTION	East of the Andes, from Venezuela south to Argentina
ADULT HABITAT	Open grassland, close to ponds, lakes, and flooded areas
LARVAL HABITAT	Ephemeral ponds
CONSERVATION STATUS	IUCN Least Concern. Has a very large range and there is no evidence of general decline. May be susceptible to chytridiomycosis

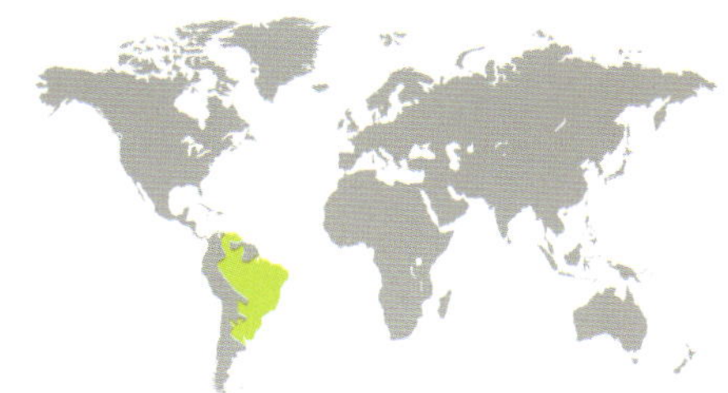

ADULT LENGTH
Male
3½–4¾ in (90–120 mm)
Female
3⅛–4⁵⁄₁₆ in (80–110 mm)

LEPTODACTYLUS LATRANS

CRIOLLA FROG

(STEFFEN, 1815)

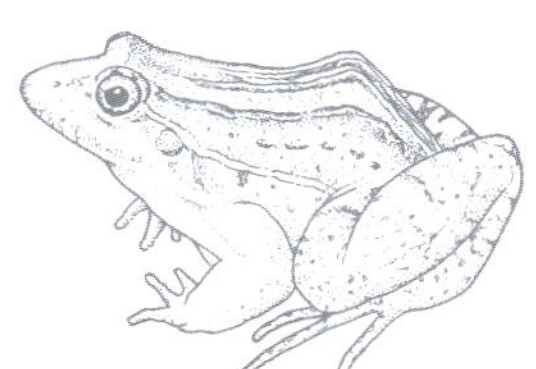

This large frog breeds from September to February. Males develop massive, muscular forearms and a pair of nuptial spines on each first finger. During amplexus, the pair make a foam nest, secreted by the female and whipped up by the male's hind legs. The nest is ring-shaped and the female sits in the middle of it. She continues to care for her young after hatching, protecting them from predators. If a bird comes close, she beats the water surface, alerting the tadpoles, which hide under her body. Criolla Frogs are predators on other frogs and are themselves eaten by people.

The Criolla Frog has prominent folds of skin along its back and sides. The back is gray, green, or reddish brown, with several white-edged dark spots. There is a large spot on the head between the eyes. The belly is white, mottled with gray. The snout is pointed and there are fringes of skin on the toes.

SIMILAR SPECIES

Also called the Criolla Frog (meaning "Creole") is *Leptodactylus chaquensis*, a large frog found in Argentina, Paraguay, Uruguay, and Bolivia. It has declined in parts of Argentina as a result of being harvested by humans for food. *Leptodactylus bolivianus* is a burrowing species found in northwestern South America.

Actual size

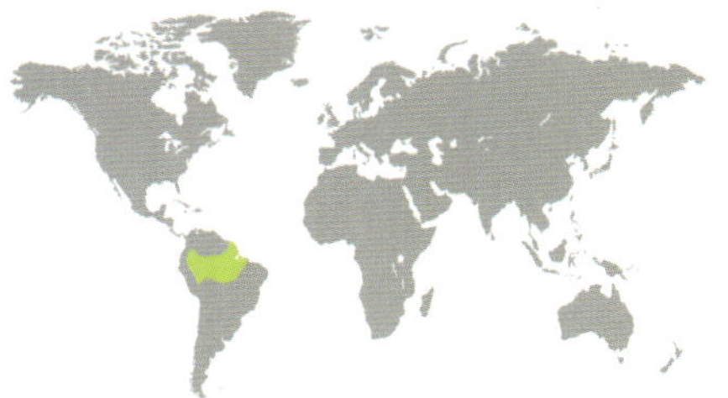

FAMILY	Leptodactylidae: Leptodactylinae
OTHER NAMES	Smokey Jungle Frog
DISTRIBUTION	Amazon Basin in Brazil, Colombia, Peru, Bolivia, and French Guiana
ADULT HABITAT	Lowland rainforest
LARVAL HABITAT	Ephemeral waterbodies
CONSERVATION STATUS	IUCN Least Concern. Has a very large range and there is no evidence of general decline

ADULT LENGTH
Male
4 1/8–7 1/8 in (106–177 mm)
Female
4 11/16–7 3/8 in (118–185 mm)

LEPTODACTYLUS PENTADACTYLUS

SOUTH AMERICAN BULLFROG

(LAURENTI, 1768)

This large and spectacular frog eats almost anything that moves, including baby birds, snakes, scorpions, and other frogs, including poison frogs. It is nocturnal, hiding by day, and breeds between May and November. The male's call is a loud "wroop." He has massive, muscular arms and spines on his chest and thumbs that help him clasp the female. Females produce around 1,000 eggs in a mass of foam contained in a cavity in the soil close to water. The tadpoles grow to a length of more than 3 in (80 mm) and are carnivorous. When molested, adults scream and produce copious quantities of toxic mucus.

SIMILAR SPECIES

The most visually striking of the 83 species in the genus *Leptodactylus* is the Coralline Frog (*L. laticeps*), from the Gran Chaco of Paraguay, Bolivia, and Argentina. It has a dramatic skin pattern of large black dots on a pale background that makes it highly prized in the international pet trade, where specimens can fetch up to US$600. It is listed as Near Threatened.

The South American Bullfrog has large eyes and prominent folds of skin along each side of its back. The skin on the upper surface of its back is smooth and is gray to reddish brown in color. There is a dark stripe on the top of the head between the eyes, and conspicuous dark spots or bars along the upper lip.

Actual size

FAMILY	Leptodactylidae: Leptodactylinae
OTHER NAMES	Pernambuco White-lipped Frog
DISTRIBUTION	Northeastern Brazil
ADULT HABITAT	Savanna grassland, sandy shores
LARVAL HABITAT	Ephemeral ponds
CONSERVATION STATUS	IUCN Least Concern. Declining where its habitat is being replaced by intensive agriculture

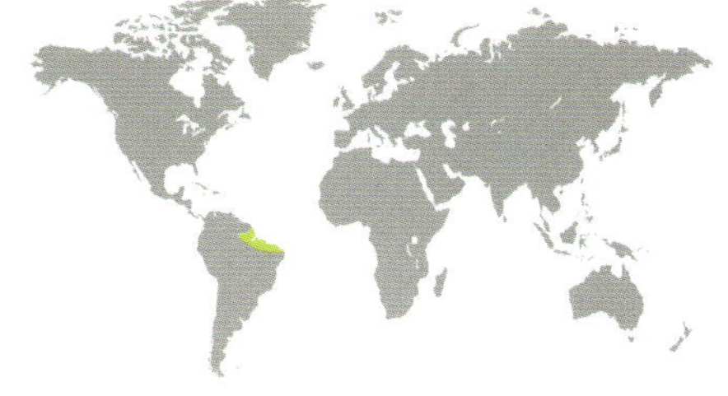

LEPTODACTYLUS TROGLODYTES

BRAZILIAN SIBILATOR FROG

LUTZ, 1926

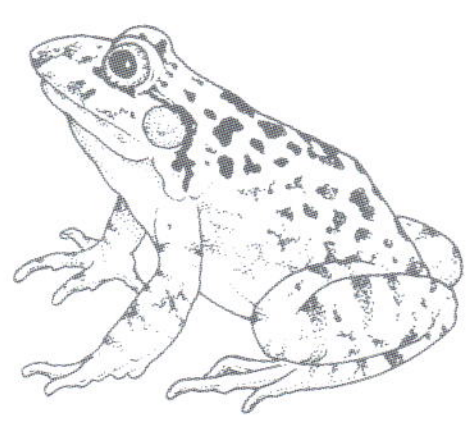

ADULT LENGTH
Male
average 1 15/16 in (49 mm)
Female
average 1 15/16 in (50 mm)

The male of this small frog has a spatula-shaped snout that he uses to dig a rather complicated burrow in which pairs mate and lay eggs. Breeding occurs between March and May, stimulated by heavy rain. The male has a simple mating call, consisting of a single note, repeated once a second. A responsive female produces a reciprocation call as she approaches him, thereby avoiding being attacked by him. The male then leads her to his burrow, which consists of a tunnel and one or more chambers, in one of which the pair mates and then the female lays about 100 eggs in a mass of foam.

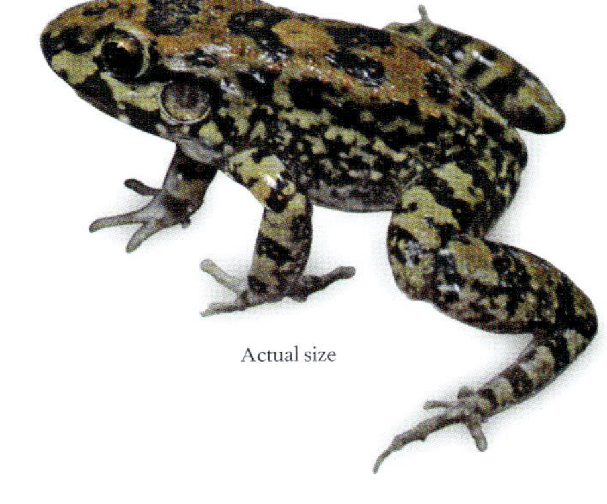

Actual size

SIMILAR SPECIES

Within this very large genus, *Leptodactylus troglodytes* belongs to the *L. fuscus* group. Sometimes known as the Rufous Frog, *L. fuscus* occurs across a huge range encompassing much of Central and South America. All the species in this group lay their eggs in foam nests in underground chambers near to water where the tadpoles develop. *Leptodactylus bufonius* lives in semiarid habitats in Argentina and neighboring countries.

The Brazilian Sibilator Frog has a pointed snout, large protruding eyes, and very distinct tympana. The hind legs are muscular and there are white tubercles on the soles of the feet. The back is pale brown with large black blotches and spots, and there are black stripes on the thighs. There is no webbing on the hands or feet.

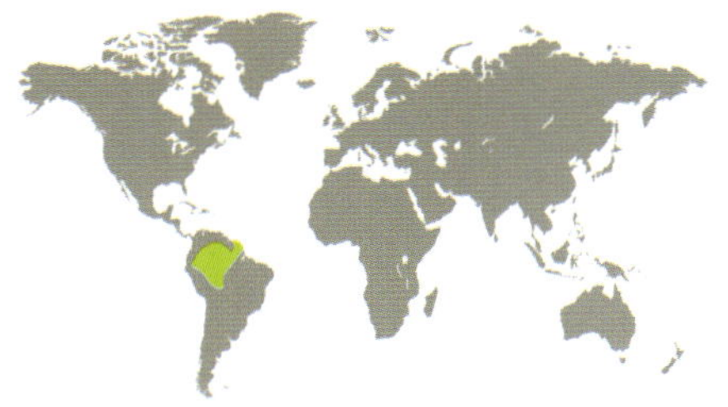

FAMILY	Leptodactylidae: Leptodactylinae
OTHER NAMES	None
DISTRIBUTION	Amazon Basin of Bolivia, Brazil, Colombia, Ecuador, French Guiana, Guyana, Peru, and Venezuela
ADULT HABITAT	Tropical forest
LARVAL HABITAT	Ephemeral pools
CONSERVATION STATUS	IUCN Least Concern. Though considered rare or uncommon, it has a huge range

ADULT LENGTH
Male
1⅜–1⅞ in (35–47 mm)
Female
1½–2 in (38–52 mm)

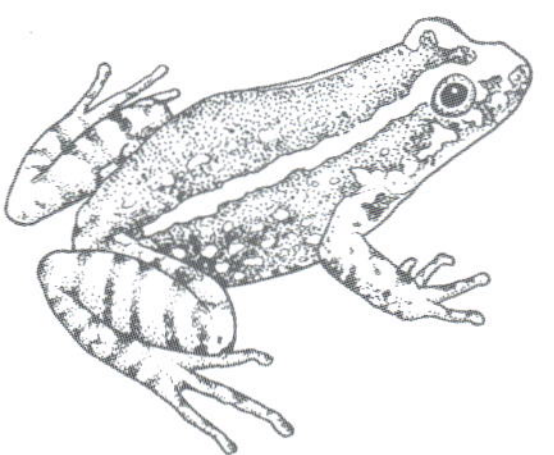

LITHODYTES LINEATUS

PAINTED ANTNEST FROG

(SCHNEIDER, 1799)

This small frog, found in leaf litter, appears to have a symbiotic relationship with the leaf-cutter ant *Atta cephalotes*. Frogs collected from ants' nests give off an aromatic odor that inhibits attack by the ants. Whether the ants benefit from this association is not known. Breeding occurs between September and February. Males call from burrows in ants' nests, producing a series of short whistles. The male and female create a foam nest at the edge of a temporary pool, in which the female lays 100–330 eggs. The female then remains close to the nest while the eggs develop.

SIMILAR SPECIES

Lithodytes lineatus is the only species in its genus. In appearance, it bears a striking resemblance to a quite unrelated species, the Brilliant-thighed Poison Frog (*Allobates femoralis*; page 356). This thus appears to be an example of mimicry, *L. lineatus* gaining protection from predators because of its resemblance to a toxic species. Another frog, also unrelated, that lives with and is unharmed by ants, is the Red Rubber Frog (*Phrynomantis microps*).

The Painted Antnest Frog has a rounded snout and a rather slender body. The back and flanks are black, and there are bright yellow stripes running down each side of the back and meeting at the tip of the nose. The limbs are tan, striped with brown, and there is a bright orange or red patch on the groin and inside of the thigh.

Actual size

FAMILY	Leptodactylidae: Paratelmatobiinae
OTHER NAMES	None
DISTRIBUTION	Southeastern Brazil
ADULT HABITAT	Bromeliads in rocky moorland at altitudes of 6,023–6,765 ft (1,836–2,062 m)
LARVAL HABITAT	Bromeliads
CONSERVATION STATUS	IUCN Critically Endangered. Threatened by bromeliad collectors

ADULT LENGTH
Male
9/16–11/16 in (14–18 mm)
Female
1/2–11/16 in (13–18 mm)

CROSSODACTYLODES ITAMBE

ITAMBE'S BROMELIAD FROG

BARATA, SANTOS, LEITE & GARCIA, 2013

Described as recently as 2013, this tiny frog completes its entire life cycle within the confines of rock-living bromeliads. It apparently occurs only within an area of 0.4 sq miles (1 sq km) at altitudes of 6,023–6,765 ft (1,836–2,062 m). Its reproductive biology has not been described. Its long-term survival is threatened by the collection of bromeliads. It is likely also that climate change, which typically has pronounced effects at high altitudes, is an additional threat.

Actual size

Itambe's Bromeliad Frog has a rounded snout, rather rough, granular skin, and adhesive disks on its fingers and toes. Males have more muscular forelimbs than females. The upper surfaces are dark brown, with a scattering of small, greenish spots.

SIMILAR SPECIES

There are five species in the genus *Crossodactylodes*, called bromeliad frogs, in southeastern Brazil. The Northern Bromeliad Frog (*C. septentionalis*) is listed as Vulnerable, while Izecksohn's Bromeliad Frog (*C. izecksohni*) is Endangered. They are under threat from destruction and degradation of the forest and, more specifically, from bromeliad collectors. In 2023 two new species, Teixeira's Bromeliad Frog (*C. teixeirai*) and the Serra Negra Bromeliad Frog (*C. serranegra*), were described from Espiritu Santo and Minas Gerais, Brazil.

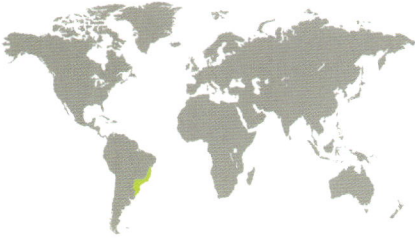

FAMILY	Leptodactylidae: Paratelmatobiinae
OTHER NAMES	None
DISTRIBUTION	Atlantic Forest, southeastern Brazil
ADULT HABITAT	Forest
LARVAL HABITAT	Small pools at the edge of ephemeral streams
CONSERVATION STATUS	IUCN Least Concern. Population declining but occurs within some protected areas

ADULT LENGTH
Male
¾–1 in (20–26 mm)
Female
1–1³⁄₁₆ in (25–30 mm)

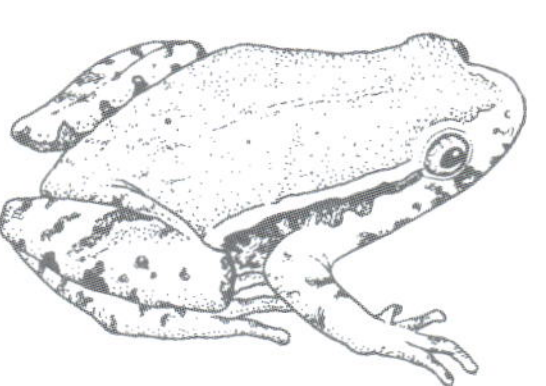

PARATELMATOBIUS POECILOGASTER

SAN ANDRÉ RAPIDS FROG

GIARETTA & CASTANHO, 1990

Actual size

The San André Rapids Frog is a plump little frog with a flat snout. The back and the legs are green or brown with small green spots, and there may be a green or blue stripe running all or part of the way down the middle of the back. A pink line running from the eye to the groin separates the back from the darker flanks. The belly is orange or red with black or white blotches.

This small, little-known frog is found on the ground near temporary streams. It breeds during the wet season, males gathering in the dried beds of streams and calling at night from pools under overhanging rocks. The eggs are laid on sloping rocks above water, so that, when they hatch, the tadpoles fall into the stream below. From time to time, heavy rain causes streams to flood, sweeping both adults and tadpoles downstream. When startled or molested, this frog flips itself over onto its back, exposing its brightly colored belly.

SIMILAR SPECIES

Although small, *Paratelmatobius poecilogaster* is the largest species in this genus, all members of which are found in the Atlantic Forest of southeastern Brazil. Seven species are known. Lutz's Rapids Frog (*P. lutzi*) and the Mantiqueira Rapids Frog (*P. mantiqueira*) are Critically Endangered, while Gaige's Rapids Frog (*P. gaigae*) is Endangered. All occur within very restricted ranges. Most records for this genus come from protected areas but their populations appear to be declining.

FAMILY	Leptodactylidae: Paratelmatobiinae
OTHER NAMES	None
DISTRIBUTION	Southeastern Brazil
ADULT HABITAT	Forest at altitudes of 2,600–3,300 ft (800–1,000 m)
LARVAL HABITAT	Ephemeral pools
CONSERVATION STATUS	IUCN Least Concern. Declining in places as a result of deforestation but remains common elsewhere

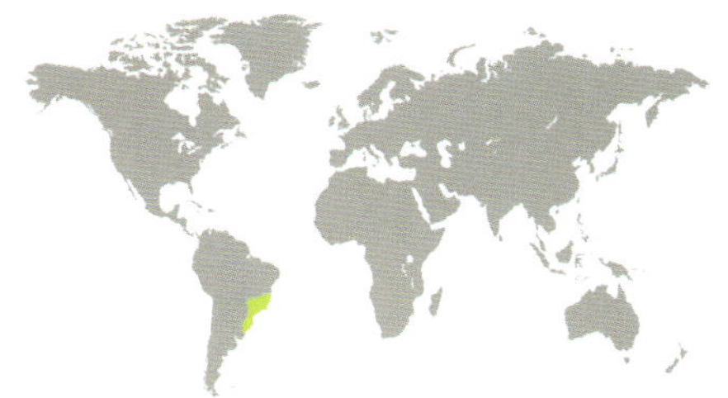

ADULT LENGTH
Male
average ⅝ in (16 mm)
Female
average 11/16 in (18 mm)

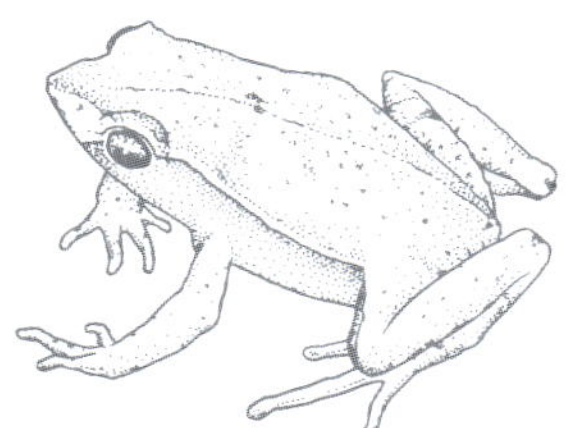

SCYTHROPHRYS SAWAYAE

BANHADO FROG

(COCHRAN, 1953)

This small, scarcely known frog possesses a variety of defenses against potential predators. The shape of its body and the colors of its skin give it a resemblance to the dead leaves among which it lives on the forest floor. It occurs in three color forms, making it more difficult for predators to learn its appearance: Around 50 percent of individuals are dark brown, 28 percent are green, and 22 percent are pale brown. Finally, when attacked or molested, it adopts a stiff-limbed posture, thrusting out its arms and legs and effectively increasing its size.

Actual size

The Banhado Frog has a wide body, a pointed snout, and pointed projections above its eyes. The legs are slender and there are small disks on the fingers and toes. The back is dark brown, green, or pale brown, and the legs are yellowish brown. A pale fold of skin separates the back from the darker flanks.

SIMILAR SPECIES

Scythrophrys sawayae is the only species in its genus. It shares with another inhabitant of Brazil's Atlantic Forest, Boie's Horned Frog (*Proceratophrys boiei*; page 254), the combination of a resemblance to dead leaves and a stiff-legged defensive posture. These two frogs are not closely related, and it seems that this combination of features has evolved independently in some forest frogs.

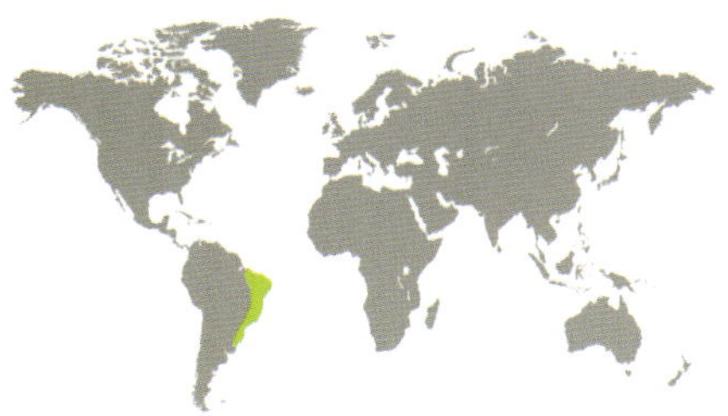

FAMILY	Odontophrynidae
OTHER NAMES	None
DISTRIBUTION	Eastern Brazil
ADULT HABITAT	Coastal forest up to 2,000 ft (600 m) altitude
LARVAL HABITAT	Ephemeral ponds
CONSERVATION STATUS	IUCN Least Concern. Declining in places but occurs within some protected areas

ADULT LENGTH
average 3⅞ in (100 mm)

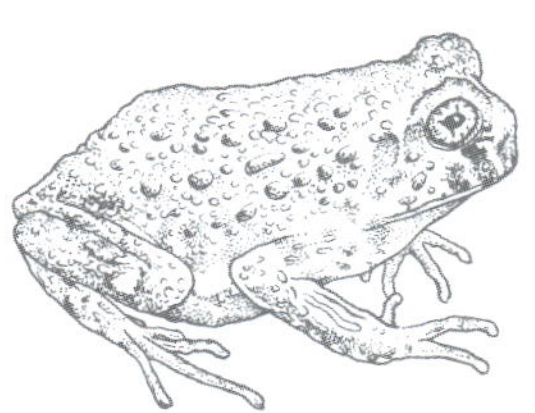

MACROGENIOGLOTTUS ALIPIOI

BAHIA FOREST FROG

CARVALHO, 1946

Males of this large frog gather around ponds that form after heavy rain and produce a chorus of calls that has been likened to the sound of foghorns. During mating, the male periodically kicks with his hind legs, perhaps to spread his sperm over the eggs. The eggs are laid in short strands attached to waterplants, and a pair lay several strands spread around a pond. While its natural habitat is the Atlantic Forest of eastern Brazil, the species is also common in cacao plantations in Espírito Santo province in the southeast of the country.

The Bahia Forest Frog has a large head with a very blunt snout and protruding eyes. There are numerous large warts on its back and flanks, and the skin on its legs is wrinkled. The fingers and toes are long. The upper surfaces are dark brown with pale brown and whitish patches. There are striking, vertical, dark brown and white stripes around the upper lip.

SIMILAR SPECIES

Macrogenioglottus alipioi is the only species in its genus. This species is most closely related to the toad-like frogs of the genus *Odontophrynus*, such as the Rio Grande Escuerzo (*O. cultripes*; page 253).

Actual size

FAMILY	Odontophrynidae
OTHER NAMES	None
DISTRIBUTION	Southern Brazil
ADULT HABITAT	Lowland forest, savanna, grassland, and swamps
LARVAL HABITAT	Ephemeral ponds
CONSERVATION STATUS	IUCN Least Concern. It occurs within several protected areas and can be common in suburban areas

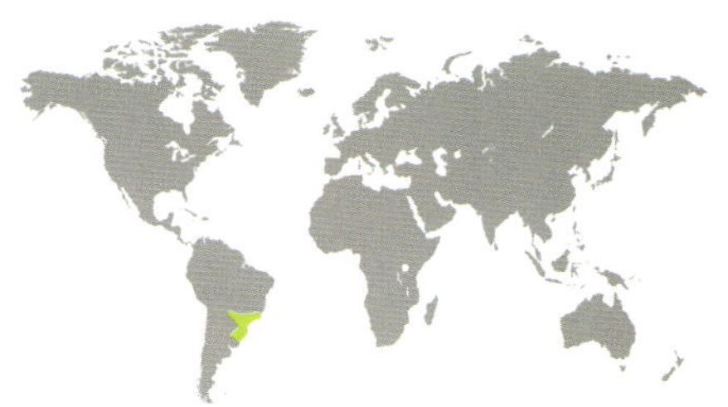

ADULT LENGTH
Male
$1\frac{15}{16}$–$2\frac{3}{8}$ in (50–60 mm)
Female
$1\frac{3}{4}$–$2\frac{13}{16}$ in (45–70 mm)

ODONTOPHRYNUS CULTRIPES

RIO GRANDE ESCUERZO

REINHARDT & LÜTKEN, 1862

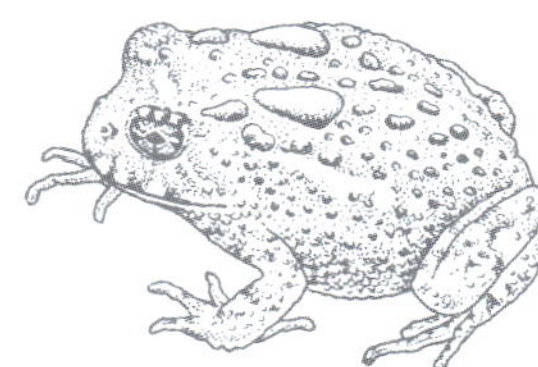

This toad-like frog is a burrowing species, spending much of its life underground. It lives in the Cerrado of Brazil, a savanna-like habitat, and in the Atlantic Forest. In some parts of its range, it breeds after rain, while in others it breeds in both dry and wet periods. It lays its eggs, in a gelatinous mass, at the bottom of ephemeral pools, where its tadpoles develop. It also breeds in a variety of manmade waterbodies and it can be very common in suburban gardens.

SIMILAR SPECIES

To date, 11 species have been described in the genus *Odontophrynus*. The Common Lesser Escuerzo (*O. americanus*) is a widespread species in Brazil, Paraguay, Uruguay, and Argentina. The Cururu Lesser Escuerzo (*O. occidentalis*) is less common and occurs in montane forest of Argentina. The Cordobo Escuerzo (*O. achalensis*) is found in montane grassland in Argentina and is listed as Vulnerable.

The Rio Grande Escuerzo has a blunt snout, large, protruding eyes, and a pair of very large, raised glands on its back, just behind the head. There are many warts on the back, colored red in some specimens. The upper surfaces are dark brown with darker and paler spots, and there is an orange or yellow band running along each flank.

Actual size

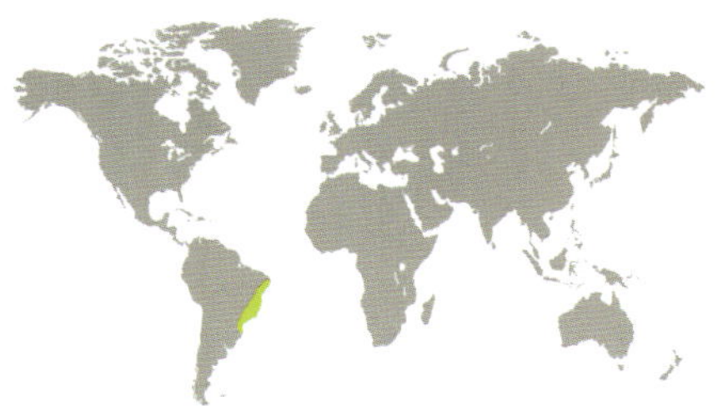

FAMILY	Odontophrynidae
OTHER NAMES	Bahia Smooth Horned Frog, Rio de Janeiro Smooth Horned Frog
DISTRIBUTION	Eastern Brazil
ADULT HABITAT	Woodland
LARVAL HABITAT	Swamps or streams
CONSERVATION STATUS	IUCN Least Concern. It is declining in places but occurs within some protected areas

ADULT LENGTH
Male
1⁹⁄₁₆–2½ in (40–62 mm)

Female
1⁹⁄₁₆–2¹⁵⁄₁₆ in (40–74 mm)

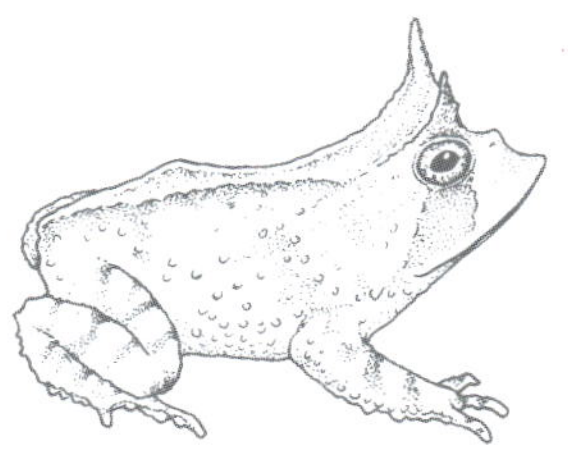

PROCERATOPHRYS BOIEI

BOIE'S HORNED FROG

(WIED-NEUWIED, 1824)

An inhabitant of Brazil's Atlantic Forest, this handsome frog is collected and exported as a pet. Its most striking feature is a pair of long triangular skin extensions above the eyes. These, together with the color pattern on its skin, make it well camouflaged among the leaf litter on the woodland floor. It feeds on a wide variety of insects and can be heard calling between September and January. When molested or attacked, it adopts a defensive posture in which it stiffens and thrusts out its arms and legs.

SIMILAR SPECIES

Forty-one species have been described in the South American genus *Proceratophrys*, known as smooth horned frogs. *Proceratophrys melanopogon* forms large breeding aggregations immediately after heavy rain. Peters' Smooth Horned Frog (*P. bigibbosa*) lives in evergreen forest in mountainous areas of northern Argentina and southern Brazil. The Plateau Smooth Horned Frog (*P. palustris*) and Araripe Horned Frog (*P. ararype*) are both Critically Endangered. See also the Botucatu Escuerzo (*P. moratoi*; page 255).

Boie's Horned Frog has a wide head, a rounded snout, and long, tapering skin extensions above each eye. The body is stout, and the skin on the back and sides is warty. The back, flanks, and legs bear a complex pattern of wide stripes in brown, red, orange, yellow, and black. A broad stripe running down the middle of the back may be brown or gray.

Actual size

FAMILY	Odontophrynidae
OTHER NAMES	None
DISTRIBUTION	Southern Brazil
ADULT HABITAT	Open areas in Cerrado woodland
LARVAL HABITAT	Slow, shallow streams
CONSERVATION STATUS	IUCN Least Concern

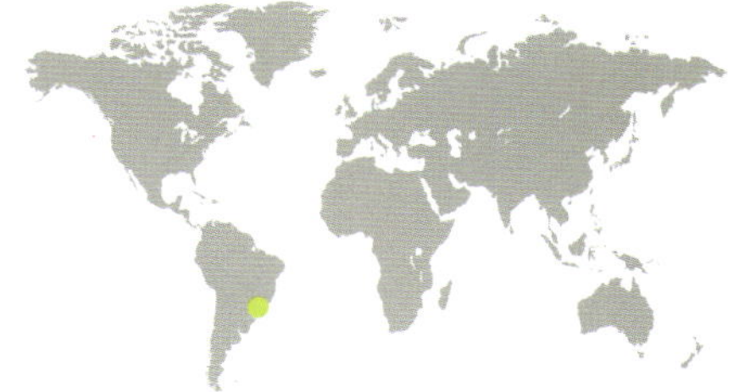

PROCERATOPHRYS MORATOI

BOTUCATU ESCUERZO

(JIM & CARAMASCHI, 1980)

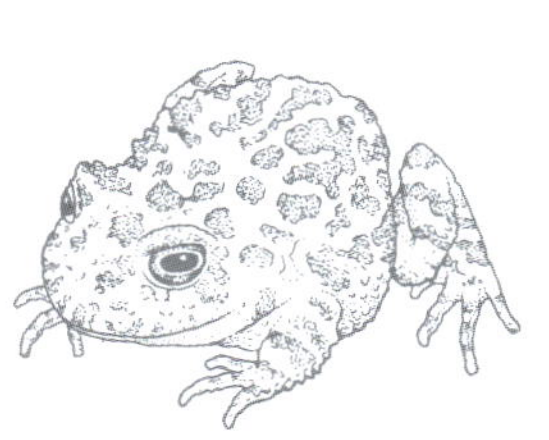

ADULT LENGTH
Male
15/16–1¼ in (24–31 mm)
Female
11/16–1 7/16 in (27–36 mm)

The conservation status of this small frog is uncertain. It was originally reported from a single location, where its habitat was subsequently destroyed and, because it had not been seen since 1990, an evaluation made in 2004 suggested that it might be extinct. However, reports in 2010 and 2011 indicated both that it still survives and that its range may be larger than previously thought. It breeds between October and February, males calling by day and night from shallow burrows at the base of plants. The eggs are laid in slow-moving streams, where the tadpoles develop.

Actual size

The Botucatu Escuerzo is a very compact little frog with short arms and legs, a short, rounded snout, large eyes, and very warty skin. Some of the warts, especially on the arms, have sharp tips. There is a complex pattern on the back of large, dark greenish-gray blotches, surrounded by broad bands of pale gray or sandy yellow.

SIMILAR SPECIES

Very little is known about the 41 species in the genus *Proceratophrys* that have been described to date. Three species—*P. kaingang*, *P. korekore*, and *P. velhochico*—were described since 2021. The Plateau Smooth Horned Frog (*P. palustris*) has a call that is similar to that of *P. moratoi*. Listed as Critically Endangered, its habitat has been adversely affected by mining, which destroys woodland and pollutes the streams in which it breeds. See also Boie's Horned Frog (*P. boiei*; page 254).

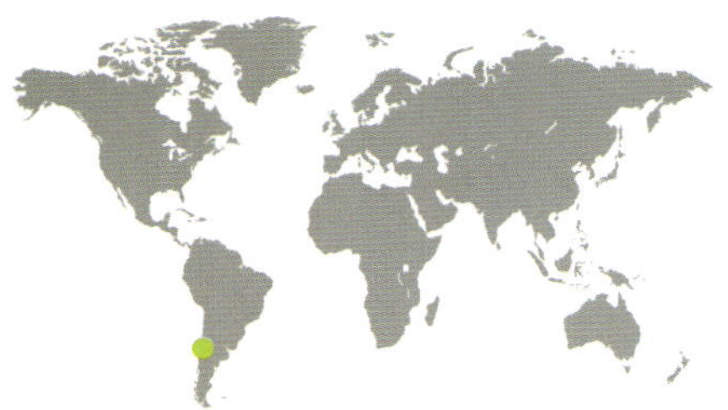

FAMILY	Rhinodermatidae
OTHER NAMES	None
DISTRIBUTION	Valdivia Province, Chile
ADULT HABITAT	Temperate forest
LARVAL HABITAT	Streams
CONSERVATION STATUS	IUCN Endangered. Has a very restricted range that is threatened by habitat destruction

ADULT LENGTH
Male
up to 1⅞ in (48 mm)
Female
up to 1$^{11}/_{16}$ in (43 mm)

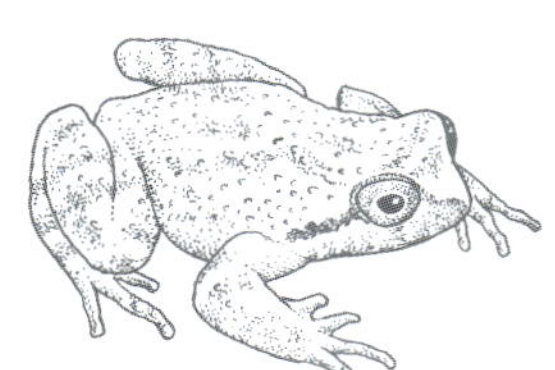

INSUETOPHRYNUS ACARPICUS

BARRIO'S FROG

BARRIO, 1970

Actual size

Barrio's Frog has very muscular arms and legs, partially webbed toes, and stubby, unwebbed fingers. The eyes are large and protruding, and have red irises. The back is warty, and is reddish brown in color with numerous small white spots. There are dark bands on the legs and the throat is pinkish yellow.

Until recently, this small frog was known from only three localities in Chile within an area of 15 sq miles (40 sq km), but in 2012 a fourth locality was found 12 miles (20 km) away. Known for its ability to leap large distances when disturbed, it hides by day under stones and is active at night. It breeds between January and May, the male developing nuptial spines on his hands and chest. These probably help him to clasp the female firmly in running water. In the cold water in which they develop, the tadpoles reach metamorphosis in 10–12 months.

SIMILAR SPECIES

Insuetophrynus acarpicus is the only species in its genus. It belongs to the small family Rhinodermatidae, along with two species of Darwin's frogs, which have a very different mode of reproduction (see page 257).

FAMILY	Rhinodermatidae
OTHER NAMES	None
DISTRIBUTION	Central and southern Chile; western Argentina
ADULT HABITAT	Near cool, wet stream banks in temperate forest
LARVAL HABITAT	Within the father's vocal sac
CONSERVATION STATUS	IUCN Endangered. Population has been greatly reduced by habitat loss and chytridiomycosis

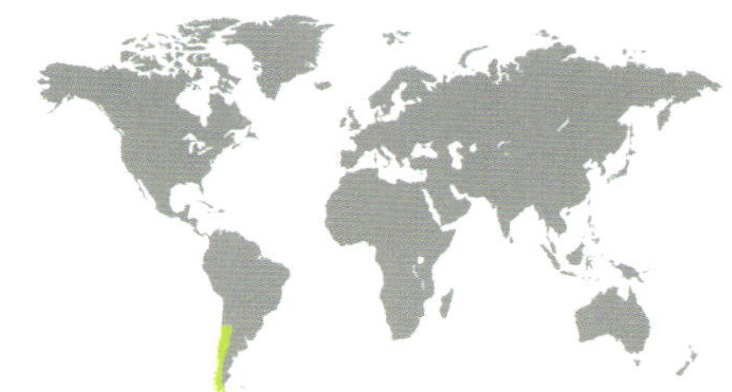

RHINODERMA DARWINII

DARWIN'S FROG

DUMÉRIL & BIBRON, 1841

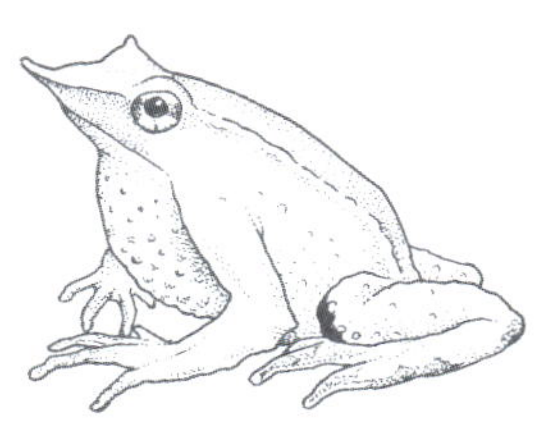

ADULT LENGTH
Male
⅞–1⅛ in (22–28 mm)
Female
1–1¼ in (25–31 mm)

This small, highly endangered frog shows unique reproductive behavior. After the male has led the female to a secluded place, she deposits up to 40 large eggs on the ground. He then guards them for 2–3 weeks until the embryos begin to move, at which point he swallows up to 19 eggs and retains them in his vocal sac for 50–70 days. There they hatch into tadpoles and develop into small frogs, the male "giving birth" to froglets from his mouth. The developing tadpoles are sustained by yolk from the eggs and a secretion from the inside of the vocal sac. The Zoological Society of London successfully bred Darwin's Frog in early 2025.

Actual size

Darwin's Frog has a long, fleshy proboscis on its snout that gives its head a triangular shape. The legs and arms are long and slender. Coloration on the back is very variable, being brown or bright green, or a mixture of both. The belly is black, often mottled with white.

SIMILAR SPECIES

Chile Darwin's Frog (*Rhinoderma rufum*) has not been found in the wild since 1980 and is thought by some researchers to have become extinct in 1982, although the IUCN list it as Critically Endangered. It differs from *R. darwinii* in that the male retains the young in his vocal sac only until the eggs have hatched into larvae, at which point he releases them into water. Both species have lost much of their habitat through deforestation, and both are infected by chytridiomycosis.

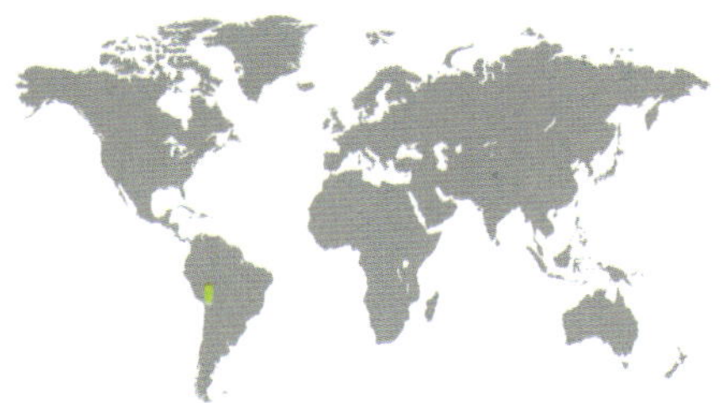

FAMILY	Telmatobiidae
OTHER NAMES	Titicaca Scrotum Frog
DISTRIBUTION	Lake Titicaca and the surrounding area, on the border between Bolivia and Peru
ADULT AND LARVAL HABITAT	Lake
CONSERVATION STATUS	IUCN Endangered. Has declined by 80 percent in the last 15 years, through overexploitation, introduced fish, and pollution

ADULT LENGTH
2¹⁵⁄₁₆–5⁷⁄₁₆ in
(75–138 mm)

TELMATOBIUS CULEUS

TITICACA WATER FROG

(GARMAN, 1875)

The Titicaca Water Frog has a large, flattened head, with a rounded snout and eyes that point forward. Being a strong swimmer, it has long, muscular hind legs and fully webbed feet. There are many folds in its skin, giving the impression that its skin is too big for its body. It is dark gray in color, with pale speckles.

This large frog is wholly aquatic and is adapted to the particular conditions in Lake Titicaca, a high-altitude lake in the Andes. Because the lake is cold and constant winds cause waves on its surface, the water contains high levels of dissolved oxygen, enabling the frog to breathe through its skin. It has very small lungs, many folds in its skin that increase its surface area, and a very high blood cell count. This enables it to swim and feed in water more than 650 ft (200 m) deep. Its recent decline is largely due to overhunting by local people, who use the frog as a food source and an ingredient in an aphrodisiac.

SIMILAR SPECIES

There are 60 species in the genus *Telmatobius*, all of them found in the Andes, from Ecuador in the north to northern Chile and northwestern Argentina in the south. Most are restricted to very small ranges, at high altitude, and, to varying degrees, are threatened by extinction. The fungus that causes chytridiomycosis has been found in two Argentinian species; this may be a threat to the entire genus.

Actual size

FAMILY	Telmatobiidae
OTHER NAMES	*Telmatobius macrostomus*, Andes Smooth Frog, Lake Junín Giant Frog
DISTRIBUTION	Lake Junín, central Peru
ADULT AND LARVAL HABITAT	Lake
CONSERVATION STATUS	IUCN Endangered. Has undergone a marked decline in the last 30 years

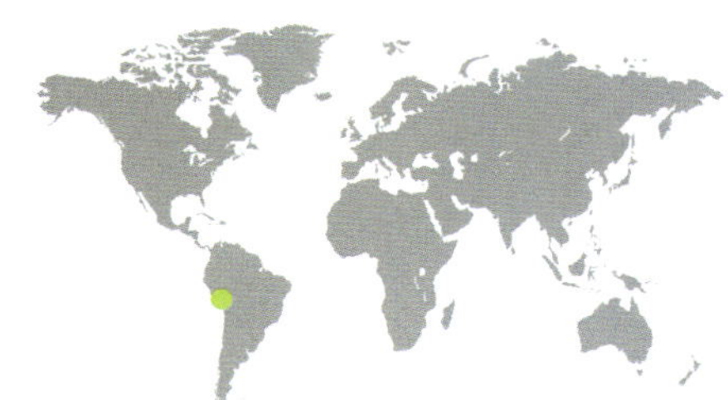

ADULT LENGTH
up to 12 in (300 mm)

TELMATOBIUS MACROSTOMUS

LAKE JUNÍN FROG

(PETERS, 1873)

One of the world's largest frogs, this wholly aquatic species lives only in Lake Junín, also known as Chinchayqucha, at an altitude of 13,412 ft (4,088 m) in the Andes of Peru. Several factors are thought to have contributed to its decline. Trout have been introduced to the lake and these prey on the tadpoles. Mining nearby produces chemical pollutants that drain into the lake, as does sewage from human settlements. Most importantly, this large frog has long been harvested by local people for food. It has been introduced to the Rio Mantaro, a long river that runs through Peru.

The Lake Junín Frog has an oval-shaped body, long legs, a pointed snout, and forward-pointing eyes. It has loose, wrinkled skin. It is dark gray to black in color.

SIMILAR SPECIES

Telmatobius macrostomus was originally in the genus *Batrachophrynus*, which contained just one other species, the Amable Maria Frog (*B. brachydactylus*, now *T. brachydactylus*), which is also Endangered. It is a semi-aquatic frog, smaller than *T. macrostomus*, that lives in and around rivers and streams that run into Lake Junín, but not in the lake itself. It is also eaten by local people and is used in traditional medicine.

Actual size

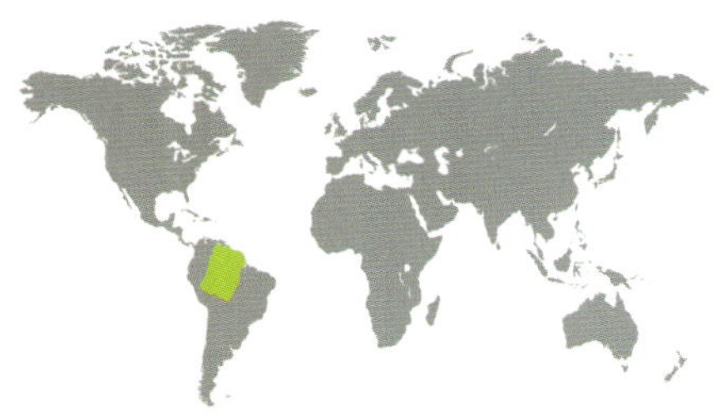

FAMILY	Allophrynidae
OTHER NAMES	None
DISTRIBUTION	Amazon Basin, including Brazil, French Guiana, Guyana, Suriname, and Venezuela
ADULT HABITAT	Tropical forest
LARVAL HABITAT	Ephemeral pools
CONSERVATION STATUS	IUCN Least Concern. It is common and has a huge range

ADULT LENGTH
Male
up to 1 in (25 mm)

Female
up to 1¹⁄₁₆ in (27 mm)

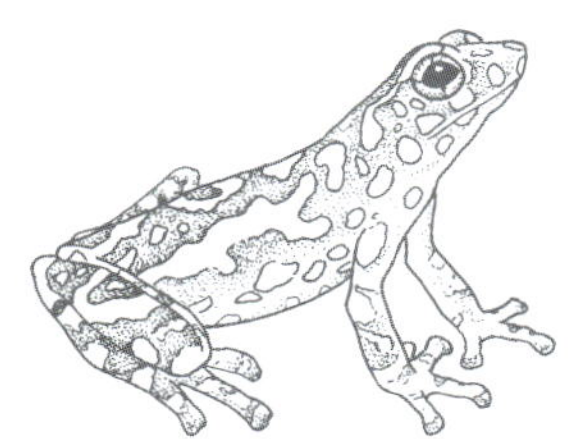

ALLOPHRYNE RUTHVENI

TUKEIT HILL FROG

GAIGE, 1926

Actual size

The Tukeit Hill Frog has a flat body and head, and well-developed adhesive disks at the tips of its fingers and toes. The skin on the back is covered in tiny pointed pustules. The back is bronze, grayish brown, gold, or yellow with a mottled black pattern (as shown here) or the reverse—black with a pale mottled pattern.

This small, scarcely known frog is described as an "explosive breeder," meaning that it gathers over a short time period in large numbers after heavy rain to mate—choruses of several hundred males have been reported. Males have a very large vocal sac and call from foliage above water-filled depressions, producing a low, raspy trill. After forming pairs, females lay around 300 eggs in the water. When not breeding, the frog is rarely seen.

SIMILAR SPECIES

Allophryne ruthveni is one of three species in its genus, which is the sole member of its family. The Resplendent Frog (*Allophryne resplendens*) is known only from a few female specimens, found in Peru, and gets its name from the iridescent yellow spots on its otherwise black back. The Relict Frog (*Allophryne relicta*), from the Atlantic Forest in eastern Brazil, was described as recently as 2013.

FAMILY	Centrolenidae: Centroleninae
OTHER NAMES	None
DISTRIBUTION	The Andes of southern Peru
ADULT HABITAT	Cloud forest
LARVAL HABITAT	Pools
CONSERVATION STATUS	IUCN Vulnerable

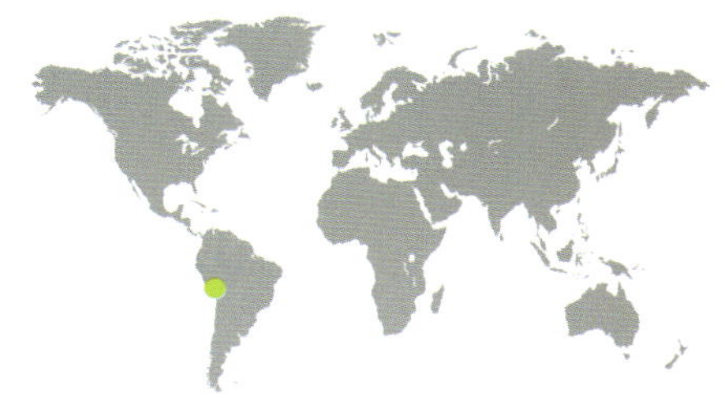

ADULT LENGTH
average 1¼ in (31 mm)

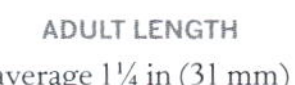

CENTROLENE SABINI

SABIN'S GLASS FROG

CATENAZZI, VON MAY, LEHR, GAGLIARDI-URRUTIA & GUAYASAMIN, 2012

This small frog, discovered as recently as 2012 in the Manú National Park of Peru, shows many of the features that give glass frogs their name, notably transparent skin on the underside, through which many of the internal organs can be seen. There is, however, a white patch concealing the heart. The predominant color is green and the frog even has green bones. Males call from foliage hanging over open water, where mating occurs and the eggs are laid. When the eggs hatch, the tadpoles wriggle out and fall into the water below.

Actual size

Sabin's Glass Frog has a bright yellow-green body, with a white stripe running from the upper lip to the pelvis. The skin on the back bears many tiny spicules. The skin on the underside is white under the chest and transparent under the belly. Males have small spines on their upper arms, and there are adhesive disks at the end of the fingers and toes.

SIMILAR SPECIES

Centrolene sabini is the southernmost member of this genus, which contains 30 other species. It is very similar to the larger *C. lemniscatum*, which occurs in northern Peru. Lynch's Glass Frog (*Centrolene lynchi*), an Endangered species from Ecuador and Colombia, is declining in many parts of its range. The cause is thought to be climate change, which is reducing the cloud cover on the mountains that is essential to maintain the moist conditions, characteristic of cloud forest, required by these frogs. See also Savage's Cochran Frog (*C. savagei*; page 262).

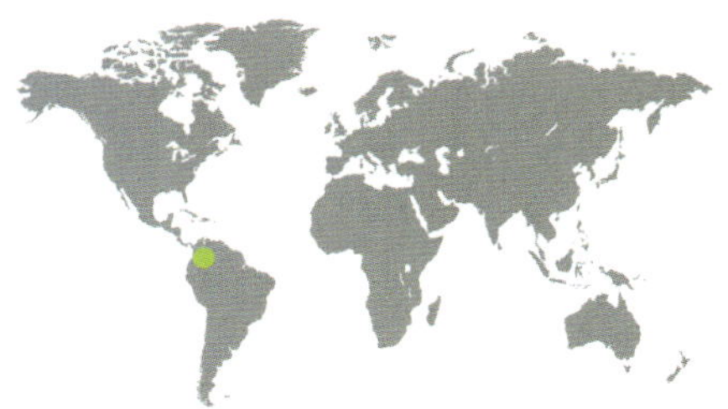

FAMILY	Centrolenidae: Centroleninae
OTHER NAMES	None
DISTRIBUTION	Central Colombia
ADULT HABITAT	Forest at altitudes of 4,600–7,900 ft (1,400–2,410 m)
LARVAL HABITAT	Streams
CONSERVATION STATUS	IUCN Least Concern, yet has a small, fragmented range that is threatened by deforestation and climate change

ADULT LENGTH
Male
$\frac{3}{4}$–$\frac{15}{16}$ in (19–23 mm)
Female
$\frac{7}{8}$–$\frac{15}{16}$ in (23–24 mm)

CENTROLENE SAVAGEI

SAVAGE'S COCHRAN FROG

RUÍZ-CARRANZA & LYNCH, 1991

Actual size

Savage's Cochran Frog has a slender body, a wide head with a rounded snout, and large, prominent eyes. There are adhesive pads on the fingers and toes. The body and legs are dappled with rounded white warts. The upper surfaces are bright green, with many small white or pale green spots, and the underside is white.

The bright green color of this delicate little frog makes it well camouflaged against leaves; even its bones are green. It is usually found in vegetation close to running water. Males call at night after daytime rain and females appear to prefer larger males as mates. During mating, females lay a clutch of around 18 large cream-colored eggs on the top of leaves, hanging 2.5–10 ft (80–300 cm) above a stream. The male remains with the eggs, guarding them until they hatch, at which point the tadpoles fall into the water below. The main predators of the eggs are insect larvae.

SIMILAR SPECIES

To date, 31 species have been described in the genus *Centrolene*. Around half of them are, to varying degrees, threatened with extinction. The Pichincha Giant Glass Frog (*C. heloderma*) is Vulnerable. An inhabitant of high-altitude cloud forest, it is now extinct in Ecuador but persists in Colombia. Climate change is reducing the extent of cloud forest, because clouds now form at higher altitudes than previously. See also Sabin's Glass Frog (*C. sabini*; page 261).

FAMILY	Centrolenidae: Centroleninae
OTHER NAMES	None
DISTRIBUTION	Ecuador and Peru
ADULT HABITAT	Cloud forest at altitudes of 4,600–6,000 ft (1,400–1,80 m)
LARVAL HABITAT	In small streams
CONSERVATION STATUS	IUCN Least Concern, yet has a restricted range that is subject to deforestation and water pollution

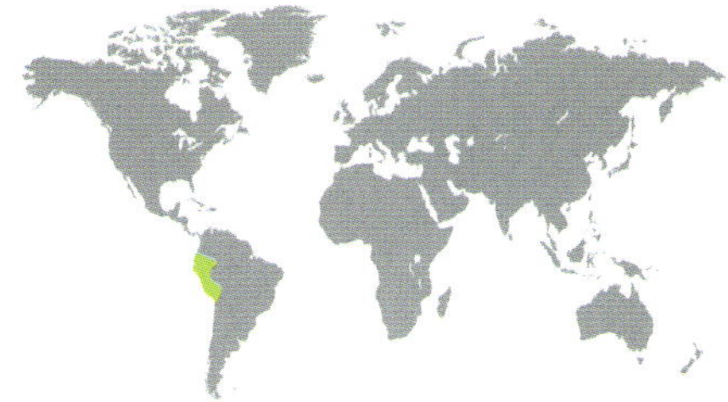

ADULT LENGTH
average ¾ in (20 mm)

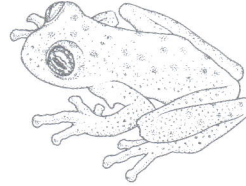

CHIMERELLA MARIAELENAE

MARIA ELENA'S GLASS FROG

(CISNEROS-HEREDIA & MCDIARMID, 2006)

Glass frogs of the family Centrolenidae get their name from a transparent patch of skin on the underside, through which the internal organs can be seen. This small, scarcely known frog lives high in the forest canopy outside the breeding season and so is rarely encountered. After rain, males descend and sit in foliage just above temporary streams. They call at night from the upper surfaces of leaves. As with other glass frogs, the eggs are probably laid on the underside of leaves, from where the tadpoles fall into the water below.

SIMILAR SPECIES

There are two other species in this genus, Corleone's Glass Frog (*C. corleone*), described in 2014, and the Surprising Glass Frog (*C. mira*), described in 2023. All three species are confined to the eastern versant of the Andes. They show similarities to the oddly named Bolivar Giant Glass Frog (*Vitreorana gorzulae*), a small species found in Venezuela and Guyana.

Actual size

Maria Elena's Glass Frog has very long, thin legs and a slender body tapering to a tiny waist. The legs are very long and thin, and there are well-developed disks on the fingers and toes. The snout is blunt and the protruding eyes are very large and point forward. The upper surfaces are pale green, dotted with small purplish spots.

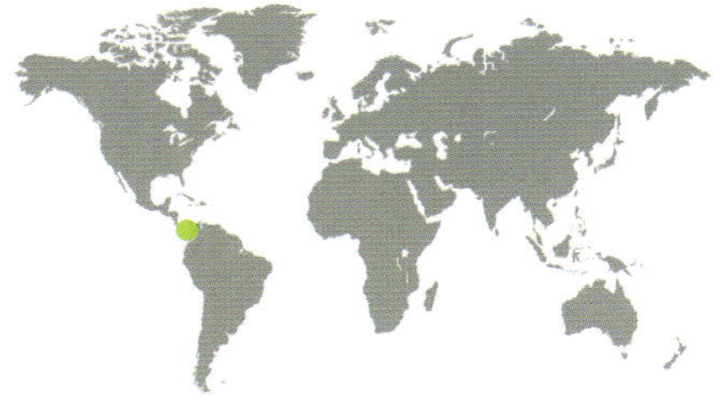

FAMILY	Centrolenidae: Centroleninae
OTHER NAMES	San José Cochran Frog
DISTRIBUTION	Costa Rica, Panama, Colombia
ADULT HABITAT	Wet and moist tropical forest up to 5,400 ft (1,650 m) altitude
LARVAL HABITAT	Fast-flowing streams
CONSERVATION STATUS	IUCN Least Concern. Declining in the northern part of its range as a result of deforestation

ADULT LENGTH
Male
⅞–1 in (21–25 mm)
Female
1–1¼ in (25–32 mm)

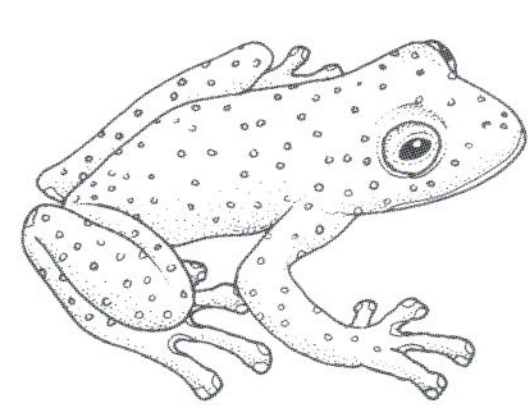

COCHRANELLA EUKNEMOS

FRINGE-LIMBED TREE FROG

(SAVAGE & STARRETT, 1967)

Actual size

The Fringe-limbed Tree Frog gets its name from the white fringes of skin that run along its lower forelimbs and hind limbs. It has a long snout and well-developed disks on the ends of the fingers and toes. The back is blue-green with conspicuous yellow spots.

This delicately formed frog inhabits montane forest up to an altitude of around 5,400 ft (1,650 m). It is becoming increasingly rare in Costa Rica and Panama, but remains common in Colombia. The main threat it faces is from deforestation. In the rainy season, from August to October, males call from trees overhanging fast-flowing mountain streams, producing a "creep... creep... creep" sound. The eggs are black and white, and are laid in large masses of jelly on leaves. Neither parent remains with them, and on hatching, the tadpoles fall into the stream below.

SIMILAR SPECIES

There are eight described species in the genus *Cochranella*, most of them having restricted ranges and are, to varying degrees, classified as under threat. For example, the Ecuadorian Blue Glass Frog (*C. mache*) has a tiny range and is considered Threatened due to deforestation. It is blue-green with yellow spots, but can rapidly change to pale blue with orange spots.

FAMILY	Centrolenidae: Centroleninae
OTHER NAMES	None
DISTRIBUTION	Eastern slopes of the Andes in Bolivia
ADULT HABITAT	Montane forest
LARVAL HABITAT	Mountain streams
CONSERVATION STATUS	IUCN Near Threatened. Population is declining, probably due to agricultural pollution

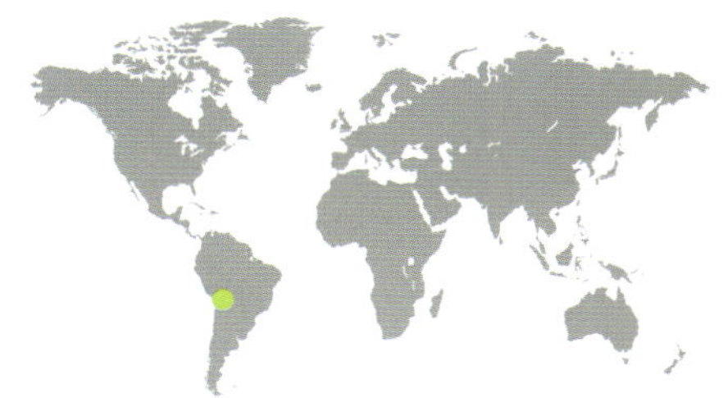

COCHRANELLA NOLA

BELL GLASS FROG

HARVEY, 1996

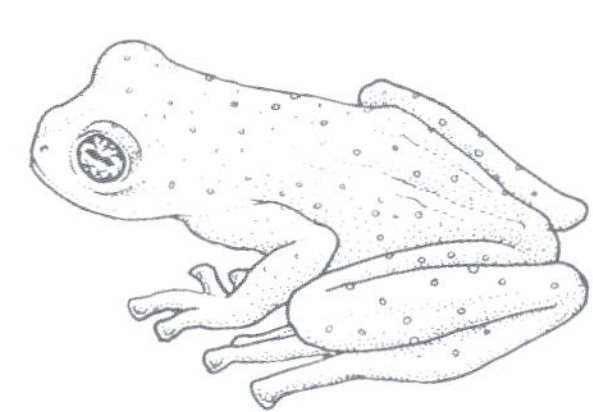

ADULT LENGTH
Male
3/4–7/8 in (20–21 mm)
Female
15/16–1 in (24–26 mm)

The specific name of this tiny frog, *nola*, means "little bell" and refers to its mating call, a high-pitched "pink." It is most commonly found along forest streams, where males form small choruses, usually of fewer than six individuals. The female is unusual among frogs of this family in laying her eggs on rocks in streams, rather than on foliage. The frog lives in drier forests than most *Cochranella* species. It is known from only one small area of the Bolivian Andes, where its population is in decline, but it is thought likely that it exists elsewhere.

Actual size

The Bell Glass Frog has large eyes and well-developed adhesive disks on the ends of its fingers and toes. It is uniformly green all over and is covered in tiny spicules, giving its skin a granular appearance. The irises are white, with a reticulated pattern of thin black lines.

SIMILAR SPECIES

Cochranella nola is similar to, and is found in the same general area as, the Bolivian Glass Frog (*Nymphargus bejaranoi*), an Endangered species. It is also similar to two species from Peru, the Data Deficient Spotted Glass Frog (*N. ocellatus*) and the Near Threatened Cuzco Glass Frog (*Rulyrana spiculata*). All were previously included in the genus *Cochranella*.

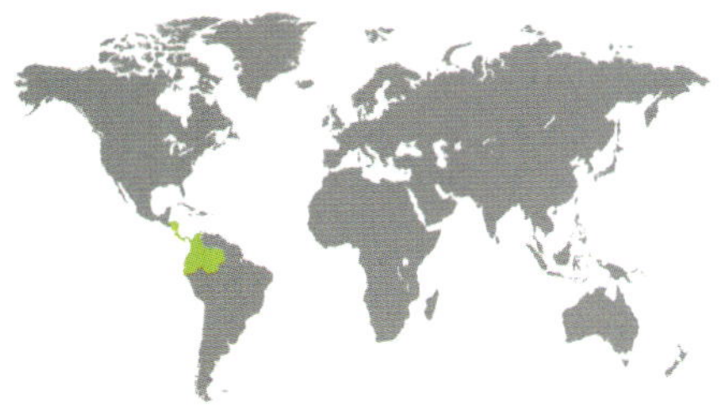

FAMILY	Centrolenidae: Centroleninae
OTHER NAMES	Nicaragua Giant Glass Frog
DISTRIBUTION	Honduras, Nicaragua, Costa Rica, Panama, Colombia, Ecuador. From sea level to 6,200 ft (1,900 m) altitude
ADULT HABITAT	Rainforest
LARVAL HABITAT	Mountain streams
CONSERVATION STATUS	IUCN Least Concern. Has declined in parts of Costa Rica but is stable elsewhere

ADULT LENGTH
Male
⅞–1⅛ in (21–28 mm)
Female
1–1¼ in (25–31 mm)

ESPADARANA PROSOBLEPON

EMERALD GLASS FROG

(BOETTGER, 1892)

Actual size

This nocturnal frog, found along the banks of fast-flowing streams, breeds between May and November. Males are strongly territorial and space themselves about 10 ft (3 m) apart. They call from leaves above a stream, producing a "dik… dik… dik" sound that warns off rival males and attracts females. If a rival enters a male's territory, a fight ensues, with the two males wrestling while hanging upside down from leaves. Such fights can last as long as 30 minutes. Mating pairs lay around 20 black eggs, covered in jelly, on a leaf. After ten days, the eggs hatch and the tadpoles fall into the stream.

SIMILAR SPECIES

There are two other species in the genus *Espadarana*. The Andes Giant Glass Frog (*E. andina*) is a common species in Colombia and Venezuela. *Espadarana callistomma* is found in Ecuador, where it is thought to be vulnerable to habitat destruction. The breeding behavior of *E. prosoblepon* is similar to that of Fleischmann's Glass Frog (*Hyalinobatrachium fleischmanni*; page 273).

The Emerald Glass Frog has a broad head with a blunt, rounded snout and large, bulging eyes. Its skin is smooth. The dorsal surfaces are emerald-green with black spots, and the ventral surface is yellow. The iris of the eye is silver with a fine black reticulated pattern.

FAMILY	Centrolenidae: Centroleninae
OTHER NAMES	Anomalous Glass Frog, Napo Cochran Frog
DISTRIBUTION	Northern Ecuador
ADULT HABITAT	Cloud forest at 5,700 ft (1,740 m) altitude
LARVAL HABITAT	Unknown; probably in streams
CONSERVATION STATUS	IUCN Endangered. Known from only one location, where much of its habitat has been destroyed by encroaching agriculture

NYMPHARGUS ANOMALUS

NAPO GLASS FROG

(LYNCH & DUELLMAN, 1973)

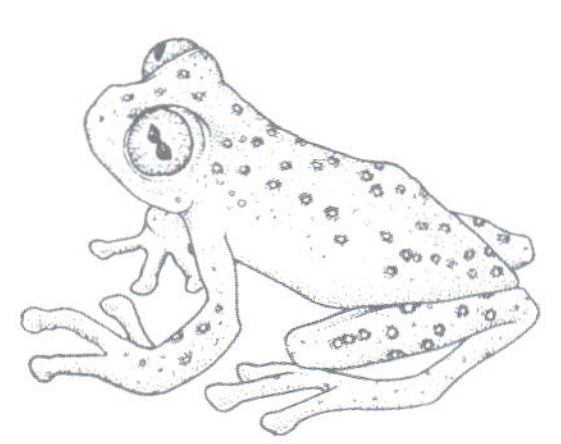

ADULT LENGTH
Male
⅞–1 in (21–25 mm)
Female
1–1¹⁄₁₆ in (25–27 mm)

The description of this frog is based on a single male specimen, found in 1971. It was not seen after that until 2009, when it was found again. Among glass frogs, which are typically bright green in color, the species is very unusual in being yellowish brown or tan. It is also unusual, but not unique, in having the dark spots on its back and legs that are arranged in rings, or ocelli. It has been found close to streams, where it is presumed to breed. Males call at night from the upper surfaces of leaves, and the eggs are laid on moss-covered twigs.

Actual size

The Napo Glass Frog has a slender body, long, slender limbs, and a wide head with a blunt snout. The eyes are large, point forward, and have yellow irises. The upper surfaces are tan, yellowish, or pink, decorated with circular ocelli of black spots that surround orange spots. The ocelli contain very small spicules.

SIMILAR SPECIES

Within the genus *Nymphargus*, two other species have spots that are arranged in the form of ocelli. Cochran's Glass Frog (*N. cochranae*) occurs on the lower Amazonian slopes of the Andes in Ecuador and Colombia. Lynch's Glass Frog (*N. ignotus*) is another yellowish-brown frog, and is found in Colombia. *Nymphargus* contains 43 species. See also the Red-spotted Glass Frog (*N. grandisonae*; page 268).

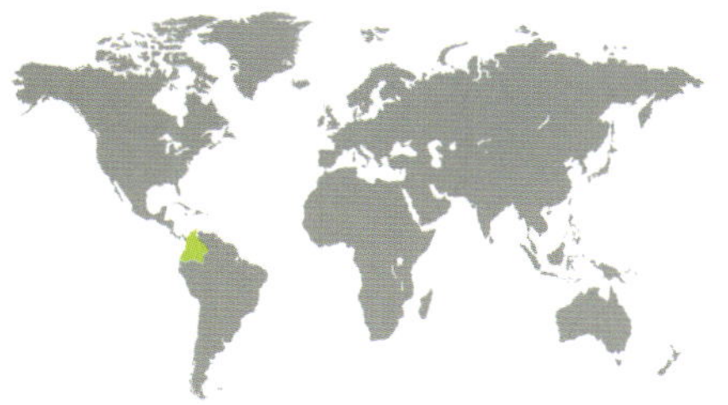

FAMILY	Centrolenidae: Centroleninae
OTHER NAMES	Giant Glass Frog
DISTRIBUTION	Colombia, northern Ecuador
ADULT HABITAT	Low montane forest and cloud forest at altitudes of 3,740–8,890 ft (1,140–2,710 m)
LARVAL HABITAT	Streams and ponds
CONSERVATION STATUS	IUCN Least Concern. Declining in places due to habitat destruction, but occurs in several protected areas

ADULT LENGTH
Male
1–1⅛ in (25–29 mm)
Female
1⅛–1¼ in (29–31 mm)

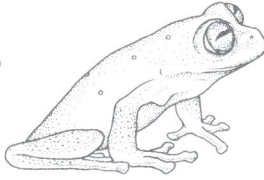

NYMPHARGUS GRANDISONAE

RED-SPOTTED GLASS FROG

(COCHRAN & GOIN, 1970)

Actual size

Despite their delicate build, males of this arboreal frog engage in protracted fights, which sometimes result in injury. They fight in defense of their calling sites, which individual males occupy for 3–92 days (on average for 36 days). Males have a small spur on their upper arm and it this that they use in fights to injure their rivals. Fights commonly occur while both protagonists are hanging from a twig by their long, slender legs. Mating and egg-laying occur on the upper surface of a leaf, the female laying 30–70 eggs. On hatching, the tadpoles fall into water below.

SIMILAR SPECIES

Of the other 42 species so far described in the genus *Nymphargus*, most similar to *N. grandisonae* is the Pepper Glass Frog (*N. griffithsi*), which gets its name from the numerous dense black speckles that decorate its green back. It has a small range in the Ecuadorian Andes. The Las Gralarias Glass Frog (*Nymphargus lasgralarias*), described in 2012 from a single location in Ecuador, has no spots at all. See also the Napo Glass Frog (*N. anomalus*; page 267).

The Red-spotted Glass Frog has a slender body, long, slender limbs, and large, protruding eyes with yellow irises. The upper surfaces are uniform pale green, adorned with small round, bright red spots. The underside of the body and legs are pale yellow to white. There are large disks on the fingers and toes, and males have a small protuberance on each upper arm.

FAMILY	Centrolenidae: Centroleninae
OTHER NAMES	None
DISTRIBUTION	Colombia
ADULT HABITAT	Forest at altitudes of 3,000–5,400 ft ((900–1,650 m)
LARVAL HABITAT	In streams. Has a small range that is severely fragmented by habitat destruction, but also occurs in protected areas
CONSERVATION STATUS	IUCN Near Threatened. Has a small range that is severely fragmented by habitat destruction, but also occurs in protected areas

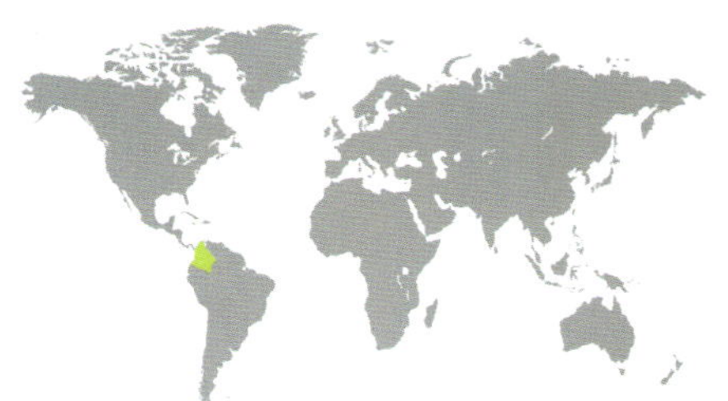

RULYRANA SUSATAMAI

SUSATAMA'S GLASS FROG

(RUÍZ-CARRANZA & LYNCH, 1995)

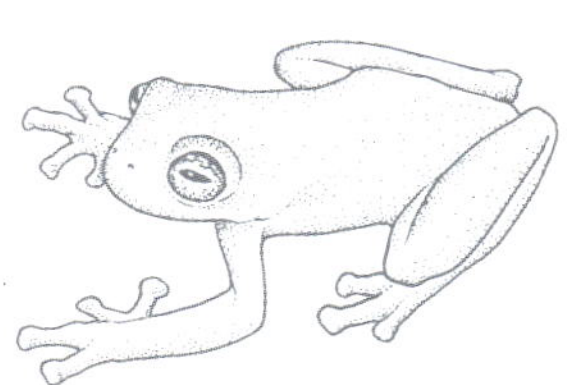

ADULT LENGTH
Male
7/8–1 in (22–26 mm)
Female
1–1 1/8 in (26–28 mm)

This beautiful glass frog has green bones and a translucent patch of skin covering the peritoneum, through which the internal organs can be seen. This species lives close to streams in the forests covering the lower eastern slopes of the Andes in central Colombia. Males call from the upper surfaces of leaves, or from a rock overhanging a stream. The eggs are laid at the male's calling site and, when they hatch, the tadpoles fall into the stream below to complete their development.

Actual size

Susatama's Glass Frog has a slender body, a wide head, and large eyes with yellow irises. There are well-developed disks on the fingers and toes. The upper surfaces are dark green, with numerous small yellow spots. The underside of the body and legs is pale yellow or white.

SIMILAR SPECIES

Rulyrana is a genus of six species of bright green glass frogs, formerly included in the genus *Cochranella*. The forests in which all species live are subject to deforestation to make way for agriculture and human settlements. The Western Glass Frog (*R. adiazeta*) occurs in Colombia and is listed as Vulnerable. The Cuzco Glass Frog (*R. spiculata*) is a Near Threatened species from Peru.

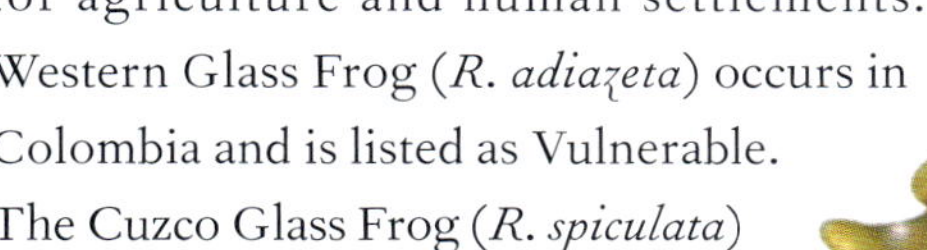

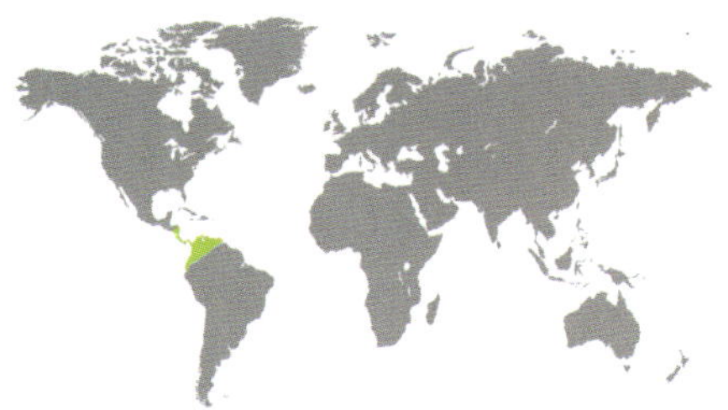

FAMILY	Centrolenidae: Centroleninae
OTHER NAMES	Limon Giant Glass Frog
DISTRIBUTION	Central and South America, from southern Nicaragua in the north to northwestern Ecuador in the south
ADULT HABITAT	Wet forest up to 4,650 ft (1,420 m) altitude
LARVAL HABITAT	Streams
CONSERVATION STATUS	IUCN Least Concern. Has declined in places but occurs within some protected areas

ADULT LENGTH
Male
$1\frac{1}{16}$–$1\frac{1}{8}$ in (27–29 mm)
Female
$1\frac{1}{8}$–$1\frac{5}{16}$ in (28–34 mm)

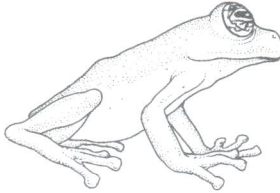

SACHATAMIA ILEX

GHOST GLASS FROG

(SAVAGE, 1967)

Actual size

The Ghost Glass Frog has a slender body, a short, rounded snout, protruding nostrils, and large, forward-pointing eyes. There are well-developed disks on the fingers and toes, and the feet are webbed. The upper surfaces are deep leaf-green, the throat and belly are white, and the irises of the eyes are silver with a blue-black reticulated pattern.

This small frog is actually the largest glass frog to be found in Central America. It is most often found in the spray zone of waterfalls and forest streams. During the day, it sleeps on top of exposed leaves, its color adjusted to match that of its background. At night, males call from the upper surfaces of leaves, usually facing the leaf's tip. Their call is a single high-pitched "click," repeated after several minutes. The eggs are black and are laid in a cluster, glued to the upper surface of a leaf overhanging a stream, into which the tadpoles fall on hatching.

SIMILAR SPECIES

Five species have been described in the genus *Sachatamia*. The Yellow-flecked Glass Frog (*S. albomaculata*) is found from Costa Rica to Ecuador. Its eggs are black and white. *Sachatamia punctulata*, from Colombia, lays white eggs. As a result of the destruction and deterioration of its forest habitat, it is listed as Vulnerable.

FAMILY	Centrolenidae: Centroleninae
OTHER NAMES	None
DISTRIBUTION	Central America, from Honduras in the north to northern Ecuador in the south
ADULT HABITAT	Lowland wet forest up to 3,150 ft (960 m) altitude
LARVAL HABITAT	Fast-moving streams
CONSERVATION STATUS	IUCN Least Concern. Declining in parts of its range, but occurs within some protected areas

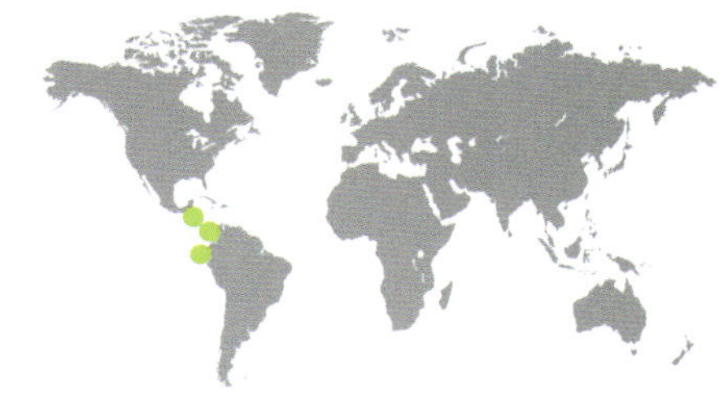

TERATOHYLA PULVERATA

CHIRIQUI GLASS FROG

(PETERS, 1873)

ADULT LENGTH
Male
⅞–1⅛ in (22–29 mm)
Female
$^{15}/_{16}$–$1^{5}/_{16}$ in (23–33 mm)

This tiny frog is common in Honduras, but is rare in other parts of its range. It breeds in the rainy season, between May and October. Males call from foliage along streams, producing a rasping "dik... dik... dik." The female lays her eggs on the upper surface of a leaf hanging over a stream so that, when they hatch, the tadpoles fall into the water below. It is not known whether one or both parents guard the eggs, as in some glass frogs. This species survives in small forest remnants, suggesting that it is less susceptible to deforestation than many other species.

Actual size

The Chiriqui Glass Frog is a slender-bodied frog with a rounded snout, and large, protruding eyes that point forward. There are well-developed disks on its fingers and toes. The upper surfaces are lime-green with small yellow or white spots, and the flanks and underside are white. The pupils of the eyes are silvery with black markings.

SIMILAR SPECIES

Five species have been described in the genus *Teratohyla*. The Spiny Glass Frog (*T. spinosa*) is another very small frog, from Central America and Colombia. It is uniform green in color and the male has spines at the base of his thumbs. The most recently described species was Amelie's Glass Frog (*T. amelie*) in 2009, from Ecuador and Peru.

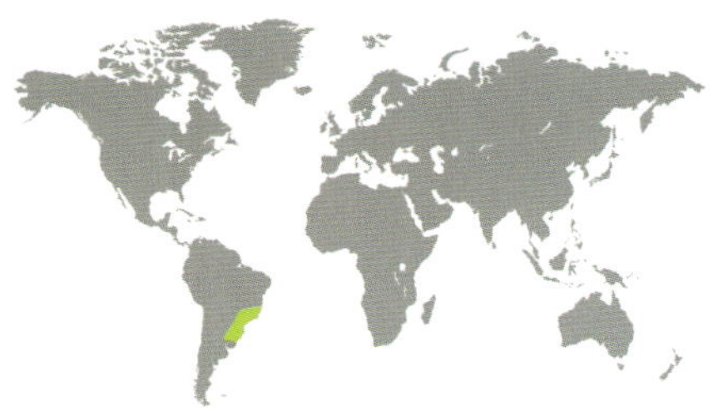

FAMILY	Centrolenidae: Centroleninae
OTHER NAMES	Includes *Vitreorana uranoscopa*, Humboldt's Glass Frog
DISTRIBUTION	Eastern and southeastern Brazil, northern Argentina 1,200 m asl
ADULT HABITAT	Forest up to 4,000 ft (1,200 m) altitude
LARVAL HABITAT	Streams
CONSERVATION STATUS	IUCN Vulnerable. Declining in places but occurs in some protected areas.

ADULT LENGTH
Male
¾–⅞ in (19–23 mm)
Female
average $^{15}/_{16}$ in (24 mm)

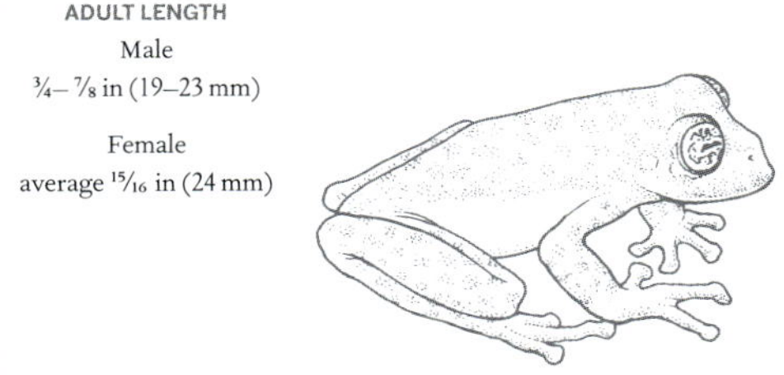

VITREORANA PARVULA

LAGES GLASS FROG

(BOULENGER, 1895)

Actual size

The Lages Glass Frog has a rounded, slightly flattened snout and eyes that are directed forward. There are well-developed disks on the fingers and toes. The upper surfaces are green, dappled with tiny white and black spots. The skin on the belly is translucent and the internal organs are visible beneath it. The irises of the eyes are gold with black markings.

Males of this small frog can be heard calling near forest streams within Brazil's Atlantic Forest between August and January, but mating activity peaks in December. They do not form choruses but space themselves out along a stream. Although active only at night, they have been reported using visual displays, such as lifting their limbs, to communicate with one another. The female lays 29–32 eggs attached to the upper surface of a leaf hanging over a stream. The tadpoles are adapted for digging in mud among dead leaves, with a long, muscular tail and small eyes.

SIMILAR SPECIES

To date, ten species have been described in the genus *Vitreorana*. As recently as 2014, the Spotted-eyed Glass Frog (*V. baliomma*) was differentiated from the Rio Glass Frog (*V. eurygnatha*), an inhabitant of the Atlantic Forest that is very similar to *V. parvula* but appears to prefer breeding in smaller streams. It is declining in many parts of its range but is listed as of Least Concern.

FAMILY	Centrolenidae: Hyalinobatrachinae
OTHER NAMES	Northern Glass Frog
DISTRIBUTION	Central and South America, from Mexico in the north to Ecuador in the south
ADULT HABITAT	Lowland forest, near streams
LARVAL HABITAT	Streams
CONSERVATION STATUS	IUCN Least Concern. Common in parts of its range and with no evidence of general decline

HYALINOBATRACHIUM FLEISCHMANNI

FLEISCHMANN'S GLASS FROG

(BOETTGER, 1893)

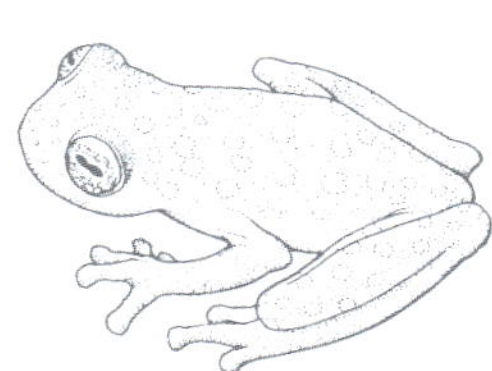

ADULT LENGTH
Male
¾–1 in (19–26 mm)
Female
⅞–1¼ in (23–32 mm)

Actual size

In this small arboreal frog, both parents look after the eggs until they hatch. During mating, the female lays 18–30 eggs on the underside of a leaf hanging over a stream. Both parents attend the eggs, sitting on them at night and staying close to them by day. The male periodically urinates on them to keep them moist. Breeding occurs between March and November, males establishing territories that they defend by calling. Their call is a "wheet," to which a "mew" is added if another male intrudes. If an intruder does not leave, a fight ensues.

SIMILAR SPECIES

To date, 35 species have been described in the genus *Hyalinobatrachium*, distributed across tropical Central and South America. The Guacharaquita Glass Frog (*H. pallidum*) is found in Venezuela and is listed as Near Threatened, while the Esmeralda Glass Frog (*H. esmeralda*), from Colombia, is Endangered. Both have very restricted ranges and have been subject to destruction of their forest habitat. See also the Reticulated Glass Frog (*H. valerioi*; page 274).

Fleischmann's Glass Frog has a flattened body, a broad head, and a round snout. The protruding eyes point forward. The upper surfaces are pale green with yellow or yellow-green spots. The underside is white and partially transparent, revealing the internal organs within a white sheath. The tips of the fingers and toes are yellow, and the irises of the eyes are gold.

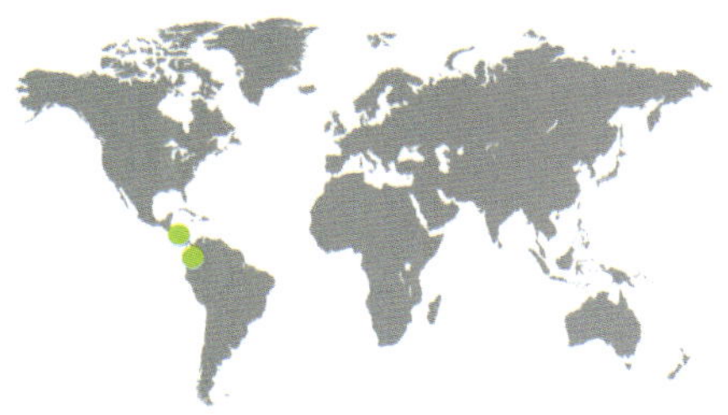

FAMILY	Centrolenidae: Hyalinobatrachinae
OTHER NAMES	La Palma Glass Frog, Valerio's Glass Frog
DISTRIBUTION	Central and South America, from Costa Rica in the north to Colombia and Ecuador in the south
ADULT HABITAT	Rainforest, near streams, up to 1,300 ft (400 m) altitude
LARVAL HABITAT	Streams
CONSERVATION STATUS	IUCN Least Concern. Not common, but has a stable population in Costa Rica; declining in Panama

ADULT LENGTH
Male
¾–^15^⁄~16~ in (19–24 mm)

Female
⅞–1 in (22–26 mm)

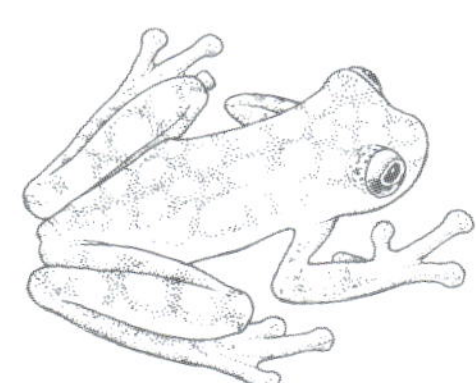

HYALINOBATRACHIUM VALERIOI

RETICULATED GLASS FROG

(DUNN, 1931)

Actual size

The blotchy skin pattern of this small frog is thought to mimic a clump of eggs, deflecting the attention of wasps that steal eggs. The female lays around 35 eggs in a sticky mass on the underside of a leaf, up to 20 ft (6 m) above a stream. She then leaves them in the care of the male, who guards them by day and night and also calls to attract other females. A male can gather as many as seven egg masses, which he keeps moist and protects against wasps. When the eggs hatch, the tadpoles fall into the stream below, where they complete their development.

SIMILAR SPECIES

The Plantation Glass Frog (*Hyalinobatrachium colymbiphyllum*) is a Central American species in which the male defends the eggs only at night; as a result, many are stolen by wasps. The Green-striped Glass Frog (*H. talamancae*) has a fragmented range along the mountains of Costa Rica. Its call is a long, low whistle. See also Fleischmann's Glass Frog (*H. fleischmanni*; page 273).

The Reticulated Glass Frog has a broad head and a blunt snout. The protruding eyes point forward. There are large adhesive disks on the fingers and toes, and the feet are webbed. The upper surfaces are green with large, yellow blotches and small black spots. The underside is transparent, revealing the red heart. The irises of the eyes are gold.

FAMILY	Centrolenidae: *incertae sedis*
OTHER NAMES	None
DISTRIBUTION	Northeastern Colombia
ADULT HABITAT	Lowland wet forest at altitudes of 3,200–5,900 ft (980–1,790 m)
LARVAL HABITAT	Streams
CONSERVATION STATUS	IUCN Vulnerable. Has a restricted range and its habitat is declining, but it occurs within a protected area

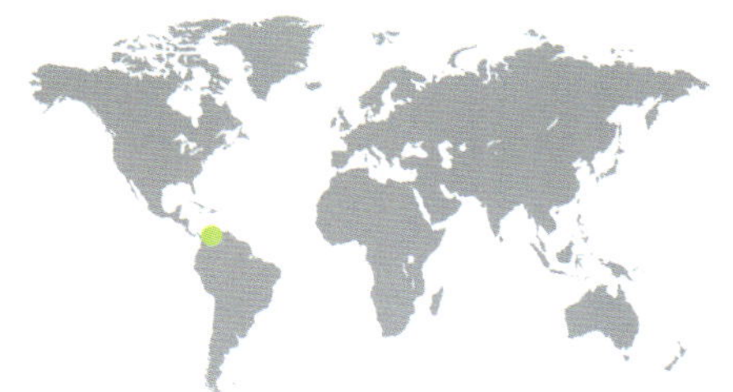

ADULT LENGTH
1⅛–1¼ in (28–31 mm)

IKAKOGI TAYRONA

MAGDALENA GIANT GLASS FROG

(RUÍZ-CARRANZA & LYNCH, 1991)

In comparison to other members of its family, this is a large glass frog. It is found at lower altitudes around the perimeter of the Sierra Nevada de Santa Marta, an isolated mountain range in Colombia. It is known from fewer than ten locations and much of its forest habitat is subject to encroaching agriculture, which leads to destruction of its forest habitat and pollution of its breeding streams by crop sprays. It is unusual among glass frogs in that the eggs may be laid on either the upper or the lower side of a leaf. The eggs are guarded by the female.

Actual size

SIMILAR SPECIES

Ikakogi tayrona was formerly included within the genus *Centrolene* (see pages 261 and 262). It was the only species in *Ikakogi* until 2019 when the Twin Glass Frog (*I. ispacue*) was described from Colombia. *Ikakogi tayrona* differs from other frogs in the family Centrolenidae in having white, not green, bones. It seems also to be unique in that it is the female, rather than the male, who cares for the eggs.

The Magdalena Giant Glass Frog has a slender body, a broad head with a rounded snout, and very large, protruding eyes. It has well-developed disks on its fingers and toes. The upper surfaces are bright green, decorated with tiny pale spots. The fingers and toes, and the underside of the body and legs, are yellow.

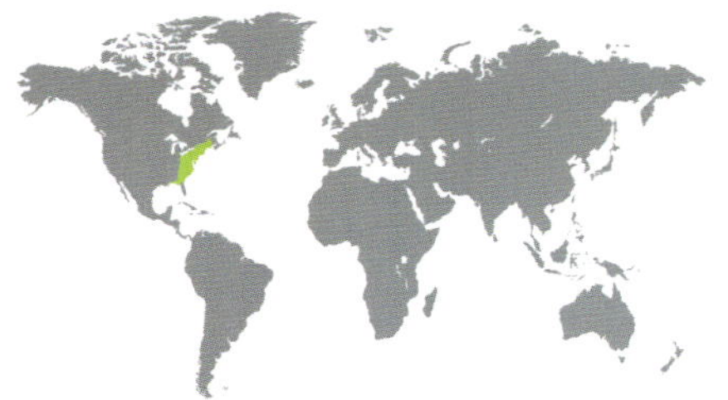

FAMILY	Hylidae: Hylinae
OTHER NAMES	None
DISTRIBUTION	Eastern USA, excluding the Florida peninsula
ADULT HABITAT	Edges of lakes and ponds
LARVAL HABITAT	Lakes and ponds
CONSERVATION STATUS	IUCN Least Concern. Declining in parts of its range. Listed as Endangered in Minnesota and Wisconsin

ADULT LENGTH
⅝–1½ in (16–38 mm); females are slightly larger than males

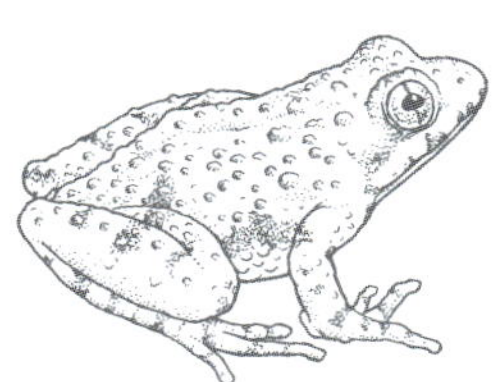

ACRIS CREPITANS

NORTHERN CRICKET FROG

BAIRD, 1854

Although it belongs to the tree frog family, this small frog does not climb. Its powerful hind legs make it capable of prodigious leaps for such a small frog. It breeds from late winter to summer, males calling mostly at night and occasionally by day from the bank of a pond or lake, or from floating vegetation. Their call is a series of clicks, like two stones being tapped together. Females tend to be preferentially attracted to the lower-pitched calls of larger males, and to the calls of males that call at the highest rate. They lay around 400 eggs, singly or in small clumps.

SIMILAR SPECIES

There are two other species in the genus *Acris*. The Southern Cricket Frog (*A. gryllus*) is found in southeastern USA. It has longer legs than the other two species and is noted for its leaping prowess. Blanchard's Cricket Frog (*A. blanchardi*), previously considered a subspecies of *A. crepitans*, occurs in central USA, Canada, and Mexico. It is slightly larger and more uniformly colored than *A. crepitans*.

The Northern Cricket Frog has warty skin and is highly variable in color. Most individuals are brown, green, or gray, and there is usually a dark triangular patch between the eyes. There is often a broad green, yellow, orange, or red stripe down the back, and there are dark spots on the flanks. The long, muscular hind limbs are marked with dark stripes and the toes are webbed.

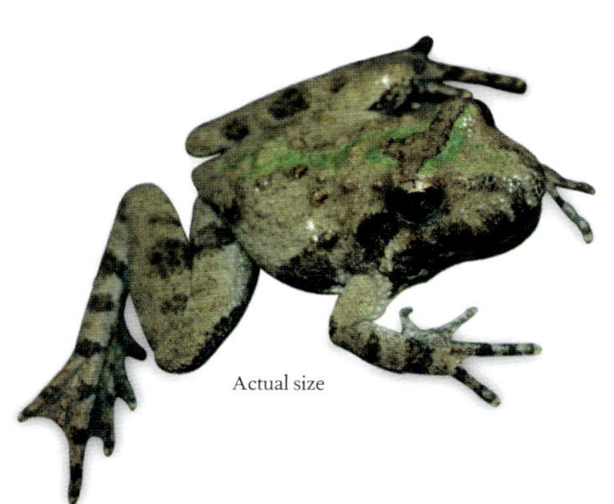

Actual size

FAMILY	Hylidae: Hylinae
OTHER NAMES	None
DISTRIBUTION	Southeastern Brazil
ADULT HABITAT	Rainforest
LARVAL HABITAT	Underground pools, then streams
CONSERVATION STATUS	IUCN Least Concern. Has remained common while other frogs living in its habitat have declined

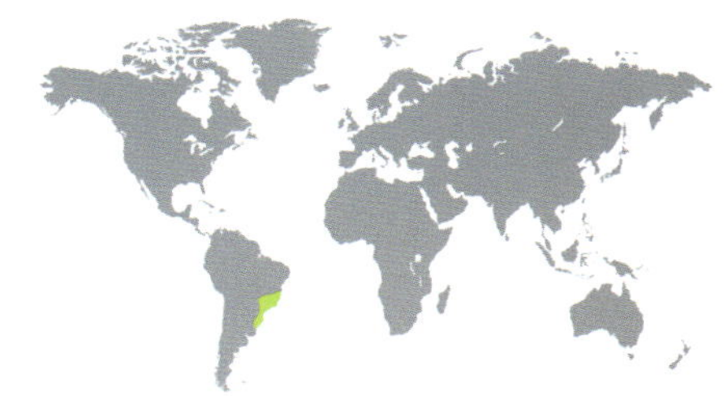

APLASTODISCUS LEUCOPYGIUS

GUINLE TREE FROG

(CRUZ & PEIXOTO, 1985)

ADULT LENGTH
Male
$1\frac{1}{2}$–$1\frac{13}{16}$ in (38–46 mm)
Female
$1\frac{11}{16}$–$1\frac{7}{8}$ in (43–47 mm)

Males of this inhabitant of Brazil's Atlantic Forest have a richer vocal repertoire than many frogs. In addition to an advertisement call, likened to the sound of a horn, they have a courtship call when interacting with females and an aggressive call when competing with rival males. They call during the rainy season, from October to March, but mating usually occurs from December to February. The male constructs an underground chamber in muddy soil, which partially fills with water. The female inspects it during an elaborate courtship ritual before laying her eggs inside. Later, rain washes the tadpoles into streams.

Actual size

The Guinle Tree Frog has a plump body, a rounded snout, large eyes, and large disks on its fingers and toes. There is a white spur on each heel. The upper surfaces are bright green with a scattering of small white spots. The lower lip is white, and the irises of the eyes are gold, tinged with orange.

SIMILAR SPECIES

The genus *Aplastodiscus* currently contains 17 species found in southern Brazil and northern Argentina. Lutz's Tree Frog (*A. albosignatus*) is a Brazilian species with very similar reproductive behavior to *A. leucopygius*. Its bright green color comes largely from its muscles and bones, the skin on the back being devoid of pigment. All species in *Aplastodiscus* are believed stable and listed as Least Concern.

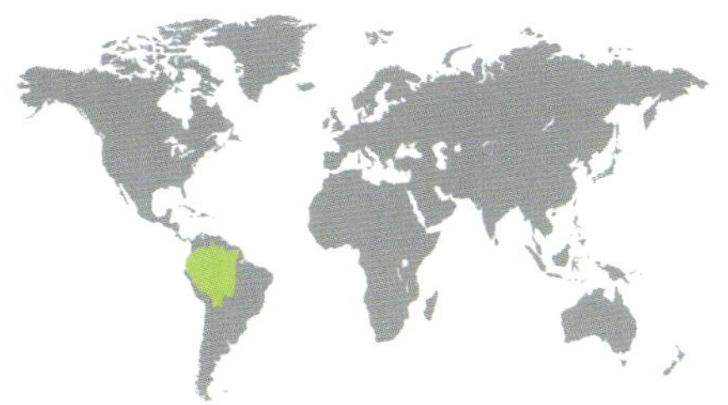

FAMILY	Hylidae: Hylinae
OTHER NAMES	Duck-footed Frog, Rusty Tree Frog
DISTRIBUTION	Amazon Basin: Brazil, Bolivia, Peru, Ecuador, Colombia, Venezuela, Guyana, Suriname, French Guiana, up to 3,300 ft (1,000 m) altitude
ADULT HABITAT	Tropical rainforest
LARVAL HABITAT	Pools, then streams
CONSERVATION STATUS	IUCN Least Concern. Abundant across most of its large range

ADULT LENGTH
Male
$3\frac{15}{16}$–$5\frac{1}{16}$ in (101–128 mm)
Female
$3\frac{9}{16}$–$4\frac{7}{8}$ in (91–123 mm)

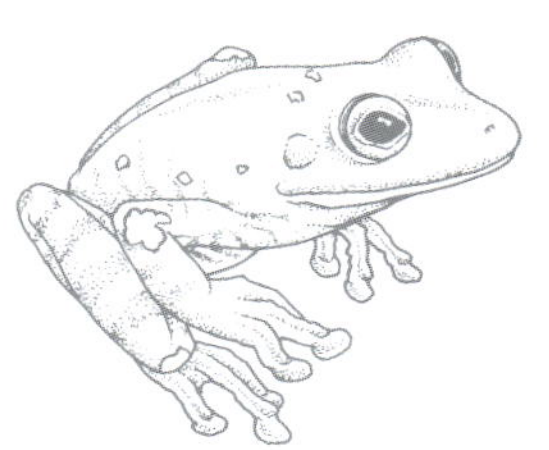

BOANA BOANS

GIANT GLADIATOR TREE FROG

(LINNAEUS, 1758)

The Giant Gladiator Tree Frog has a slender body, large eyes, webbed hands and feet, and large adhesive disks on its fingers and toes. The upper surfaces are brown in males and orange-brown in females, and there may be delicate dark stripes on the flanks and legs. Some individuals are spotted. The underside is cream or white.

This large tree frog is active at night and is most common close to streams. It breeds in the dry season, between July and December. Males call from trees, but mating occurs at small pools close to, and often linked to, streams. If there are no natural pools available, the male digs one. The female lays 1,300–3,000 eggs, which form a film on the water's surface. The male defends his pool against other males, using sharp spikes on his hands to inflict wounds on opponents. As they grow, the tadpoles, which are unpalatable to fish, make their way to streams, where they complete their development.

SIMILAR SPECIES

Only a few of the 99 species in the genus *Boana* are of conservation concern. The Campo Grande Tree Frog (*B. cymbalum*) was found only near São Paulo, Brazil. Because of destruction of its habitat, it was listed as Critically Endangered; it is now listed as Extinct. The Los Bracitos Tree Frog (*B. heilprini*) occurs on the island of Hispaniola, where its habitat has been seriously degraded; it is listed as Vulnerable. See also Map Tree Frog (*B. geographica*; page 279) and Rosenberg's Gladiator Frog (*B. rosenbergi*; page 280).

Actual size

FAMILY	Hylidae: Hylinae
OTHER NAMES	None
DISTRIBUTION	Lowland tropical regions east of the Andes in South America; Trinidad
ADULT HABITAT	Forest, woodland, grassland, and savanna up to 1,600 ft (500 m) altitude (3,900 ft/1,200 m in Ecuador)
LARVAL HABITAT	Rivers and pools
CONSERVATION STATUS	IUCN Least Concern. Abundant throughout most of its enormous range

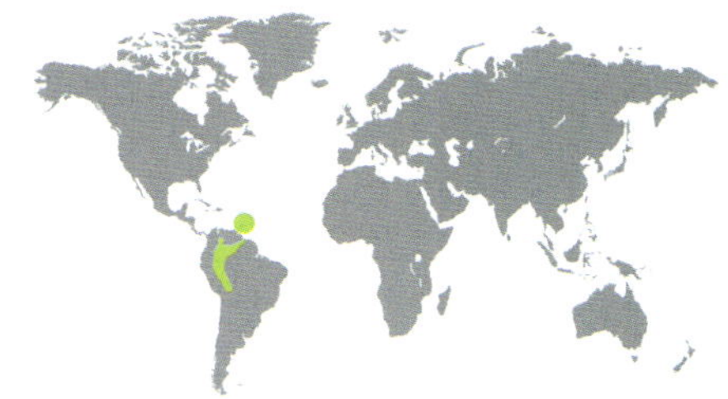

ADULT LENGTH
Male
up to 2⅜ in (60 mm)

Female
up to 3 in (76 mm)

BOANA GEOGRAPHICA

MAP TREE FROG

(SPIX, 1824)

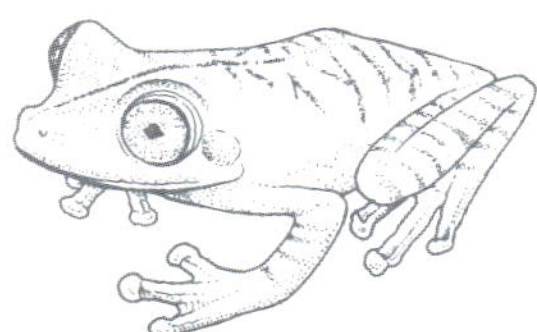

This common tree frog has a variety of defensive behavior patterns, including feigning death, emitting a foul odor, and a "boo" display, in which it spreads its hands out on either side of its head. Breeding occurs after rain, males calling from branches above water. Their call has been described as a moan interrupted by chuckles. The female lays up to 2,000 eggs. Unlike the tadpoles of many frogs, those of this species are unpalatable to fish, enabling them to survive in most waterbodies. However, they are palatable to dragonfly larvae, and the tadpoles reduce their risk of being eaten by moving about in very compact shoals.

SIMILAR SPECIES

There are currently 99 species recognized in this genus, most of them previously assigned to the genus *Hyla*. The Blue-flanked Tree Frog (*Boana calcarata*) and Gunther's Banded Tree Frog (*B. fasciata*) both occur across much of the Amazon Basin and have similar habits to *B. geographica*. See also the Giant Gladiator Tree Frog (*B. boans*; page 278)and Rosenberg's Gladiator Frog (*B. rosenbergi*; page 280).

The Map Tree Frog has a slender body, long, slender limbs, a wide head, and huge eyes. There are characteristic white spurs on the heels of the hind legs, and the hands and feet are partially webbed. It is highly variable in color, varying both according to age and location. Young individuals are creamy white with black markings; adults may be green, brown, or yellowish.

Actual size

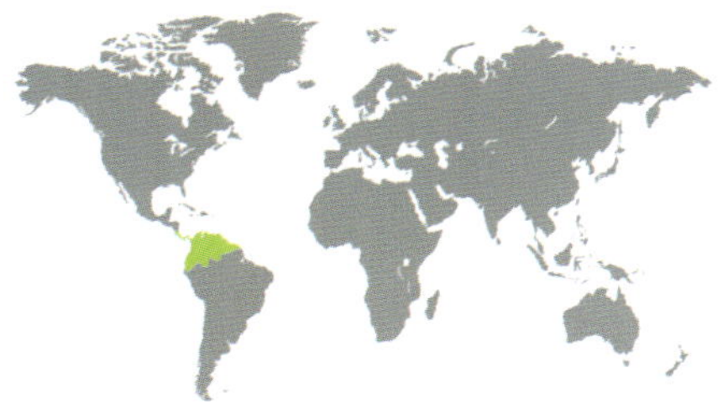

FAMILY	Hylidae: Hylinae
OTHER NAMES	Rosenberg's Gladiator Tree Frog, Rosenberg's Tree Frog
DISTRIBUTION	Costa Rica, Panama, Colombia, Venezuela, Ecuador
ADULT HABITAT	Tropical forest
LARVAL HABITAT	Pools made by their father
CONSERVATION STATUS	IUCN Least Concern. It is declining in places but has a large range and lives in some altered habitats

ADULT LENGTH
Male
$2\frac{3}{8}$–$3\frac{9}{16}$ in (60–91 mm)
Female
$2\frac{3}{8}$–$3\frac{11}{16}$ in (60–95 mm)

BOANA ROSENBERGI

ROSENBERG'S GLADIATOR FROG

(BOULENGER, 1898)

Males of this frog possess a sharp spine on each forelimb. This is normally withdrawn into a sheath but, during fights, it is unsheathed and is used to puncture the eyes or eardrums of the opponent. Such fights can result in blindness, deafness, or death. Inter-male aggression in this species is in defense of water-filled pools that males construct at the start of the breeding season. They call from the edge of their nests, warning off rival males and attracting females to mate and lay their eggs in the nest. Intruding males are usually seen off by aggressive calling or chases, and fighting is engaged in only as a last resort.

Rosenberg's Gladiator Frog is a large tree frog with a long snout, adhesive disks on its fingers and toes, and webbing on its hands and feet. The back is tan or reddish brown with darker markings. There are narrow, wavy dark stripes on the flanks and often a dark stripe running down the middle of the head.

SIMILAR SPECIES

The Blacksmith Tree Frog (*Boana faber*) is found in tropical forests in Brazil, Argentina, and Paraguay. Males build and defend muddy pools in which their eggs and larvae develop. Fights consist of escalating phases, in which one of the males, usually the intruder, may move away at any phase, bringing the contest to an end. Fights occur only when the intruder has not moved away during any of the early phases. See also the Giant Gladiator Tree Frog (*B. boans*; page 278) and the Map Tree Frog (*B. geographica*; page 279).

Actual size

FAMILY	Hylidae: Hylinae
OTHER NAMES	None
DISTRIBUTION	Nicaragua, Costa Rica, Panama
ADULT HABITAT	Wet lowland forest up to 2,100 ft (650 m) altitude
LARVAL HABITAT	In swamps
CONSERVATION STATUS	IUCN Least Concern. Has a large range that includes several protected areas and its population is stable

ADULT LENGTH
Male
1 9/16–1 15/16 in (39–49 mm)
Female
1 13/16–2 1/8 in (46–55 mm)

BOANA RUFITELA

CANAL ZONE TREE FROG

(FOUQUETTE, 1961)

This very beautiful tree frog is distinguished from many similar species by the bright red webbing between its fingers and toes. It breeds in swamps within lowland forest. Males call throughout the year, producing a series of high-pitched clucks, but breeding mostly occurs on rainy nights between August and October. It underwent a population explosion in 1980 on the island of Barro Colorado in the Panama Canal Zone, following unusually heavy rain. By 1983, however, the population had returned to its normal level. Males have a small spine on each hand, which they are thought to use during fighting over females.

Actual size

The Canal Zone Tree Frog has large, protruding eyes, large disks on its fingers and toes, and a small projection on each heel. The upper surfaces are lime-green or bluish green with scattered black, blue, or white spots. The flanks are yellow-green and there may be a yellow or red longitudinal stripe. The webbing on the hands and feet is bright red.

SIMILAR SPECIES

Central and South America are home to 99 species of *Boana*, many of them very common. The White-edged Tree Frog (*B. albomarginata*) occurs in Brazil, Gunther's Banded Tree Frog (*B. fasciata*) occurs in Ecuador and Peru, and the Montevideo Tree Frog (*B. pulchella*) is found in Uruguay and southern Brazil.

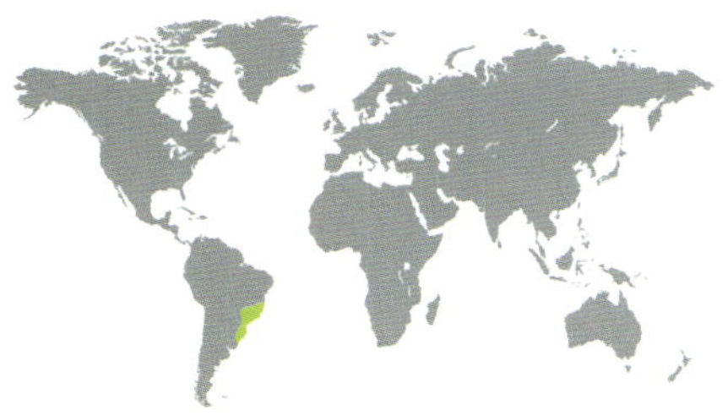

FAMILY	Hylidae: Hylinae
OTHER NAMES	None
DISTRIBUTION	Southeastern Brazil
ADULT HABITAT	Forest
LARVAL HABITAT	Ephemeral pools
CONSERVATION STATUS	IUCN Critically Endangered. Its habitat has been largely destroyed and it may be extinct

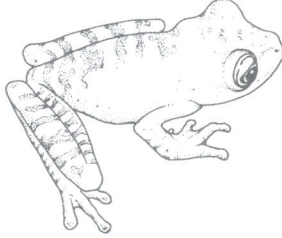

ADULT LENGTH
average 1¾ in (45 mm)

BOKERMANNOHYLA IZECKSOHNI

IZECKSOHN'S TREE FROG

(JIM & CARAMASCHI, 1979)

Actual size

Izecksohn's Tree Frog has a slender body, a rounded snout, and large, protruding eyes. There are well-developed disks on its fingers and toes. The back and the upper surfaces of the legs are brown with darker brown transverse stripes. The flanks are yellow, the underside is cream, and the irises of the eyes are gold.

This small, scarcely known tree frog is one of more than 200 amphibian species listed in 2012 as "lost," meaning that it is possibly extinct. An inhabitant of Brazil's Atlantic Forest, it is known from only four specimens. Following its description in 1979, it was not seen until 2006, when three individuals were discovered at a new location. The forest at its original location has been destroyed to make way for agriculture and human settlements. Males lack vocal sacs, suggesting that they do not call to attract females, but they do produce a distress call when molested.

SIMILAR SPECIES

To date, 31 species have been described in the genus *Bokermannohyla* (formerly *Hyla*), all of which are found in Brazil. Bahia Spiny-handed Tree Frog (*Bokermannohyla juiju*), from Bahia state in northeastern Brazil, was described in 2009. Males of this and some other *Bokermannohyla* species have a gland underneath their chin that may be similar to the mental gland that plays an important part in sexual interactions in some salamanders. See also the Reservoir Tree Frog (*B. luctuosa*; page 283).

FAMILY	Hylidae: Hylinae
OTHER NAMES	None
DISTRIBUTION	Southeastern Brazil
ADULT HABITAT	Forest
LARVAL HABITAT	Ephemeral pools
CONSERVATION STATUS	IUCN Least Concern. Declining in places but occurs within some protected areas

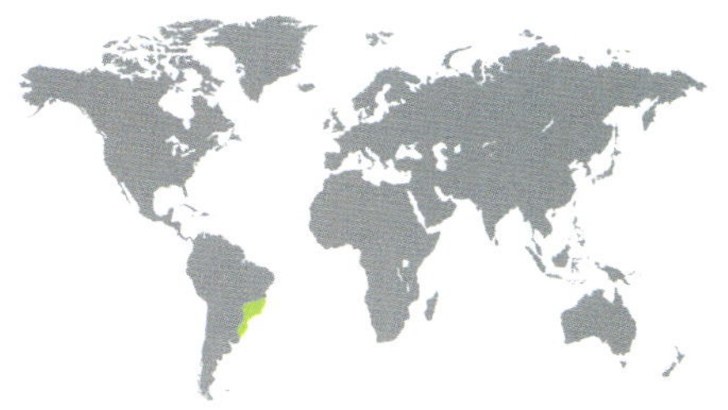

BOKERMANNOHYLA LUCTUOSA

RESERVOIR TREE FROG

(POMBAL & HADDAD, 1993)

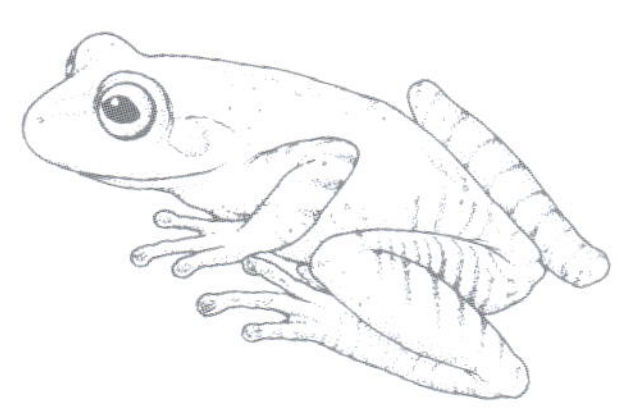

ADULT LENGTH
Male
2⅛–2⅜ in (55–61 mm)
Female
up to 2 13/16 in (70 mm)

The specific name of this large tree frog comes from the Latin *luctuosus*, meaning "sad," and refers to the mournful calls the males make late at night between November and March. Females lay their rather large eggs in small temporary pools and ponds on the floor of the species' Atlantic Forest home. Males have thicker arms than females and a spine on each hand, enabling them to maintain a firm grip on females during mating. When disturbed, young frogs have been seen to open their mouths wide and emit distress calls.

SIMILAR SPECIES

There are 31 species in the genus *Bokermannohyla* (formerly *Hyla*), all found in Brazil. The Espirito Santo Tree Frog (*B. circumdata*) is found in the mountains of southeastern Brazil. The Atlantic Forest Tree Frog (*B. hylax*) occurs in the states of Paraná and São Paulo; it has tested positive for the fungus that causes the disease chytridiomycosis. See also Izecksohn's Tree Frog (*B. izecksohni*; page 282).

The Reservoir Tree Frog has a slender body, a rounded snout, and large, protruding eyes. There are well-developed disks on its fingers and toes. The back and the upper surfaces of the legs are brown with darker brown transverse stripes. The flanks are yellow, the underside is cream, and the irises of the eyes are yellow.

Actual size

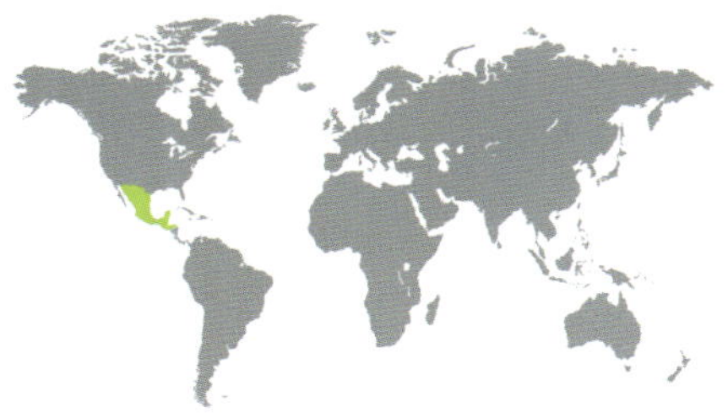

FAMILY	Hylidae: Hylinae
OTHER NAMES	None
DISTRIBUTION	Guatemala, Belize, Honduras, Mexico
ADULT HABITAT	Cloud forest up to 5,870 ft (1,790 m) altitude
LARVAL HABITAT	Very small pools in bromeliads
CONSERVATION STATUS	IUCN Least Concern, but has declined dramatically as a result of habitat destruction and, probably, chytridiomycosis

ADULT LENGTH
Male
15/16–1 3/16 in (24–30 mm)
Female
average 1 5/16 in (33 mm)

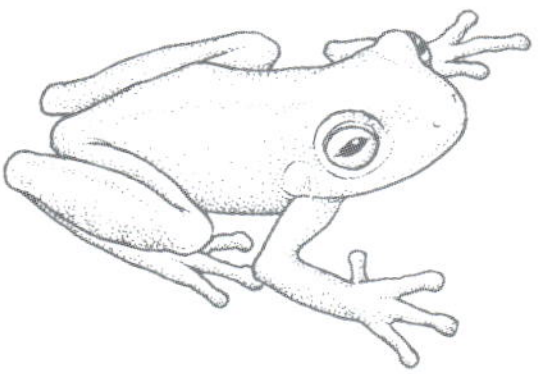

BROMELIOHYLA BROMELIACIA

BROMELIAD TREE FROG

(SCHMIDT, 1933)

Actual size

The Bromeliad Tree Frog has a slender body, a rounded snout, and rather short legs. There are well-developed disks on each of its fingers and toes. The upper surfaces are pale brown or yellow, there may be a pink tinge on the flanks, and the underside is whitish. The irises of the eyes are gold.

An inhabitant of cloud forest, this small tree frog lives around bromeliads and the leaf sheaths of banana trees. It is thought to breed year-round, depositing its eggs in the tiny pools that form in the axils of plants in the damp atmosphere that characterizes cloud forest. The tadpoles develop in these pools, using their long, muscular tails to wriggle around on the surface of leaves to move from pool to pool. It is estimated that this species declined by 50 percent between 2000 and 2010. This decline was most marked at higher altitudes in Guatemala, suggesting that it was largely due to chytridiomycosis.

SIMILAR SPECIES

There are two other species of *Bromeliohyla*. The Greater Bromeliad Tree Frog (*Bromeliohyla dendroscarta*), which is slightly larger than *B. bromeliacia*, is pale yellow in color and has bright red eyes. It also lives in bromeliads in cloud forest, in the states of Veracruz and Oaxaca in Mexico. The other species is Lesser Bromeliad Tree Frog (*B. melacaena*) from Honduras. Both species are listed as Endangered and thought to be victims of chytridiomycosis.

FAMILY	Hylidae: Hylinae
OTHER NAMES	None
DISTRIBUTION	Northeastern Brazil
ADULT HABITAT	Semiarid savanna
LARVAL HABITAT	Ephemeral streams
CONSERVATION STATUS	IUCN Least Concern. Its population is believed to be stable

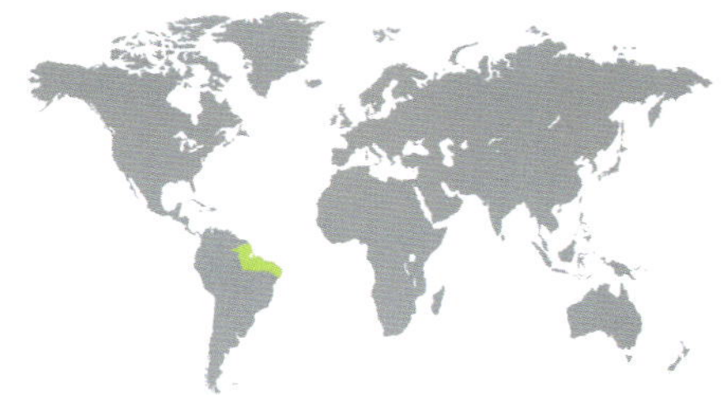

CORYTHOMANTIS GREENINGI

GREENING'S TREE FROG

BOULENGER, 1896

ADULT LENGTH
Male
average 2⅞ in (73 mm)
Female
average 3⅜ in (87 mm)

This unusual frog inhabits the dry Caatinga grassland of northeastern Brazil. It is most often found hiding in damp places, such as holes in trees, crevices in rocks, and bromeliads. It uses its curiously shaped head to block the entrance of its hiding place to protect itself against predators and to reduce water loss from its body. After sufficient rain has fallen to allow streams to flow, males defend small territories by calling and, if necessary, by wrestling with intruders. Pairs form on land, and the female then carries the male to water, where she attaches around 700 eggs to a rock.

SIMILAR SPECIES

Corythomantis greeningi was long thought to be the only species in its genus but, in 2012, a new species, the Bahia Helmeted Tree Frog (*C. galeata*), was reported from Bahia state in Brazil. An Endangered species, it was subsequently transferred to genus *Nyctimantis*, but in 2021 another new species was described, the Botoque Tree Frog (*C. botoque*), from Bahia and Minas Gerais.

Greening's Tree Frog has a slender body, a long, narrow head with bony crests behind the eyes, and a long, flattened snout covered in small spicules. The skin is rough and warty, and there are well-developed disks on the fingers and toes. The upper surfaces are gray or light brown with dark brown and red patches. Females are typically darker than males.

Actual size

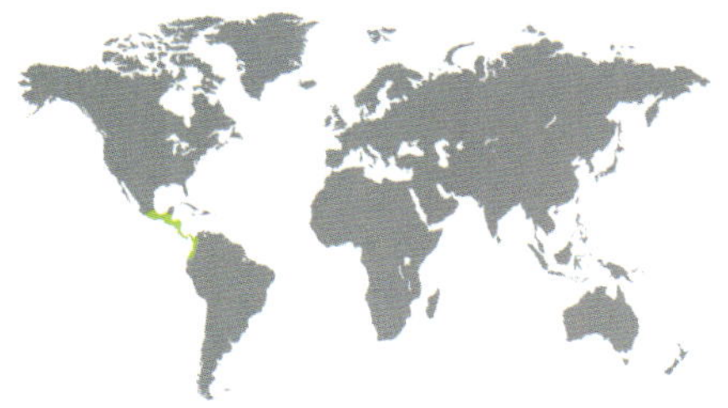

FAMILY	Hylidae: Hylinae
OTHER NAMES	Pantless Tree Frog
DISTRIBUTION	Central and South America, from Mexico to Ecuador
ADULT HABITAT	Humid tropical forest
LARVAL HABITAT	Ephemeral and permanent pools
CONSERVATION STATUS	IUCN Least Concern. Has a large range and there is no evidence of general decline

ADULT LENGTH
Male
7/8–1 1/16 in (23–27 mm)
Female
1 3/16–1 3/8 in (30–35 mm)

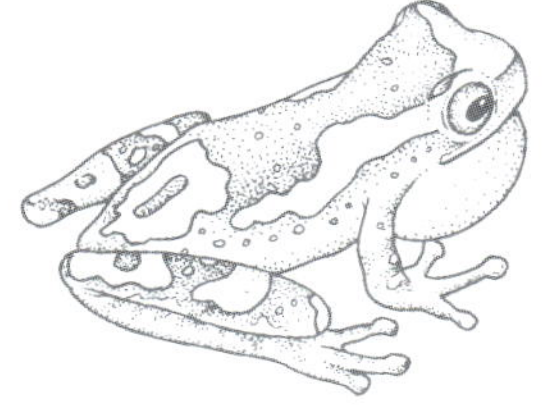

DENDROPSOPHUS EBRACCATUS

HOURGLASS TREE FROG

(COPE, 1874)

Actual size

The Hourglass Tree Frog gets its name from the shape of a brown patch on its otherwise yellow head and back. The legs are yellow or orange with brown patches, and there is a bright yellow stripe along the upper lip. Its coloration is generally brighter at night than it is during the day. The feet are webbed and there are adhesive disks on the fingers and toes.

Males of this brightly colored little frog gather in choruses, defending their territories by calling and, occasionally, by fighting. The call consists of a buzz note, followed by one to four clicks. Females show a preference for calls with more clicks. As in many frogs that form choruses, some males adopt a silent satellite strategy, attempting to intercept females moving toward callers. The eggs may be laid on leaves hanging over water, or in the water itself: If the habitat is well shaded, the female chooses a leaf; if there is little shade, she chooses the water.

SIMILAR SPECIES

Genus *Dendropsophus* contains 107 species. *Dendropsophus ebraccatus* often breeds in the same pools as two other small, predominantly yellow, tree frogs, the Yellow Tree Frog (*D. microcephalus*; page 287) and the San Carlos Tree Frog (*D. phlebodes*; page 288). Another yellow tree frog, the White-leaf Frog or Beireis' Tree Frog (*D. leucophyllatus*), is found in the Amazon Basin. It occurs in many color morphs, including one called "giraffe," which is brown with a bright yellow reticulated pattern.

FAMILY	Hylidae: Hylinae
OTHER NAMES	Small-headed Tree Frog, Yellow Cricket Tree Frog
DISTRIBUTION	Central and South America, from Mexico to Brazil; Trinidad and Tobago
ADULT HABITAT	Pastures, forest edges, and disturbed forest
LARVAL HABITAT	Ephemeral pools
CONSERVATION STATUS	IUCN Least Concern. Has a large range and its population is thought to be generally expanding

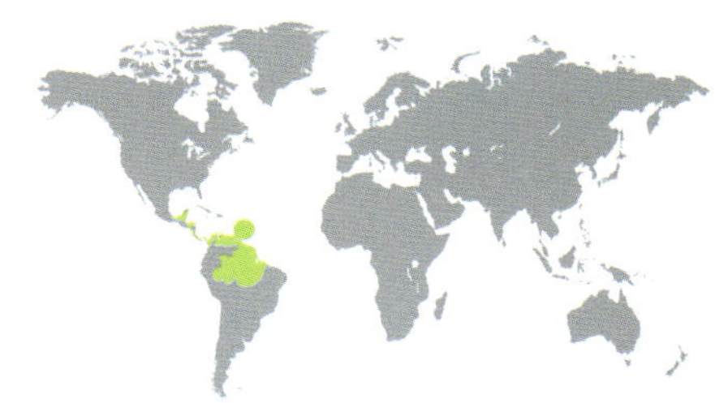

DENDROPSOPHUS MICROCEPHALUS

YELLOW TREE FROG

(COPE, 1886)

ADULT LENGTH
Male
⅞–1$\frac{1}{16}$ in (23–27 mm)
Female
1–1¼ in (26–31 mm)

This small tree frog thrives in disturbed habitats and, as a result, is apparently becoming more common. Males gather around temporary waterbodies, often in enormous numbers. Their call is an insect-like "creek… eek… eek." Females show a preference for males that produce more complex calls with additional notes, which require them to expend more energy. As a result, males respond to the calls of other males by increasing the complexity of their calls. They also exchange aggressive calls to maintain their spacing. In Panama, where many frogs have been devastated by chytridiomycosis, this species has remained unscathed.

Actual size

The Yellow Tree Frog gets its name from its bright yellow coloration which, in many individuals, is dappled with brown spots or patches. It is able to change color and is paler at night. It has a short snout, large golden eyes, and webbed toes. There are well-developed adhesive disks on its fingers and toes.

SIMILAR SPECIES

Currently, 107 species are included in the genus *Dendropsophus*; most of them were previously included in *Hyla*. Only a few of them are of conservation concern. *Dendropsophus amicorum*, listed as Critically Endangered, is known from just one specimen, which is found in cloud forest in northwestern Venezuela. The Pacific Lowland Tree Frog (*D. gryllatus*), a lowland forest species from Ecuador, is listed as Endangered. See also the Hourglass Tree Frog (*D. ebraccatus*; page 286) and the San Carlos Tree Frog (*D. phlebodes*; page 288).

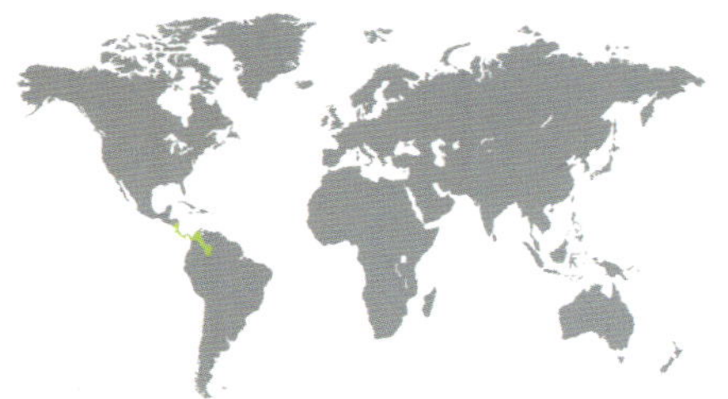

FAMILY	Hylidae: Hylinae
OTHER NAMES	None
DISTRIBUTION	Nicaragua, Costa Rica, Panama, eastern Colombia
ADULT HABITAT	Lowland forest
LARVAL HABITAT	Ephemeral pools
CONSERVATION STATUS	IUCN Least Concern. Has a large range and its population is thought to be stable

ADULT LENGTH
Male
up to 15/16 in (24 mm)
Female
up to 1 1/8 in (28 mm)

DENDROPSOPHUS PHLEBODES

SAN CARLOS TREE FROG

(STEJNEGER, 1906)

Actual size

The San Carlos Tree Frog has a short, pointed snout, large eyes, and larger adhesive disks on its fingers than on its toes. The upper surfaces are yellow or tan with brown markings, usually in the form of a network of thin lines. The flanks are yellow and the underside is white.

This small tree frog breeds after heavy rain, males calling from tall grasses around temporary ponds. Their call has a long primary note, followed by several short notes: "creek... eek... eek." As more males call in a chorus, each male synchronizes his calls with those of his neighbors, increases his call rate, and adds extra short notes to each call. In Panama, this species and the related Hourglass Tree Frog (*D. ebraccatus*; page 286) and Yellow Tree Frog (*D. microcephalus*; page 287) are sometimes found breeding together in the same pond. They each have slightly different mating calls, reducing the risk that females will choose males of the wrong species.

SIMILAR SPECIES

Like *Dendropsophus phlebodes*, *D. ebraccatus* (page 286) and *D. microcephalus* (page 287), females of most members of this genus lay their eggs in water. Some species, however, lay them attached to leaves overhanging water, so that the tadpoles fall into the water on hatching. These include the Upper Amazon Tree Frog (*D. bifurcus*), a common species in the western Amazon Basin, and Ruschi's Tree Frog (*D. ruschii*), a little-known species from the Atlantic Forest of eastern Brazil.

FAMILY	Hylidae: Hylinae
OTHER NAMES	None
DISTRIBUTION	Colombia, Ecuador, Peru, Bolivia, western Brazil
ADULT HABITAT	Rainforest at altitudes of 650–4,000 ft (200–1,200 m)
LARVAL HABITAT	Ephemeral and permanent pools
CONSERVATION STATUS	IUCN Least Concern. A common species with a stable population. Has a large range that includes several protected areas

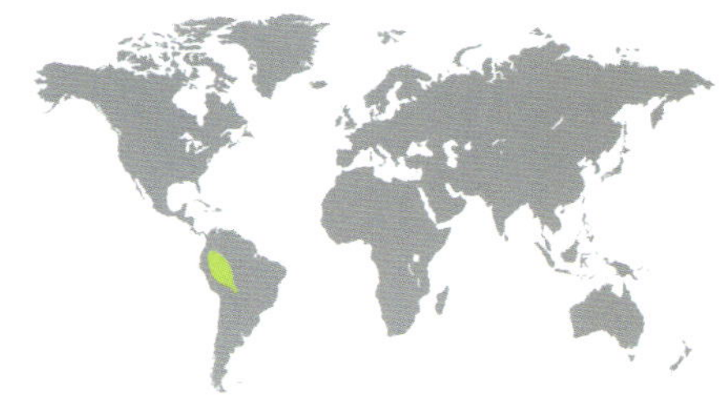

ADULT LENGTH
Male
$\frac{3}{4}$–$\frac{15}{16}$ in (20–24 mm)
Female
1–$1\frac{1}{8}$ in (26–29 mm)

DENDROPSOPHUS RHODOPEPLUS

RED-SKIRTED TREE FROG

(GÜNTHER, 1858)

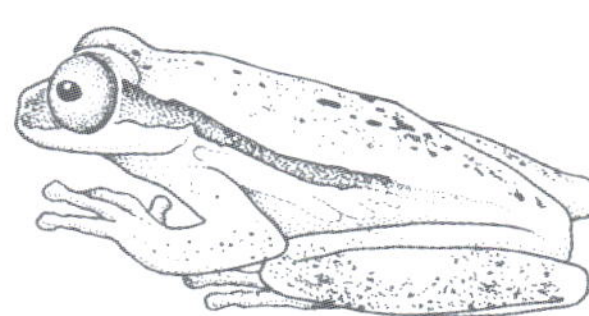

This colorful little tree frog occurs in dense forest in the upper Amazon Basin to the east of the Andes. It breeds after heavy rain, males gathering to call at night around ponds and pools. Their call consists of a pair of high-pitched notes, repeated about once every second. Females produce 140–380 eggs, depending on their size, which they deposit in clumps in the water. The tadpoles are orange with brown flecks. The species persists in forest that has been somewhat affected by human activity, suggesting that it is not wholly dependent on pristine habitat.

Actual size

The Red-skirted Tree Frog has a rounded snout, large eyes, and rather small disks on its fingers and toes. The back is yellow or tan, fading to white during the day, and is decorated with red flecks. There is a reddish-brown stripe along the flank and a yellow stripe along the upper lip. The arms and legs are yellow, and there are red stripes on the legs.

SIMILAR SPECIES

Dendropsophus rhodopeplus is one of a number of small, predominantly yellow, tree frogs that are found in the upper Amazon Basin. Koechlin's Tree Frog (*D. koechlini*) is a common species in Bolivia, Brazil, and Peru. The Yellow-toed Tree Frog (*Dendropsophus leali*) has a similar range and is also found in Colombia. See also the Hourglass Tree Frog (*D. ebraccatus*; page 286), the Yellow Tree Frog (*D. microcephalus*; page 287), and the San Carlos Tree Frog (*D. phlebodes*; page 288).

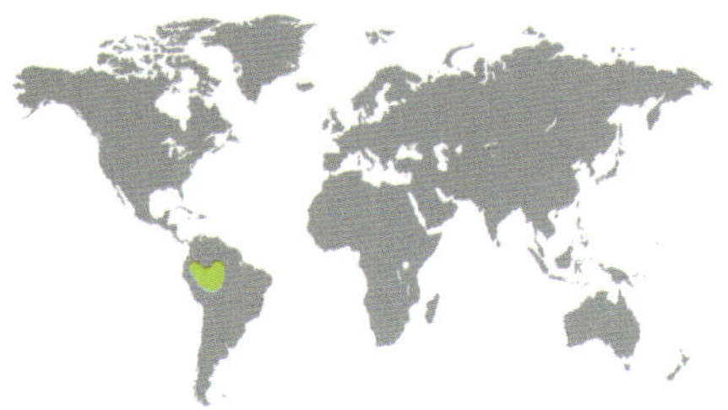

FAMILY	Hylidae: Hylinae
OTHER NAMES	Shreve's Sarayacu Tree Frog
DISTRIBUTION	Colombia, Ecuador, Peru, Bolivia, Venezuela, western Brazil
ADULT HABITAT	Rainforest up to 5,500 ft (1,700 m) altitude
LARVAL HABITAT	Ephemeral and permanent pools
CONSERVATION STATUS	IUCN Least Concern. A common species with a stable population. Has a large range that includes several protected areas

ADULT LENGTH
Male
15/16–1 1/8 in (24–29 mm)
Female
1 5/16–1 7/16 in (34–37 mm)

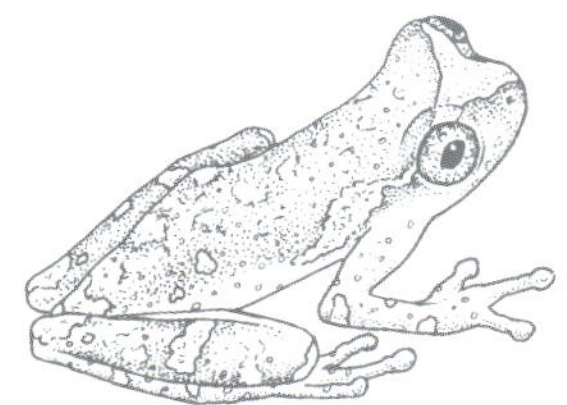

DENDROPSOPHUS SARAYACUENSIS

SARAYACU TREE FROG

(SHREVE, 1935)

Actual size

The Sarayacu Tree Frog has a slender body, a rounded snout, large eyes, and toes that are partially webbed. The upper surfaces are mottled dark and pale brown with pale patches on the back and pale stripes on the legs. There is a conspicuous pale triangular patch on the top of its snout. The hands, feet, and hidden parts of the limbs are orange.

This common tree frog inhabits the western part of the Amazon Basin and is unusual in giving off a distinctive herbal odor. It is nocturnal and generally found in low vegetation. It breeds during the rainy season, males gathering in choruses of five or ten individuals. The female lays 70–170 eggs, attaching them in several batches to leaves, moss-covered roots, or tree trunks overhanging a pool. After 10–13 days, the eggs hatch and the tadpoles fall into the water.

SIMILAR SPECIES

Living in much the same region as *Dendropsophus sarayacuensis* are the Triangle Tree Frog (*D. triangulum*) and Ross Allen's Tree Frog (*D. rossalleni*), the latter found only in low-lying areas. The Elegant Forest Tree Frog (*D. elegans*) is a common species in the Atlantic Forest that runs along the southeastern coast of Brazil. See also pages 286–289.

FAMILY	Hylidae: Hylinae
OTHER NAMES	None
DISTRIBUTION	Southern USA, especially Louisiana, Mississippi, Alabama, and Georgia
ADULT HABITAT	Hardwood swamps and forested floodplains
LARVAL HABITAT	Pools in swamps
CONSERVATION STATUS	IUCN Least Concern. Common across most of its range

DRYOPHYTES AVIVOCA

BIRD-VOICED TREE FROG

(VIOSCA, 1928)

ADULT LENGTH
Male
$1\frac{1}{8}$–$1\frac{9}{16}$ in (28–39 mm)
Female
$1\frac{1}{4}$–$2\frac{1}{16}$ in (32–53 mm)

Sometimes drab in appearance, this small tree frog has a whistling call that has been described as "hauntingly beautiful" as it echoes through cypress swamps. With water always available in its habitat, the breeding season is long, lasting from April to August. Males call at night, initially from treetops, before descending to meet females in the lower branches, calling all the while. The female then carries the male to water, where she lays 400–800 eggs. More territorial than most American tree frogs, males of this species defend their calling perches by producing aggressive calls and, occasionally, by wrestling with an intruder.

Actual size

The Bird-voiced Tree Frog has a warty skin and large adhesive disks on its fingers and toes. It is very variable in color, being green, gray, or nearly black. There is a yellowish spot below the eye and there are dark cross-bars on the legs. There is a pale greenish patch on the inside of its hind legs.

SIMILAR SPECIES

Dryophytes avivoca is similar to, but rather smaller than, Cope's Gray Tree Frog (*D. chrysoscelis*; page 292) and the Eastern Gray Tree Frog (*D. versicolor*; page 294). The species is dull in appearance in contrast to the vibrantly colored Arizona Tree Frog (*D. wrightorum*), which breeds "explosively" after heavy thunderstorms in ephemeral pools. Dull gray in color, the Canyon Tree Frog (*D. arenicolor*) lives in rocky areas, including the Grand Canyon.

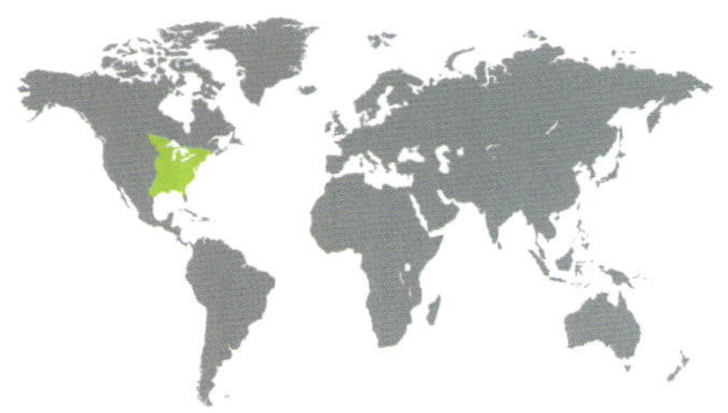

FAMILY	Hylidae: Hylinae
OTHER NAMES	None
DISTRIBUTION	Southeastern states and the lower Midwest, USA; extreme southern Canada
ADULT HABITAT	Woodland
LARVAL HABITAT	Semipermanent ponds
CONSERVATION STATUS	IUCN Least Concern. Has a very large range and there is no evidence of general decline

ADULT LENGTH
1¼–2⅜ in (32–60 mm)

DRYOPHYTES CHRYSOSCELIS

COPE'S GRAY TREE FROG

(COPE, 1880)

Breeding in this tree frog occurs between March and August, depending on locality. Males begin to call at dusk, initially from high up in trees and then moving down until they are calling from the ground, very close to a pond. They do not call at any pond, avoiding those that contain sunfish, which feed on their tadpoles. Females also avoid ponds containing predatory fish and salamanders that eat tadpoles. A clutch of 30–40 eggs is laid, deposited on the water's surface or on submerged vegetation. Females may breed up to three times in a season.

SIMILAR SPECIES

In appearance, *Dryophytes chrysoscelis* is identical to the Eastern Gray Tree Frog (*D. versicolor*; page 294), but it has half as many chromosomes. Though often found together, hybrids are very rare, females of the two species ensuring correct species identification by responding selectively to the male mating call. In both species, the call is a trill, with that of *D. chrysoscelis* being faster than the trill of *D. versicolor*. *Dryophytes* contains 20 species in North America and East Asia.

Cope's Gray Tree Frog has warty skin and large adhesive disks on its fingers and toes. It is gray or green, or a combination of both, in a pattern that resembles the bark of a tree; it is able to vary the color of its skin to match its background. There are orange or yellow patches on the inside of its hind legs.

Actual size

FAMILY	Hylidae: Hylinae
OTHER NAMES	None
DISTRIBUTION	Southeastern and Gulf Coast states, USA
ADULT HABITAT	Swamps and surrounding woodland
LARVAL HABITAT	Lakes, ponds, swamps, ditches
CONSERVATION STATUS	IUCN Least Concern. A common species with a large range

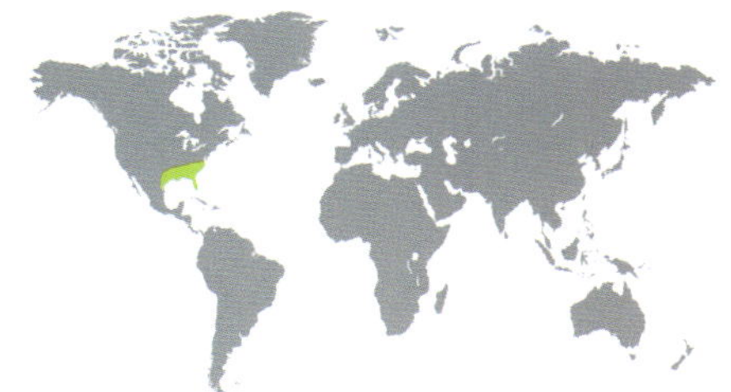

DRYOPHYTES CINEREUS

AMERICAN GREEN TREE FROG

(SCHNEIDER, 1799)

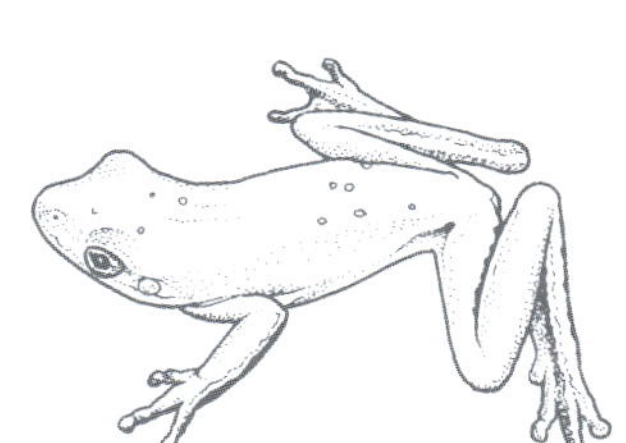

ADULT LENGTH
1¼–2$^{9}/_{16}$ in (32–64 mm)

This arboreal frog breeds from March to August, forming choruses after warm rains. Calling and mating occur at night, males and females spending the day in trees nearby. The male's call is a nasal "quonk... quonk... quonk." Males often call alternately with their nearest neighbors, and switch to a rapid, rattling, aggressive call if rivals come too close. Some males adopt a silent "satellite" strategy, attempting to intercept females approaching calling males. Males vary in the number of nights for which they attend a chorus; those that attend most often have the highest mating success.

Actual size

The American Green Tree Frog has a flat head and a pointed snout. There are large adhesive disks on the fingers and toes, and the hind feet are webbed. The skin is smooth and is green on the back, often with yellowish spots. In many populations there is a white stripe, bordered in black, running along each flank. The frog can change color, being yellowish when calling and gray when cold.

SIMILAR SPECIES

Three other *Dryophytes* species occur in southeastern USA and are sometimes called rain frogs. The call of the Barking Tree Frog (*D. gratiosus*) is a resonant "tonk." It occasionally hybridizes with *D. cinereus*. The Pine Barrens Tree Frog (*D. andersonii*) has a lavender stripe along its flanks. The Squirrel Tree Frog (*D. squirellus*) is green only while breeding, being otherwise brown. Females of these four species identify males of their own species on the basis of subtle differences in their calls.

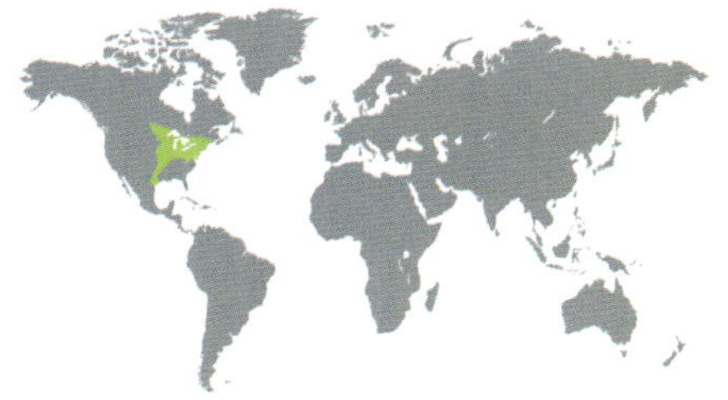

FAMILY	Hylidae: Hylinae
OTHER NAMES	Morse-code Frog, Pinewoods Tree Frog
DISTRIBUTION	Eastern states of USA, extreme southern Canada
ADULT HABITAT	Woodland
LARVAL HABITAT	Ephemeral and semipermanent ponds
CONSERVATION STATUS	IUCN Least Concern. Has a very large range and there is no evidence of a general decline

ADULT LENGTH
Male
$1\frac{1}{4}$–2 in (32–51 mm)
Female
$1\frac{5}{16}$–$2\frac{3}{8}$ in (33–60 mm)

DRYOPHYTES VERSICOLOR

EASTERN GRAY TREE FROG

(LECONTE, 1825)

This plump frog is noted for its ability to change color to match its background. It can do this within 30 minutes of moving to a new perch. It is quite often found in and around houses and farm buildings. It breeds between April and August, gathering around ponds and pools that do not contain fish. The male's call is a melodious trill and the female lays 30–40 eggs. The tadpoles of this species have been the subject of a great deal of research into the harmful effects of pesticides, such as carbaryl, on their development and survival.

SIMILAR SPECIES

In appearance, *Dryophytes versicolor* is identical to Cope's Gray Tree Frog (*D. chrysoscelis*; page 292), but it has a different call and twice as many chromosomes. The Pine Woods Tree Frog (*D. femoralis*) occurs in the coastal plains of the southeastern states of the USA. Its call sounds like the tapping of a telegraph key, giving it the alternative common name of Morse-code Frog.

The Eastern Gray Tree Frog has warty skin and large adhesive disks on its fingers and toes. It is very variable in color, being gray, brown, or green, or a combination of these that resembles the bark of a tree. There is a white spot below the eye, more marked in males than females, and there are orange or yellow patches on the inside of its hind legs.

Actual size

FAMILY	Hylidae: Hylinae
OTHER NAMES	None
DISTRIBUTION	Northern Honduras
ADULT HABITAT	Wet forest up to 4,600 ft (1,400 m) altitude
LARVAL HABITAT	Pools or slow-moving streams
CONSERVATION STATUS	IUCN Endangered. Has a very small, fragmented range that is subject to habitat destruction

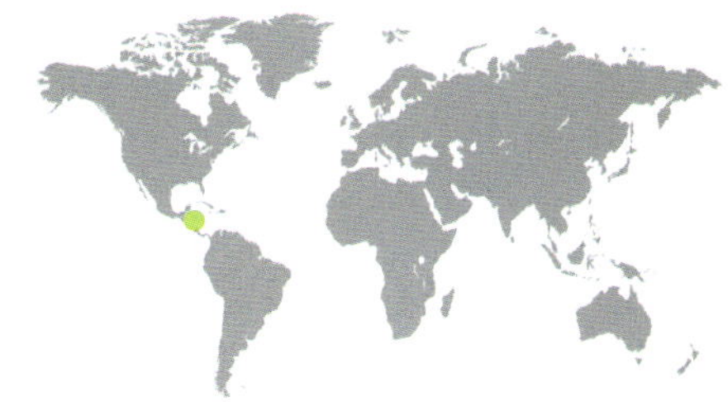

DUELLMANOHYLA SALVAVIDA

HONDURAN BROOK FROG

(MCCRANIE & WILSON, 1986)

ADULT LENGTH
Male
1–1⅛ in (25–28 mm)

Female
1$\frac{5}{16}$ in (34 mm);
based on one specimen

So rare is this small tree frog that detailed description of it is based on just three males and one female. It is named after a brand of Honduran beer called SalvaVida, meaning "lifesaver." Males call along slow-moving forest streams and around pools, perched on low vegetation overhanging the water. The female lays around 100 eggs, attached to leaves so that, on hatching, the tadpoles fall into the water below. Tadpoles have been observed under leaves at the bottom of streams. The species is potentially threatened by the disease chytridiomycosis.

Actual size

The Honduran Brook Frog has a short snout and well-developed disks on its fingers and toes. The back is dull green and the underside is yellow. There is a white stripe along the upper lip, which merges into an indistinct stripe along the flanks. The irises of the eyes are deep red.

SIMILAR SPECIES

Of the ten species currently contained in the genus *Duellmanohyla*, five are listed as Endangered, one as Vulnerable, and three as Near Threatened. Only the Rufous-eyed Brook Frog (*D. rufioculis*) is listed as of Least Concern. Found in the montane forests of Costa Rica, it breeds in streams and apparently is not affected by chytridiomycosis.

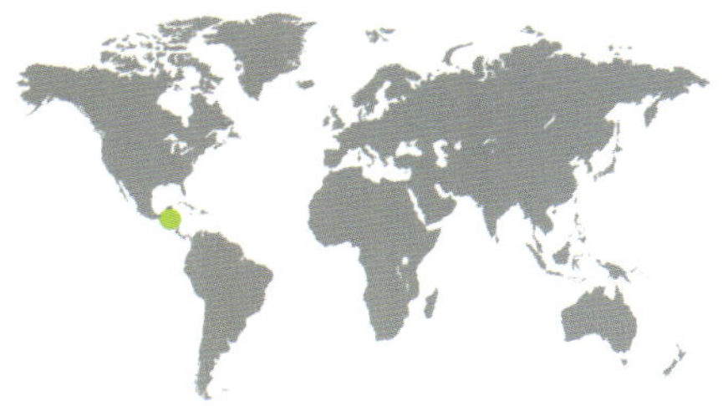

FAMILY	Hylidae: Hylinae
OTHER NAMES	None
DISTRIBUTION	Northwestern Honduras, northeastern Guatemala
ADULT HABITAT	Wet forest up to 5,150 ft (1,570 m) altitude
LARVAL HABITAT	Pools in mountain streams
CONSERVATION STATUS	IUCN Endangered. Threatened by encroaching agriculture, which is leading to deforestation and water pollution. Also heavily infected by chytridiomycosis

ADULT LENGTH
Male
1–1¼ in (25–32 mm)

Female
up to 1½ in (38 mm)

DUELLMANOHYLA SORALIA

COPAN BROOK FROG

(WILSON & MCCRANIE, 1985)

Actual size

The Copan Brook Frog has a short snout and well-developed disks on its fingers and toes. The back is dark brown, decorated with complex bright green markings that resemble lichen. The underside is yellow. The irises of the large eyes are bright red.

This tiny red-eyed frog is typically found in low vegetation close to mountain streams. Males produce a series of low-pitched "peeps" at intervals of 20–30 seconds. The tadpoles, found in quiet pools in streams, are long and slender, growing to a length of 1$^{11}/_{16}$ in (43 mm). They have large funnel-shaped mouths and are unusual in being covered in iridescent green spots. Although found in three national parks, the species' populations are declining due to habitat destruction and the disease chytridiomycosis. Along with other Honduran amphibians, it is the subject of a rescue program involving captive breeding.

SIMILAR SPECIES

The Red-eyed Stream Frog (*Duellmanohyla uranochroa*) is another Endangered frog, found in Costa Rica and Panama. It is one of 24 out of 53 native species that declined dramatically at higher altitudes in Costa Rica in the early 1990s. The Chamula Mountain Brook Frog (*D. chamulae*), also Endangered, breeds in fast-flowing streams at high altitudes in Mexico. See also the Honduran Brook Frog (*D. salvavida*; page 295).

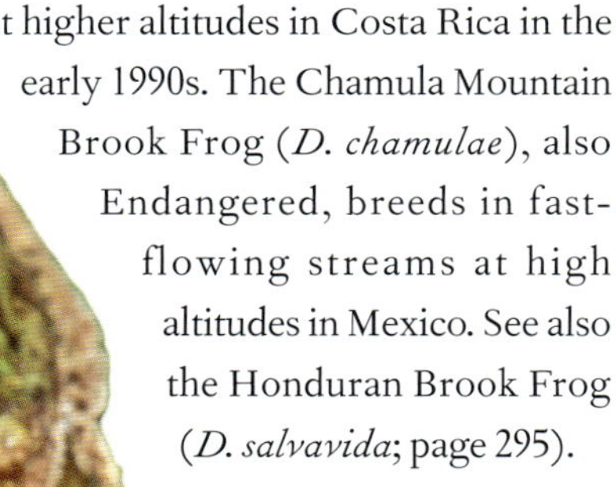

FAMILY	Hylidae: Hylinae
OTHER NAMES	None
DISTRIBUTION	Central Panama
ADULT HABITAT	Cloud forest
LARVAL HABITAT	Pools in tree-holes
CONSERVATION STATUS	IUCN Critically Endangered. Almost certainly extinct in the wild since 2007, but a captive population survives

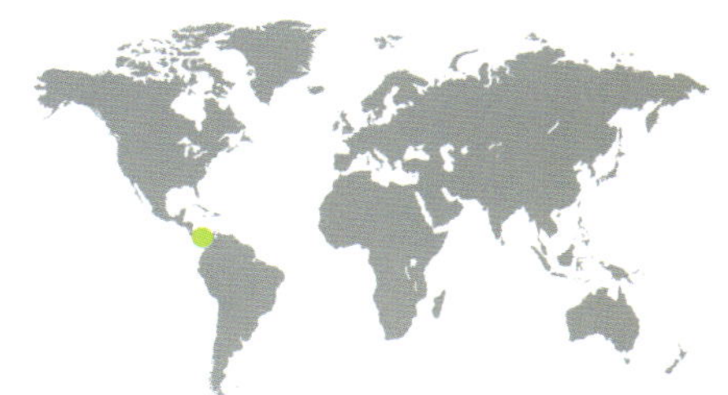

ECNOMIOHYLA RABBORUM

RABB'S FRINGE-LIMBED TREE FROG

MENDELSON, SAVAGE, GRIFFITH, ROSS, KUBICKI & GAGLIARDO, 2008

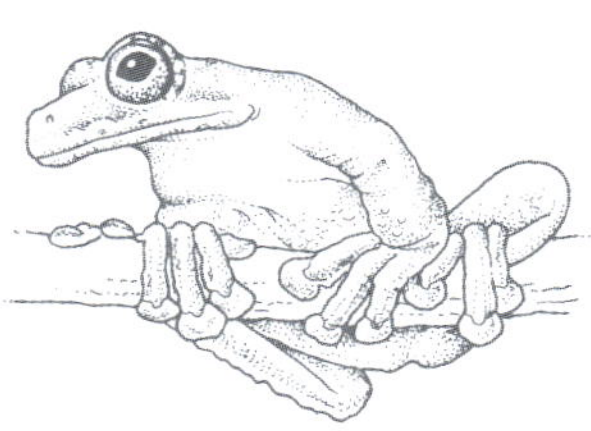

ADULT LENGTH
Male
2½–3$^{13}/_{16}$ in (62–97 mm)
Female
2⅜–3⅞ in (61–100 mm)

Listed as Critically Endangered, this unusual frog is probably extinct, a testament to the destructive power of the disease chytridiomycosis, which reached its range in 2006. Found in only a small region of Panama, it is a canopy-dweller, capable of gliding from tree to tree. Mating occurs from March to May. The male calls from close to a water-filled tree hollow, in which the female lays 60–200 eggs, leaving the male to defend them. He feeds the tadpoles in a remarkable way, sitting in the water while they nibble at his skin. Males also call outside the breeding season.

SIMILAR SPECIES

There are 12 species in this genus, distributed across Central America and northern South America. At least half of them are threatened by extinction. The Heredia Tree Frog (*Ecnomiohyla fimbrimembra*) is a large brown gliding frog from Costa Rica and Panama. Even larger is Cope's Brown Tree Frog (*E. miliaria*), found from Costa Rica to Nicaragua; it is listed as Vulnerable. This species is unusual among frogs in that the male is larger than the female.

Rabb's Fringe-limbed Tree Frog has massive forearms, equipped with fringes of skin that are covered in spines. The feet and hands are large and fully webbed, and when spread wide, enable it to glide from tree to tree. There are large adhesive disks on its fingers and toes. The frog is brown, or brown and green.

Actual size

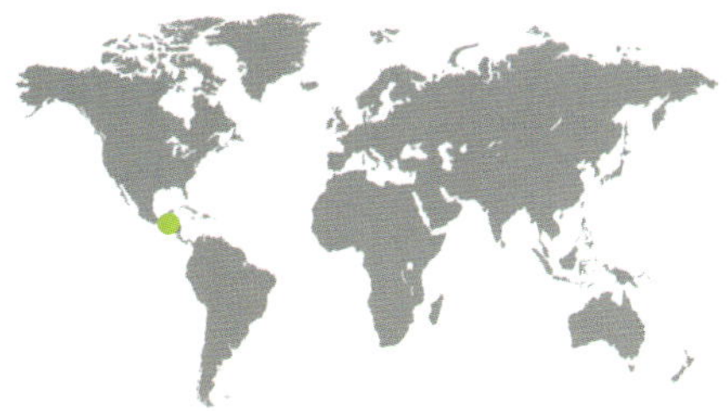

FAMILY	Hylidae: Hylinae
OTHER NAMES	None
DISTRIBUTION	Guatemala
ADULT HABITAT	Subtropical rainforest at altitudes of 3,400–3,500 ft (1,050–1,080 m)
LARVAL HABITAT	Ephemeral pools and slow-moving streams
CONSERVATION STATUS	IUCN Endangered. Its very restricted range is subject to encroachment by agriculture and human settlements, and it does not occur in any protected areas

ADULT LENGTH
Male
unknown

Female
average 1⅜ in (35 mm)
based on two specimens

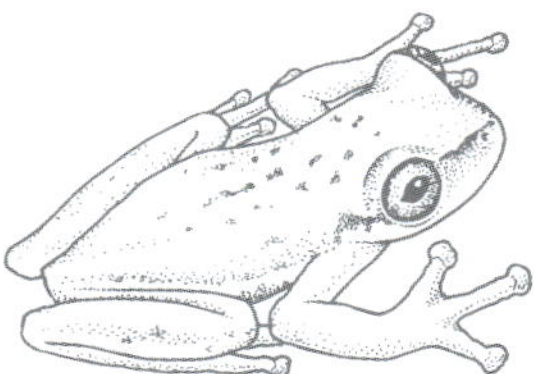

EXERODONTA PERKINSI

PERKINS' TREE FROG

(CAMPBELL & BRODIE, 1992)

Actual size

Perkins' Tree Frog has a slender body, a wide head, and large eyes. There are large disks on the fingers and toes, and the feet are partially webbed. The upper surfaces are pale brown or dull green, with small, dark brown patches. A pale stripe runs from behind the eye to the groin, and there is a white stripe above the eye.

The description of this very rare tree frog is based on just two females, found hiding during the day in the axils of elephant-ear plants. It is known only from an area of very wet forest on the northern slopes of the Sierra de los Cuchumatanes and just 20 specimens have been collected. This is an area of widespread population declines among frogs, especially those breeding in streams, probably caused by the disease chytridiomycosis.

SIMILAR SPECIES

The eight members of the genus *Exerodonta*, known as highlands tree frogs, are found in Mexico, Guatemala, and Honduras. The Río Aloapan Tree Frog (*E. abdivita*) is a small brown, Near Threatened, frog found in the Oaxaca region of Mexico. Tadpoles of the Black-eyed Tree Frog (*E. melanomma*), from Mexico, are suspected of being infected with chytridiomycosis. The species is listed as Vulnerable, as is the Puebla Tree Frog (*E. xera*), which occurs in desert habitat in Mexico.

FAMILY	Hylidae: Hylinae
OTHER NAMES	None
DISTRIBUTION	Continental Europe, except for southern Spain, southern France, and Italy
ADULT HABITAT	Open forests, gardens, vineyards, orchards, parks
LARVAL HABITAT	Lakes, ponds, swamps, reservoirs, and ditches
CONSERVATION STATUS	IUCN Least Concern. Declining in parts of western and central Europe

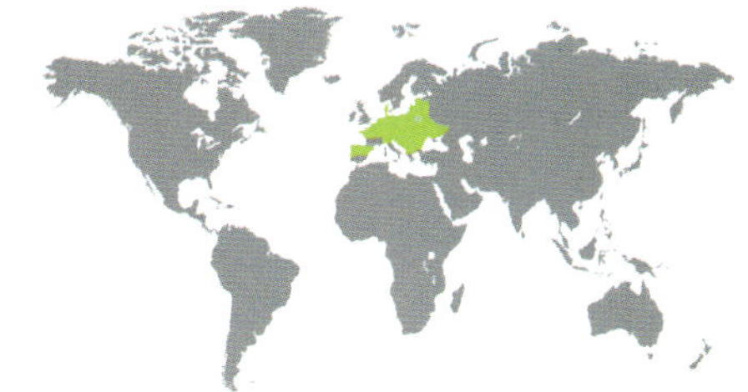

HYLA ARBOREA

EUROPEAN TREE FROG

(LINNAEUS, 1758)

ADULT LENGTH
Male
1¼–1¹¹⁄₁₆ in (32–43 mm)
Female
1⁹⁄₁₆–1¹⁵⁄₁₆ in (40–50 mm)

Found in very diverse habitats, although not in dense forests, this frog has a call that is regarded in many places as a harbinger of spring. The date at which it breeds varies across Europe, but is generally from March to June. The male's call is very loud, sounding like "krak… krak… krak"—choruses can be heard half a mile (1 km) away. Females are more likely to respond positively to the lower-pitched calls of larger males. Although mating occurs at night, visual cues are also important, with females preferring to mate with males that have more brightly colored vocal sacs.

SIMILAR SPECIES

The Stripeless Tree Frog (*Hyla meridionalis*), as its name implies, lacks the prominent stripes along the flanks characteristic of *H. arborea*. Slightly larger than *H. arborea*, it is found in eastern Spain and southern France. It has declined in many places along the Mediterranean coast, as a result of the widespread use of insecticides to control mosquitoes. The Italian Tree Frog (*H. intermedia*; page 300) occurs only in Italy.

The European Tree Frog has long legs with webbed feet, and well-developed adhesive disks on its fingers and toes. The back and legs are dark or light green but, depending on the temperature and the background, may also be yellow, brown, or gray. The underside is white or yellow. There is a white-edged dark stripe running down each flank from eye to groin.

Actual size

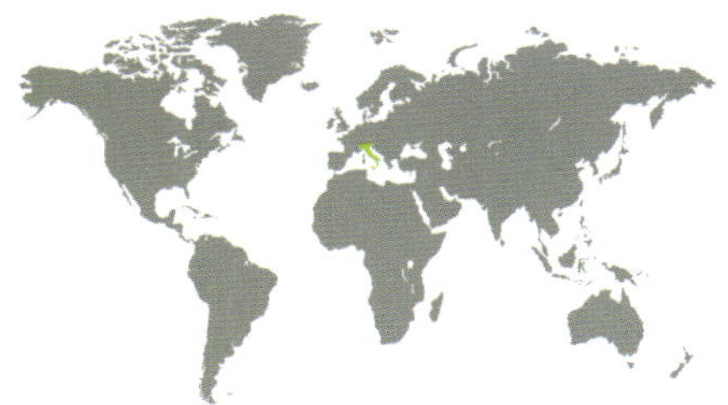

FAMILY	Hylidae: Hylinae
OTHER NAMES	None
DISTRIBUTION	Italy, including Sicily, parts of Slovenia, southern Switzerland
ADULT HABITAT	Woodlands, hedgerows, wetlands
LARVAL HABITAT	Lakes, ponds, swamps, reservoirs, and ditches
CONSERVATION STATUS	IUCN Least Concern. Declining in parts of Italy due to predation by introduced crayfish

ADULT LENGTH
Male
1¼–1¹¹⁄₁₆ in (32–43 mm)
Female
1⁹⁄₁₆–1¹⁵⁄₁₆ in (40–50 mm)

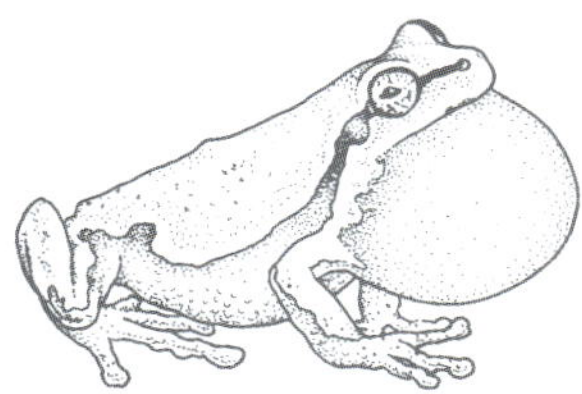

HYLA INTERMEDIA

ITALIAN TREE FROG

BOULENGER, 1882

Actual size

The Italian Tree Frog has long legs with webbed feet, and well-developed adhesive disks on its fingers and toes. The back and legs are dark or light green but, depending on the temperature and the background, may be yellow, brown, or gray. The underside is white or yellow. There is a white-edged dark stripe running down each flank from eye to groin.

Mating in this frog occurs at a lek, which is a gathering of calling males at which females can choose a mate. Females mate preferentially with those males that put most energy into calling on a given night, but the most important determinant of male mating success is the number of nights on which they attend a chorus. Females don't always get to mate with the male of their choice, because non-calling satellite males gather around the most attractive males and try to intercept approaching females. This species is declining in parts of its range due to predation by the freshwater crayfish *Procambarus clarkii*, which has been introduced from the USA.

SIMILAR SPECIES

Formerly considered to be a subspecies of the European Tree Frog (*H. arborea*; page 299), this frog is very similar in appearance but is genetically distinct. It is sometimes found in human-created habitats, such as rice fields. Also similar in appearance is the Tyrrhenian Tree Frog (*H. sarda*), found on Corsica, Sardinia, and Elba, as well as some other Mediterranean islands. Genus *Hyla* now contains 16 species.

FAMILY	Hylidae: Hylinae
OTHER NAMES	None
DISTRIBUTION	Southern Peru, Bolivia
ADULT HABITAT	Cloud forest at altitudes of 5,600–7,900 ft (1,700–2,400 m)
LARVAL HABITAT	Fast-flowing streams
CONSERVATION STATUS	IUCN Near Threatened. Declining in parts of its range due to habitat destruction but it occurs in some protected areas

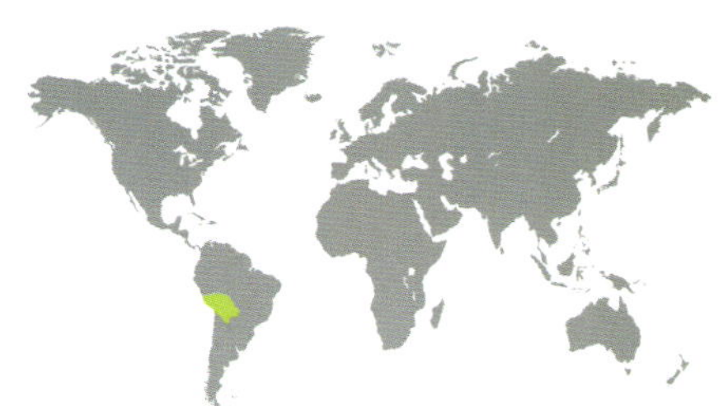

HYLOSCIRTUS ARMATUS

ARMED TREE FROG

(BOULENGER, 1902)

ADULT LENGTH
Male
1⅞–2¾ in (47–69 mm)
Female
1⅞–2¹⁵⁄₁₆ in (47–75 mm)

This large tree frog, found on the eastern slopes of the Andes, is notable for the very unusual anatomy of the male's forearms. They are much thicker than the female's, due to their greater muscular development, and they have two arrays of sharp spines, one on each thumb and one on each forearm. It is not known if these are important during mating, during fighting, or both. A very common species in some places, it breeds in mountain streams. The tadpoles have a very large sucker around the mouth, enabling them to cling to rocks and grow to a very large size (3¹⁄₁₆ in/78 mm).

SIMILAR SPECIES

The unique nature of the male's forearms set *Hyloscirtus armatus* apart from other tree frogs. It has been suggested that it is, in fact, a complex of at least three species.

The Armed Tree Frog has a blunt snout, large eyes, and large disks on its fingers and toes. The upper surfaces may be uniform gray or pale brown, or yellow-brown with a dark brown reticulated pattern. The forelimbs of the male are greatly enlarged and bear two clusters of sharp spines.

Actual size

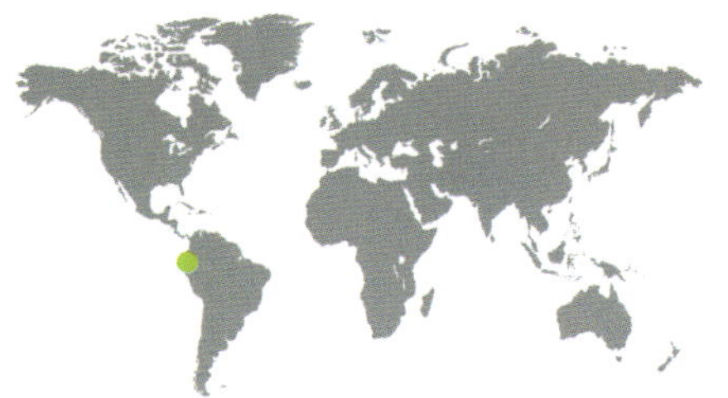

FAMILY	Hylidae: Hylinae
OTHER NAMES	None
DISTRIBUTION	Ecuador
ADULT HABITAT	Cloud forest
LARVAL HABITAT	Streams
CONSERVATION STATUS	IUCN Critically Endangered

ADULT LENGTH
Male
$2\frac{3}{4}$–$2\frac{13}{16}$ in (68–71 mm)

Female
unknown

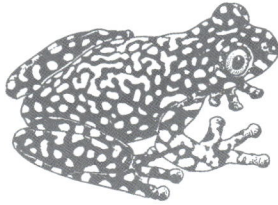

HYLOSCIRTUS PRINCECHARLESI

PRINCE CHARLES' STREAM TREE FROG

COLOMA, CARVAJAL-ENDARA, DUEÑAS, PAREDES-RECALDE, MORALES-MITE, ALMEIDA-REINOSO, TAPIA, HUTTER, TORAL-CONTRERAS & GUAYASAMIN, 2012

This unusually colored tree frog is named after the UK's Prince Charles, now King Charles III, in recognition of his work to raise awareness of deforestation and its consequences for biodiversity. It is known from just three males, found at a single location in the mountains of Ecuador. It was discovered in 2008 and described in 2012, and has not yet been assessed for its conservation status. Its future must be precarious, both because of its tiny range and because it lives in cloud forest, a habitat that is under serious threat from deforestation and climate change.

The Prince Charles' Stream Tree Frog has very long arms and legs, glossy skin, and large, protruding eyes. The adhesive disks on its fingers and toes are not expanded. The upper surface is black, decorated with numerous orange spots. The underside is black with a marbled gray pattern.

SIMILAR SPECIES

To date, 41 species have been described in the genus *Hyloscirtus*, most of them formerly assigned to the genus *Hyla*. Several of them are threatened with extinction. For example, *H. colymba* occurs in the cloud forest of Costa Rica and Panama, and is listed as Endangered. *Hyloscirtus larinopygion* comes from Ecuador and Colombia, and now categorized as Least Concern. Its Latin species name means "fat buttocks," which is a reference to its distinctive swollen cloacal region.

Actual size

FAMILY	Hylidae: Hylinae
OTHER NAMES	Gunther's Costa Rican Tree Frog
DISTRIBUTION	Costa Rica, Panama
ADULT HABITAT	Rainforest
LARVAL HABITAT	Ephemeral pools and roadside ditches
CONSERVATION STATUS	IUCN Least Concern. Has recovered from past declines and occurs within some protected areas

ISTHMOHYLA PSEUDOPUMA

MEADOW TREE FROG

(GÜNTHER, 1901)

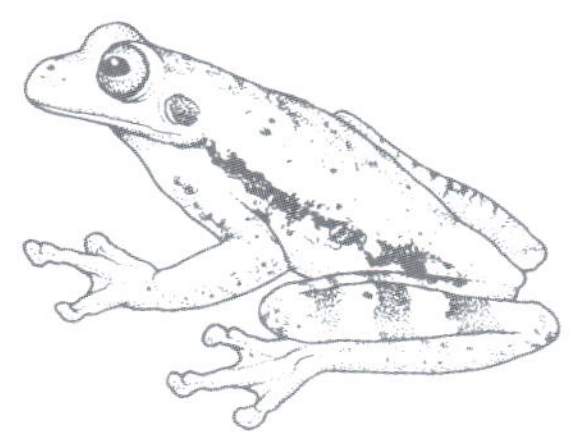

ADULT LENGTH
Male
$1\frac{7}{16}$–$1\frac{3}{4}$ in (37–45 mm)
Female
$1\frac{5}{8}$–2 in (41–52 mm)

Males of this usually rather drab tree frog develop a bright yellow back during the breeding season. The species is an "explosive breeder," all mating activity occurring within a 24-hour period following heavy rain. Males call from water, producing a series of short, low-pitched notes. They greatly outnumber females and "mating balls" of several males sometimes develop around unfortunate single females. The female lays 1,800–2,500 eggs, in several clumps of less than 500, scattered around a pond. Larger tadpoles become cannibalistic, eating the eggs and small larvae of their own species. Adults sometimes fall prey to land crabs.

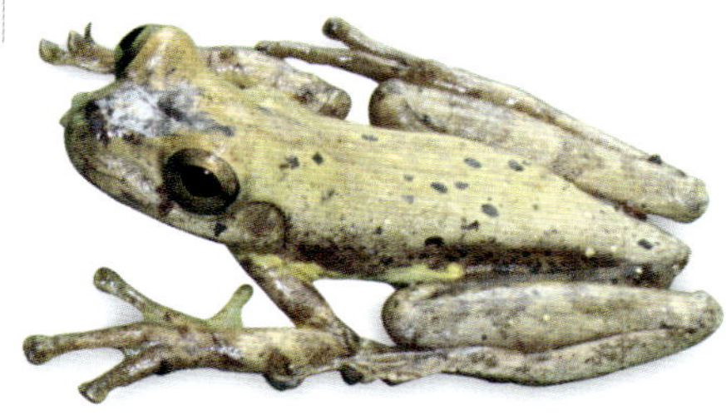

Actual size

The Meadow Tree Frog has a slender body, a blunt snout, and large eyes. The arms and legs are slender, and the fingers and toes are tipped with well-developed adhesive disks. It is generally dull yellow or pale brown in color, with small dark spots. The back of males becomes bright yellow during the breeding season.

SIMILAR SPECIES

There are 13 species in the genus *Isthmohyla*, found in Honduras, Costa Rica, and Panama. The majority of them are threatened by extinction to varying degrees. The Narrow-lined Tree Frog (*I. angustilineata*), from Panama and Costa Rica, has declined dramatically and is listed as Critically Endangered. Its decline is probably due to a combination of chytridiomycosis and climate change. See also Starrett's Tree Frog (*I. tica*; page 304).

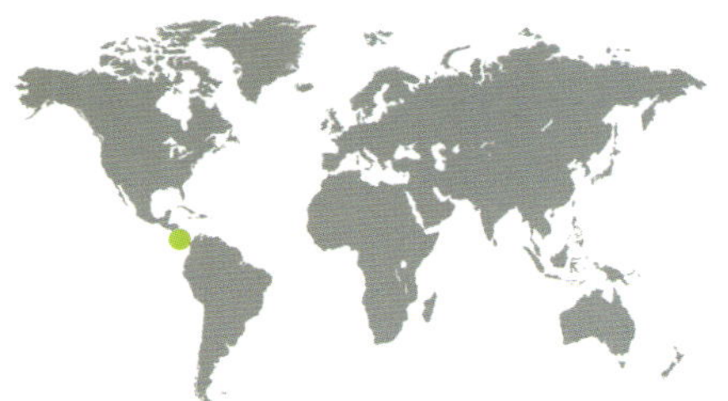

FAMILY	Hylidae: Hylinae
OTHER NAMES	None
DISTRIBUTION	Costa Rica, northern Panama
ADULT HABITAT	Rainforest at altitudes of 3,600–5,400 ft (1,100–1,650 m)
LARVAL HABITAT	Fast-flowing streams
CONSERVATION STATUS	IUCN Critically Endangered. Has virtually disappeared since 1992, most probably because of chytridiomycosis

ADULT LENGTH
Male
$1\frac{1}{16}$–$1\frac{5}{16}$ in (27–34 mm)
Female
$1\frac{5}{16}$–$1\frac{5}{8}$ in (33–42 mm)

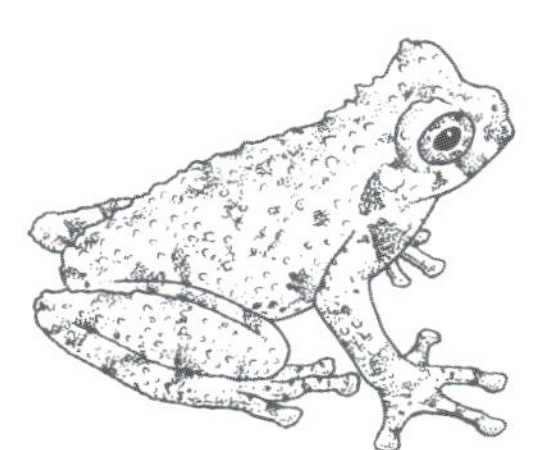

ISTHMOHYLA TICA

STARRETT'S TREE FROG

(STARRETT, 1966)

Actual size

Starrett's Tree Frog has a short, rounded snout, large, protruding eyes, well-developed adhesive disks on its fingers and toes, and many small warts on its back. The head, back, and upper surfaces of the legs bear a mottled pattern of green and brown. The underside is white.

This small tree frog is one of several frog species that have disappeared in Central America as a result of chytridiomycosis. First detected in northern Costa Rica in 1987, the disease has spread inexorably southward at the rate of 12–15 miles (20–25 km) per year, reaching central Panama in 2006. Most severely affected have been frogs breeding in streams at higher altitudes. This species breeds in the dry season, between February and April. Males call from vegetation above fast-flowing streams, producing cricket-like chirps. The tadpoles are adapted for life in fast-flowing water, having very long tails and sucker-like mouths.

SIMILAR SPECIES

Another victim of chytridiomycosis in Central America is *Isthmohyla calypsa*. Bright metallic green in color, this small tree frog is unique among Costa Rican frogs in having prominent spines on its back. Males call along streams and show remarkable site fidelity, returning to the same spot in successive years. Abundant in the 1990s, the species declined dramatically from 1992, when dead animals were found. See also the Meadow Tree Frog (*I. pseudopuma*; page 303).

FAMILY	Hylidae: Hylinae
OTHER NAMES	None
DISTRIBUTION	Eastern Brazil, northeastern Argentina, eastern Paraguay
ADULT HABITAT	Rainforest
LARVAL HABITAT	Permanent and ephemeral pools
CONSERVATION STATUS	IUCN Least Concern. Declining in places but occurs within several protected areas

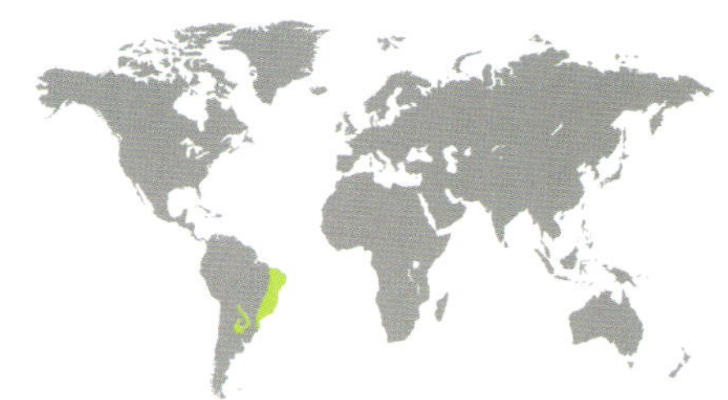

ITAPOTIHYLA LANGSDORFFII

OCELLATED TREE FROG

(DUMÉRIL & BIBRON, 1841)

ADULT LENGTH
Male
2⅝–3⅛ in (65–81 mm)
Female
3⅜–4 in (87–103 mm)

This large tree frog is notable for its coloration, which resembles mosses and lichen, making it well camouflaged in its arboreal habitat. It occurs in the Atlantic Forest region of eastern Brazil, an area of unusually high frog diversity. It breeds "explosively" after rain, with large numbers of animals found gathering around temporary or permanent waterbodies. Females lay 700–3,000 small eggs, which float on the water's surface. The tadpoles have been reported being preyed on by spiders.

SIMILAR SPECIES

Itapotihyla langsdorffii is the only species in its genus. In common with some other tree frogs, it shows fusion (or co-ossification) of the skin and the bones on the top of the skull—see also Bruno's Casque-headed Tree Frog (*Nyctimantis brunoi*; page 307). When such frogs hide in tree-holes, with their head closing the entrance, this acts as a defense against predators and may also reduce water loss from the top of the head.

The Ocellated Tree Frog has a long, slender body and long limbs. There are well-developed disks on the fingers and toes, and both hands and feet are webbed. There are bony crests on the top of the head. The upper surfaces are green, tan, and brown, with dark patches on the back and stripes on the legs. The belly is yellow or orange.

Actual size

FAMILY	Hylidae: Hylinae
OTHER NAMES	None
DISTRIBUTION	Central Veracruz state, Mexico
ADULT HABITAT	Cloud forest
LARVAL HABITAT	Streams
CONSERVATION STATUS	IUCN Critically Endangered. Has a very small range and its habitat is being destroyed

ADULT LENGTH
Male
$1\frac{1}{4}$–$1\frac{7}{16}$ in (32–37 mm)
Female
average $1\frac{7}{16}$ in (37 mm)

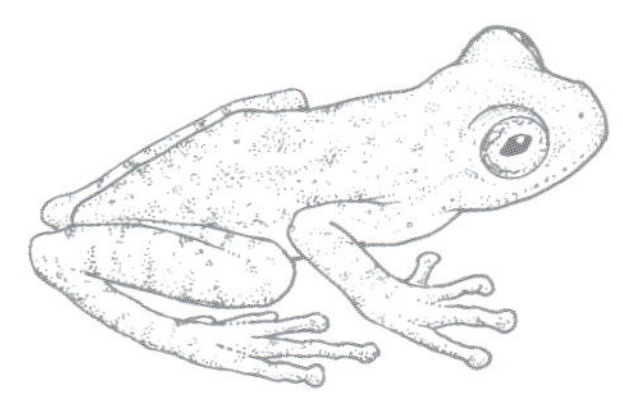

MEGASTOMATOHYLA NUBICOLA

CLOUD FOREST TREE FROG

(DUELLMAN, 1964)

Actual size

The Cloud Forest Tree Frog has long, slender arms and legs, large eyes, and a rounded snout. The feet are fully webbed, and the adhesive disks on the fingers are larger than those on the toes. The upper surfaces are reddish brown or tan, with darker blotches on the back and stripes on the legs. The underside is white.

Cloud forest is found where altitude and climate combine to produce conditions of persistent fog. It is characterized by an abundance of mosses and other moisture-loving plants. Although only one percent of the world's woodland is cloud forest, it is an important habitat because of the large number of species that are adapted to it and that live nowhere else, including many frogs. The cloud forest of Mexico is threatened by climate change and by deforestation to make way for coffee plantations. This small and little-known tree frog lives in an isolated and shrinking area of cloud forest.

SIMILAR SPECIES

There are just four species in the genus *Megastomatohyla*, all found in areas of cloud forest in Mexico. They are differentiated from other tree frogs by their large-mouthed tadpoles. The Variegated Tree Frog (*M. mixomaculata*) is listed as Endangered. The Mixe Tree Frog (*M. mixe*) and the Oaxacan Tree Frog (*M. pellita*) are both listed as Critically Endangered.

FAMILY	Hylidae: Hylinae
OTHER NAMES	None
DISTRIBUTION	Atlantic Coast, southeastern Brazil
ADULT HABITAT	Coastal tropical forest
LARVAL HABITAT	Probably water-filled cavities in bromeliads and other plants
CONSERVATION STATUS	IUCN Least Concern. Has a restricted range that is potentially threatened by habitat destruction

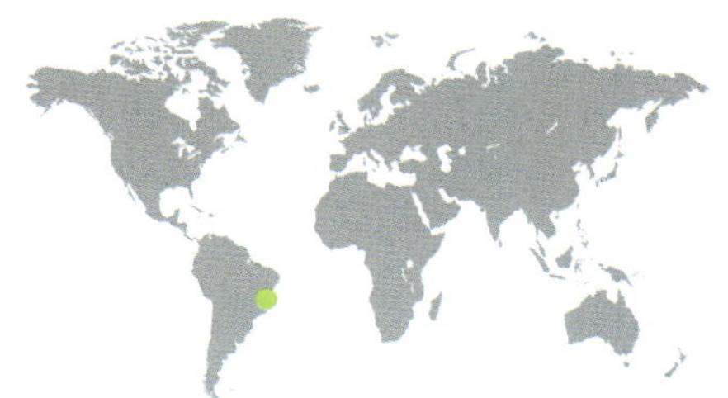

ADULT LENGTH
Male
1[15]/16–2½ in (49–62 mm)
Female
2[3]/16–3⅛ in (56–81 mm)

NYCTIMANTIS BRUNOI

BRUNO'S CASQUE-HEADED TREE FROG

(MIRANDA-RIBEIRO, 1920)

This rather odd-looking frog gets its name from the form of its head, on the top of which the skin is fused to the skull, a phenomenon called co-ossification. This is related to its habit of hiding by day in a bromeliad or a hole in a tree with the top of its head sealing off the entrance. This behavior, called phragmosis, probably serves a dual function, protecting the frog from predators and reducing the amount of water it loses by evaporation from its skin. Individuals vary considerably in size and they appear to choose hiding places that very closely match their size.

SIMILAR SPECIES

This species used to be in the genus *Aparasphenodon*, with three other poorly known Brazilian species, including the Bahia Broad-snout Casque-headed Tree Frog (*N. arapapa*), a small frog, described in 2009, from coastal tropical forest in northeastern Brazil. There is anecdotal evidence that it feeds its tadpoles on unfertilized eggs. Two other species transferred from *Aparasphenodon* to *Nyctimantis* are Pomba River Casque-headed Frog (*N. pomba*) and Bokermann's Casque-headed Frog (*N. bokermanni*).

Bruno's Casque-headed Tree Frog has a large, flattened head with a pointed snout. The eyes are large and protruding, and point forward. The arms and legs are long and slender, and there are adhesive disks on the fingers and toes. The frog is pale brown in color, with irregular dark brown patches.

Actual size

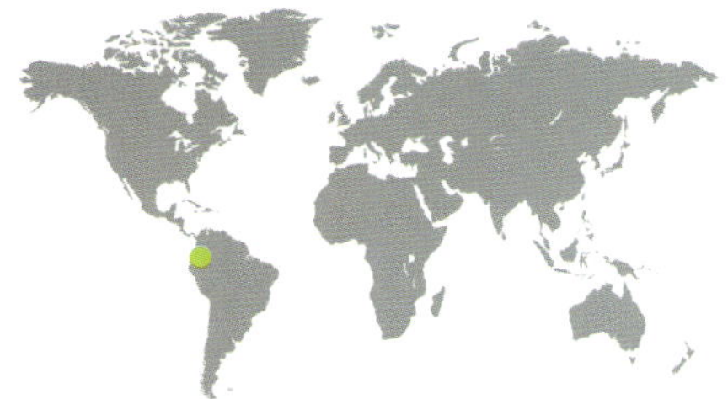

FAMILY	Hylidae: Hylinae
OTHER NAMES	Chocolate Tree Frog
DISTRIBUTION	Ecuador, Peru, Colombia
ADULT HABITAT	Lowland rainforest up to 3,900 ft (1,200 m) altitude
LARVAL HABITAT	Small pools in trees and bamboo
CONSERVATION STATUS	IUCN Least Concern. The population of this little-known species is believed to be stable

ADULT LENGTH
Male
$2^{1}/_{8}$–$2^{3}/_{4}$ in (55–68 mm)
Female
$2^{1}/_{4}$–$2^{1}/_{2}$ in (59–63 mm)

NYCTIMANTIS RUGICEPS

BROWN-EYED TREE FROG

BOULENGER, 1882

This strikingly colored tree frog is typically found close to stands of bamboo. It breeds in the tiny pools of water that form in broken bamboo stems, as well as in other holes in trees. The male calls from a hole, up to 30 ft (10 m) above ground, producing a "knock… knock… knock" sound. The female lays her eggs in a water-filled hollow and it is believed that she periodically visits the tadpoles in order to feed them on unfertilized eggs. The tadpoles have not been observed. When molested or attacked, this frog rolls itself up into a ball.

Actual size

The Brown-eyed Tree Frog has a slender body, large eyes, and very large hands. There are well-developed adhesive disks on the fingers and toes. The back is cream, contrasting strongly with the flanks, which are black with large lemon-yellow spots. The arms and legs are yellow where they join the body, otherwise cream, and the fingers and toes are black.

SIMILAR SPECIES

Nyctimantis rugiceps is one of seven species in genus *Nyctimantis*. It differs from other frogs in the family Hylidae by having pupils that are vertical, rather than horizontal slits. It is thought to be most closely related to the genus *Triprion* (see pages 330–332).

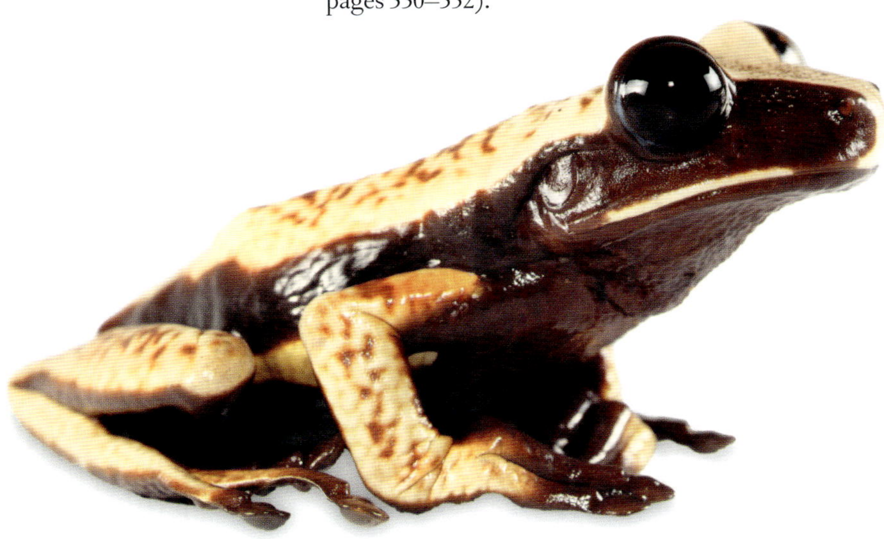

FAMILY	Hylidae: Hylinae
OTHER NAMES	None
DISTRIBUTION	Northern Argentina, Uruguay
ADULT HABITAT	Lowland forest
LARVAL HABITAT	Ephemeral pools
CONSERVATION STATUS	IUCN Least Concern, but declining due to habitat destruction. Occurs in some protected areas

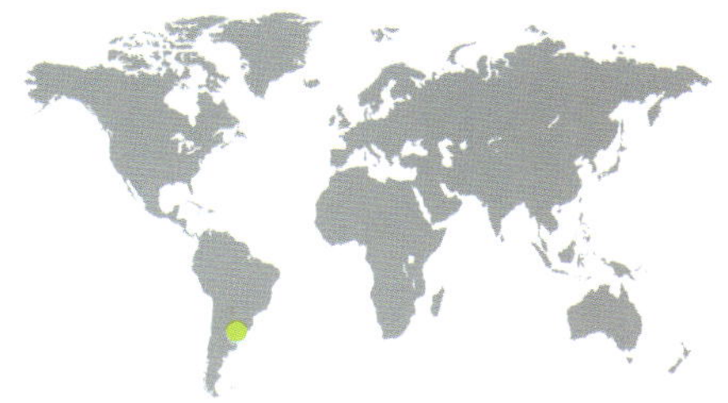

NYCTIMANTIS SIEMERSI

RED-SPOTTED ARGENTINA FROG

(MERTENS, 1937)

ADULT LENGTH
Male
2⅜–3 in (60–77 mm)
Female
2 15/16–3¼ in (74–83 mm)

This large, colorful frog is typically found in and around bromeliads, which provide a safe refuge when it is inactive. It is an "explosive breeder," mating and egg-laying being completed within three days. The male, which has paired vocal sacs, produces a short, frequently repeated call while floating in a temporary pool. The female lays 2,500–9,000 eggs in clumps among submerged vegetation. In the early stages of their development, the black tadpoles are adorned with red spots. The wet lowland habitat of this species is threatened by expanding agriculture and the construction of a huge ship canal.

SIMILAR SPECIES

This was the only species in the genus *Argenteohyla* before being transferred to *Nyctimantis*. It is most closely related to other frogs that have the skin of their head fused to their skull, such as the Pepper Tree Frog (*Trachycephalus typhonius*; page 329).

The Red-spotted Argentina Frog has long, slender legs, long fingers and toes, a pointed snout, and large, protruding eyes. The upper surfaces are greenish brown with a pattern of black reticulated blotches and stripes. There are reddish spots on the thighs and the underside is pale violet.

Actual size

FAMILY	Hylidae: Hylinae
OTHER NAMES	None
DISTRIBUTION	Northern Bolivia, southern Peru
ADULT HABITAT	Lowland rainforest up to 890 ft (270 m) altitude
LARVAL HABITAT	Tiny pools in Brazil-nut shells and palm bracts
CONSERVATION STATUS	IUCN Least Concern. A common species showing no evidence of decline

ADULT LENGTH
Male
$1\frac{7}{8}$–2 in (47–52 mm)
Female
$1\frac{7}{8}$–$2\frac{9}{16}$ in (47–64 mm)

OSTEOCEPHALUS CASTANEICOLA

BRAZIL-NUT TREE FROG

MORAVEC, APARICIO, GUERRERO-REINHARD, CALDERÓN, JUNGFER & GVOZDÍK, 2009

Actual size

This elegant tree frog lives in tropical rainforest where Brazil-nut trees (*Bertholletia excelsa*) are common. The nuts are eaten by humans and agoutis, and the empty cup-shaped shells are discarded on the forest floor. They retain water longer than holes in the ground do and thus provide an ideal breeding habitat for frogs and a variety of insects. This frog also lays its eggs in the water-filled bracts of palms. The developing tadpoles are fed by the female, which provides them with unfertilized eggs.

SIMILAR SPECIES

To date, 28 species have been described in the genus *Osteocephalus*, the slender-legged tree frogs of South America. None of them is of conservation concern. *Osteocephalus planiceps* is a common, large species found in Colombia, Ecuador, and Peru. The female Oophagous Slender-legged Tree Frog (*O. oophagus*) visits its tadpoles every five days in order to feed them on eggs. Strangely, the male remains in amplexus on her back throughout this process.

The Brazil-nut Tree Frog has very long legs, webbed feet, large eyes, and well-developed adhesive pads on its fingers and toes. The back varies in color from tan to purplish brown, and the legs are striped in dark and pale brown. The underside is creamy white and there is a white stripe along the upper lip.

FAMILY	Hylidae: Hylinae
OTHER NAMES	None
DISTRIBUTION	Hispaniola
ADULT HABITAT	Forest and plantations
LARVAL HABITAT	Mountain streams
CONSERVATION STATUS	IUCN Vulnerable. Reduced to a few scattered populations by deforestation and mining

OSTEOPILUS VASTUS

HISPANIOLAN GIANT TREE FROG

(COPE, 1871)

ADULT LENGTH
Male
up to 4¼ in (109 mm)

Female
up to 5⅝ in (142 mm)

It is common among frogs for the female to be larger than the male, but this large tree frog takes this sexual dimorphism to an extreme, with females up to 50 percent larger than males. The male calls from trees overhanging running water, producing an "ook… ook… ook" sound. The eggs are deposited in small, muddy pools close to streams and, as the tadpoles develop, they make their way into clearer, running water. Deforestation and mining have destroyed much of the frog's habitat, and this has had the effect of reducing this once widespread species to a few remnant populations.

SIMILAR SPECIES

There are eight species in the genus *Osteopilus*, the West Indian tree frogs. The Hispaniolan Yellow Tree Frog (*O. pulchrilineatus*) is a small, Vulnerable tree frog found in the Dominican Republic. As Vulnerable, it is being bred in captivity. The Cuban Tree Frog (*O. septentrionalis*) is a large frog that eats smaller frogs, not least in Florida, where it has been introduced and where efforts are now being made to eradicate it.

Actual size

The Hispaniolan Giant Tree Frog has a warty back, large eyes, and a rounded snout. Its hands and feet are large, the toes are fully webbed, and the fingers and toes have large adhesive pads. The upper surfaces are green or gray with dark gray or black markings, and there is a reddish triangle on the head. The legs are striped green and black.

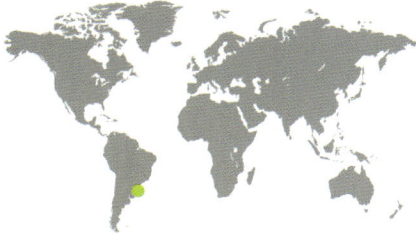

FAMILY	Hylidae: Hylinae
OTHER NAMES	None
DISTRIBUTION	Southeastern Brazil, in coastal areas up to 2,100 ft (650 m) altitude
ADULT HABITAT	Restinga scrubland
LARVAL HABITAT	Tiny pools in the leaf axils of bromeliads
CONSERVATION STATUS	IUCN Least Concern. Declining in places due to loss of habitat, but occurs within some protected areas

ADULT LENGTH
Male
5/8–15/16 in (16–24 mm)
Female
9/16–15/16 in (14–24 mm)

PHYLLODYTES LUTEOLUS

YELLOW HEART-TONGUED FROG

(WIED-NEUWIED, 1824)

Actual size

The Yellow Heart-tongued Frog is a small, slender-bodied frog with a blunt snout. It is uniformly pale green with a yellow tinge on the hands and feet. Many individuals have a narrow brown stripe running from the eye to the armpit. The irises of the eyes are gold.

This very small frog is totally dependent, throughout its life history, on bromeliads. Where the density of frogs is high in relation to the availability of bromeliad plants, males fight one another to establish possession of plants, using a pair of sharp tusks in their lower jaw to wound one another. Males call to attract females, which lay their eggs in the water-filled leaf axils, most commonly just one egg per axil. The tadpoles complete their development and growth within their tiny pools. New frog populations have become established within the Rio de Janeiro municipality, carried in bromeliads collected from the wild and planted in gardens.

SIMILAR SPECIES

Sixteen species have been described in the genus *Phyllodytes*, all dependent on bromeliads and most of them found in Brazil. An exception is the Endangered Trinidad Heart-tongued Frog (*P. auratus*), which is found at the top of a single mountain on Trinidad. It lives on only one species of giant bromeliad, which is subject to collection.

FAMILY	Hylidae: Hylinae
OTHER NAMES	None
DISTRIBUTION	Northern Honduras at altitudes of 3,050–5,085 ft (930–1,550 m)
ADULT HABITAT	Wet montane forest
LARVAL HABITAT	Mountain streams
CONSERVATION STATUS	IUCN Critically Endangered. Has declined dramatically throughout its restricted range

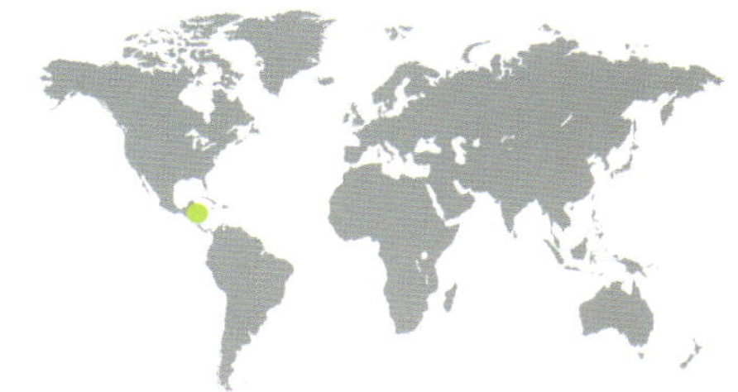

PLECTROHYLA CHRYSOPLEURA

GOLDEN-SIDED TREE FROG

WILSON, MCCRANIE & CRUZ-DÍAZ, 1994

ADULT LENGTH
Male
2³⁄₁₆–2¹¹⁄₁₆ in (56–66 mm)
Female
2½–2¹¹⁄₁₆ in (63–66 mm)

This large tree frog gets its common name from bright yellow patches on each flank, largely hidden in the armpit and the groin when it is resting. Males have been observed at night, sitting on boulders in the splash zone of mountain streams. Originally known from only one locality, its numbers declined by an estimated 80 percent between 1994 and 2004. This decline was due partly to destruction of its forest habitat; in addition, the presence of tadpoles with deformed mouths indicates infection by chytridiomycosis. The species was reported from a new locality in 2011.

SIMILAR SPECIES

There are 21 species in the Central American genus *Plectrohyla*, known as spike-thumbed frogs. As many species again have been transferred to the genus *Sarcohyla*. Within *Plectrohyla*, eight species each are listed as Critically Endangered and Endangered. The Cave Spikethumb Tree Frog (*P. tecunumani*) and the Alta Verapaz Spikethumb Tree Frog (*P. teuchestes*) are both from Guatemala and Critically Endangered. The Puebla Tree Frog (*P. charadricola*) is listed as Endangered. It lives in montane forest in Mexico where, like other species in this genus, it is threatened by deforestation and chytridiomycosis. See also the Honduras Spike-thumb Frog (*P. dasypus*; page 314).

The Golden-sided Tree Frog has a rounded snout and well-developed disks on its fingers and toes. Its upper surfaces are pale gray to brown, with indistinct darker markings and an overall bronze hue. There are bright yellow patches in the armpit, the groin, and on concealed parts of the legs. The iris is gold with a black reticulated pattern.

Actual size

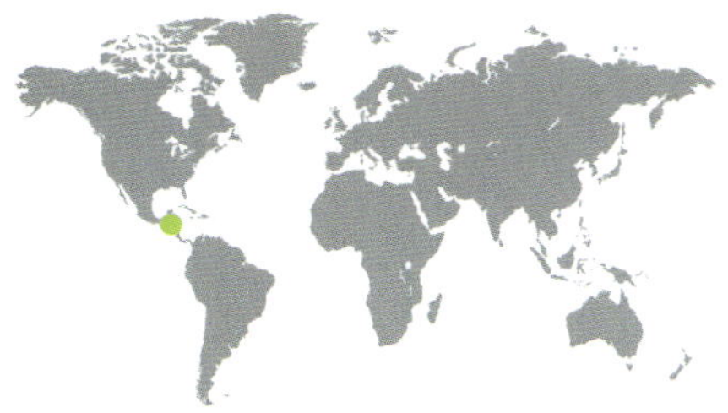

FAMILY	Hylidae: Hylinae
OTHER NAMES	None
DISTRIBUTION	Northwestern Honduras at altitudes of 4,635–6,530 ft (1,410–1,990 m)
ADULT HABITAT	Cloud forest
LARVAL HABITAT	Mountain streams
CONSERVATION STATUS	IUCN Critically Endangered. Has declined dramatically throughout its restricted range

ADULT LENGTH
Male
1¼–1⅝ in (31–42 mm)
Female
1¼–1¾ in (31–44 mm)

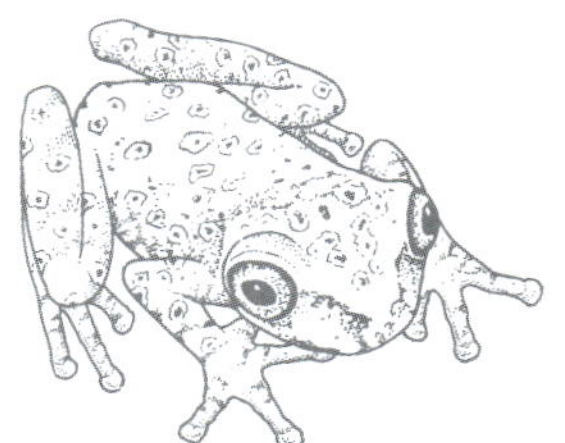

PLECTROHYLA DASYPUS

HONDURAS SPIKE-THUMB FROG

MCCRANIE & WILSON, 1981

Actual size

The Honduras Spike-thumb Frog has a slender body, a short snout, and long fingers and toes ending in well-developed disks that have a yellowish tinge. The upper surfaces are brown or bronze with scattered black spots, and the underside is gray. The iris is copper with a black reticulated pattern.

This small tree frog gets its common name from a short, hard spur on each hand, called a prepollical spine. A feature of other species in this genus, the spine appears to be used in combat between males—some museum specimens of males have scars on their heads. Males also have more muscular forearms than females. The entire range of this very rare frog falls within the Parque Nacional Cusuco in northern Honduras. Surveys for the disease chytridiomycosis have found very high levels of infection in the stream-living tadpoles of this species.

SIMILAR SPECIES

The Exquisite Tree Frog (*P. exquisitus*) is named for its beautiful coloration, which exists in two color phases, a partot green and a buff phase. It occurs at moderate elevations in northwestern Honduras, and its tadpoles are often found with those of *P. dasypus*, and like *P. dasypus* it is Critically Endangered. Many stream-breeding frogs in Central America have been very badly affected by chytridiomycosis. See also the Golden-sided Tree Frog (*P. chrysopleura*; page 313).

FAMILY	Hylidae: Hylinae
OTHER NAMES	None
DISTRIBUTION	Eastern USA and eastern Canada
ADULT HABITAT	Woodlands near ephemeral and permanent waterbodies
LARVAL HABITAT	Ephemeral and permanent waterbodies
CONSERVATION STATUS	IUCN Least Concern. Has a very large range and faces no serious threats other than draining of its wetland habitats

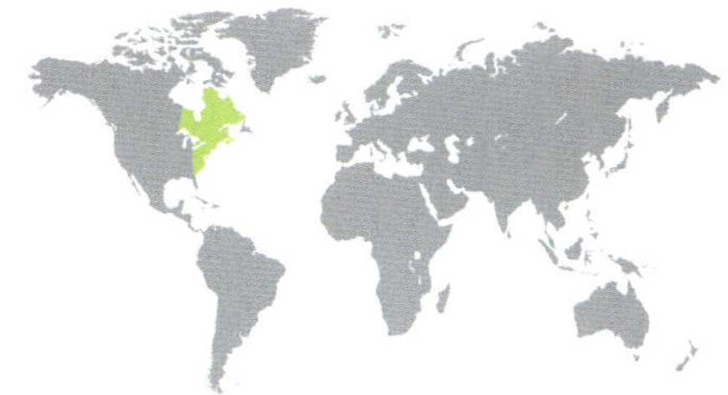

PSEUDACRIS CRUCIFER

SPRING PEEPER

(WIED-NEUWIED, 1838)

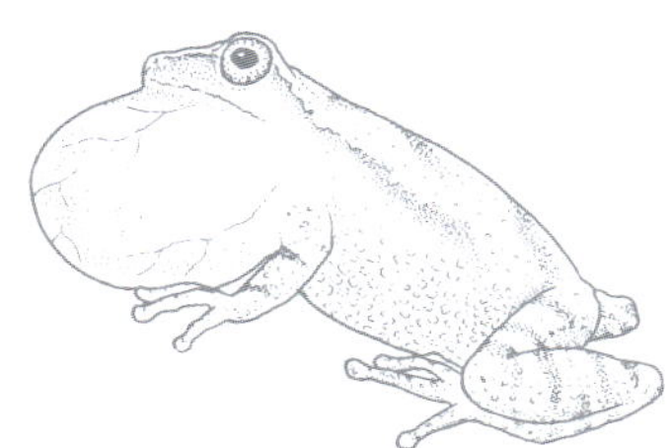

ADULT LENGTH
Male
$^{11}/_{16}$–1$^{1}/_{8}$ in (18–28 mm)

Female
$^{7}/_{8}$–1$^{1}/_{4}$ in (23–32 mm)

The Spring Peeper provides one of the first sounds of spring, the male's distinctive "peep" being heard by day and night as long as the temperature is above freezing. From a distance, choruses of males have been likened to distant sleigh bells. Within a chorus, adjacent males often form duets and trios, alternating their calls with one another. If a rival male gets too close, they switch to an aggressive call, which is a stuttering trill. At the north end of its range the species is able to survive the long, severe winter without being frozen, thanks to glucose in its cells, which acts as an antifreeze.

Actual size

The Spring Peeper is a rather plump little frog with a pointed snout and adhesive disks on its fingers and toes. It varies in color, being straw, brown, gray, olive, or reddish. There is usually a distinct, dark "X" mark on its back and dark bands on its legs, and there is a dark line on the head between the eyes and a dark stripe through the eye.

SIMILAR SPECIES

The Little Grass Frog (*Pseudacris ocularis*) is found in southeastern USA and is North America's smallest frog. The Ornate Chorus Frog (*P. ornata*), also from southeastern USA, is larger than *P. crucifer* and breeds in the winter. The largest of the chorus frogs is Strecker's Chorus Frog (*P. streckeri*), which is found mostly in Texas and Oklahoma. This species also breeds in winter and survives the heat in summer by burrowing headfirst into the ground. *Pseudacris* contains 17 species.

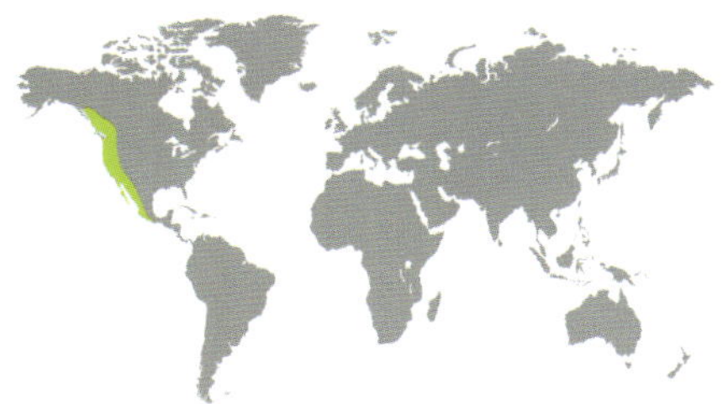

FAMILY	Hylidae: Hylinae
OTHER NAMES	Pacific Chorus Frog
DISTRIBUTION	Western USA, western Canada, Mexico
ADULT HABITAT	Low vegetation near water
LARVAL HABITAT	Ephemeral and permanent waterbodies
CONSERVATION STATUS	IUCN Least Concern. Has a very large range and shows no evidence of widespread decline

ADULT LENGTH
Male
1–1⅞ in (25–48 mm)

Female
1–1⅞ in (25–47 mm)

PSEUDACRIS REGILLA

PACIFIC TREE FROG

(BAIRD & GIRARD, 1852)

Actual size

The Pacific Tree Frog is a plump little frog with adhesive disks on its fingers and toes. It varies in color, being green, brown, beige, copper, tan, rust, and even black. Individuals can change color within a few minutes. There is a distinctive broad black or brown stripe through the eye.

Sometimes known as the Hollywood Frog because its distinctive "rib-bit" call is used as background noise in many films, the Pacific Tree Frog is variable in color. Experiments with individuals that are green or brown have shown that they choose a background color that matches their own. The timing of breeding varies considerably, depending on latitude and altitude. Calling males space themselves out and avoid calling at the same time. Some studies have found that larger males are more likely to attract females, but they are also more likely to be eaten by predatory bugs.

SIMILAR SPECIES

The Boreal Chorus Frog (*Pseudacris maculata*) has a huge range, covering most of central Canada and the USA. It may occasionally hybridize with *P. regilla* at the extreme west of its range. The California Tree Frog (*P. cadaverina*) is found in southern California and Mexico. In color it resembles the pale rocks on which it lives; its Latin species name means "corpse-like."

FAMILY	Hylidae: Hylinae
OTHER NAMES	Shrinking Frog
DISTRIBUTION	South America, east of the Andes, from Venezuela south to Argentina; Trinidad
ADULT AND LARVAL HABITAT	Ponds, lakes, and swamps with floating vegetation
CONSERVATION STATUS	IUCN Least Concern. Has a very large range and shows no evidence of general decline

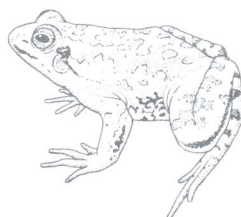

ADULT LENGTH
1¾–2¹⁵⁄₁₆ in (45–75 mm)

PSEUDIS PARADOXA

PARADOXICAL FROG

(LINNAEUS, 1758)

The name of this frog refers to the fact that its tadpole can grow to be three to four times larger (up to 8½ in/220 mm long) than the adult into which it metamorphoses. This occurs only in permanent ponds, tadpoles developing in ephemeral ponds metamorphosing while still small before their pond dries up. Giant tadpoles are unusual in that they start to develop adult organs, such as the gonads, intestine, and lungs. Adults are wholly aquatic, are active by day and night, and feed on insects and small frogs. They lay their eggs in a floating foam nest at the edge of a pond.

SIMILAR SPECIES

There is a total of seven species in the genus *Pseudis*, all with a similar life history. All are found in northern South America and none is of conservation concern. *Pseudis cardosoi* is a smaller species than *P. paradoxa* and inhabits open, seasonally flooded grassland in southern Brazil. Its call is likened to the grunting of pigs. The skin of pseudid frogs contains medicinal compounds, including antimicrobials and a treatment for type 2 diabetes.

The Paradoxical Frog has smooth, slimy skin, a short, paunchy body, and a small head. The protruding eyes and the nostrils point upward so that, when floating in water, they break the surface. The hind limbs are long and muscular, and the toes are webbed but the fingers are not. The frog is green and brown in color, and there is a black patch behind the eye.

Actual size

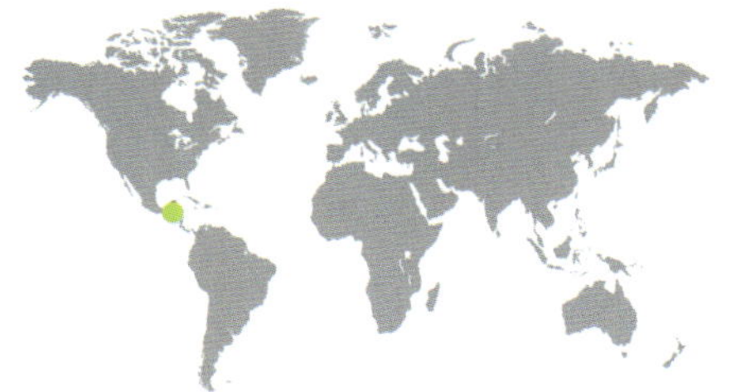

FAMILY	Hylidae: Hylinae
OTHER NAMES	None
DISTRIBUTION	Eastern Guatemala
ADULT HABITAT	Cloud forest
LARVAL HABITAT	Streams
CONSERVATION STATUS	IUCN Endangered. Its very restricted range is subject to deforestation and chytridiomycosis is a potential threat

ADULT LENGTH
Male
1⅛–1⁵⁄₁₆ in (28–33 mm)

Female
1⅝–2 in (41–51 mm)

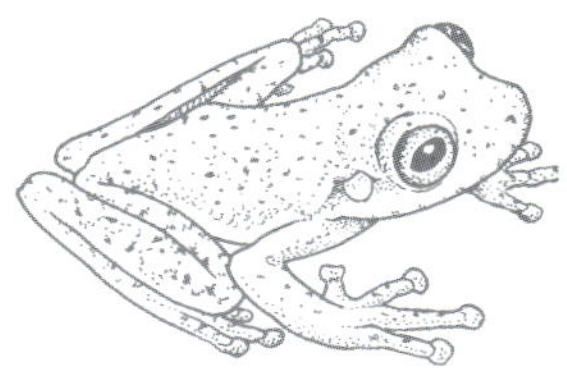

QUILTICOHYLA SANCTAECRUCIS

CHINAMOCOCH STREAM FROG

(CAMPBELL & SMITH, 1992)

Actual size

The Chinamococh Stream Frog has a slender body, and long, thin arms and legs. It has well-developed disks on its fingers and toes. The hind feet are partially webbed. The upper surfaces are lime-green with yellow blotches and small black dots. The underside of the body, head, and legs is white, and the irises of the eyes are brown.

This small, nocturnally active frog breeds in February, during the dry season. Breeding males develop nuptial pads, consisting of many tiny spines, on their thumbs. These enable them to maintain a firm grip on the female during mating. They also have a large glandular patch on the chest, the function of which is unknown. The male's mating call is a soft, low-pitched "wraack." This species has lost much of its habitat to expanding agriculture, wood extraction, and human settlement. Chytridiomycosis has caused population declines in other species of this genus in Guatemala.

SIMILAR SPECIES

Quilticohyla sanctaecrucis and three other species comprise the southern Mexican and Guatemalan genus *Quilticohyla*, known as mountain stream frogs. They were previously in *Ptychohyla*. Worldwide, frogs that breed in high-altitude streams are threatened by habitat loss, climate change, and chytridiomycosis. The Chinamococh Stream Frog and the Guerreran Stream Frog (*Q. erythromma*) are Endangered while the Warty Mountain Stream Frog (*Q. acrochorda*) and the Zoque Stream Frog (*Q. zoque*) are Critically Endangered.

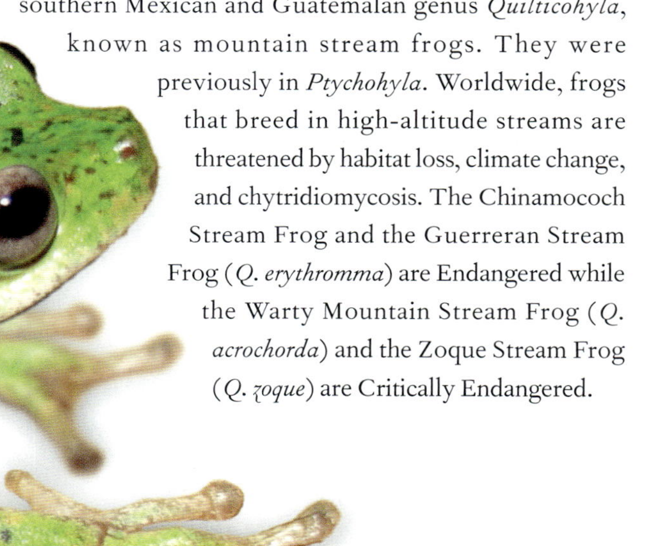

FAMILY	Hylidae: Hylinae
OTHER NAMES	None
DISTRIBUTION	Venezuela, Colombia, Trinidad and Tobago
ADULT HABITAT	Open areas, including flooded grassland and swamps
LARVAL HABITAT	Ponds and swamps
CONSERVATION STATUS	IUCN Least Concern. Very common and appears to be expanding its range

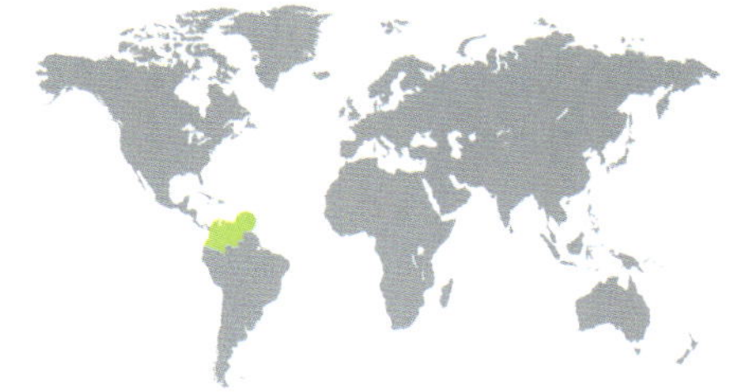

SCARTHYLA VIGILANS

MARACAIBO BASIN TREE FROG

(SOLANO, 1971)

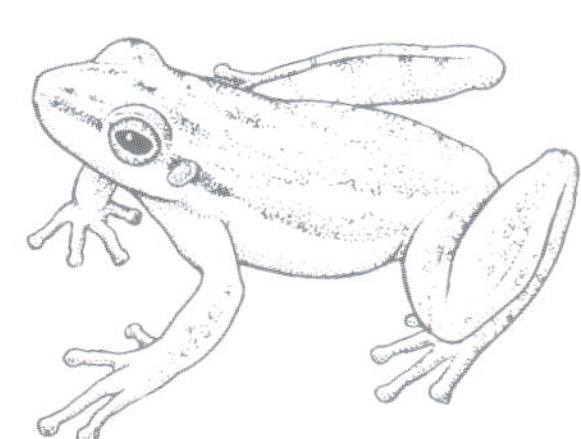

ADULT LENGTH
Male
$^{11}/_{16}$–$^{7}/_{8}$ in (17–21 mm)
Female
$^{3}/_{4}$–$^{7}/_{8}$ in (19–23 mm)

This very small frog changes color depending on the time of day. At night, when it is active, it is a uniform lime-green, whereas by day it is brown with dark longitudinal stripes. The male's call is like a cricket's chirp and is so quiet that it is often drowned out by the calls of other frogs. When disturbed, the frog jumps away athletically and skitters across the water's surface. It has recently been found in some new locations in Venezuela, suggesting that its range is expanding. It is thought that it may be dispersed by floating on rafts of Water Hyacinths (*Eichhornia crassipes*).

Actual size

The Maracaibo Basin Tree Frog has a slender body, long hind legs, and a pointed snout. The feet are webbed and the hands are not. There are small disks on the fingers and toes. By day, the upper surfaces are brown, the underside is white, and there is a dark longitudinal stripe along the flanks. By night the frog is uniform lime-green.

SIMILAR SPECIES

There is just one other species in the genus *Scarthyla*. The Tarauaca Snouted Tree Frog (*S. goinorum*) is a common species in the upper Amazon Basin, in Bolivia, Peru, Colombia, and Brazil. Its call is a low whistle. It has a very unusual tadpole, which swims just below the water's surface, feeding on duckweed. Its very muscular tail enables it to leap out of the water and skim across the surface.

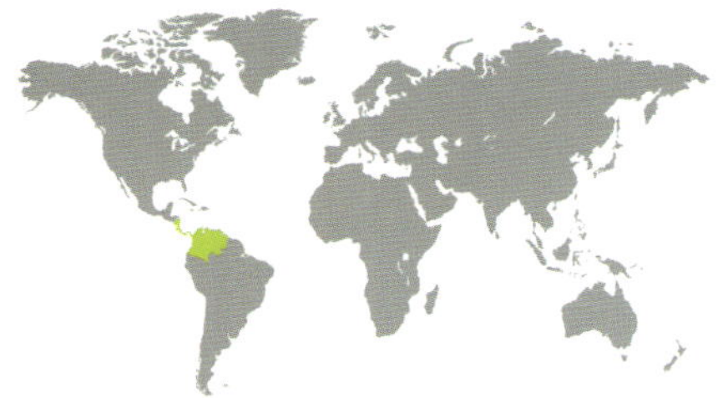

FAMILY	Hylidae: Hylinae
OTHER NAMES	Boulenger's Snouted Tree Frog
DISTRIBUTION	Central and South America, from Nicaragua in the north to Colombia and Venezuela in the south
ADULT HABITAT	Lowland tropical forest
LARVAL HABITAT	Permanent pools and ponds
CONSERVATION STATUS	IUCN Least Concern. A common species with a stable population

ADULT LENGTH
Male
1 7/16–1 15/16 in (36–49 mm)
Female
1 5/8–2 1/16 in (42–53 mm)

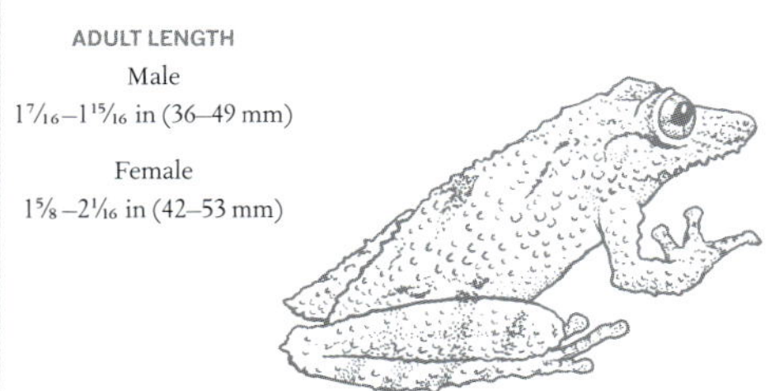

SCINAX BOULENGERI

BOULENGER'S TREE FROG

(COPE, 1887)

Actual size

Boulenger's Tree Frog has a flattened body, a long, pointed snout, and protruding nostrils. There are triangular pads on the fingers and toes. The skin on the back and limbs is covered in numerous, sharp-pointed warts. The upper surfaces are brown or green with darker markings, taking the form of bold cross-stripes on the hind legs. The irises of the eyes are bronze.

This common tree frog has a prolonged breeding season, with peaks of mating and egg-laying activity in May, June, and August. Males establish small territories in hidden positions close to ponds, which they then defend against their rivals. Their call is a single low-pitched, guttural note. Most males attend a chorus for between one and ten nights, but exceptional individuals attend for 50 nights. Their persistence reflects their physical condition. While attending and calling in a chorus, males lose weight as they use up their energy reserves. Females lay 600–700 eggs in the water.

SIMILAR SPECIES

There are around 78 species in the genus *Scinax*, found in South and Central America and the West Indies. The name of the genus comes from the Greek *skinos*, meaning "quick" or "active." Most species are Least Concern, but three species, including Joly's Snouted Tree Frog (*S. jolyi*) from the Guianas, are Endangered, one each are Vulnerable and Near Threatened, and three are Data Deficient. See also the Red Snouted Tree Frog (*S. ruber*; page 321).

FAMILY	Hylidae: Hylinae
OTHER NAMES	Allen's Snouted Tree Frog
DISTRIBUTION	Amazon Basin and the surrounding countries; Trinidad and Tobago. Introduced to Martinique, Puerto Rico, St. Lucia
ADULT HABITAT	Forest, open habitats, parks and gardens
LARVAL HABITAT	Ephemeral pools
CONSERVATION STATUS	IUCN Least Concern. A common species with a huge range and a stable population

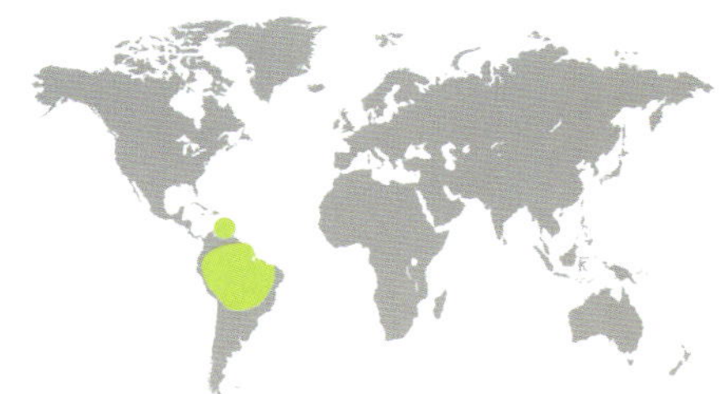

ADULT LENGTH
Male
1¼–1⁷⁄₁₆ in (31–37 mm)
Female
1⁹⁄₁₆–1⅝ in (40–42 mm)

SCINAX RUBER

RED SNOUTED TREE FROG

(LAURENTI, 1768)

This very common tree frog is an opportunistic, "explosive breeder," meaning that heavy rain triggers the rapid formation of dense choruses of males that last for just a few nights. Most males spend only one night at a chorus. Females enter a chorus and move from male to male. Their preference is for a male about 20 percent smaller than themselves. This size ratio brings male and female cloacas close together, maximizing the proportion of eggs that are fertilized. However, the female's preference is often overridden by larger males, which will attack and displace the male from her back.

Actual size

SIMILAR SPECIES

Given its enormous range, across which there is much genetic variation, this is probably more than one species. Even though a number of species were transferred to the genus *Ololygon*, the genus *Scinax* is still expanding with 12 new species being described since the first edition of this book was published. The Olive Tree Frog (*S. elaeochroa*) is found in Costa Rica, Nicaragua, and Panama. It is noted for its ability to sit, head down, on the surface of a vertically hanging leaf. See also Boulenger's Tree Frog (*S. boulengeri*; page 320).

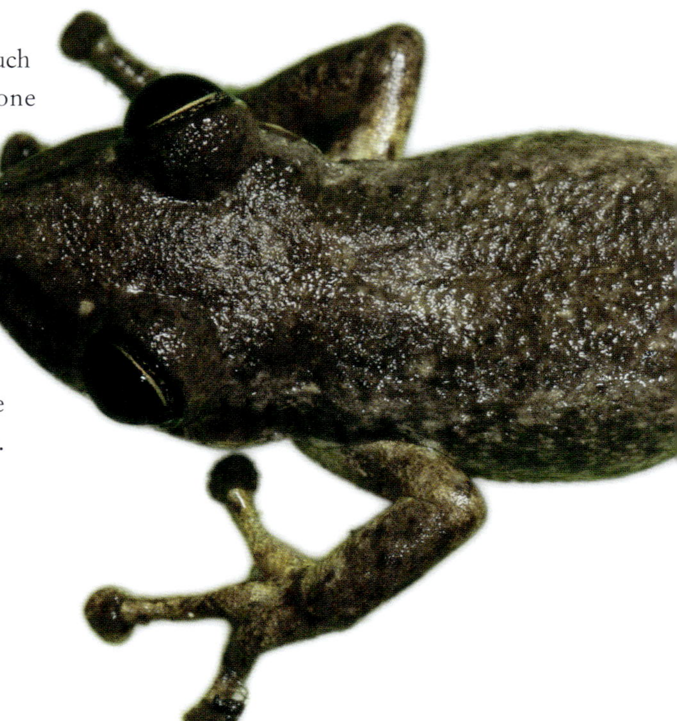

The Red Snouted Tree Frog has a slender, flattened body, a pointed snout, long legs, and large disks on its fingers and toes. The feet are webbed. The upper surfaces are green, brown, or yellow, with pale patches on the back and pale stripes along the flanks. There are yellow or orange spots in the groin and the irises of the eyes are bronze.

FAMILY	Hylidae: Hylinae
OTHER NAMES	Baudin's Tree Frog, Mexican Smilisca, Tooter
DISTRIBUTION	Central America, from Mexico to Costa Rica; extreme southeastern corner of Texas, USA
ADULT HABITAT	Humid, semi-humid, and semiarid lowlands, up to 5,280 m (1,610 m) altitude. Also suburban gardens with ponds
LARVAL HABITAT	Ephemeral ponds
CONSERVATION STATUS	IUCN Least Concern. Widespread and abundant, and thrives in human-altered habitats

ADULT LENGTH
Male
1 7/8–2 15/16 in (47–75 mm)

Female
2 3/16–3 1/2 in (56–90 mm)

SMILISCA BAUDINII

MEXICAN TREE FROG

(DUMÉRIL & BIBRON, 1841)

A remarkable ability to change color, like a chameleon, is a feature of this large frog, which can be brown, gray, green, tan, or yellow. Its habitat is very dry for much of the year and it survives periods of drought by enclosing itself in a cocoon made up of up to 40 layers of shed skin, interspersed with dry mucus. It hides in tree-holes, under bark, and in the leaf axils of plants. It can breed at any time of year, in response to heavy rain. The male's call, a nasal "heck... heck... heck," has been likened to the sound of an old car starting up.

The Mexican Tree Frog has a rather chubby body, short legs, and round pads on its fingers and toes. The color of the upper surfaces varies greatly, to match its background. There are darker markings on the back and legs, a dark stripe through the eye, and a pale spot beneath the eye. The sides and belly are cream with dark spots.

SIMILAR SPECIES

There are nine species in the genus *Smilisca*, which are found mostly in Central America. The Lowland Burrowing Tree Frog (*S. fodiens*) survives dry periods by burrowing underground and enclosing itself in a cocoon. It occurs in western Mexico and southern Arizona. The Upland Burrowing Tree Frog (*S. dendata*), also from Mexico, has lost much of its grassland habitat to agriculture and is the only species listed as Endangered.

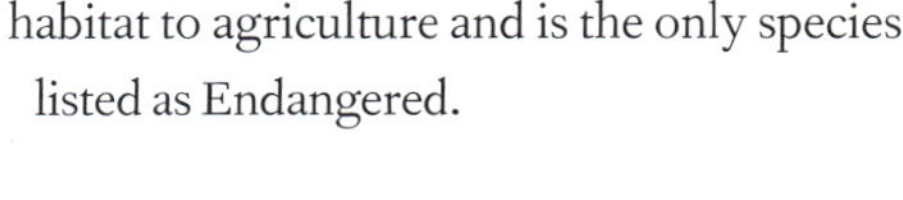

Actual size

FAMILY	Hylidae: Hylinae
OTHER NAMES	Veragua Cross-banded Tree Frog
DISTRIBUTION	Central America, from Honduras in the north to Panama in the south. One record from Colombia
ADULT HABITAT	Forest up to 5,000 ft (1,525 m) altitude
LARVAL HABITAT	Streams or basins created by parents
CONSERVATION STATUS	IUCN Least Concern. Widespread and abundant, and thrives in human-altered habitats

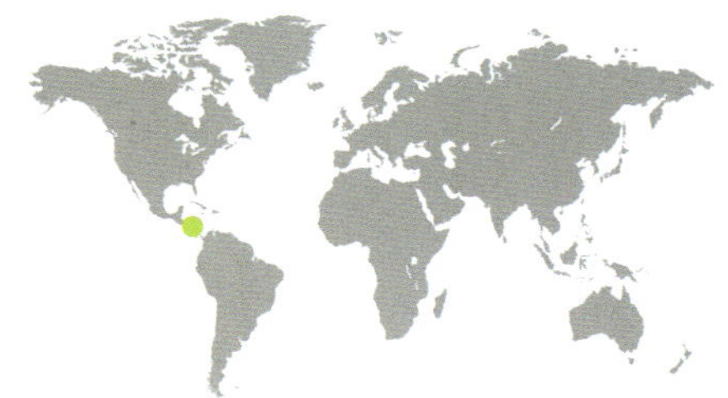

SMILISCA SORDIDA

DRAB TREE FROG

(PETERS, 1863)

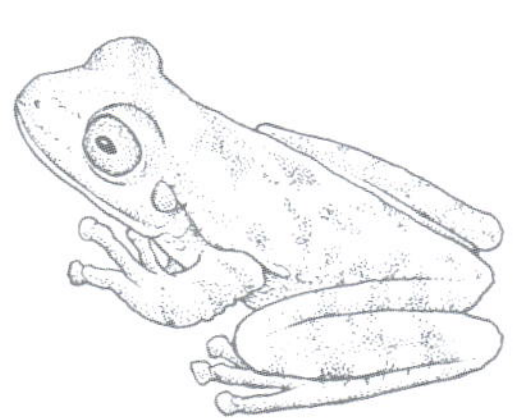

ADULT LENGTH
Male
1¼–2⅛ in (32–54 mm)
Female
2³⁄₁₆–2⁹⁄₁₆ in (56–64 mm)

This dull brown tree frog lives near creeks and rivers, and is often found hiding inside bromeliads. It breeds in forest streams when these become shallow and clear during the dry season. While in amplexus, a pair may dig a water-filled basin at the edge of a stream in which they lay their eggs, or they may lay their eggs directly in the stream. A basin provides some protection against egg predators and appears to be the method of choice when such predators are numerous. The tadpoles have large adhesive disks around their mouths with which they can anchor themselves to rocks in streams.

SIMILAR SPECIES

The Panama Cross-banded Tree Frog (*Smilisca sila*) is similar to *S. sordida* in laying its eggs in a specially prepared basin next to a stream. In contrast, the Nicaragua Cross-banded Tree Frog (*S. puma*) lays its eggs in open water, where they float on the surface. See also the Mexican Tree Frog (*S. baudinii*; page 322).

The Drab Tree Frog has webbed feet and large adhesive disks on its fingers and toes. The upper surfaces are grayish brown or tan, with indistinct darker patches on the back and faint cross-stripes on the legs. The underside is white and there is a purple patch in the groin. The irises of the eyes are yellow with black netlike veins.

Actual size

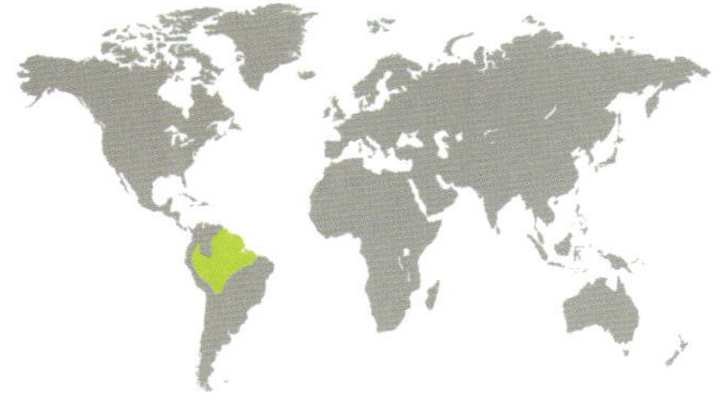

FAMILY	Hylidae: Hylinae
OTHER NAMES	Greater Hatchet-faced Tree Frog, Slope-headed Tree Frog
DISTRIBUTION	South America, from Peru in the east, to Bolivia in the south; Trinidad and Tobago
ADULT HABITAT	Permanently flooded areas, lagoons with floating vegetation
LARVAL HABITAT	Permanent ponds and lagoons
CONSERVATION STATUS	IUCN Least Concern. A common species that occurs within some protected areas

ADULT LENGTH
Male
1–1⅝ in (25–42 mm)
Female
1½–1$^{13}/_{16}$ in (38–46 mm)

SPHAENORHYNCHUS LACTEUS

ORINOCO LIME TREE FROG

(DAUDIN, 1800)

Actual size

The Orinoco Lime Tree Frog has a streamlined body shape, a very pointed snout, fully webbed feet, and partially webbed hands. The upper surfaces are lime green, with a blue tinge in places. There is a thin brown stripe between the eyes and the snout. The underside is white, and the irises of the eyes are gold or bronze.

An inhabitant of the Amazon and Orinoco river basins, this semi-aquatic frog feeds primarily on ants. Breeding occurs in the wet season, males and females congregating around ponds at night. Males call from emergent or floating vegetation, producing a series of clacking notes and inflating a large, bright blue vocal sac. The female lays around 60 eggs, in the water or attached to a plant. While being nocturnal protects the frog from daytime predators, at night it falls prey to spiders.

SIMILAR SPECIES

Sphaenorhynchus lacteus is one of the larger of the 15 species described to date in the genus *Sphaenorhynchus*, known as lime tree frogs. The Bahia Lime Tree Frog (*S. bromelicola*) has a limited range in eastern Brazil. It lives in bromeliads, breeds in ponds and is Critically Endangered. *Sphaenorhynchus caramaschii*, from upland areas of eastern Brazil, was described in 2007, since when four other new species have been described.

FAMILY	Hylidae: Hylinae
OTHER NAMES	*Osteocephalus exophthalma*
DISTRIBUTION	Western Guyana, eastern Venezuela
ADULT HABITAT	Montane forest up to 4,700 ft (1,430 m) altitude
LARVAL HABITAT	Probably in pools
CONSERVATION STATUS	IUCN Least Concern. Lives in a highly inaccessible habitat and faces no apparent threats

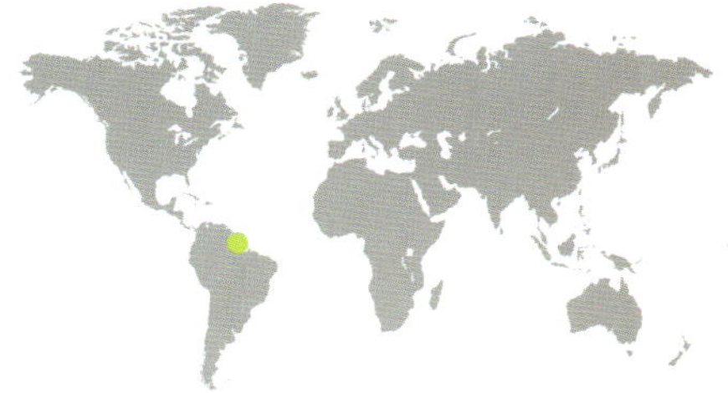

TEPUIHYLA EXOPHTHALMA

BIG-EYED SLENDER-LEGGED TREE FROG

(SMITH & NOONAN, 2001)

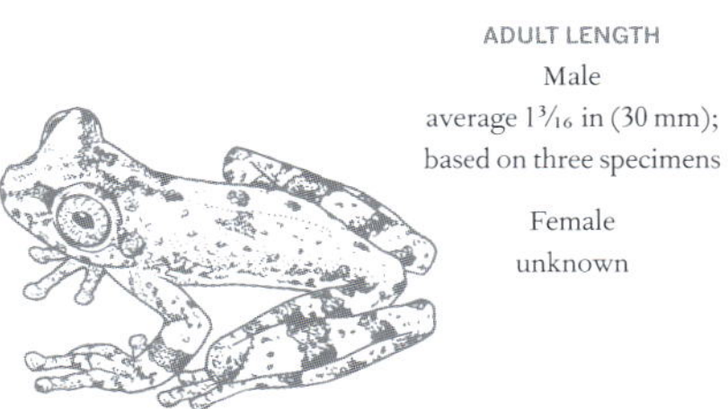

ADULT LENGTH
Male
average 1³⁄₁₆ in (30 mm);
based on three specimens

Female
unknown

This striking tree frog is known from only three specimens, all male, found in forest on the steep slopes of the tepuis, or flat-topped mountains, that are found in Venezuela and Guyana. This is an inaccessible and poorly explored region that is home to many relatively unknown frogs. It is likely that, like other species in its genus, this frog hides by day in bromeliads, calls from branches above water at night, and lays its eggs in pools. Its scientific species name, *exophthalma*, refers to its very large, protuberant eyes.

Actual size

The Big-eyed Slender-legged Tree Frog has a slender body, a broad head, and enormous, protruding eyes. The arms and legs are long and slender, and there are well-developed adhesive disks on the fingers and toes. The upper surfaces are brown, with black patches on the back and black stripes on the legs. The underside is yellow with black spots.

SIMILAR SPECIES

Tepuihyla exophthalma is the smallest of the nine species described to date in its genus. Warren's Tepui Tree Frog (*T. warreni*) and the Chimantá Tepui Tree Frog (*T. obscura*) are listed as Endangered, and three others are Near Threatened. All are found in Venezuela and Guyana. While some species are described as being common, their distribution, natural history, and conservation status are scarcely known, reflecting the difficulty of studying frogs in this remote mountainous region.

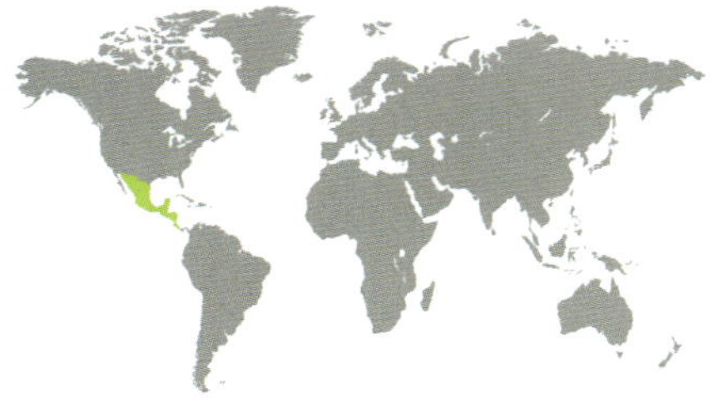

FAMILY	Hylidae: Hylinae
OTHER NAMES	Loquacious Tree Frog
DISTRIBUTION	Central America, from Mexico to Costa Rica
ADULT HABITAT	Forest up to 3,300 ft (1,000 m) altitude
LARVAL HABITAT	Ephemeral and permanent ponds
CONSERVATION STATUS	IUCN Least Concern. Tolerates alteration of its habitat and occurs within some protected areas

ADULT LENGTH
Male
1⁵⁄₁₆–1¾ in (33–45 mm)
Female
1½–1⅞ in (38–47 mm)

TLALOCOHYLA LOQUAX

MAHOGANY TREE FROG

(GAIGE & STUART, 1934)

Actual size

The Mahogany Tree Frog has a rounded, blunt snout and large eyes. Depending on the time of day, its upper surfaces are light gray, yellow, or reddish brown, while the underside is yellow. There are bright red patches in the armpits, the groin, and the back of the thighs, and the webbing on the hind feet is red.

A nocturnally active species, this uncommon frog changes color according to the time of day. When resting by day, it is very pale, being yellow, gray, or white, but at night it darkens to deep yellow or reddish brown. It breeds in the middle of the rainy season, in July and August. Males call from floating leaves in the middle of a pond, producing a sound like the honking of geese. The female lays a clutch of around 250 eggs that she attaches to submerged vegetation. The tadpoles live in the deepest parts of a pond.

SIMILAR SPECIES

There are four other species in the genus *Tlalocohyla*, all found in Mexico. The Dwarf Mexican Tree Frog (*T. smithii*) and the Painted Tree Frog (*T. picta*) are very common species. Godman's Tree Frog (*T. godmani*) breeds in temporary streams in montane forest in eastern Mexico, but is Vulnerable due to habitat loss to agriculture. The recently described Tapir Valley Tree Fog (*T. celeste*) is Critically Endangered.

FAMILY	Hylidae: Hylinae
OTHER NAMES	Surinam Casque-headed Frog
DISTRIBUTION	Eastern Brazil, Ecuador, Peru, Bolivia. Separate population in northern Brazil, Guyana, French Guiana, Suriname
ADULT HABITAT	Lowland rainforest
LARVAL HABITAT	Ephemeral and permanent pools
CONSERVATION STATUS	IUCN Least Concern. Has a huge range but is threatened in places by deforestation

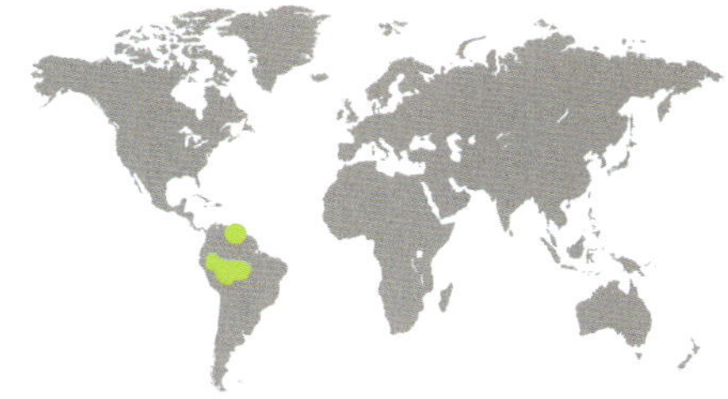

ADULT LENGTH
Male
2–2⅜ in (52–60 mm)
Female
2–2⅝ in (52–65 mm)

TRACHYCEPHALUS CORIACEUS

SURINAM GOLDEN-EYE TREE FROG

(PETERS, 1867)

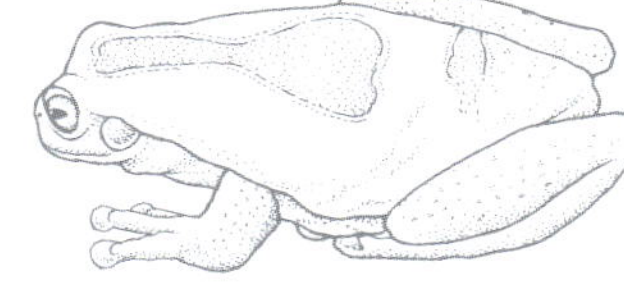

A large, nocturnally active tree frog, this species is found in two distinct parts of the Amazon Basin, typically in areas where the forest becomes flooded in the wet season. Males call while floating in the water, producing a loud growl. They have enormous paired vocal sacs that almost meet above the head when fully inflated. The eggs float on the surface of the water. When handled, the frog inflates its body, making itself much larger and causing the large black spots on each flank to be visible from the front. This is probably a defense mechanism against predators. The species is eaten in parts of its range by humans.

SIMILAR SPECIES

The Black-spotted Casque-headed Tree Frog (*Trachycephalus nigromaculatus*) occurs in coastal areas of southern Brazil. It is often found inside bromeliads, a habit that has resulted in it being accidentally distributed by plant traders to areas outside its natural range. The Rio Golden-eye Tree Frog (*T. imitatrix*) occurs in mountainous areas of southern and southeastern Brazil. See also the Amazon Milk Frog (*T. resinifictrix*; page 328) and the Pepper Tree Frog (*T. typhonius*; page 329).

The Surinam Golden-eye Tree Frog has a wide head, a blunt, rounded snout, and large disks on its rather short fingers and toes. The upper surfaces are brown, reddish brown, or tan in color, and there is a large black patch on each flank, just behind the armpit. The webbing between the toes is often bright red.

Actual size

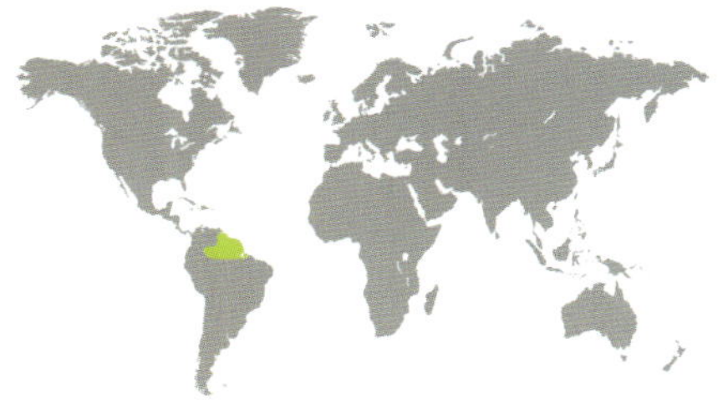

FAMILY	Hylidae: Hylinae
OTHER NAMES	Blue Milk Frog, Boatman Frog, Mission Golden-eyed Tree Frog, Wife Toad
DISTRIBUTION	Northern Brazil, Guyana, French Guiana, Suriname
ADULT HABITAT	Tropical forest
LARVAL HABITAT	Water-filled holes in trees
CONSERVATION STATUS	IUCN Least Concern. Has a huge range that includes several protected areas

ADULT LENGTH
Male
average 3 in (77 mm)

Female
average 3⁷⁄₁₆ in (88 mm)

TRACHYCEPHALUS RESINIFICTRIX

AMAZON MILK FROG

(GOELDI, 1907)

This striking, intricately patterned inhabitant of the Amazon Basin is arboreal and is active at night. It breeds between November and May. Males call on dry, cloudless nights from water-filled holes in trees 6–100 ft (2–30 m) above the ground. Their call has been likened to the sound of a paddle being tapped against the side of a canoe. Having located a male, the female lays a gelatinous mass of around 2,500 eggs in the water in his tree-hole. The tadpoles feed on unhatched eggs and plant matter, eventually emerging from their arboreal nursery as tiny froglets. When handled or attacked, this frog secretes a milky fluid from its skin.

SIMILAR SPECIES

Trachycephalus resinifictrix was once thought to occur across much of northern South America but, in 2013, it was found to be two separate species. The newly described species, the slightly smaller *T. cunauaru*, occurs to the west of the Amazon Basin as far as Ecuador. The related *T. mambaiensis* is found in the Cerrado, a savanna ecoregion in southern Brazil.

The Amazon Milk Frog has conspicuous alternating bands of brown and cream, often with a bluish tinge, across its body and limbs. The irises of the eyes are gold with a black Maltese cross centered on the pupil. It has a wide head and a blunt, rounded snout. There are large disks on its fingers and toes and large, pale warts on its back.

Actual size

FAMILY	Hylidae: Hylinae
OTHER NAMES	None
DISTRIBUTION	Central and South America, from Mexico in the north to Argentina in the south
ADULT HABITAT	Lowland forest up to 935 ft (285 m) altitude
LARVAL HABITAT	Ephemeral pools
CONSERVATION STATUS	IUCN Least Concern. A common species with a huge range and a stable population

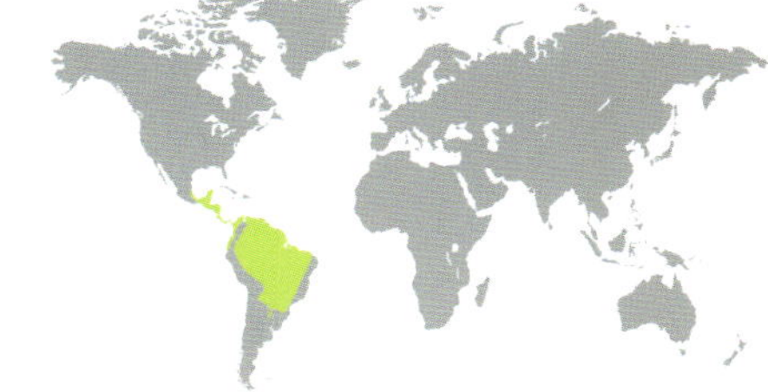

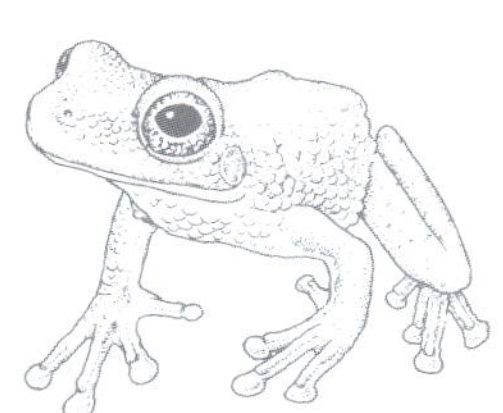

ADULT LENGTH
Male
$2^{13}/_{16}$–$3^{15}/_{16}$ in (70–101 mm)
Female
$3^{5}/_{8}$–$4^{1}/_{2}$ in (93–114 mm)

TRACHYCEPHALUS TYPHONIUS
PEPPER TREE FROG
(LAURENTI, 1768)

This large tree frog is so named because its skin secretion causes humans to sneeze. When attacked, glands on its head produce a copious sticky white secretion that deters predators, including snakes. It breeds over just one night, after the first rains have fallen, gathering around temporary ponds. Males call from the water, producing an extremely loud growl, amplified by a pair of very large vocal sacs. The female lays her eggs in a thin, floating layer on the water's surface. Active at night, the frog is rarely seen outside the context of breeding. Its webbed feet enable it to parachute from one tree to another.

The Pepper Tree Frog has a wide head and a blunt, rounded snout. Its skin is thick and glandular, and there are large disks on its fingers and toes. The feet are fully webbed, the hands partially so. The upper surfaces are yellow, tan, brown, or gray, and the underside is cream. There is a bold, dark pattern on the back.

SIMILAR SPECIES

There are 18 species in the genus *Trachycephalus*. Called casque-headed frogs, they hide in tree-holes and use their bony heads to seal the entrance. They are distributed across Central and South America and none is of conservation concern. The Amazon Milk Frog (*T. resinifictrix*; page 328) breeds in holes high up in trees and is very striking in appearance. It has alternating bands of brown and cream with a bluish tinge.

Actual size

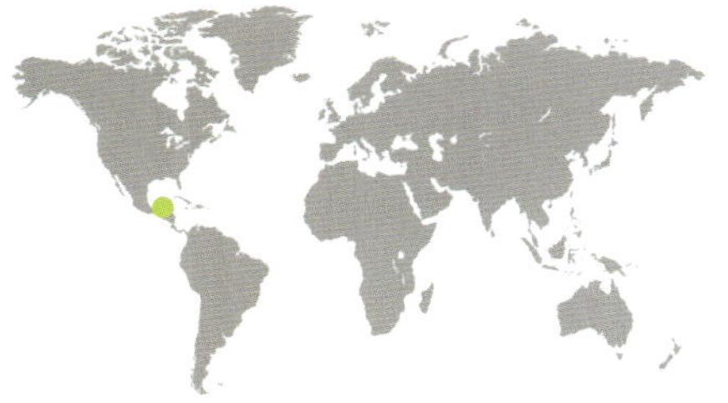

FAMILY	Hylidae: Hylinae
OTHER NAMES	Duck-billed Tree Frog, Yucatan Casque-headed Tree Frog
DISTRIBUTION	Yucatán peninsula, Mexico; northern Belize, northern Guatemala
ADULT HABITAT	Grassland with patches of shrubs and trees
LARVAL HABITAT	Pools
CONSERVATION STATUS	IUCN Least Concern. Declining in some places but its population is generally stable

ADULT LENGTH
Male
$1\frac{7}{8}$–$2\frac{3}{8}$ in (48–61 mm)
Female
$2\frac{5}{8}$–$2\frac{15}{16}$ in (65–74 mm)

TRIPRION PETASATUS

YUCATAN SHOVEL-HEADED TREE FROG

(COPE, 1865)

Actual size

The Latin species name of this curious-looking frog, *petasatus*, means "with a hat on." This refers to the hard, shield-like structure on its head, which continues forward as a hard, flat upper lip that extends beyond the lower jaw. When hiding in a tree-hole, either during the day or throughout the dry season, the frog uses its head to block the entrance. Breeding occurs after rain, most commonly in July. Males call from trees at night, making a sound that is like the quacking of a duck. Amplexed pairs form in trees and then descend to pools, where the eggs are laid in clumps.

The Yucatan Shovel-headed Tree Frog has a flattened, duck-like snout and a saddle-like flap behind the eyes. The disks on the fingers and toes are large, and the hind feet are webbed. The upper surfaces are olive-green in males, and tan or olive-brown in females, both with darker markings on the back and dark stripes on the legs. The belly is white.

SIMILAR SPECIES

Triprion petasatus was the only species in its genus, until the Mexican Shovel-headed Tree Frog, which was in the genus *Daiglena* and has a similarly strange snout, was transferred to *Triprion*, as *Triprion spatulatus* (page 331).

FAMILY	Hylidae: Hylinae
OTHER NAMES	*Triprion spatulatus*, Shovel-nosed Tree Frog
DISTRIBUTION	Western Mexico
ADULT HABITAT	Coastal lowland forest up to 1,640 ft (500 m) altitude
LARVAL HABITAT	Ephemeral streams and ponds
CONSERVATION STATUS	IUCN Least Concern. There is no evidence of general decline and it occurs within at least one protected area

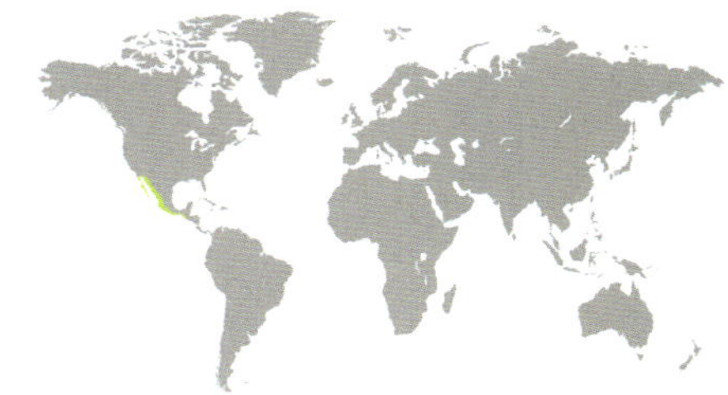

ADULT LENGTH
Male
$2\frac{3}{8}$–$3\frac{7}{16}$ in (61–88 mm)
Female
$2\frac{15}{16}$–$3\frac{15}{16}$ in (75–102 mm)

TRIPRION SPATULATUS

MEXICAN SHOVEL-HEADED TREE FROG

GÜNTHER, 1882

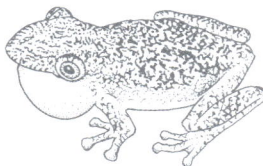

This large tree frog gets its name from its very unusual snout, which is spoon- or spatula-shaped and extends some way beyond the mouth. The function of this curious structure is unknown, but it may be that this frog uses it to close the entrance to a tree-hole when it has retreated inside it during dry weather. It breeds during the rainy season, between June and November. Males call from muddy or rocky banks close to the temporary pools or streams where the eggs are laid, producing a single low-pitched "braaaa" note.

SIMILAR SPECIES

Originally the sole species in the genus *Diaglena* this species was moved to the genus *Triprion*, which already contained another species with a strange flattened snout, the Yucatan Shovel-headed Tree Frog (*Triprion petasatus*; page 330).

The Mexican Shovel-headed Tree Frog has a slender body and head, a flat, shovel-shaped snout, and large, widely spaced eyes that are directed partially forward. It has well-developed disks on its fingers and toes. The upper surfaces are green with brown markings. The frog is very variable in color, and individuals toward the south of its range are generally darker.

Actual size

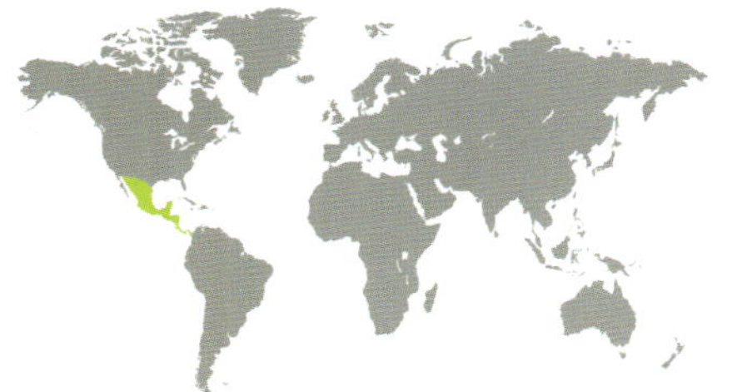

FAMILY	Hylidae: Hylinae
OTHER NAMES	Spiny-headed Tree Frog
DISTRIBUTION	Mexico, Honduras, Costa Rica, Panama
ADULT HABITAT	Rainforest
LARVAL HABITAT	Water-filled cavities in bromeliads and other plants
CONSERVATION STATUS	IUCN Near Threatened. Has become very rare in Mexico and Honduras

ADULT LENGTH
Male
2¼–2¾ in (59–69 mm)
Female
2¼–3⅛ in (58–80 mm)

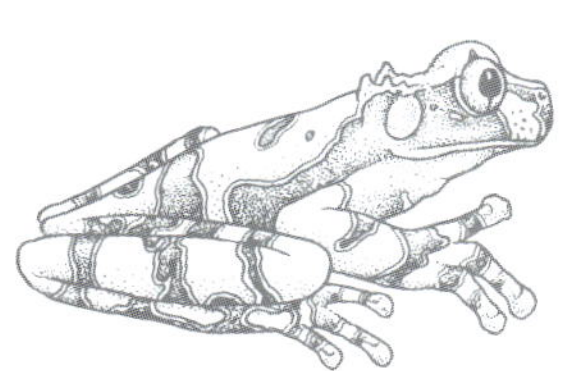

TRIPRION SPINOSUS

CORONATED TREE FROG

(STEINDACHNER, 1864)

Quite often heard but rarely seen, this large frog is arboreal and nocturnal, and is most often found on bromeliads and banana plants. Males produce a loud "boop... boop... boop" call from water-filled cavities in plants. The eggs are laid just above the waterline and the tadpoles develop in the pool. Pairs may mate and lay eggs again on subsequent nights, in the same or in different pools. The female periodically visits pools where she has laid eggs, and the tadpoles nibble at her to elicit the production of the unfertilized eggs on which they feed. A single pool can produce 1–16 froglets.

The Coronated Tree Frog gets its name from an array of sharp, pointed spines on its head. The skin on its head is fused to its skull and it has a very large tympanum. It has long legs and adhesive disks on the ends of its fingers and toes. The back is pale brown or gray and the flanks are dark brown or black, and they are separated by a white border.

SIMILAR SPECIES

Originally in the genus *Anotheca*, this spiny-necked frog was moved to *Triprion*, which contains the Yucatan and Mexican Shovel-snouted Tree Frogs (*T. petasatus* and *T. spatulatus*; pages 330–331). In its reproductive habits, such as breeding in water-filled holes in plants and feeding its young on unfertilized eggs, it is very similar to frogs of other families, such as the Strawberry Poison-dart Frog (*Oophaga pumilio*; page 382) and the Imitating Poison Frog (*Ranitomeya imitator*; page 386).

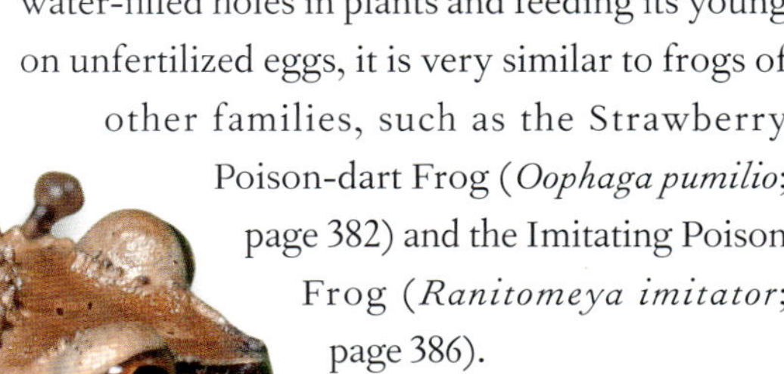

Actual size

FAMILY	Hylidae: Hylinae
OTHER NAMES	None
DISTRIBUTION	Rio de Janeiro state, Brazil
ADULT HABITAT	Coastal woodlands up to 160 ft (50 m) altitude
LARVAL HABITAT	Ephemeral pools
CONSERVATION STATUS	IUCN Vulnerable. Has a restricted range that is subject to habitat destruction resulting from urban development

ADULT LENGTH
average 1$^{9}/_{16}$ in (39 mm)

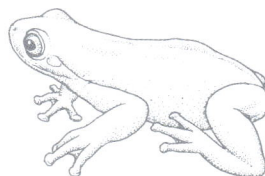

XENOHYLA TRUNCATA

IZECKSOHN'S BRAZILIAN TREE FROG

(IZECKSOHN, 1959)

Uniquely among frogs, whole fruits are a major component of the diet of this species. It lives in the restinga habitat, consisting largely of shrubs and small trees, that is characteristic of low-lying coastal areas of Brazil. It hides in bromeliads and breeds in temporary pools, and it feeds on insects and spiders as well as fruits. Significantly, it defecates viable seeds from the fruit that it has digested and, as it frequents damp areas where seeds will germinate, it is probably a very effective dispersal agent for the trees. When attacked, it thrusts its hind legs out to the side and inflates its lungs with air.

Actual size

Izecksohn's Brazilian Tree Frog has a plump body, a small head, and a pointed snout. There are disks at the ends of its fingers and toes. It is uniformly reddish brown in color and the irises of its eyes are red.

SIMILAR SPECIES

The genus *Xenohyla* contains only one other species, *X. eugenioi*, which occurs in low-lying coastal areas of Bahia state, Brazil, to the north of *X. truncata*. It is not known if it also feeds on fruit. It differs from *X. truncata* in having pale longitudinal stripes on its back. It appears to be rare, but its conservation status has not yet been determined and it is listed as Least Concern.

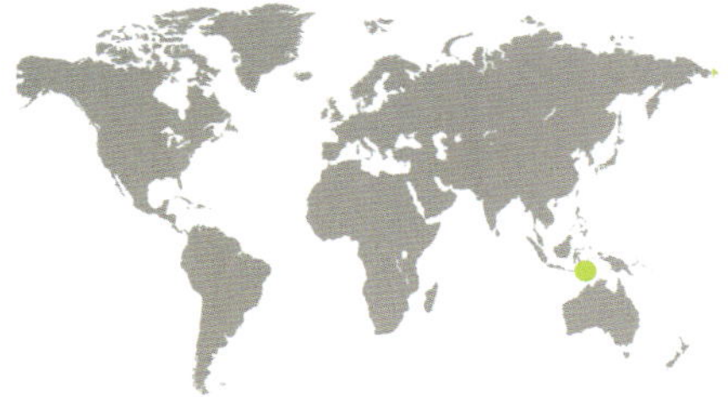

FAMILY	Hylidae: Pelodryadinae
OTHER NAMES	Everett's Treefrog
DISTRIBUTION	West Timor, Alor, Sumba, and Savu, Indonesia; Timor-Leste
ADULT HABITAT	Tropical and subtropical riverine forest, dry woodland, marshes, and perianthropic habitats such as banana groves up to 5,000 ft (1,500 m) altitude
LARVAL HABITAT	Unknown
CONSERVATION STATUS	IUCN Least Concern. No specific threats have been identified

ADULT LENGTH
Male
$1\frac{3}{8}$–$1\frac{11}{16}$ in (35–43 mm)
Female
$1\frac{9}{16}$–$1\frac{15}{16}$ in (40–49 mm)

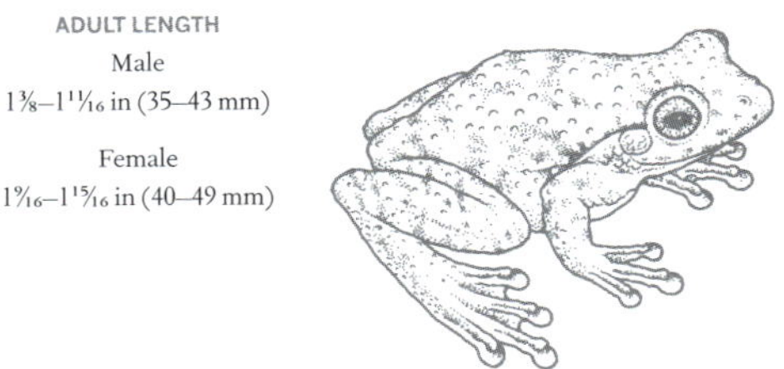

LITORIA EVERETTI

TIMOR TREEFROG

(BOULENGER, 1897)

The Timor Treefrog is a moderately large species confined to the Lesser Sunda Islands of southeastern Indonesia and Timor-Leste, where it demonstrates a rather punctuated distribution. It does not appear to be particularly common, but it may be found in pristine forest streams and rivers, often perched on boulders from where it will leap into the water and hide on the bottom if disturbed. It is also encountered in man-mediated habitats such as banana groves, where it shelters in the water-filled leaf axils. Alfred Hart Everett (1848–1898) was a British civil servant and naturalist who collected extensively in Borneo and the Lesser Sunda Islands, collecting the type series on Sumba and Savu.

SIMILAR SPECIES

Although *Litoria* is a large genus with 117 species, they are mostly distributed through New Guinea and Australia. Only one other species occurs in the Wallacea region, the Ambon Treefrog (*L. amboinensis*), on Ambon and Seram.

The Timor Treefrog is pale unicolor brown or cyan, and its large eyes, with horizontal pupils, are at least twice the size of its tympana. It has powerful limbs, with well-developed adhesive disks on its fingers and toes and minimal webbing between the digits.

Actual size

FAMILY	Hylidae: Pelodryadinae
OTHER NAMES	White-lipped Tree Frog
DISTRIBUTION	Northeastern coast of Queensland, Australia; Papua New Guinea, including the Bismarck Archipelago; eastern Indonesia
ADULT HABITAT	Diverse habitats, including rainforest and gardens
LARVAL HABITAT	Ponds and pools
CONSERVATION STATUS	IUCN Least Concern. Shows no evidence of population decline

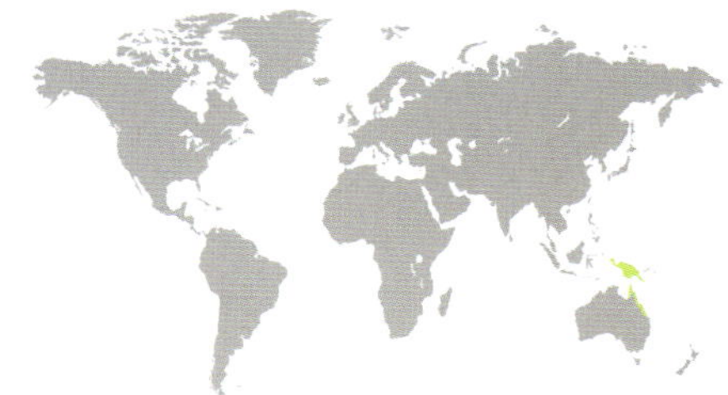

NYCTIMYSTES INFRAFRENATUS

GIANT TREE FROG

(GÜNTHER, 1867)

ADULT LENGTH
Male
$2\frac{1}{2}$–$3\frac{15}{16}$ in (62–102 mm)
Female
$2\frac{7}{8}$–$5\frac{5}{16}$ in (73–135 mm)

Despite being the world's largest tree frog, and often being found around human habitation, rather little is known about this spectacular creature. The male's call is like the bark of a dog, the species breeds in ponds and pools after heavy rain, and the female lays 200–400 eggs. It is also known that the frog first breeds at age 3 or 4 years, and that it lives to be at least 10 years old in the wild. It sometimes turns up, well outside its natural range, in boxes of bananas.

The Giant Tree Frog has long, muscular legs and very large adhesive disks on its fingers and toes. Its upperparts are uniform green or bronze in color, and there is a conspicuous white stripe on its lower lip. There is a white stripe on the lower part of each leg, which becomes salmon-pink in breeding males. The iris of the eyes is yellow, orange, or red.

SIMILAR SPECIES

A white stripe on its lower lip distinguishes *Nyctimystes infrafrenatus* from the other large green tree frog found in both Australia and Papua New Guinea, White's Tree Frog (*Ranoidea caerulea*; page 339). The frog fauna of the islands north of Australia is scarcely known and many new discoveries are currently being made there. The Northern New Guinea Tree Frog (*N. gramineus*), from Indonesia and Papua New Guinea, was described in 1905, but other large green tree frogs, such as the Huon Peninsula Green Tree Frog (*N. dux*) and Sauron's Eye Tree Frog (*N. sauroni*), were described as recently as 2006.

Actual size

FAMILY	Hylidae: Pelodryadinae
OTHER NAMES	Golden Bell Frog, Green Frog, Green and Golden Swamp Frog
DISTRIBUTION	Victoria and New South Wales, Australia. Introduced to New Zealand, New Caledonia, and Vanuatu
ADULT AND LARVAL HABITAT	Large permanent ponds
CONSERVATION STATUS	IUCN Near Threatened. Has declined severely in Australia but is flourishing in New Zealand and on the other islands where it has been introduced

ADULT LENGTH
Male
$2\frac{3}{16}$–$2\frac{3}{4}$ in (57–69 mm)

Female
$2\frac{5}{8}$–$4\frac{1}{4}$ in (65–108 mm)

RANOIDEA AUREA

GREEN AND GOLDEN BELL FROG

(LESSON, 1830)

The Green and Golden Bell Frog has smooth skin, which on its back is green with irregular gold or bronze patches. There is a creamy-white fold of skin running along each side of the body from eye to groin, bordered below by a black stripe. The flanks are brown with cream spots, and there is bright blue coloration on the back of the thighs and in the groin.

Once abundant in southeastern Australia, this colorful frog has declined rapidly in the last 30 years. This is thought to be due to fragmentation of its habitat and the introduction of Mosquitofish (*Gambusia affinis*) to control mosquito larvae. Unfortunately, the fish also eat tadpoles. Although a member of the tree frog family, the species is mostly found on or close to the ground. It breeds from August to March, males producing a call likened to the sound of a distant motorbike. The female lays 3,000–10,000 eggs in a gelatinous mass that floats for a few hours before sinking to the bottom of the pond.

SIMILAR SPECIES

There are six *Ranoidea* species that are closely related to *R. aurea*, one in northern Australia, two in the southwest, and three in the southeast. Dahl's Aquatic Frog (*R. dahlii*) is found on the large floodplains of northern Australia. The Spotted-thighed Frog (*R. cyclorhynchus*) and Moore's Frog (*R. moorei*) both occur in western Australia and occasionally hybridize.

Actual size

FAMILY	Hylidae: Pelodryadinae
OTHER NAMES	Northern Snapping Frog, Round Frog
DISTRIBUTION	Northern Australia
ADULT HABITAT	Grassland and open woodland
LARVAL HABITAT	Temporary ponds
CONSERVATION STATUS	IUCN Least Concern. Has a very large range and shows no evidence of a general decline

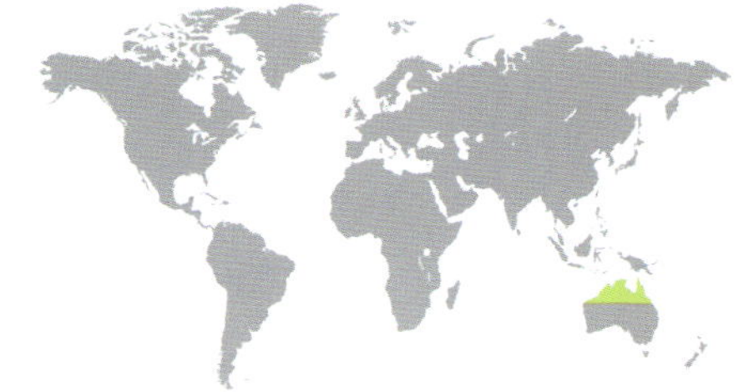

ADULT LENGTH
Male
2¹³⁄₁₆–3¹⁄₁₆ in (71–79 mm)
Female
2¹³⁄₁₆–4¹⁄₈ in (71–105 mm)

RANOIDEA AUSTRALIS

GIANT FROG

(GRAY, 1842)

A common species, the Giant Frog can frequently be seen in the wet season basking by day beside temporary pools. It breeds between December and February, males producing a loud, repeated "unk." Females lay up to 7,000 eggs in a clump that initially floats, then sinks to the bottom. The tadpoles often gather together in shoals and can tolerate water temperatures as high as 109°F (43°C). During the dry season, the frog burrows into the ground, its bladder filled with water, and surrounds itself with a waterproof cocoon. It can survive in this way for up to six months.

SIMILAR SPECIES

There are 72 species in the genus *Ranoidea*. Those previously in the genus *Cyclorana* are all burrowers. The Long-footed Frog (*R. longipes*) occupies the same habitat as *R. australis* and their ranges overlap. It is a smaller frog with a call likened to the lowing of cattle. The Daly Waters Frog (*R. maculosa*), also found in northern Australia, is noted for its coloration, which is yellow with irregular brown patches. See also the New Holland Frog (*R. novaehollandiae*; page 343).

The Giant Frog has a rotund body and a large, broad head. Its color is very variable, being gray, brown, or green, or occasionally pink. There is a dark stripe running from the snout, through the eye, and over the tympanum. There are two prominent folds of skin running along each side of the body.

Actual size

FAMILY	Hylidae: Pelodryadinae
OTHER NAMES	None
DISTRIBUTION	Great Dividing Range, eastern New South Wales, Australia
ADULT AND LARVAL HABITAT	Fast-flowing rocky streams
CONSERVATION STATUS	IUCN Endangered. Has disappeared from most of its range since the 1980s

ADULT LENGTH
Male
1 7/16–1 5/8 in (36–42 mm)

Female
1 7/8–2 1/8 in (48–54 mm)

RANOIDEA BOOROOLONGENSIS

BOOROOLONG FROG

(MOORE, 1961)

Actual size

The Booroolong Frog has a slightly warty skin and adhesive disks on its fingers and toes. It is variable in color, being gray, olive, or reddish brown with black reticulations on the back. The underside is cream. The throat is white in females and dark in males.

A victim of the chytridiomycosis epidemic that caused declines and extinctions among frogs living in the uplands of eastern Australia in the 1980s, the Booroolong Frog is now confined to an area of less than 4 sq miles (10 sq km). It has also been adversely affected by introduced fish and habitat degradation of the streams in which it breeds. Most active at night, it breeds in August, males producing a soft, purring call from rocks in or near streams. The females lay a clump of 700–1,800 eggs, attached to a rock.

SIMILAR SPECIES

Related to Lesueur's Frog (*Ranoidea lesueuri*), a pale brown frog with a narrow but prominent stripe running from the nostril to the eye, after which it broadens out and extends on the body. Like *R. booroolongensis*, it breeds in rocky streams. It occurs from New South Wales into Victoria and is listed as Least Concern.

FAMILY	Hylidae: Pelodryadinae
OTHER NAMES	Dunny Frog, Green Tree Frog
DISTRIBUTION	Northern and eastern Australia, Papua New Guinea
ADULT HABITAT	Woodland near streams and swamps
LARVAL HABITAT	Temporary waterbodies
CONSERVATION STATUS	IUCN Least Concern. Common over a very large range

RANOIDEA CAERULEA

WHITE'S TREE FROG

(WHITE, 1790)

ADULT LENGTH
Male
$2^{11}/_{16}$–3 in (67–77 mm)
Female
$2^{3}/_{8}$–$4^{5}/_{16}$ in (60–110 mm)

Sometimes known as the Dunny Frog, this large tree frog is often found in toilet bowls, water cisterns, and mailboxes. Relaxed in the presence of humans, it is very popular as a pet. It breeds from November to February. Stimulated by heavy rain, males call initially from trees, then from the ground closer to water, making a loud "craack… craack" sound. Females lay between 200 and 2,000 eggs that spread out over the water's surface. White's Tree Frog is extremely tolerant of high temperatures and more resistant to desiccation than most frogs. Its skin contains compounds that have antibacterial properties.

White's Tree Frog has a smooth skin with many pores. Its fingers and toes end in large adhesive disks and the toes are partially webbed. There are large glands on the back of the head and a fleshy fold on the side of the head that partially overhangs the tympanum. The upper surfaces are bright green, and some individuals have a few white spots on the back. The underside is white.

SIMILAR SPECIES

The Centralian Tree Frog (*Ranoidea gilleni*) is a smaller, bright green frog with large glands on its head, and is found in a restricted area around Alice Springs. The Cave-dwelling Frog (*R. cavernicola*) lacks the fold of skin on the side of the head and is found in limestone caves in northern Western Australia. See also the Giant Tree Frog (*Nyctimystes infrafrenatus*; page 335) and the Splendid Tree Frog (*R. splendida*; page 345).

Actual size

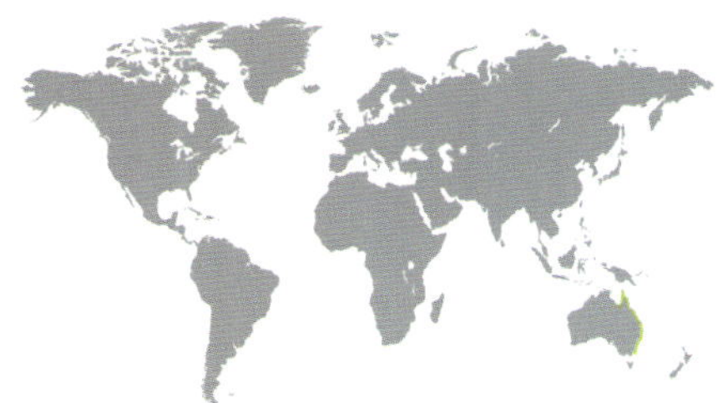

FAMILY	Hylidae: Pelodryadinae
OTHER NAMES	Australian Red-eyed Tree Frog
DISTRIBUTION	Coastal Queensland and New South Wales, Australia
ADULT HABITAT	Coastal rainforest
LARVAL HABITAT	Ephemeral pools and mountain streams
CONSERVATION STATUS	IUCN Vulnerable. Shows no evidence of population decline

ADULT LENGTH

Male
2⅛–2½ in (54–62 mm)

Female
2¼–2¾ in (58–68 mm)

RANOIDEA CHLORIS

RED-EYED GREEN TREE FROG

(BOULENGER, 1892)

Actual size

This highly photogenic frog is rarely seen outside the breeding season because it is very arboreal and lives high up in trees. In spring and summer, heavy rain stimulates it to descend and aggregate around flooded areas and streams. It is an "explosive breeder," mating and egg-laying being completed within a few days. One study found that smaller males mated with more females than larger ones, probably because they spent more nights at the breeding site. The species is of interest to the pharmaceutical industry because its skin secretes peptides with antibiotic and anticancer properties.

SIMILAR SPECIES

The Orange-thighed Frog (*Ranoidea xanthomera*) is a very similar species that lives along the norther Queensland coast. Its orange legs distinguish it from *R. chloris*. The Dainty Green Tree Frog (*R. gracilenta*) is smaller than *R. chloris* and occurs along the northeastern coast of Australia and in Papua New Guinea. It is quite often found in fruit plantations and, as a result, has been translocated from place to place in fruit packages.

The Red-eyed Green Tree Frog has a uniformly green back and a bright yellow underside. There is a purple patch on the back of each thigh and the hands and feet are largely yellow. It gets its name from the bright red irises of its eyes. There are large disks on the ends of the fingers and toes.

FAMILY	Hylidae: Pelodryadinae
OTHER NAMES	*Nyctimystes dayi*, Day's Big-eyed Tree Frog, Lace-eyed Tree Frog
DISTRIBUTION	Northeastern Queensland, Australia
ADULT HABITAT	Rainforest up to 4,000 ft (1,200 m) altitude
LARVAL HABITAT	Fast-flowing rocky streams
CONSERVATION STATUS	IUCN Vulnerable. Has declined over much of its restricted range, especially at higher altitudes

RANOIDEA DAYI

AUSTRALIAN LACE-LID

(GÜNTHER, 1897)

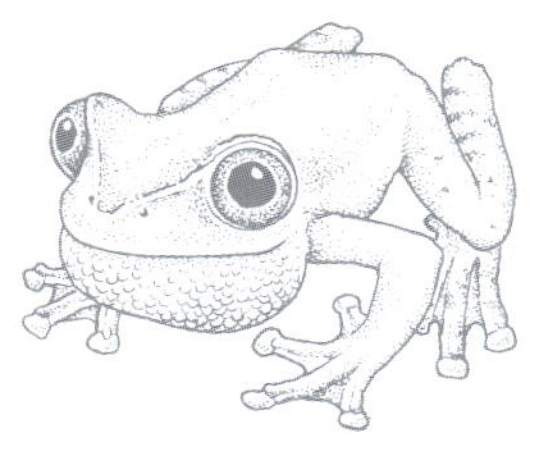

ADULT LENGTH
Male
1³⁄₁₆–1⅝ in (30–42 mm)
Female
1¾–2⅛ in (45–55 mm)

This beautiful frog gets its name from its lacy white lower eyelids, which it draws across its large, dark eyes when it is resting. It breeds between October and April, males calling at night from rocks and foliage beside streams. Males space themselves out, at least 3 ft (1 m) apart, and produce a short, sharp "ee" every 5 or 6 seconds. The eggs, laid in the water, hatch into tadpoles that have sucker-like mouths and muscular tails, adaptations for living in fast-flowing water. Because this frog breeds in streams and has vanished from higher altitudes, it is very likely that its decline is due to chytridiomycosis.

SIMILAR SPECIES

Until 2016 *Ranoidea dayi* was included in the genus *Nyctimystes*, which still contains the White-lipped Tree Frog (*N. infrafrenatus*; page 335), the Sandy Big-eyed Tree Frog (*N. kubori*), which lives near rainforest streams at altitudes of 3,300–6,600 ft (1,000–2,000 m) in Indonesia and Papua New Guinea, and 45 other species from New Guinea.

The Australian Lace-lid has a slender body and legs, very large eyes, and a flattened snout. There is webbing on the hands and feet, and large disks on the fingers and toes. The upper surfaces are variable in color but are often reddish brown, sometimes with white spots. The flanks are yellow and the underside is creamy white.

Actual size

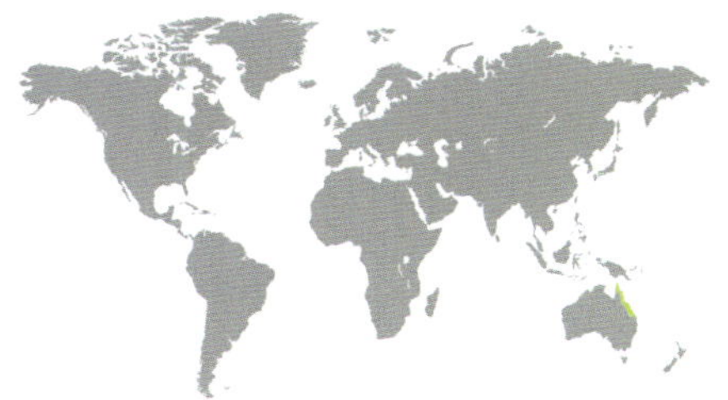

FAMILY	Hylidae: Pelodryadinae
OTHER NAMES	Waterfall Frog
DISTRIBUTION	Northeastern Queensland, Australia
ADULT AND LARVAL HABITAT	Mountain streams and waterfalls in coastal rainforest
CONSERVATION STATUS	IUCN Least Concern, although suffered severe declines at higher altitudes in the 1980s

ADULT LENGTH
Male
$1^{9}/_{16}$–$1^{7}/_{8}$ in (40–48 mm)
Female
$1^{15}/_{16}$–$2^{1}/_{8}$ in (49–55 mm)

RANOIDEA NANNOTIS

TORRENT TREE FROG

(ANDERSSON, 1916)

While many forest-living frogs enter streams only to breed, the Torrent Tree Frog lives in fast-flowing streams and waterfalls for its entire life. It appears to breed at any time of year. Males have spiny nuptial pads on their hands and patches of spines on their chests; these enable them to grip females firmly. Females lay 130–220 eggs in a gelatinous mass underneath rocks. The tadpoles show features typical of stream-living species, such as a streamlined body, a very muscular tail, and a large mouth that acts as a sucker.

SIMILAR SPECIES

Ranoidea nannotis is the largest of four *Ranoidea* species that are known as torrent frogs. All are found in the coastal forests of Queensland and all are of conservation concern. The Creek Frog (*R. rheocola*) is listed as Near Threatened, the Armoured Frog (*R. lorica*) as Critically Endangered, and the Nyakala Frog (*R. nyakalensis*) is believed Extinct. Torrent frogs have declined most dramatically at higher altitudes, where colder conditions allow the fungus that causes chytridiomycosis to thrive.

The Torrent Tree Frog has a broad head with large eyes and a rounded snout. The back is covered with warts and has a marbled pattern of gray and green with black markings. The underside is pale. There are well-developed adhesive disks on the fingers and toes, and the toes are fully webbed.

Actual size

FAMILY	Hylidae: Pelodryadinae
OTHER NAMES	Eastern Snapping Frog, Wide-mouthed Frog
DISTRIBUTION	Queensland and northern New South Wales, Australia
ADULT HABITAT	Varied semiarid habitats but not at high altitudes or in forest
LARVAL HABITAT	Still or slow-moving water
CONSERVATION STATUS	IUCN Least Concern. Has a large range and shows no evidence of a general decline

RANOIDEA NOVAEHOLLANDIAE

NEW HOLLAND FROG

(STEINDACHNER, 1867)

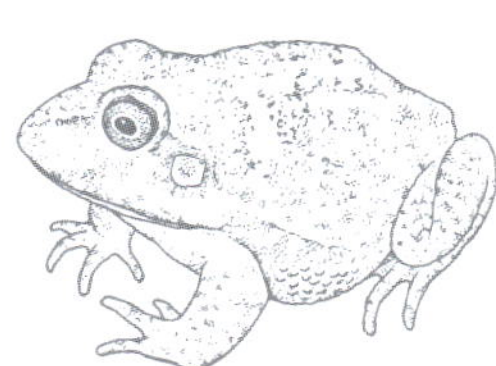

ADULT LENGTH
Male
$2\frac{3}{8}$–$3\frac{1}{8}$ in (61–81 mm)
Female
$2\frac{13}{16}$–$3\frac{15}{16}$ in (71–101 mm)

Described as a "mean machine," this burrowing frog has a voracious appetite and a very large mouth. Heavy rain stimulates its emergence from the burrow in which it spends the long dry season, males calling from dams and ditches, and producing a short, explosive "unk." Females lay around 1,000 eggs in clumps in shallow water; these float for a while and then sink. The New Holland Frog takes a wide variety of prey, from small insects and worms to other frogs. Depending on the size of its prey and the way it moves, the frog lunges at it open-mouthed or flicks out its tongue to catch it.

SIMILAR SPECIES

The Short-footed Frog (*Ranoidea brevipes*) is a smaller burrowing frog, found in Queensland and northern New South Wales. The Eastern Water-holding Frog (*R. platycephala*) inhabits semidesert areas and scrubland in New South Wales; it differs from other *Ranoidea* species in having small eyes. See also the Giant Frog (*R. australis*; page 337).

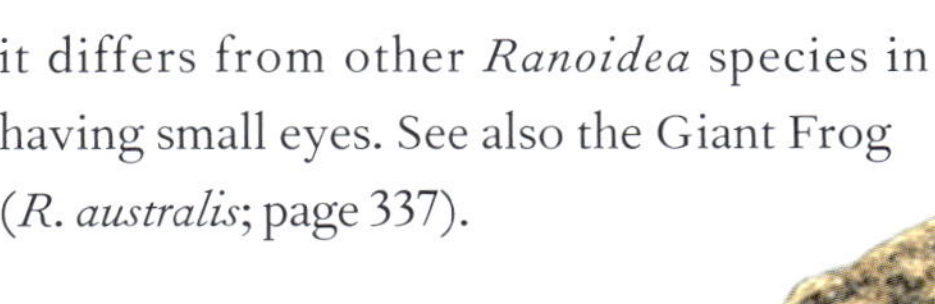

The New Holland Frog is a large, muscular frog with a massive head and a plump body. The large eyes are positioned high on the head. There are two folds of skin running along the flanks. It is usually gray or brown in color, often with a pale stripe down the middle of the back. The belly is white.

Actual size

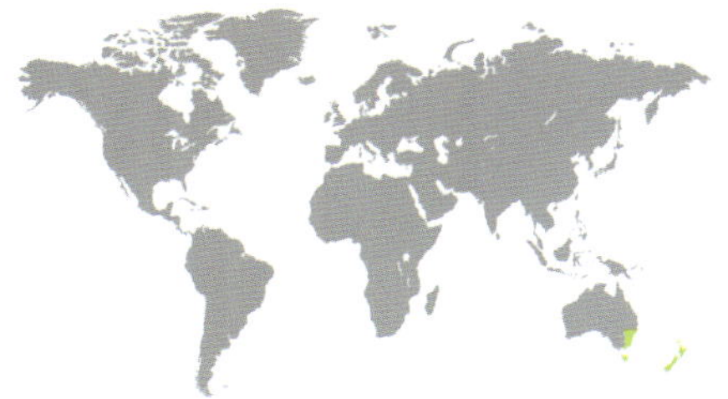

FAMILY	Hylidae: Pelodryadinae
OTHER NAMES	Growling Grass Frog, Warty Bell Frog
DISTRIBUTION	Southeastern Australia and Tasmania. Introduced to New Zealand
ADULT HABITAT	Woodland and pasture near large permanent ponds
LARVAL HABITAT	Large permanent ponds
CONSERVATION STATUS	IUCN Vulnerable. Has declined severely in Australia but is flourishing in New Zealand

ADULT LENGTH
Male
2⅛–2⅝ in (55–65 mm)

Female
2⅜–4 in (60–104 mm)

RANOIDEA RANIFORMIS

SOUTHERN BELL FROG

(KEFERSTEIN, 1867)

The Southern Bell Frog has large eyes, a warty back, and long, muscular legs with webbed feet. The upper surfaces are green with irregular gold or bronze patches, and there is often a pale green stripe down the middle of the back. The flanks are brown with cream spots, and there is bright blue coloration on the back of the thighs and in the groin.

This large, colorful frog gets one of its common names from its mating call, a long growling sound produced by males floating in open water. Breeding occurs between August and April, with the female producing a loose clump of around 1,700 eggs. The frog is a voracious predator, sometimes eating other frogs, including small members of its own species. The causes of its dramatic decline in Australia are thought to be habitat loss, introduced Mosquitofish (*Gambusia affinis*), and the disease chytridiomycosis. It was introduced to New Zealand in 1860 and has become a common species there.

SIMILAR SPECIES

Ranoidea raniformis is very similar in appearance and habits to the Green and Golden Bell Frog (*R. aurea*; page 336). Both have declined in Australia but have successfully invaded New Zealand. *Ranoidea raniformis* is more warty and has less gold coloration than *R. aurea*, and the two species can also be distinguished by their calls.

Actual size

FAMILY	Hylidae: Pelodryadinae
OTHER NAMES	Magnificent Tree Frog
DISTRIBUTION	Kimberley region, northwestern Australia
ADULT HABITAT	Moist forest; also found in caves, gorges, and buildings
LARVAL HABITAT	Temporary waterbodies
CONSERVATION STATUS	IUCN Least Concern. Faces no apparent threats and shows no evidence of population decline

RANOIDEA SPLENDIDA

SPLENDID TREE FROG

(TYLER, DAVIES & MARTIN, 1977)

ADULT LENGTH
Male
$3^{7}/_{16}$–4 in (88–104 mm)
Female
$3^{5}/_{8}$–$4^{1}/_{8}$ in (94–106 mm)

This large frog appears to be similar in behavior and ecology to White's Tree Frog (*Ranoidea caerulea*; page 339). It is distinguished by a huge parotoid gland on its head; this secretes a poison that deters predators such as snakes and birds. It breeds at the onset of the rainy season, in December and January, the male producing a deep, barking call. The female lays 6,500 eggs that form a layer on the water's surface. Male frogs typically attract females by calling, sometimes by means of visual displays. Males of this species are highly unusual in secreting a waterborne sexual attractant, called splendiferin, from its skin.

The Splendid Tree Frog has a huge gland on its head that, in large animals, hangs over the large tympanum. The back is green, with scattered yellow or white spots. The backs of the thighs are yellow or orange, and the belly is white. The fingers and toes have large adhesive disks, and the toes are fully webbed.

SIMILAR SPECIES

In recent years much research has centered on the various compounds that *Ranoidea caerulea* and its close relatives produce in their skin, which include peptides with antimicrobial properties and others that are effective anticancer agents. Those species with the most active peptides, such as the Red-eyed Green Tree Frog (*R. chloris*; page 340) and *R. caerulea*, seem to be more resistant to chytridiomycosis than those with less active peptides, such as the Torrent Tree Frog (*R. nannotis*; page 342).

Actual size

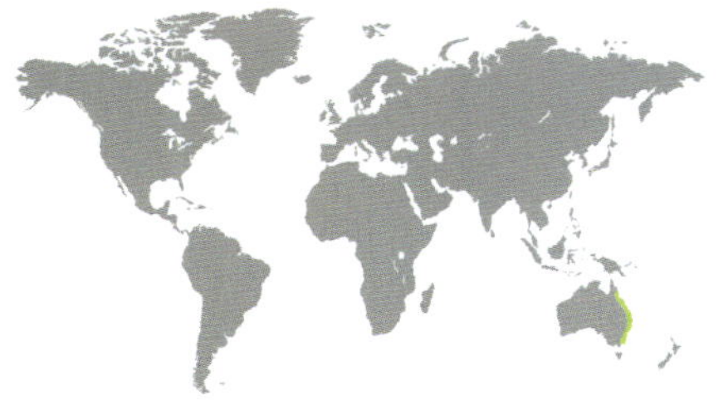

FAMILY	Hylidae: Pelodryadinae
OTHER NAMES	Wilcox's Frog
DISTRIBUTION	Coastal hills and mountains of eastern Australia, from northern Queensland to New South Wales
ADULT AND LARVAL HABITAT	Fast-flowing rocky streams
CONSERVATION STATUS	IUCN Least Concern. Has a large range and there is no evidence of a general decline

ADULT LENGTH
Male
1⅜–1⅞ in (35–48 mm)

Female
1$\frac{9}{16}$–2¾ in (39–69 mm)

RANOIDEA WILCOXII

STONY CREEK FROG

(GÜNTHER, 1864)

Males of this small terrestrial frog are noted for their ability to change color very rapidly. Within five minutes of going into amplexus with a female, a male's back changes from dull brown to bright lemon-yellow. The biological function of this change is not known. The frog breeds in streams, females laying eggs in clumps attached to stones. It is a common species, despite the fact that it is extensively infected by the fungus that causes chytridiomycosis. It appears that the Stony Creek Frog is resistant to the disease, and may thus be a reservoir host from which the disease can be passed to susceptible stream-breeding species.

SIMILAR SPECIES

The Stony Creek Frog has only recently been separated, on the basis of genetic differences, from Lesueur's Frog (*Ranoidea lesueurii*), which also lives in the eastern highlands of Australia, as far south as the state of Victoria. It is also similar to the Booroolong Frog (*R. booroolongensis*; page 338), an Endangered species that is highly susceptible to chytridiomycosis.

The Stony Creek Frog has a slender body, a pointed snout, and long, muscular hind legs. There are disks on the tips of the fingers and toes, and the feet are webbed. The upper surfaces are pale brown or fawn, becoming bright yellow in males during mating. The underside is white and there is usually a dark stripe running through the eye.

Actual size

FAMILY	Hylidae: Phyllomedusinae
OTHER NAMES	Gaudy Leaf Frog, Red-eyed Leaf Frog
DISTRIBUTION	Central America, from southern Mexico to Panama
ADULT HABITAT	Humid forest up to 4,100 ft (1,250 m) altitude
LARVAL HABITAT	Pools and ponds
CONSERVATION STATUS	IUCN Least Concern. Declining in places due to habitat destruction and collection for the pet trade

AGALYCHNIS CALLIDRYAS

RED-EYED TREE FROG

(COPE, 1862)

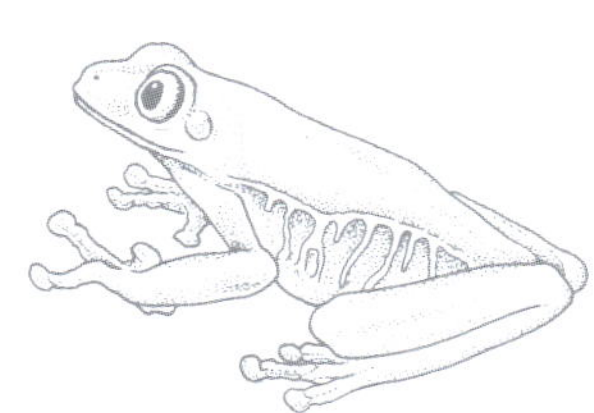

ADULT LENGTH
Male
1$\frac{3}{16}$–2$\frac{1}{4}$ in (30–59 mm)
Female
1$\frac{7}{8}$–3 in (48–77 mm)

This large frog rests in the open during the day, adopting a posture that conceals the brightly colored parts of its body. At night, several frogs gather in a tree overhanging a pond and males call, initially from elevated positions and then from the pond edge. The male climbs onto the female's back and she enters the pond to take up water through her skin and store it in her bladder, and she then climbs back up into the tree. She lays a batch of around 50 eggs on a leaf and then, still carrying the male, returns to water before laying another batch. The eggs hatch after about five days, the tadpoles falling into the pond.

SIMILAR SPECIES

Among the 14 species in the genus *Agalychnis*, the Lemur Tree Frog (*A. lemur*) is one of the rarest, being listed as Critically Endangered. It is distributed from Costa Rica and Panama, into Colombia, and, like other species, is impacted by the destruction of its forest habitat, collection for the global pet trade, and infection by the disease chytridiomycosis. See also the Gliding Leaf Frog (*A. spurrelli*; page 349).

The Red-eyed Tree Frog has a slender body, long, delicate limbs, well-developed toe pads on the fingers and toes, and large, prominent eyes. The upper surfaces are green, with blue and white stripes on the flanks, red and yellow on the inside of the legs, and yellow, orange, or red fingers and toes. The irises of the eyes are bright red.

Actual size

FAMILY	Hylidae: Phyllomedusinae
OTHER NAMES	*Agalychnis dacnicolor*, Mexican Giant Leaf Frog
DISTRIBUTION	Along the Pacific coast of Mexico
ADULT HABITAT	Dry lowlands and deciduous woodland
LARVAL HABITAT	Ponds and swamps
CONSERVATION STATUS	IUCN Least Concern. Its population is stable and it occurs within some protected areas

ADULT LENGTH
Male
2⅜–3⅛ in (60–80 mm)
Female
2¹³⁄₁₆–3⅞ in (70–100 mm)

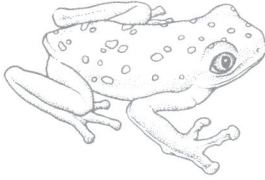

AGALYCHNIS DACNICOLOR

MEXICAN LEAF FROG

(COPE, 1864)

This large frog lives in areas where there is a prolonged dry season, when it hides in rodent burrows and other damp places. Breeding usually begins in May, males gathering around pools and establishing territories from which they call. They occupy the same territories night after night. Mating and egg-laying take 5–6 hours because the female lays her eggs in batches, on leaves hanging over the water, and returns to the pool between each batch to take on water by absorption through the skin and into the bladder, from where the water is transferred to the jelly surrounding the eggs. During this process, the male is sometimes displaced by a rival.

The Mexican Leaf Frog has a slender body, long, thin arms and legs, and long fingers and toes. The upper surfaces are green, with scattered small white spots. There is a white band along the flanks and the underside, and the fingers and toes are orange. The irises of the eyes are white with delicate black markings.

SIMILAR SPECIES

Before being transferred to *Agalychnis*, this species was included in the genus *Pachymedusa*, which is very similar to tree frogs in the genera *Callimedusa* (page 350), *Pithecopus* (page 355), and *Phyllomedusa*. *Agalychnis* species lay their eggs on leaves over a pool so that, when they hatch, the tadpoles fall into the water below.

Actual size

FAMILY	Hylidae: Phyllomedusinae
OTHER NAMES	Gliding Tree Frog, Spurrell's Leaf Frog
DISTRIBUTION	Central and South America, from Costa Rica to Colombia and Ecuador
ADULT HABITAT	Humid forest up to 2,460 ft (750 m) altitude
LARVAL HABITAT	Ephemeral pools and ponds
CONSERVATION STATUS	IUCN Least Concern. Declining in places due to habitat destruction but occurs within some protected areas

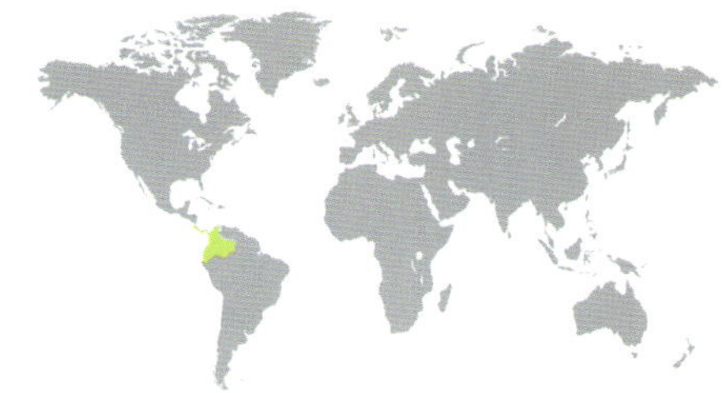

AGALYCHNIS SPURRELLI

GLIDING LEAF FROG

BOULENGER, 1913

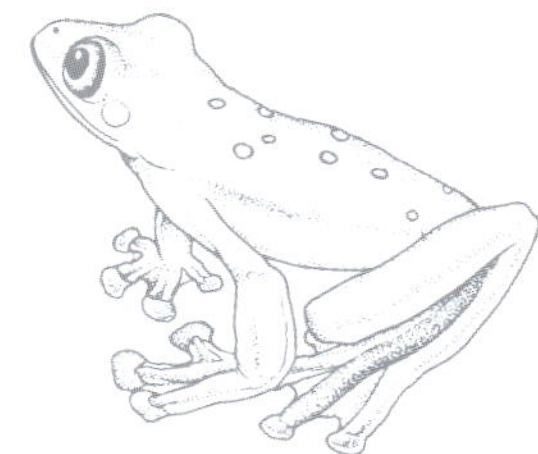

ADULT LENGTH
Male
1⅞–3 in (48–76 mm)

Female
2⅜–3⅝ in (60–93 mm)

This large frog varies considerably in size across its range, females reaching 2⅞ in (72 mm) in Costa Rica, 3⅜ in (87 mm) in Panama, and 3⅝ in (93 mm) in Colombia. It moves through foliage and climbs up vines by means of a hand-over-hand movement. It glides from one tree to another, at an angle of about 45 degrees, by spreading out its webbed feet and hands to slow its descent. It breeds in the rainy season, between May and October. The male's call is a low-pitched moan. The female lays her eggs on the upper side of a leaf hanging over a pool. Yellowy green in color by day, the frog turns dark green at night.

SIMILAR SPECIES

The Parachuting Red-eyed Tree Frog or Misfit Leaf Frog (*Agalychnis saltator*) moves from tree to tree in a similar manner to *A. spurrelli*, but has less well-developed webbing between its fingers and toes. The use of large webbed hands and feet to glide also occurs in the Southeast Asian Dring's Flying Frog (*Feihyla kajau*; page 605) and the Green Flying Frog (*Rhacophorus reinwardtii*; page 619). See also the Red-eyed Tree Frog (*A. callidryas*; page 347).

The Gliding Leaf Frog has very well-developed pads on, and webbing between, its fingers and toes. The upper arms are thin and the forearms are robust. The back is green, often with black-edged white spots, the flanks are orange, the underside is white, and the hands and feet are yellow. The irises of the very large eyes are dark red.

Actual size

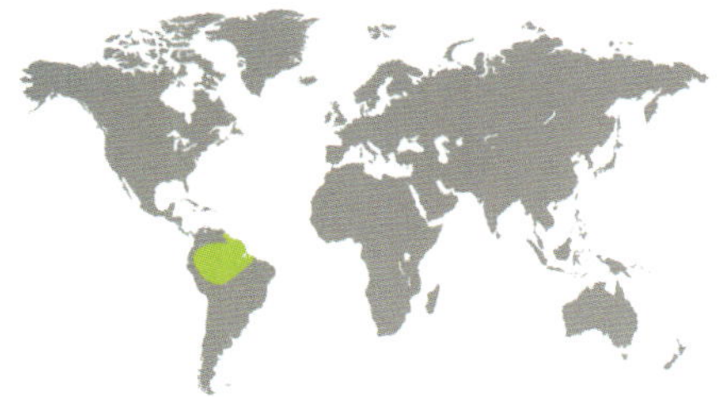

FAMILY	Hylidae: Phyllomedusinae
OTHER NAMES	Barred Leaf Frog, Monkey Frog
DISTRIBUTION	Northern South America, east of the Andes, including northern Brazil, Bolivia, Peru, Ecuador, Colombia, Venezuela, Guyana, Suriname, and French Guiana
ADULT HABITAT	Pristine rainforest up to 1,640 ft (500 m) altitude
LARVAL HABITAT	Ephemeral pools
CONSERVATION STATUS	IUCN Least Concern. Has a very large range and shows no evidence of a general decline

ADULT LENGTH
Male
1¾–2⅛ in (44–54 mm)

Female
average 2⅜ in (60 mm)

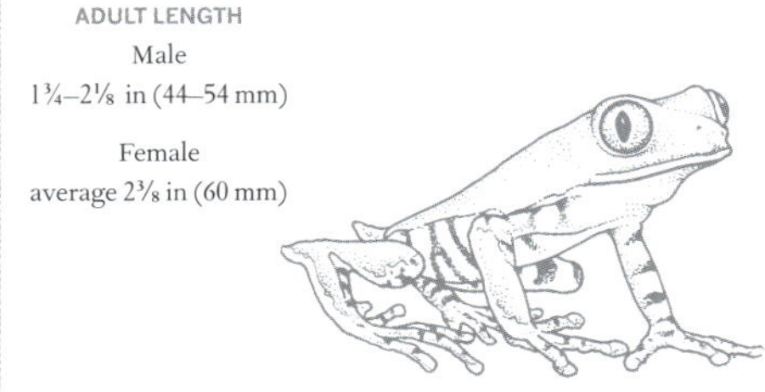

CALLIMEDUSA TOMOPTERNA

TIGER-STRIPED LEAF FROG

(COPE, 1868)

Most common in Suriname, this arboreal frog is active only at night. It breeds in the wet season, between December and May, gathering in trees and bushes around pools. Pairs lay around 70 eggs in a gelatinous mass, wrapped up in folded leaves above the pool. When the eggs hatch, the tadpoles fall into the water below. More than 50 percent of the eggs are lost to predators before they have hatched, the most common predators being beetles and flies. The frog is found only in pristine rainforest and is therefore very vulnerable to deforestation.

SIMILAR SPECIES

The Agua Rica Leaf Frog (*Phyllomedusa ecuatoriana*) is a slightly larger frog, found in high elevation Ecuadorian cloud forests. It does not have the orange and black ventral coloration of *P. tomopterna*, but is orange with red reticulations. It is an uncommon species and listed as Vulnerable. The Warty Leaf Frog (*P. atelopoides*), found in Brazil and Peru, differs from other species in this genus by being brown, and by living mostly on the ground. *Callimedusa* contains six species.

The Tiger-striped Leaf Frog has a slender body, large, prominent eyes, and very long, slender limbs. There are well-developed adhesive disks on the fingers and toes. The upper surfaces of the body and legs are bright, pale green. The flanks, inner surfaces of the legs, hands, and feet are orange with vertical black stripes. The underside is white or orange.

Actual size

FAMILY	Hylidae: Phyllomedusinae
OTHER NAMES	Splendid Tree Frog
DISTRIBUTION	Central and South America, from Nicaragua to Ecuador
ADULT HABITAT	Lowland wet and moist forest up to 560 ft (170 m) altitude
LARVAL HABITAT	Water-filled hollows at the base of trees
CONSERVATION STATUS	IUCN Least Concern. Not a common species, but shows no evidence of a general decline

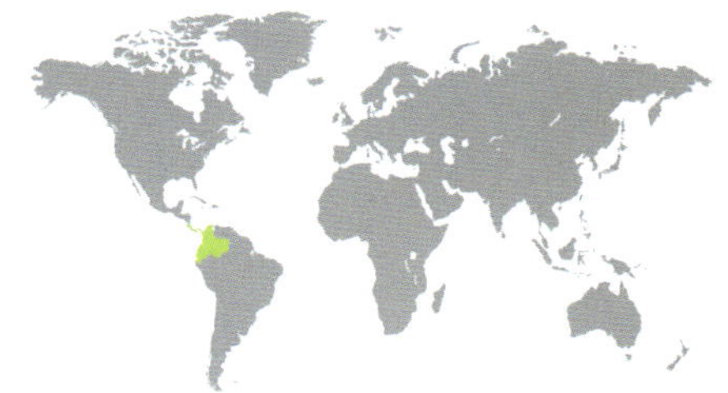

CRUZIOHYLA CALCARIFER

SPLENDID LEAF FROG

(BOULENGER, 1902)

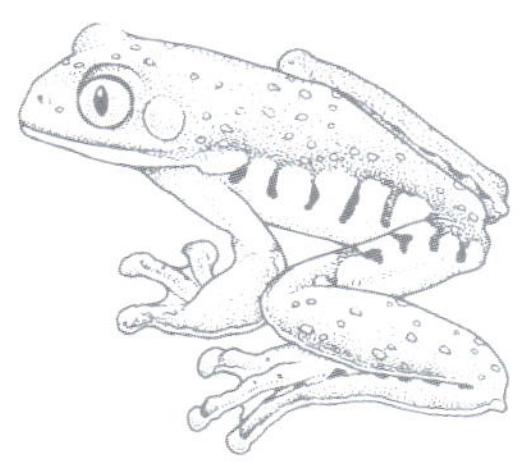

ADULT LENGTH
Male
2–3⅛ in (51–81 mm)
Female
2⅜–3⅜ in (61–87 mm)

Active at night, and living at the top of tall trees, this colorful frog is rarely seen. During the rainy season, from March to October, it comes closer to the ground and males call to attract females. Once in amplexus, the female enters a pool to fill her bladder and then climbs up to lay a clutch of 10–54 eggs on a tree trunk, branch, or foliage above shallow, water-filled cavities among the roots of trees. As they develop, the tadpoles are washed by rain into the pool below. Adults are able to move between trees by gliding, using their spread hands and feet to slow their descent.

The Splendid Leaf Frog has a broad head, large eyes, and a large and distinct tympanum. There are well-developed adhesive disks on the fingers and toes, which are fully webbed. The back is dark green, the flanks and thighs are orange with black or purple stripes, and the underside is yellow or orange. The hands and feet are bright orange and there are small whitish spurs on the heels.

SIMILAR SPECIES

There are two other species in this genus, the Fringe Leaf Frog (*Cruziohyla craspedopus*), previously included within *Agalychnis*, is very similar in appearance to *C. calcarifer* and inhabits Colombia, Ecuador, Peru, and Brazil, and may possibly also occur in Bolivia. Sylvia's Leaf Frog (*C. sylviae*) was described in 2018 from Nicaragua, Costa Rica, and Panama.

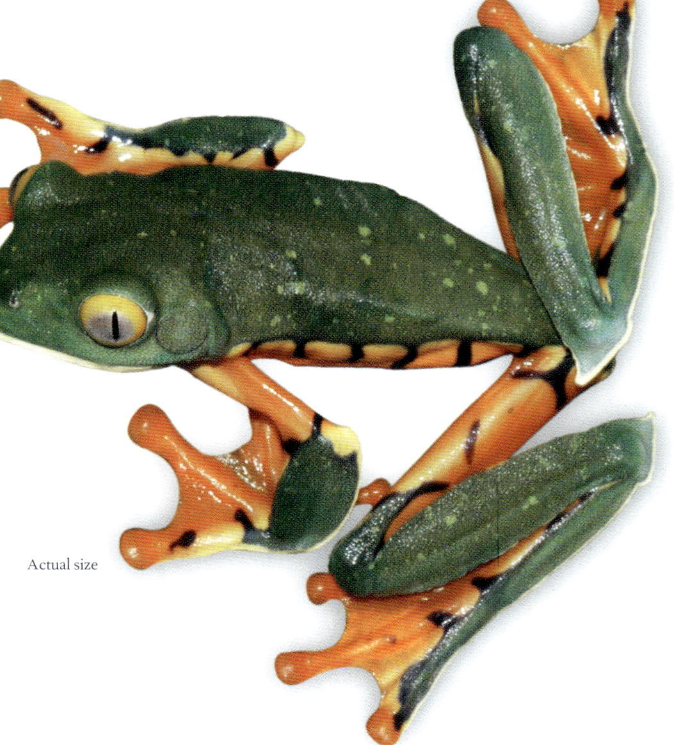

Actual size

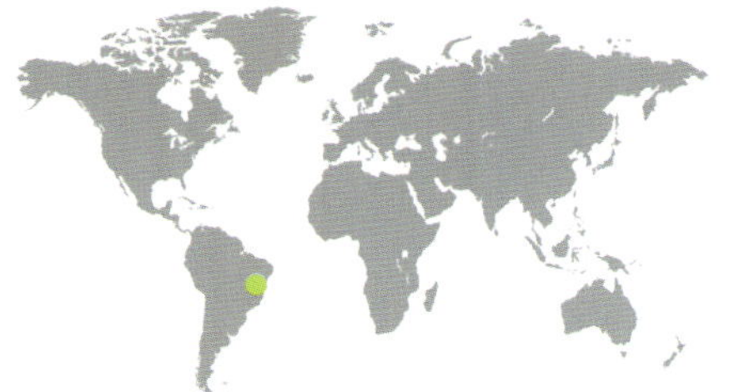

FAMILY	Hylidae: Phyllomedusinae
OTHER NAMES	*Agalychnis aspera*
DISTRIBUTION	Southern Bahia state, Brazil
ADULT HABITAT	Lowland forest up to 160 ft (50 m) altitude
LARVAL HABITAT	Ephemeral pools
CONSERVATION STATUS	IUCN Least Concern. Has a restricted range but its population is stable

ADULT LENGTH
Male
1½–1¹³⁄₁₆ in (38–46 mm)
Female
1¾–1⅞ in (45–48 mm)

HYLOMANTIS ASPERA

ROUGH LEAF FROG

PETERS, 1873

Actual size

This handsome frog gets its name from the rough texture of its skin. It lives close to the sea along the Atlantic coast of Brazil. It is found in forest or on the edge of forest, close to swamps and pools, and is abundant in places. The male's call consists of three or four short pulses. The female lays around 60 eggs on leaves hanging over water. When the eggs hatch, the tadpoles fall into the water, where they complete their development.

SIMILAR SPECIES

The genus *Hylomantis* contains only one other species; the Granular Leaf Frog (*H. granulosa*) is only found in northeastern Brazil. Other species formerly included in *Hylomantis* were transferred to the genus *Agalychnis* (see pages 347–349).

The Rough Leaf Frog has a long, slender body, very long, slender limbs, and large, protruding eyes. The skin on the head, back, and legs has a very rough texture. The upper surfaces are bright green, with an irregular pattern of reddish-brown and white blotches. The flanks, the underside of the limbs, and the fingers and toes are orange.

FAMILY	Hylidae: Phyllomedusinae
OTHER NAMES	None
DISTRIBUTION	Southeastern Brazil
ADULT HABITAT	Forest up to 2,620 ft (800 m) altitude
LARVAL HABITAT	Fast-moving streams
CONSERVATION STATUS	IUCN Least Concern. Declining in places due to deforestation, but occurs within some protected areas

ADULT LENGTH
Male
$1\frac{1}{8}$–$1\frac{7}{16}$ in (28–37 mm)
Female
$1\frac{5}{8}$–$1\frac{13}{16}$ in (41–46 mm)

PHASMAHYLA COCHRANAE

CHOCOLATEFOOT LEAF FROG

(BOKERMANN, 1966)

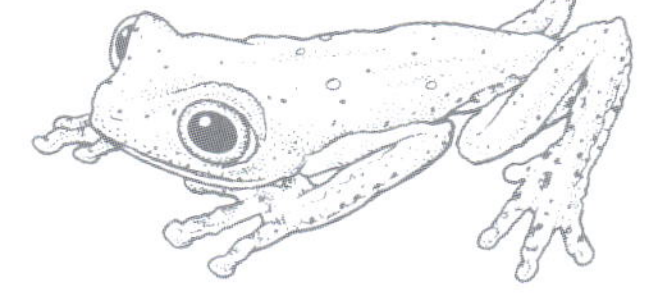

This small frog gets its name from its reddish-brown toes. It moves mostly by walking, when bright yellow and brown coloration on its flanks and legs is revealed. These colors do not change, unlike that on the back, which can change rapidly from bright green to brown. Males are generally both paler and smaller than females. During the breeding season, from October to April, pairs deposit around 32 eggs, enclosed in a gelatinous mess and folded into a leaf hanging over a stream. On hatching, the tadpoles fall into the stream below, where they are swept downstream to calmer pools.

SIMILAR SPECIES

Eight species have been described in the genus *Phasmahyla*. Known as shiny leaf frogs, they are found only in Brazil and their biology is little known. The Spotted Leaf Frog (*P. guttata*) is found in the coastal mountains along Brazil's southeastern coast. Lis and Bella's Leaf Frog (*Phasmahyla lisbella*), described in 2018, has a very limited range in southeastern Brazil.

Actual size

The Chocolatefoot Leaf Frog has a slender body, a rounded snout, long, slender limbs, and large eyes that are directed slightly forward. The upper surfaces are green, brown, or reddish brown with scattered, small, white and black spots. The flanks, the undersides of the legs, and the feet are yellow with large chocolate-brown or purple spots. The eyes have vertical pupils and silver irises.

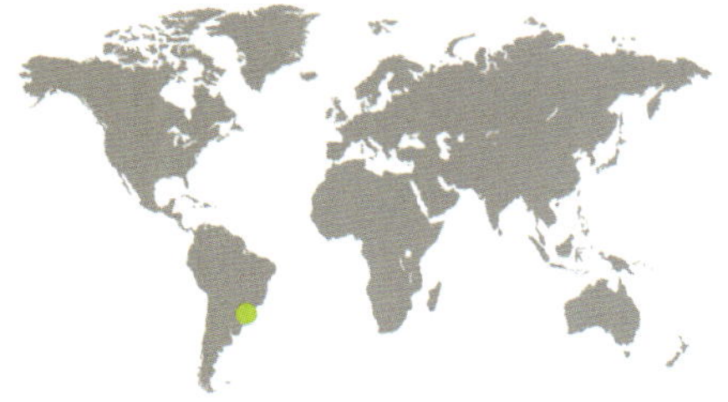

FAMILY	Hylidae: Phyllomedusinae
OTHER NAMES	None
DISTRIBUTION	Santa Catarina and São Paulo states, Brazil
ADULT HABITAT	Pristine rainforest with a closed canopy at altitudes of 330–3,300 ft (100–1,000 m)
LARVAL HABITAT	Oxbow lakes, blackwater ponds, and other water in rainforest habitats
CONSERVATION STATUS	IUCN Critically Endangered. Threats include chytridiomycosis, air pollution, and loss of habitat to coal mining and agriculture

ADULT LENGTH
Male
1¼–1⅜ in (32–35 mm)

Female
1¼–1⁷⁄₁₆ in (32–36 mm)

PHRYNOMEDUSA APPENDICULATA

SANTA CATARINA LEAF FROG

(LUTZ, 1925)

The Santa Catarina Leaf Frog was known to inhabit the Atlantic Coastal Forest ecoregion of southeastern Brazil. It was believed to be extinct, having not been seen since the early 1970s until being rediscovered in São Paulo state in 2011. Fewer than 50 specimens are thought to exist, and it may still be extinct in Santa Catarina, where it was first described, although authors also comment that this is a remarkably difficult tree frog to find due to its secretive habits. This leaf frog prefers pristine rainforest habitats with a closed canopy. Females lay their eggs on vegetation overhanging standing water, avoiding loss to egg-eating aquatic predators, the hatching tadpoles wriggling free and dropping into the water below.

SIMILAR SPECIES

The genus *Phrynomedusa* contains five further specimens, all endemic to the Atlantic Coastal Forest of southeastern Brazil and occurring as far north as Espirito Santo state (Bicolored Leaf Frog, *P. marginata*) to as far south as Santa Catarina state. *Phrynomedusa* is most closely related to the leaf frog genus *Cruziohyla*, which has three species, including the Splendid Leaf Frog (*C. calcarifer*, page 351).

The Santa Catarina Leaf Frog is unicolor green above and pale green to off-white underneath, with contrasting red pigment on the thighs and the undersides of the limbs. The limbs are long and slender, with large adhesive disks on the fingers and toes and minimal webbing between the toes. The eyes are large, with vertical pupils. The ankle on each foot bears a green and white spike-like protuberance called a calcar.

Actual size

FAMILY	Hylidae: Phyllomedusinae
OTHER NAMES	Orange-legged Tree Frog, Tiger-legged Monkey Frog
DISTRIBUTION	Northern South America, east of the Andes, from Venezuela south to Argentina
ADULT HABITAT	A very wide range of habitats, including arid grassland, open areas in forests, and gardens
LARVAL HABITAT	Ephemeral pools
CONSERVATION STATUS	IUCN Least Concern. Has a very large range and shows no evidence of a general decline

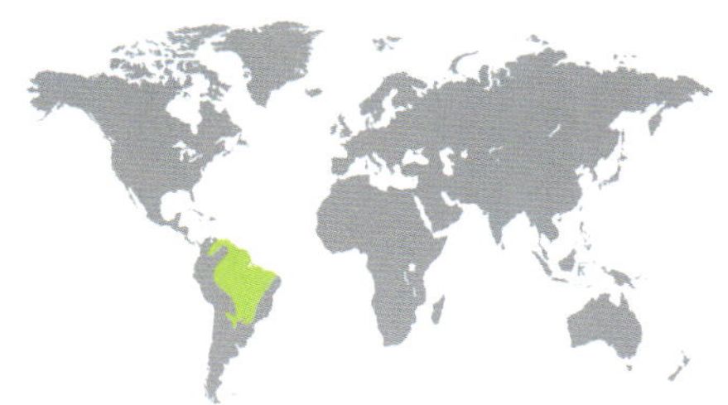

ADULT LENGTH
Male
average 1½ in (38 mm)

Female
average 2 in (51 mm)

PITHECOPUS HYPOCHONDRIALIS

ORANGE-SIDED LEAF FROG

(DAUDIN, 1800)

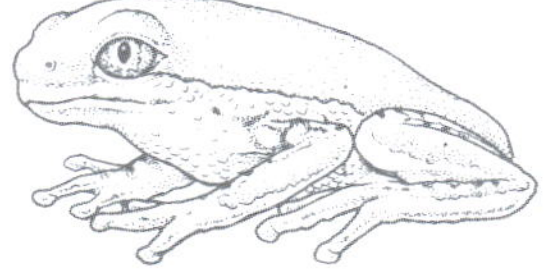

This nocturnal frog is able to live in arid habitats because of its ability to reduce water loss. During the heat of the day it wipes a waxy secretion, produced by special glands in its skin, all over its body. When attacked, it feigns death, rolling onto its back with its legs tucked up, and produces an unpleasant odor. After heavy rain, males call from bushes over temporary pools. The eggs are wrapped in leaves hanging over water, the female surrounding them with fluid-filled capsules that provide extra water for the developing embryos in dry weather.

Actual size

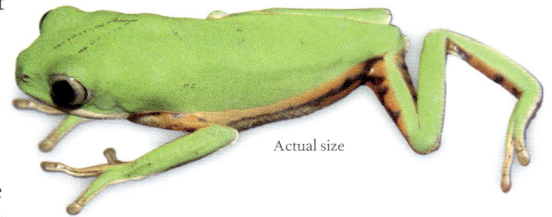

The Orange-sided Leaf Frog has a slender body, large, prominent, forward-pointing eyes, and very long, slender limbs. There are well-developed adhesive disks on the fingers and toes. The upper surfaces of the body and legs are bright, pale green. The undersides of the limbs are orange with black stripes, and the belly and the chin are white.

SIMILAR SPECIES

There are currently 12 described species in the genus *Pithecopus*, which are distributed across Central and South America. *Pithecopus azureus* is so similar to *P. hypochondrialis* that, until recently, it was considered to be the southern form of the same species. Males of the species, which is found in Bolivia, Brazil, Argentina, and Paraguay, have been observed grappling with one another in defense of their calling sites.

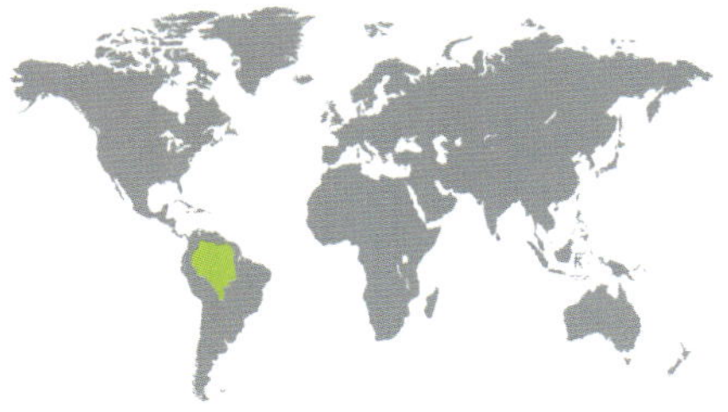

FAMILY	Aromobatidae: Allobatinae
OTHER NAMES	None
DISTRIBUTION	Amazon Basin region of Brazil, Bolivia, Peru, Ecuador, Colombia, Guyana, Suriname, and French Guiana
ADULT HABITAT	Lowland forest
LARVAL HABITAT	Tiny pools in leaf litter and fruit husks
CONSERVATION STATUS	IUCN Least Concern. Has a very large range and shows no evidence of a general decline

ADULT LENGTH
Male
$1\frac{1}{8}$–$1\frac{5}{16}$ in (28–33 mm)
Female
$1\frac{5}{16}$–$1\frac{3}{8}$ in (33–35 mm)

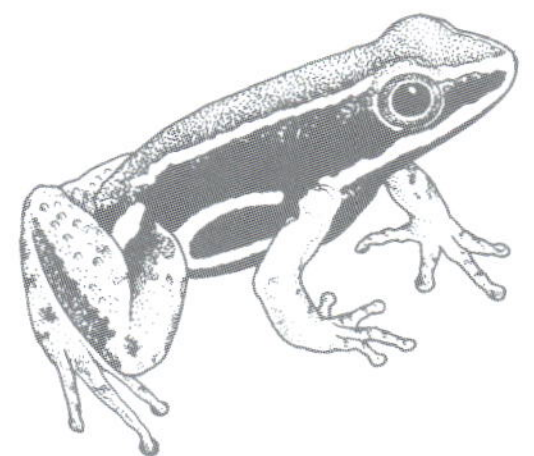

ALLOBATES FEMORALIS

BRILLIANT-THIGHED POISON FROG

(BOULENGER, 1884)

Actual size

The Brilliant-thighed Poison Frog has a somewhat pointed snout and granular skin. The body is dark brown or black, with lines of white, blue, or pale brown running from the snout to the base of the hind legs. There is a crescent-shaped orange patch at the base of the thighs, and an orange or yellow patch behind the forelimbs. The legs are brown, the throat black, and the belly white with black markings.

Males of this small diurnal frog species establish territories on the forest floor, which they then defend against other males by calling, and occasionally by fighting. Those males that call most often have the largest territories. Females are not attacked and move about among males before choosing one as a mate. Between 8 and 20 eggs are laid among the leaf litter and are defended by the male. When they hatch, the male carries the tadpoles—up to eight at a time—to tiny pools; the female occasionally assists him. Experiments have shown that males can home in directly on their territory after being displaced up to 230 ft (70 m) away.

SIMILAR SPECIES

It is likely that this wide-ranging species is actually more than a single species. It differs in male call characteristics from one location to another. It can be confused with *Lithodytes lineatus* (page 248), an edible frog that mimics its color patterns. The Martinique Volcano Frog (*A. chalcopis*) is found only on the West Indian island of Martinique, where it lives in grassy areas among lava flows. Listed as Critically Endangered, it is threatened by volcanic eruptions.

FAMILY	Aromobatidae: Allobatinae
OTHER NAMES	Striped Rocket Frog
DISTRIBUTION	Costa Rica, Panama, Colombia, Ecuador
ADULT HABITAT	Lowland moist and wet forest
LARVAL HABITAT	Streams
CONSERVATION STATUS	IUCN Least Concern. Has a large range and shows no evidence of a general decline

ALLOBATES TALAMANCAE

TALAMANCA ROCKET FROG

(COPE, 1875)

ADULT LENGTH
Male
11/16–15/16 in (17–24 mm)

Female
5/8–1 in (16–25 mm)

This small frog gets its name from its defensive behavior. When disturbed, it makes a huge leap, often into water. Despite not being toxic, it is active by day and is most commonly found near streams. Males call from the leaf litter, producing a high-pitched "peet… peet… peet" sound. Mating and egg-laying also occur on the ground. When the eggs hatch, both parents carry the tadpoles, in batches, to a stream. It is reported that this species comes together in quite large numbers, possibly as a defense against predators.

SIMILAR SPECIES

Allobates contains 63 species. The Sanguine Poison Frog (*Allobates zaparo*), also known as Zaparo's Poison Frog, is similar to *A. talamancae* except that, despite being toxic, it is rather secretive in its habits. It is found in Ecuador and Peru. Another similar species is the Three-striped Rocket Frog (*A. trilineatus*) from Bolivia, Ecuador, and Colombia. Neither species is of conservation concern.

Actual size

The Talamanca Rocket Frog has tiny rounded tubercles on its back, which is chocolate-brown in color. There is a pale stripe running the full length of the body from the snout, with a broader black stripe beneath it. The underside is pale and the legs are brown with dark markings. The throat of the male is black, while the female's is white, cream, or yellow.

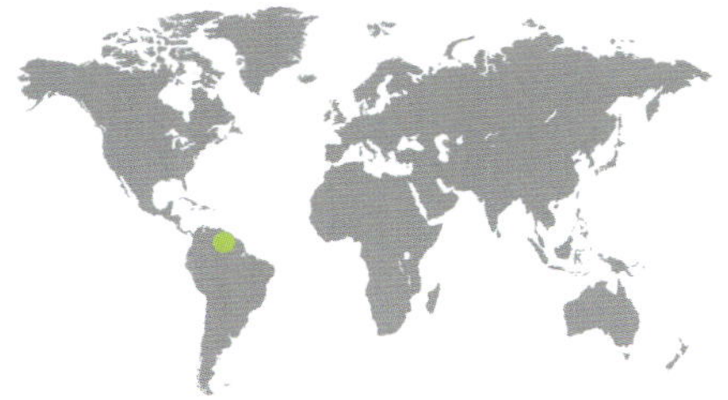

FAMILY	Aromobatidae: Anomaloglossinae
OTHER NAMES	None
DISTRIBUTION	Venezuela, Guyana, possibly Brazil
ADULT HABITAT	Scrub and meadow at altitudes of 6,100–8,900 ft (1,860–2,700 m)
LARVAL HABITAT	Tiny pools in bromeliads
CONSERVATION STATUS	IUCN Endangered. Has a very small range and its habitat is affected by tourism

ADULT LENGTH
⅝–¾ in (16–19 mm)

ANOMALOGLOSSUS RORAIMA

RORAIMA ROCKET FROG

(LA MARCA, 1997)

Actual size

The Roraima Rocket Frog has large eyes, and small disks on the tips of its fingers and toes that are decorated with pale blue spots. The body and legs are various shades of brown, marked with scattered black spots. There is a dark brown stripe running through the eye.

Mount Roraima is a large tepui, or flat-topped mountain, situated where the borders of Venezuela, Brazil, and Guyana meet. Discovered by Sir Walter Raleigh in 1596, it is separated from the surrounding land by 1,300 ft-high (400 m) cliffs. These present an insuperable barrier for many plants and animals, and Roraima is home to many unique species, including this tiny frog. Males attract females with a mating call consisting of a single high-pitched note. Eggs are laid in clutches of about five on the leaves of bromeliads. It is not known how the tadpoles reach pools in the bromeliad leaf axils.

SIMILAR SPECIES

There are around 32 species in the South American genus *Anomaloglossus*. *Anomaloglossus breweri* is found on Aprada-tepui in Venezuela and is listed as Near Threatened. Stephen's Rocket Frog (*A. stepheni*) is a very small, common, and widespread species found in Brazil. Unrelated to *A. roraima* but, like it, confined to the same mountain, is the Endangered Roraima Bush Toad (*Oreophrynella quelchii*).

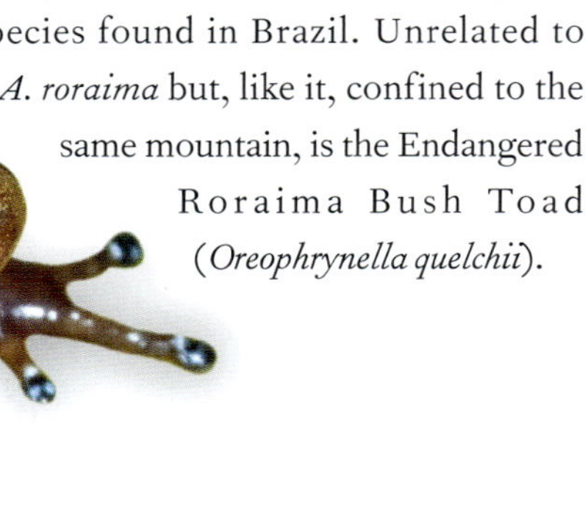

FAMILY	Aromobatidae: Anomaloglossinae
OTHER NAMES	None
DISTRIBUTION	Central Colombia
ADULT HABITAT	Cloud forest and tropical rainforest
LARVAL HABITAT	Streams
CONSERVATION STATUS	IUCN Least Concern. A common species with a stable population

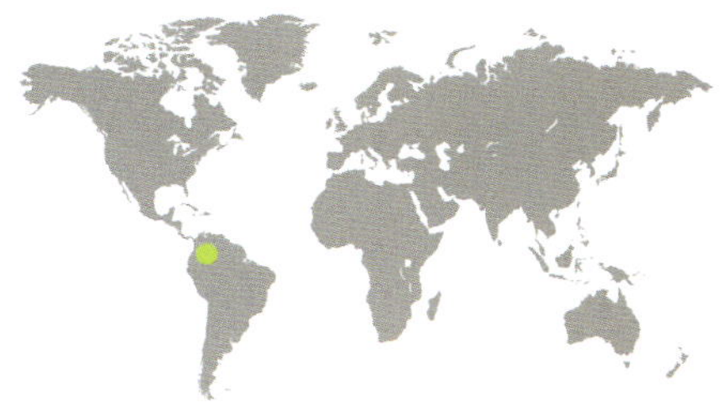

RHEOBATES PALMATUS

PALM ROCKET FROG

(WERNER, 1899)

ADULT LENGTH
Male
$^{5}/_{16}$–1$^{7}/_{16}$ in (23–37 mm)
Female
1$^{1}/_{8}$–1$^{9}/_{16}$ in (28–39 mm)

This diminutive frog uses small, damp caves and crevices for shelter and for depositing its eggs. Males call during daylight hours, producing a melodic whistle. They jump up and down as they call and turn black in color. If a female is receptive, she approaches a male's inflated vocal sac and slides under his chin. He then leads her to an egg-laying site. The male stays with the eggs until they hatch, when he carries them, 20–30 at a time, to a stream. The tadpoles sometimes stay on their father's back for up to seven days.

Actual size

The Palm Rocket Frog is a small, rather plump frog with a blunt snout. There are disks on its fingers and toes. The upper surfaces are dark brown with black spots and the underside is pale gray-brown. There is a dark stripe through the eye and there may be pale spots on the flanks.

SIMILAR SPECIES

There are only two species in the genus *Rheobates*, both previously included within *Colostethus*. The False Palm Rocket Frog (*Rheobates pseudopalmatus*) is known from only one small area in central Colombia. Its conservation status is the same as its congeneric, Least Concern.

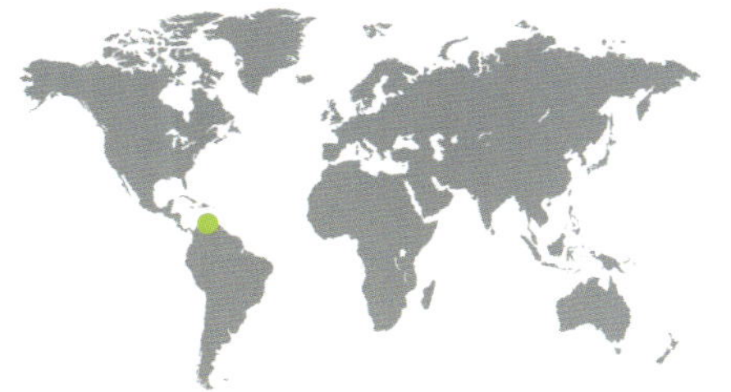

FAMILY	Aromobatidae: Aromobatinae
OTHER NAMES	None
DISTRIBUTION	Northwestern Venezuela
ADULT HABITAT	Cloud forest
LARVAL HABITAT	Cold-water streams
CONSERVATION STATUS	IUCN Critically Endangered. Has a very small range that is subject to habitat destruction

ADULT LENGTH
Male
1¾–2 in (45–52 mm)

Female
2$^{1}/_{16}$–2½ in (53–62 mm)

AROMOBATES NOCTURNUS

VENEZUELAN SKUNK FROG

MYERS, PAOLILLO & DALY, 1991

This little-known frog gets its name from the fact that it releases an odor like a skunk when handled. Its skin secretion seems to be repellent rather than toxic. It belongs to the poison frog family, but unlike other poison frogs it is active at night and appears to live almost entirely in streams. It is known only from an area of 4 sq miles (10 sq km) of forest that has been extensively felled to make way for agriculture and roads. Its population declined by 80 percent between 2000 and 2010, and in 2012 it was listed as "lost."

SIMILAR SPECIES

There are 18 species in the genus *Aromobates*, all found in the northern Andes, in Venezuela, Colombia, and Bolivia. Very little is known about them, other than that, unlike many poison frogs, they are cryptically colored. Several species have not been seen for many years. The Merida Rocket Frog *(A. meridensis)*, from Venezuela and Bolivia, has tested positive for the fungus that causes chytridiomycosis and is also listed as Critically Endangered.

The Venezuelan Skunk Frog is a compact, muscular frog with webbed hind feet and adhesive pads on its fingers and toes. It is olive-green in color, often with paler markings.

Actual size

FAMILY	Aromobatidae: Aromobatinae
OTHER NAMES	None
DISTRIBUTION	Northern Venezuela
ADULT HABITAT	Forest up to 5,280 ft (1,610 m) altitude
LARVAL HABITAT	Streams
CONSERVATION STATUS	IUCN Near Threatened. Has a small range and its population is declining as a result of habitat destruction

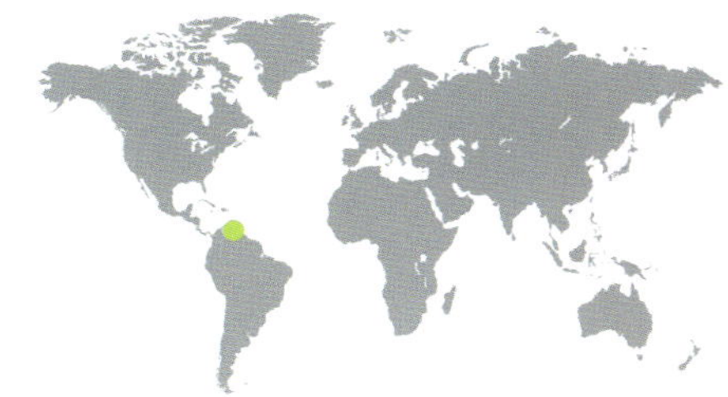

MANNOPHRYNE HERMINAE

HERMINA'S POISON FROG

(BOETTGER, 1893)

ADULT LENGTH
Male
¾–15⁄16 in (19–24 mm)
Female
⅞–1⅛ in (21–28 mm)

Unlike most poison frogs, this small species is not brightly colored. Its skin patterns provide good, effective camouflage and it seems to prefer darker parts of the forest, where it is usually found close to streams. Males call by day to attract females and the eggs are laid on the ground. The male carries the tadpoles on his back, in groups of 4–15, to streams, where they complete their development. The species lives near Venezuela's northern coast and is threatened by forest fires and encroaching agriculture. It has tested positive for the fungus that causes chytridiomycosis.

Actual size

Hermina's Poison Frog is a small, compact frog with a pointed snout and large eyes. The upper surfaces are olive-green, brown, or tan with dark patches on the back and dark stripes on the legs. There is a dark stripe running through the eye and, above it, a pale stripe running the length of the body.

SIMILAR SPECIES

Rivero's Poison Frog (*Mannophryne riveroi*) also lives along the northern coast of Venezuela. Much of its forest habitat has been destroyed and it is listed as Endangered. The same is true of the Collared Poison Frog (*M. collaris*), which occurs farther west, in the Venezuelan Andes. See also the Yellow-throated Frog (*M. trinitatis*; page 362).

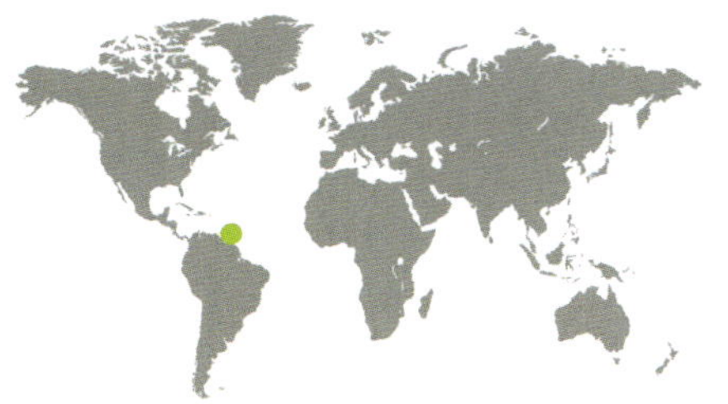

FAMILY	Aromobatidae: Aromobatinae
OTHER NAMES	Trinidad Poison Frog, Trinidadian Stream Frog
DISTRIBUTION	Trinidad
ADULT HABITAT	Moist forest
LARVAL HABITAT	Streams
CONSERVATION STATUS	IUCN Least Concern. Has a small range and its population is declining as a result of water pollution and habitat destruction

ADULT LENGTH
Male
$^{11}/_{16}$–$^{3}/_{4}$ in (17–20 mm)
Female
$^{11}/_{16}$–$^{7}/_{8}$ in (18–22 mm)

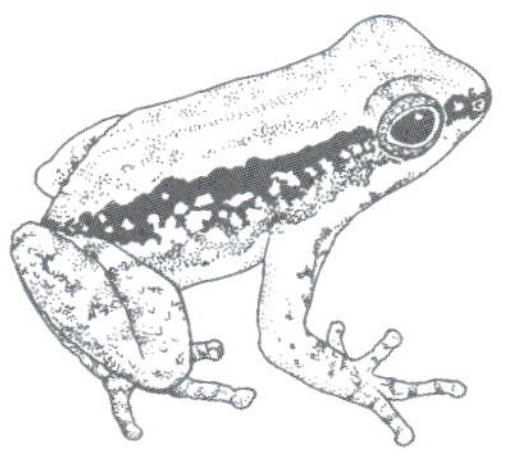

MANNOPHRYNE TRINITATIS

YELLOW-THROATED FROG

(GARMAN, 1888)

This small frog gets its name from the bright yellow throat of the female; the male's is gray. It lives along clear streams, and mating follows calling by the male and an elaborate courtship ritual, during which he changes color from brown to black. A clutch of 2–12 eggs is laid in a rock crevice or among leaves and is tended for 21 days by the male. He then carries the tadpoles on his back to pools within streams, taking considerable care to inspect pools and to avoid those containing predatory fish or shrimps.

SIMILAR SPECIES

Most of the 21 species in the genus *Mannophryne* are listed as Near Threatened, Vulnerable, Endangered, or Critically Endangered. In the Vulnerable category is the Bloody Bay Poison Frog (*M. olmonae*), which is found in both Trinidad and Tobago. It has tested positive for the pathogen that causes chytridiomycosis but it is not yet clear whether it is seriously affected by the disease. See also Hermina's Poison Frog (*M. herminae*; page 361).

The Yellow-throated Frog is a small, compact frog with a pointed snout, large eyes, and disks on its fingers and toes. The head and back are pale brown, and the flanks are black with white spots. The legs are pale brown with dark brown stripes. A white stripe runs below the eye.

Actual size

FAMILY	Dendrobatidae: Colostethinae
OTHER NAMES	Ruby Poison-arrow Frog
DISTRIBUTION	Southern Ecuador, northern Peru, Colombia
ADULT HABITAT	Tropical forest
LARVAL HABITAT	Pools or small streams
CONSERVATION STATUS	IUCN Least Concern. Declining in places due to deforestation, but occurs within some protected areas

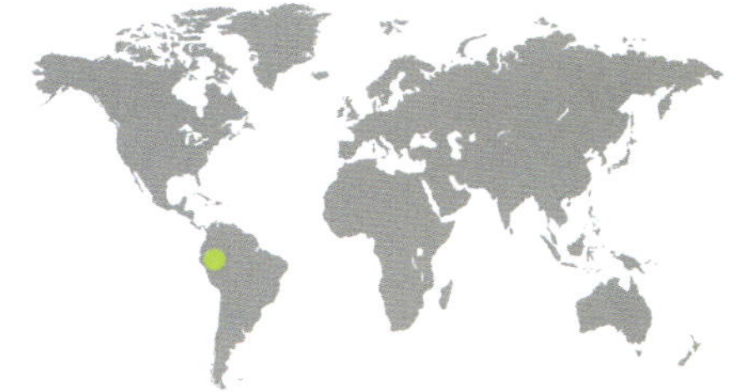

AMEEREGA PARVULA

RUBY POISON FROG

(BOULENGER, 1882)

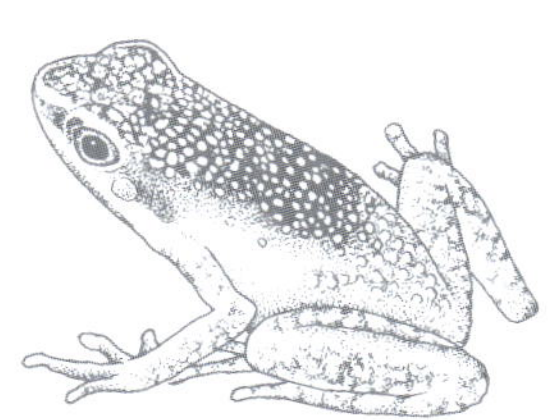

ADULT LENGTH
Male
11/16–15/16 in (17–24 mm)
Female
3/4–15/16 in (19–24 mm)

This very small, colorful frog is an inhabitant of the upper Amazon Basin. It is active by day, hopping about among the leaf litter. Its bright coloration warns potential predators that its skin contains toxic compounds, which it derives from the ants on which it feeds. The eggs are laid in nests on the ground and, when they hatch, the male carries the tadpoles to small puddles or slow-moving streams, where they complete their development. It is reported that this species performs foot-waving displays, but it is not clear if these are used in the context of aggression or courtship, or both.

Actual size

SIMILAR SPECIES

To date, 29 species have been described in the genus *Ameerega*, some of them very recently. All are found in South and Central America. The Ecuador Poison Frog (*A. bilinguis*) is similar in appearance to *A. parvula* and is found in Ecuador and Colombia. The Cainarachi Poison Frog (*A. cainarachi*), from Peru, has lost much of its forest habitat to agriculture and is listed as Endangered. See also the Three-striped Poison Frog (*A. trivittata*; page 365) and Silverstone's Poison Frog (*A. silverstonei*; page 364).

The Ruby Poison Frog has a pointed snout and large, dark eyes. The background color is black or brown. There are many raised red spots on the back that increase in density toward the head, and raised blue spots on the legs. A pale blue stripe runs along the upper lip and the flank. The underside is blue and black.

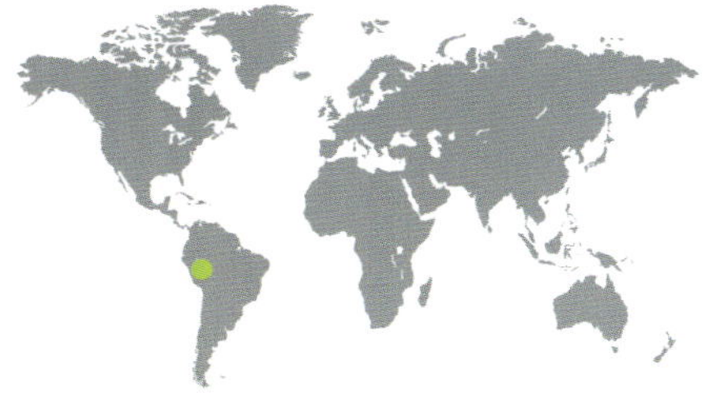

FAMILY	Dendrobatidae: Colostethinae
OTHER NAMES	None
DISTRIBUTION	Central Peru
ADULT HABITAT	Montane rainforest
LARVAL HABITAT	Streams
CONSERVATION STATUS	IUCN Endangered. Has a very restricted range that is subject to deforestation. Also smuggled illegally around the world

ADULT LENGTH
Male
up to 1½ in (38 mm)

Female
up to $1\frac{11}{16}$ in (43 mm)

AMEEREGA SILVERSTONEI

SILVERSTONE'S POISON FROG

(MYERS & DALY, 1979)

Actual size

Larger than many poison frogs, this rare species has striking coloration that makes it a target for illegal collection and distribution in the international pet trade. It is found only in one mountain range, the Cordillera Azul on the eastern side of the Andes. Its habitat is very wet and it probably breeds throughout the year. Males produce a trill-like call to attract females, which lay clutches of around 30 eggs on the ground. Males guard the eggs and carry the tadpoles to streams. Its skin secretions are not as powerful as those of other poison frogs, and although they are very distasteful they are not lethal to snakes.

SIMILAR SPECIES

The Pleasing Poison Frog (*Ameerega bassleri*), a Vulnerable species from eastern Peru, is very variable in color, being black with bold yellow, orange, green, or blue stripes. Another Peruvian species, the Pongo de Aguirre Poison Frog (*A. pongoensis*), is known from only one location and is listed as Vulnerable. It lives at lower altitudes near streams. See also the Ruby Poison Frog (*A. parvula*; page 363) and the Three-striped Poison Frog (*A. trivittata*; page 365).

Silverstone's Poison Frog has a pointed snout and coarse granular skin. The head, body, and thighs are bright red, orange, or yellow, often decorated with dark spots. The long hind legs are black, sometimes with red on the thighs. The upper part of the iris is gold.

FAMILY	Dendrobatidae: Colostethinae
OTHER NAMES	Three-striped Poison-arrow Frog
DISTRIBUTION	Amazon Basin, from Peru in the west to Suriname in the east and south to Bolivia
ADULT HABITAT	Tropical rainforest
LARVAL HABITAT	Small pools
CONSERVATION STATUS	IUCN Least Concern. Declining in places due to deforestation, but occurs within some protected areas

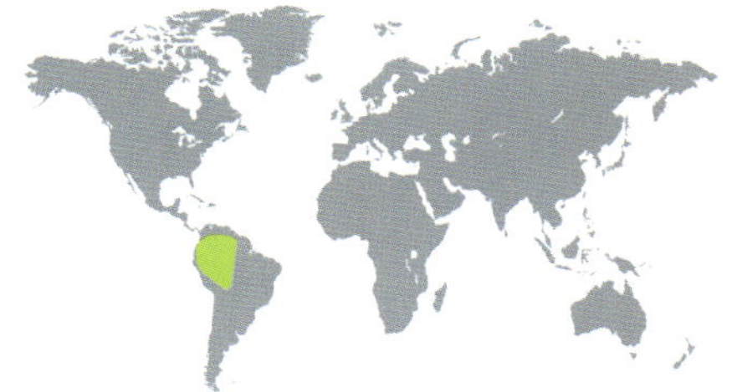

ADULT LENGTH
Male
up to 1⅝ in (42 mm)
Female
up to 1 15/16 in (50 mm)

AMEEREGA TRIVITTATA

THREE-STRIPED POISON FROG

(SPIX, 1824)

One of the larger poison frogs, this species has a very extensive range. During the rainy season, from May to October, males establish territories measuring 43–1,680 sq ft (4–156 sq m), which they defend against other males by calling and fighting. Females move about among the males' territories and are courted, not attacked, by resident males. Females prefer to mate with those males that hold larger territories. The eggs are laid under leaves and, when they hatch, the male carries the tadpoles on his back to small pools, where they complete their development.

SIMILAR SPECIES

Ameerega trivittata is very similar in appearance to the Spot-legged Poison Frog (*A. picta*). Both species are toxic and may have evolved to look alike, thus increasing the effectiveness of their defense against predators—a strategy called Müllerian mimicry. See also the Ruby Poison Frog (*A. parvula*; page 363) and Silverstone's Poison Frog (*A. silverstonei*; page 364).

The Three-striped Poison Frog has a pointed snout and large, dark eyes. The skin has a granular texture on the back and is smooth on the flanks. The back and flanks are black with green stripes, or green with black stripes. A pale green stripe runs below the eye. The legs may be yellow, green, or brown.

Actual size

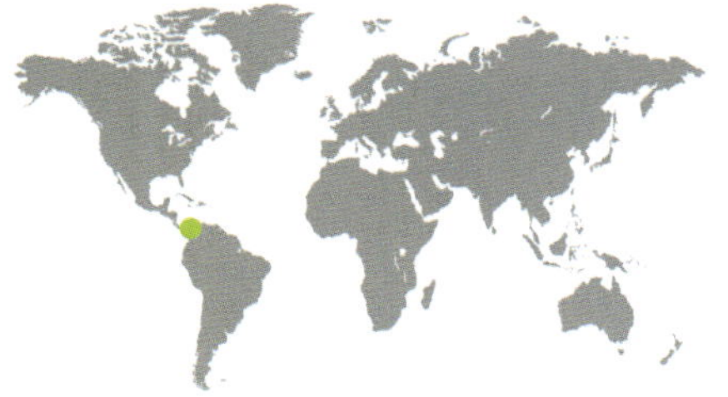

FAMILY	Dendrobatidae: Colostethinae
OTHER NAMES	None
DISTRIBUTION	Panama, northern Colombia
ADULT HABITAT	Lowland forest up to 2,600 ft (800 m) altitude
LARVAL HABITAT	The mother's back, then small pools
CONSERVATION STATUS	IUCN Least Concern. Declining in places, probably due to infection by chytridiomycosis

ADULT LENGTH
Male
up to 1 1/16 in (27 mm)
Female
up to 1 1/8 in (28 mm)

COLOSTETHUS PANAMANSIS

PANAMA ROCKET FROG

(DUNN, 1933)

Actual size

The Panama Rocket Frog is a small, compact frog with a blunt snout. The upper surfaces of the body and legs are brown, the flanks are darker brown, and the underside is creamy white. A narrow white or yellow stripe runs along the posterior half of each flank. There are yellow patches on the armpits, the groins, and the underside of the legs. Shown here is a female carrying tadpoles on her back.

In this small poison frog, the female plays a major role in the care of the young. The eggs are laid in leaf litter and, when they hatch, they wriggle onto the female's back. She carries them around, 20–35 at a time, for up to nine days, before depositing them in a pool of water. They grow while she is carrying them, feeding either on yolk or on food gathered when their mother enters a pool. Breeding activity is most intense in May and June, at the start of the wet season. Both sexes are very aggressive, engaging in fights to repel intruders from their territories.

SIMILAR SPECIES

Most of the 12 species of *Colostethus*, known as rocket frogs, are found in Panama, though some occur in northern South America. The Common Rocket Frog (*C. inguinalis*) occurs only in northern Colombia and is very similar to *C. panamansis*. Pratt's Rocket Frog (*C. pratti*) has a range that extensively overlaps that of *C. panamansis*. Individuals of the two species compete directly for the same spaces and respond to each other's aggressive calls.

FAMILY	Dendrobatidae: Colostethinae
OTHER NAMES	None
DISTRIBUTION	Southwestern Ecuador, northwestern Peru
ADULT HABITAT	Tropical forest at altitudes of 500–4,600 ft (150–1,400 m)
LARVAL HABITAT	Pools in streams
CONSERVATION STATUS	IUCN Near Threatened. Has a small range and its habitat is subject to degradation and chemical contamination

EPIPEDOBATES ANTHONYI

ANTHONY'S POISON ARROW FROG

(NOBLE, 1921)

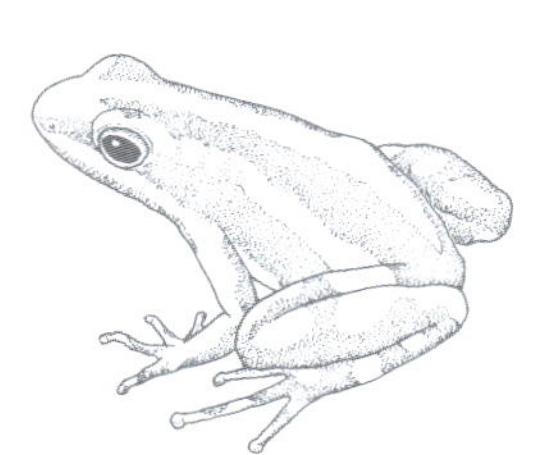

ADULT LENGTH
Male
3/4–1 in (19–25 mm)

Female
7/8–1 1/16 in (21–27 mm)

Males of this small, colorful frog species establish territories in the rainy season, producing a trill-like call to deter intruders and attract females. It is most active early in the morning and late in the afternoon, and is usually found close to streams. During mating, the male produces a series of croaks. The female lays a clutch of 15–40 eggs in a nest in the leaf litter within a male's territory. He protects the eggs for up to two weeks, keeping them moist and seeing off intruders. When they hatch, he carries them to a pool in a stream, where they complete their development.

Actual size

Anthony's Poison Arrow Frog has smooth skin, a pointed snout, and short, muscular hind legs. It is dark red to brown in color, with three white or yellowish longitudinal stripes on the back and white spots on the legs. The stripes and spots often have a bluish tinge. The underside has a marbled pattern in red or brown and white.

SIMILAR SPECIES

Epipedobates anthonyi was recently separated from the Phantasmal Poison Frog (*E. tricolor*; page 368), a Vulnerable species found in Ecuador. The two species look alike and both produce a toxin in their skin called epibatidine. As well as being very toxic, this compound is a powerful analgesic, 200 times more effective than morphine and synthetic versions of the toxin are used to make non-lethal painkillers.

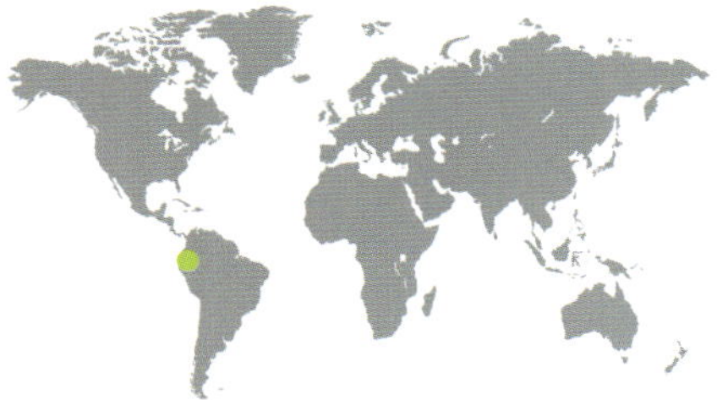

FAMILY	Dendrobatidae: Colostethinae
OTHER NAMES	Phantasmal Poison-arrow Frog
DISTRIBUTION	Central Ecuador
ADULT HABITAT	Forest at altitudes of 3,300–5,803 ft (1,000–1,769 m)
LARVAL HABITAT	Pools or slow-moving streams
CONSERVATION STATUS	IUCN Vulnerable. Known from just seven locations within a very small range; subject to habitat loss and collecting, and may be susceptible to chytridiomycosis

ADULT LENGTH
Male
average ¾ in (20 mm)
Female
average ⅞ in (22 mm)

EPIPEDOBATES TRICOLOR

PHANTASMAL POISON FROG

(BOULENGER, 1899)

Actual size

The Phantasmal Poison Frog has smooth skin and a pointed snout. It is dark red to brown in color, with three longitudinal white or yellowish stripes on the back and white spots on the legs. The stripes are often broken into short bands. The underside has a marbled pattern in red or brown and white.

The striking color patterns of this small frog are very variable, making it possible for field researchers to recognize individuals. Males are very aggressive and are the main providers of parental care. Each male defends a territory by calling and, if necessary, by grappling with an intruder and trying to press him down to the ground. The female lays a clutch of about ten eggs; the male guards these and, when they hatch, carries the tadpoles to water. Females prefer to mate with males that make more elaborate calls. Such males are older, and father offspring that are more likely to survive.

SIMILAR SPECIES

There are seven species in the genus *Epipedobates*, all found to the west of the Andes and on their slower slopes, in Ecuador and Colombia. *Epipedobates machalilla* is a coffee-colored little frog found at lower altitudes in Ecuador. The male cares for the eggs and carries the tadpoles to water. The Marbled Poison Frog (*E. boulengeri*), from Colombia and Ecuador, thrives in altered habitats such as gardens.

FAMILY	Dendrobatidae: Colostethinae
OTHER NAMES	None
DISTRIBUTION	Central and South America, from southern Costa Rica to northern Colombia
ADULT HABITAT	Rainforest up to 5,250 ft (1,600 m) altitude
LARVAL HABITAT	Small streams
CONSERVATION STATUS	IUCN Vulnerable. Declining in many parts of its range due to chytridiomycosis

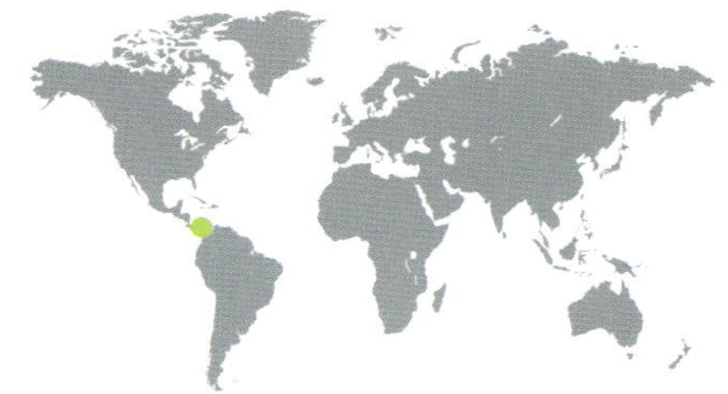

SILVERSTONEIA NUBICOLA

BOQUETE ROCKET FROG

(DUNN, 1924)

ADULT LENGTH
Male
5/8–3/4 in (16–20 mm)
Female
11/16–7/8 in (17–23 mm)

This tiny frog appears to breed at any time of year. It lives in leaf litter on the forest floor and is most commonly found close to rocky sections of streams. Males call during the day, producing a high-pitched "peet… peet… peet." The eggs are laid in the leaf litter and are attended by the male, which carries them on his back, 1–11 at a time, to small streams, where they complete their development. In 2004, several dead specimens of this species were discovered in El Cope National Park in Panama; they were found to be infected with chytridiomycosis.

Actual size

The Boquete Rocket Frog has a dark brown back and black flanks. The legs are reddish brown with black spots. A conspicuous white stripe runs from its snout, along the flank to the groin. A paler stripe runs from the snout along the length of the body. The underside is yellow.

SIMILAR SPECIES

To date, eight species have been described in this genus, five of them as recently as 2013. The Rainforest Rocket Frog (*Silverstoneia flotator*) is a very common species found in Costa Rica and Panama. It is very similar in appearance and habits to *S. nubicola*. The tadpoles of both species have an unusual upward-pointing funnel around the mouth, which enables them to feed on food items floating at the water's surface.

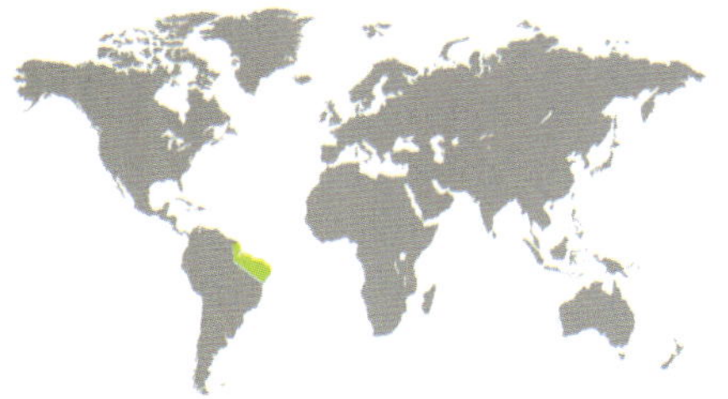

FAMILY	Dendrobatidae: Dendrobatinae
OTHER NAMES	None
DISTRIBUTION	Northeastern Brazil
ADULT HABITAT	Lowland tropical forest
LARVAL HABITAT	Water-filled hollows, notably those in Brazil-nut capsules
CONSERVATION STATUS	IUCN Least Concern. Shows no evidence of decline, but may be threatened by the pet trade

ADULT LENGTH
$\frac{11}{16}$–$\frac{7}{8}$ in (18–23 mm), females slightly larger than males

ADELPHOBATES CASTANEOTICUS

BRAZIL-NUT POISON FROG

(CALDWELL & MYERS, 1990)

Actual size

The Brazil-nut Poison Frog has glossy black skin on its back, decorated with white or yellow spots, which are sometimes arranged in rows. The arms and legs are brown, and there are larger yellow or orange spots where they join the body.

When the nuts of Brazil-nut trees (*Bertholletia excelsa*) fall to the forest floor, their contents are eaten by mammals such as agoutis, leaving behind their outer capsules, which soon fill with water. These tiny pools make ideal nurseries for the larvae of frogs, toads, and many insects. The larvae compete with, or feed on, one another, and the species that gets there first is usually the one that succeeds. The male of this tiny, colorful frog protects the eggs until they hatch, when he carries the tadpoles to the Brazil-nut husks and other tiny pools. The tadpoles are large and aggressive, and feed on insect larvae.

SIMILAR SPECIES

There are two other, very colorful species in this genus. The Splash-backed Poison Frog (*A. galactonotus*) from eastern Brazil lives in the leaf litter and is black with a bright yellow or orange back. The Rio Madeira Poison Frog (*A. quinquevittatus*), from Brazil and Peru, has a black body with white longitudinal stripes and yellow arms and legs with black spots. Currently, the most serious threat to all three species is their coloration, which makes them popular in the pet trade.

FAMILY	Dendrobatidae: Dendrobatinae
OTHER NAMES	None
DISTRIBUTION	Colombia
ADULT HABITAT	Cloud forest and dry forest at altitudes of 5,200–6,900 ft (1,580–2,100 m)
LARVAL HABITAT	Water-filled bromeliads
CONSERVATION STATUS	IUCN Vulnerable. Has a very fragmented range. Threatened by habitat destruction, removal of bromeliads, and collecting for the pet trade

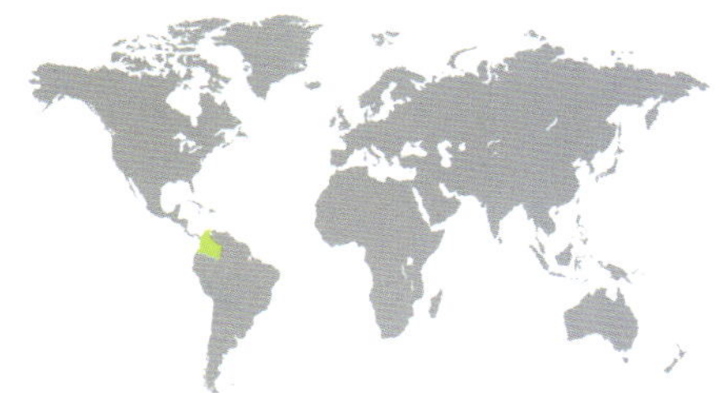

ANDINOBATES BOMBETES

CAUCA POISON FROG

(MYERS & DALY, 1980)

ADULT LENGTH
Male
⅝–¾ in (16–19 mm)
Female
$^{11}/_{16}$–¾ in (18–20 mm)

This rare frog gets its scientific species name, *bombetes*, from the male's call, which is like the buzzing of a bee. It lives in the Colombian Andes, among the leaf litter on the forest floor, and is active by day. The eggs are laid on the ground and, when they hatch, they are carried by their father, one or two at a time and glued to his back with mucus, to a bromeliad, where they complete their development in a water-filled leaf axil. Recent research has shown that males living near streams produce higher-pitched calls than those living further away from streams (high-pitched sounds are more audible against the sound of rushing water).

Actual size

The Cauca Poison Frog has a slender body, a narrow head, and a rounded snout. The fingers and toes have square tips. The upper surfaces are black or dark brown, with red or orange stripes converging toward the snout. The flanks and underside are pale green, blue-green, or yellow, mottled with black, and the upper arms are red.

SIMILAR SPECIES

Also found in the Colombian Andes is the Santander Poison Frog (*Andinobates virolinensis*). Largely bright red in color, its habitat, and that of another vivid red poison frog, the recently described Chocó Poison Frog (*A. cassidyhornae*), has been extensively destroyed by logging, encroaching agriculture, and pollution by agrochemicals. In addition, both of these species are illegally collected for the pet trade, and both species are listed as Vulnerable.

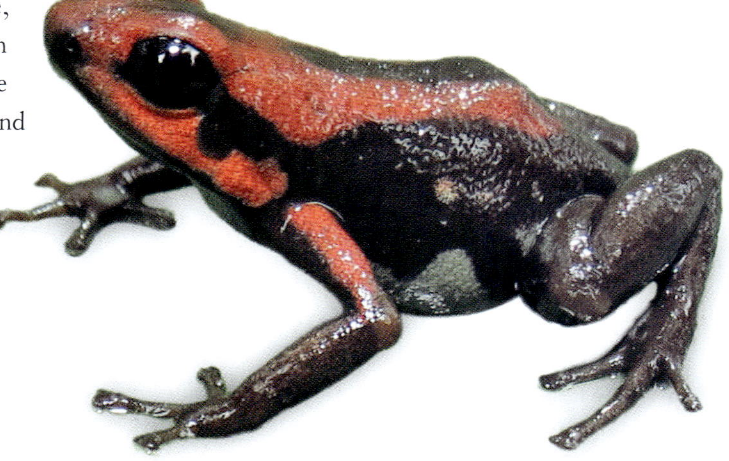

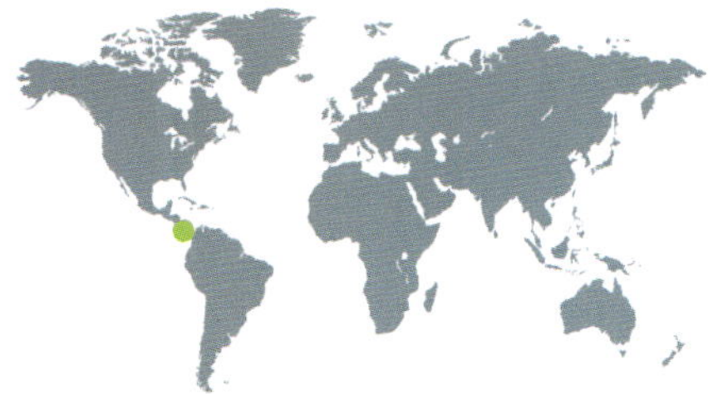

FAMILY	Dendrobatidae: Dendrobatinae
OTHER NAMES	None
DISTRIBUTION	Central Panama
ADULT HABITAT	Lowland forest at altitudes of around 450 ft (136 m)
LARVAL HABITAT	Unknown
CONSERVATION STATUS	IUCN Critically Endangered. Quite common within a restricted area, but this is threatened by deforestation and copper mining

ADULT LENGTH
Male
7⁄16–9⁄16 in (11–14 mm)
Female
average ½ in (13 mm)

ANDINOBATES GEMINISAE

GEMINIS' DART FROG

BATISTA, JARAMILLO, PONCE & CRAWFORD, 2014

Actual size

Geminis' Dart Frog has a slender body, a rounded snout, and small disks at the end of its fingers and toes. Its skin has a granular texture. It is uniformly bright chrome-orange in color, sometimes with small brown patches. Its fingers and toes are bluish gray, and the eyes are black.

Discovered in 2011, this tiny poison frog occurs only in undisturbed forest, where it is found under logs and rocks toward the tops of ridges. It is active by day, hopping about on the forest floor, where its vivid coloration makes it very visible. Males call mostly in the morning, producing around two buzz-like calls per minute. The fact that one individual has been observed carrying a single tadpole on its back indicates that the species performs parental care. It is thought that its tadpoles develop in water-filled cupped leaves, tree hollows, and pitcher plants.

SIMILAR SPECIES

Geminis' Dart Frog is closely related to the Bluebelly Poison Frog (*A. minutus*), a tiny frog from Panama and Colombia that has a striking gold and black pattern on its back. It lays its eggs on the ground and carries its tadpoles to tiny pools in the leaf axils of bromeliads. Geminis' Dart Frog is similar in appearance to the Andean Poison Frog (*A. opisthomelas*; page 373), which can be found in Colombia.

FAMILY	Dendrobatidae: Dendrobatinae
OTHER NAMES	None
DISTRIBUTION	Colombia
ADULT HABITAT	Montane forest at altitudes of 3,800–7,200 ft (1,160–2,200 m)
LARVAL HABITAT	Water-filled bromeliads
CONSERVATION STATUS	IUCN Vulnerable. Has a very small range, and is threatened by habitat loss and collecting for the pet trade

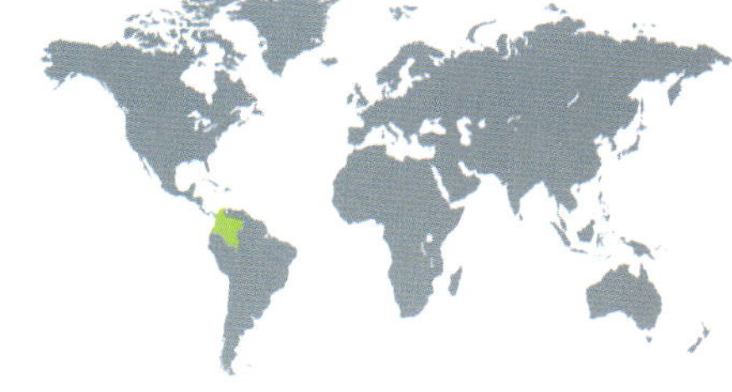

ADULT LENGTH
$^9/_{16}$–$^3/_4$ in (14–20 mm)

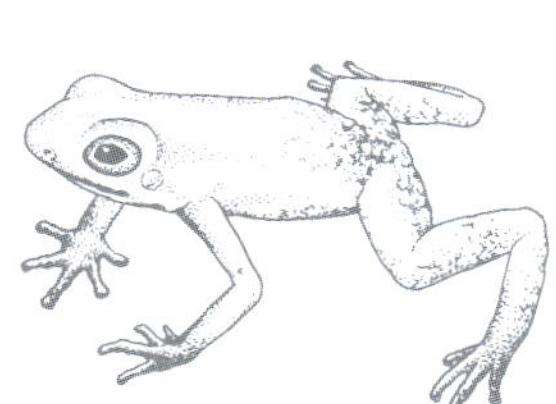

ANDINOBATES OPISTHOMELAS

ANDEAN POISON FROG

(BOULENGER, 1899)

Very little is known about this small, colorful frog. It lives mostly in the leaf litter on the forest floor, where it lays its eggs. The male cares for the young, carrying the tadpoles one at a time to pools in the leaf axils of bromeliads. The species occurs in a very restricted range on the eastern slopes of the Andes, where it is threatened by deforestation as land is cleared for agriculture. Its bright coloration has resulted in it being heavily collected for the pet trade.

Actual size

The Andean Poison Frog has a slender body and protruding black eyes. The fingers and toes have blunt tips. The body and legs are bright red, sometimes with a few black spots. In some individuals the lower arms and legs are brown or black.

SIMILAR SPECIES

Of the 16 species in this genus, two are categorized as Critically Endangered, three are Endangered, and six are Vulnerable. Most are found only in Colombia. The Yellow-bellied Poison Frog (*Andinobates fulguritus*) is not threatened and is found in Panama as well as Colombia. At mating, the male does not clasp the female; instead, he deposits sperm on the ground and the female lays one to five eggs on top of it.

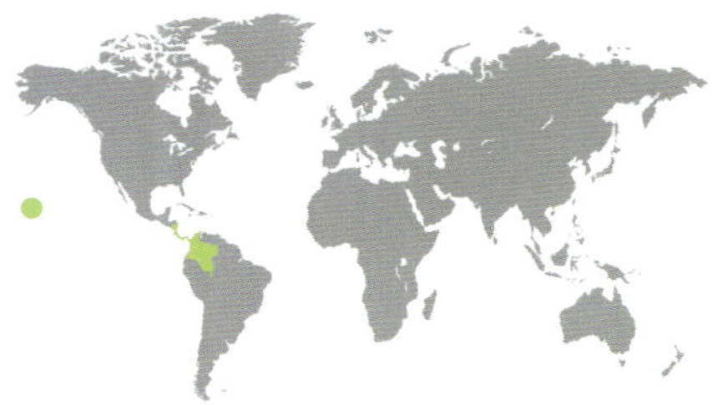

FAMILY	Dendrobatidae: Dendrobatinae
OTHER NAMES	Green and Black Dart-poison Frog
DISTRIBUTION	Nicaragua, Costa Rica, Panama, Colombia. Introduced to Oahu, Hawaii
ADULT HABITAT	Lowland rainforest
LARVAL HABITAT	Water-filled hollows in trees and bromeliads
CONSERVATION STATUS	IUCN Least Concern. Shows no evidence of a general decline

ADULT LENGTH
Male
1–1$\frac{9}{16}$ in (25–40 mm)

Female
1$\frac{1}{16}$–1$\frac{5}{8}$ in (27–42 mm)

DENDROBATES AURATUS

GREEN POISON FROG

(GIRARD, 1855)

Actual size

The Green Poison Frog has a pointed head and glossy skin. It is typically black with brilliant green patches, but there is much variation in color: the black can be replaced by brown or bronze, and the green by yellow or blue. The skin of some individuals has a golden sheen. It is a good climber and has small adhesive pads on the tips of its fingers and toes.

This colorful frog is active by day, the male defending a territory containing a nest in the leaf litter. The female is more proactive than the male in soliciting mating, approaching him and beating on his back with her hind feet. Females sometimes fight one another for access to a preferred male. Males guard up to six clutches of eggs, produced by different females. He rotates the eggs, keeps them wet, and removes fungus from them. When they hatch, he carries the tadpoles on his back, one or two at a time, to small pools in bromeliads or tree-holes.

SIMILAR SPECIES

There is no other species like the Green Poison Frog in appearance, and it is unusual in terms of its behavior: The active role played by females in mating appears to be unique among frogs. There are five species in the genus *Dendrobates*. All are active by day, are brightly colored, and possess toxic secretions in their skin, derived from ants and other poisonous insects on which they feed. See also the Black and Yellow Poison Frog (*D. leucomelas*; page 375) and the Dyeing Poison Frog (*D. tinctorius*; page 376).

FAMILY	Dendrobatidae: Dendrobatinae
OTHER NAMES	Bumblebee Poison Dart Frog, Yellow-headed Dart-poison Frog
DISTRIBUTION	Venezuela, Guyana, Brazil, Colombia, Bolivia
ADULT HABITAT	Lowland rainforest
LARVAL HABITAT	Water-filled hollows in trees
CONSERVATION STATUS	IUCN Least Concern. Shows no evidence of a general decline

ADULT LENGTH
$1\frac{1}{4}$–$1\frac{15}{16}$ in (31–50 mm)

DENDROBATES LEUCOMELAS

BLACK AND YELLOW POISON FROG

STEINDACHNER, 1864

An inhabitant of very hot, wet forest, this colorful frog is active by day and feeds largely on ants, the source of the toxins in its skin. The male is the sole provider of parental care, the female's role being simply to produce eggs. In captivity, females produce 2–12 eggs in a clutch, but over a breeding season each can produce 100–1,000 eggs. The female's choice of male to look after a clutch of eggs is based on his call, those with more elaborate calls being preferred. The male looks after the eggs in a nest on the ground and then carries the tadpoles to water-filled holes in trees.

Actual size

The Black and Yellow Poison Frog is larger than other *Dendrobates* species. Its skin is glossy and black, with three cross-bands of yellow or orange, which often contain black spots. The limbs are patterned in black and yellow or orange. There are adhesive pads on the fingers and toes.

SIMILAR SPECIES

Dendrobates leucomelas is unlike any other frog in appearance. While it has proved relatively easy to breed in captivity, a related species from Colombia, the Yellow-striped Poison Frog (*D. truncatus*), does not breed in captivity, and this has led to it being protected by the CITES regulations that control trade in endangered species. See also the Green Poison Frog (*D. auratus*; page 374) and the Dyeing Poison Frog (*D. tinctorius*; page 376).

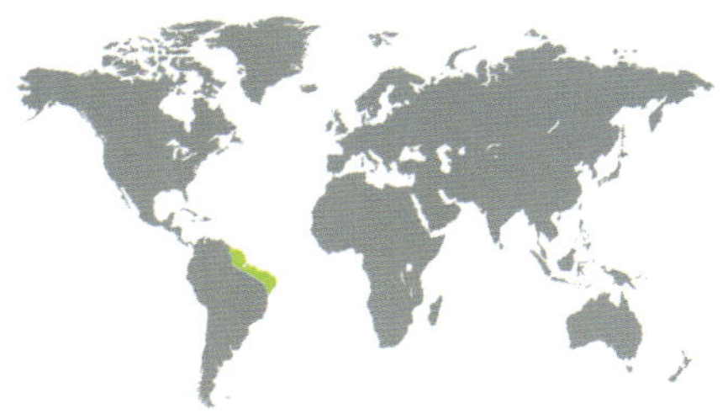

FAMILY	Dendrobatidae: Dendrobatinae
OTHER NAMES	Dyeing Poison-arrow Frog, Tinging Frog
DISTRIBUTION	Guyana, Suriname, French Guiana, northeastern Brazil
ADULT HABITAT	Lowland forest up to 2,000 ft (600 m) altitude
LARVAL HABITAT	Tiny pools in plants
CONSERVATION STATUS	IUCN Least Concern. Has a stable population. It is popular in the pet trade but breeds well in captivity

ADULT LENGTH
1⁵⁄₁₆–1¹⁵⁄₁₆ in (34–50 mm)

DENDROBATES TINCTORIUS

DYEING POISON FROG

(CUVIER, 1797)

This colorful frog gets its name from the curious way it is used by indigenous tribes. Feathers are plucked from the back of young parrots and a Dyeing Poison Frog is rubbed on their exposed skin. When the feathers regrow, the frog's toxin makes them yellow or red rather than green. The species occurs in a number of different color forms; one of two common forms was previously considered to be a distinct species, *Dendrobates azureus*. It is brilliant blue with black and white spots. The female Dyeing Poison Frog is the more active partner during mating, approaching the male and stroking his snout and back with her hands.

The Dyeing Poison Frog has long, thin arms, legs, fingers, and toes. There are well-developed adhesive pads on the fingers and toes; these are circular in females and heart-shaped in males. Color is highly variable. The form shown here has a pale blue body, decorated with darker spots, and dark blue arms and legs. Other forms have varying amounts of bright yellow on their head and back, and blue or black legs. Rare forms are much more yellow.

SIMILAR SPECIES

The Yellow-striped Poison Frog (*Dendrobates truncatus*) occurs in forest on the western slopes of the Andes in Colombia. It is very popular in the pet trade. Another poison frog that has many different color forms is the Splash-backed Poison Frog (*Adelphobates galactonotus*). This species is found in the forests of eastern Brazil, to the south of the Amazon. See also the Green Poison Frog (*D. auratus*; page 374) and the Black and Yellow Poison Frog (*D. leucomelas*; page 375).

Actual size

FAMILY	Dendrobatidae: Dendrobatinae
OTHER NAMES	None
DISTRIBUTION	Northern Peru
ADULT HABITAT	Forest at altitudes of 3,000–3,600 ft (900–1,100 m)
LARVAL HABITAT	Tiny pools in the leaf axils of bromeliads
CONSERVATION STATUS	IUCN Endangered. Habitat destruction has reduced its range to a few forest fragments, and it is also subject to collection for the pet trade

ADULT LENGTH
$1\frac{1}{16}$–$1\frac{1}{8}$ in (27–29 mm)

EXCIDOBATES MYSTERIOSUS

MARAÑÓN POISON FROG

(MYERS, 1982)

This distinctively colored little frog lives in close association with bromeliads. Males produce a rattle-like buzz to attract females and the eggs are laid on a bromeliad bract. The male then carries them to tiny pools in the leaf axils of the plant, where the tadpoles develop. The species' forest habitat has been extensively destroyed and replaced by coffee plantations, and it now survives in just five or six small patches of forest. It has also been subject to collection for the international pet trade. It was rediscovered in 1989, having not been seen for 60 years.

Actual size

The Marañón Poison Frog has a distinctive coloration: black or dark brown with white polka dots on its back, head, limbs, and underside. In some individuals, the spots have a pale blue tinge. The size and distribution of the spots is variable but there is always a single spot under the chin.

SIMILAR SPECIES

There are three species in the genus *Excidobates*. All have oval spots on the underside of their thighs. The Santiago Poison Frog (*E. captivus*) is a very small species, not exceeding $\frac{5}{8}$ in (17 mm) in length. Known from only two locations, it was not seen in the wild between 1929 and 1960. Its tadpoles develop in pools of water that form in *Heliconia* plants. *Excidobates condor*, from Ecuador, was described in 2012.

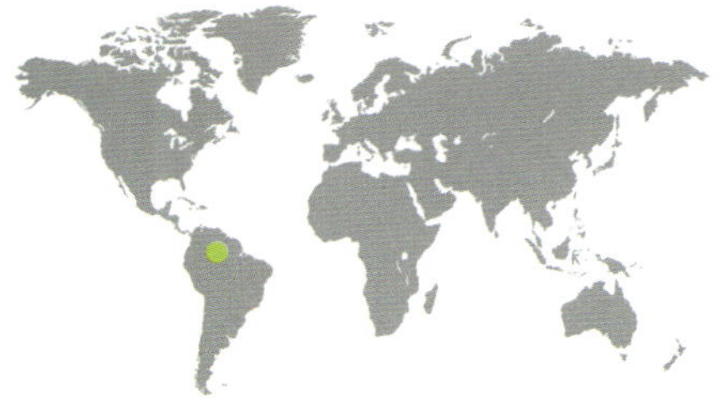

FAMILY	Dendrobatidae: Dendrobatinae
OTHER NAMES	Demonic Poison-arrow Frog, Yapacana's Little Red Frog
DISTRIBUTION	Southern Venezuela
ADULT HABITAT	Rainforest at altitudes of 2,000–4,000 ft (600–1,200 m)
LARVAL HABITAT	Tiny pools in the leaf axils of bromeliads and other plants
CONSERVATION STATUS	IUCN Critically Endangered. Has a very restricted range and is subject to habitat destruction and illegal collection

ADULT LENGTH
½–¾ in (12–19 mm)

MINYOBATES STEYERMARKI

DEMONIC POISON FROG

(RIVERO, 1971)

Actual size

The Demonic Poison Frog has long, thin arms and smooth skin on its back. There are disks on its fingers, which are larger than those on the toes. The upper and lower surfaces are red-brown, dark red, or scarlet with black spots, which are more numerous on the underside. The limbs are sometimes salmon-pink.

Cerro Yapacana is a tepui, or flat-topped mountain, in southern Venezuela. The only home of this small frog, it is also an area of intense gold mining and its attendant pollution and habitat destruction. This frog is found in mossy areas near rocks. The male attracts a female with a series of soft calls and then looks after their clutch of three to nine eggs, keeping them moist until they hatch after 10–14 days. The male then carries the tadpoles to tiny pools in the leaf axils of bromeliads and other plants, where it takes them up to seven weeks to reach metamorphosis.

SIMILAR SPECIES

Minyobates steyermarki is the only species in its genus. While it is similar to other poison frogs in being brightly colored, active by day, and having toxic skin secretions, it differs from them in two ways. First, during mating, the male clasps the female around her head, rather than around her body. And second, its skin secretions, which are relatively mild, are chemically different from those of other poison frogs.

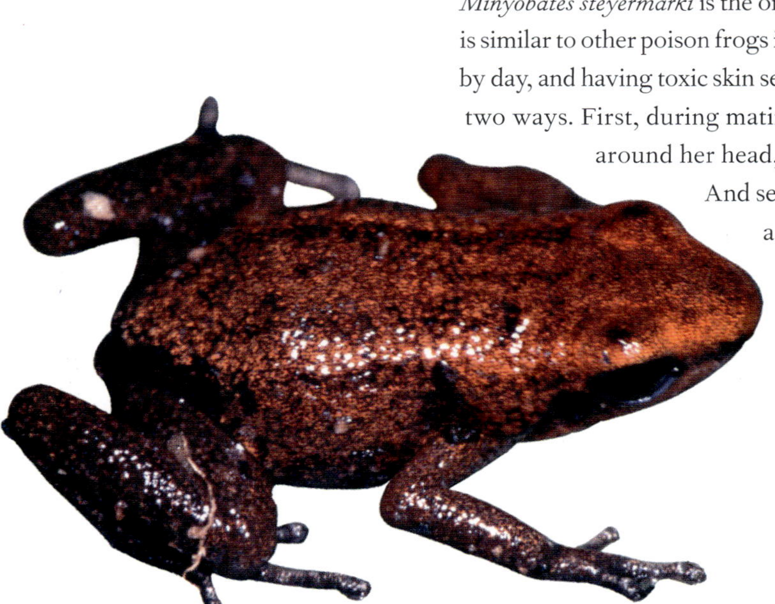

FAMILY	Dendrobatidae: Dendrobatinae
OTHER NAMES	None
DISTRIBUTION	Western Panama
ADULT HABITAT	Forest, from sea level to 3,675 ft (1,120 m) altitude
LARVAL HABITAT	Tiny pools in the leaf axils of bromeliads
CONSERVATION STATUS	IUCN Critically Endangered. Has a very small range that is subject to habitat destruction

OOPHAGA ARBOREA

POLKADOT POISON FROG

(MYERS, DALY & MARTÍNEZ, 1984)

ADULT LENGTH
¾–⅞ in (20–22 mm)

This small frog is highly toxic and is generally arboreal in its habits. The male calls from a bromeliad leaf to attract a female, which approaches him, drums the leaf with her legs, and touches his snout. There is no amplexus, the pair instead mating in a vent-to-vent position. Mating yields a clutch of four to eight eggs, which is defended very aggressively by the male. It is not clear which parent then carries the resulting tadpoles to pools in bromeliads, where they are fed unfertilized eggs by the female.

SIMILAR SPECIES

In two other species in this genus, the Harlequin Poison Frog (*Oophaga histrionica*) and the Splendid Poison Frog (*O. speciosa*), all aspects of parental care, including guarding the eggs, are carried out by the female. Harlequin Poison Frog females have been observed eating other females' eggs. The Splendid Poison Frog is a bright red frog that occupies a very small range in Panama and is listed as Endangered. See also the Granular Poison Frog (*O. granulifera*; page 380), Lehmann's Frog (*O. lehmanni*; page 381), and the Strawberry Poison-dart Frog (*O. pumilio*; page 382).

Actual size

The Polkadot Poison Frog gets its common name from the array of circular white or yellow spots on a black or brown background. It is unusual in that the spots tend to be slightly raised above the surrounding skin. It has rather short hind legs and well-developed adhesive pads on its fingers and toes.

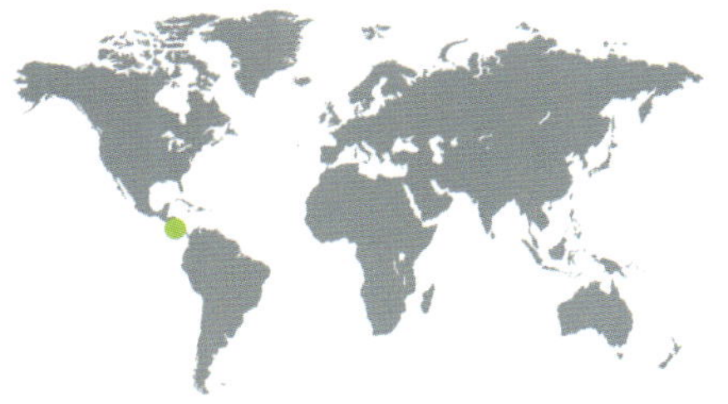

FAMILY	Dendrobatidae: Dendrobatinae
OTHER NAMES	Granular Poison-arrow Frog, Granulated Poison-dart Frog
DISTRIBUTION	Costa Rica, along the Pacific (western) coast and in a small area near the east coast
ADULT HABITAT	Humid lowland forest up to 330 ft (100 m) altitude
LARVAL HABITAT	Tiny pools in the leaf axils of plants
CONSERVATION STATUS	IUCN Vulnerable. Declining due to habitat loss and collecting

ADULT LENGTH
11/16–7/8 in (18–22 mm)

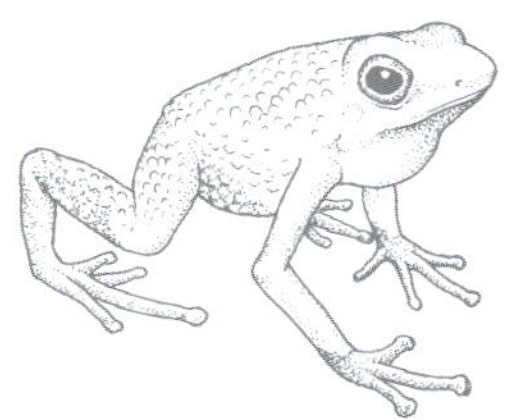

OOPHAGA GRANULIFERA

GRANULAR POISON FROG

(TAYLOR, 1958)

Actual size

The Granular Poison Frog is so called because of the granular skin on its back. Disks are larger on the fingers than the toes. Typical coloration is orange on the back, head, and upper arms, and green on the belly, lower arms, and hind legs. In individuals living in the Quepos region on Costa Rica's Pacific coast, however, the back, head, and upper arms are olive-green rather than red.

This small frog is most commonly found on the forest floor near streams. Males are territorial, defending egg-deposition sites by calling and fighting with intruders. Their call, a series of chirps, is heard most often early and late in the day. When a female enters his territory, the male leads her to a curled-up leaf, where she lays three or four eggs. The male guards them and keeps them wet by urinating on them. The female continues to lay batches of eggs until the first batch hatches. She then carries them, one or two at a time, to the water-filled leaf axils of plants.

SIMILAR SPECIES

In all 12 species of *Oophaga*, the female feeds the tadpoles on unfertilized eggs. The tadpoles, which are usually left singly in tiny pools, solicit feeding from their mother by swimming in a characteristic manner. In addition to nutrients, the eggs contain poisons, so that the developing tadpoles acquire the toxic properties of their mother. See also the Polkadot Poison Frog (*O. arborea*; page 379), Lehmann's Frog (*O. lehmanni*; page 381), and the Strawberry Poison-dart Frog (*O. pumilio*; page 382).

FAMILY	Dendrobatidae: Dendrobatinae
OTHER NAMES	Red-banded Poison Frog
DISTRIBUTION	Eastern Colombia
ADULT HABITAT	Montane forest
LARVAL HABITAT	Tiny pools in the leaf axils of plants, tree-holes, etc.
CONSERVATION STATUS	IUCN Critically Endangered. Known from only two locations, where the habitat is being degraded; also collected for the pet trade

ADULT LENGTH
$1\frac{1}{4}$–$1\frac{7}{16}$ in (31–36 mm)

OOPHAGA LEHMANNI

LEHMANN'S FROG

(MYERS & DALY, 1976)

The striking coloration of this small and very rare frog has made it a target for the international pet trade. Unfortunately, it is a very delicate species and rarely survives for long in captivity. Both parents care for the young. The male guards the eggs and, from time to time, rotates them to ensure that they are adequately oxygenated. The female carries the tadpoles to tiny pools in bromeliads, hollows in trees, and bamboo stems, and then visits them periodically to feed them on unfertilized eggs. The tadpoles are cannibalistic, so she usually places only one tadpole in each pool.

Actual size

Lehmann's Frog has smooth, shiny skin. The fingers and toes are long and end in adhesive disks; these are silver in males. There are three color forms, with two broad red, orange, or yellow bands around the body, against a black or brown background. The arms and legs are also brown or black, with broad colored bands around them.

SIMILAR SPECIES

Oophaga lehmanni has bred successfully, in captivity, with the Harlequin Poison Frog (*O. histrionica*), raising doubt as to whether the two are really distinct species. The Harlequin Poison Frog is extremely variable in color, including forms that have blue, red, and yellow markings, and is found in Colombia and Ecuador. See also the Polkadot Poison Frog (*O. arborea*; page 379), the Granular Poison Frog (*O. granulifera*; page 380), and the Strawberry Poison-dart Frog (*O. pumilio*; page 382).

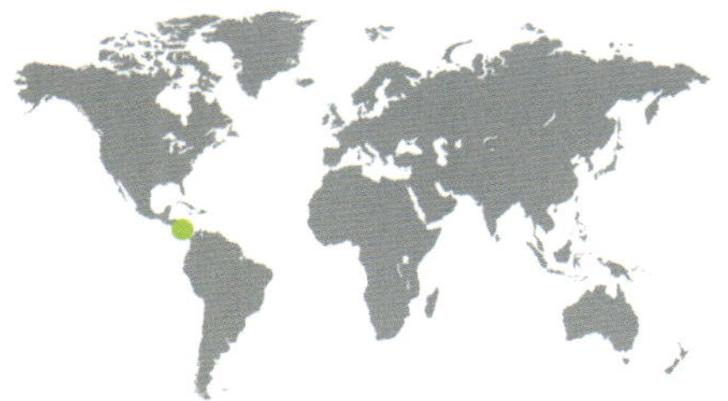

FAMILY	Dendrobatidae: Dendrobatinae
OTHER NAMES	Formerly *Dendrobates pumilio*, Flaming Poison Frog, Red-and-blue Poison Frog, Strawberry Poison Frog
DISTRIBUTION	Costa Rica, Nicaragua, Panama
ADULT HABITAT	Rainforests of the Caribbean coast, up to 3,150 ft (960 m) altitude
LARVAL HABITAT	Tiny water pools in the leaf axils of plants
CONSERVATION STATUS	IUCN Least Concern. Intensively collected for the pet trade

ADULT LENGTH
up to ¹⁵⁄₁₆ in (24 mm)

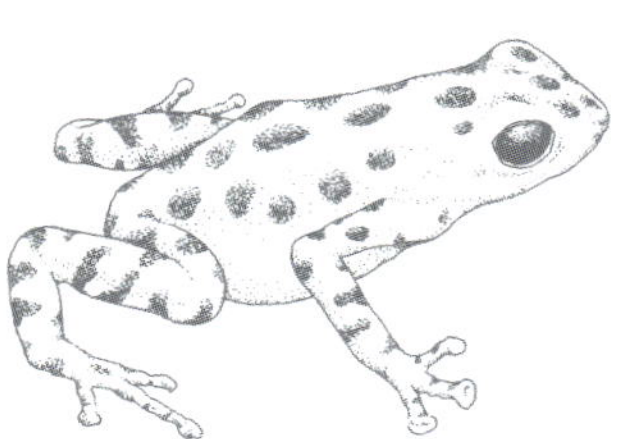

OOPHAGA PUMILIO

STRAWBERRY POISON-DART FROG

(SCHMIDT, 1857)

Actual size

The Strawberry Poison-dart Frog is highly variable in color: no fewer than 30 distinct color morphs have been described. The morph shown here is known as "blue jeans" and is one of the more common forms. Other morphs are all-red, all-blue, and all-green in color. This species acquires its skin poison by eating ants and other toxic insects; captive animals that are not fed on such food are not toxic.

Both sexes of this species play a role in the parental care of their young. Males call from the forest floor to defend a small territory from intrusion by other males and to attract females. A clutch of 3–17 eggs is laid in a hollow in the soil within a male's territory, and he defends them, urinating on them to keep them moist, until they hatch. The female then carries the tadpoles on her back, one or two at a time, to tiny pools of water in the leaf axils of bromeliads and other plants. She then visits them periodically, feeding them with unfertilized eggs.

SIMILAR SPECIES

There are 12 species in the genus *Oophaga*, which means "egg-eating" and refers to the fact that the tadpoles feed exclusively on unfertilized eggs provided by their mother. Should their mother die or forget where she has left them, her tadpoles will starve to death. All species of *Oophaga* are brightly colored, active by day, and are found in Central and South America. See also the Polkadot Poison Frog (*O. arborea*; page 379), the Granular Poison Frog (*O. granulifera*; page 380), and Lehmann's Frog (*O. lehmanni*; page 381).

FAMILY	Dendrobatidae: Dendrobatinae
OTHER NAMES	Black-legged Poison Dart Frog, Two-toned Arrow-poison Frog
DISTRIBUTION	Eastern Colombia up to 5,000 ft (1,500 m) altitude
ADULT HABITAT	Tropical rainforest
LARVAL HABITAT	Pools in streams
CONSERVATION STATUS	IUCN Endangered. Its limited range is subject to deforestation and pollution from illegal crop-spraying

PHYLLOBATES BICOLOR

BLACK-LEGGED POISON FROG

DUMÉRIL & BIBRON, 1841

ADULT LENGTH

Male

1¼–1$^{9}/_{16}$ in (32–40 mm)

Female

1⅜–1$^{11}/_{16}$ in (35–43 mm)

This brightly patterned frog changes color during its life. Juveniles are dark brown or black with two longitudinal yellow stripes, whereas adults have a uniformly yellow back. The species lives on the forest floor, close to streams. The female lays a clutch of 12–20 eggs in a covered nest in the leaf litter. The male guards them and keeps them moist, and, when they hatch, he carries the tadpoles to pools in streams. Found only in Colombia's Cordillera Occidental, the species is threatened by the destruction of its forest habitat to make way for agriculture and this frog occurs in only one protected area.

Actual size

The Black-legged Poison Frog is very variable in color, but typically has a bright yellow or orange back, and black or dark blue flanks and legs, often with yellow spots. The skin is smooth, and the tips of the fingers and toes are slightly widened. The eyes are black.

SIMILAR SPECIES

Phyllobates bicolor is one of only three species that have been documented as being used by humans to poison their hunting darts; the others are the Golden Poison Frog (*P. terribilis*; page 384) and the Kokoe Poison Frog (*P. aurotaenia*). Of the four species in this genus, two are from Costa Rica and Panama: the Lovely Poison Frog (*P. lugubris*) and the Golfodulcean Poison Frog (*P. vittatus*).

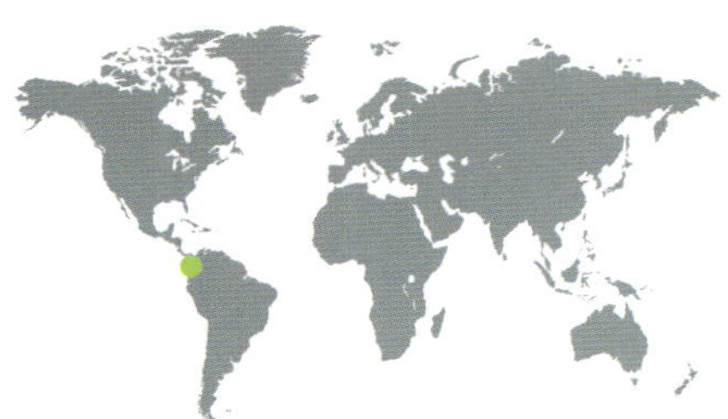

FAMILY	Dendrobatidae: Dendrobatinae
OTHER NAMES	Golden Frog, Golden Poison Dart Frog
DISTRIBUTION	A small area near the Pacific coast of Colombia
ADULT HABITAT	Rainforest
LARVAL HABITAT	Water-filled bromeliad leaf axils and tree-holes
CONSERVATION STATUS	IUCN Endangered. Its very small range is subject to deforestation and pollution from illegal crop-spraying

ADULT LENGTH
Male
1⁷⁄₁₆–1¾ in (37–45 mm)

Female
1⁹⁄₁₆–1⅞ in (40–47 mm)

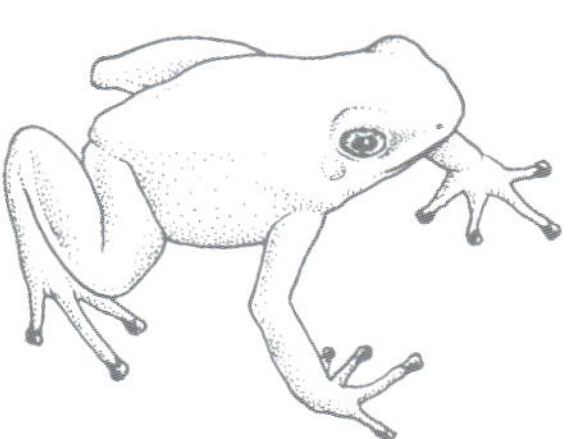

PHYLLOBATES TERRIBILIS

GOLDEN POISON FROG

MYERS, DALY & MALKIN, 1978

Actual size

The Golden Poison Frog is uniformly bright yellow, golden orange, pale metallic green, or white. Individuals in the same locality tend to be similarly colored. The tips of the fingers and toes are black and slightly widened. The eyes are black and some individuals have black patches on one or more parts of their body.

The Golden Poison Frog is generally considered to be the world's most poisonous frog, one individual containing sufficient poison in its skin to kill around 22,000 mice, or ten people. It has long been used by Colombian Indians to make their hunting darts toxic. It derives its poison from its natural insect prey and loses its toxicity when kept in captivity. It is active by day, hopping about on the ground and making no attempt to hide when disturbed. Its only natural enemy is a snake that is immune to its toxin. As in other poison frogs, males care for the eggs and carry tadpoles to tree-holes.

SIMILAR SPECIES

There are six other species in the genus *Phyllobates*, all with a similar life history. The Black-legged Poison Frog (*P. bicolor*; page 383) is a smaller species. It and the Kokoe Poison Frog (*P. aurotaenia*) are less toxic than *P. terribilis* but are also used by Colombian Indians for their poisons. Two new species from Colombia were described in 2024; Samper's Poison Frog (*P. samperi*), and Bezos' Poison Frog (*P. bezosi*), named for the owner of Amazon.

FAMILY	Dendrobatidae: Dendrobatinae
OTHER NAMES	None
DISTRIBUTION	Peru
ADULT HABITAT	Lowland rainforest
LARVAL HABITAT	Water-filled leaf axils of bromeliads
CONSERVATION STATUS	IUCN Vulnerable. Subject to collecting for the pet trade and destruction of its habitat

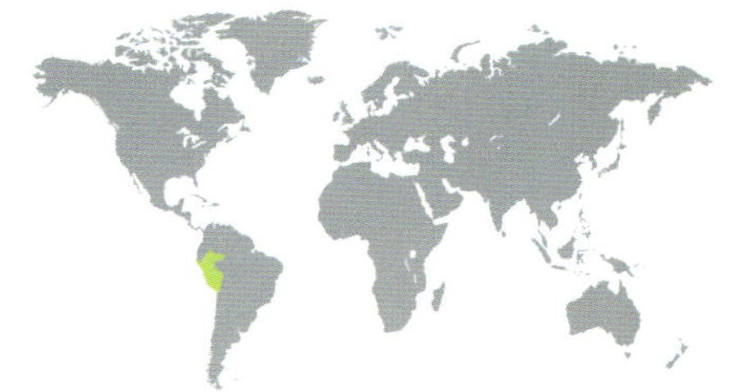

RANITOMEYA BENEDICTA

BLESSED POISON FROG

BROWN, TWOMEY, PEPPER & SANCHEZ RODRIGUEZ, 2008

ADULT LENGTH
Male
$\frac{9}{16}$–$\frac{11}{16}$ in (15–18 mm)
Female
$\frac{11}{16}$–$\frac{3}{4}$ in (17–20 mm)
both based on very small samples

This small, strikingly colored frog is mainly arboreal in its habits, but descends to the forest floor to breed. It is most commonly found where fallen trees have created an open area in the forest. The male attracts females with a frequently repeated buzz-like call. The female lays four to six eggs in the leaf litter and the male guards them. When they hatch, he carries the tadpoles to the water-filled leaf axils of large bromeliads. Recognized as a species as recently as 2008, it is one of many poison frogs living in South America that have been described in the last 20 years.

Actual size

The Blessed Poison Frog has a distinctive bright red head and neck, often with black spots. The body and legs are black or dark blue, often with a reticulated pattern in brilliant pale blue. The pads at the end of the fingers and toes are triangular in shape.

SIMILAR SPECIES

Ranitomeya benedicta was formerly regarded as a color variant of the Crowned Poison Frog (*R. fantastica*). So was Summers' Poison Frog (*R. summersi*), also found in Peru, which is black with a striking pattern of yellow stripes. It is listed as Endangered because, like *R. benedicta*, it is being collected for the international pet trade and its rainforest habitat is being degraded. See also the Imitating Poison Frog (*R. imitator*; page 386).

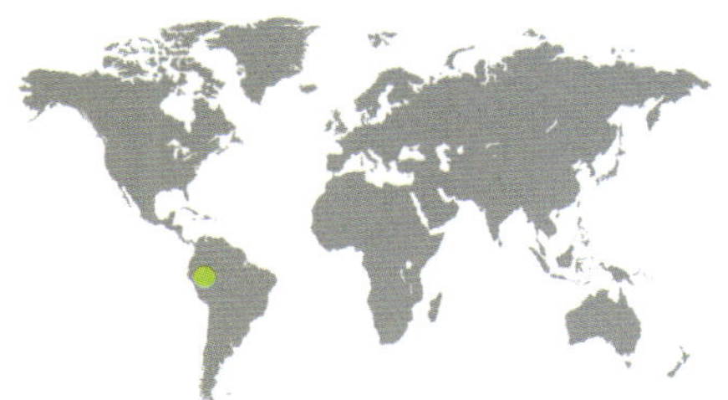

FAMILY	Dendrobatidae: Dendrobatinae
OTHER NAMES	Mimic Poison Frog
DISTRIBUTION	Eastern Peru, in the foothills of the Andes up to 3,300 ft (1,000 m) altitude
ADULT HABITAT	Forest
LARVAL HABITAT	Tiny pools in the leaf axils of plants
CONSERVATION STATUS	IUCN Least Concern. Popular in the pet trade, but this controlled under CITES

ADULT LENGTH
11/16–7/8 in (17–22 mm)

RANITOMEYA IMITATOR

IMITATING POISON FROG

(SCHULTE, 1986)

Actual size

The Imitating Poison Frog is a very small frog with large eyes. It is an excellent jumper and climber, having long, slender legs and expanded tips to its fingers and toes. The skin has a granular look and is very glossy. Coloration is very variable—this specimen has a bright orange body with blue legs and belly—but most commonly consists of yellow or green patterns on a black background.

This small frog is remarkable in two ways. First, it is monogamous, males and females forming stable pair-bonds and together looking after their offspring. Second, it occurs in at least three distinct color morphs in different parts of its range, each of which mimics the coloration of another poison frog species that lives in the same locality. The male's contribution to parental care consists of guarding the eggs until they hatch, carrying tadpoles, one at a time, to tiny pools in the leaf axils of plants, and guarding the tadpoles. The female's role is to feed the developing tadpoles by providing them with unfertilized eggs.

SIMILAR SPECIES

Depending on where it is, *Ranitomeya imitator* almost exactly resembles one of three related species, the Variable Poison Frog (*R. variabilis*), the Crowned Poison Frog (*R. fantastica*), and the Amazonian Poison Frog (*R. ventrimaculata*), but can be distinguished from each by its call. This is an example of Müllerian mimicry, a phenomenon in which all species involved have similar bright coloration that warns potential predators that their skin contains toxic secretions. See also the Blessed Poison Frog (*R. benedicta*; page 385).

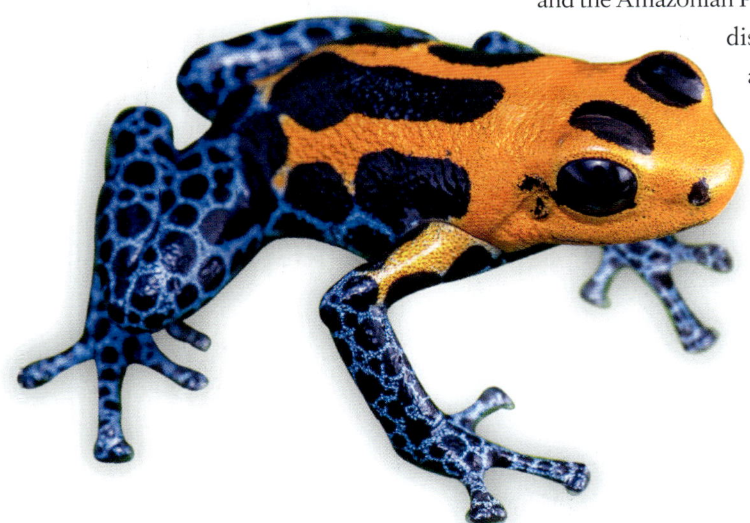

FAMILY	Dendrobatidae: Dendrobatinae
OTHER NAMES	Reticulated Poison Frog
DISTRIBUTION	Eastern Peru, eastern Ecuador
ADULT HABITAT	Lowland rainforest up to 660 ft (200 m) altitude
LARVAL HABITAT	Tiny pools in the leaf axils of plants
CONSERVATION STATUS	IUCN Least Concern. Has a large range, is reasonably common, and is legally protected against being collected

RANITOMEYA RETICULATA

RED-BACKED POISON FROG

(BOULENGER, 1884)

ADULT LENGTH
Male
up to ½ in (12 mm)
Female
up to ¾ in (20 mm)

This colorful frog lives on the ground in the upper Amazon Basin, occasionally climbing up tree trunks. Males attract females by calling, and then court them by stroking and licking them. If a female is ready to mate, she stamps her feet. The eggs are laid on the forest floor and, when they hatch, the male carries the tadpoles to tiny pools, called phytotelmata, in the leaf axils of bromeliads and other plants. The female visits the tadpoles at intervals to feed them with unfertilized eggs. The frog's bright coloration signals to potential predators that its skin contains toxic secretions.

Actual size

The Red-backed Poison Frog has a slender body, large black eyes, and well-developed adhesive disks on its fingers and toes. The lower back, flanks, arms, and legs are black, decorated with a white or sky-blue mesh-like pattern. The head and the front part of the back are vivid orange or red.

SIMILAR SPECIES

Several of the 16 species in the genus *Ranitomeya* described to date are brightly colored and toxic. Zimmermann's Poison Frog (*R. variabilis*), from Peru and Ecuador, is a very toxic species with a complex green, yellow, and black skin pattern. The Spotted Poison Frog (*R. vanzolinii;* page 388), from Brazil and Peru, is black with yellow spots and blue legs. See also the Blessed Poison Frog (*R. benedicta*; page 385) and the Imitating Poison Frog (*R. imitator*; page 386).

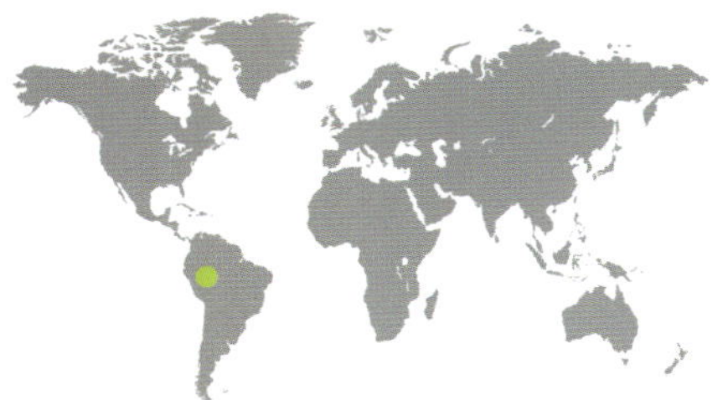

FAMILY	Dendrobatidae: Dendrobatinae
OTHER NAMES	Brazilian Poison Frog
DISTRIBUTION	Eastern Peru, southwestern Brazil
ADULT HABITAT	Lowland rainforest up to 4,300 ft (1,300 m) altitude
LARVAL HABITAT	Tiny pools in tree-holes
CONSERVATION STATUS	IUCN Least Concern. Abundant in many places but there is concern about the potential impact on the species of collecting for the pet trade

ADULT LENGTH
Male
average 11/16 in (17 mm)

Female
average ¾ in (19 mm)

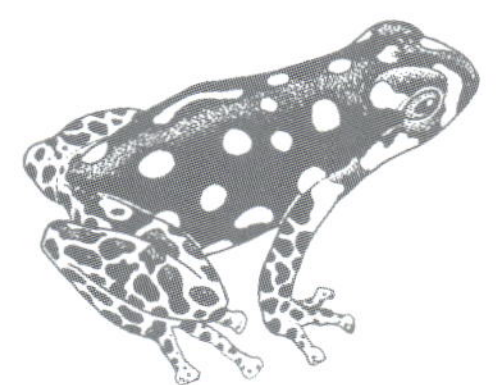

RANITOMEYA VANZOLINII

SPOTTED POISON FROG

(MYERS, 1982)

Actual size

The Spotted Poison Frog has a narrow head, very smooth and shiny skin, and well-developed adhesive disks on its fingers and toes. The back is black, decorated with pale yellow to orange round or elongated spots. The long hind limbs and the arms are black, decorated with a complex blue reticulated pattern.

This small inhabitant of the Amazon rainforest is the first frog found to show a monogamous mating pattern in which both sexes care for the young. Males defend territories by calling and, during breeding, females remain within their mate's territory. The pair mate in a tiny water-filled tree-hole, usually in a sapling or vine, producing one or a small number of eggs. When the eggs hatch, the male carries each tadpole to a new tree-hole, so that only one tadpole occupies each hole. The female visits the tadpoles, on average every five days, laying one or two unfertilized eggs on which the tadpoles feed.

SIMILAR SPECIES

The striking spotted pattern on its back distinguishes *Ranitomeya vanzolinii* from other members of the genus. Several little-known species occur in Peru, including *R. amazonica*, which has a complex pattern of yellow, orange, or red stripes on its black back. See also the Blessed Poison Frog (*R. benedicta*; page 385), the Imitating Poison Frog (*R. imitator*; page 386), and the Red-backed Poison Frog (*R. reticulata*; page 387).

FAMILY	Dendrobatidae: Hyloxalinae
OTHER NAMES	None
DISTRIBUTION	Peru
ADULT HABITAT	Lowland rainforest
LARVAL HABITAT	Water-filled hollows, then streams
CONSERVATION STATUS	IUCN Endangered. Declining in its restricted range due to habitat loss and illegal collecting

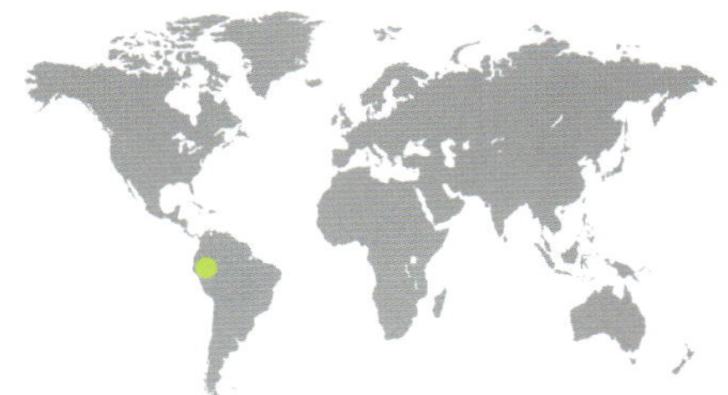

ADULT LENGTH
up to 1 1/16 in (27 mm)

HYLOXALUS AZUREIVENTRIS

SKY BLUE POISON DART FROG

(KNELLER & HENLE, 1985)

This small, secretive poison frog lives in a very restricted area on the eastern slopes of the Peruvian Andes, where its forest habitat is being destroyed to make way for agriculture. A ground-living species, it typically breeds among piles of rocks. The male attracts a female by calling "vee… vee… vit" and then grasping her around the head. The female lays 12–16 eggs in water-filled coconut shells, hollows in leaf litter, or bromeliads. The male cares for the eggs for around two weeks before carrying the tadpoles to streams.

SIMILAR SPECIES

There are 64 species in the genus *Hyloxalus*; all are small frogs with a similar life history. The Los Tayos Rocket Frog (*H. nexipus;* page 390) lives at intermediate elevations in the mountains of Peru and Ecuador, and the Forest Rocket Frog (*H. sylvaticus*) in the high-altitude cloud forest of Peru. The Santiago Rocket Frog (*H. shuar*), from Ecuador, has tested positive for the fungus that causes chytridiomycosis, which may be a threat to species in this genus.

Actual size

The Sky Blue Poison Dart Frog has glossy black skin. The body is decorated with longitudinal stripes in yellow or orange, and the legs with a complex pattern of dots in bright sky-blue or turquoise-green. The fingers and toes have slightly broadened tips.

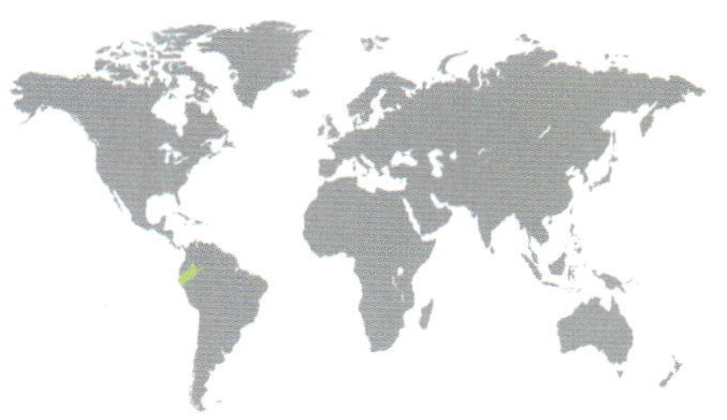

FAMILY	Dendrobatidae: Hyloxalinae
OTHER NAMES	Limon Rocket Frog
DISTRIBUTION	Southern Ecuador, northern Peru
ADULT HABITAT	Rainforest at altitudes of 1,600–5,100 ft (500–1,550 m)
LARVAL HABITAT	Streams
CONSERVATION STATUS	IUCN Least Concern. A common species with a stable population and no known threats

ADULT LENGTH
Male
up to 1 3/16 in (30 mm)
Female
up to 1 3/8 in (35 mm)

HYLOXALUS NEXIPUS

LOS TAYOS ROCKET FROG

(FROST, 1986)

Actual size

The Los Tayos Rocket Frog has a slender body, a pointed snout, and well-developed disks on its fingers and toes. The back and flanks are dark brown or black, and there is a red, orange, or yellow stripe running from the tip of the snout and along the flank. The arms and legs are blue-gray with dark gray stripes. Shown here is a male carrying tadpoles on his back.

This colorful frog is found on the forested eastern slopes of the Andes and in the westernmost parts of the Amazon Basin. It is active by day and lives in close association with rocky streams. Males defend territories in streams, jumping onto the top of rocks to call. Rival males sometimes jump up and wrestle with resident males, trying to displace them. The eggs are laid on the ground near a stream. When they hatch, the male carries the tadpoles, several at a time, to a quiet pool within the stream, where they complete their development.

SIMILAR SPECIES

Once a common species in southern Ecuador, the South American Rocket Frog (*Hyloxalus anthracinus*) has not been seen since 1991 and is listed as Critically Endangered. Lehmann's Rocket Frog (*H. lehmanni*) is a high-altitude species from Colombia and Ecuador, and is categorized as Near Threatened. It is thought that both species have been affected by the disease chytridiomycosis. See also the Sky Blue Poison Dart Frog (*H. azureiventris*; page 389).

FAMILY	Brachycephalidae
OTHER NAMES	None
DISTRIBUTION	Southeastern Brazil
ADULT HABITAT	Montane coastal forest at altitudes of 2,460–3,940 ft (750–1,200 m)
LARVAL HABITAT	Within the egg
CONSERVATION STATUS	IUCN Least Concern. Has a large range, including extensive protected areas

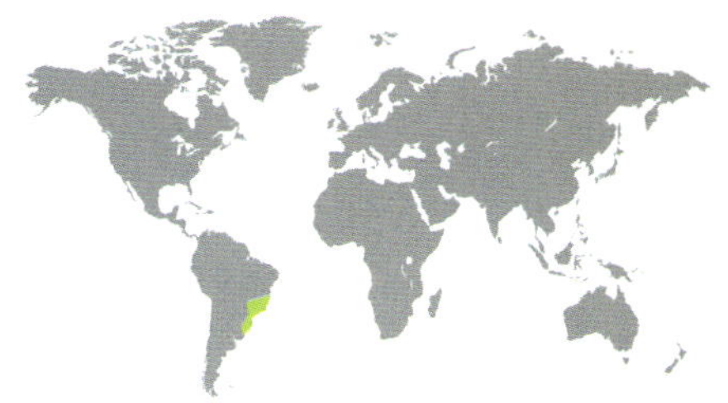

ADULT LENGTH
½–¾ in (12–20 mm)

BRACHYCEPHALUS EPHIPPIUM

PUMPKIN TOADLET

(SPIX, 1824)

This plump little frog lives on the forest floor and becomes active by day in the rainy season. Males defend territories by calling (a rather quiet buzz), and by waving a hand up and down in front of their face while adopting an upright posture. Territorial disputes occasionally escalate to wrestling. During mating, the female carries the male to a spot in the leaf litter, where she lays up to five eggs. After the male has fertilized them, she rolls them in the dirt so as to camouflage them and then leaves them to develop. Development is direct, the eggs hatching to release tiny froglets.

Actual size

The Pumpkin Toadlet is so called because of its spherical shape and bright yellow or orange color. It has a stout body and very short legs, and only three fingers and three toes. The iris is black.

SIMILAR SPECIES

There are 42 species in the genus *Brachycephalus*, all occurring in Brazil. They are also called saddle-back toads because they have a bony shield above the vertebrae. Most are bright red, orange, or yellow, and are probably toxic, as *B. ephippium* is; its skin contains tetrodotoxin, a toxin it shares with puffer fish and the Californian Newt (*Taricha torosa*). The Flea-frog (*B. didactylus*) is a candidate for the title of world's smallest frog, being only ⅜ in (10 mm) in length.

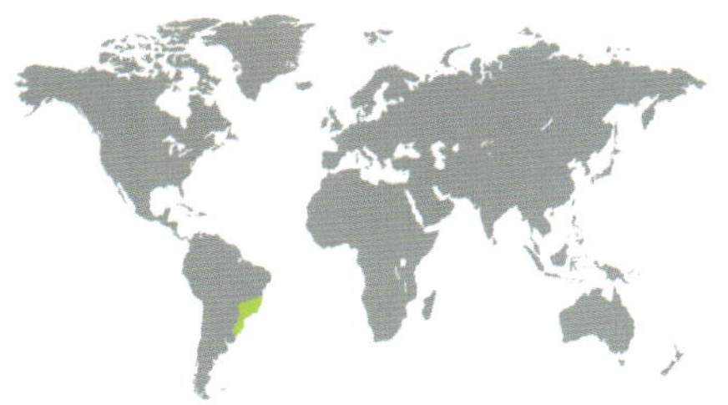

FAMILY	Brachycephalidae
OTHER NAMES	None
DISTRIBUTION	Southeastern Brazil
ADULT HABITAT	Rainforest at altitudes above 4,000 ft (1,200 m)
LARVAL HABITAT	Within the egg
CONSERVATION STATUS	IUCN Near Threatened. Has a large range and shows no evidence of a general decline

ADULT LENGTH
Male
⅝–1⁵⁄₁₆ in (16–34 mm)

Female
1⅛–1¹³⁄₁₆ in (28–46 mm)

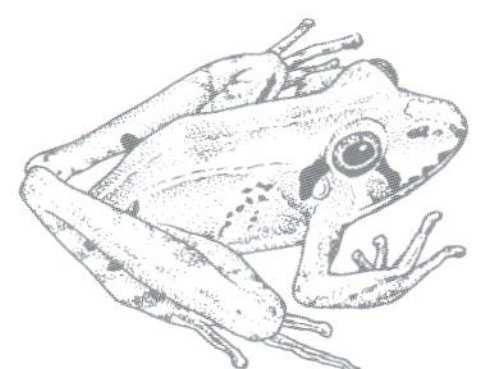

ISCHNOCNEMA GUENTHERI

STEINDACHNER'S ROBBER FROG

(STEINDACHNER, 1864)

Actual size

Found in Brazil's Atlantic Forest, a biodiversity hotspot, this small frog lives on the forest floor, hopping about among the leaf litter. Males call sporadically to attract females, their call starting quietly and becoming louder. The female lays a clutch of 20–30 eggs in a small cave-like hollow in the ground. There are no free-living tadpoles as development is direct, the eggs hatching to release tiny froglets. When this frog is disturbed, its response is to feign death.

SIMILAR SPECIES

One of 39 species described to date in this genus, and found in eastern Brazil and northeastern Argentina, *Ischnocnema guentheri* is thought to be a complex of several species. Little is known about the biology of the genus as a whole, and it may well be that there are several new species yet to be described.

Steindachner's Robber Frog has a long snout and large, prominent eyes. It has long, slender hind legs, and toes tipped with large disks. The tips of the fingers are slightly dilated. The back is brown, brick-red, green, or cream with darker patches. There are dark stripes on the legs and there is often a broken, dark eye-stripe.

FAMILY	Ceuthomantidae
OTHER NAMES	None
DISTRIBUTION	Guyana
ADULT HABITAT	Cloud forest at altitudes of 4,890–5,050 ft (1,490–1,540 m)
LARVAL HABITAT	Unknown
CONSERVATION STATUS	IUCN Vulnerable

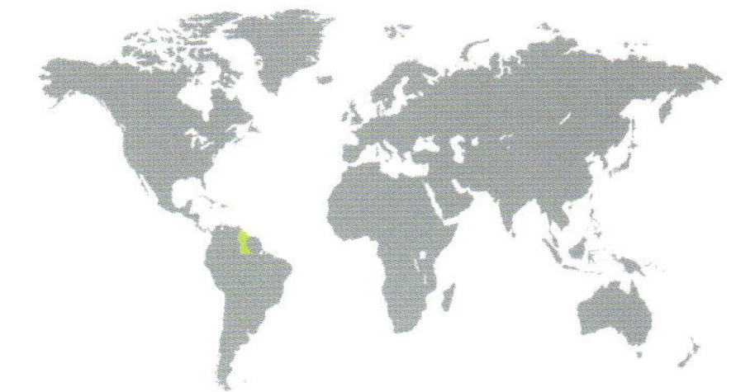

ADULT LENGTH
around ¾ in (20 mm); based on a single specimen of each sex

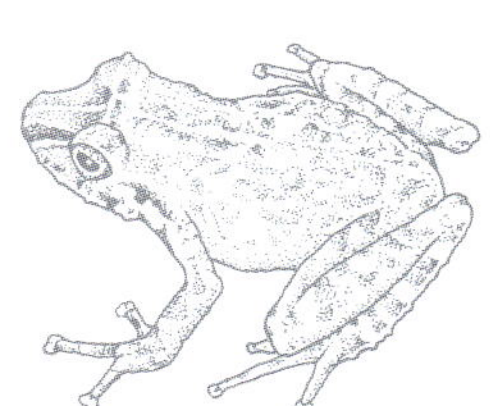

CEUTHOMANTIS SMARAGDINUS

KAMANA FALLS EMERALD-BARRED FROG

HEINICKE, DUELLMAN, TRUEB, MEANS, MACCOLLOCH & HEDGES, 2009

This unusual species represents a small family of frogs that was "discovered," using genetic techniques, as recently as 2009. The Guiana Shield is an ancient geological formation in northeastern South America, the higher mountains of which are known as the Guiana Highlands. These stretch from eastern Venezuela to northern Brazil and western Guyana, and are where the distinctive table-like mountains known as tepuis are found. These "islands in the sky" are home to a number of plants and animals of ancient origin that have remained isolated for millions of years, including this species. It is assumed that, like its closest relatives, this frog has direct development, the eggs hatching to produce tiny froglets.

Actual size

The Kamana Falls Emerald-barred Frog has a long snout, a spiny wart on each eyelid, and distinctive notched disks on its fingers. There are paired protrusions of unknown function on the shoulders and lower back. The upper surfaces are olive-green with black blotches and bright green stripes, one of them between the eyes.

SIMILAR SPECIES

There are three other species in this recently described genus, all of which can be found in the Guiana Highlands: Cerro Aracamuni Emerald-barred Frog (*Ceuthomantis aracamuni*) and Duellman's Emerald-barred Frog (*C. duellmani*), both from Venezuela, and the Cave-singing Frog (*Ceuthomantis cavernibardus*), from Brazil and Venezuela, the only species not listed as Vulnerable or Near Threatened.

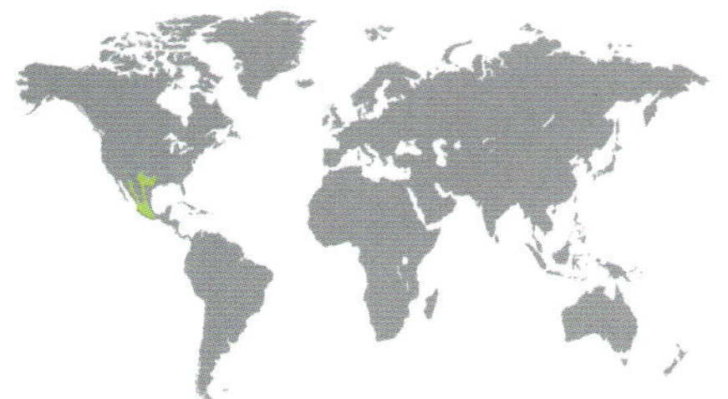

FAMILY	Craugastoridae
OTHER NAMES	None
DISTRIBUTION	Mexico; Arizona and Texas, USA
ADULT HABITAT	Woodland, bushland, near cliffs in desert, caves, rocky outcrops
LARVAL HABITAT	Within the egg
CONSERVATION STATUS	IUCN Least Concern. Shows no evidence of decline in any part of its range

ADULT LENGTH
Male
1⅞–3⅜ in (47–87 mm)
Female
2⅛–3⅝ in (54–94 mm)

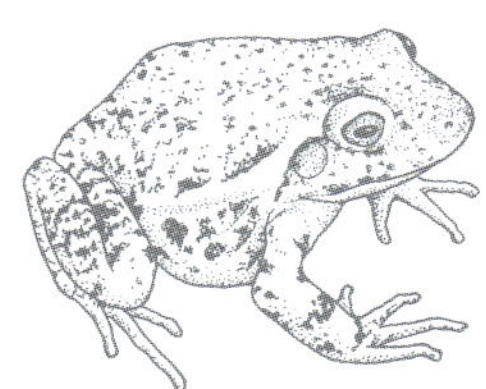

CRAUGASTOR AUGUSTI

BARKING FROG

(DUGÈS, 1879)

This terrestrial frog gets its name from the male's call, likened to the barking of a dog. A secretive species that is hard to find, it breeds during periods of rain any time between February and August, but most commonly in April and May. Males call from rock crevices and lead females to rain-filled cracks and other damp places, where 50–76 eggs are laid. This species shows direct development, the larval stage being completed within the eggs, which hatch to release tiny froglets. It is thought that females remain with their eggs, keeping them moist by urinating on them while they develop.

The Barking Frog has a plump body and large, protuberant eyes. A fold of skin runs across the back of the head and extends backward part of the way along the flanks. There is no webbing between the fingers or toes. The back is yellow, gray, green, or brown with dark brown patches. There are large tubercles on the soles of the feet.

SIMILAR SPECIES

There are 122 species in the genus *Craugastor*, known as northern rain frogs. The Guatemala Robber Frog (*C. brocchi*), found in Guatemala and Mexico, is listed as Vulnerable. McCranie's Robber Frog (*C. chrysozetetes*), once found in Honduras, is listed as Critically Endangered and may be Extinct, mainly as a result of deforestation. See also Fitzinger's Robber Frog (*C. fitzingeri*; page 395) and the Broad-headed Rain Frog (*C. megacephalus*; page 396).

Actual size

FAMILY	Craugastoridae
OTHER NAMES	None
DISTRIBUTION	Central and South America, from Honduras in the north to Colombia in the south
ADULT HABITAT	Forest from sea level to 3,900 ft (1,200 m) altitude
LARVAL HABITAT	Within the egg
CONSERVATION STATUS	IUCN Least Concern. Shows no evidence of decline in any part of its range

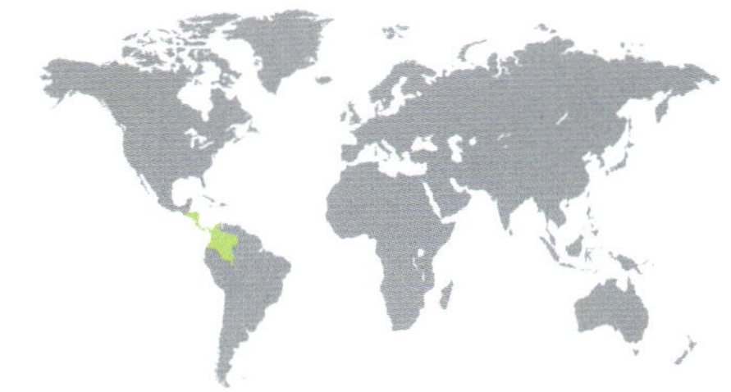

CRAUGASTOR FITZINGERI

FITZINGER'S ROBBER FROG

(SCHMIDT, 1857)

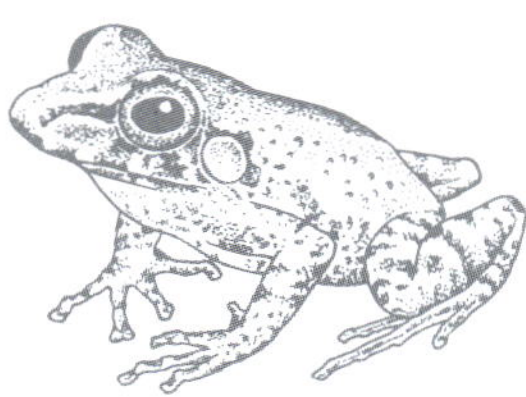

ADULT LENGTH
Male
1⅞–3⅜ in (47–87 mm)
Female
2⅛–3⅝ in (54–94 mm)

Like other species in this family, this frog does not have free-swimming tadpoles. Development is direct, being completed within the egg, from which the young emerge as tiny froglets. The female lays 20–85 eggs in a cavity in the ground and then guards them until they hatch. An abundant and widespread species, the frog is active by night, in bushes and trees, and hides by day in the leaf litter. The call made by males sounds like two stones being struck together; because it attracts the attentions of the frog-eating Fringe-lipped Bat (*Trachops cirrhosus*), males call only sporadically.

SIMILAR SPECIES

The Isla Bonita Robber Frog (*Craugastor crassidigitus*) is found in Costa Rica, Panama, and Colombia. It is a common species and is found in coffee plantations, indicating that it is able to thrive in altered habitats. See also the Barking Frog (*C. augusti*; page 394) and the Broad-headed Rain Frog (*C. megacephalus*; page 396).

Fitzinger's Robber Frog is a slender and long-legged frog, with partially webbed feet and adhesive disks on its fingers and toes. The upper surfaces are brown or tan with scattered black spots, and there are brown or yellow spots on the back of the thighs. Many individuals have a broad, pale stripe running down the middle of the head and back.

Actual size

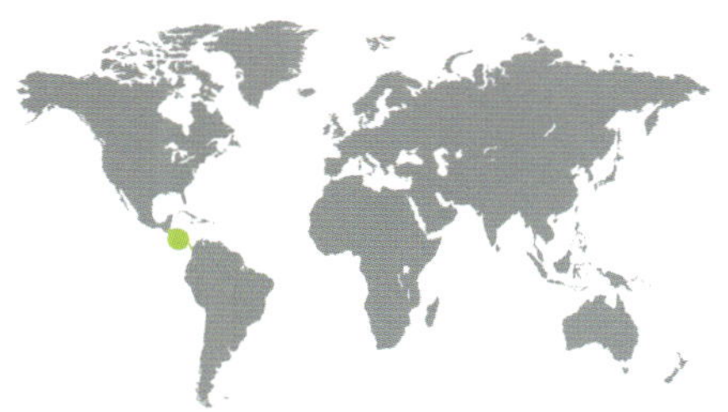

FAMILY	Craugastoridae
OTHER NAMES	None
DISTRIBUTION	Central America, from Honduras in the north to Panama in the south
ADULT HABITAT	Forest, from sea level to 4,000 ft (1,200 m) altitude
LARVAL HABITAT	Within the egg
CONSERVATION STATUS	IUCN Least Concern. Has declined in parts of its range due to chytridiomycosis

ADULT LENGTH
Male
1³⁄₁₆–1¹¹⁄₁₆ in (30–43 mm)
Female
1¹⁵⁄₁₆–2¹³⁄₁₆ in (50–70 mm)

CRAUGASTOR MEGACEPHALUS

BROAD-HEADED RAIN FROG

(COPE, 1875)

This frog lives on the ground, among leaf litter. It spends the day within a burrow made by another animal and at night sits at the entrance to its burrow, waiting to ambush passing prey. It feeds mostly on insects, but larger individuals can also eat small frogs and lizards. Males are apparently mute and the species' mating behavior has not been described. The fungus causing chytridiomycosis has been found in this species and it is likely that marked declines in parts of its range are due to the disease.

SIMILAR SPECIES

The Veragua Robber Frog (*C. rugosus*) is found in Costa Rica and Panama. It is a large frog with very rough, wrinkled skin. *Craugastor opimus* is a large, rather rare frog found in Panama and Colombia. See also the Barking Frog (*C. augusti*; page 394) and Fitzinger's Robber Frog (*C. fitzingeri*; page 395).

Actual size

The Broad-headed Rain Frog gets its name from its relatively large, broad head. There are ridges on its back that make an hourglass pattern. Its upper surfaces are gray, tan, or olive in color, sometimes tinged with salmon-pink. The underside of the body and the legs is red or orange with a complex, dark brown pattern.

FAMILY	Craugastoridae
OTHER NAMES	None
DISTRIBUTION	Western and northwestern Honduras
ADULT HABITAT	Montane forest at altitudes of 3,450–5,650 ft (1,050–1,720 m)
LARVAL HABITAT	Within the egg
CONSERVATION STATUS	IUCN Critically Endangered. Much of its forest habitat has been destroyed and it has declined in pristine areas due to chytridiomycosis

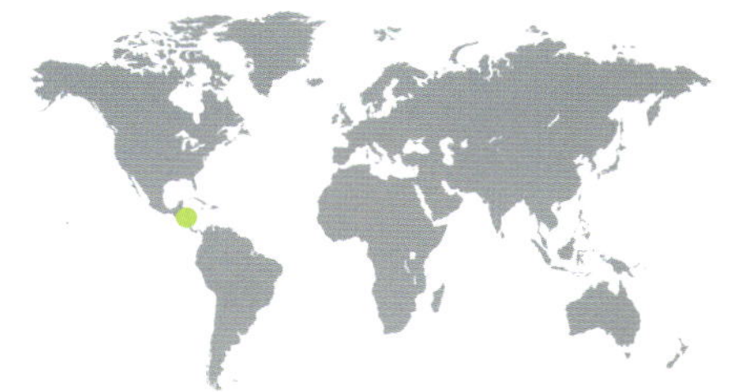

CRAUGASTOR MILESI

MILES' ROBBER FROG

(SCHMIDT, 1933)

ADULT LENGTH
Male
¾–1 in (19–26 mm)
Female
1–1 7/16 in (25–37 mm)

This small frog is typically found in leaf litter close to streams. Once a common species, its numbers declined dramatically in the 1980s. It was declared extinct in 2004, having not been seen since 1983 but, in 2008, a single living specimen was discovered in the Parque Nacional Cusuco. It is one of many species in Central America, especially frogs breeding in high-altitude streams, that have declined in apparently pristine habitats as a result of the spread of the disease chytridiomycosis.

Actual size

SIMILAR SPECIES

The Rio Viejo Robber Frog (*Craugastor aurilegulus*) is a Vulnerable species, found beside streams in cloud forest in northern Honduras. It has largely disappeared from higher, but not lower, altitudes, a pattern common among species affected by chytridiomycosis. See also the Barking Frog (*C. augusti*; page 394), Fitzinger's Robber Frog (*C. fitzingeri*; page 395) and the Broad-headed Rain Frog (*C. megacephalus*; page 396).

Miles' Robber Frog has a rounded snout and small disks at the tips of its long fingers and toes. Its upper surfaces are olive-green in color, with irregular darker markings, and there are dark cross-stripes on the hind legs. The underside of the body and legs is yellow with darker flecks, and there is an orange patch in the groin and inside of the thighs.

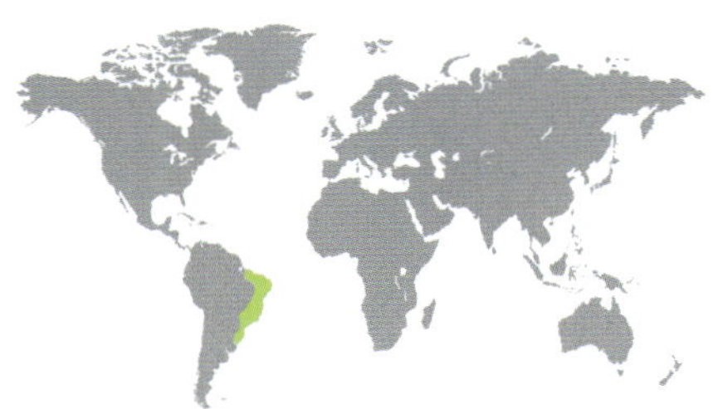

FAMILY	Craugastoridae
OTHER NAMES	None
DISTRIBUTION	Eastern and southeastern Brazil
ADULT HABITAT	Forest
LARVAL HABITAT	Within the egg
CONSERVATION STATUS	IUCN Least Concern. Show no evidence of decline across its large range

ADULT LENGTH
Male
1¼–1¾ in (32–45 mm)
Female
1¾–2⁹⁄₁₆ in (44–64 mm)

HADDADUS BINOTATUS

CLAY ROBBER FROG

(SPIX, 1824)

Actual size

Found in Brazil's Atlantic Forest, this species is common but little known. It is usually found in leaf litter on the forest floor but moves up into low foliage at night. It breeds at the start of the rainy season, males calling from bushes and producing a short, low-pitched call. Like other species in the Craugastoridae family, its eggs develop directly, hatching to release tiny froglets. It feeds on spiders and insects but, unlike many similar frogs, appears to avoid ants.

SIMILAR SPECIES

There are currently three species in the genus *Haddadus*, all of which are found in the Atlantic Forest in eastern Brazil. They are little known and it is possible that *H. binotatus* is a complex of several species. The Iguarasse Robber Frog (*H. plicifer*) was last collected 100 years ago; as there is no available information on its population status it is listed as Data Deficient. The recently described Giant Robber Frog (*H. aramunha*), from Bahia, Brazil, is Vulnerable.

The Clay Robber Frog has a somewhat pointed snout and long, slender legs. Its hands and feet are not webbed but do have tubercles on their undersides. It is yellow, ocher, reddish brown, or brown in color, with variable numbers of dark spots. Many individuals have a dark stripe through the eye.

FAMILY	Eleutherodactylidae: Eleutherodactylinae
OTHER NAMES	Yellow Dink Frog
DISTRIBUTION	Western Panama
ADULT HABITAT	Forest at altitudes of 2,230–2,590 ft (680–790 m)
LARVAL HABITAT	Within the egg
CONSERVATION STATUS	IUCN Least Concern

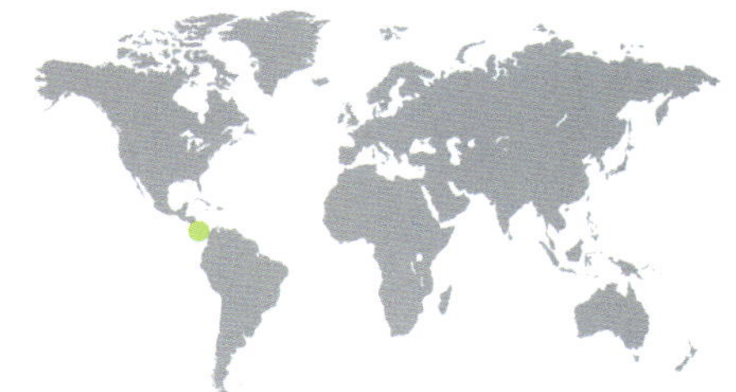

DIASPORUS CITRINOBAPHEUS

YELLOW DYER RAIN FROG

HERTZ, HAUENSCHILD, LOTZKAT & KÖHLER, 2012

ADULT LENGTH
Male
$^{11}/_{16}$–$^{3}/_{4}$ in (17–20 mm)

Female
$^{7}/_{8}$ in (22 mm);
based on one specimen

This tiny brightly colored frog gets its name from the fact that it leaves a yellow stain on the fingers when handled. The function of this skin secretion is unknown; it appears not to be toxic. This species was discovered as recently as 2012, and the first clue to its existence came from its mating call, which is distinctive. It hides in very dense vegetation and is thus rarely seen. It is presumed that, like other frogs in the family Eleutherodactylidae, its eggs develop directly, hatching to produce tiny froglets. Found in only a few locations, it is likely to be under serious threat from deforestation.

Actual size

The Yellow Dyer Rain Frog is one of the world's smallest frogs. It has large eyes and adhesive disks on its fingers and toes. It is yellow or orange in color, and some individuals have brown markings, particularly in the form of a stripe on the top of the head between the eyes.

SIMILAR SPECIES

There are 17 species described to date in the genus *Diasporus*, all found in Central America and northern South America. Very little is known about them. The Esmeraldas Robber Frog (*D. gularis*) is common in western Colombia and northwestern Ecuador. The Tigrillo Dink Frog (*Diasporus tigrillo*), known only from two male specimens, has been found in just one valley in Costa Rica, and listed as Near Threatened.

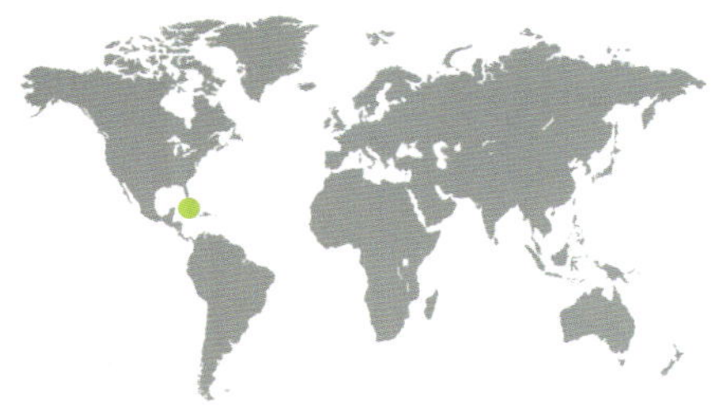

FAMILY	Eleutherodactylidae: Eleutherodactylinae
OTHER NAMES	Cuban Red-rumped Frog
DISTRIBUTION	Eastern Cuba
ADULT HABITAT	Humid forest at altitudes of 100–3,800 ft (30–1,150 m)
LARVAL HABITAT	Within the egg
CONSERVATION STATUS	IUCN Critically Endangered. A rare species, threatened by habitat degradation, although it does occur in some protected areas

ADULT LENGTH
Male
up to ⅞ in (21 mm)
Female
up to 1 in (25 mm)

ELEUTHERODACTYLUS ACMONIS

EL YUNQUE ROBBER FROG

SCHWARTZ, 1960

Actual size

The El Yunque Robber Frog has a slender body, long, slender limbs, and large eyes. The anterior part of the body and the arms are yellow or tan, decorated with a complex brown or black marbled pattern. The posterior body and the hind legs are orange or reddish brown. The hands and feet are pale gray and the underside of the body is brown.

This small, colorful frog is found in two separate areas in Cuba's Guantánamo province. It is an uncommon frog, even in areas of suitable habitat. It is found on the ground, in rock crevices, and occasionally on low vegetation. The male's mating call is a series of trills. The frog breeds by direct development, with its eggs, laid on the ground, hatching to release young frogs. Its forest habitat is threatened by destruction and degradation through charcoal manufacture, subsistence farming, and expanding tourism.

SIMILAR SPECIES

Eleutherodactylus acmonis is one of 58 species in this genus that occur in Cuba, many of which are threatened with extinction due to the destruction of the forests on the island. Ricord's Robber Frog (*E. ricordii*) and the Rio Yumuri Robber Frog (*E. bresslerae*) both occur in eastern Cuba. The former is listed as Vulnerable and the latter as Critically Endangered.

FAMILY	Eleutherodactylidae: Eleutherodactylinae
OTHER NAMES	Puerto Rican Rock Frog
DISTRIBUTION	Southeastern Puerto Rico
ADULT HABITAT	Crevices and caves in rocky areas
LARVAL HABITAT	Within the egg
CONSERVATION STATUS	IUCN Endangered. Has a very restricted range that is subject to habitat destruction

ELEUTHERODACTYLUS COOKI

COOK'S ROBBER FROG

GRANT, 1932

ADULT LENGTH
Male
average $1^{11}/_{16}$ in (43 mm)
Female
average 2 in (51 mm)

This very rare frog has enormous eyes, an adaptation that enables it to see in its cave habitat. Males guard the developing eggs, which are laid in clutches of around 16 on damp rock surfaces. While many males have just a single clutch to defend, some have two or three, and a very few have four. Males vary in the extent of yellow coloration under their chins, and it appears that females prefer to mate with, and leave their eggs in the care of, yellower males. The species' habitat is very restricted and is under threat from urban development.

SIMILAR SPECIES

There are 221 species currently listed in the genus *Eleutherodactylus*, many of which occur on Caribbean islands. The Martinique Robber Frog (*E. martinicensis*), for example, occurs on a number of islands. It is listed as Near Threatened. It has become extinct on St. Lucia through habitat loss and predation by introduced rats, cats, and mongooses.

Cook's Robber Frog has a slender body, long, slender limbs, and large, protuberant eyes. The upper surfaces are uniformly pale reddish brown or greenish brown. There are large adhesive disks on the fingers and toes. The underside is whitish, with varying amounts of yellow under the chin and chest in males.

Actual size

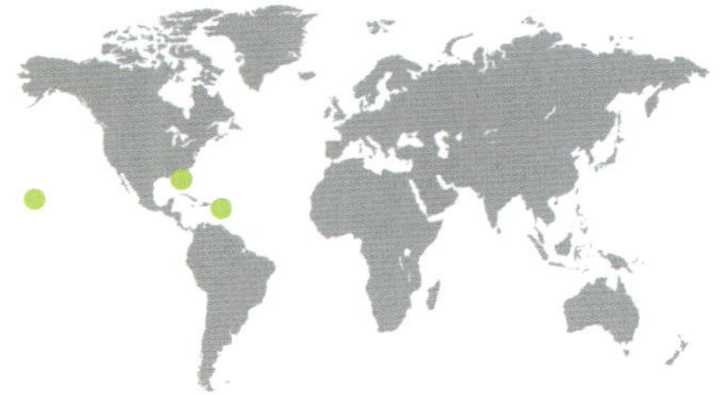

FAMILY	Eleutherodactylidae: Eleutherodactylinae
OTHER NAMES	Coqui Frog
DISTRIBUTION	Puerto Rico. Introduced to the Virgin Islands, Dominica; and Florida and Hawaii, USA
ADULT HABITAT	All kinds of habitat, including urban areas
LARVAL HABITAT	Within the egg
CONSERVATION STATUS	IUCN Least Concern. Common in its native Puerto Rico and an invasive species elsewhere

ADULT LENGTH
Male
1³⁄₁₆–1⁷⁄₁₆ in (30–37 mm)
Female
1⁷⁄₁₆–2 in (36–52 mm)

ELEUTHERODACTYLUS COQUI

PUERTO RICAN COQUI

THOMAS, 1966

Actual size

This frog gets its name from the male's call, which consists of a low-pitched "co," followed by a high-pitched "qui" (pronounced "kee"). The sensitivities of male and female ears differ, so that males hear the "co," a territorial signal, and females hear the "qui," a sexual signal. The male leads the female to a suitable egg-laying site, such as a rolled-up leaf. This species is highly unusual in having internal fertilization. During mating, a pair adopt a reverse hind-leg clasp, in which the male's cloaca is pressed against the female's. Males defend the eggs until they hatch to produce tiny froglets, which are just ¼ in (6 mm) long.

SIMILAR SPECIES

Johnstone's Whistling Frog (*Eleutherodactylus johnstonei*) occurs on many West Indian islands. It is unusual in that either sex, but not both, may care for the eggs. Like *E. coqui*, it has been introduced to new locations as a "stowaway," most likely in consignments of plants. It has not, however, become a serious pest, as *E. coqui* has on the four main Hawaiian islands.

The Puerto Rican Coqui has a rather plump body, and fingers and toes that are not webbed but are tipped with adhesive disks. Its eyes are large and the tympana are conspicuous. The upper surfaces are uniformly gray or grayish brown, and there is a dark stripe through the eye.

FAMILY	Eleutherodactylidae: Eleutherodactylinae
OTHER NAMES	Jamaican Rock Frog
DISTRIBUTION	Western Jamaica
ADULT HABITAT	Forest up to 2,100 ft (635 m) altitude
LARVAL HABITAT	Within the egg
CONSERVATION STATUS	IUCN Vulnerable. Breeds in limestone caves that are subject to mining, quarrying, and tourism

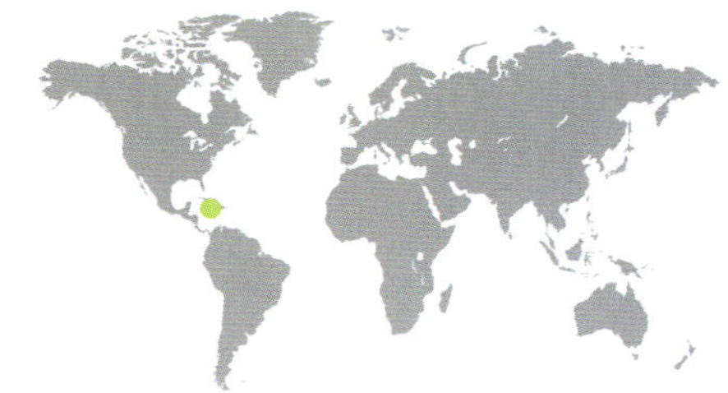

ELEUTHERODACTYLUS CUNDALLI

CUNDALL'S ROBBER FROG

DUNN, 1926

ADULT LENGTH
Male
$^{7}/_{16}$–$1^{7}/_{16}$ in (11–36 mm)
Female
average $2^{3}/_{8}$ in (60 mm)

This frog has unique reproductive habits. A forest-dwelling species, it moves into deep limestone caves to breed, going as far as 285 ft (87 m) from the cave entrance. Males are territorial and call from rock outcrops. The eggs are laid on the cave floor and are attended by the female until they hatch into tiny froglets after more than 30 days. The froglets then climb onto their mother's back and she carries them, as many as 72 at a time, out into the forest. Caves provide a stable environment for egg development in terms of temperature and humidity, but the hatched young need parental help to get above ground into the forest.

SIMILAR SPECIES

There are 17 *Eleutherodactylus* species in Jamaica, several of them threatened with extinction. The Portland Ridge Frog (*E. cavernicola*) is a cave-dwelling species found in only one location on Jamaica's southern coast. It is listed as Critically Endangered, as is the Leaf Mimic Frog (*E. sisyphodemus*), which is known from a single forest reserve in the northwest of the island.

Cundall's Robber Frog has a pointed snout, long, slender limbs, and large eyes. There are well-developed disks on the long, thin fingers and toes. The upper surfaces are various shades of brown with irregular, darker spots and blotches, and there is a black stripe running through the eye. The photo below shows a female carrying many froglets on her back.

Actual size

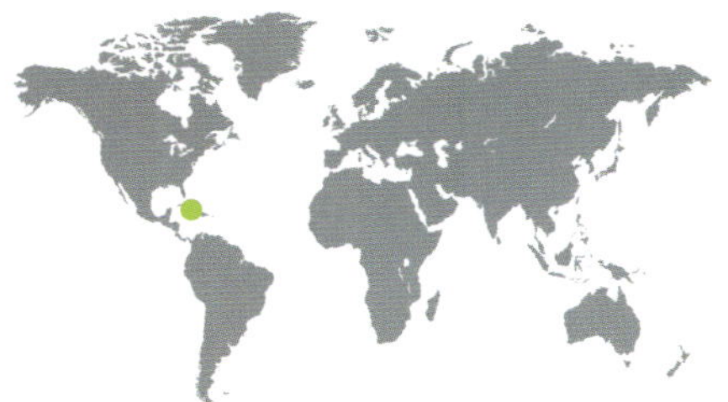

FAMILY	Eleutherodactylidae: Eleutherodactylinae
OTHER NAMES	Monte Iberia Dwarf Eleuth
DISTRIBUTION	Eastern Cuba
ADULT HABITAT	Humid forest up to 2,000 ft (600 m) altitude
LARVAL HABITAT	Within the egg
CONSERVATION STATUS	IUCN Vulnerable. Has a very restricted range that is threatened by deforestation and mining

ADULT LENGTH
Male
average 3/8 in (10 mm)

Female
average 7/16 in (11 mm)

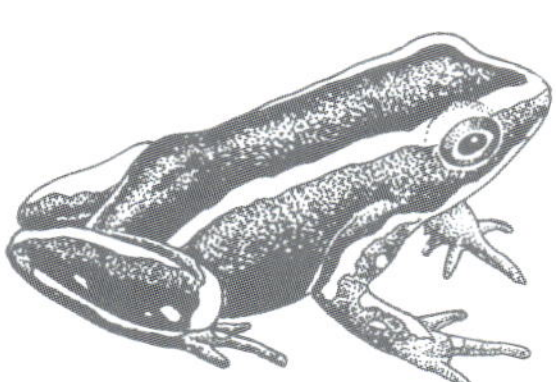

ELEUTHERODACTYLUS IBERIA

MONTE IBERIA ELEUTH

ESTRADA & HEDGES, 1996

Actual size

The Monte Iberia Eleuth has a pointed snout and glossy skin. The body and legs are black or dark brown. A stripe runs down each side from the tip of the snout, where it is yellow or copper, to the groin, where it is white. There are also white stripes running down the arms and hind legs.

A candidate for the title of the world's smallest frog, this tiny species lives on Mt. Iberia in Cuba. It feeds mostly on mites, from which it derives the compounds that it uses to make toxic skin secretions. Its bright coloration warns potential predators that it is toxic, an association of characters that it shares with South American poison frogs, to which it is not related. The male's call consists of irregular high-pitched chirps. Females lay just one large egg at a time. The tadpole phase of its life cycle is completed within the egg, which hatches to produce a tiny froglet.

SIMILAR SPECIES

Three other *Eleutherodactylus* species living in Cuba share the characteristics of very small size and laying a single egg. The Habana Robber Frog (*E. limbatus*) is 1/2 in (12 mm) long. It has a restricted range in southern Cuba and is listed as Critically Endangered. *Eleutherodactylus orientalis* is 1/2 in (13 mm) long, lives in eastern Cuba, and is Critically Endangered. The Cuban Robber Frog (*E. cubanus*), which is 9/16 in (14 mm) long and is found in scattered locations across the island, is Endangered.

FAMILY	Eleutherodactylidae: Eleutherodactylinae
OTHER NAMES	Norton's Robber Frog, Spiny Green Frog
DISTRIBUTION	Southern Haiti, southeastern Dominica
ADULT HABITAT	Caves in forests
LARVAL HABITAT	Within the egg
CONSERVATION STATUS	IUCN Critically Endangered. Has a very small range and its habitat is being destroyed

ADULT LENGTH
2½–2¹¹⁄₁₆ in
(62–66 mm)

ELEUTHERODACTYLUS NORTONI

SPINY GIANT FROG

SCHWARTZ, 1976

This little-known frog has not been seen since 2006. Its habitat has been largely destroyed by deforestation, for the manufacture of charcoal and to create space for agriculture, and by mining. Its restricted range includes some protected areas, but the management of these is ineffective and its habitat continues to be degraded. It lives in sinkhole caves in wooded areas and males call from trees or rocks. The eggs, laid on the ground, develop directly into frogs, the tadpole stage being completed within the confines of the egg.

SIMILAR SPECIES

The Diquini Robber Frog (*Eleutherodactylus inoptatus*) is a large Near Threatened species that also lives in both Haiti and Dominica. The False Green Robber Frog (*E. chlorophenax*) is found only in Haiti, where its upland forest habitat has been largely destroyed. It is listed as Critically Endangered, as is Karl's Robber Frog (*E. karlschmidti*) of Puerto Rico, which inhabits mountain streams. Its decline is thought to be due to the disease chytridiomycosis, and it is listed as Critically Endangered, and possibly extinct.

The Spiny Giant Frog gets its name from small spines scattered on its back and legs, most obviously above its eyes. It has large eyes and triangular pads at the end of its fingers and toes. The head, back, and legs are pale gray, with irregular green patches on the back and green stripes on the legs.

Actual size

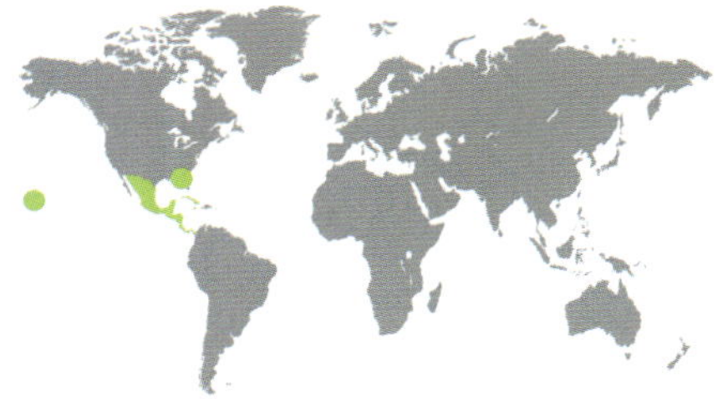

FAMILY	Eleutherodactylidae: Eleutherodactylinae
OTHER NAMES	None
DISTRIBUTION	Cuba, Bahamas, Turks and Caicos Islands, Cayman Islands, USA (Florida, Louisiana and Alabama). Introduced to Jamaica, Guam and Hawaii, Central America (Mexico, Honduras, Panama)
ADULT HABITAT	Wooded habitats, gardens, urban areas
LARVAL HABITAT	Within the egg
CONSERVATION STATUS	IUCN Endangered. It has a very large range, occurs in a wide variety of habitats, and has a stable population

ADULT LENGTH
Male
1³⁄₁₆–1⅜ in (30–35 mm)
Female
1⁹⁄₁₆–1¹⁵⁄₁₆ in (40–50 mm)

ELEUTHERODACTYLUS PLANIROSTRIS

GREENHOUSE FROG

(COPE, 1862)

This small frog is so called because it thrives in gardens. It has invaded Florida and, from there, expanded its range into other southern states of the USA. It was long assumed that it was introduced to Florida from the West Indies by humans in the late 1800s, but a recent genetic analysis suggests that it arrived 70–400 million years ago. In the Everglades it breeds between April and September. Males produce a soft, cricket-like chirp and the female lays 2–26 eggs in moist soil. The eggs hatch directly into tiny frogs, just ³⁄₁₆ in (5 mm) long, still with the remains of their larval tail.

The Greenhouse Frog has a pointed snout, large eyes, and long, slender fingers and toes with flattened tips. The back is brown or tan with red patches and darker markings, and the tip of the snout is orange or red. Some individuals have two pale stripes running along the back. The legs are striped and the underside is white.

SIMILAR SPECIES

Of the 211 species in the genus *Eleutherodactylus*, three occur in Texas. The Rio Grande Chirping Frog (*E. cystignathoides*) occurs in southeastern Texas and Mexico, and is very common in gardens. The Cliff Chirping Frog (*E. marnockii*) and the Spotted Chirping Frog (*E. guttilatus*) both occur in rocky areas, where they hide in crevices.

Actual size

FAMILY	Eleutherodactylidae: Eleutherodactylinae
OTHER NAMES	None
DISTRIBUTION	Eastern Cuba
ADULT HABITAT	Montane rainforest at altitudes of 1,500–6,000 ft (450–1,830 m)
LARVAL HABITAT	Within the egg
CONSERVATION STATUS	IUCN Critically Endangered. Has a range of less than 4 sq miles (10 sq km) and its forest habitat is deteriorating

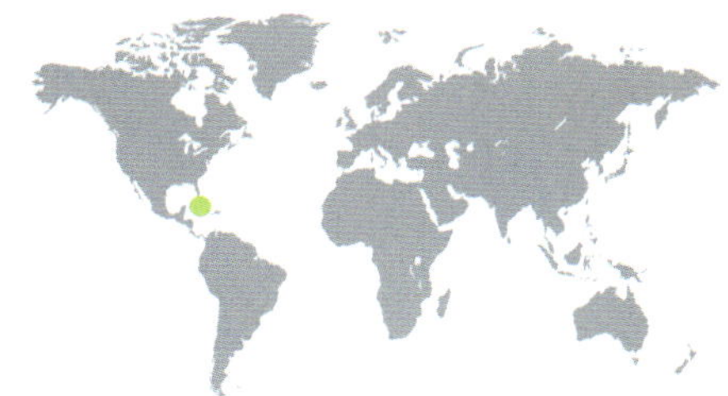

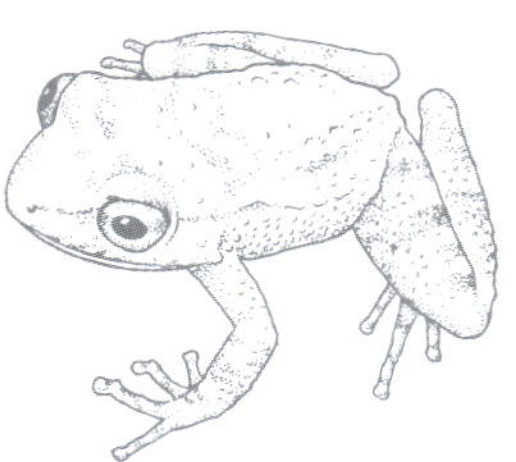

ADULT LENGTH
Male
up to 1 7/16 in (37 mm)
Female
up to 2 1/16 in (53 mm)

ELEUTHERODACTYLUS TURQUINENSIS

TURQUINO ROBBER FROG

BARBOUR & SHREVE, 1937

This very rare frog is known only from the Sierra Maestra mountains of eastern Cuba, where it is typically found in or very close to mountain streams. When disturbed, it dives quickly into deep water. Males call at night from crevices between rocks, producing a series of chirps. It is presumed that, like other species in this genus, the eggs are laid on land and develop directly into tiny frogs instead of tadpoles. The species' cloud forest habitat is being destroyed to make way for agriculture and human settlements, and it is likely that it is susceptible to chytridiomycosis.

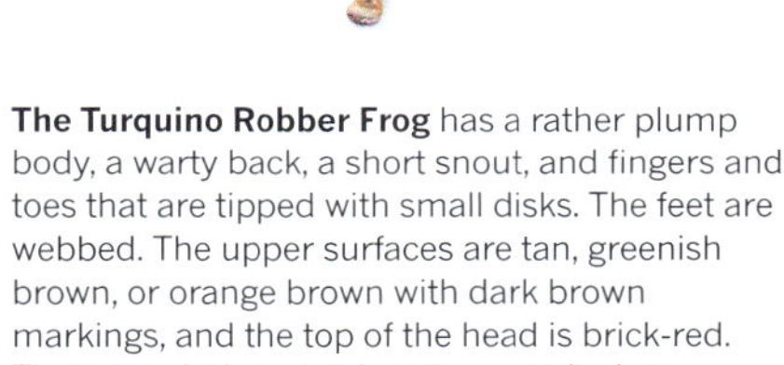

Actual size

The Turquino Robber Frog has a rather plump body, a warty back, a short snout, and fingers and toes that are tipped with small disks. The feet are webbed. The upper surfaces are tan, greenish brown, or orange brown with dark brown markings, and the top of the head is brick-red. There are dark and pale stripes on the legs.

SIMILAR SPECIES

Like *Eleutherodactylus turquinensis*, *E. rivularis* lives close to streams in the Sierra Maestra mountains. Known from only three locations, it is also listed as Vulnerable. The Juventud Robber Frog (*E. cuneatus*) is a common species, found in the mountains of southeastern Cuba. It lays its eggs under rocks close to the edge of streams. *Eleutherodactylus riparius*, also common, is found in a variety of habitats across most of Cuba.

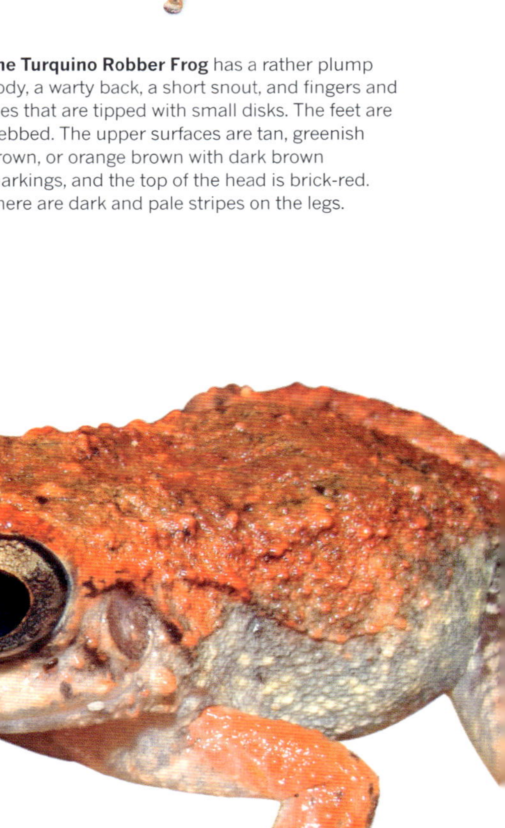

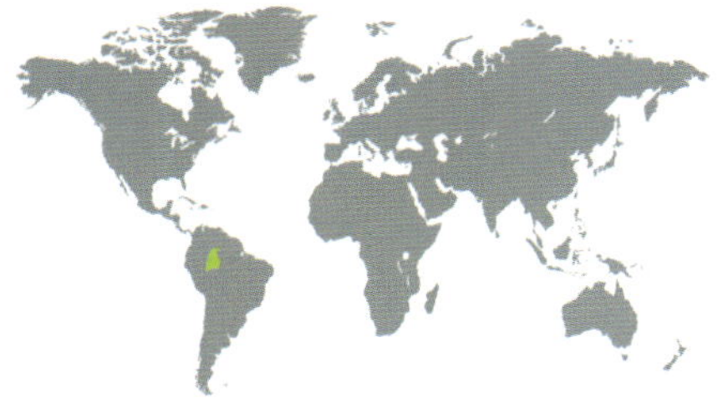

FAMILY	Eleutherodactylidae: Phyzelaphryninae
OTHER NAMES	None
DISTRIBUTION	Colombia, eastern Ecuador, northeastern Peru, western Brazil
ADULT HABITAT	Lowland tropical forest up to 660 ft (200 m) altitude
LARVAL HABITAT	Assumed to be within the egg
CONSERVATION STATUS	IUCN Least Concern. Its population is thought to be stable and it occurs in a number of protected areas

ADULT LENGTH
½–⁹⁄₁₆ in (13–14 mm)

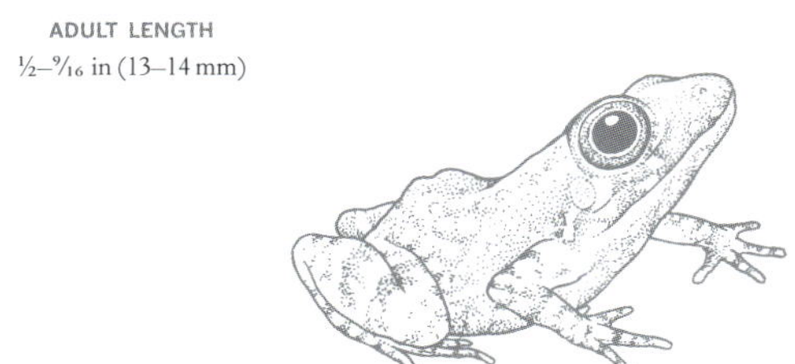

ADELOPHRYNE ADIASTOLA

YAPIMA SHIELD FROG

HOOGMOED & LESCURE, 1984

Actual size

The Yapima Shield Frog has a rounded snout, a very short fourth finger, and sharp points at the tips of its fingers and toes. The upper surfaces are brown with darker spots and there are dark stripes on the legs. The throat and chest are pale with dark spots.

The deep leaf litter that accumulates on the floor of tropical forests is an ideal habitat for small frogs, providing them with excellent cover and a rich supply of creatures on which to feed. Unfortunately, such frogs are very difficult to observe, especially if, like this species, they are cryptically colored. As a result, their diversity and abundance are very likely to be underestimated. This frog is one of several species of leaf litter frogs that have been described recently. It is very secretive in its habits. Males are sometimes heard calling, but mating and egg-laying have not been observed.

SIMILAR SPECIES

Twelve species have been included to date in the genus *Adelophryne*, all found in northern South America. Their fingers and toes have sharp tips and they have small fourth fingers. The Guiana Shield Frog (*A. gutturosa*) occurs in western Venezuela, Guyana, and Brazil. It is known that the female lays a single egg, which is likely to be true also of other species in this genus given their very small size.

FAMILY	Strabomantidae: Holoadeninae
OTHER NAMES	None
DISTRIBUTION	Southern Peru
ADULT HABITAT	Wet grassland and shrubland at altitudes of 11,150–12,140 ft (3,400–3,700 m)
LARVAL HABITAT	Within the egg
CONSERVATION STATUS	IUCN Endangered. Has a small range and its habitat is threatened by encroaching agriculture

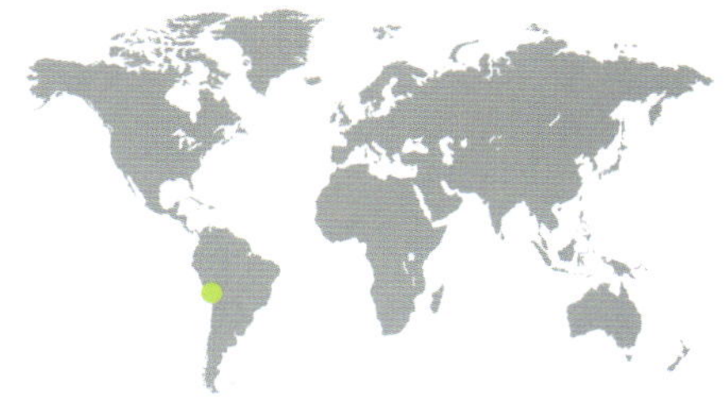

BRYOPHRYNE COPHITES

CUSCO ANDES FROG

(LYNCH, 1975)

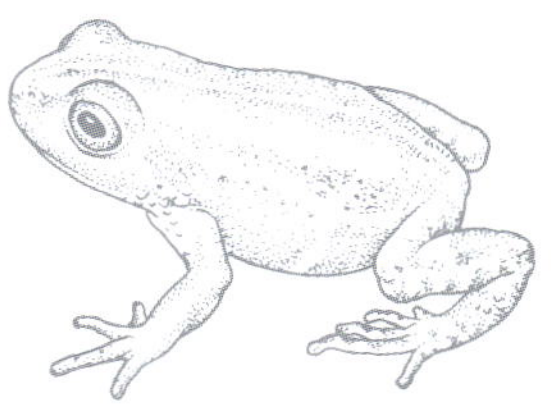

ADULT LENGTH
Male
up to ⅞ in (23 mm)

Female
up to 1⅛ in (29 mm)

This plump little frog can neither hear nor call. Males lack all the vocal apparatus that is found in most frogs, and both sexes lack eardrums and internal ear structures. The species' reproduction involves direct development, the tadpole stage being completed within the egg. The female produces around 20 large eggs in a nest of moss and attends them until they hatch into froglets, ¼ in (6–7 mm) long. The frog is known only from Manú National Park in southern Peru, but it may exist elsewhere. It tested positive for the disease chytridiomycosis in 2009.

Actual size

The Cusco Andes Frog has a plump body, a narrow head, large eyes, and a rounded snout. The upper surfaces are gray, brown, or tan, often with a very thin, pale line running down the middle of the back. The underside is gray with a dark, mottled pattern.

SIMILAR SPECIES

Bryophryne cophites was the first of 11 species to be described in this genus, whose name means "moss toad." All are found in southern Peru, where their grassland habitat is subject to destruction to make way for agriculture. The Umasbamba Moss Toad (*B. bustamantei*) is listed as Least Concern, but other species have been described recently and their conservation status has yet to be assessed and is listed as Data Deficient.

FAMILY	Strabomantidae: Holoadeninae
OTHER NAMES	None
DISTRIBUTION	Rio de Janeiro state, southeastern Brazil
ADULT HABITAT	Forest
LARVAL HABITAT	Presumed to be within the egg
CONSERVATION STATUS	IUCN Least Concern. Declining in places where there is deforestation, but occurs in some protected areas

ADULT LENGTH
9/16–11/16 in (15–18 mm)

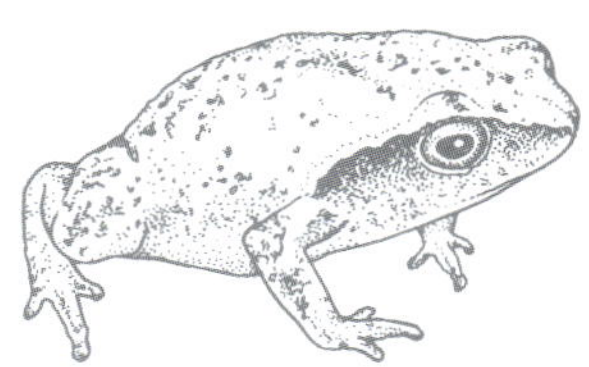

EUPARKERELLA COCHRANAE

COCHRAN'S GUANABARA FROG

IZECKSOHN, 1988

Actual size

Cochran's Guanabara Frog has a narrow head, a pointed snout, and marked ridges running from the tip of the snout to just behind the ear. The upper surfaces are reddish brown, with darker blotches on the back and transverse stripes on the legs.

An inhabitant of the Atlantic Forest ecoregion, this small frog lives in leaf litter on the forest floor. It is found in Rio de Janeiro state (formerly Guanabara, hence the species' common name). It is active at night, males produce a single-note call, and, when attacked, it adopts a stiff-legged defensive posture. It is an example of miniaturization, an evolutionary trend that has occurred in a number of frog families. This is associated with living on the ground, among leaf litter, where there is an abundance of very small invertebrates on which to feed. Like other miniaturized frogs, this species has reduced numbers of bones in its fingers and toes.

SIMILAR SPECIES

There are four other species in the genus *Euparkerella*, all small leaf litter frogs living in the Atlantic Forest. The Brazilian Guanabara Frog (*E. brasiliensis*) also lives in Rio de Janeiro state. It is a common frog and occurs in urban gardens where there are trees. The Critically Endangered Izecksohn's Guanabara Frog (*E. robusta*) and the Three-toed Guanabara Frog (*E. tridactyla*) both have rather small ranges in Espírito Santo state. The recently described Cryptic Guanabara Frog (*E. cryptica*) is also Critically Endangered.

FAMILY	Strabomantidae: Holoadeninae
OTHER NAMES	None
DISTRIBUTION	Southeastern Brazil
ADULT HABITAT	Montane rainforest up to 4,000 ft (1,200 m) altitude
LARVAL HABITAT	Presumed to be within the egg
CONSERVATION STATUS	Data Deficient. Its habitat is threatened by encroaching agriculture and urbanization

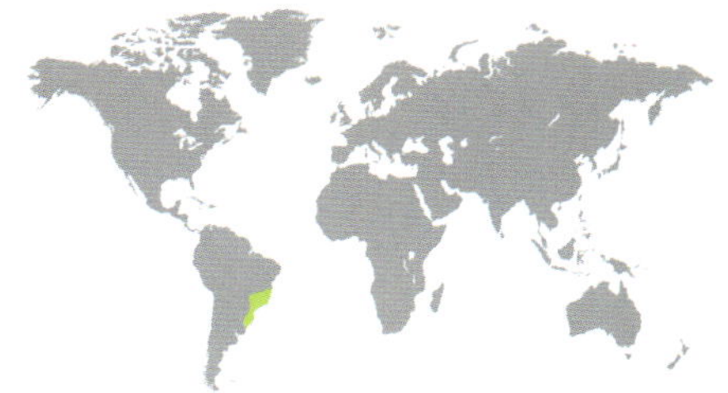

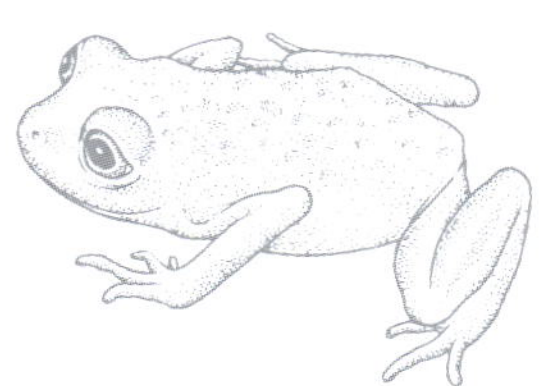

ADULT LENGTH
Male
average 1⅝ in (41 mm)
Female
1¾–1⅞ in (44–48 mm)

HOLOADEN PHOLETER

HOLE-DWELLING HIGHLAND FROG

POMBAL, SIQUEIRA, DORIGO, VRCIBRADIC & ROCHA, 2008

This recently described frog is very difficult to find, largely because it spends much of its time underground. Its specific name, *pholeter*, is Greek for "one who lives in a hole." It has been found in shallow burrows in the soil, with the ground inside and near the hole apparently swept clear of leaves. When attacked, it releases a sticky, colorless secretion of unknown chemical composition. Like other frogs in its family, it very probably lays large eggs that hatch directly into tiny frogs. Another species in this genus (*Holoaden bradei*) has been reported to show parental care of its eggs.

Actual size

The Hole-dwelling Highland Frog has a large head with a slightly rounded snout, large, forward-pointing eyes, and long, slender limbs. The back is covered with many small, bulging glands. The back and flanks are purplish brown to black, the limbs are dark brown, and there is a white blotch on the belly.

SIMILAR SPECIES

There are three other species in the genus *Holoaden*, known as highland frogs, all found in southeastern Brazil. The Itatiaia Highland Frog (*H. bradei*) has a tiny range in the Itatiaia Mountains. Its habitat has been badly degraded and it is also thought to have been badly affected by extreme frosts; it is listed as Critically Endangered. The other species are Lüderwaldt's Highland Frog (*H. luederwaldti*) and Suarez' Highland Frog (*H. suarezi*), from São Paulo state. This last species was described in 2013.

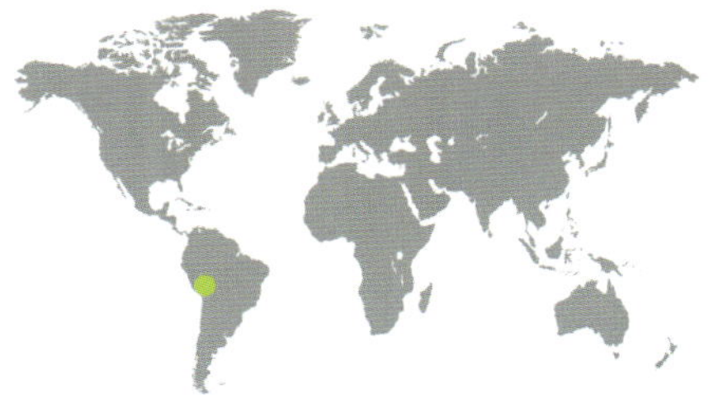

FAMILY	Strabomantidae: Holoadeninae
OTHER NAMES	None
DISTRIBUTION	Southern Peru
ADULT HABITAT	Cloud forest at altitudes of 6,000–9,000 ft (1,830–2,740 m)
LARVAL HABITAT	Within the egg
CONSERVATION STATUS	IUCN Critically Endangered. Has a small range that is fragmented by encroachment of small-scale agriculture

ADULT LENGTH
Male
½–⅝ in (13–16 mm)

Female
⁹⁄₁₆–¾ in (14–19 mm)

NOBLELLA BAGRECITO

BAGRECITO ANDES FROG

(LYNCH, 1986)

Actual size

The Bagrecito Andes Frog has a plump body and a narrow head with a pointed snout. The back and limbs are striped in various shades of brown, and the sides of the body and head are dark brown. The underside is white or cream, with a mottled brown pattern.

This small, plump frog is found on the Amazonian slopes of the Andes in Peru and Ecuador. It lives at high altitudes, where conditions are cold for most of the year, and is typically found in the leaf litter that covers the ground in cloud forest. It is thought to breed between August and October. Although its reproduction has not been reported, it is assumed that, like other members of its family, it has direct development. This means that there is no free-living larval stage and that the frog lays large eggs that hatch directly into tiny froglets.

SIMILAR SPECIES

To date, 12 species have been included in the genus *Noblella*. All are found at high altitudes in the Andes. The Acjanaco Puna Frog (*N. usurpator*) is a Near Threatened species found in high-altitude scrubland in Peru. It has lost much of its habitat to agriculture, particularly the cultivation of potatoes. The Madre Selva Frog (*N. madreselva*) from Peru is Critically Endangered.

FAMILY	Strabomantidae: Holoadeninae
OTHER NAMES	Previously *Noblella coloma*
DISTRIBUTION	Northern Ecuador
ADULT HABITAT	Cloud forest at altitudes of 4,875–7,267 ft (1,486–2,215 m)
LARVAL HABITAT	Unknown. Presumed to be within the egg
CONSERVATION STATUS	IUCN Critically Endangered. Very rare; occurs in a small area that includes two protected areas

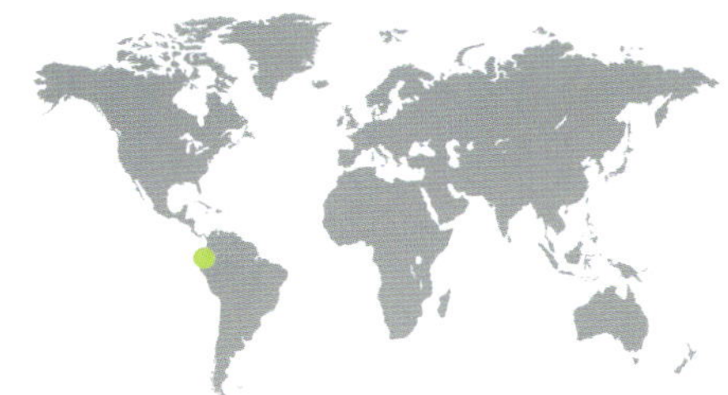

ADULT LENGTH
Male
average 9/16 in (15 mm)
Female
average 5/8 in (16 mm)

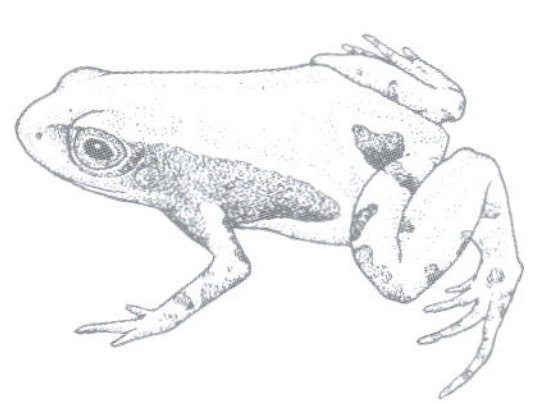

PHYLLONASTES COLOMA

COLOMA'S RAIN FROG

(GUAYASAMIN & TERÁN-VALDEZ, 2009)

This extremely rare frog from the western slopes of the Andes lives in leaf litter in dense cloud forest and is active during the day and at night. It was discovered in 1994 and has been seen only 11 times since. Nothing is known about its reproduction but it is assumed that, like other members of its family, it has direct development. This means that it does not have a free-living larval stage but instead lays large eggs that hatch directly into tiny froglets.

Actual size

Coloma's Rain Frog is a small frog with a rounded snout and pointed fingers and toes. The back is brown, the flanks and sides of the head are black, and the underside is bright orange or red. There are some darker patches in the groin and on the arms and legs.

SIMILAR SPECIES

The bright orange belly of *Phyllonastes coloma* differentiates it from the 12 species in the genus *Noblella*, in which it was previously included, or 11 species in *Phyllonastes*, its current genus. The Loreto Leaf Frog (*Phyllonastes myrmecoides*) is a relatively common species from western Peru, eastern Brazil, and northern Bolivia. Its scientific species name, meaning "ant-like," refers to its small size.

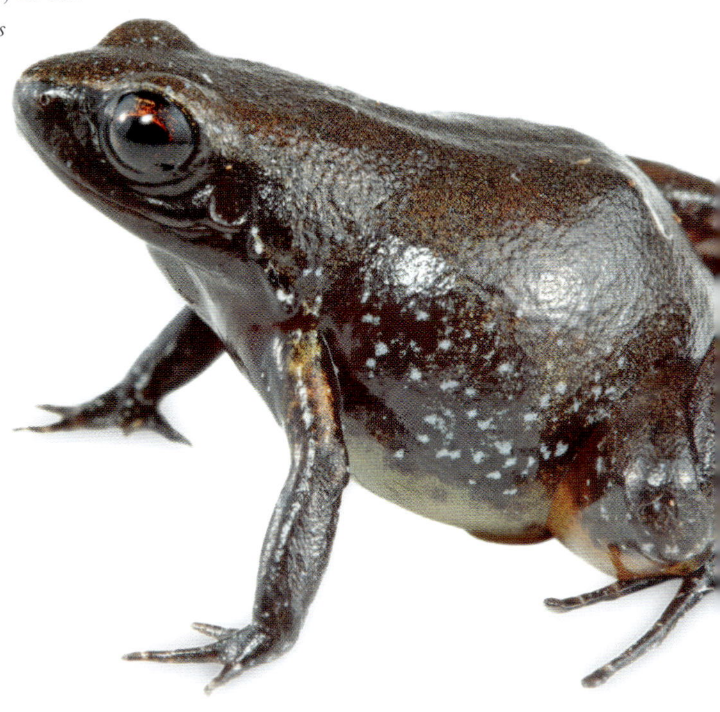

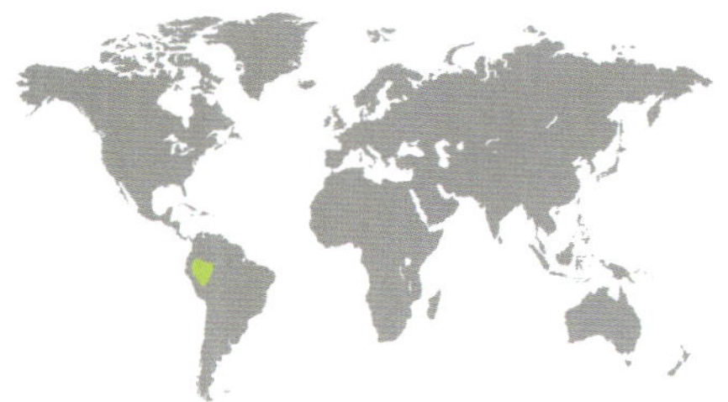

FAMILY	Strabomantidae: Hypodactylinae
OTHER NAMES	*Hypodactylus nigrovittatus*
DISTRIBUTION	Peru, Colombia, Ecuador, western Brazil
ADULT HABITAT	Rainforest and cloud forest at altitudes of 330–6,350 ft (100–1,935 m)
LARVAL HABITAT	Within the egg
CONSERVATION STATUS	IUCN Least Concern. Has a very large range that includes several protected areas

ADULT LENGTH
Male
11⁄16–1 in (17–25 mm)
Female
1–1¼ in (25–31 mm)

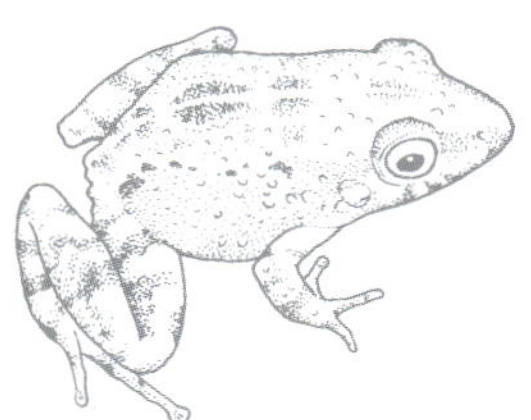

NICEFORONIA NIGROVITTATA

BLACK-BANDED ROBBER FROG

(ANDERSSON, 1945)

Actual size

The Black-banded Robber Frog has a plump body, a pointed snout, and long fingers and toes that have pointed tips. The upper surfaces are brown, with darker patches on the back and dark stripes on the legs. There is a conspicuous black patch on the back near the top of each leg. The belly is yellow and the throat is white.

Found across a large area that includes the eastern slopes of the Andes and the upper Amazon Basin, this common frog is active on the ground by day, living in the leaf litter. At night it has occasionally been found in vegetation above the ground. Its snout has a fleshy projection that is believed to be used for burrowing. In common with other frogs in its family, it lays relatively large eggs on the ground. There is no free-living tadpole but the eggs instead hatch directly into tiny frogs.

SIMILAR SPECIES

The genus *Niceforonia* currently contains 15 species, found in the Andes of Colombia, Ecuador, and Peru. The Bright Robber Frog (*N. lucida*), from Peru, and the Carchi Andes Toad (*N. brunnea*), from Ecuador, are listed as Endangered. Both have lost much of their habitat as the result of the clearance of forest to make way for agriculture.

FAMILY	Strabomantidae: Pristimantinae
OTHER NAMES	None
DISTRIBUTION	Northern Argentina, southern Bolivia
ADULT HABITAT	Humid montane forest at altitudes of 2,300–7,200 ft (700–2,200 m)
LARVAL HABITAT	Within the egg
CONSERVATION STATUS	Data Deficient. Declining in places due to deforestation, but occurs in some protected areas

ADULT LENGTH
Male
1–1¼ in (25–31 mm)

Female
1⅛–1⁹⁄₁₆ in (29–40 mm)

OREOBATES DISCOIDALIS

TUCUMAN ROBBER FROG

(PERACCA, 1895)

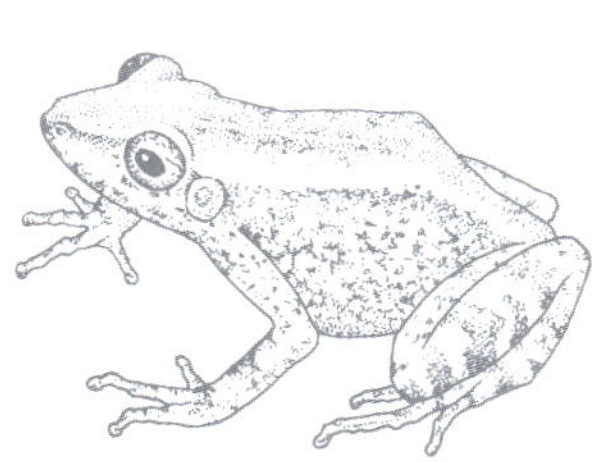

This ground-living frog is found in leaf litter in forests covering the eastern flanks of the Andes. Males are strongly territorial and have a richer vocal repertoire than many frogs, consisting of three distinct calls. The advertisement call serves both to attract females and to warn off other males. The territorial call is produced when a rival male is heard fairly close to a male's territory. The aggressive call is produced when a male has come very close and a fight has ensued. The eggs are laid under logs, where they hatch directly into tiny froglets.

Actual size

The Tucuman Robber Frog has a plump body, a pointed snout, and large eyes. The fingers and toes are long and slender. It is variable in color, being pale to dark brown above, with cross-stripes on its arms and legs. There is a dark stripe running through the eye, the iris of which is gold.

SIMILAR SPECIES

The genus *Oreobates* currently contains 26 species, some described very recently. *Oreobates barituensis* is very similar to *O. discoidalis*, but can be distinguished by its advertisement call. It occurs in Argentina. The La Paz Robber Frog (*O. cruralis*) is found in humid rainforest in Bolivia and Peru. *Oreobates ibischi* is a common species in Bolivia, where it is found in gardens as well as in forest.

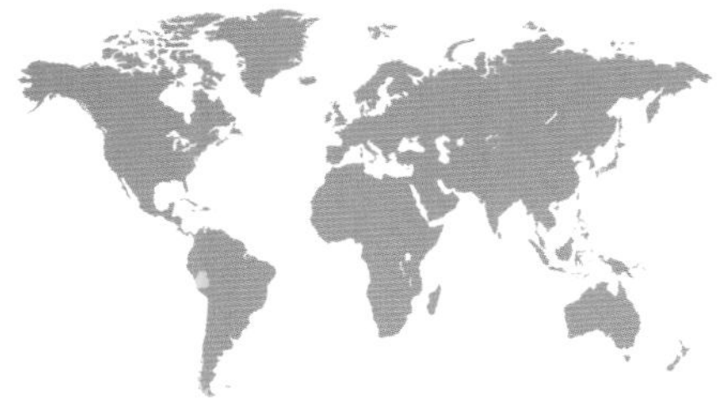

FAMILY	Strabomantidae: Pristimantinae
OTHER NAMES	None
DISTRIBUTION	Central Peru
ADULT HABITAT	Cloud forest at altitudes of 7,500–8,900 ft (2,300–2,700 m)
LARVAL HABITAT	Within the egg
CONSERVATION STATUS	Data Deficient. Has a very restricted range; while this lies largely within a protected area, it is threatened by expanding agriculture

ADULT LENGTH
Male
⁹⁄₁₆–⅝ in (15–16 mm)
Female
average ¾ in (20 mm)

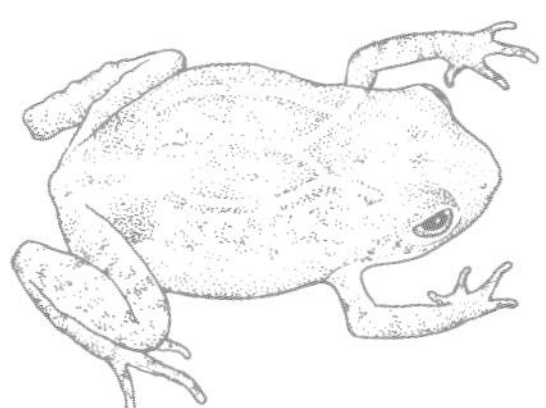

PHRYNOPUS BRACKI

BRACK'S ANDES FROG

HEDGES, 1990

Actual size

Brack's Andes Frog has a plump body, large eyes, and rather short arms and legs with short fingers and toes. The skin is smooth and shiny, and is predominantly brown with some indistinct darker markings.

This small, secretive frog is known from only one locality, in the Cordillera Yanachaga on the eastern side of the Peruvian Andes. A terrestrial species, it is typically found in leaf litter on the forest floor. Its eggs, which are laid on the ground, hatch directly into tiny frogs. Although much of its range falls within a protected area, its future is uncertain because of pressure to cut down the forest to create space for chili pepper plantations.

SIMILAR SPECIES

Of the 37 species currently included in the genus *Phrynopus*, most are found in the Andes, many at high altitude and occupying very small ranges. Three are listed as Critically Endangered, including Dagmar's Robber Frog (*P. dagmarae*), known from three mountain peaks; while eight are Endangered, including Heims' Robber Frog (*P. heimorum*), known from a single location. Both are threatened by the destruction of their habitat to make space to grow potatoes.

FAMILY	Strabomantidae: Pristimantinae
OTHER NAMES	Pacific Robber Frog
DISTRIBUTION	Ecuador, southern Colombia
ADULT HABITAT	Cloud forest at altitudes of 4,800–9,200 ft (1,460–2,800 m)
LARVAL HABITAT	Within the egg
CONSERVATION STATUS	IUCN Vulnerable. It is, however, vulnerable to habitat loss through deforestation and/or climate change

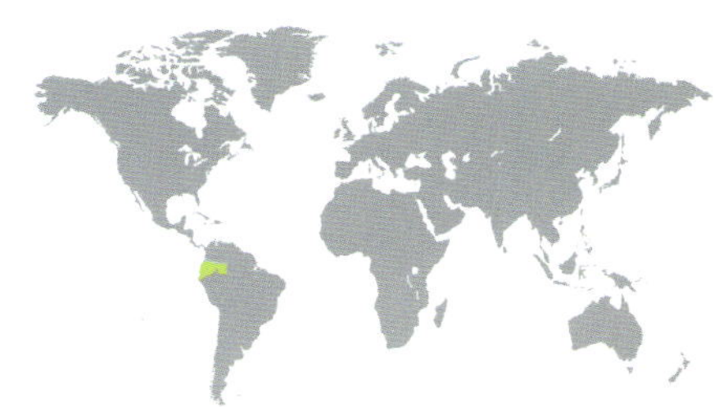

PRISTIMANTIS APPENDICULATUS

PINOCCHIO RAIN FROG

(WERNER, 1894)

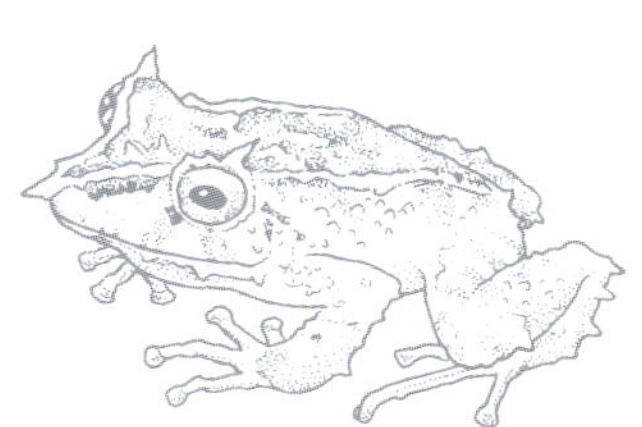

ADULT LENGTH
Male
¾–⅞ in (19–21 mm)
Female
1³⁄₁₆–1⅜ in (30–35 mm)

This distinctive frog gets its name from the fleshy tubercle at the tip of its snout, which looks like a long nose. Found on the Pacific slopes of the Andes, the species prefers low vegetation in the denser parts of the forest. Adults are active only at night, but juveniles are also active by day. The frog reproduces in the wet season, the male producing a call that is barely audible to human ears. The eggs, which are laid in leaf litter, hatch directly into tiny froglets. This species is notable for the very large size difference between males and females.

SIMILAR SPECIES

The genus *Pristimantis*, currently containing 612 species, is the largest among vertebrates. The Spurred Rain Frog (*P. calcarulatus*) is a locally common species in cloud forest in Ecuador and Colombia. It has prominent spurs on its heels and is listed as Vulnerable because, like *P. appendiculatus*, the long-term future of its habitat is precarious.

Actual size

The Pinocchio Rain Frog has pointed tubercles above its eyes, on its heels, and along the outside of its legs, in addition to the one on its snout. The eyes are large and have silver irises. The fingers and toes are long, thin, and tipped by triangular pads. The skin coloration consists of patches of green, yellow, and various shades of brown.

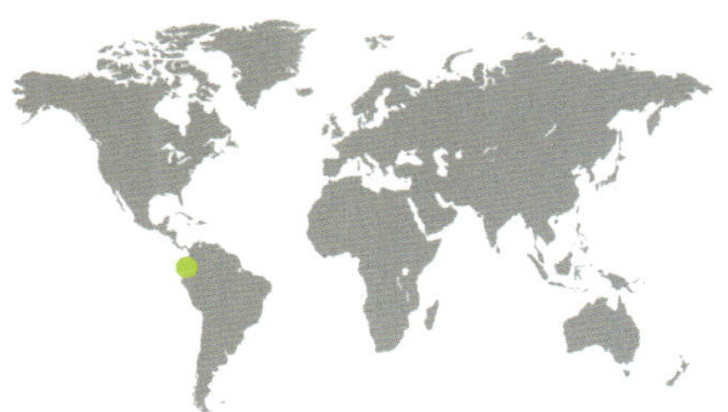

FAMILY	Strabomantidae: Pristimantinae
OTHER NAMES	Valley Robber Frog
DISTRIBUTION	Colombia, Ecuador
ADULT HABITAT	Humid tropical forest up to 6,300 ft (1,910 m) altitude
LARVAL HABITAT	Within the egg
CONSERVATION STATUS	IUCN Least Concern. It is, however, declining over much of its range

ADULT LENGTH
Male
$^{11}/_{16}$–1$^{1}/_{16}$ in (18–27 mm)
Female
1$^{1}/_{8}$–1$^{1}/_{4}$ in (28–31 mm)

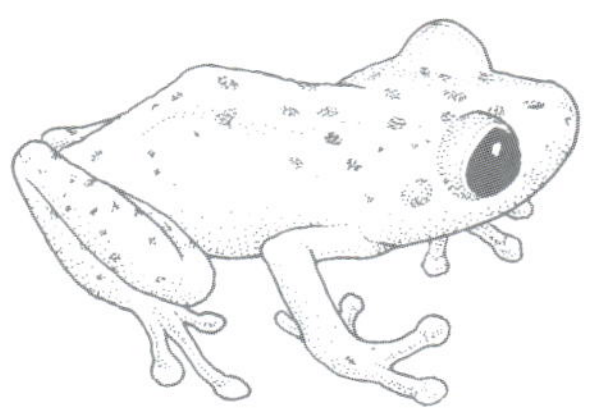

PRISTIMANTIS CHALCEUS

COPPER RAIN FROG

(PETERS, 1873)

Actual size

The Copper Rain Frog has a slender body, a blunt snout, and large, protruding eyes with black irises. There are well-developed disks on the fingers and toes. The upper surfaces are rusty pink to rosy white, the underside is white, and the hands and feet are yellow.

The most distinctive feature of this arboreal frog is the texture of the skin on its back, head, and legs, which is covered in tiny bumps. It lives on the lower Pacific slopes of the Andes, sheltering by day in bromeliads or in the leaf axils of elephant-ear plants. At night, it is found climbing in vegetation 2–5 ft (60–150 cm) above the ground. It has been found in banana plantations, as well as in natural forest. Formerly considered a common species, the frog has not been found during recent surveys, suggesting that it has declined, possibly as a result of habitat loss, climate change, or disease.

SIMILAR SPECIES

The closest relative of *Pristimantis chalceus* is the Ricuarte Robber Frog (*P. scolodiscus*), a Vulnerable species that occurs on the western slopes of the Andes in Colombia and Ecuador. The major threat to this species is thought to be habitat loss caused by agricultural development, including the cultivation of illegal crops. It may also have been adversely affected by disease.

FAMILY	Strabomantidae: Pristimantinae
OTHER NAMES	None
DISTRIBUTION	Southeastern Costa Rica, Panama
ADULT HABITAT	Forest at altitudes of 1,500–4,800 ft (450–1,450 m)
LARVAL HABITAT	Within the egg
CONSERVATION STATUS	Not yet assessed

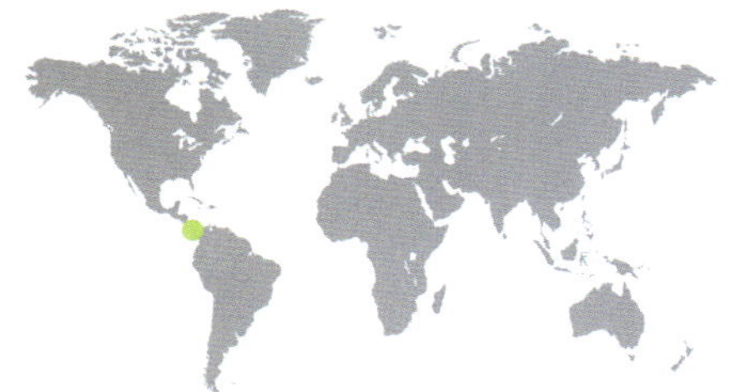

PRISTIMANTIS EDUCATORIS

EDUCATOR'S ROBBER FROG

RYAN, LIPS & GIERMAKOWSKI, 2010

ADULT LENGTH
Male
¾–⅞ in (19–21 mm)
Female
15/16–1½ in (23–38 mm)

This rain frog lives in forest where there is dense vegetation near the ground. The male's call is a soft "pweep." The female lays 15–20 eggs on the upper surface of a palm leaf. She then remains with them for at least 28 days, until they hatch into tiny froglets. By day, she sits on top of the eggs; at night she sits beside them. Her presence protects the eggs from predators, and covering them by day probably prevents them from drying out. This species seems to have been unaffected by the wave of chytridiomycosis that passed through Central America in the late 1980s and early 1990s.

Actual size

Educator's Robber Frog has a slender body, a large head with an elongated snout, and long limbs. The upper surfaces are pale tan or brown, and the underside is white. On the back there are four or five chevron-shaped stripes of brown spots and the legs are striped. Some individuals have large white or yellow spots on the flanks.

SIMILAR SPECIES

The La Loma Robber Frog (*Pristimantis caryophyllaceus*) is found in Costa Rica, Panama, and Colombia. Smaller than *P. educatoris*, it is listed as Near Threatened. Much of its forest habitat has been destroyed and, in some places, it is infected with chytridiomycosis. The Rio San Juan Robber Frog (*P. ridens*) is a common species found from Honduras to Colombia.

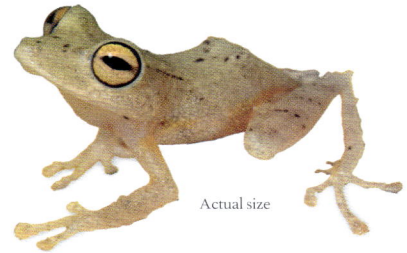

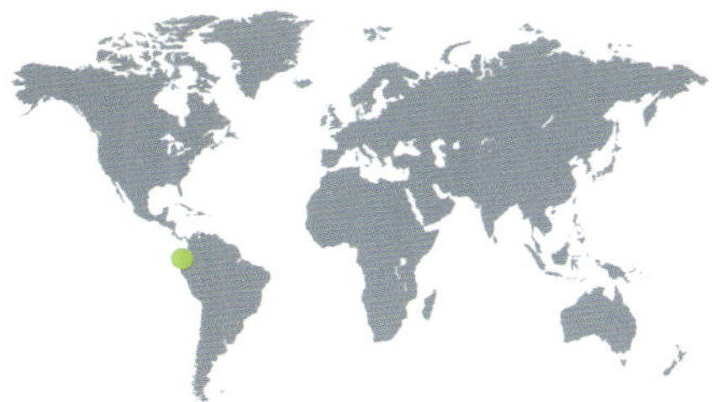

FAMILY	Strabomantidae: Pristimantinae
OTHER NAMES	None
DISTRIBUTION	Northern Ecuador
ADULT HABITAT	Forest at altitudes of 6,070–6,768 ft (1,850–2,063 m)
LARVAL HABITAT	Within the egg
CONSERVATION STATUS	IUCN Endangered. Known from only three localities

ADULT LENGTH
Male
average 11/16 in (17 mm)

Female
3/4–15/16 in (20–24 mm)

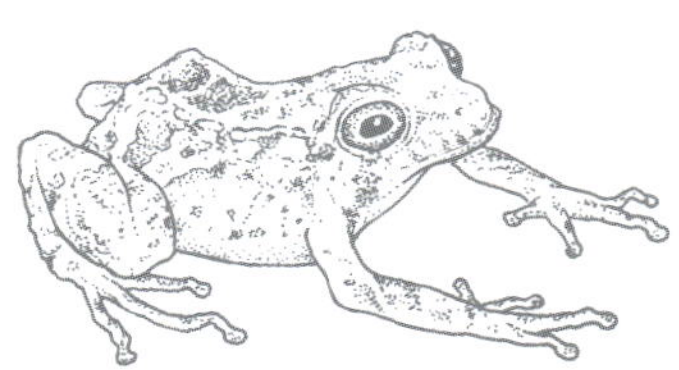

PRISTIMANTIS MUTABILIS

MUTABLE RAIN FROG

GUAYASAMIN, KRYNAK, KRYNAK, CULEBRAS & HUTTER, 2015

Actual size

The Mutable Rain Frog has a short, rounded snout, large eyes, and long, thin fingers and toes that are tipped with round adhesive pads. The upper surfaces are pale brown to grayish green, patterned with dark brown patches, and there is a thin orange line along each side of the back. Females have a patch of red in the groin, extended along the inner thigh.

This very unusual frog, discovered in 2013, gets its name from its ability to change the texture of its skin both radically and rapidly. Over a period of a few minutes, it goes from being very smooth to being covered in numerous spine-like tubercles. The function of this transformation is not known, but it may serve to enhance the frog's ability to camouflage itself against the moss-covered branches of trees. An arboreal species, the Mutable Rain Frog has a repertoire of at least three distinct, high-pitched calls.

SIMILAR SPECIES

The closest relative of this species is the Zacualtipan Robber Frog (*Pristimantis verecundus*), a Near Threatened inhabitant of high-altitude forest in Colombia and Ecuador. Unlike *P. mutabilis*, it cannot change its skin. The Sobetes Robber Frog (*P. sobetes*) can change its skin texture and is found at high altitudes in Ecuador. It is categorized as Vulnerable.

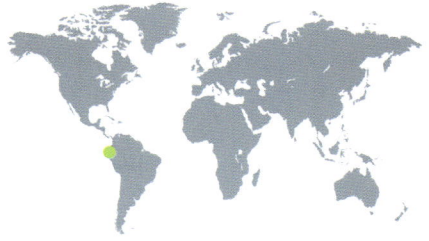

FAMILY	Strabomantidae: Pristimantinae
OTHER NAMES	Ornate Robber Frog
DISTRIBUTION	Ecuador
ADULT HABITAT	Humid forest at altitudes of 1,300–5,900 ft (400–1,800 m)
LARVAL HABITAT	Within the egg
CONSERVATION STATUS	IUCN Endangered. Has lost much of its habitat to agriculture and logging but occurs in some protected areas

PRISTIMANTIS ORNATISSIMUS

ORNATE RAIN FROG

(DESPAX, 1911)

ADULT LENGTH
Male
7/8–1 1/16 in (22–27 mm)
Female
1 1/4–1 9/16 in (31–40 mm)

This very beautiful rain frog is found on the lower Pacific slopes of the Andes. It is active at night, perching on large leaves 2.5–10 ft (80–300 cm) above the ground. By day, it sleeps on leaves or in the axils of elephant-ear plants. It has also been found in banana plantations. When handled, it produces an unpleasant-smelling secretion. The species has become much less common than it was in the past. The causes of its decline include the loss and fragmentation of its forest habitat, and may also involve irregular rainfall patterns and disease.

Actual size

The Ornate Rain Frog has a pointed snout, large eyes, and long fingers and toes that end in triangular adhesive pads. The upper surfaces are yellow or bright green, with a complex pattern of black blotches and stripes on the back and transverse stripes on the arms and legs. The fingers and toes are orange, and the underside is yellow or white.

SIMILAR SPECIES

Among the approximately 612 species of *Pristimantis*, the closest relative of *P. ornatissimus* is the Striped Robber Frog (*P. unistrigatus*). This lives in valleys within the Andes of Colombia and Ecuador. It is a common species that thrives in human-altered habitats such as farmland and gardens, and is commonly heard after heavy rain in the city of Quito. It lays its eggs in a burrow.

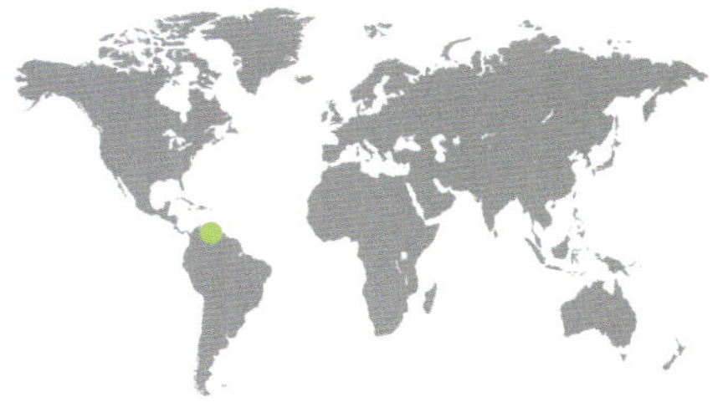

FAMILY	Strabomantidae: Strabomantinae
OTHER NAMES	None
DISTRIBUTION	Northern Venezuela
ADULT HABITAT	Cloud forest at altitudes of 800–5,250 ft (250–1,600 m)
LARVAL HABITAT	Within the egg
CONSERVATION STATUS	IUCN Least Concern. Its numbers are declining due to clearing of its forest habitat for agriculture

ADULT LENGTH
Male
1³⁄₁₆–1½ in (30–38 mm)
Female
1¹⁵⁄₁₆–2¹⁵⁄₁₆ in (49–74 mm)

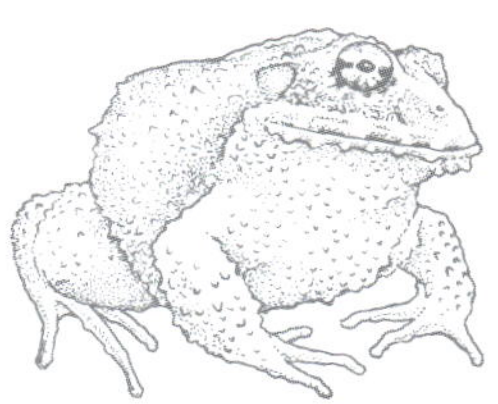

STRABOMANTIS BIPORCATUS

PUERTO CABELLO ROBBER FROG

PETERS, 1863

This ground-living frog is notable for its very wide head. It is active at night and appears to be very sedentary in its habits, individuals being found in the same hiding place over several days. One female was found in a cavity under a thick layer of leaf litter, attending a clutch of 45 eggs. The species has direct development, the tadpole stage being completed within the eggs, which hatch into very small frogs. When disturbed, females defending eggs inflate themselves with air, rise up on extended limbs, and emit a series of loud shrieks.

SIMILAR SPECIES

The 16 species in the genus *Strabomantis*, sometimes known as broad-headed frogs, are found in Central and South America. The Warty Ground Frog (*S. helonotus*) is a very rough-skinned species from Ecuador. Listed as Critically Endangered, it may be extinct. The Mindo Robber Frog (*S. necerus*), also from Ecuador and also Critically Endangered, has not been seen since 1995 and may have been affected by the disease chytridiomycosis.

The Puerto Cabello Robber Frog has a very large, broad head, a plump body, and powerful hind legs. There are numerous warts on the head, back, and legs, and there are conspicuous folds of skin on the back. The upper surfaces are brown with irregular black markings, and the underside is tan or white.

Actual size

FAMILY	Neblinaphrynidae
OTHER NAMES	None
DISTRIBUTION	Pico da Neblina, Cerro de la Neblina, Amazonas state, Brazil
ADULT AND LARVAL HABITAT	Burrower, under rocks in high-elevation grassy areas with scattered short trees and steep rocky slopes at altitudes of 6,604–9,826 ft (2,013–2,995 m)
CONSERVATION STATUS	Not yet assessed but its small range leaves it vulnerable to chytridiomycosis and climate change. It may yet be listed as Critically Endangered

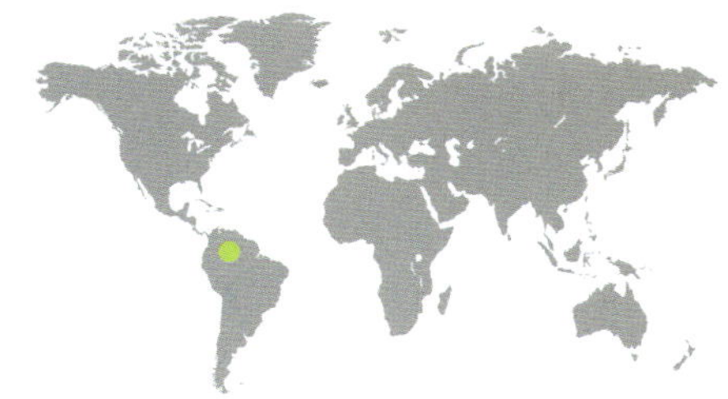

NEBLINAPHRYNE MAYERI

MAYER'S NEBLINA FROG

FOUQUET, KOK, RECODER, PRATES, CAMACHO, MARQUES-SOUZA, GHELLERE, MCDIARMID & RODRIGUES, 2023

ADULT LENGTH
Male
$\frac{9}{16}$–$\frac{3}{4}$ in (14–19 mm)

Female
$\frac{5}{8}$–$1\frac{5}{16}$ in (16–33 mm)

This recently described frog is endemic to the Pico da Neblina tepui, Brazil's highest mountain in Amazonas state, where it has been recorded at the summit, with the majority of specimens collected above 8,500 ft (2,600 m). Its discovery, together with that of *Caligophryne doylei* (page 424), both described in monotypic families, supports the argument that the tabletop tepuis of the Pantepui region are early Cenozoic refugia. This secretive species was found in small groups under rocks in grassy areas and is not known to vocalize. That it may be a burrowing frog was suggested based on the large metatarsal tubercles on its feet. *Neblinaphryne mayeri* is believed to have direct development, in common with other frogs in the Brachycephaloidea. The species was named for Brazilian Army General Sinclair James Mayer, who provided logistical support to the expedition that discovered this frog.

Actual size

Mayer's Neblina Frog is a small, dark and light brown, smooth-skinned frog with short limbs that terminate in slender, tapering, webless digits without disks. It has four fingers and five toes. It has moderately large eyes, but tympana are absent.

SIMILAR SPECIES

Neblinaphrynidae is one of seven families in the superfamily Brachycephaloidea, which also contains the large family Stabomantidae (815 spp.). Neblinaphrynidae was monotypic until a second species, the Serra do Imeri Frog (*Neblinaphryne imeri*) was described by the same authors in 2024, demonstrating the importance of the Brazilian tepuis as centers of endemism.

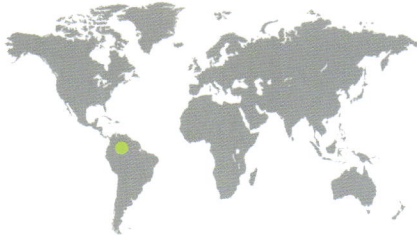

FAMILY	Caligophrynidae
OTHER NAMES	None
DISTRIBUTION	Pico da Neblina, Cerro de la Neblina, Amazonas states, Brazil and Venezuela
ADULT AND LARVAL HABITAT	Open grassy areas with scattered short trees and steep rocky slopes at altitudes of 6,600–8,500 ft (2,000–2,600 m)
CONSERVATION STATUS	Not yet assessed but its small range leaves it vulnerable to chytridiomycosis and climate change. It may yet be listed as Critically Endangered

ADULT LENGTH
Male
$\frac{3}{4}$–$1\frac{3}{16}$ in (19–30 mm)
Female
$\frac{3}{4}$–$1\frac{5}{16}$ in (19–34 mm)

CALIGOPHRYNE DOYLEI

CONAN DOYLE'S NEBLINA FROG

FOUQUET, KOK, RECODER, PRATES, CAMACHO, MARQUES-SOUZA, GHELLERE, MCDIARMID & RODRIGUES, 2023

Actual size

Conan Doyle's Neblina Frog is a stocky, red-brown frog, cream or mottled brown below, with tuberculate skin, a pointed snout, moderately large eyes, small tympana, stout limbs, and long, slender, tapering digits. It has four fingers and five toes.

The towering, flat-topped tepuis of Brazil, Venezuela, and Guyana are believed to have been the inspiration behind Conan Doyle's *The Lost World*, and so the describers of this unique frog species named it in the author's honor. *Caligophryne doylei* is endemic to the Cerro de la Neblina. It is usually found at elevations around 6,600 ft (2,000 m) but has also been collected at 8,500 ft (2,600 m). This is in contrast to *Neblinaphryne mayeri* (page 423), which was described by the same authors from the same massif, but in the family Neblinaphrynidae, and which primarily occurs above 8,500 ft (2,600 m). A female *Caligophryne doylei* collected on the Venezuelan slope of the Pico da Neblina was found "seemingly guarding" a clutch of 11 eggs. Frogs in the Brachycephaloidea, the superfamily containing both Caligophrynidae and Neblinaphrynidae, have direct development. Males are not known to vocalize.

SIMILAR SPECIES

Caligophrynidae is a monotypic (one species) family in the superfamily Brachycephaloidea, which also contains Neblinaphrynidae. These two tiny families demonstrate the significance of the Brazilian tepuis as remnants of the Cenozoic era and centers of endemism, and highlight the importance of their conservation.

FAMILY	Arthroleptidae: Arthroleptinae
OTHER NAMES	None
DISTRIBUTION	Mt. Nimba, at the border between Guinea, Côte d'Ivoire, and Liberia
ADULT HABITAT	Montane grassland, marshes, and forest at altitudes of 1,640–5,400 ft (500–1,650 m)
LARVAL HABITAT	Within the egg
CONSERVATION STATUS	IUCN Near Threatened. Has a very limited range that is threatened by habitat destruction through mining and agriculture

ADULT LENGTH
average ¾ in (20 mm)

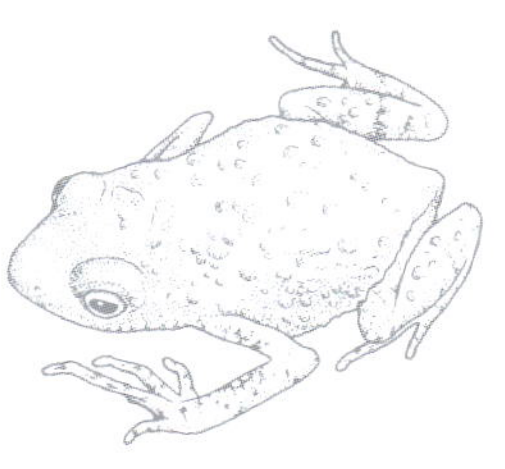

ARTHROLEPTIS CRUSCULUM

GUINEA SCREECHING FROG

ANGEL, 1950

Although it is a World Heritage Site, Mt. Nimba is being extensively mined for iron ore and bauxite. In addition, the natural habitat is being destroyed to provide agricultural land. Like other *Arthroleptis* species, known in southern Africa as "squeakers," the Guinea Screeching Frog is much easier to hear than it is to see. It lays large yolk-rich eggs in cavities in the soil. Development is direct, the eggs hatching to release tiny frogs, the tadpole stage having been completed within the egg. Although recorded only in the Guinea sector of Mt. Nimba, it may also occur in the Liberian and Côte d'Ivoire sectors.

Actual size

The Guinea Screeching Frog has a wide head and short limbs. There are warts on its back, snout, and thighs. In common with other *Arthroleptis* species, the third finger is very long, especially in males. This frog is reddish brown in color, often with dark brown patches.

SIMILAR SPECIES

There are 49 species in the genus *Arthroleptis*, distributed across much of sub-Saharan Africa. The Mottled Squeaker (*Arthroleptis poecilonotus*) has a huge range, extending from Guinea-Bissau in the west to Uganda in the east. By contrast, the Cave Squeaker (*A. troglodytes*) is restricted to a high-altitude area in Zimbabwe, where it lives in caves; it is listed as Critically Endangered. See also the Problem Squeaker Frog (*A. palava*; page 426) and the Common Squeaker (*A. stenodactylus*; page 427).

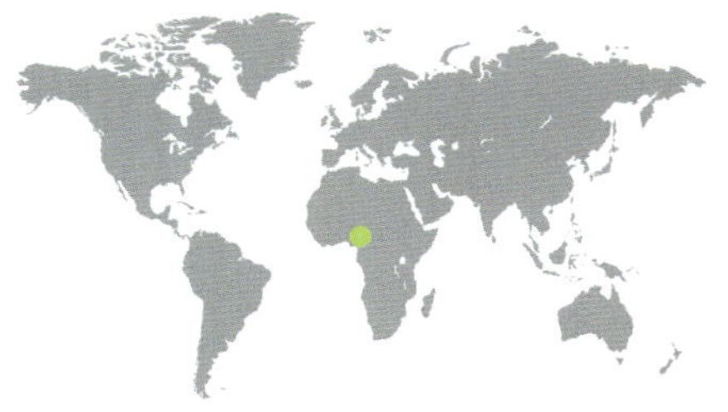

FAMILY	Arthroleptidae: Arthroleptinae
OTHER NAMES	None
DISTRIBUTION	Mountains of Cameroon and Nigeria at altitudes of 3,300–6,200 ft (1,000–1,900 m)
ADULT HABITAT	Forest, *Eucalyptus* plantations, farmland, grassland
LARVAL HABITAT	Within the egg
CONSERVATION STATUS	IUCN Least Concern. Appears to thrive in habitats altered by humans

ADULT LENGTH
Male
⅞–¹⁵⁄₁₆ in (22–24 mm)
Female
¹⁵⁄₁₆–1⅛ in (24–29 mm)
both based on very few specimens

ARTHROLEPTIS PALAVA

PROBLEM SQUEAKER FROG

BLACKBURN, GVOŽDÍK & LEACHÉ, 2010

Actual size

The Problem Squeaker Frog has a globular body, stout hind limbs, and a triangular-shaped head. The fingers and toes are long and are not webbed. The skin is smooth and is pale brown, with darker brown and black markings on the back. There is often a narrow, pale stripe down the middle of the back.

Palava means "problem" in Central and West African pidgin English, and the name of this frog refers to the fact that, until 2010, it was not clear if it was a separate species or not. Previously, it was confused with *Arthroleptis poecilonotus*, a species ranging from Guinea-Bissau in the west to Uganda in the east. Nothing is known of its natural history and it is assumed that, like other squeakers, it lives in leaf litter and lays eggs that hatch directly into tiny frogs.

SIMILAR SPECIES

Arthroleptis palava is one of several species in the genus found in West and Central Africa. The Granular Squeaker (*Arthroleptis adelphus*) is fairly common in Cameroon, Equatorial Guinea, and Gabon, but the Krokosua Squeaking Frog (*A. krokosua*), in Ghana, and Perret's Squeaking Frog (*A. perreti*), in Cameroon, are listed as Critically Endangered and Endangered respectively as their forest habitat is being degraded and destroyed by expanding human settlements. See also the Guinea Screeching Frog (*A. crusculum*; page 425) and the Common Squeaker (*A. stenodactylus*; page 427).

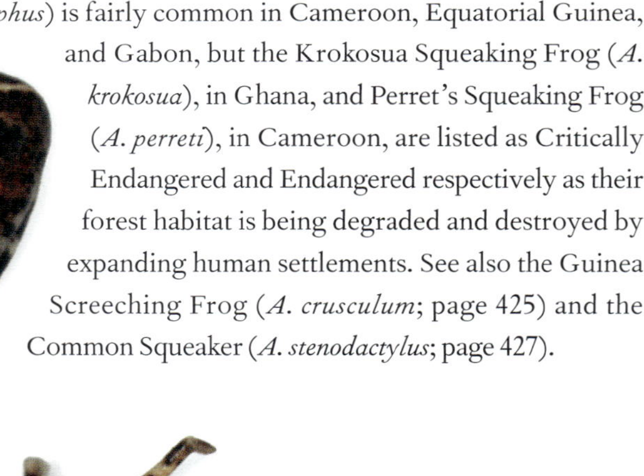

FAMILY	Arthroleptidae: Arthroleptinae
OTHER NAMES	Dune Squeaker, Savannah Squeaking Frog, Shovel-footed Squeaker
DISTRIBUTION	Eastern and southern Africa, from Kenya in the north to northeastern South Africa in the south, and west to Angola
ADULT HABITAT	Dry forest and coastal woodland. Also suburban gardens
LARVAL HABITAT	Within the egg
CONSERVATION STATUS	IUCN Least Concern. Has a very large range and shows no evidence of a general decline

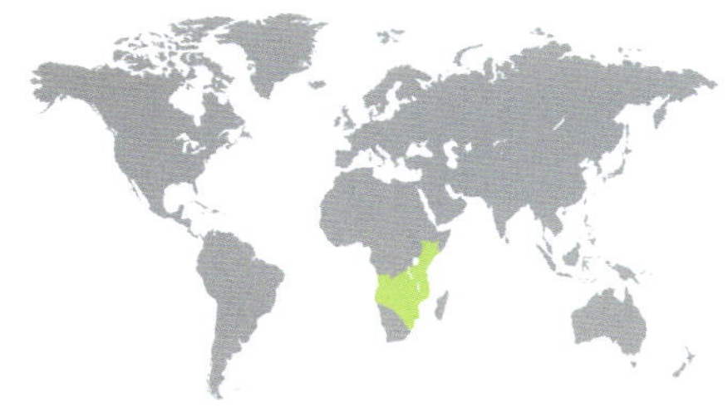

ARTHROLEPTIS STENODACTYLUS

COMMON SQUEAKER

PFEFFER, 1893

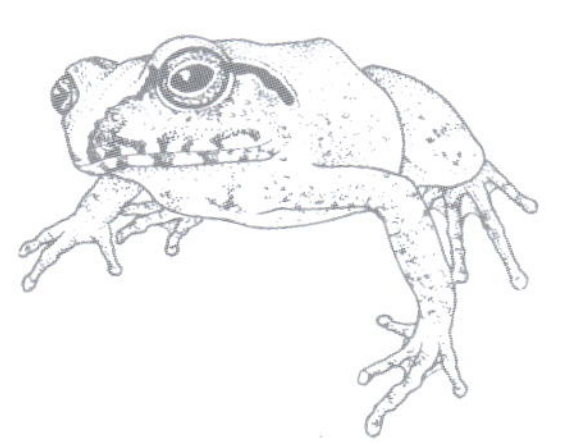

ADULT LENGTH
Male
1⅛–1⅜ in (28–35 mm)
Female
1⅛–1¾ in (28–44 mm)

Their high-pitched metallic squeaks make these small frogs easy to hear but they are much more difficult to see. The color and small size of these frogs make them well camouflaged among the leaf litter in which they live. Breeding is stimulated by summer rain and, in South Africa, occurs between December and February. Males call from the ground and pairs lay their eggs, in clutches of 33–80, in shallow burrows. The clutch is not tended by the parents and development is direct, the eggs hatching to produce tiny froglets.

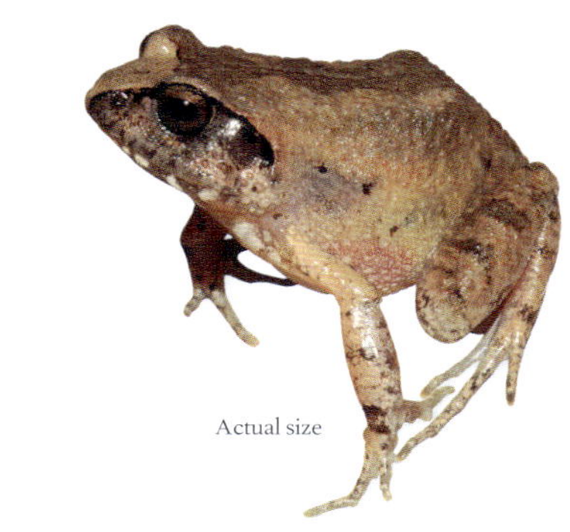

Actual size

SIMILAR SPECIES

In contrast to the huge geographic range of *Arthroleptis stenodactylus*, the other member of this genus to occur in South Africa, the Bush Squeaker (*A. wahlbergii*), has a very restricted range in the northeast of the country. A smaller frog, it produces fewer eggs: clutches number 11–30 eggs. Although this species is listed as of Least Concern, it lives in an area of expanding sugar cane production and urban development. See also the Guinea Screeching Frog (*A. crusculum*; page 425) and the Problem Squeaker Frog (*A. palava*; page 426).

The Common Squeaker has short hind legs and a hard tubercle on each foot, used for digging. The third finger is very long, especially in males, and neither fingers nor toes are webbed. The back is light or dark brown and there is a dark stripe running through the eye.

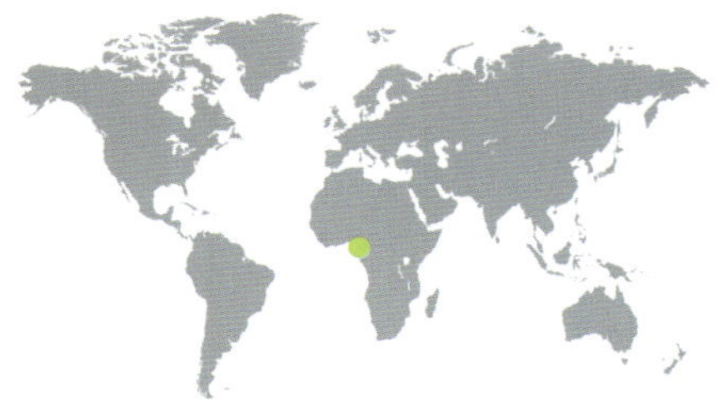

FAMILY	Arthroleptidae: Arthroleptinae
OTHER NAMES	None
DISTRIBUTION	Western Cameroon
ADULT HABITAT	Montane forest close to rivers at altitudes of 5,600–6,900 ft (1,700–2,100 m)
LARVAL HABITAT	Streams
CONSERVATION STATUS	IUCN Endangered. Has a very restricted range that is threatened by deforestation

ADULT LENGTH
Male
15/16–1 3/16 in (24–30 mm)
Female
1–1 5/16 in (25–34 mm)

CARDIOGLOSSA ALSCO

ALSCO LONG-FINGERED FROG

HERRMANN, HERRMANN, SCHMITZ & BÖHME, 2004

Actual size

The Alsco Long-fingered Frog has a plump body and a pointed snout. The back is tan with large, dark brown patches and the flanks are pink. The legs are tan with dark brown cross-stripes. A black stripe runs from the snout, through the eye, to the groin. The undersides of the legs and body are blue.

This small frog gets its name from the American Linen Supply Company (ALSCO), which funded the expedition that discovered it in 2000. It is known only from forest on the southern slopes of Mt. Tchabal Mbaba in western Cameroon, an area where there is extensive deforestation. The frog is associated with streams and is typically found hiding under stones. It has been heard calling in the dry season. The eggs are laid under stones in streams, where the tadpoles develop.

SIMILAR SPECIES

Most of the 18 species in the genus *Cardioglossa* described to date are found in West Africa, where many are threatened by the loss of their forest habitat. The Black Long-fingered Frog (*C. pulchra*) occurs in Cameroon and Nigeria and is listed as Endangered. The Nsoung Long-fingered Frog (*C. trifasciata*) occurs in Cameroon and is categorized as Critically Endangered.

FAMILY	Arthroleptidae: Astylosterninae
OTHER NAMES	None
DISTRIBUTION	Western and southwestern Cameroon
ADULT HABITAT	Upland forest
LARVAL HABITAT	Forest streams
CONSERVATION STATUS	IUCN Least Concern. Its habitat is being destroyed by logging and smallholder farming

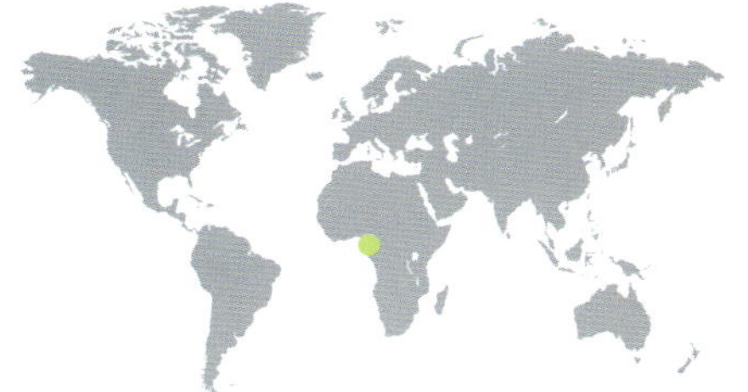

ADULT LENGTH
average 2 in (50 mm)

ASTYLOSTERNUS DIADEMATUS

VICTORIA NIGHT FROG

WERNER, 1898

The Latin species name of this frog means "wearing a crown" and refers to the stripe on its head between the eyes. Very little is known about the natural history of the genus, found in West and Central Africa and called night frogs. Unlike their close relatives the squeakers (genus *Arthroleptis*), the eggs of *Astylosternus* species hatch into free-swimming tadpoles that live in streams. Night frogs are very unusual in having concealed claws in their toes. These are actually the last bones in the toe and they seem to be used for defense—although only in extreme circumstances—the claw cutting its way out through the skin and underlying tissues.

Actual size

The Victoria Night Frog is a smooth-skinned frog with a broad head, a rounded snout, and large, prominent eyes. The fingers and toes are long and thin. The back is dark brown with darker spots and the hind legs have dark brown cross-bands. On the top of the head, between the eyes, there is a dark band bounded with yellow. The irises of the eyes are bright red.

SIMILAR SPECIES

The Benito River Night Frog (*Astylosternus batesi*) is a relatively common species found over a large area of Central Africa covering Gabon and neighboring countries. The Nganha Night Frog (*A. nganhanus*) is known from only five specimens, found at high altitude in Cameroon. It is listed as Critically Endangered. Genus *Astylosternus* contains 13 species.

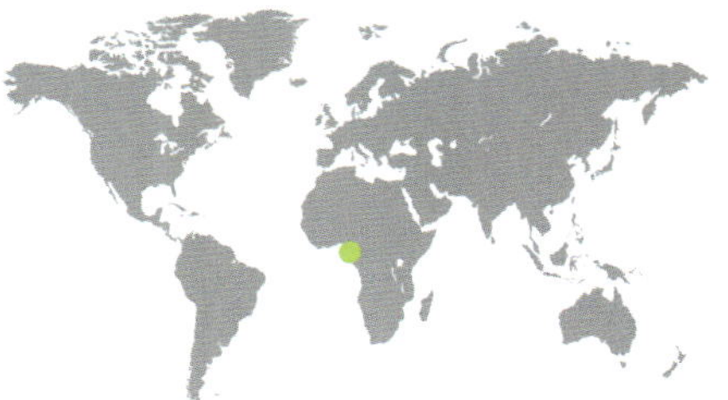

FAMILY	Arthroleptidae: Astylosterninae
OTHER NAMES	Previously *Trichobatrachus robustus*
DISTRIBUTION	Coastal Central Africa, from southeastern Nigeria to the Democratic Republic of the Congo
ADULT HABITAT	Lowland rainforest
LARVAL HABITAT	Fast-flowing rivers
CONSERVATION STATUS	IUCN Least Concern. Rare near villages as result of being harvested for food

ADULT LENGTH
Male
3¹³⁄₁₆–5⅛ in (98–130 mm)
Female
3⅛–4⅜ in (80–113 mm)

ASTYLOSTERNUS ROBUSTUS

HAIRY FROG

(BOULENGER, 1900)

This strange frog appears to be unique in two ways. First, males develop long, hairlike skin extensions on their flanks and thighs in the breeding season. Breeding has not been observed, but it is known that males protect the eggs, which are attached to submerged rocks in streams, by sitting on them. It is assumed that the "hairs," which have a good blood supply, enable the male to obtain oxygen from the water, obviating the need to breathe air. The second unique feature is that this frog can inflict quite serious wounds with sharp "claws" on its hind feet. These are not true claws but the broken ends of bones pushed out through the skin.

SIMILAR SPECIES

Astylosternus robustus was previously the only species in the genus *Trichobatrachus* before being transferred to *Astylosternus*. In Cameroon and other countries in West and Central Africa, it is collected for food, both as adults and as tadpoles, along with other frogs such as the Goliath Frog (*Conraua goliath*; page 527) and members of the genus *Astylosternus* (*A. diadematus*; page 429). Frogs in this part of the world are a significant source of food for local people.

The Hairy Frog has a short, robust body, a broad head with a rounded snout, and powerful, muscular legs. The large eyes have vertical pupils and the hind feet are partially webbed. It is generally olive-green or brown in color, and there is often a broad, darker patch down the middle of the back.

Actual size

FAMILY	Arthroleptidae: Astylosterninae
OTHER NAMES	None
DISTRIBUTION	Central and western Cameroon
ADULT HABITAT	Montane forest at altitudes of 4,000–8,700 ft (1,200–2,650 m)
LARVAL HABITAT	Streams
CONSERVATION STATUS	IUCN Endangered. Has a restricted range that is subject to habitat destruction

ADULT LENGTH
Male
average ⅞ in (21 mm)

Female
average 1 in (25 mm)

LEPTODACTYLODON PERRETI

PERRET'S EGG FROG

AMIET, 1971

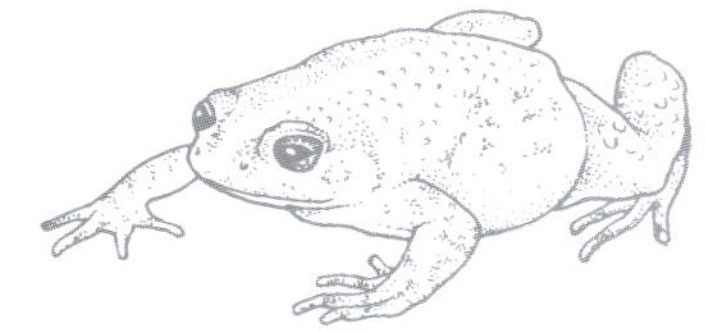

Compared with frogs from southern and eastern Africa, those of West Africa have been very little studied. There are a total of 15 species in the genus *Leptodactylodon*, many of them commonly known as "egg frogs" because of their shape. They are found in eastern Nigeria, Cameroon, Equatorial Guinea, and Gabon. Males of this species call from holes in the ground close to streams. The tadpoles, which live in streams, have umbrella-shaped mouths that point upward, and they hang in the water with their mouths at the surface. It is not clear whether this posture is related to feeding or to breathing.

Actual size

Perret's Egg Frog has a round, plump body, long, slender toes, and no webbing on its hands or feet. The skin on the back is orange to brown with numerous black spots. The limbs are darker brown with black markings, and the irises of the eyes are black.

SIMILAR SPECIES

The Redbelly Egg Frog (*Leptodactylodon erythrogaster*), listed as Critically Endangered, is found only on the southeastern slopes of Mt. Manengouba in western Cameroon. Its tadpoles are long and thin, an adaptation for wriggling into the crevices between stones, and are able to swim backward as well as forward. Mertens' Egg Frog (*L. mertensi*), also from Mt. Manengouba, is listed as Endangered, while Boulenger's Egg Frog (*L. boulengeri*), from the mountains of western Cameroon, is categorized as Near Threatened.

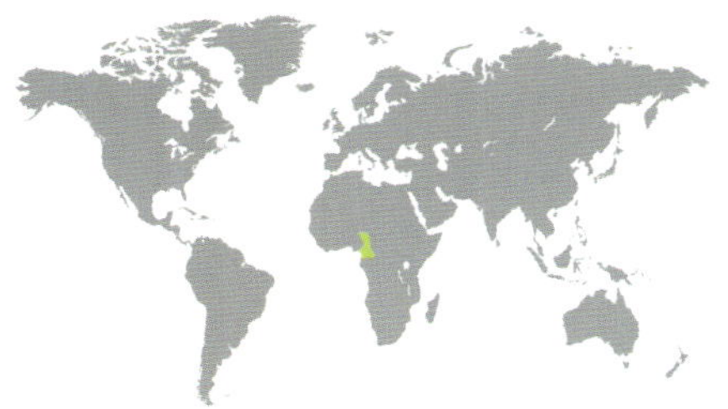

FAMILY	Arthroleptidae: Astylosterninae
OTHER NAMES	None
DISTRIBUTION	Western Africa, from southeastern Nigeria through Cameroon to northern Equatorial Guinea
ADULT HABITAT	Lowland rainforest up to 3,000 ft (900 m) altitude
LARVAL HABITAT	Streams
CONSERVATION STATUS	IUCN Least Concern. May be declining in places due to deforestation, but has a very large range that includes some protected areas

ADULT LENGTH
average 2¹⁄₁₆ in (53 mm)

NYCTIBATES CORRUGATUS

SOUTHERN NIGHT FROG

BOULENGER, 1904

Actual size

The Southern Night Frog has a large head, large eyes with vertical pupils, and numerous folds of skin on its back. The limbs are slender, and the fingers and toes are long with slightly swollen tips. The upper surfaces are brown and there are narrow darker stripes on the legs.

This little-studied frog gets its scientific species name, *corrugatus*, from the texture of the skin on its back, which contains many small chevron-shaped folds. A common species, it lives in lowland areas, where it breeds in fast-flowing rocky streams containing clean water. Males call from the ground close to streams. The tadpoles have long, muscular tails, enabling them to swim against the current.

SIMILAR SPECIES

Nyctibates corrugatus is the only species in its genus. It is most similar to two other West African frogs, the Gaboon Forest Frog (*Scotobleps gabonicus*; page 433) and the Hairy Frog (*Astylosternus robustus*; page 430).

FAMILY	Arthroleptidae: Astylosterninae
OTHER NAMES	None
DISTRIBUTION	Western Africa, from Nigeria in the north to Angola in the south
ADULT HABITAT	Lowland rainforest
LARVAL HABITAT	Streams
CONSERVATION STATUS	IUCN Least Concern. Has a very large range but is declining in many places due to deforestation

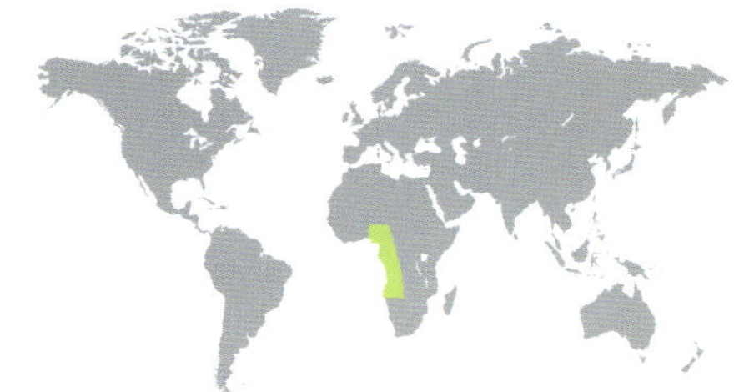

ADULT LENGTH
average 2$^{3}/_{16}$ in (57 mm)

SCOTOBLEPS GABONICUS

GABOON FOREST FROG

BOULENGER, 1900

Although this frog is widespread and common, almost all we know about it dates from the original 1900 description of a single specimen by the Belgian biologist George Albert Boulenger, who described more than 2,000 animal species, including many frogs. It is found in lowland rainforest and breeds in streams, apparently preferring those that are wide and shallow with sandy banks. It has been heard calling in April and its eggs have been found in May. A survey conducted in 2009 found several individuals infected with the pathogen that causes chytridiomycosis, but it is not known how badly it may be affected by the disease.

Actual size

SIMILAR SPECIES

Scotobleps gabonicus is the only species in its genus. It is most similar to two other West African frogs, the Southern Night Frog (*Nyctibates corrugatus*; page 432) and the Hairy Frog (*Astylosternus robustus*; page 430).

The Gaboon Forest Frog has a plump body, a large head, large eyes with vertical pupils, and numerous warts on its back. The fingers and toes are long and have slightly swollen tips, and the hind feet are partially webbed. The upper surfaces are brown, with small black spots on the back and stripes on the legs. There is a dark patch on the head between the eyes.

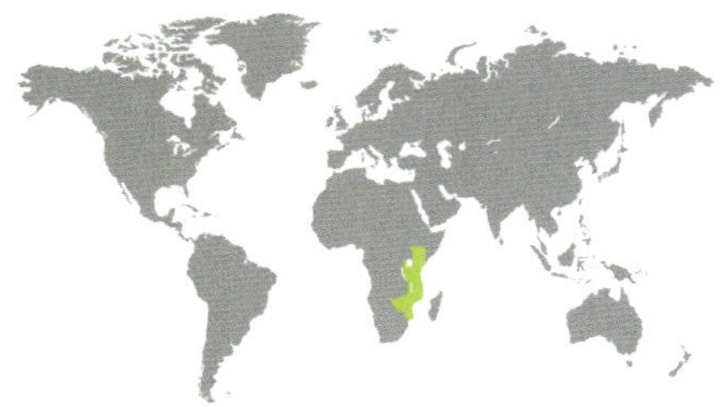

FAMILY	Arthroleptidae: Leptopelinae
OTHER NAMES	Brown Forest Tree Frog, Johnston's Tree Frog
DISTRIBUTION	Kenya, Tanzania, Malawi, Mozambique, Zimbabwe
ADULT HABITAT	Dry lowland forest
LARVAL HABITAT	Ponds and pools
CONSERVATION STATUS	IUCN Least Concern. Has a very large range, but is threatened in places by habitat destruction

ADULT LENGTH
Male
1¾–1^15^⁄16 in (44–50 mm)
Female
2⅜–2^13^⁄16 in (60–70 mm)

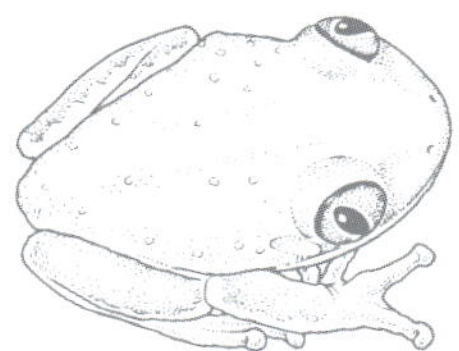

LEPTOPELIS FLAVOMACULATUS

AFRICAN YELLOW-SPOTTED TREE FROG

(GÜNTHER, 1864)

This large arboreal frog is very variable in color and only some individuals live up to its name. In fact, only young individuals reliably have yellow spots. Males call from sunset to dawn, from vegetation up to 13 ft (4 m) above ground, or from a burrow in the ground, producing a soft, drawn-out cry that sounds more distant than it actually is. The eggs are laid in a shallow pit close to a pond or pool. From there, the tadpoles find their way to the water.

SIMILAR SPECIES

There are 46 species in the genus *Leptopelis*, found in Africa's forests. The Amani Tree Frog (*L. vermiculatus*) is very similar in appearance to brown individuals of *L. flavomaculatus*. Its habitat in the East Usambara Mountains of Tanzania has recently come under serious threat from the activities of illegal gold miners, and the species is listed as Endangered. See also the Natal Tree Frog (*L. natalensis*; page 435) and the Uluguru Tree Frog (*L. uluguruensis*; page 436).

The African Yellow-spotted Tree Frog is a robust frog with short limbs and large eyes. There are two color phases: Juveniles and some adult males are bright green with yellow spots; and adult females and all other adult males are gray-brown with a dark brown triangular patch on the back. The toes are webbed, and there are large adhesive disks on the fingers and toes.

Actual size

FAMILY	Arthroleptidae: Leptopelinae
OTHER NAMES	Forest Tree Frog, Raucous Tree Frog
DISTRIBUTION	Coastal KwaZulu-Natal, South Africa
ADULT HABITAT	Coastal forest and swamps
LARVAL HABITAT	Stagnant pools and swamps
CONSERVATION STATUS	IUCN Least Concern. Declining in places as a result of loss of habitat through water extraction

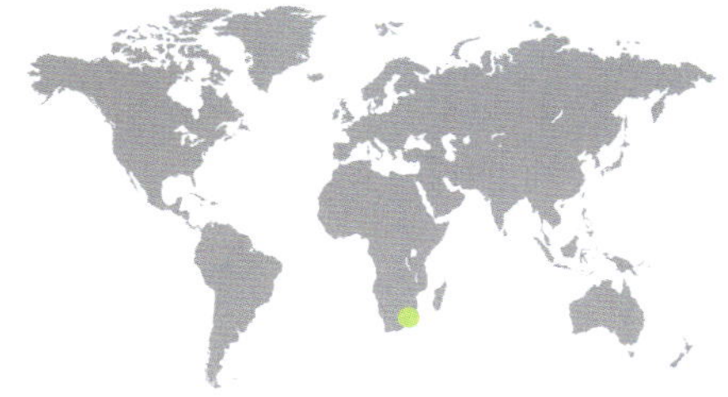

LEPTOPELIS NATALENSIS

NATAL TREE FROG

(SMITH, 1849)

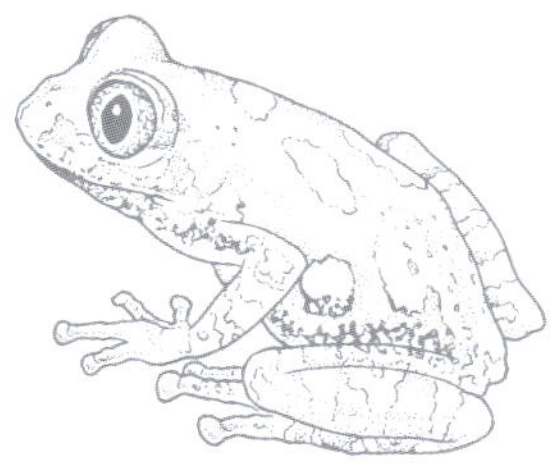

ADULT LENGTH
Male
average $1^{15}/_{16}$ in (49 mm)
Female
average $2^{3}/_{8}$ in (60 mm);
up to $2^{5}/_{8}$ in (65 mm)

This handsome frog breeds in the summer rainy season. Males call from high up in trees, producing a loud "yack… yack," preceded by a quiet buzz. The female climbs up to find the male, and the pair, in amplexus, then descend to the ground, where the female digs a shallow nest in the soil or among decaying leaves close to water. She lays a clutch of nearly 200 eggs, which she covers with leaves. On hatching, the tadpoles make their own way to water. Unusually, they are able to wriggle over land and can even jump by flicking their tails.

SIMILAR SPECIES

The Mozambique Tree Frog (*Leptopelis mossambicus*) has a similar natural history to, and a range overlapping that of, *L. natalensis*. It is uniformly brown. The Long-toed Tree Frog (*L. xenodactylus*) is lime-green in color. It is found in a small area in the highlands of KwaZulu-Natal, and has long fingers and toes that lack adhesive disks. Its habitat is being degraded and destroyed, and it is listed as Endangered. See also the African Yellow-spotted Tree Frog (*L. flavomaculatus*; page 434) and the Uluguru Tree Frog (*L. uluguruensis*; page 436).

The Natal Tree Frog has huge forward-pointing eyes with red or gold irises. It is highly variable in color, being green, pale brown, or cream. Some individuals are beautifully mottled in a combination of these colors. There are well-developed disks on its fingers and toes.

Actual size

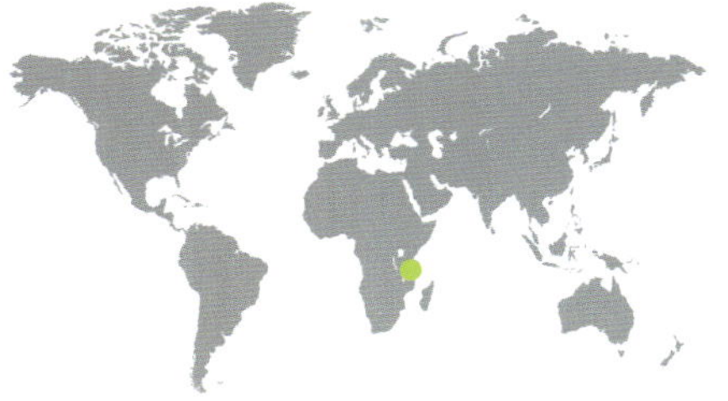

FAMILY	Arthroleptidae: Leptopelinae
OTHER NAMES	Uluguru Forest Tree Frog
DISTRIBUTION	Eastern Arc Mountains, Tanzania
ADULT HABITAT	Rainforest
LARVAL HABITAT	Forest pools
CONSERVATION STATUS	IUCN Near Threatened. Threatened by habitat loss due to deforestation and illegal mining

ADULT LENGTH
Male
$1\frac{1}{8}$–$1\frac{1}{2}$ in (28–38 mm)
Female
$1\frac{13}{16}$–$1\frac{7}{8}$ in (46–48 mm)

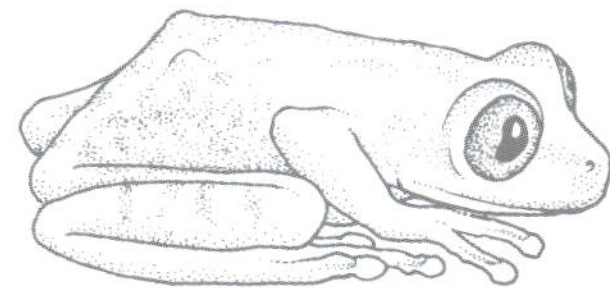

LEPTOPELIS ULUGURUENSIS

ULUGURU TREE FROG

BARBOUR & LOVERIDGE, 1928

Actual size

The name of this frog refers to the Uluguru Mountains, one of several ranges that make up Tanzania's Eastern Arc Mountains. These are ancient highlands, covered in forest and grassland, that support a very high diversity of plants and animals found nowhere else. This makes them one of 24 biodiversity hotspots that have been identified across the world. The Uluguru Tree Frog is one of many tree frogs that are found in these mountains. Very little is known about it other than that the male's call is a brief "clack."

SIMILAR SPECIES

Forests in the Eastern Arc Mountains are under threat from logging and clearance to make room for agriculture. The Eastern Usambara range is also threatened by illegal gold mining, which leads to streams being polluted by mercury. The Amani Tree Frog (*Leptopelis vermiculatus*) occurs in green and brown color morphs and is listed as Endangered. Susan's Tree Frog (*L. susanae*), from Ethiopia, is also categorized as Endangered. See also the African Yellow-spotted Tree Frog (*L. flavomaculatus*; page 434) and the Natal Tree Frog (*L. natalensis*; page 435).

The Uluguru Tree Frog has a broad head, a blunt snout, and huge eyes with vertical pupils and brown or silver irises. The feet are partially webbed, and there are large adhesive disks on the fingers and toes. The upper surfaces are bluish green or yellowish brown, often with white or yellow spots and rings.

FAMILY	Hyperoliidae: Acanthixalinae
OTHER NAMES	None
DISTRIBUTION	West and Central Africa, from southeastern Nigeria in the west, through Equatorial Guinea and Gabon, to northeastern Democratic Republic of the Congo in the east
ADULT HABITAT	Lowland rainforest
LARVAL HABITAT	Water-filled tree-holes
CONSERVATION STATUS	IUCN Least Concern. Declining in places and threatened by deforestation

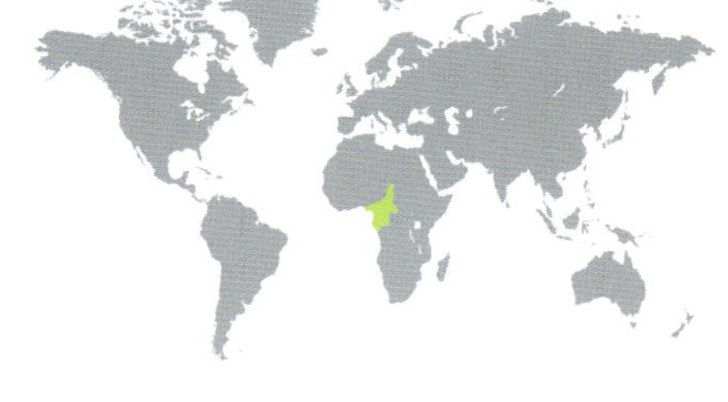

ADULT LENGTH
up to 1 7/16 in (36 mm)

ACANTHIXALUS SPINOSUS

AFRICAN WART FROG

(BUCHHOLZ & PETERS, 1875)

This very secretive frog gets its name from numerous large warts on its head, back, and legs. It spends the day in water-filled holes in trees, with only the tip of its nose out of water. No mating call has been reported and males appear to be mute. It has been suggested that males and females locate one another by smell. Eggs are laid in clutches of eight to ten above the waterline in tree-holes. On hatching, the tadpoles fall into the water, where they take around three months to reach metamorphosis. Young frogs are orange and purple in color.

SIMILAR SPECIES

There is one other species in this genus, the Ivory Coast Wart Frog (*Acanthixalus sonjae*). It has a wider head than *A. spinosus* and is found in Côte d'Ivoire and Ghana. Its population is declining and it is listed as Vulnerable. Both species are dependent on large, mature trees for the water-filled tree-holes in which they live and breed, but such trees are becoming rarer and rarer.

Actual size

The African Wart Frog has a long snout, large, protruding eyes, and many warts on its head, back, and legs. There is an array of spines on the lower part of each hind leg. There are well-developed disks on the fingers and toes; these are larger in males than in females. The upper surfaces are olive-green or brown with black blotches.

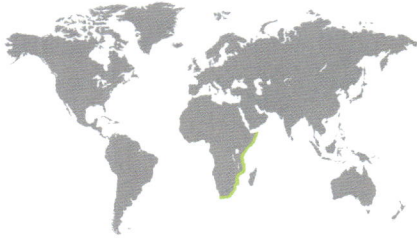

FAMILY	Hyperoliidae: Hyperoliinae
OTHER NAMES	Delicate Leaf-folding Frog, Pickersgill's Banana Frog
DISTRIBUTION	Coastal lowlands of East Africa, from Somalia in the north to South Africa in the south
ADULT HABITAT	Coastal woodland, bushland, and grassland
LARVAL HABITAT	Permanent marshes and pools
CONSERVATION STATUS	IUCN Least Concern. Has a very large range and lives in a broad range of habitats, including human-altered ones

ADULT LENGTH
Male
9/16–7/8 in (15–22 mm)
Female
5/8–15/16 in (16–24 mm)

AFRIXALUS DELICATUS

DELICATE SPINY REED FROG

PICKERSGILL, 1984

Actual size

The Delicate Spiny Reed Frog has a slender, elongated body, large, protruding eyes, and a pointed snout. There is webbing between its toes and adhesive disks on both the fingers and toes. It is very variable in color, the upper surfaces being yellow-brown or silvery with dark longitudinal stripes, two of which merge between the eyes.

This small, slender frog breeds at the edge of pools, gluing together the leaves of long-leaved plants to form a protective chamber for the eggs. The eggs are laid just above or just below the water's surface. Occasionally a female will lay her eggs in a series of batches up to three days apart, these being fertilized by different males. She can mate with up to three males in this process. Prior to mating, males gather in small groups, each producing a buzz-like call lasting up to 22 seconds. The tadpoles are long and streamlined in shape, and grow to 1½ in (38 mm) long.

SIMILAR SPECIES

Afrixalus delicatus has a more pointed snout than the otherwise similar Natal Spiny Reed Frog (*A. spinifrons*; page 440). The Snoring Leaf-folding Frog (*A. crotalus*) is another small reed frog that breeds in ephemeral ponds in the grasslands of Malawi, Mozambique, and Zimbabwe. The male's call is a series of indistinct rattles and clicks.

FAMILY	Hyperoliidae: Hyperoliinae
OTHER NAMES	Fornasini's Spiny Reed Frog
DISTRIBUTION	Eastern and southern Africa, from Kenya to South Africa
ADULT HABITAT	Savanna, bushveld, and grassland, in vegetation around pools and swamps
LARVAL HABITAT	Pools and swamps
CONSERVATION STATUS	IUCN Least Concern. Has a very large range and shows no evidence of a general decline

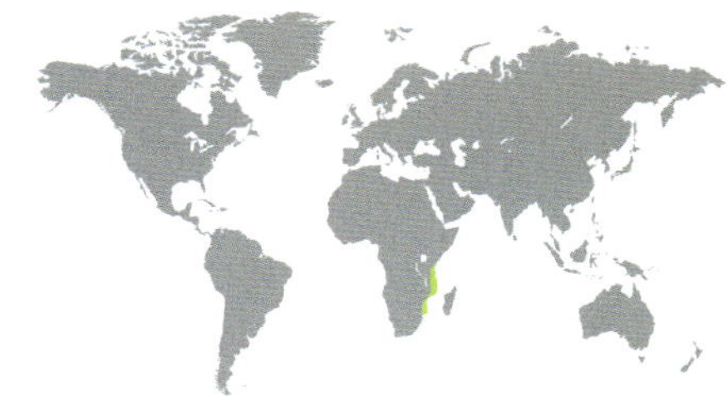

AFRIXALUS FORNASINI

GREATER LEAF-FOLDING FROG

(BIANCONI, 1849)

ADULT LENGTH
Male
1⅛–1½ in (29–38 mm)

Female
1³⁄₁₆–1⁹⁄₁₆ in (30–40 mm)

Between late September and June, males of this species call from elevated positions on emergent aquatic vegetation. They defend their call sites against other males by calling, chasing, and, occasionally, fighting. When a female finds a male, the pair move to a position closer to the pond and lay 20–50 eggs on a leaf overhanging the water. The male uses his hind legs to fold the leaf around the eggs, holding the leaf until it is firmly glued by a secretion produced by the female. When the eggs hatch, the tadpoles fall into the water.

Actual size

The Greater Leaf-folding Frog has an elongated body, a long snout, and long, slender limbs. There are adhesive disks and webbing on both fingers and toes. The back is pale, often with a conspicuous dark brown longitudinal stripe down the middle. The back, flanks, and legs are covered with small black-tipped spines, each positioned on a small white spot.

SIMILAR SPECIES

There are 37 species in this African genus. These frogs often perch in very exposed positions under a hot sun and have a very high resistance to water loss from the skin. The Golden Leaf-folding Frog (*Afrixalus aureus*) is a common, small species from Mozambique, South Africa, and Swaziland. Another member of the genus, the Knysna Leaf-folding Frog (*A. knysnae*), has a very restricted range and is listed as Endangered. See also the Natal Spiny Reed Frog (*A. spinifrons*; page 440).

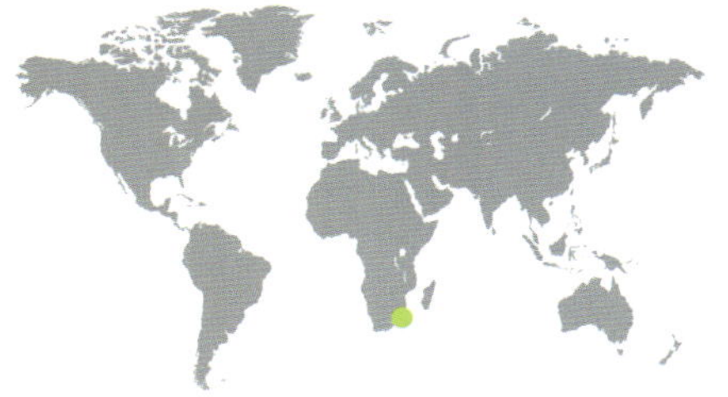

FAMILY	Hyperoliidae: Hyperoliinae
OTHER NAMES	Golden Spiny Reed Frog, Natal Banana Frog, Natal Leaf-folding Frog
DISTRIBUTION	KwaZulu-Natal and Eastern Cape coast, South Africa
ADULT HABITAT	Shrubland and dry forest
LARVAL HABITAT	Ephemeral ponds and dams
CONSERVATION STATUS	IUCN Least Concern. Declining due to loss of habitat to agriculture and urbanization

ADULT LENGTH
Male
up to ¾ in (20 mm)

Female
up to 1 in (25 mm)

AFRIXALUS SPINIFRONS

NATAL SPINY REED FROG

(COPE, 1862)

Actual size

The Natal Spiny Reed Frog gets its name from the numerous small black spines on its back, head, and snout. It has a rounded snout and large, prominent eyes. There are well-developed adhesive disks on its fingers and toes. The upper surfaces are yellow or brown, and there may be one or two dark stripes running the length of the back.

This small frog breeds from spring to midsummer, sometimes gathering in large aggregations, in which males form choruses. The rather quiet call has two parts: "zip… trill." The first part is an aggressive signal directed at other males, while the second part of the call is responded to by females. In large choruses, around five percent of males are silent "satellite" males that seek to intercept females as they approach the callers. A clutch of 10–50 eggs is laid on leaves just above the water's surface and folded into a leaf during mating. The tadpoles fall into the water on hatching.

SIMILAR SPECIES

The wide-ranging Delicate Spiny Reed Frog (*Afrixalus delicatus*; page 438) has a more pointed snout than *A. spinifrons*. The Uluguru Banana Frog (*A. uluguruensis*) occurs only in the Eastern Arc Mountains of Tanzania, where it is threatened by deforestation and gold mining. It is listed as Vulnerable. See also the Greater Leaf-folding Frog (*A. fornasini*; page 439).

FAMILY	Hyperoliidae: Hyperoliinae
OTHER NAMES	None
DISTRIBUTION	Cameroon, Equatorial Guinea, Gabon, Republic of the Congo, Democratic Republic of the Congo
ADULT HABITAT	Tropical forest
LARVAL HABITAT	Still and flowing water
CONSERVATION STATUS	IUCN Least Concern. A common species that seems able to adapt to habitat change

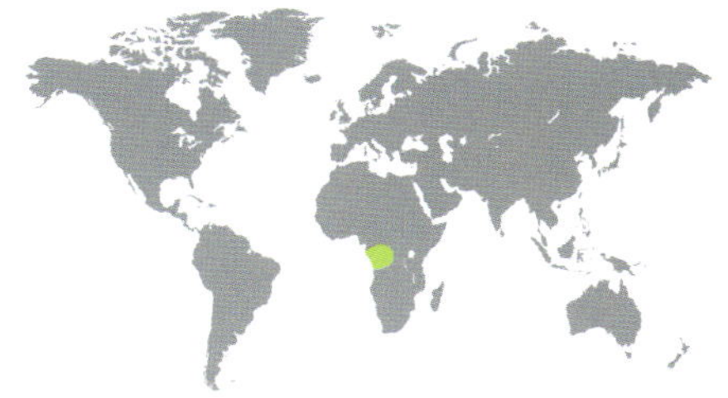

CRYPTOTHYLAX GRESHOFFII

GRESHOFF'S WAX FROG

(SCHILTHUIS, 1889)

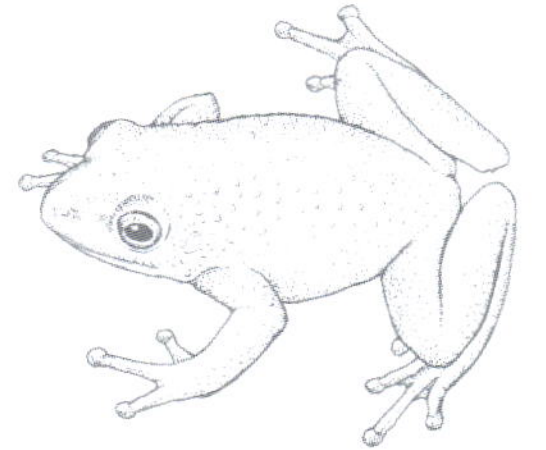

ADULT LENGTH
Male
1$^{9}/_{16}$–2$^{1}/_{8}$ in (39–54 mm)

Female
1$^{7}/_{8}$–2$^{1}/_{4}$ in (48–58 mm)

This common frog is typically found in forest clearings, but occurs in a variety of habitats, including open country. It is strongly associated with swamps, lakes, streams, and large rivers, where it breeds. The male calls from low branches near water, producing a sound likened to knocking on wood. The female lays 11–19 large eggs in a mass of jelly on a leaf hanging over water. On hatching, the tadpoles fall into the water below. Females are more colorful than males, being generally redder and having bright orange hands and feet.

SIMILAR SPECIES

The only other species in the genus *Cryptothylax* is the Tiny Wax Frog (*C. minutus*). Smaller than *C. greshoffii*, it is known only from Lake Tumba in the Democratic Republic of the Congo. It has not been recorded for many years and is included on a list of 200 amphibians that are described as "lost."

Greshoff's Wax Frog has large eyes with diamond-shaped pupils and golden-green irises. The feet are webbed and there are disks on the fingers and toes. The upper surfaces range from tawny brown to red, and the underside is pink to white. The female is generally redder than the male.

Actual size

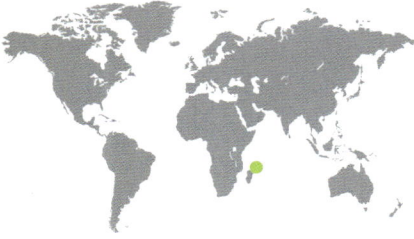

FAMILY	Hyperoliidae: Hyperoliinae
OTHER NAMES	Blue-back Reed Frog
DISTRIBUTION	Northeastern Madagascar
ADULT HABITAT	Edges of rainforest, dry forest, savanna, dunes. Also farmland, villages, urban areas
LARVAL HABITAT	Ephemeral and permanent pools
CONSERVATION STATUS	IUCN Least Concern. Has a very large range and shows no evidence of a general decline

ADULT LENGTH
Male
up to 1⅜ in (35 mm)

Female
up to 1 9/16 in (40 mm)

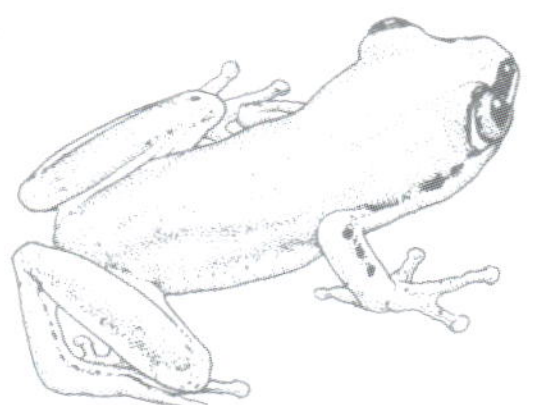

HETERIXALUS MADAGASCARIENSIS

MADAGASCAR REED FROG

(DUMÉRIL & BIBRON, 1841)

Actual size

The Madagascar Reed Frog has large, protuberant eyes with gold irises. There is a dark stripe between the nostril and the eye, and the hands, feet, and undersides of the legs are orange. The color on the back is highly variable, being green, brown, blue, or white, and it changes with temperature, tending toward brown when cold and white when hot.

This small frog is a member of the genus *Heterixalus*, endemic to the island of Madagascar, which because of its isolation is inhabited by many frogs found nowhere else. Widespread destruction of Madagascar's forest has had an adverse effect on numerous frogs, but these reed frogs seem to have been less affected than most, probably because they thrive in human-created habitats such as rice fields and sugar-cane plantations. The species appears to breed throughout the year, males calling at night, after rain, from sites close to ponds and swamps. By day, it hides in the water if disturbed.

SIMILAR SPECIES

There are 11 species in the genus *Heterixalus*, all with a similar life history. Only two are of conservation concern, being listed as Near Threatened. Carbone's Reed Frog (*Heterixalus carbonei*) is never seen outside forest and seems not to have adapted to human-dominated habitats. Rutenberg's Reed Frog (*H. rutenbergi*) is a habitat specialist, living in moorland at high altitude and breeding only in acidic water.

FAMILY	Hyperoliidae: Hyperoliinae
OTHER NAMES	Argus Sedge Frog, Yellow-spotted Reed Frog
DISTRIBUTION	Coastal areas of East Africa, from Somalia in the north to South Africa in the south
ADULT HABITAT	Near water in savanna
LARVAL HABITAT	Ephemeral or permanent shallow pools and marshes
CONSERVATION STATUS	IUCN Least Concern. Has a very large range but is subject to habitat loss in South Africa

ADULT LENGTH

$1\frac{1}{16}$–$1\frac{5}{16}$ in (27–34 mm)

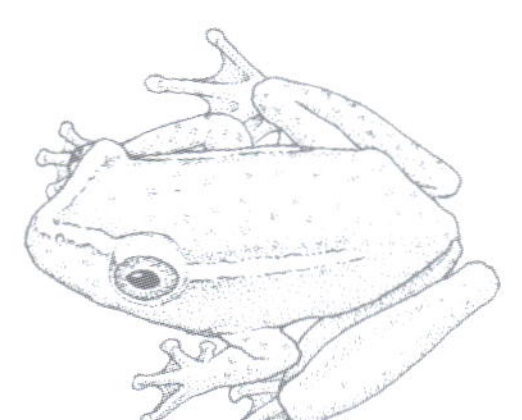

HYPEROLIUS ARGUS

ARGUS REED FROG

PETERS, 1854

This pretty little frog shows striking sexual dimorphism in appearance, males and females differing in both color and skin pattern. Breeding occurs after spring rains, males calling from emergent aquatic vegetation and producing a rapidly repeated, low-pitched clucking sound. The female lays 200 eggs in clutches of 30. These are typically found underwater, either because they are laid there or because they are laid as the water is rising. The color difference between the sexes has been used to study the effect of chemical pollutants on sexual development. Young males exposed to such chemicals are feminized, developing female coloration.

Actual size

The Argus Reed Frog is a wide-bodied reed frog with webbed hands and feet that have well-developed adhesive disks. Males are typically green above, with a pale midline stripe, and white below. Females are light brown or reddish above, orange below, and with red limbs and feet. Their back is often decorated with large round cream spots, edged in black.

SIMILAR SPECIES

Differentiating *Hyperolius argus* from related species on the basis of appearance alone is problematic, as there is variation within each sex, and from one part of the species' range to another. In South Africa, it occurs alongside the Water Lily Reed Frog (*H. pusillus*; page 448), which is considerably smaller. And in West Africa it is similar in size and appearance to the Dotted Reed Frog (*H. guttulatus*), and there are another 142 *Hyperolius* species across Sub-Saharan Africa.

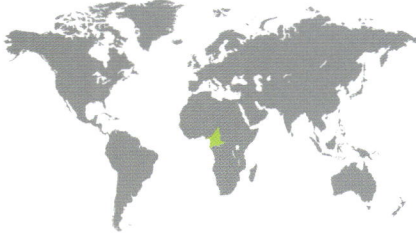

FAMILY	Hyperoliidae: Hyperoliinae
OTHER NAMES	*Chlorolius koehleri*
DISTRIBUTION	Cameroon, southeastern Nigeria, Equatorial Guinea, northern Gabon
ADULT HABITAT	Forest and grassy meadows up to 5,900 ft (1,800 m) altitude
LARVAL HABITAT	Streams
CONSERVATION STATUS	IUCN Least Concern. Its population status is known but is likely to be declining as a result of deforestation

ADULT LENGTH
Male
1–1¹⁄₁₆ in (26–27 mm)

Female
unknown

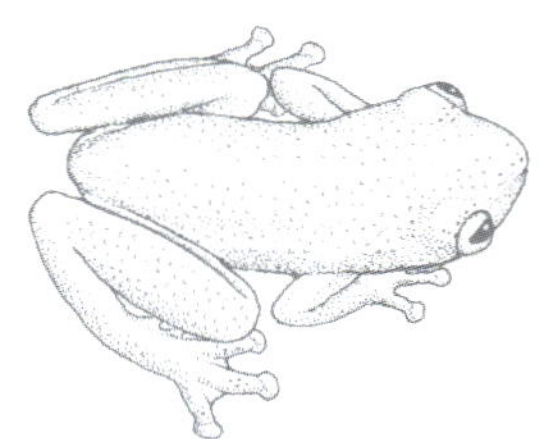

HYPEROLIUS KOEHLERI

KÖHLER'S GREEN REED FROG

MERTENS, 1940

Actual size

Köhler's Green Reed Frog has a flattened head, large, protruding eyes, and large disks on its fingers and toes. The upper surfaces are grass-green with diffuse reddish-brown spots. The legs, hands, and feet are yellowish green and the underside is pale green. The throat is turquoise-blue.

Males of this little-known frog have a very quiet call, making them difficult to find. They have black spines on their flanks and on the underside of their limbs and feet. These probably serve to help the male grasp the female in the fast-flowing streams where they breed. The species lives in lowland and montane forest and grassy meadows, and is also found in coffee plantations and in forest that has been degraded by logging and other human activities. It lays its eggs in streams, where the tadpoles develop.

SIMILAR SPECIES

Hyperolius koehleri used to be the only species in the genus *Chlorolius*. It differs from other *Hyperolius* in that males lack a vocal sac and have spines on their flanks and on the underside of their hind legs.

FAMILY	Hyperoliidae: Hyperoliinae
OTHER NAMES	Marbled Reed Frog
DISTRIBUTION	East Africa, from northern Mozambique to South Africa
ADULT HABITAT	Varied habitats, close to temporary or permanent ponds, dams, marshes, etc.
LARVAL HABITAT	Temporary or permanent ponds, dams, and marshes
CONSERVATION STATUS	IUCN Least Concern. Has a very large range and shows no evidence of a general decline. Expanding its range in South Africa

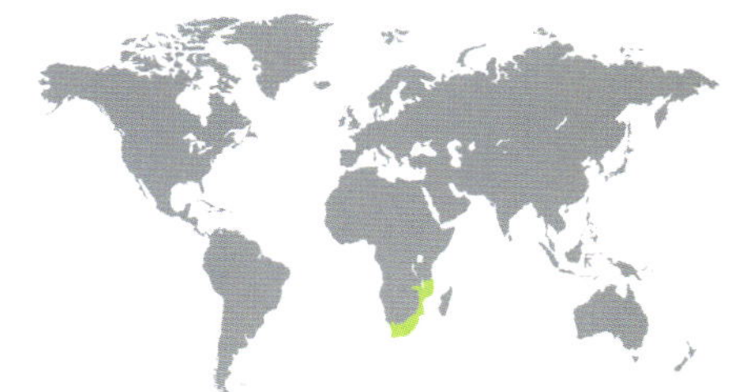

HYPEROLIUS MARMORATUS

PAINTED REED FROG

RAPP, 1842

ADULT LENGTH
Male
15⁄16–1¼ in (25–31 mm)
Female
1 1⁄16–1 5⁄16 in (27–33 mm)

Despite their very small size, males of this common species produce one of the loudest calls of any frog: a very short "whipp" sound. Calling is energetically very expensive and males typically call for just a few nights, then leave the chorus for a few nights to build up their energy reserves, before returning. Females show a preference for the lower-pitched calls of larger males, but the most important determinant of male mating success is the number of nights during a season that they attend a chorus and call. Females lay 150–650 eggs in water, in clumps of around 20.

Actual size

The Painted Reed Frog has a rounded snout and large eyes. There are well-developed adhesive disks on the fingers and toes. Broadly speaking, there are three patterns of marking on the back: spotted, striped (shown here), and marbled, with some individuals having combinations of these. Color is even more variable, the most consistent feature being that the hands and feet are bright red.

SIMILAR SPECIES

The status of *Hyperolius marmoratus* is a subject of considerable debate. While some biologists regard it as a separate species, others regard it as a subspecies of the Common Reed Frog (*H. viridiflavus*), which occurs over much of sub-Saharan Africa. Appearance is of little use in this context, as at least 40 different color patterns have been described across the *H. viridiflavus* complex.

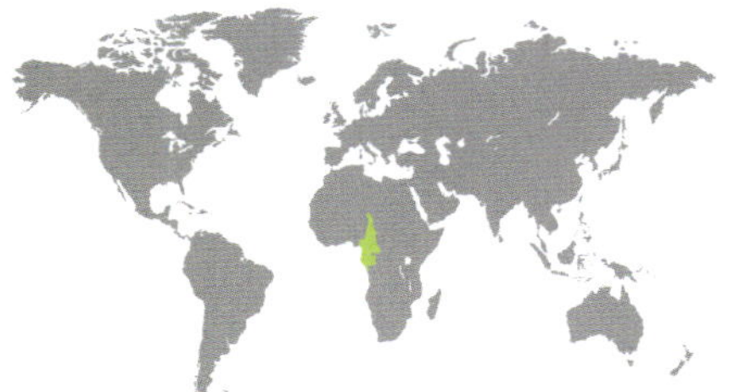

FAMILY	Hyperoliidae: Hyperoliinae
OTHER NAMES	Previously *Alexteroon obstetricans*
DISTRIBUTION	Cameroon and Gabon
ADULT HABITAT	Dense tropical forest
LARVAL HABITAT	Streams
CONSERVATION STATUS	IUCN Least Concern. Shows no evidence of decline but is vulnerable to deforestation

ADULT LENGTH
1–1¼ in (26–31 mm)

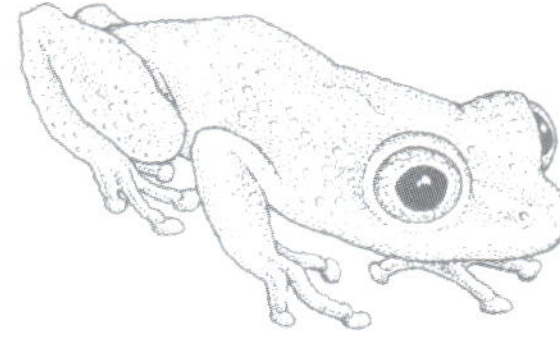

HYPEROLIUS OBSTETRICANS

COMMON EGG-GUARDING FROG

(AHL, 1931)

Actual size

The Common Egg-guarding Frog has a flattened body, large adhesive disks on its fingers and toes, and huge eyes. There are small frills of skin along the edges of its limbs. The skin on the back is bright green, dotted with small, slightly raised white spots. The underside of the limbs is turquoise and the skin on the belly is translucent, showing the organs beneath.

This attractive little frog is apparently not gregarious during mating, males spacing themselves far apart when calling from foliage above streams. Their call is a metallic "toc," repeated five or six times. A pair produce a clutch of 40–50 eggs encased in a mass of clear jelly, which the female then guards. The jelly apparently does not dissolve as it does in many species, and the female has to help the tadpoles to wriggle free from it. They fall into the water below, where they feed on plants.

SIMILAR SPECIES

Hyperolius obstetricans and two other Central African species, the Cross-banded Egg-guarding Frog (*H. hypsiphonus*) and the Smooth Egg-guarding Frog (*H. jynx*), were until recently included in the genus *Alexteroon*. Both inhabit dense forests, with *H. hypsiphonus* a widespread species in Cameroon, Gabon, and the Congo, where it calls high up in trees, but lays its eggs low down over streams. *Hyperolius jynx* is very small, and listed as Critically Endangered because it is found at only one locality in Cameroon.

FAMILY	Hyperoliidae: Hyperoliinae
OTHER NAMES	Avoca Reed Frog
DISTRIBUTION	Coast of KwaZulu-Natal, South Africa; Swaziland
ADULT HABITAT	Coastal grassland and bushveld
LARVAL HABITAT	Ponds and marshes
CONSERVATION STATUS	IUCN Endangered. Has a very restricted range, subject to habitat destruction and pollution

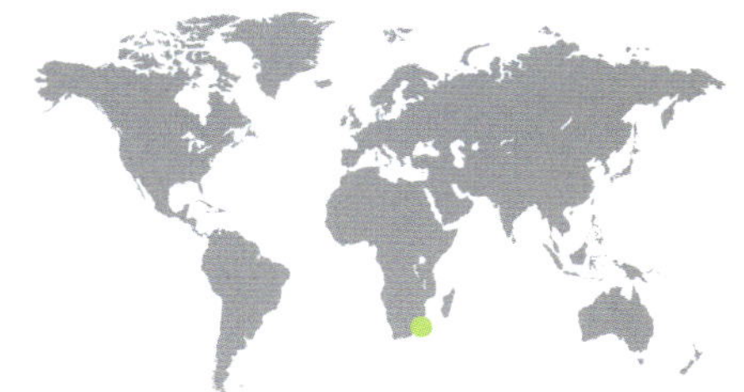

HYPEROLIUS PICKERSGILLI

PICKERSGILL'S REED FROG

RAW, 1982

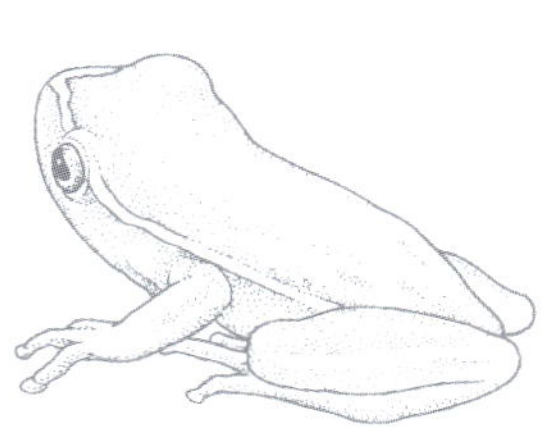

ADULT LENGTH
Male
up to ⅞ in (22 mm)

Female
up to 1⅛ in (29 mm)

This very secretive little frog breeds in stagnant water in coastal ponds and marshes. Males produce soft, cricket-like calls from concealed positions in dense vegetation near water. The eggs are laid in a gelatinous mass attached to vegetation above the water, the tadpoles falling into the water when the eggs hatch. This species is sexually dimorphic in color (see caption). It has a very small range and is threatened by the loss of its habitat to agriculture and urbanization. It has also been adversely affected by DDT, used to control malaria-carrying mosquitoes.

Actual size

Pickersgill's Reed Frog has two color forms. Juveniles and many mature males are pale brown with a dark-edged white or yellow stripe running along each side from the snout, through the eye, to the groin. All females and some mature males become bright yellowish green, the longitudinal stripes fading as they get older.

SIMILAR SPECIES

The range of *Hyperolius pickersgilli* overlaps that of the Yellow-striped Reed Frog (*H. semidiscus*; page 449), a common species found in Swaziland and South Africa. The Long-nosed Reed Frog (*H. nasutus*) has a large range across sub-Saharan Africa. It has an elongated body and a pointed snout and the male's call is a high-pitched chirp.

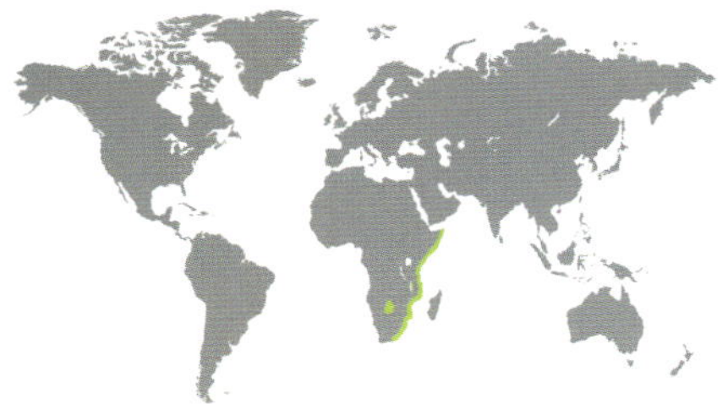

FAMILY	Hyperoliidae: Hyperoliinae
OTHER NAMES	Dwarf Reed Frog, Lily Pad Frog, Translucent Tree Frog
DISTRIBUTION	Coastal plains in East Africa, from Somalia in the north to South Africa in the south; also inland in Malawi and Botswana
ADULT HABITAT	Open swamps in savanna and shrubland
LARVAL HABITAT	Ponds
CONSERVATION STATUS	IUCN Least Concern. A common species in many places with a range that includes some protected areas

ADULT LENGTH
Male
⅝–¾ in (16–20 mm)

Female
$^{11}/_{16}$–1 in (18–25 mm)

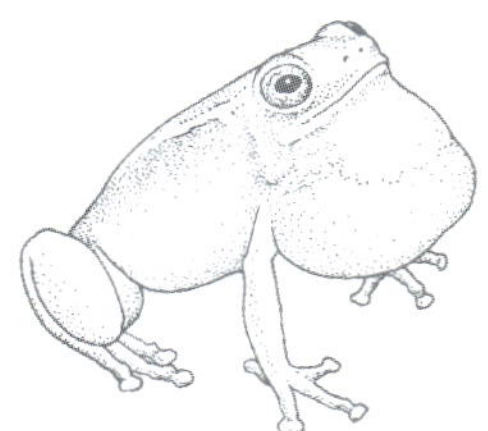

HYPEROLIUS PUSILLUS

WATER LILY REED FROG

COPE, 1862

Actual size

The Water Lily Reed Frog has a blunt snout, webbed feet, and well-developed disks on its fingers and toes. The upper surfaces are translucent green, sometimes decorated with small, dark spots. The fingers and toes may be green, yellow, or orange. The belly has a translucent patch through which the frog's internal organs can be seen.

As its name implies, this tiny frog is associated with water lilies. Heavy rain triggers the formation of large choruses of males at lily ponds. They call while sitting on lily leaves, producing a series of high-pitched clicks. Their vocal sac is unusual in that it has two subsidiary lobes. Males are very aggressive in defense of their calling sites and butt one another with their inflated vocal sacs. The female lays up to 500 eggs, in batches of 20–120, deposited in a single layer between the leaves of floating vegetation. The eggs and newly hatched tadpoles are green.

SIMILAR SPECIES

The Green Reed Frog (*Hyperolius viridis*) is another small, entirely green reed frog that occurs only in western Tanzania. It differs from *H. pusillus* in calling from grass around the edge of ponds. The Sharp-nosed Reed Frog (*H. acuticeps*) also calls from elevated positions around ponds, producing insect-like chirps. It has a large range, from Ethiopia to Zimbabwe and South Africa.

FAMILY	Hyperoliidae: Hyperoliinae
OTHER NAMES	Arum Frog, Hewitt's Reed Frog, Yellow-flanked Reed Frog
DISTRIBUTION	Eastern coast of South Africa, Swaziland
ADULT HABITAT	Dense reedbeds close to deep water
LARVAL HABITAT	Deep ponds, lakes, and rivers
CONSERVATION STATUS	IUCN Least Concern. Not a common species, and is threatened in parts of its range by habitat destruction

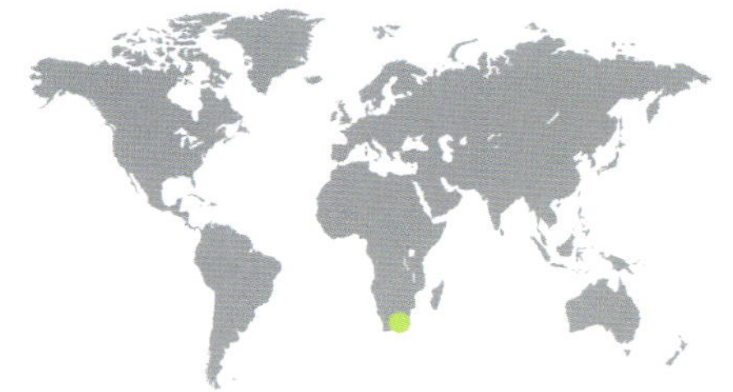

HYPEROLIUS SEMIDISCUS

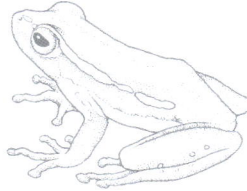

YELLOW-STRIPED REED FROG

HEWITT, 1927

ADULT LENGTH
Male
average 1 in (25 mm)

Female
average 1⅛ in (28 mm);
up to 1⅜ in (35 mm)

This small frog is found close to deep bodies of water, where its tadpoles grow to a large size, reaching 1⅞ in (48 mm) in length. The male's call is a harsh croak followed by a short creak, uttered from floating vegetation such as water-lilies, or from elevated positions on reeds or trees close to the water's edge. Females lay up to 200 eggs, deposited on submerged vegetation in batches of around 30. This species is noted for its ability to change color, from green to brown and back again.

Actual size

SIMILAR SPECIES

The range of *Hyperolius semidiscus* overlaps that of Pickersgill's Reed Frog (*H. pickersgilli*; page 447), a very rare species that is also found in Swaziland and South Africa, but is distinguished from that species by the red coloration on its limbs. In having a longitudinal yellow stripe, *H. semidiscus* is similar to the Argus Reed Frog (*H. argus*; page 443).

The Yellow-striped Reed Frog gets its name from the bold yellow stripe, usually edged in black, that runs from its eye or its snout and along the flanks. The upper surfaces are green or brown, the underside is cream or yellow, and the concealed surfaces of the limbs, the fingers, and the toes are orange or red. There are often yellow spots on the thighs.

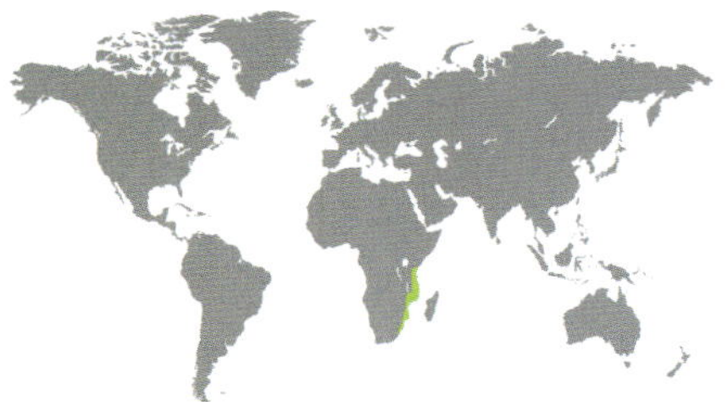

FAMILY	Hyperoliidae: Hyperoliinae
OTHER NAMES	Green Reed Frog, Smith's Reed Frog, Straw-or-green Sedge Frog, Yellow-green Reed Frog
DISTRIBUTION	East Africa, from Kenya in the north to South Africa in the south
ADULT HABITAT	Savanna, open woodland, bushland, and coastal areas
LARVAL HABITAT	Reedbeds around rivers, ephemeral ponds, dams, etc.
CONSERVATION STATUS	IUCN Least Concern. A common species that is not declining

ADULT LENGTH
Male
1–1⁵⁄₁₆ in (25–33 mm)

Female
1³⁄₁₆–1⁹⁄₁₆ in (30–40 mm)

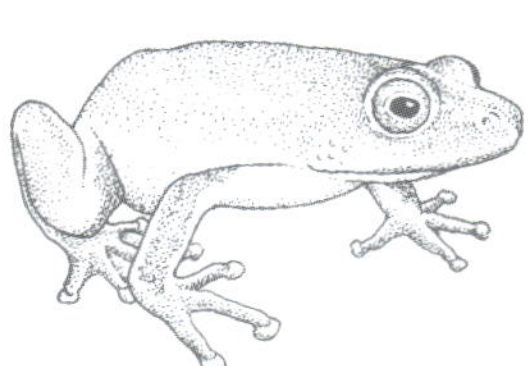

HYPEROLIUS TUBERILINGUIS

TINKER REED FROG

SMITH, 1849

Actual size

The Tinker Reed Frog is distinguished from other reed frogs by its pointed snout. It occurs in two color phases: All juveniles and some males are brown to green with yellow longitudinal stripes, and all females and some males are uniformly bright green or yellow. The eyes are large and there are disks on the fingers and toes.

The call of this common frog consists of a variable number of staccato notes—usually two or three, but can be as many as six. The number depends on the competitive sound environment in which a male finds himself, with males close to one another in a chorus matching each other in the number of notes in their call. Females are more likely to approach calls containing more notes, probably because they are easier to locate. Males call from dawn to dusk, from plants emerging from the water. A sticky mass of 300–400 eggs is laid, attached to vegetation just above the water.

SIMILAR SPECIES

Hyperolius tuberilinguis is one of four very similar species whose ranges span sub-Saharan Africa. In West Africa, the Variable Reed Frog (*H. concolor*) ranges from Sierra Leone to Cameroon. Balfour's Reed Frog (*H. balfouri*) occurs from Cameroon to Kenya, while the range of the Kivu Reed Frog (*H. kivuensis*) includes Angola, Zambia, Tanzania, Uganda, South Sudan, and Ethiopia.

FAMILY	Hyperoliidae: Hyperoliinae
OTHER NAMES	Witte's Running Frog
DISTRIBUTION	Southern Democratic Republic of the Congo, western and northern Zambia, probably Angola
ADULT HABITAT	Moist savanna
LARVAL HABITAT	Unknown; probably flooded grassland
CONSERVATION STATUS	IUCN Least Concern. Nothing is known of its current status

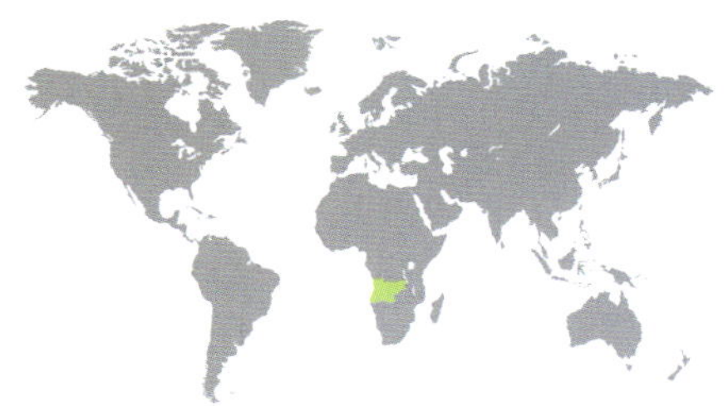

ADULT LENGTH
$^{11}/_{16}$–$^{7}/_{8}$ in (17–22 mm); females on average very slightly larger than males

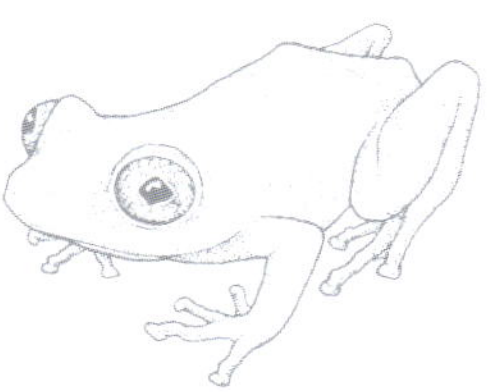

KASSINULA WITTEI

DE WITTE'S CLICKING FROG

LAURENT, 1940

This very small frog is secretive in its habits and is rarely seen. It occurs in open savanna grassland in a high-altitude region of Central Africa. It breeds in the wet season, between November and April. Males call from the base of flooded tufts of grass, producing a series of double metallic clicks. Although smaller than members of the genus *Kassina* (see pages 455–456), it is very similar to them in shape and habits.

Actual size

SIMILAR SPECIES

Kassinula wittei is the only species in its genus. It is closely related to frogs in the genus *Kassina* but is characterized by being far more diminutive, having many small differences in its skeleton, and by having a very different mating call.

De Witte's Clicking Frog is a small, compact frog with a pointed snout and large eyes. It has small disks on its fingers and toes. The upper surfaces are black or dark brown, with a complex pattern of interlocking gold longitudinal stripes. The underside is white.

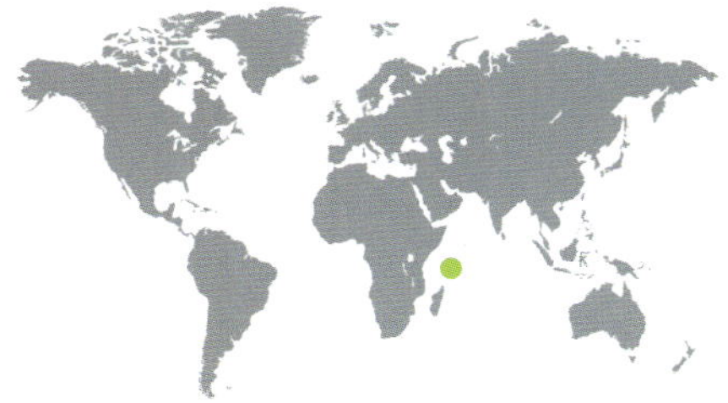

FAMILY	Hyperoliidae: Hyperoliinae
OTHER NAMES	Seychelles Island Frog
DISTRIBUTION	Mahé, Praslin, Silhouette, and La Digue islands, Seychelles
ADULT HABITAT	Remnants of rainforest, plantations
LARVAL HABITAT	Pools
CONSERVATION STATUS	IUCN Least Concern. Has a very small range but shows no evidence of decline

ADULT LENGTH
Male
up to 2 in (51 mm)
Female
up to 3 in (76 mm)

TACHYCNEMIS SEYCHELLENSIS

SEYCHELLES TREE FROG

(DUMÉRIL & BIBRON, 1841)

This large, handsome tree frog is rarely seen, being nocturnal in its habits and spending the day sleeping high up on the leaves of trees and the fronds of palms. It descends to breed, laying 100–500 eggs in pools and slow-flowing streams. Although much of its natural rainforest habitat has been destroyed, it has remained reasonably common, thriving in plantations.

SIMILAR SPECIES

Tachycnemis seychellensis is the only species in its genus. Its closest relatives are the tree frogs of the genus *Heterixalus* (see page 442), found in Madagascar. They diverged from one another between 10 and 35 million years ago when the former supercontinent Gondwanaland began to break up.

The Seychelles Tree Frog has large adhesive disks on its fingers and smaller ones on its toes. Its color varies among the different islands on which it occurs. On Mahé and Praslin, males are brown and females are green, while on Silhouette and La Digue, both sexes are green. In all color forms, the frog is paler on the flanks and belly than on the back.

Actual size

FAMILY	Hyperoliidae: Kassininae
OTHER NAMES	Previously *Phlyctimantis leonardi*
DISTRIBUTION	Congo Basin, west-central Africa
ADULT HABITAT	Rainforest and adjacent open grassland
LARVAL HABITAT	Ponds and large pools
CONSERVATION STATUS	IUCN Least Concern. A common species with a large range and a stable population

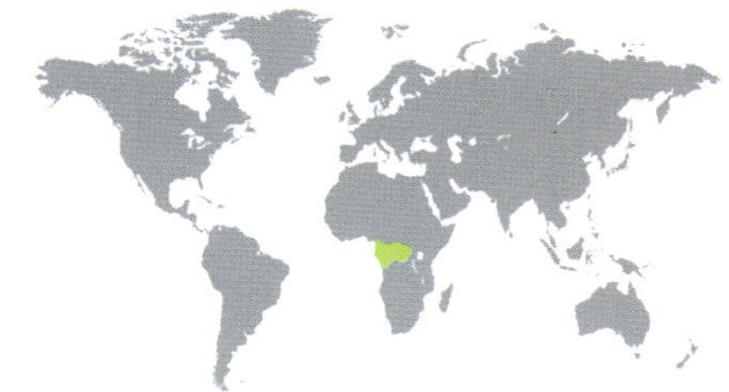

ADULT LENGTH
1¾–2¼ in (45–59 mm)

HYLAMBATES LEONARDI

LEONARD'S WOT-WOT

BOULENGER, 1906

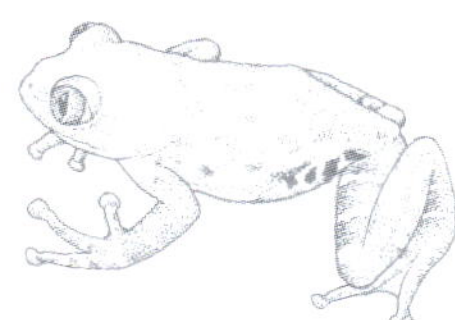

This species belongs to a small genus of large frogs associated with forests in central Africa. Their common name comes from their call which comprises a fast series of wot-wots. Leonard's Wot-wot emerges to breed in open areas, often flooded grassland and artificial ponds in farmland. Males call from bushes near the water's edge, and jump into the water if they are disturbed. The bright colors on the frog's limbs and belly are probably warning signals. When attacked or molested, other species in this genus adopt postures that expose their bright color patterns, startling their attacker.

Actual size

Leonard's Wot-wot has a rounded snout, large, prominent eyes, slender arms and legs, and small disks on its fingers and toes. The back is cream, yellowish brown, brown, or gray, sometimes with scattered darker markings. The inner surfaces of the legs are striped in black and yellow. The underside is pale cream or gray.

SIMILAR SPECIES

Hylambates leonardi and four other species used to be placed in the genus *Phlyctimantis*. In the west, from Liberia to Cameroon, Boulenger's Wot-wot (*H. boulengeri*) is a common species. Its defensive posture involves putting its head between its arms. Further east, in Uganda and Rwanda, the Warty Wot-wot (*H. verrucosus*) is noted for producing a foul-smelling secretion when handled. Keith's Wot-wot (*H. keithae*), from Tanzania, is Endangered.

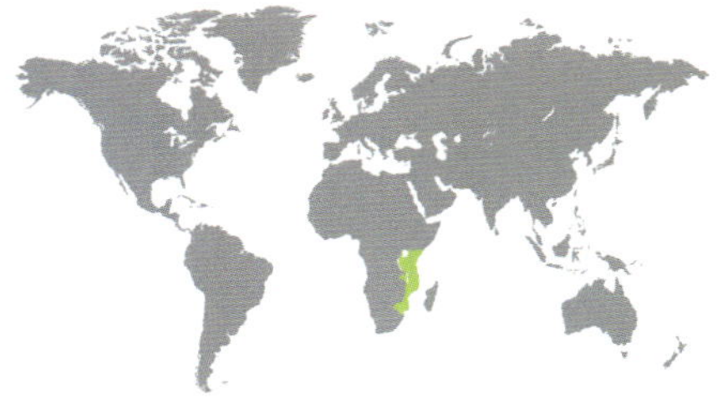

FAMILY	Hyperoliidae: Kassininae
OTHER NAMES	Previously *Kassina maculata*, Brown-spotted Tree Frog, Spotted Running Frog, Vlei Frog
DISTRIBUTION	East Africa, from southern Kenya in the north to northeastern South Africa in the south
ADULT HABITAT	Savanna and grassland, near water; also found in agricultural areas
LARVAL HABITAT	Large ephemeral and permanent ponds, pools, and dams
CONSERVATION STATUS	IUCN Least Concern. Has a very large range and shows no evidence of a general decline

ADULT LENGTH
Male
1¾–2⅝ in (45–65 mm)
Female
1$^{15}/_{16}$–2¾ in (50–68 mm)

HYLAMBATES MACULATUS

EASTERN WOT-WOT

DUMÉRIL, 1853

The red patches that give this frog its name are largely hidden when it is resting but are exposed when it is attacked. They signal that its skin contains toxic secretions capable of causing severe sickness in mammalian predators. Although generally aquatic, the frog can climb up to shelter in the leaf axils of vegetation. The male calls while floating in the water, the sound likened to that of bursting bubbles. The eggs are laid singly or in clusters of four or five, attached to submerged vegetation. The tadpoles can take as long as ten months to reach metamorphosis, sometimes attaining a length of 5⅛ in (130 mm).

SIMILAR SPECIES

The red coloration of *Hylambates maculatus* distinguishes it from the similar-looking Bubbling Kassina (*K. senegalensis*; page 456). See also Leonard's Wot-wot (*H. leonardi*; page 453).

The Eastern Wot-wot has a plump body, large, protruding eyes, and smooth skin. There are round disks on its fingers and toes. The upper surfaces are gray with large, white-edged black spots. The groin, armpits, and undersides of the legs are bright red with black spots.
The belly is white.

Actual size

FAMILY	Hyperoliidae: Kassininae
OTHER NAMES	None
DISTRIBUTION	Côte d'Ivoire, Ghana
ADULT HABITAT	Forest
LARVAL HABITAT	Large, well-vegetated ponds
CONSERVATION STATUS	IUCN Vulnerable. Its forest habitat is being destroyed to make way for agriculture and human settlements

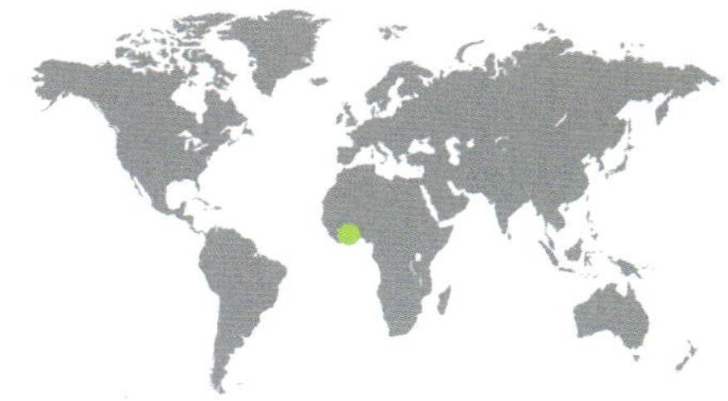

KASSINA ARBORICOLA

IVORY COAST RUNNING FROG

PERRET, 1985

ADULT LENGTH
Male
1$^{7}/_{16}$–1$^{9}/_{16}$ in (37–40 mm)

Female
unknown

Like other kassinas, this small frog does not hop or jump but runs across the ground, its hind legs moving alternately. Males call to attract females, either from the ground or from elevated perches in vegetation. Their call is a popping sound. The eggs are attached to submerged vegetation. The tadpoles, which feed on plant material, have a very deep tail fin and can grow up to 2¼ in (58 mm) long. They move through the water in a slow and graceful manner.

SIMILAR SPECIES

Also found in West Africa, from Sierra Leone in the west to southern Ghana in the east, is Cochran's Running Frog (*Kassina cochranae*). It is very similar in appearance and habits to *K. arboricola*. Schiøtz's Running Frog (*K. schioetzi*) is a recently described species, also found in Côte d'Ivoire.

The Ivory Coast Running Frog has a slender body, arms, and legs with large, protruding eyes and smooth skin. The body and legs are gray, with many large, white-edged black spots. In some individuals the spots are arranged in lines. The belly is dark gray, and there are yellow patches on the upper arms, thighs, and feet.

Actual size

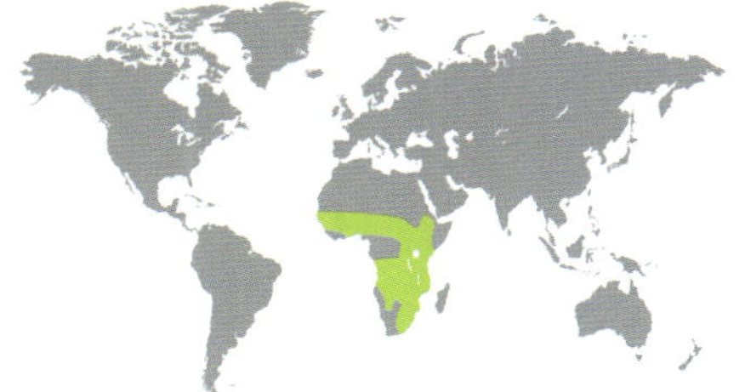

FAMILY	Hyperoliidae: Kassininae
OTHER NAMES	Running Frog, Senegal Kassina
DISTRIBUTION	Sub-Saharan Africa
ADULT HABITAT	Near water in tropical savanna and grassland
LARVAL HABITAT	Ephemeral and permanent ponds and pools
CONSERVATION STATUS	IUCN Least Concern. Has a very large range and shows no evidence of a general decline

ADULT LENGTH
1–1[15]/16 in (25–49 mm)

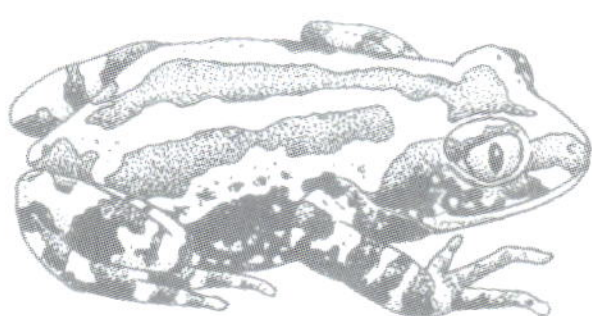

KASSINA SENEGALENSIS

BUBBLING KASSINA

(DUMÉRIL & BIBRON, 1841)

The common name of this frog refers to the sound made by many males in a chorus, each responding to one another by producing a "boip" call. It is one of the commonest sounds in the African savanna, especially after rain. The female lays 260–400 eggs attached to submerged vegetation. As in other kassinas, the tadpoles have a characteristic tail that is very deep and reddish in color. They swim slowly and gracefully, and can reach up to 3⅛ in (80 mm) in length. In dry weather, adults are often found in ants' nests and the burrows made by scorpions, mole rats, and golden moles.

SIMILAR SPECIES

All 15 *Kassina* species have a similar natural history. Parker's Running Frog (*Kassina maculifer*) is a common species found in very dry savanna in Kenya, Somalia, and Ethiopia. *Kassina schioetzi*, a forest species from Côte d'Ivoire and Guinea, is olive-green or beige with large black spots. *Kassina jozani* is found in a very restricted area—the Jozani Forest of Zanzibar—and is listed as Endangered.

The Bubbling Kassina has a bullet-shaped body, large, protruding eyes, and smooth skin. It walks rather than hops and has comparatively small hind legs. Across its large range it is very variable in color. In South Africa, it is gray or ocher with a large brown or black stripe down the middle of the back, as well as other stripes and spots.

Actual size

FAMILY	Hyperoliidae: Kassininae
OTHER NAMES	None
DISTRIBUTION	Southern Côte d'Ivoire
ADULT HABITAT	Rainforest
LARVAL HABITAT	Streams and rivers
CONSERVATION STATUS	IUCN Vulnerable. Currently known from only one, restricted location

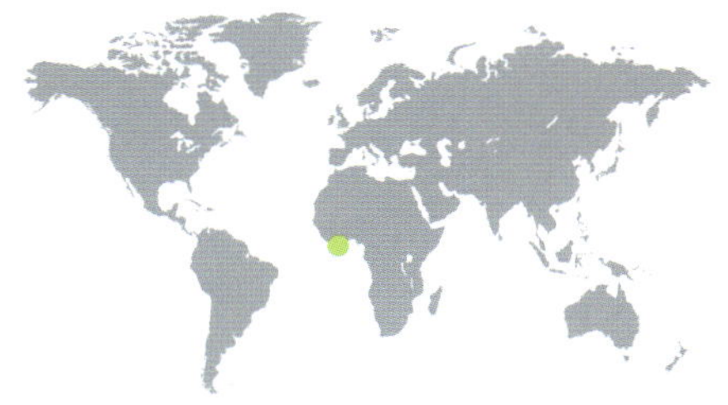

MORERELLA CYANOPTHALMA

BLUE-EYED FROG

RÖDEL, ASSEMIAN, KOUAMÉ, TOHÉ & PERRET, 2009

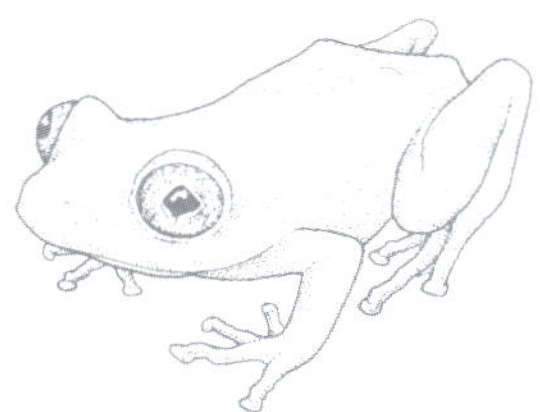

ADULT LENGTH
Male
$\frac{7}{8}$–$1\frac{5}{16}$ in (23–34 mm)

Female
$1\frac{1}{8}$–$1\frac{3}{8}$ in (28–35 mm)

Described in 2009, this small frog is unusual in that males and females differ strikingly in appearance, the female being the more colorful sex. It is known only from the Banco National Park, close to the city of Abidjan, but it is thought that it occurs elsewhere in Côte d'Ivoire. It is active at night, males calling from foliage just above streams and small rivers. The mating call is a single note, repeated two or three times. The eggs are laid in groups of 30–144 attached to foliage overhanging water. Adults have been seen close by, suggesting that there may be some form of parental care.

Actual size

The Blue-eyed Frog has a slender body, large, protruding eyes, and disks on its fingers and toes. The female is reddish brown, bright red, or orange, with gray or bright blue irises. The male is reddish brown by day and becomes yellowish at night. The irises of his eyes are brown. The underside of both sexes is white to yellow.

SIMILAR SPECIES

Morerella cyanopthalma is the only species in its genus. In addition to its very unusual coloration, it differs from other frogs in its family in a number of skeletal features and is genetically distinct.

FAMILY	Hyperoliidae: Kassininae
OTHER NAMES	None
DISTRIBUTION	Nigeria, Cameroon, Equatorial Guinea, Gabon, Democratic Republic of the Congo
ADULT HABITAT	Lowland tropical forest
LARVAL HABITAT	Slow-flowing streams
CONSERVATION STATUS	IUCN Least Concern. Declining in places due to deforestation, but occurs within some protected areas

ADULT LENGTH
1³⁄₁₆–1⁵⁄₁₆ in (30–33 mm)

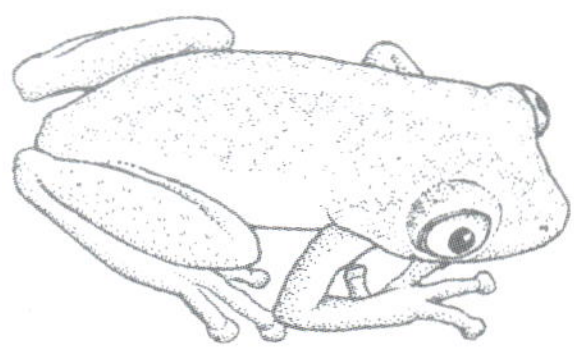

OPISTHOTHYLAX IMMACULATUS

GRAY-EYED FROG

(BOULENGER, 1903)

Actual size

During mating in this small frog, male and female cooperate to protect their eggs in a foam nest wrapped in a leaf. Males call from high up in trees overhanging streams. Their call is a deep, nasal "doet… doet." The female lays 6–10 large, yolk-filled eggs on a leaf. The male then slides backward to clasp the female around her waist. While she uses her hind legs to beat a sticky secretion into a foam, he uses his to fold the end of the leaf around the eggs. After 2–3 weeks the eggs hatch and the tadpoles fall into the stream far below.

SIMILAR SPECIES

Opisthothylax immaculatus is the only species in its genus. It is also the only frog in the African family Hyperoliidae that makes a foam nest.

The Gray-eyed Frog has a slender body, warty skin, a pointed snout, large, protruding eyes, and disks on its fingers and toes. The upper surfaces are variable in color, ranging from reddish brown, through orange to yellow. The undersides of the body and limbs are yellow. The iris of the eye is gray or yellow.

FAMILY	Hyperoliidae: Kassininae
OTHER NAMES	Mocquard's Mountain Kassina
DISTRIBUTION	Ethiopia
ADULT HABITAT	Montane grassland at altitudes of 6,500–10,500 ft (1,980–3,200 m)
LARVAL HABITAT	Ephemeral or permanent pools
CONSERVATION STATUS	IUCN Vulnerable. Known from only a very small area, where its habitat is declining

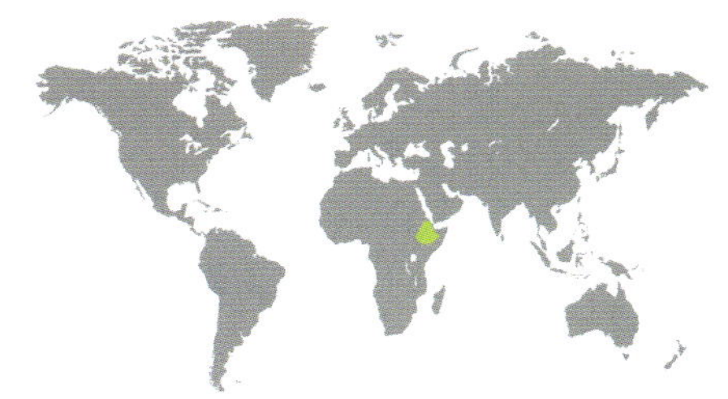

ADULT LENGTH
Male
1⅜–1⅞ in (35–47 mm)
Female
1$^{9}/_{16}$–1$^{15}/_{16}$ in (39–49 mm)

PARACASSINA KOUNHIENSIS

KOUNI VALLEY STRIPED FROG

(MOCQUARD, 1905)

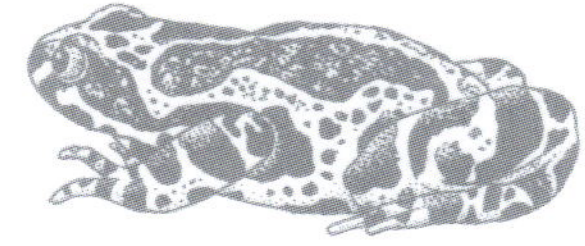

This is one of two species found in the Ethiopian Highlands that are unique in being specialized to feed on terrestrial snails. The skull and jaws are adapted such that the mouth has an enormous gape, enabling the frog to swallow a snail whole, with its shell intact. Males utter a single-note "popping" call, which they produce while floating in water supported by waterplants. They often call during the day. The species is found to the east of the Great Rift Valley, where its range includes the Bale Mountains National Park. It is reported to have curved teeth that enable it to eat slugs as well as snails.

Actual size

The Kouni Valley Striped Frog has a plump body, a blunt snout, rather short hind limbs, and long fingers and toes. The upper surfaces are golden brown, decorated with large rounded, pale-edged black spots. There is a conspicuous yellow patch in each groin. The underside is mottled in pale and dark gray.

SIMILAR SPECIES

The only other species in this genus is the Ethiopia Striped Frog (*Paracassina obscura*). This is similar in behavior, appearance, and diet, but is found to the west of the Great Rift Valley, up to altitudes of 10,000 ft (3,000 m). It sometimes occurs in rural gardens and urban areas.

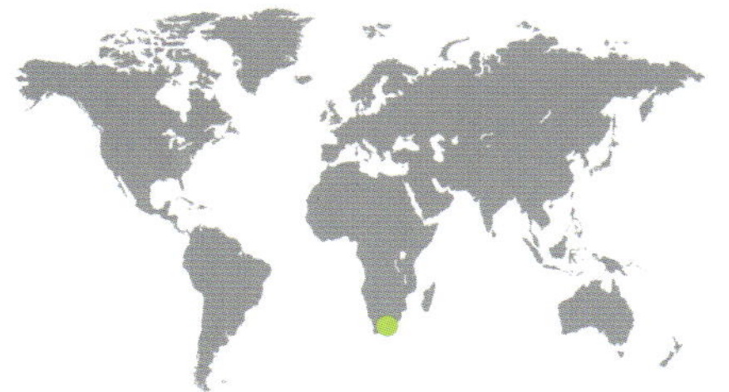

FAMILY	Hyperoliidae: Kassininae
OTHER NAMES	Rattling Frog
DISTRIBUTION	South Africa, Lesotho, Swaziland
ADULT HABITAT	Grassland and fynbos heath
LARVAL HABITAT	Ephemeral and permanent ponds
CONSERVATION STATUS	IUCN Least Concern. Shows no evidence of decline and occurs in some human-altered habitats

ADULT LENGTH
Male
average $1^{5}/_{16}$ in (33 mm);
up to $1^{3}/_{4}$ in (44 mm)

Female
average $1^{3}/_{8}$ in (35 mm)

SEMNODACTYLUS WEALII

WEALE'S RUNNING FROG

(BOULENGER, 1882)

An agile climber, this frog has an unusual arrangement of fingers, which are divided into two, widely separated pairs, enhancing its ability to grasp foliage. On the ground, it tends to walk rather than hop. The male's call is a loud, coarse rattle, lasting about half a second. Pairs form some distance from water, then walk to a pond to lay 100–500 eggs on submerged vegetation. When molested, the frog feigns death, rolling onto its back with its legs tucked in. Its skin contains toxins that make any animal eating it violently sick.

Weale's Running Frog has an elliptical body, long arms and legs, no disks on its fingers or toes, and large, protruding eyes. The upper surfaces are gray, olive-green, or yellow with black longitudinal stripes. The undersides of the legs and arms, the hands, and the feet are bright yellow.

SIMILAR SPECIES

Semnodactylus wealii is the only species in its genus, which is closely related to the kassinas. In particular, it is rather similar in appearance to the Bubbling Kassina (*Kassina senegalensis*; page 456).

Actual size

FAMILY	Hyperoliidae *incertae sedis*
OTHER NAMES	None
DISTRIBUTION	Eastern Democratic Republic of the Congo
ADULT HABITAT	Grassland at altitudes of 7,900–9,350 ft (2,400–2,850 m)
LARVAL HABITAT	Flooded grassland
CONSERVATION STATUS	IUCN Endangered. A very rare species with a restricted range

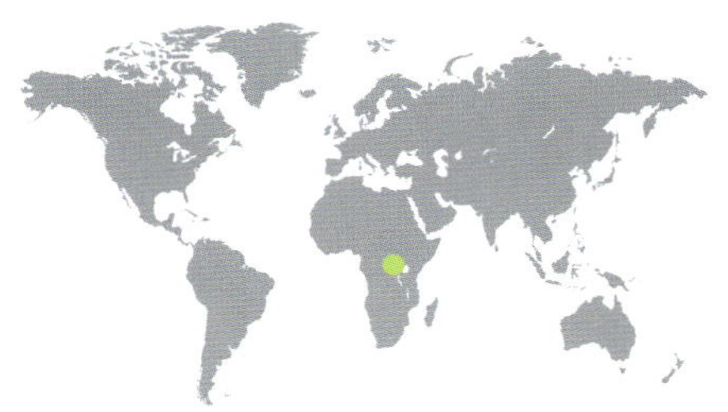

ADULT LENGTH
Male
¾–15/16 in (19–24 mm)
Female
1 1/16–1 5/16 in (27–34 mm)

CHRYSOBATRACHUS CUPREONITENS

ITOMBWE GOLDEN FROG

LAURENT, 1951

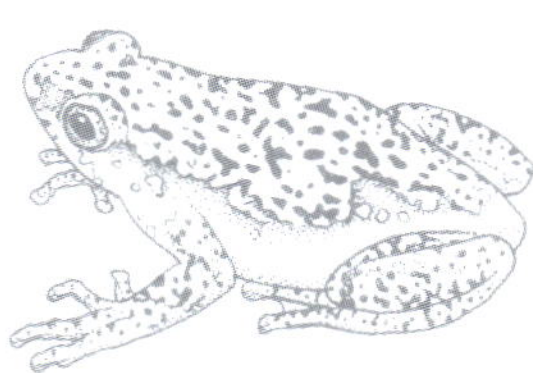

The Itombwe Mountains are forested highlands to the west of Lake Tanganyika. They are home to gorillas, chimpanzees, and African Bush Elephants (*Loxodonta africana*), as well as several frogs that occur nowhere else, and were made a protected area in 2006. The colorful Itombwe Golden Frog had not been seen since its discovery in 1951, but an expedition in 2011 "rediscovered" it, along with three other frog species. It is unusual in that females are very much larger than males. Very little is known about its natural history or behavior apart from the fact that, during mating, the male clasps the female around her lumbar region, not under her armpits.

Actual size

The Itombwe Golden Frog has short hind legs and disks on its fingers and toes. The upper surfaces of the body and legs are bright green and brown with black dots, and there is a golden metallic sheen. The flanks and undersides of the body and legs are white or pink.

SIMILAR SPECIES

Chrysobatrachus cupreonitens is the only species in its genus. It differs from all other species in its family in a number of anatomical features and in having lumbar amplexus. It is therefore considered *incertae sedis* (of unknown placement) within the Hyperoliidae.

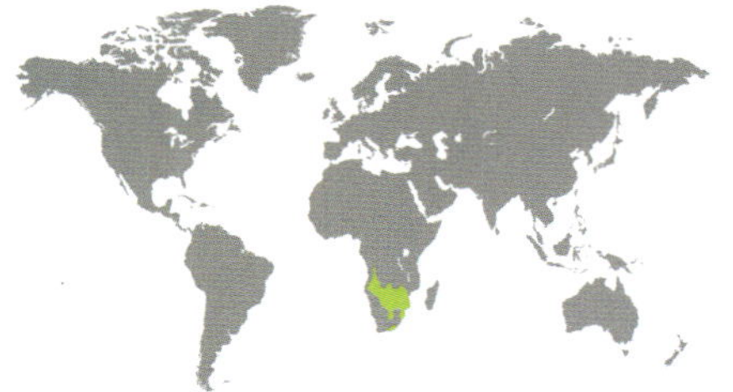

FAMILY	Brevicipitidae
OTHER NAMES	Common Rain Frog
DISTRIBUTION	Southern Africa, from Angola in the west to Mozambique in the east
ADULT HABITAT	Bushveld, woodland with sandy soil
LARVAL HABITAT	Within the egg
CONSERVATION STATUS	IUCN Least Concern. Has a large range and occurs in a wide range of habitats, including suburban environments

ADULT LENGTH
Male
average $1\frac{7}{16}$ in (37 mm)

Female
average $1\frac{15}{16}$ in (49 in);
up to $2\frac{3}{8}$ in (60 mm)

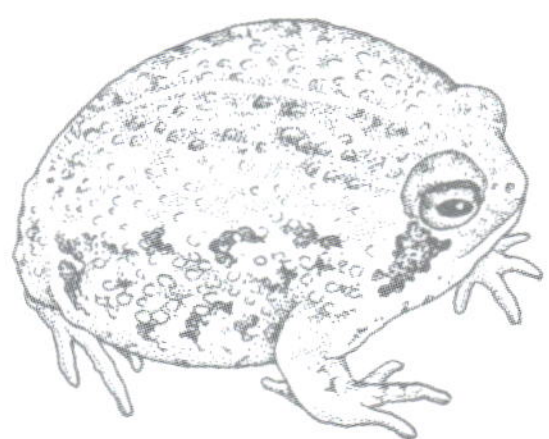

BREVICEPS ADSPERSUS

BUSHVELD RAIN FROG

PETERS, 1882

A burrowing species, this small frog lives mostly underground, emerging to feed and mate only after rain. It digs backward into the soil, using its hind feet. The male is so small, relative to the female, and his limbs are so short, that he cannot clasp her during mating as most frogs do. Instead, the female produces a secretion from her back that acts as a glue, holding the mating pair together. Around 45 eggs are laid underground in compact, sticky balls. These eggs hatch directly into small froglets. If disturbed underground, the frog can inflate its body, lodging itself firmly in its burrow.

The Bushveld Rain Frog has a stout, globular body, short, stout limbs, a flattened face, a small mouth, and forward-pointing eyes. There are horny tubercles on the hind feet, which are used for digging. The back is light or dark brown, with rows of lighter, dark-bordered yellowish or orange patches. There is a mask-like dark stripe running from the eye to the armpit.

SIMILAR SPECIES

Breviceps adspersus is most similar to the Mozambique Rain Frog (*B. mossambicus*), from eastern South Africa, Mozambique, and as far north as Tanzania. The two species occasionally hybridize. Bilbo's Rain Frog (*B. bagginsi*) is found in only a few locations in KwaZulu-Natal, South Africa. It is named after the title character in J. R. R. Tolkien's *The Hobbit* and is listed as Near Threatened. See also the Desert Rain Frog (*B. macrops*; page 463) and the Plaintive Rain Frog (*B. verrucosus*; page 464). *Breviceps* contains 20 species.

Actual size

FAMILY	Brevicipitidae
OTHER NAMES	Boulenger's Short-headed Frog, Web-footed Rain Frog
DISTRIBUTION	Namaqualand coast, Namibia and South Africa
ADULT HABITAT	Coastal sand dunes
LARVAL HABITAT	Within the egg
CONSERVATION STATUS	IUCN Near Threatened. Threatened by habitat loss due to diamond mining

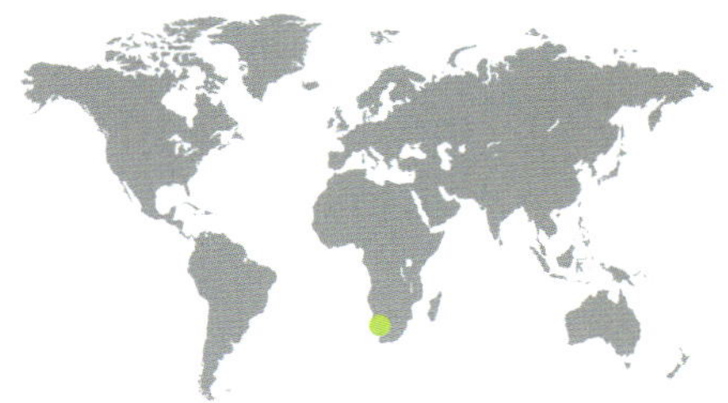

BREVICEPS MACROPS

DESERT RAIN FROG

BOULENGER, 1907

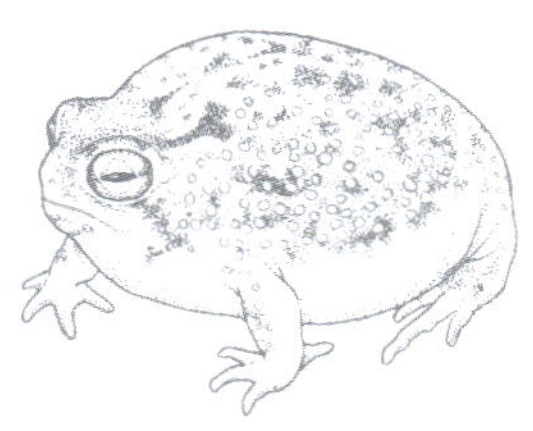

ADULT LENGTH
Male
average 1¼ in (31 mm)
Female
average 1⁹⁄₁₆ in (40 mm);
up to 1¹⁵⁄₁₆ in (50 mm)

The arid coast of southwestern Africa is an unlikely place to find a frog, as the only source of water is thick fog that occasionally blows in from the sea. This is sufficient for this small, burrowing frog, whose scientific name means "short-headed, large-eyed frog." It has webbed feet that enable it to walk on, and dig into, soft sand, and it ventures onto the surface only after dark. There it looks for animal dung, where it may find beetles and other insects. Mating has not been observed, but its call—a long, drawn-out whistle—is heard during and after foggy periods from June to October.

Actual size

The Desert Rain Frog has a stout, round body, short legs, and very large eyes. The hands and feet are paddle-like and there is fleshy webbing between the toes. The back is white or yellowish, often with a marbled, dark-brown pattern. The underside is white and there is a "window" of transparent skin over the belly.

SIMILAR SPECIES

Like the Desert Rain Frog, the Namaqua Rain Frog (*Breviceps namaquensis*) has large eyes, but it differs in that it does not have webbed feet. It also occurs in southwestern Africa, but in the less arid Karoo biome. The Cape Mountain Rain Frog (*B. montanus*) has smaller eyes and is found in fynbos vegetation on Table Mountain and other ranges near South Africa's Cape Peninsula. See also the Bushveld Rain Frog (*B. adspersus*; page 462) and the Plaintive Rain Frog (*B. verrucosus*; page 464).

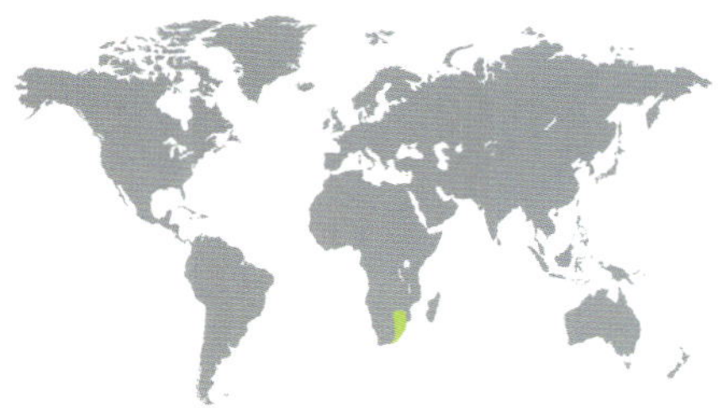

FAMILY	Brevicipitidae
OTHER NAMES	Natal Short-headed Frog, Rough Rain Frog
DISTRIBUTION	Eastern South Africa, Lesotho, Swaziland
ADULT HABITAT	Forest and adjacent grassland; coastal forest in KwaZulu-Natal
LARVAL HABITAT	Within the egg
CONSERVATION STATUS	IUCN Least Concern. Threatened by deforestation in places; thrives in suburban gardens

ADULT LENGTH
Male
average 1⁵⁄₁₆ in (33 mm)

Female
average 1⅝ in (42 mm);
up to 2¹⁄₁₆ in (53 mm)

BREVICEPS VERRUCOSUS

PLAINTIVE RAIN FROG

RAPP, 1842

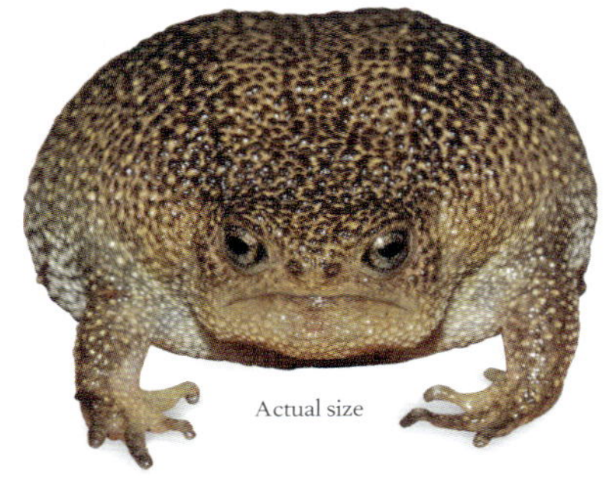

Actual size

The Plaintive Rain Frog has a very stout, round body, a very small head, short legs, and tiny eyes. Its upper surfaces are tan or dark brown with black markings. A black face mask is present in some individuals, as is a pale line down the middle of the back. The underside is generally pale, with males having a darker throat.

The common English name of this species refers to the male's mating call, a long, mournful whistle. Its scientific name, *verrucosus*, refers to its skin, which is warty and coarse. During the rainy season, between August and November, quite large choruses of calling males can form. Males respond to one another, so that waves of calling spread through a population. As in other rain frogs, the male is only a third of the weight of the female and has to become glued to her back during mating (see page 462). The eggs are laid in damp soil deep underground and hatch directly into tiny frogs.

SIMILAR SPECIES

The Northern Forest Rain Frog (*Breviceps sylvestris*) is found in wooded mountains in northern South Africa. Males build a network of shallow tunnels and underground chambers. Its habitat is subject to deforestation and it is listed as Near Threatened. The Whistling Rain Frog (*B. sopranus*) is a very small species whose mating call is a high-pitched whistle. It was described as recently as 2003. See also the Bushveld Rain Frog (*B. adspersus*; page 462) and the Desert Rain Frog (*B. macrops*; page 463).

FAMILY	Brevicipitidae
OTHER NAMES	None
DISTRIBUTION	North Pare Mountains, near Mt. Kilimanjaro, Tanzania
ADULT HABITAT	Humid montane forest at altitudes of 5,680–6,600 ft (1,730–2,000 m)
LARVAL HABITAT	Presumed to be within the egg
CONSERVATION STATUS	IUCN Critically Endangered. Has a very small range and its habitat is at risk from deforestation

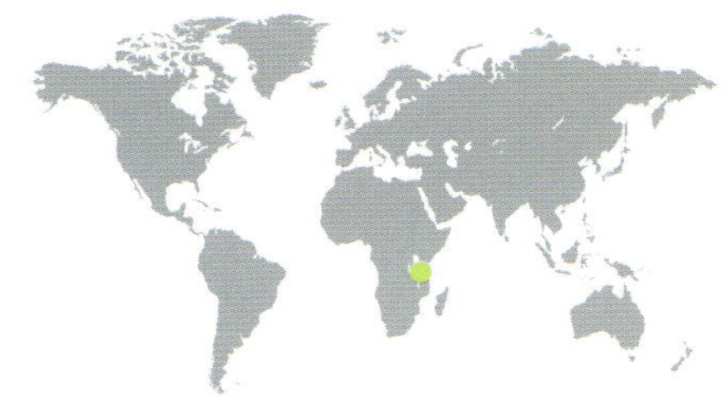

CALLULINA LAPHAMI

LAPHAM'S WARTY FROG

LOADER, GOWER, NGALASON & MENEGON, 2010

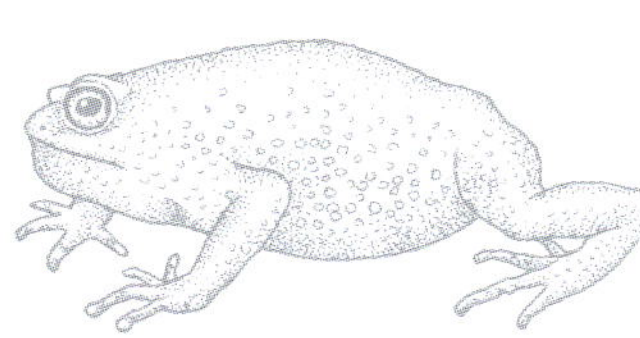

ADULT LENGTH
Male
7/8–1 1/8 in (23–29 mm)

Female
1 5/16–1 3/4 in (33–45 mm)

Actual size

Lapham's Warty Frog is a stout, robust frog with a globular body and a flattened face. The legs are slender, and the hands and feet lack webbing or adhesive disks. The upper surfaces are dark brown with small white speckles. The underside is paler. There is a red, or occasionally green, stripe on the head between the eyes.

Discovered recently, this small frog has a very uncertain future. Its habitat has been reduced by deforestation to an area of only 6.4 sq miles (16.5 sq km), and even this is under threat. It occurs at altitudes of 5,680–6,600 ft (1,730–2,000 m), and has been found by day under damp rocks, and at night in bushes and small trees. Males call from low branches, producing a series of trills. The mode of reproduction of this species and other members of the genus is not known, but it is presumed that eggs are laid underground and hatch directly into small frogs.

SIMILAR SPECIES

There are nine species in the genus *Callulina*, all found in the Eastern Arc Mountains of Tanzania and Kenya. Until recently, only one species, Krefft's Secret Frog (*C. kreffti*), was known, but detailed research has shown that several of the isolated mountain ranges in the region have endemic species. For example, the Taita Warty Frog (*C. dawida*) is found only in the Taita Hills of Kenya, and Critically Endangered.

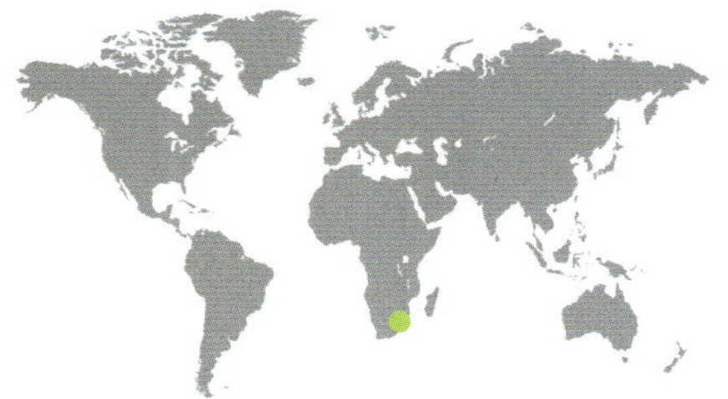

FAMILY	Hemisotidae
OTHER NAMES	Spotted Shovel-nosed Frog, Spotted Snout-burrower
DISTRIBUTION	Eastern South Africa
ADULT HABITAT	Grassland and savanna
LARVAL HABITAT	Seasonal waterbodies
CONSERVATION STATUS	IUCN Near Threatened. Has a rather small range, fragmented by habitat destruction. Its habitat is also affected by invasive plants and agrochemical pollution

ADULT LENGTH
Male
average 1⅝ in (41 mm);
up to 2 in (51 mm)

Female
average 2⅝ in (65 mm);
up to 3⅛ in (80 mm)

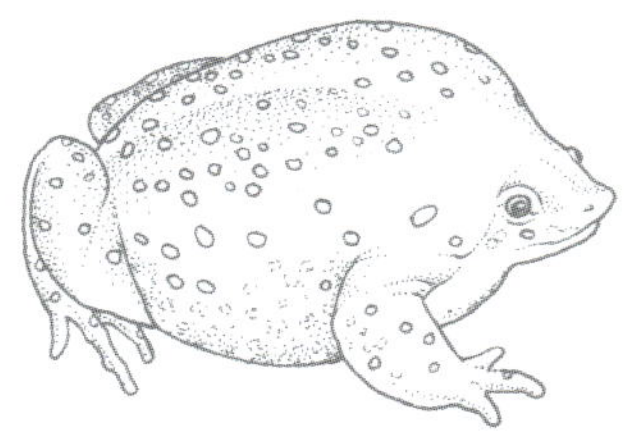

HEMISUS GUTTATUS

SPOTTED BURROWING FROG

(RAPP, 1842)

This large frog breeds at the edges of swamps and rivers during rain, between October and December. Males produce a cricket-like trill from hidden positions and are notoriously hard to find. When a pair forms, the male clasps the female around her abdomen and she carries him while she digs a burrow in the mud. She lays around 200 eggs in the burrow, covering them with a protective layer of empty egg capsules. She stays with the eggs until they hatch, and then she helps the tadpoles out of the burrow into open water, carrying them on her back.

The Spotted Burrowing Frog has a small head, tiny eyes, and a pointed snout with a hard tip. The body is globular and the skin is smooth and shiny. The legs and arms are muscular and the fingers are thick. Its back is olive-green, dark brown, or purple in color, with many yellow spots; the underside is white.

SIMILAR SPECIES

The nine species in this genus are all forward-burrowers, using their pointed snouts to push into soft soil and their powerful forelimbs to clear away the spoil. They are found in wet habitats across eastern and southern Africa. The Lake Zwai Shovelnose Frog (*Hemisus microscaphus*) is found in Ethiopia, while the Olive Snout-burrower (*H. olivaceous*) is found in the Congo Basin. See also the Mottled Shovel-nosed Frog (*H. marmoratus*; page 467).

Actual size

FAMILY	Hemisotidae
OTHER NAMES	Marbled Snout-burrower, Shovel-nosed Frog
DISTRIBUTION	Much of sub-Saharan Africa
ADULT HABITAT	Grassland and savanna
LARVAL HABITAT	Seasonal waterbodies
CONSERVATION STATUS	IUCN Least Concern. Has a huge range and shows no evidence of a general decline

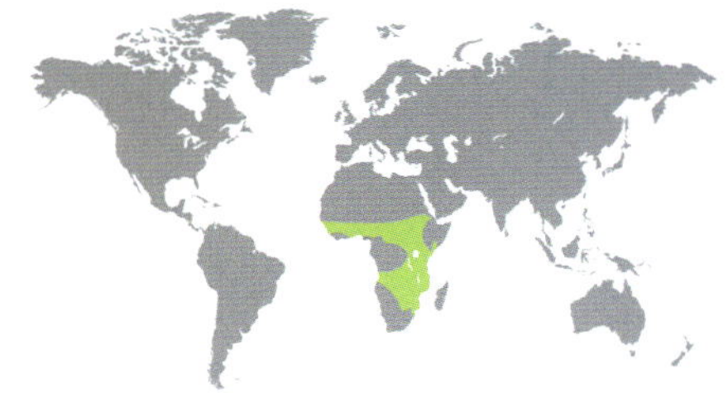

HEMISUS MARMORATUS

MOTTLED SHOVEL-NOSED FROG

(PETERS, 1854)

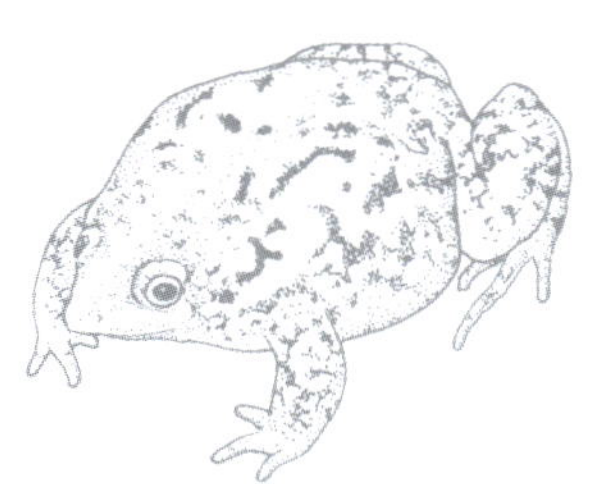

ADULT LENGTH
Male
⅞–1⅜ in (22–35 mm)

Female
1⅛–2⅛ in (29–55 mm)

This frog spends much of its life underground, emerging onto the surface when conditions are wet to feed on ants and termites, which it catches with its long, sticky tongue. In the dry season it becomes inactive deep down, encased in mud. After heavy rain, males call from concealed positions, producing a buzzing sound that lasts several seconds. A clutch of 80–250 eggs is laid in a burrow under a log or stone close to water. Further rain raises the water level, flooding the nest and allowing the tadpoles to escape into open water, helped by the mother, which digs a channel for them.

SIMILAR SPECIES

While most *Hemisus* species are found in savanna and grassland, Perret's Snout-burrower (*H. perreti*) is found in the tropical forest of the Congo Basin. The Pig-nosed Frog (*H. guineensis*) is found in grassland near temporary waterbodies across a huge range, from Angola in the west to Mozambique in the east. See also the Spotted Burrowing Frog (*H. guttatus*; page 466).

The Mottled Shovel-nosed Frog has a plump, globular body, a small head with a sharp, hardened snout, and a transverse fold in the skin just above the small eyes. The forelimbs are muscular. The skin is pale brown, typically with a marbled pattern in dark brown or dark gray, although some individuals are uniform in color.

Actual size

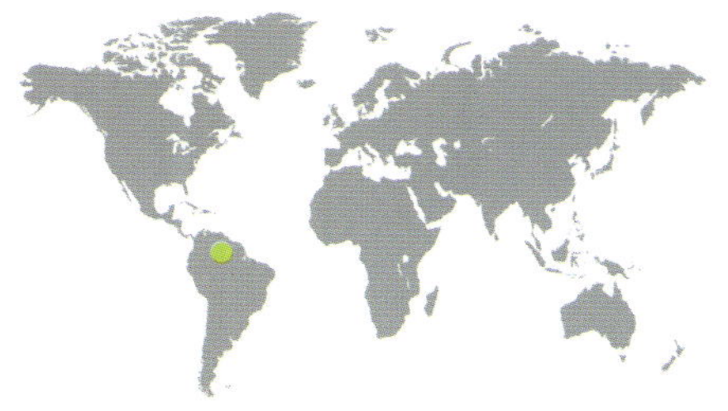

FAMILY	Microhylidae: Adelastinae
OTHER NAMES	Neblina Frog
DISTRIBUTION	Cerro de la Neblina and Rio Negro, Amazonas states, Brazil and Venezuela; plateau above Meamu River, Guyana
ADULT HABITAT	Rainforest on a sandy substrate and along streams with temporary pools at altitudes of 230–460 ft (70–140 m)
LARVAL HABITAT	Unknown
CONSERVATION STATUS	IUCN Least Concern. Probably threatened by habitat loss

ADULT LENGTH
Male
⅞–1⅛ in (22–29 mm)
Female
unknown

ADELASTES HYLONOMOS

NEBLINA FOREST FROG

ZWEIFEL, 1986

Actual size

The Neblina Forest Frog is a small, smooth-scaled, dark brown species, pale brown below, possessing small, black eyes with horizontal pupils and lacking tympana. Its limbs are well developed and its four fingers and five toes are unwebbed.

The Neblina Forest Frog is the sole species in the microhylid subfamily Adelastinae. It was considered endemic to the Cerro de la Neblina but has also been collected from the Rio Negro, Brazil, and from a plateau above the Meamu River, Guyana. The Cerro de la Neblina is a Cenozoic refugia and center of endemism for frogs, being the type locality of other rare frogs, e.g. *Neblinaphryne mayeri* (page 423) and *Caligophryne doylei* (page 424). Unlike these two high-elevation grassland species, *Adelastes hylonomos* is a lowland rainforest frog that hides and calls at night from within the leaf litter. No females or tadpoles have been collected, and reproductive strategy is unknown.

SIMILAR SPECIES

The Neblina Forest Frog belongs to a monotypic subfamily of the Microhylidae and therefore has no close relatives. A recent molecular phylogenetic study, which erected the subfamily Adelastinae, also proposed that *Adelastes* is closest to the Gastrophryninae (pages 492–501) or the Otophryninae (pages 516–517), but this is not surprising as these are the only other American Microhylidae families, and both occur in the Guiana Shield region where the Cerro de la Neblina is located.

FAMILY	Microhylidae: Asterophryinae
OTHER NAMES	Previously *Pseudocallulops pullifer*
DISTRIBUTION	Papua province, Indonesia
ADULT HABITAT	Rainforest at altitudes of 1,150–2,800 ft (350–850 m)
LARVAL HABITAT	Within the egg
CONSERVATION STATUS	Data Deficient. There is no apparent threat to the species and it occurs in at least one protected area

ASTEROPHRYS PULLIFER

WANDAMMEN BUSH FROG

(GÜNTHER, 2006)

ADULT LENGTH
Male
1–$1\frac{5}{16}$ in (26–33 mm)
Female
$1\frac{1}{16}$–$1\frac{5}{16}$ in (27–34 mm)

Described in 2006, this frog is known from only a single location, on the rocky eastern slope of the Wondiwoi Mountains, but it is likely that it occurs elsewhere on New Guinea. It can be common: One researcher counted 100 calling males during a 2.5-mile (4-km) walk. The male calls from a rock crevice, producing a series of loud croaks in the first few hours after nightfall. The eggs are presumably guarded by the male because, when they hatch, he is there for the tiny young frogs to climb onto his back. The male then carries them for a few days.

Actual size

The Wandammen Bush Frog has a slender body, long, slender limbs, and large eyes. The disks on the fingers are larger than those on the toes. The upper surfaces are brown with irregular, darker spots and blotches, and the underside of the body and legs is cream. There is a pale spot below the knee on each hind leg. This male is carrying three froglets on his back.

SIMILAR SPECIES

This species, and another, the Onin Bush Frog (*A. eurydactylus*), were previously placed in the genus *Pseudocallulops*. *Asterophrys eurydactylus* is a slightly larger frog of similar habits, which is also found in western New Guinea. As in the Horned Land Frog (*Sphenophryne cornuta*; page 482), carrying young frogs probably serves to disperse the young as widely as possible.

FAMILY	Microhylidae: Asterophryinae
OTHER NAMES	Star-eyed Bush Frog
DISTRIBUTION	New Guinea
ADULT HABITAT	Lowland and foothill rainforest up to 3,300 ft (1,000 m) altitude; also found in urban gardens
LARVAL HABITAT	Unknown
CONSERVATION STATUS	IUCN Least Concern. Has a large range and its population is stable

ADULT LENGTH
up to 2⅝ in (65 mm)

ASTEROPHRYS TURPICOLA

NEW GUINEA BUSH FROG

(SCHLEGEL, 1837)

This rather rotund frog is noted for its aggressive behavior and appearance. It has many sharp spines on its body and legs, most notably on its eyelids and lower jaw. Its large mouth and wide gape enable it to feed on lizards, insects, and other frogs. When disturbed, it opens its mouth wide and protrudes a bright blue tongue. It lives on the forest floor, from where males produce a call which is likened to the mew of a kitten. The species tolerates habitat disturbance and has been found in urban gardens.

SIMILAR SPECIES

This was the only species in the genus *Asterophrys*, until 1994 when the White-footed Bush Frog (*A. leucopus*) was described from northern New Guinea. Smaller than *A. turpicola*, it is spiny like *A. turpicola*, but its tongue is not blue. More recently another six species have been transferred to *Asterophrys* from other genera such as *Pseudocallulops*. See also the Wandammen Bush Frog (*A. pullifer*; page 469).

The New Guinea Bush Frog has a broad body and head, short legs, and many sharply pointed warts on its head, body, and legs. A spiny crest runs from the eardrum to the armpit, and there are disks on its fingers and toes. The upper surfaces of the frog are yellowish brown with large black blotches. The flanks and underside are pale brown.

Actual size

FAMILY	Microhylidae: Asterophryinae
OTHER NAMES	Golden Land Frog, Shrill Chirper
DISTRIBUTION	Northern Queensland, Australia; Papua New Guinea
ADULT HABITAT	Grassy woodland and lowland rainforest
LARVAL HABITAT	Unknown
CONSERVATION STATUS	IUCN Least Concern. Its population is stable and its habitat is not under threat

AUSTROCHAPERINA GRACILIPES

SLENDER LAND FROG

FRY, 1912

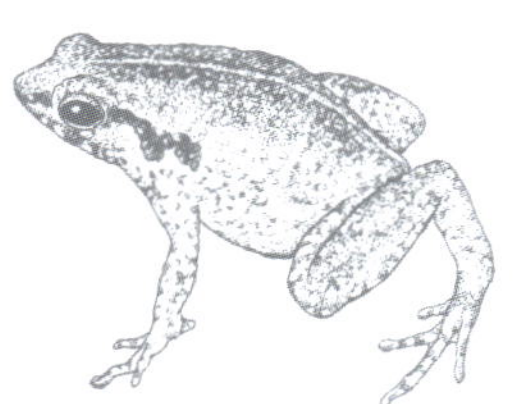

ADULT LENGTH

Male
$^{11}/_{16}$–$^{3}/_{4}$ in (17–20 mm)

Female
$^{3}/_{4}$–$^{7}/_{8}$ in (19–23 mm)

This secretive little frog is found mostly in leaf litter, close to ephemeral and permanent streams. The male calls from a perch about 1½ ft (50 cm) above the ground, producing a series of high-pitched pips lasting 10–20 seconds. Other aspects of its reproductive biology are unknown, but it is likely that it lays large terrestrial eggs that hatch directly into small frogs. It has a restricted range at the tip of Australia's Cape York and is more widely distributed in Papua New Guinea.

Actual size

The Slender Land Frog has a squat body, a pointed snout, and small eyes. There are small disks on the fingers, and larger ones on the toes. The upper surfaces are grayish brown with darker markings and a thin, pale line down the middle of the back. The flanks are black and there are orange patches in the armpits, groin, and inside of the thighs.

SIMILAR SPECIES

Twenty-nine species have been described to date in the genus *Austrochaperina*. Only five of them occur in Australia, most being discovered in Papua New Guinea, where new species are still being found. The Northern Territory Frog (*A. adelphe*) lives in the extreme north of Australia's Arnhem Land. The Owen Stanley Land Frog (*A. brevipes*) has a tiny range in Papua New Guinea; males of this species have been observed guarding eggs.

FAMILY	Microhylidae: Asterophryinae
OTHER NAMES	None
DISTRIBUTION	West Papua province, Indonesia
ADULT HABITAT	Rainforest at altitudes of 1,600–2,500 ft (500–750 m)
LARVAL HABITAT	Within the egg
CONSERVATION STATUS	IUCN Least Concern. It seems to be common within a restricted area

ADULT LENGTH
Male
1 15/16–2 3/16 in (49–57 mm)
Female
2 3/16–2½ in (56–62 mm)

CALLULOPS WONDIWOIENSIS

WONDIWOI LAND FROG

GÜNTHER, STELBRINK & VON RINTELEN, 2012

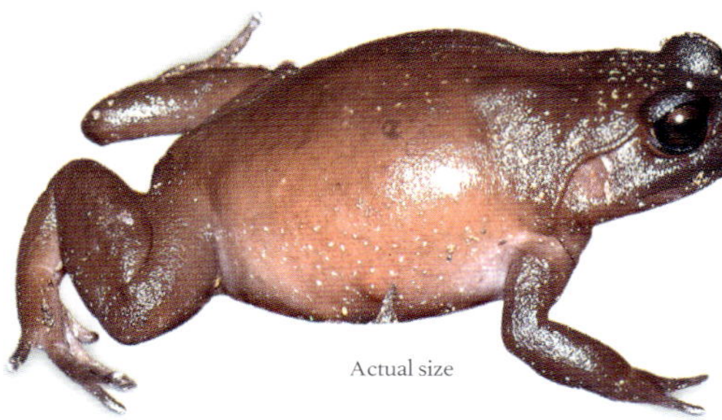

Actual size

The Wondiwoi Land Frog has a plump body, short hind legs, a rounded snout, and large eyes. There are small disks on the tips of the fingers and toes. The upper surfaces are grayish or reddish brown with small white spots, most numerous on the flanks. The underside is pale brown to white.

This recently discovered frog is named after the Wondiwoi Mountains in western New Guinea, where it is found in burrows in soft soil on the forest floor. Males call during the early part of the night, from inside or close to their burrows, producing six to nine loud croaks that sound like the quacking of a Mallard duck. Calling males maintain a distance between one another of at least 65 ft (20 m). The juveniles of this species are usually found close to waterfalls. They have very large eyes and, unlike their brown parents, are black with white spots and yellow blotches.

SIMILAR SPECIES

The genus *Callulops* currently contains 29 species, several of them described very recently. Fakfak Land Frog (*Callulops valvifer*), found in the Fakfak Mountains of western New Guinea, has a call that is so loud it drowns out the calls of other frogs. It was originally described in 1910 on the basis of a single juvenile specimen found in the stomach of a snake.

FAMILY	Microhylidae: Asterophryinae
OTHER NAMES	Previously *Albericus siegfriedi*
DISTRIBUTION	Eastern Papua New Guinea
ADULT HABITAT	Montane rainforest at altitudes of 7,900–8,200 ft (2,400–2,500 m)
LARVAL HABITAT	Not reported, but probably within the egg
CONSERVATION STATUS	IUCN Critically Endangered. Declining within its very restricted range

ADULT LENGTH
9/16–7/8 in (14–21 mm)

CHOEROPHRYNE SIEGFRIEDI

MOUNT ELIMBARI WARTY BUZZING FROG

(MENZIES, 1999)

This small frog is found only on the slopes of Mt. Elimbari in Papua New Guinea. It is typically found climbing in low vegetation at night, when males produce a buzz-like call. The species' reproductive biology has not been observed, but it is assumed that it has direct development, the tadpole stage being completed within the eggs, which hatch into tiny frogs. The forest around the mountain where it lives has been largely destroyed and its habitat is subject to increasingly frequent bush fires.

Actual size

The Mount Elimbari Warty Buzzing Frog has a plump body; short, thin legs: a pointed snout; and numerous sharply pointed warts on its back, head, and legs. There are disks on its fingers and toes. The upper surfaces are pale brown with diffuse dark brown markings.

SIMILAR SPECIES

This small frog originally belonged to the genus *Albericus*, which was named after the dwarf king Alberich, in the mythological Germanic poem *Nibelungenlied*. The species name *siegfriedi* was for the hero of the poem. In 2016 *Albericus* was synonymized with *Choerophryne*, which now contains 39 species, endemic to New Guinea. Most are rather drab in color, but the Mount Simpson Buzzing Frog (*C. sanguinopictus*) is pale blue or pale green, decorated with red spots. It occurs only on Mt. Simpson in the southeast of the island. Most species have buzz-like calls, but the Mouse Frog (*C. murritus*), found in the central highlands, produces single peeps.

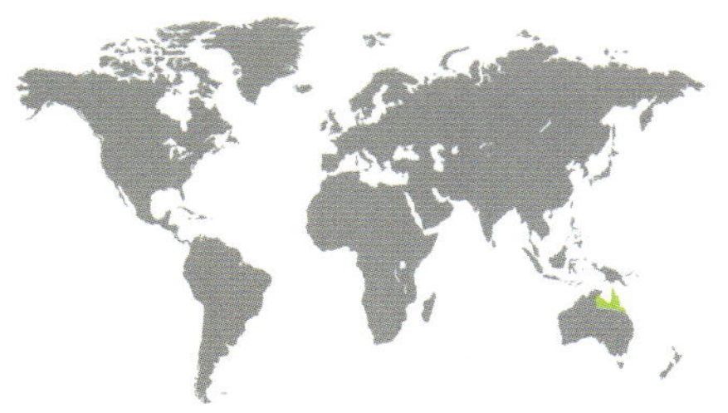

FAMILY	Microhylidae: Asterophryinae
OTHER NAMES	Buzzing Nursery Frog, Buzzing Rainforest Frog
DISTRIBUTION	Northern Queensland, Australia
ADULT HABITAT	Wet forest at altitudes of 3,000–4,300 ft (900–1,300 m)
LARVAL HABITAT	Within the egg
CONSERVATION STATUS	IUCN Least Concern. Has a very small range and its habitat is threatened by water extraction

ADULT LENGTH
Male
½–⁹⁄₁₆ in (12–15 mm)
Female
½–¹¹⁄₁₆ in (13–17 mm)

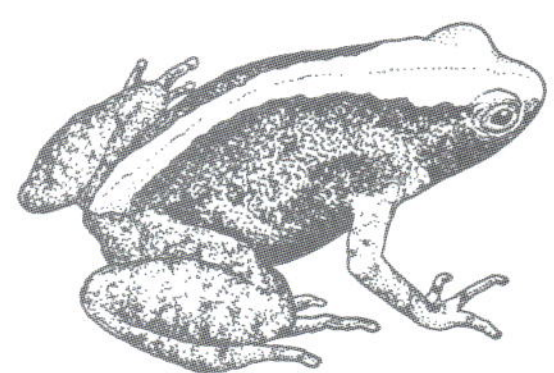

COPHIXALUS BOMBIENS

WINDSOR BUZZING FROG

ZWEIFEL, 1985

Actual size

The Windsor Buzzing Frog has a plump body and a pointed snout. The upper surfaces are dull reddish brown with darker markings. There is often a paler brown stripe along the middle of the back that gets broader toward the head. The underside is dark with gray flecks.

This tiny frog gets its name from the male's mating call, an insect-like buzzing sound, which it makes close to the ground. Other aspects of the species' breeding biology are unknown, but it is presumed that it has direct development, the tadpole stage being completed within the egg. It occurs in three small areas in the Windsor Tableland, in a habitat known as vine forest, which is reliant on abundant heavy rain. Where it occurs, it is common and seems to be relatively unaffected by selective logging.

SIMILAR SPECIES

To date, 63 species have been described in the genus *Cophixalus*, of which 19 occur in Australia. The Elegant Frog (*C. concinnus*) is a Critically Endangered species found in northeastern Queensland. It is known from only a single location, less than 4 sq miles (10 sq km) in area, in montane rainforest, which is threatened by climate change. See also the Rattling Frog (*C. crepitans*; page 475).

FAMILY	Microhylidae: Asterophryinae
OTHER NAMES	Northern Nursery Frog, Rattling Nursery Frog, Rusty Rainforest Frog
DISTRIBUTION	Northern Queensland, Australia
ADULT HABITAT	Tropical rainforest
LARVAL HABITAT	Within the egg
CONSERVATION STATUS	IUCN Least Concern. Has a very small range that lies within a protected area

COPHIXALUS CREPITANS

RATTLING FROG

ZWEIFEL, 1985

ADULT LENGTH
Male
½–⁹⁄₁₆ in (12–14 mm)
Female
average ⁹⁄₁₆ in (14 mm)

This very small frog gets its name from the male's mating call, a high-pitched rattle lasting for around two seconds. Details of its breeding biology are unknown, but it is presumed that it has direct development, the tadpole stage being completed within the egg. It lives in a very small area of forest close to the town of Coen in northern Queensland, where logging used to threaten its future existence, but it is now fully protected and the only threat it faces is from tourism.

Actual size

The Rattling Frog has a plump body, large eyes, and a blunt snout. The legs are relatively long, and there are disks on the fingers and toes. The upper surfaces are brown with yellow or orange patches and black markings, including a W-shaped mark between the shoulders. The underside is orange, yellow, or yellow-green.

SIMILAR SPECIES

The Cape York Frog (*Cophixalus peninsularis*) is found in the same area as *C. crepitans*. Known from only two specimens, it is slightly larger. The Inelegant Frog (*C. infacetus*) is found in rainforest to the south of Cairns, Queensland. It lays 8–13 eggs in a string that is left in a damp place, where the eggs hatch directly into tiny frogs. See also the Windsor Buzzing Frog (*C. bombiens*; page 474).

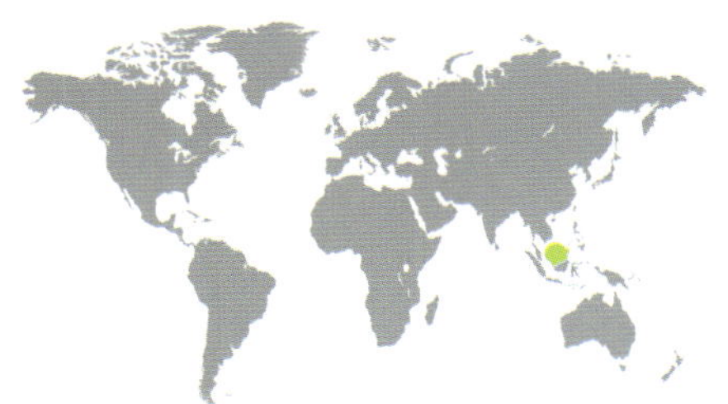

FAMILY	Microhylidae: Asterophryinae
OTHER NAMES	Long Snouted Frog
DISTRIBUTION	Northern Borneo
ADULT HABITAT	Lowland forest below 1,640 ft (500 m) altitude
LARVAL HABITAT	Unknown
CONSERVATION STATUS	IUCN Least Concern. Its habitat is severely threatened by deforestation

ADULT LENGTH
Male
unknown

Female
$1\frac{3}{16}$–$1\frac{5}{8}$ in (30–41 mm)

GASTROPHRYNOIDES BORNEENSIS

BORNEO NARROWMOUTH TOAD

(BOULENGER, 1897)

Actual size

Although called a toad, there is nothing toad-like about this frog. It has a slender, elegant shape and very smooth skin. It is rarely encountered, and has been found under dead leaves, in water-filled tree-holes, and within a dead log. Its body shape, pointed snout, and powerful hind legs suggest that it is a burrowing species, and its small eyes suggest that it rarely emerges above ground. The lowland forests of Borneo, where it lives, are subject to widespread clear-cutting, leading to destruction and fragmentation of its habitat. It is not found in any protected areas.

SIMILAR SPECIES

For a long time, *Gastrophrynoides borneensis* was thought to be the only species in its genus but, in 2008, a very similar frog was found on an isolated mountain-top in Peninsular Malaysia. Tung's Narrow-mouthed Frog (*G. immaculatus*) is very similar in appearance, but has a longer snout and is not spotted. Its natural history and breeding biology are as mysterious as those of *G. borneensis*. It is listed as Vulnerable.

The Borneo Narrowmouth Toad has a cone-shaped head, short arms, stubby fingers and toes, and small, bulbous eyes. The skin is smooth and shiny. The upper surfaces are gray-brown, decorated with numerous scattered, small white spots. The underside is white.

FAMILY	Microhylidae: Asterophryinae
OTHER NAMES	None
DISTRIBUTION	Throughout Papua New Guinea and some small offshore islands
ADULT HABITAT	Lowland forest up to 11,700 ft (3,570 m) altitude
LARVAL HABITAT	Within the egg
CONSERVATION STATUS	Data Deficient. A common species with a stable population

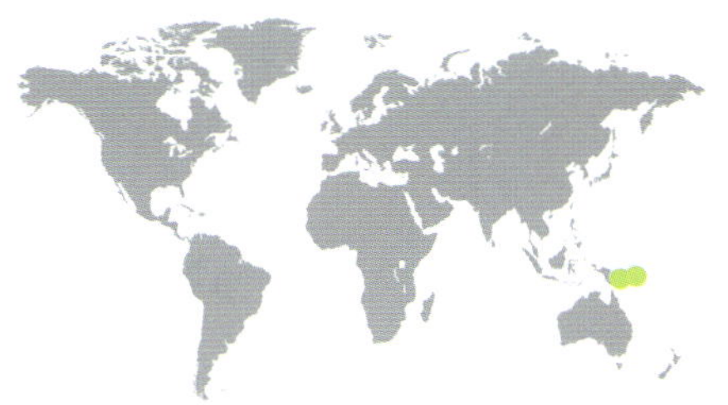

HYLOPHORBUS RUFESCENS

RED MAWATTA FROG

MACLEAY, 1878

ADULT LENGTH
Male
1⅛–1¼ in (28–32 mm)
Female
1$\frac{9}{16}$–1⅝ in (39–42 mm)

This small frog, described as a "walker-hopper," is active on the forest floor at night. The male attracts females with a call consisting of a series of loud notes. He makes a cup-shaped nest in the ground, beneath the leaf litter, and the female lays around 13 large eggs in it. Experiments have shown that the male guards the eggs constantly and aggressively, inflating his body and lunging when under threat, and eating intruding ants. If a guarding male is removed, the eggs invariably die, either through predation or fungal attack. The eggs hatch directly into tiny frogs.

Actual size

The Red Mawatta Frog has long arms and legs, and long fingers and toes. The disks on the toes are larger than those on the fingers. The snout is pointed, the eyes are large, and there are warts on the back. It is dark olive-brown or reddish in color, with a darker, mottled pattern.

SIMILAR SPECIES

Seventeen species have been described in the little-known genus *Hylophorbus*, many of them recently. The smallest of them, at ⅞ in (23 mm) long, is *H. richardsi*, known from only two localities in the mountains of central Papua New Guinea. The largest, at 1⅜ in (35 mm) long, is *H. wondiwoi*, found only on the Wandammen Peninsula in Indonesia's Papua province in western New Guinea. It lives in an area that is not protected from logging.

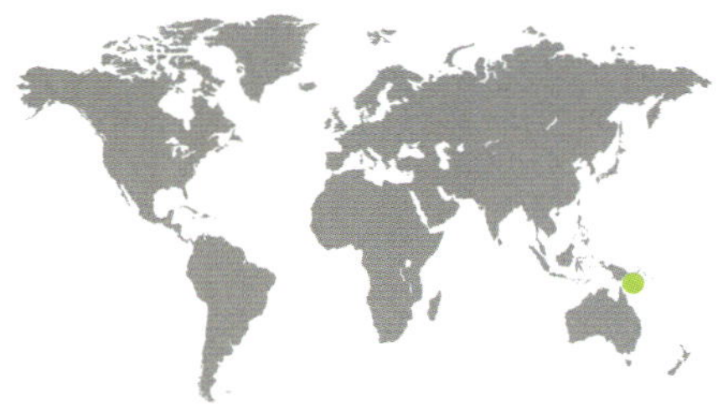

FAMILY	Microhylidae: Asterophryinae
OTHER NAMES	None
DISTRIBUTION	Southeastern Papua New Guinea
ADULT HABITAT	Cloud forest at altitudes of 2,100–2,620 ft (630–800 m)
LARVAL HABITAT	Not reported, but probably within the egg
CONSERVATION STATUS	IUCN Critically Endangered. Has a very restricted range

ADULT LENGTH
Male
⅞–1 in (22–26 mm)

Female
average 1 in (25 mm)

OREOPHRYNE EZRA

MOUNT RIO CROSS FROG

KRAUS & ALLISON, 2009

Actual size

The Mount Rio Cross Frog is a small frog with a rounded snout, large, protruding eyes, and well-developed smaller disks on the toes and larger ones on the fingers. The toes are webbed. The upper surfaces are black with yellow spots in juveniles and peach in adults. The underside is pale blue in juveniles, and yellow in adults. The iris of the eyes is bright, pale blue.

Many frogs change color as they grow older, but very few do so as dramatically as this small frog. Juveniles are black with yellow spots, whereas adults are bright peach with brilliant blue eyes. Both color patterns may warn potential predators that this frog is distasteful or toxic, but this hypothesis has not been tested. The species feeds on ants and lives in cloud forest on a single mountain, Mt. Riu on Sudest Island in the Louisiade Archipelago, Papua New Guinea. It exemplifies a category of frogs whose future is precarious: those that are confined to the tops of mountains. If climate change adversely alters their habitat, they simply have nowhere else to go.

SIMILAR SPECIES

Oreophryne ezra is probably unique in its genus, most members of which are tan, brown, or gray throughout their lives. However, this genus is not well known and there may well be similar species still to be discovered. Changes in color with age are common among African reed frogs of the genus *Hyperolius*. For example, the Painted Reed Frog (*H. marmoratus*; page 445) starts life uniformly brown but develops into a variety of color forms.

FAMILY	Microhylidae: Asterophryinae
OTHER NAMES	Montane Chorus Frog
DISTRIBUTION	Lombok and Bali, Indonesia
ADULT HABITAT	Montane forest above 3,300 ft (1,000 m) altitude
LARVAL HABITAT	Presumed to be within the egg
CONSERVATION STATUS	IUCN Endangered. Has not been reported for many years

ADULT LENGTH
average 1⅛ in (29 mm)

OREOPHRYNE MONTICOLA

LOMBOK CROSS FROG

(BOULENGER, 1897)

Very little is known about this small frog, which has not been collected since the 1930s. An arboreal species found in the forested highlands of Lombok and Bali, it is presumed to breed by direct development, laying eggs that develop into tiny frogs without going through a free-living tadpole stage. In Bali, the forest is protected, but with human recreation rather than biodiversity protection in mind. It is thought that the species may occur in a protected area of Lombok, and it has recently been reported from lowland swamps in Bali, but this has yet to be confirmed.

Actual size

SIMILAR SPECIES

The Southeast Asian—Melanesian genus *Oreophryne* currently contains 73 species, many described very recently. Known as cross frogs, their natural history is not well understood. At just ½ in (12 mm) long, *O. minuta* is the smallest species. It is known from a single locality in Papua New Guinea. At the other end of the scale, Anthony's Cross Frog (*O. anthonyi*) is 1 15/16 in (50 mm) long. It is common in Papua New Guinea, where it occurs up to an altitude of 9,200 ft (2,800 m).

The Lombok Cross Frog has a short snout, short legs, and larger disks on its fingers than on its toes. There are numerous warts on the head, back, and legs. The upper surfaces are various shades of brown, sometimes with one or more pale longitudinal stripes. The concealed parts of the thighs are bright pink and the underside is pale with a speckled brown pattern.

FAMILY	Microhylidae: Asterophryinae
OTHER NAMES	None
DISTRIBUTION	Southern Papua New Guinea
ADULT HABITAT	Rainforest up to 3,300 ft (1,000 m) altitude
LARVAL HABITAT	Within the egg
CONSERVATION STATUS	IUCN Least Concern

ADULT LENGTH
Male
11/16–7/8 in (18–22 mm)

Female
7/8–1 1/16 in (21–27 mm)

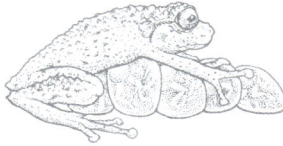

OREOPHRYNE OVIPROTECTOR

EGG-GUARDING CROSS FROG

GÜNTHER, RICHARDS, BICKFORD & JOHNSTON, 2012

Actual size

The Egg-guarding Cross Frog is a small frog with numerous small warts on its back, head, and legs, and well-developed adhesive disks, larger on the fingers than the toes. The upper surfaces are brown with a bright green patch between the eyes and bright green stripes along the flanks. The underside is white.

This small frog gets its name from the way it protects its eggs. The eggs are large and are laid, in clutches of around eight, on the underside of leaves 3⅞ in–11½ ft (10–350 cm) above the ground. At night, one of the parents—almost always the father but occasionally the mother—sits on the eggs. If the carer is removed, the eggs die of desiccation. The eggs develop directly, hatching into tiny froglets. The male calls at night, producing a loud rattle. Each male responds to the call of the male nearest him, so that their combined sound moves through the forest like a wave.

SIMILAR SPECIES

To date, 73 species in the genus *Oreophryne* have been described, some of them very recently. On the basis of their calls, they can be divided into "peepers" and "rattlers," like *O. oviprotector*. Other "rattlers" are the Alpine Cross Frog (*O. alticola*), found in high-altitude grassland in Papua New Guinea, and the Idenburg Cross Frog (*O. idenburgensis*), a large frog found in rainforest, also in Papua New Guinea.

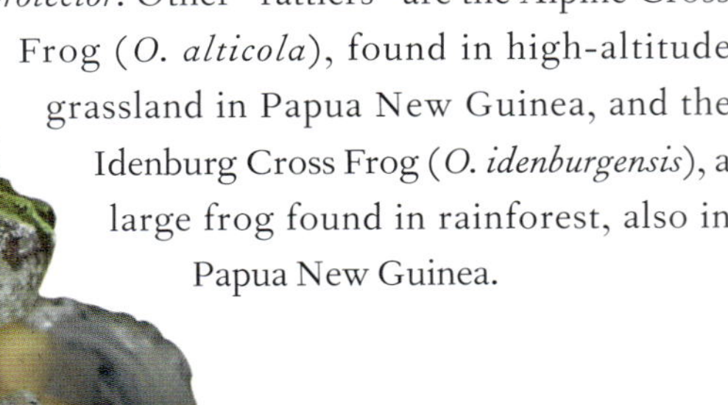

FAMILY	Microhylidae: Asterophryinae
OTHER NAMES	None
DISTRIBUTION	Southeastern tip of Papua New Guinea
ADULT HABITAT	Forest
LARVAL HABITAT	Probably within the egg
CONSERVATION STATUS	IUCN Least Concern

ADULT LENGTH
¼–⁵⁄₁₆ in (7–8 mm)

PAEDOPHRYNE AMAUENSIS

AMAU MINIATURE FROG

RITTMEYER, ALLISON, GRÜNDLER, THOMPSON & AUSTIN, 2012

Discovered in 2009, this tiny frog is believed to be the world's smallest vertebrate. So far known from only one locality, it lives among leaf litter and is capable of jumping 30 times its body length. It is active at night and males call at dawn and at dusk, producing a series of very high-pitched notes that sound like a stridulating insect. They call continuously for 1 or 2 minutes and then take a rest. Mating has not been observed and it is assumed that, like other members of its family, the species breeds by direct development, laying eggs that hatch directly into frogs.

Actual size

The Amau Miniature Frog is a tiny frog with a broad head and proportionally long legs. The upper surfaces are dark brown with various irregular tan or rusty-brown patches. The sides and belly are brown or gray with a speckled bluish-white pattern.

SIMILAR SPECIES

To date, seven species have been described in the genus *Paedophryne*, all since 2010. The Kamilai Miniature Frog (*Paedophryne swiftorum*), at ⁵⁄₁₆–⅜ in (8–9 mm) long, is slightly larger than *P. amauensis* and is very similar in its habits. It can be distinguished by its slightly lower-pitched call. Both species show features typical of amphibians that have undergone miniaturization during their evolution. For example, their first toe and finger are reduced to stumps, and their other digits have a reduced number of bones. The Mount Simpson Miniature Frog (*P. kathismaphlox*) is listed as Critically Endangered.

FAMILY	Microhylidae: Asterophryinae
OTHER NAMES	None
DISTRIBUTION	Papua New Guinea
ADULT HABITAT	Rainforest up to 4,900 ft (1,500 m) altitude; also occurs in gardens and degraded forest
LARVAL HABITAT	Within the egg
CONSERVATION STATUS	IUCN Least Concern. A common species with a large range

ADULT LENGTH
average 1⅝ in (42 mm)

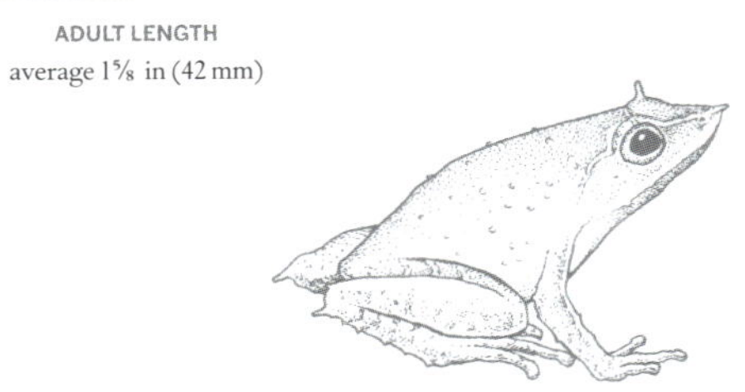

SPHENOPHRYNE CORNUTA

HORNED LAND FROG

PETERS & DORIA, 1878

Actual size

Parental care takes an unusual form in this nocturnally active frog. The female lays 20–30 eggs that hatch directly into tiny frogs. When they emerge from the eggs, these climb onto their father's back. He hides by day, but for 3–9 nights he moves around the forest floor, up to 160 ft (50 m) or so from where the eggs were laid. Each night, a few of the young frogs jump off their father and disappear into the leaf litter. This behavior ensures that the young are widely dispersed, reducing competition between them for food.

SIMILAR SPECIES

Until recently *Sphenophryne cornuta* was the only species in its genus, but several species have been transferred from other genera, making a total of 15 species. It is closely related to another Papuan frog in which newly hatched froglets are similarly carried around by their father, Schlaginhaufen's Land Frog (*S. schlaginhaufeni*). See also Wandammen Bush Frog (*Asterophrys pullifer*; page 469).

The Horned Land Frog has a pointed, projecting snout, a prominent, soft spine above each eye, and long legs with spines on their outer edge. The fingers and toes are long and end in disks. The upper surfaces are dull gray or brown, the flanks are paler, and the underside is gray, reddish brown, or red with white spots.

FAMILY	Microhylidae: Asterophryinae
OTHER NAMES	None
DISTRIBUTION	Southeastern Papua New Guinea
ADULT HABITAT	Montane wet forest at altitudes of 2,700–3,430 ft (830–1,045 m)
LARVAL HABITAT	Within the egg
CONSERVATION STATUS	IUCN Least Concern. It appears to have a very restricted distribution

ADULT LENGTH
Male
1¼–1¾ in (32–44 mm)

Female
$1\frac{11}{16}$–$1\frac{13}{16}$ in (43–46 mm)

SPHENOPHRYNE MINIAFIA

MOUNT TRAFALGAR LAND FROG

(KRAUS, 2014)

Papua New Guinea is home to many frog species that have been described recently, and many more that have yet to be described. There are huge areas of the country that have still not been thoroughly explored, so the chance of finding new species, especially in mountainous regions, is very high. This recently described frog is known from only a single, isolated mountain group, of volcanic origin, near Cape Nelson. It appears to be a burrowing species and has been heard calling from a burrow at night. Its call consists of a single note.

Actual size

SIMILAR SPECIES

Sphenophryne miniafia is the most recently described of 15 species included in the genus *Sphenophryne*. The Owen Stanley Land Frog (*S. rhododactyla*) gets its name from the mountain range in eastern Papua New Guinea where it occurs. Its call is like the meow of a cat. The Alotau Land Frog (*S. dentata*) comes from the southeastern tip of Papua New Guinea. Males respond to one another's calls in such a way that a wave of sound passes through the forest.

The Mount Trafalgar Land Frog has a wide head, long legs, and long fingers and toes. A fold of skin runs from the tip of the pointed snout, over the eye, and along the flank. On the head, this separates a dark brown face mask from the paler crown. The upper surfaces vary between green, brown, and ocher, and there are faint dark stripes on the legs.

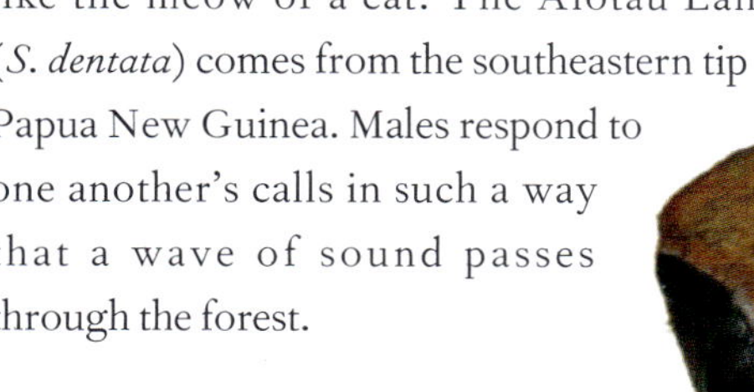

FAMILY	Microhylidae: Asterophryinae
OTHER NAMES	None
DISTRIBUTION	Northwestern Papua New Guinea
ADULT HABITAT	Rainforest at altitudes of 3,000–4,600 ft (900–1,400 m)
LARVAL HABITAT	Within the egg
CONSERVATION STATUS	Least Concern. Appears to be a common species and occurs in remote areas where there are no apparent threats

ADULT LENGTH
Male
$1\frac{7}{16}$–$1\frac{13}{16}$ in (37–46 mm)

Female
unknown

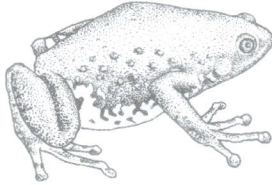

XENORHINA ARBORICOLA

ARBOREAL SNOUTED FROG

ALLISON & KRAUS, 2000

Known from only a very few male specimens, this unusual frog has many anatomical features typical of burrowing frogs but lives high up in trees. It does, indeed, bury itself in leaf litter but, rather than on the forest floor, it is found among orchids, ferns, and other epiphytes that live on trees in wet tropical forest. One specimen was found buried in the leaf litter that had gathered in an old bird's nest. Males call mostly at night from high in the trees, producing a sequence of 12–14 melodious notes, and one male was found attending a string of 11 eggs.

SIMILAR SPECIES

The genus *Xenorhina* currently contains 41 species. All are burrowing frogs that lay eggs that hatch directly into tiny frogs. *Xenorhina arboricola* differs from other species in having very large disks on its fingers and toes, a reflection of its arboreal habits. The Sharp Snouted Frog (*X. oxycephala*) is a common species from western and northern Papua New Guinea. Males call from beneath the soil, producing a sound similar to that made by *X. arboricola*.

The Arboreal Snouted Frog has a plump body, long, slender limbs, a pointed snout, and small eyes. The disks on the long fingers and toes are well developed. The skin is smooth with scattered white tubercles. The upper surfaces are grayish brown, the underside is gray, and there are irregular blotches on the flanks, legs, fingers, and toes.

Actual size

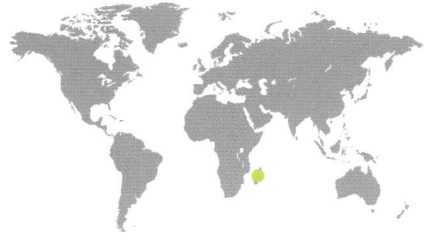

FAMILY	Microhylidae: Cophylinae
OTHER NAMES	Previously *Stumpffia helenae*, *Rhombophryne helenae*
DISTRIBUTION	Central Madagascar
ADULT HABITAT	Montane forest
LARVAL HABITAT	Unknown
CONSERVATION STATUS	IUCN Critically Endangered. Exists in two small habitat fragments in a forest reserve that is not well protected

ANILANY HELENAE

HELENA'S STUMP-TOED FROG

(VALLAN, 2000)

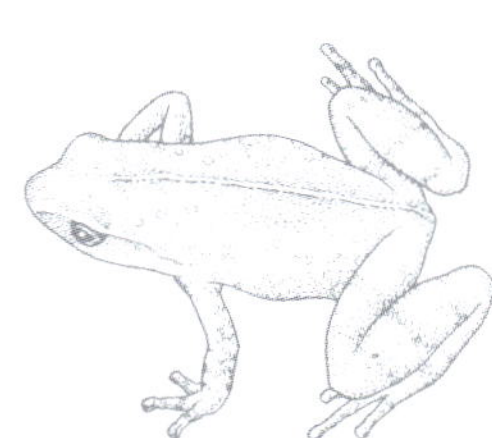

ADULT LENGTH
Male
average 9/16 in (14 mm)
Female
average 9/16 in (15 mm)

Actual size

The stump-toed frogs of Madagascar are noted for the slow, deliberate way in which they move, though they can jump away quickly when alarmed. This small frog is active by day and is usually found on the ground among leaf litter. On its lower back are two dark spots that resemble eyes. It is thought that these deflect attacks by predators away from its head. The male's call is a regularly repeated, high-pitched chirp. Otherwise, the species' reproductive behavior has not been described, but it is thought that it lays its eggs in a terrestrial foam nest.

Helena's Stump-toed Frog has short fingers and toes, and slightly enlarged fingertips. It has a pointed snout, and is gray in color with two black spots on the lower part of its back, which are concealed by its thighs when it is at rest. There is often a narrow orange stripe down the middle of the back.

SIMILAR SPECIES

Anilany helenae used to be included in the stump-toed frog genera *Stumpffia* and *Rhombophryne*, which currently contain 44 and 20 species respectively. *Anilany* is a monotypic genus. Within *Stumpffia*, the Giant Stump-toed Frog (*S. grandis*), although only around 1 in (25 mm) long, is the largest species and is found in eastern and northeastern Madagascar. The smallest species, the Andoany Stump-toed Frog (*S. pygmaea*), is only 7/16 in (11 mm) long and is confined to the small islands of Nosy Be and Nosy Komba, just off the Madagascan coast.

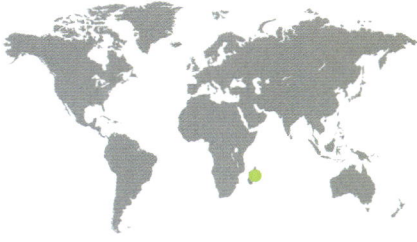

FAMILY	Microhylidae: Cophylinae
OTHER NAMES	None
DISTRIBUTION	Eastern Madagascar
ADULT HABITAT	Forest up to 4,300 ft (1,300 m) altitude
LARVAL HABITAT	Water-filled holes in trees and other plants
CONSERVATION STATUS	IUCN Near Threatened. Declining where its habitat is being destroyed, but it occurs within some protected areas

ADULT LENGTH
Male
11/16–7/8 in (17–22 mm)
Female
11/16–7/8 in (18–21 mm)

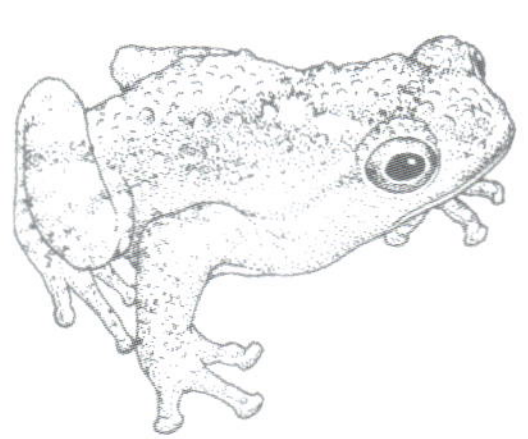

ANODONTHYLA BOULENGERII

BOULENGER'S CLIMBING FROG

MÜLLER, 1892

Actual size

Boulenger's Climbing Frog has a blunt snout, large eyes, well-developed disks on its fingers, and many sharp tubercles on its head, back, and legs. Its coloration is highly variable, but the upper surfaces are typically brown with white spots. The underside is pale gray.

Males of this small, common frog call at night from perches on tree trunks 3–6.5 ft (1–2 m) above the ground. Their call is a single melodious note, repeated 140–175 times per minute. The female lays 23–30 large eggs in a water-filled tree-hole or leaf axil on a palm tree. The male remains close to the eggs and the developing tadpoles, guarding them against predators. The tadpoles are endotrophic, meaning that they do not have mouths but complete their entire development by absorbing the egg yolk provided by their mother.

SIMILAR SPECIES

Eleven species have been described in the genus *Anodonthyla*, several of them as recently as 2010. All are found in Madagascar. The Ranomafana Climbing Frog (*Anodonthyla moramora*) is a small frog that lives in a tiny area within a national park in the east of the island. The Black-throated Climbing Frog (*A. nigrigularis*) is found in south and central Madagascar. Both species are listed as Endangered.

FAMILY	Microhylidae: Cophylinae
OTHER NAMES	None
DISTRIBUTION	Northwestern Madagascar
ADULT HABITAT	Forest
LARVAL HABITAT	Water-filled holes in trees and other plants
CONSERVATION STATUS	IUCN Endangered. Has a very small range

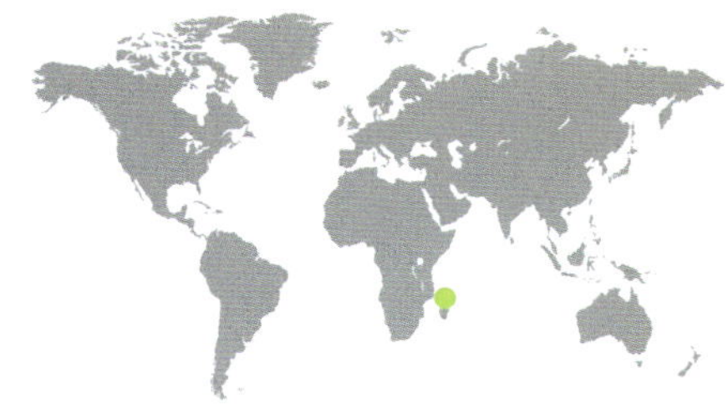

COPHYLA BERARA

BERARA CLIMBING FROG

VENCES, ANDREONE & GLAW, 2005

ADULT LENGTH
Male
⅞–1 in (23–26 mm)
Female
unknown

No females of this very rare species have been collected. The male calls at night from leaves 3–6.5 ft (1–2 m) above the ground. His call is a single melodious note, lasting just under a second in duration. It is presumed that, like closely related species, the frog lays its eggs in water-filled tree-holes and that the tadpoles are sustained entirely on maternal yolk. It has not been found outside the Anabohazo Forest, on the Sahamalaza Peninsula in northwestern Madagascar. This forest has recently been declared a protected area.

Actual size

The Berara Climbing Frog has a slender body, a short snout, and large eyes. There are well-developed disks on the fingers. The upper surfaces are pale brown with darker markings, and there is a dark stripe running through the eye to the armpit. The underside is cream to white.

SIMILAR SPECIES

Cophyla berara is the most recently described of three species in the genus. The Nosy Be Climbing Frog (*Cophyla occultans*) is a smaller, little-known species. The Whistling Tree Frog (*C. phyllodactyla*) is a slightly larger frog, known since 1880, which differs from *C. berara* in having a shorter call. It appears able to survive in some altered habitats and is quite common in northern Madagascar.

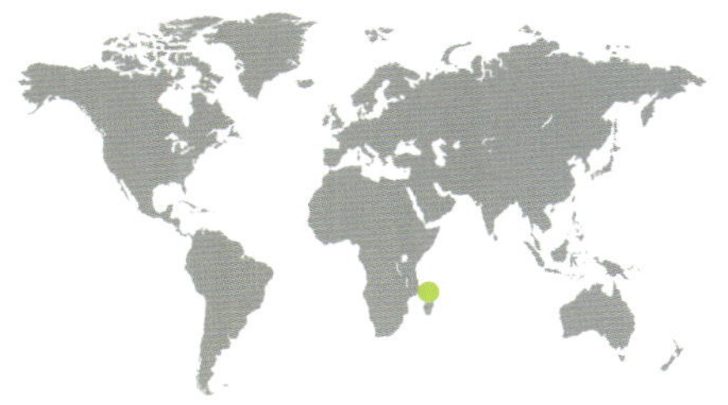

FAMILY	Microhylidae: Cophylinae
OTHER NAMES	None
DISTRIBUTION	Northwestern Madagascar, including the island of Nosy Be
ADULT HABITAT	Forest
LARVAL HABITAT	Probably small pools in leaf axils
CONSERVATION STATUS	IUCN Endangered. Has a very restricted range and is threatened by destruction of its forest habitat

ADULT LENGTH
1–1³⁄₁₆ in (25–30 mm)

PLATYPELIS MILLOTI

MILLOT'S CLIMBING FROG

GUIBÉ, 1950

Actual size

Millot's Climbing Frog has a pointed snout and large eyes. There are well-developed adhesive pads on its fingers and toes. The upper surfaces are chocolate-brown, with a striking pattern of black, yellow, and red patches, some with white borders. There is a broad, yellow transverse band between the eyes, and a narrow yellow line down the middle of the back. The belly and the underside of the legs are bright red.

This small, colorful frog is typically found in the leaf axils of plants, particularly screw pines (*Pandanus* spp.). Males call at night, producing a short, harmonious note that they repeat rapidly and regularly. The species' breeding behavior has not been observed but it is likely that, in common with its close relatives, it lays its eggs in pools in plant axils and that its tadpoles do not feed, lacking a mouth, but depend on yolk from the egg to sustain their growth. Like many forest frogs in Madagascar, its habitat is being destroyed for timber extraction, charcoal burning, grazing by livestock, fires, and encroachment by human settlements.

SIMILAR SPECIES

Seven species are currently recognized in the genus *Platypelis*, the giant tree frogs of Madagascar. Despite their name, most are very small. Barbour's Climbing Frog (*P. barbouri*) is tiny and is both common and widespread in eastern Madagascar. Its tadpoles develop in water-filled tree-holes. Boulenger's Giant Climbing Frog (*P. grandis*) is a large frog, up to 4⅛ in (105 mm) in length; males of the species have been observed guarding the eggs.

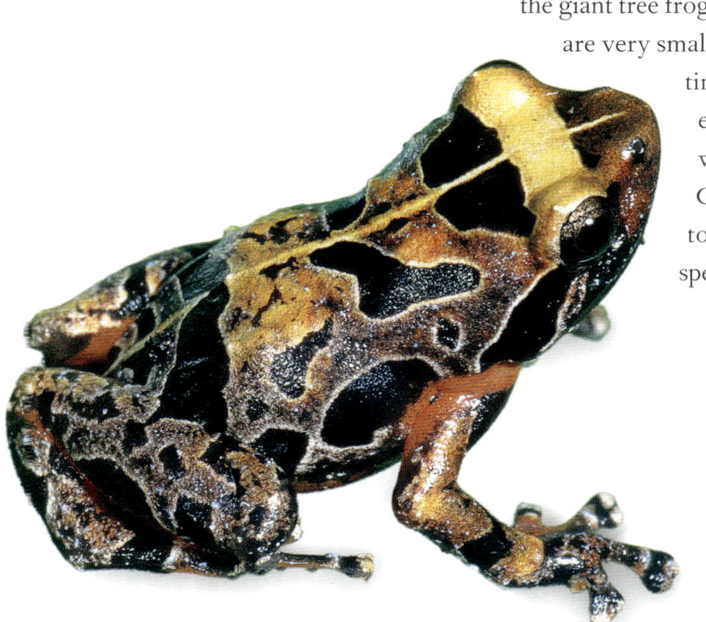

FAMILY	Microhylidae: Cophylinae
OTHER NAMES	Previously *Rhombophryne alluaudi*, Sakana Digging Frog
DISTRIBUTION	Eastern Madagascar
ADULT HABITAT	Forest up to 4,900 m (1,500 m) altitude
LARVAL HABITAT	Inside a burrow
CONSERVATION STATUS	IUCN Least Concern. Declining in places but occurs within some protected areas

ADULT LENGTH
1⁹⁄₁₆–2⅜ in (40–60 mm)

PLETHODONTOHYLA ALLUAUDI

FORT DAUPHIN DIGGING FROG

(MOCQUARD, 1901)

This squat, plump frog is rarely seen as it spends most of its time in a burrow, under a fallen log, or in deep leaf litter. It is difficult to find, but its mating call—a single note produced at long intervals—is sometimes heard coming from the forest floor. The eggs are laid in a burrow and are large enough to provide the tadpoles with sufficient yolk to complete their development, so that they do not need to find food for themselves. The forests of Madagascar are widely subject to degradation, through encroaching agriculture, timber extraction, charcoal collection, and the presence of livestock.

Actual size

SIMILAR SPECIES

Eleven species have been described in the genus *Plethodontohyla*. This and other *Plethodontohyla* species were previously included in the genus *Rhombophryne*, which contains 20 species, including the Nosy Be Burrowing Frog (*R. testudo*) which emerges above ground only after heavy rain. It has very small eyes, and a distinctive defensive posture when disturbed, in which it extends its arms and legs and makes its back concave. Known from only three localities, the species is listed as Endangered. See Malagasy Climbing Rain Frog (*P. mihanika*; page 490).

The Fort Dauphin Digging Frog has a very wide body and head, and smooth skin. There is a fold of skin running from the eye to the armpit. The upper surfaces are brown, with a variable, dark brown pattern on the back and dark brown stripes on the legs. The belly is yellow.

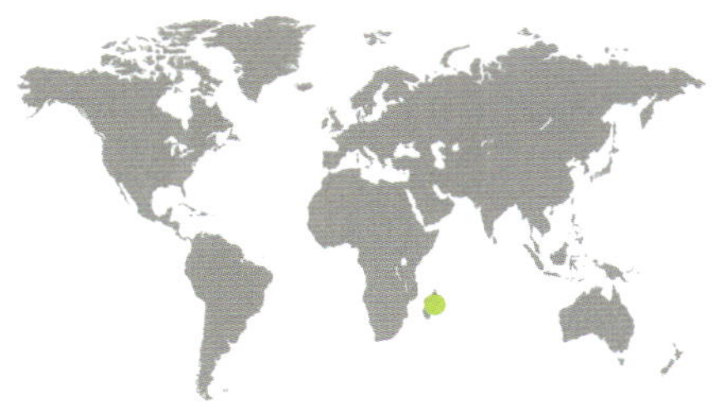

FAMILY	Microhylidae: Cophylinae
OTHER NAMES	None
DISTRIBUTION	Central and eastern Madagascar
ADULT HABITAT	Forest
LARVAL HABITAT	Thought to be water-filled tree-holes
CONSERVATION STATUS	IUCN Least Concern. Declining in places due to deforestation, but occurs within some protected areas

ADULT LENGTH
Male
1–1³⁄₁₆ in (26–30 mm)
Female
1⅛–1¼ in (29–31 mm)

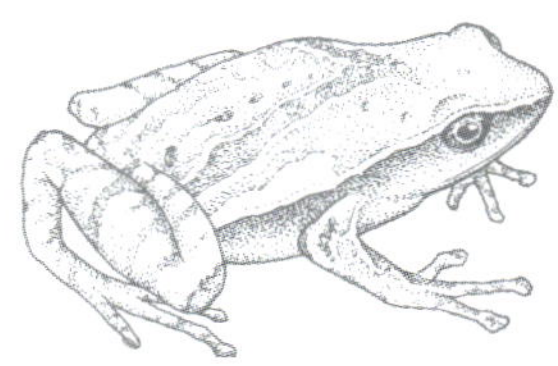

PLETHODONTOHYLA MIHANIKA

MALAGASY CLIMBING RAIN FROG

VENCES, RAXWORTHY, NUSSBAUM & GLAW, 2003

Actual size

Adults, juveniles, and tadpoles of this small frog have been found together in water-filled tree-holes, suggesting that this is where it breeds and raising the possibility that adults care for their young. Males call from tree trunks, producing melodious notes at the rate of around 11 per minute. The eggs are laid in strings. The species' forest habitat is being destroyed as a result of expanding agriculture, logging, charcoal manufacture, the invasion and spread of alien eucalyptus trees, grazing by livestock, fire, and expanding human settlements.

SIMILAR SPECIES

To date, 11 species have been described in the Madagascan genus *Plethodontohyla*. Some, such as the Port Dauphin Digging Frog (*P. alluaudi*; page 489) and the Ocellated Digging Frog (*P. ocellata*), are terrestrial and dig into the soil. This species has two large black spots, edged in white, on its lower back. Others, such as the Mahanoro Digging Frog (*P. notosticta*), are, like *P. mihanika*, arboreal in their habits.

The Malagasy Climbing Rain Frog has a pointed snout and enlarged adhesive disks on its fingers and toes. A pale stripe running from snout to groin forms a clear border between the paler back and the darker flanks. The back is pale brown with darker markings, often including an inverted "V" and two black spots on the lower back.

FAMILY	Microhylidae: Dyscophiinae
OTHER NAMES	None
DISTRIBUTION	Northeastern Madagascar
ADULT HABITAT	Coastal forest and scrub; also found in urban gardens
LARVAL HABITAT	Pools or slow-moving streams
CONSERVATION STATUS	IUCN Least Concern. Has declined in much of its range but is less subject to collecting than it once was

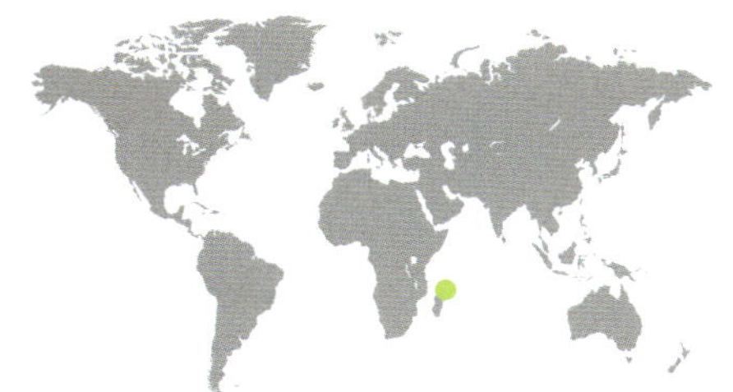

ADULT LENGTH
Male
2⅜–2⅝ in (60–65 mm)
Female
3⅛–4¾ in (80–120 mm)

DYSCOPHUS ANTONGILII

TOMATO FROG

GRANDIDIER, 1877

The vivid color of this aptly named frog warns potential enemies that its skin produces a defensive secretion when it is attacked or handled. Against predators such as snakes, this white substance acts as glue, sticking their jaws together; in humans, it induces an allergic skin reaction. The Tomato Frog breeds after heavy rain at any time of year, males producing a series of low-pitched notes. Between 1,000 and 1,500 small black eggs are laid in pools or slow-moving streams. Once seriously threatened by collecting for the international pet trade, the frog is now protected and is a "flagship species" for conservation in Madagascar.

SIMILAR SPECIES

Dyscophus antongilii is one of three species in its genus. The Sambava Tomato Frog (*D. guineti*) is generally yellower in color and is found along forest streams in eastern Madagascar. It sometimes hybridizes with *D. antongilii*. The Antsouhy Tomato Frog (*D. insularis*) is smaller than the other two species, is gray-brown in color, and occurs in western Madagascar.

Actual size

The Tomato Frog has a very plump body, a short snout, and large, protruding eyes with golden irises. The male is yellow to orange in color, the female orange to dark red or sometimes brown. The underside is white in both sexes. Some individuals also have black markings along their flanks.

FAMILY	Microhylidae: Gastrophryninae
OTHER NAMES	None
DISTRIBUTION	Southeastern Brazil
ADULT HABITAT	Lowland forest up to 650 ft (200 m) altitude
LARVAL HABITAT	Ephemeral pools
CONSERVATION STATUS	IUCN Least Concern. Declining in places but occurs within some protected areas

ADULT LENGTH
Male
average ¾ in (20 mm)
Female
average 1 in (26 mm)

ARCOVOMER PASSARELLII

PASSARELLI'S FROG

CARVALHO, 1954

Actual size

Passarelli's Frog has a pointed snout, rather small eyes, and long legs, arms, fingers, and toes. It is brown or gray in color, with a darker, broad, wavy-edged band down the middle of the back. There is a white stripe running from the eye to the armpit, and a V-shaped white stripe on the top of the snout.

This small, little-known frog is one of many frog species found only in the Atlantic Forest that runs along Brazil's southeastern coast. It is usually found in the leaf litter on the forest floor. The male's call is a short, sharp whistle and the female lays 70–100 eggs in temporary pools that form after heavy rain. Other aspects of its reproductive biology have not been described. When molested, it adopts a characteristic stiff-legged defensive posture, with all four limbs stretched out in a star pattern.

SIMILAR SPECIES

Arcovomer passarellii is the only species in its genus to have been described. It has been suggested that the somewhat smaller individuals found in the state of Espírito Santo may be a separate species.

FAMILY	Microhylidae: Gastrophryninae
OTHER NAMES	None
DISTRIBUTION	Southeastern Brazil
ADULT HABITAT	Forest up to 2,600 ft (800 m) altitude
LARVAL HABITAT	Ponds
CONSERVATION STATUS	IUCN Least Concern. A very common species, with an apparently stable population

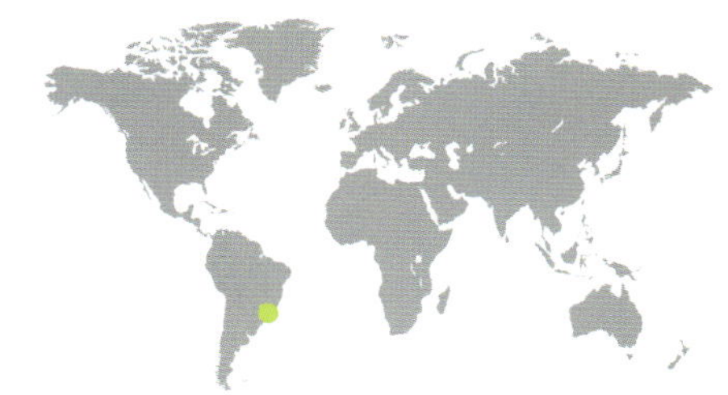

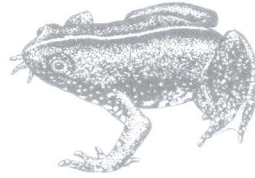

ADULT LENGTH
Male
average ⅞ in (21 mm)
Female
average 1 in (26 mm)

CHIASMOCLEIS LEUCOSTICTA

SANTA CATARINA HUMMING FROG

(BOULENGER, 1888)

An inhabitant of Brazil's Atlantic Forest, this small frog has an unusual mode of reproduction. Large groups of calling males gather around ponds after heavy rain. When pairs form, they float in open water, the male glued to the female's back. She dips her head into the water and releases one to four eggs, which the male then fertilizes. This is repeated around 70 times until a floating mass of around 200 fertilized eggs has formed. The pair, still in amplexus, then dive beneath the eggs and blow air bubbles out of their nostrils; these provide support for the eggs from below.

Actual size

The Santa Catarina Humming Frog has a plump body, a small head with a pointed snout, and small eyes. The skin is smooth and the upper surfaces are black or brown in color, dotted with small, pale blue spots. The belly is mottled in black and white, and there may be orange patches on the arms and legs.

SIMILAR SPECIES

To date, 29 species have been described in the genus *Chiasmocleis*. Known as humming frogs, they are found in Panama and tropical South America. Eleven species have been described since the first edition of this book, including Jake Dalton's Humming Frog (*C. jacki*) from French Guiana and Amapa, Brazil, and ABOFOA's Humming Frog (*C. abofoa*) from Peru, both in 2022.

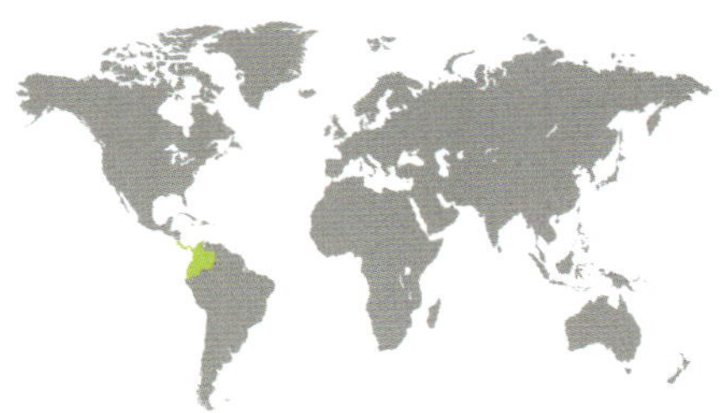

FAMILY	Microhylidae: Gastrophryninae
OTHER NAMES	Costa Rica Nelson Frog
DISTRIBUTION	Costa Rica, Panama, Colombia, Ecuador
ADULT HABITAT	Lowland forest up to 5,250 ft (1,600 m) altitude
LARVAL HABITAT	Ephemeral pools and swamps
CONSERVATION STATUS	IUCN Least Concern. Threatened in parts of its range by deforestation

ADULT LENGTH
Male
1⁹⁄₁₆–2⅜ in (40–60 mm)
Female
1¹⁵⁄₁₆–2¹³⁄₁₆ in (50–70 mm)

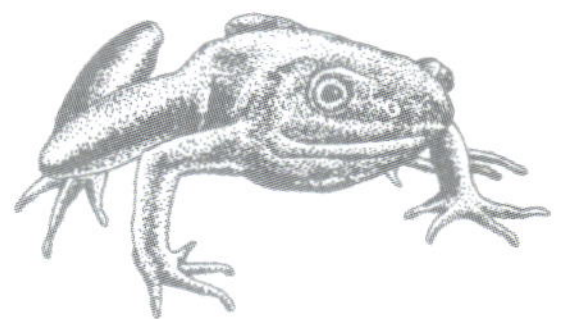

CTENOPHRYNE ATERRIMA

BLACK NARROW-MOUTHED FROG

(GÜNTHER, 1901)

This plump frog is very secretive in its habits, being usually found in leaf litter, under fallen logs, and occasionally deep underground. It becomes active during rainfall at night, and breeds between June and August. It appears that males do not call to attract females, and that eggs are laid in shallow pools and swamps, where the rather large tadpoles are found. The innermost toe of each foot on the frog bears a large, spade-like tubercle, suggesting that it sometimes burrows into the soil.

SIMILAR SPECIES

Six species in the genus *Ctenophryne* have been described to date. Known as egg frogs, they are found in Central and South America. The Brown Egg Frog (*C. geayi*) is a common species that lives in lowland habitats in the Amazon Basin. The Carpish Cloud Forest Frog (*Ctenophryne carpish*) is found at high altitude in the Peruvian Andes and is listed as Endangered. Two other species are also Endangered, the Equatorial Egg Frog (*C. aequatorialis*) from Ecuador and the Bearded Egg Frog (*C. barbatula*) from Peru.

The Black Narrow-mouthed Frog has a globular body, a short, narrow head, and small eyes. Its arms are short, its legs are long, and its toes are webbed. There is a transverse fold of skin just behind the eyes. The upper surfaces are uniformly dark gray, brown, or black, and the underside is brown.

Actual size

FAMILY	Microhylidae: Gastrophryninae
OTHER NAMES	None
DISTRIBUTION	Argentina, Bolivia, Brazil, Paraguay
ADULT HABITAT	Soft, wet soil in open areas
LARVAL HABITAT	Ephemeral ponds
CONSERVATION STATUS	IUCN Least Concern. Although rarely seen, it is very common in suitable habitat

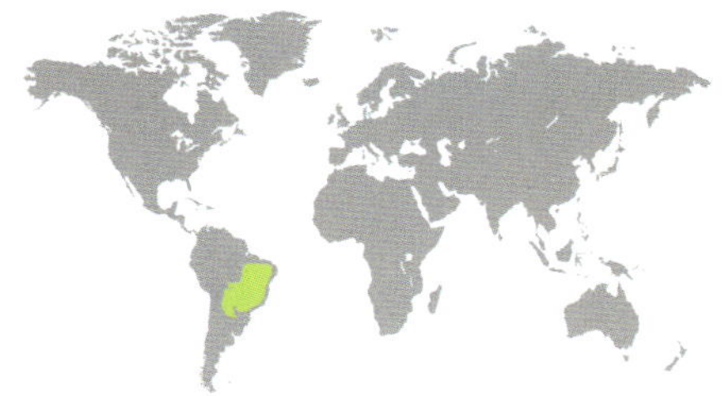

ADULT LENGTH
1⁹⁄₁₆–1¹⁵⁄₁₆ in (40–50 mm); female is larger than male

DERMATONOTUS MUELLERI

MULLER'S TERMITE FROG

(BOETTGER, 1885)

This unusual frog is very abundant in places, but is rarely seen because it lives deep underground. Using its forelimbs, it digs headfirst into soft, damp soil and creates an underground chamber where it lives during the dry season. It feeds primarily on termites. It emerges between September and February, often in large numbers, to breed in ephemeral ponds. Described as an "explosive breeder," it completes mating and egg-laying within five days. Its skin produces a white secretion that is of interest to the pharmaceutical industry because of its antiparasitic and antifungal properties.

Actual size

Muller's Termite Frog has a stout body, a very small head, and short arms and legs. The eyes are small and point forward, and there is a fold in the skin behind the eyes. The fingers and toes have blunt, rounded tips. The skin is smooth and is colored olive-green to brown on the back, with irregular black markings on the flanks and legs.

SIMILAR SPECIES

Dermatonotus muelleri is the only species in its genus, and is very distinctive in its morphology and habits. It provides a remarkable example of convergent evolution, sharing with the burrowing frogs of the African genus *Hemisus* (see pages 466–467) the habit of burrowing headfirst. (Most burrowing frogs burrow backward, using their hind feet.) Moreover, *D. muelleri* and *Hemisus* species, alone among frogs, can bend their heads downward, by as much as 90 degrees.

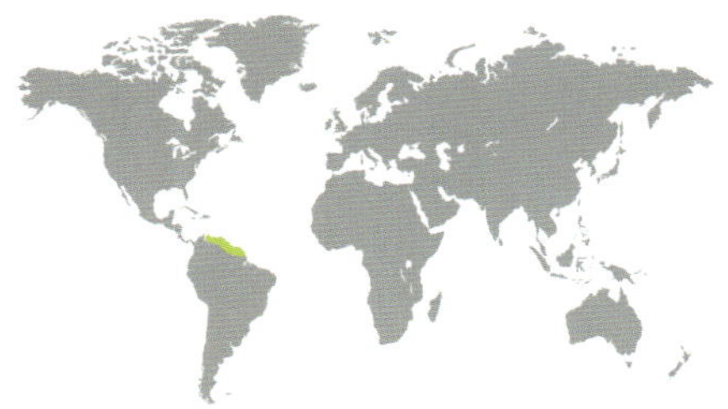

FAMILY	Microhylidae: Gastrophryninae
OTHER NAMES	Northeastern Oval Frog
DISTRIBUTION	Coastal Venezuela, Guyana, Suriname, French Guiana, northern Brazil, Trinidad
ADULT HABITAT	Lowland tropical forest, wet savanna, and freshwater marshes up to 1,650 ft (500 m) altitude
LARVAL HABITAT	Ephemeral pools
CONSERVATION STATUS	IUCN Least Concern. Widely distributed and relatively common but locally may be threatened by habitat loss

ADULT LENGTH
Male
1 5/16–1 3/8 in (33–35 mm)
Female
1 7/16–1 7/8 in (36–48 mm)

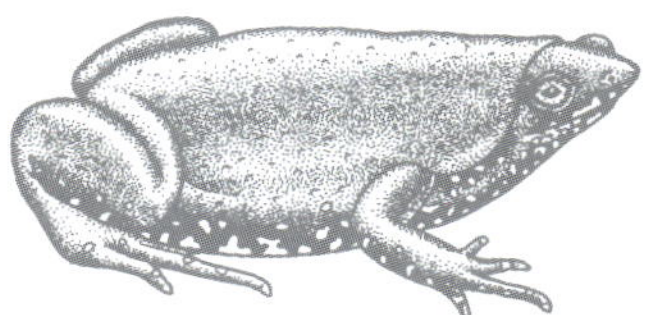

ELACHISTOCLEIS SURINAMENSIS

SURINAME OVAL FROG

(DAUDIN, 1802)

Actual size

The Suriname Oval Frog is more pear-shaped than oval. Its body is stout but its head is small, with a pointed snout and small, black eyes. Its limbs are short and its four fingers and five toes are long and unwebbed. It is drab gray above, usually with a fine, light stripe down the middle of its back, but while other species exhibit undersides blotched with yellow, orange, or red, *Elachistocleis surinamensis* has a brown belly blotched with gray.

The Suriname Oval Frog is a secretive frog that lives in leaf litter in lowland tropical forests. In dry weather, oval frogs estivate underground, but during heavy rain, the males gather in flooded areas and call to attract females. During amplexus, males of some species (e.g. Bicolored Oval Frog, *Elachistocleis bicolor*) become glued to the female's back so they don't fall off. Eggs are laid in ephemeral ponds and hatch in two days, the tadpoles then metamorphosing within eight weeks, before the ponds dry out. The narrow, pointed snouts of oval frogs suggest they feed on small invertebrates, and they have been documented feeding on ants and termites that abound in their leaf-litter habitats.

SIMILAR SPECIES

Elachistocleis currently contains 17 species, distributed from Costa Rica to Argentina. The often-cited name *E. ovalis*, in which most populations were once "lumped," cannot confidently be applied to any specific population. *Elachistocleis surinamensis* is the only species within its range, but a further eight species occur in Brazil. In recent years the genus has received quite a lot of attention, with 10 species being described since 2000, the most recent being *E. sikuani* and *E. tinigua*, both from Colombia, in 2022.

FAMILY	Microhylidae: Gastrophryninae
OTHER NAMES	Texas Narrow-mouthed Toad, Western Narrow-mouthed Toad
DISTRIBUTION	Central and southern USA, Mexico
ADULT HABITAT	Grassland, desert scrub, and pine–oak woodland
LARVAL HABITAT	Permanent and ephemeral ponds, and other small waterbodies
CONSERVATION STATUS	IUCN Least Concern. Common in many places and there is no evidence of a general decline

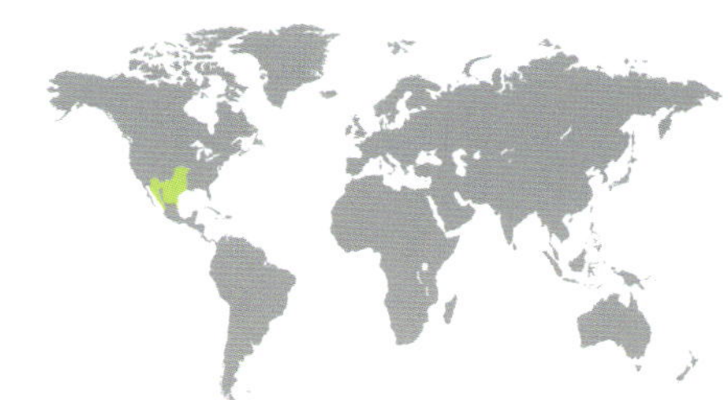

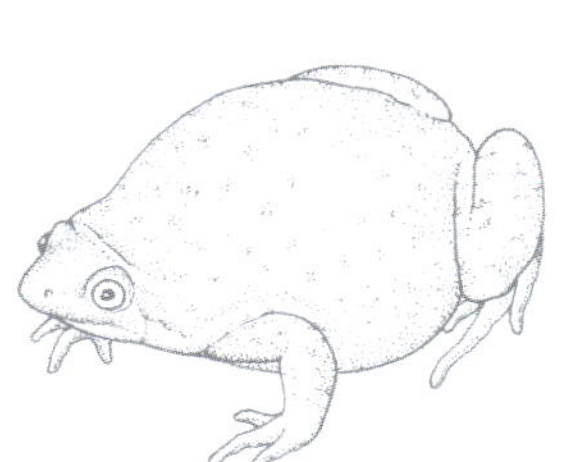

ADULT LENGTH
¾–1⅝ in (19–42 mm)
female is larger than male

GASTROPHRYNE OLIVACEA

GREAT PLAINS NARROW-MOUTHED TOAD

(HALLOWELL, 1856)

This secretive frog is hard to find except when it is breeding. It lives in any habitat that provides moisture and somewhere to hide. Heavy rain stimulates males to gather and form choruses in ponds. They call from the pond edge, with only their heads sticking out of the water. Their call is a sharp "peep" followed by a sound like that made by an angry bee. When a male grasps a female, he secretes a sticky substance from glands on his chest to glue himself to the female's back. The female lays up to 2,000 eggs that float on the water's surface.

Actual size

The Great Plains Narrow-mouthed Toad has a plump body, a small head with a pointed snout, and small eyes. The arms and legs are short, and there is no webbing between the toes. The skin is smooth. The back is gray, olive-green, or tan, and may be decorated with small black spots. The belly is pale and lacks markings.

SIMILAR SPECIES

There are four species in the genus *Gastrophryne*, known as American narrow-mouthed toads. The Eastern Narrow-mouthed Toad (*G. carolinensis*) is found in the southern states of the USA and is similar in appearance and behavior to *G. olivacea*. The Elegant Narrow-mouthed Toad (*G. elegans*) occurs in Central America and Mexico.

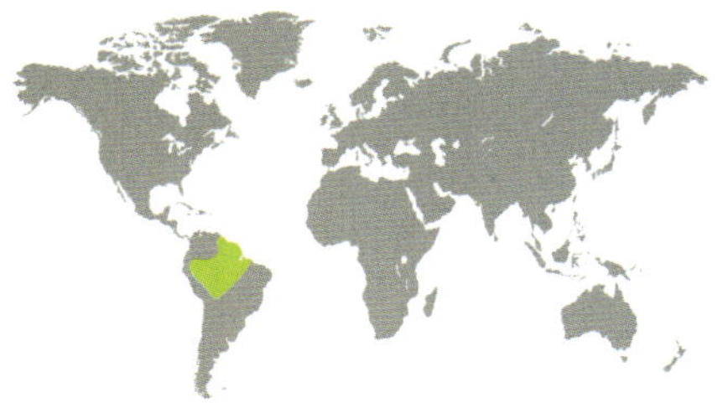

FAMILY	Microhylidae: Gastrophryninae
OTHER NAMES	Amazon Sheep Frog
DISTRIBUTION	Northern and western sides of the Amazon Basin in South America
ADULT HABITAT	Forest
LARVAL HABITAT	Ponds
CONSERVATION STATUS	IUCN Least Concern. Common in parts of its range and rare in others, but shows no evidence of decline

ADULT LENGTH
Male
$1\frac{5}{16}$–$1\frac{9}{16}$ in (34–39 mm)
Female
$1\frac{9}{16}$–$1\frac{3}{4}$ in (39–44 mm)

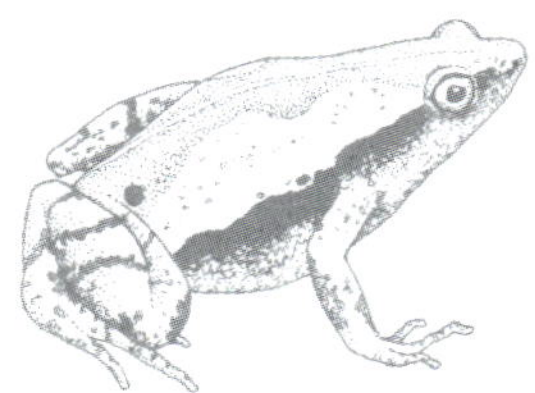

HAMPTOPHRYNE BOLIVIANA

BOLIVIAN BLEATING FROG

(PARKER, 1927)

Actual size

This aquatic frog gets its name from the male's mating call, which has been likened to the bleating of sheep. It is an "explosive breeder," gathering in large numbers after rain to form choruses of calling males around ponds. When the eggs are laid, they spread out to form a thin film on the water's surface. Outside the breeding season, adults are usually found among leaf litter on the forest floor. This species is rare over much of its large range, but is common in Bolivia and Peru.

SIMILAR SPECIES

There is one other species in this genus. The Wing-mouthed Frog (*Hamptophryne alios*) is named for a pair of enlarged structures on either side of the tadpole's oral disc, that resemble wings. It occurs in Peru and is widespread in Bolivia, but the full extent of its range is not known. Its conservation status has been assessed as Least Concern.

The Bolivian Bleating Frog has a plump body, a small head with a pointed snout, and rather small eyes. The upper surfaces are brown or reddish brown, with darker patches. There is a very thin, pale stripe down the middle of the back, and there is a dark band running along the flanks from the snout, through the eye, to the groin.

FAMILY	Microhylidae: Gastrophryninae
OTHER NAMES	Mexican Narrow-mouthed Toad
DISTRIBUTION	Texas, USA; Central America, from Mexico south to Costa Rica
ADULT HABITAT	Thorn scrub, savanna
LARVAL HABITAT	Ephemeral and permanent ponds
CONSERVATION STATUS	IUCN Least Concern. Shows no evidence of decline in much of its range but is listed as Threatened in Texas

HYPOPACHUS VARIOLOSUS

SHEEP FROG

(COPE, 1866)

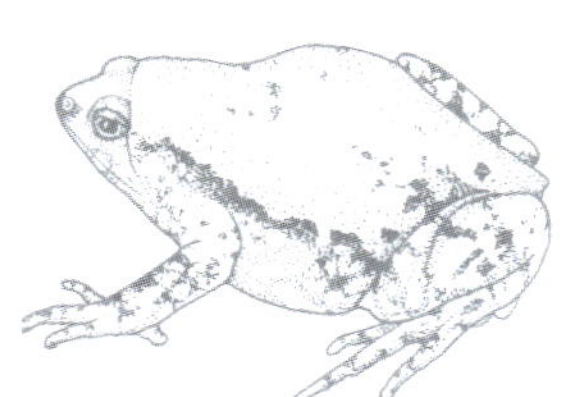

ADULT LENGTH
Male
1–1½ in (25–38 mm)
Female
1⅛–1⅝ in (29–41 mm)

This small frog is a specialist feeder, eating ants and termites. It has horny, spade-like tubercles on its hind feet that it uses to dig backward into soft soil. It hides by day, in a burrow or under a log. After heavy rain, sheep frogs migrate to permanent or temporary ponds to breed, males calling from the pond edge or while floating in open water, producing a sound like bleating sheep. The female lays around 700 eggs. Destruction of its habitat has led to this species being afforded protection in Texas, but elsewhere in its range it remains common.

Actual size

The Sheep Frog has a rotund body, a pointed snout, and a transverse fold in the skin across its head, just behind the eyes. There is no webbing between the long toes. The back is brown with dark spots, and there is a narrow yellow line down its middle. A broad black band runs from behind the eye along the flank.

SIMILAR SPECIES

There are three other species of sheep frog in the genus *Hypopachus*, all found in Central America. The Two-spaded Narrow-mouthed Toad (*H. ustum*) differs from the others in having a pair of horny tubercles on each hind foot rather than one. Barber's Sheep Frog (*H. barberi*) is listed as Near Threatened owing to the widespread destruction of its pine–oak forest habitat. The Lenca Sheep Frog (*H. guancasco*), from Honduras, is Endangered.

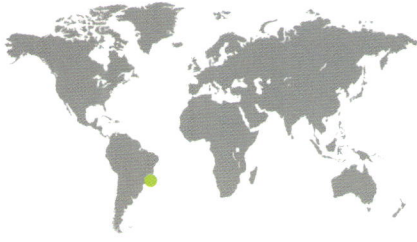

FAMILY	Microhylidae: Gastrophryninae
OTHER NAMES	None
DISTRIBUTION	Southeastern Brazil
ADULT HABITAT	Forest
LARVAL HABITAT	Within the egg
CONSERVATION STATUS	IUCN Least Concern. Its population is stable and its range includes several protected areas

ADULT LENGTH
9/16–1 9/16 in (15–40 mm)

MYERSIELLA MICROPS

RIO ELONGATED FROG

(DUMÉRIL & BIBRON, 1841)

Actual size

The Rio Elongated Frog has a globular body, a small, triangular head, and a pointed snout that projects well beyond the mouth. The arms are short and stout, the eyes are very small, and the skin is smooth. The upper surfaces are dark brown with small white spots, and the underside is light brown.

This inhabitant of Brazil's Atlantic Forest is very hard to find because it lives underground, using its distinctive pointed snout to burrow headfirst into soft soil. It becomes active on the surface in the rainy season, when males call from the ground, producing a long, clear note. The male grasps the female around her pelvis and she drags him underground as she excavates a chamber in which to lay the eggs. The eggs are large and she lays only around 14 of them. They hatch directly into tiny frogs over the course of about 19 days. One report describes a female apparently guarding her eggs.

SIMILAR SPECIES

The only species in its genus, *Myersiella microps* is similar in appearance and behavior to the disc frogs of the genus *Synapturanus* (see page 517). In common with some other frogs that dig headfirst, such as those in the African genus *Hemisus* (see pages 466–467), it has a strengthened arrangement of the bones of the shoulder.

FAMILY	Microhylidae: Gastrophryninae
OTHER NAMES	None
DISTRIBUTION	Southeastern Brazil
ADULT HABITAT	Forest
LARVAL HABITAT	Small ephemeral ponds
CONSERVATION STATUS	IUCN Least Concern. Its population is stable and its range includes several protected areas

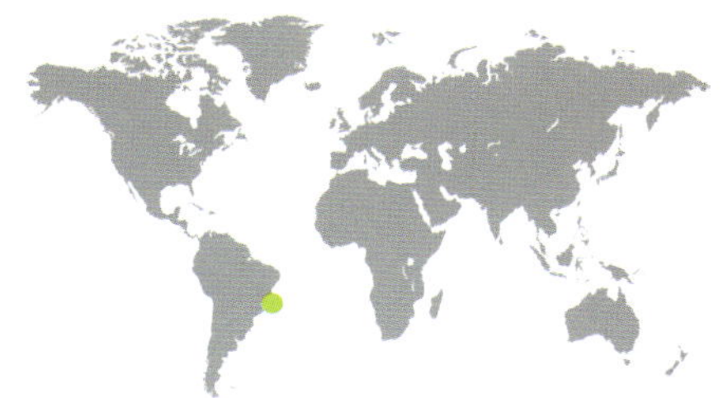

STEREOCYCLOPS INCRASSATUS

BRAZILIAN DUMPY FROG

COPE, 1870

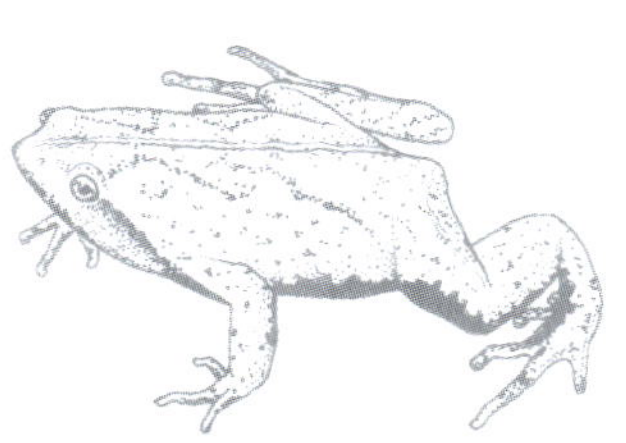

ADULT LENGTH
Male
1 7/16–1 13/16 in (37–46 mm)

Female
1 7/16–1 3/4 in (37–45 mm)

This small, scarcely known species is found in the Atlantic Forest that runs along the southeastern coast of Brazil. Although this frog is quite common, it is difficult to find, being generally secretive in its habits and active at night. Occasionally it is discovered in leaf litter on the forest floor. It eats a wide variety of prey, particularly ants and beetles. The species is an "explosive breeder," gathering in large numbers at forest ponds after the first heavy rain of the year. The eggs and tadpoles develop in water.

Actual size

The Brazilian Dumpy Frog has a rather flattened body, a small head, small front legs, and very muscular hind legs. The eyes are small. The upper surfaces are pale brown or gray with a narrow, pale stripe running down the middle of the back. There are delicate darker patterns on the back and the flanks are dark brown.

SIMILAR SPECIES

Three other species have been described in the Brazilian genus *Stereocyclops*. Parker's Dumpy Frog (*S. parkeri*), from southeastern Brazil, is most similar to *S. incrassatus*. The Bahia Yellow Frog (*S. histrio*) is found in eastern Brazil. Its conservation status has not yet been determined. The same is true of the Web-footed Dumpy Frog (*S. palmipes*), which was described in 2012.

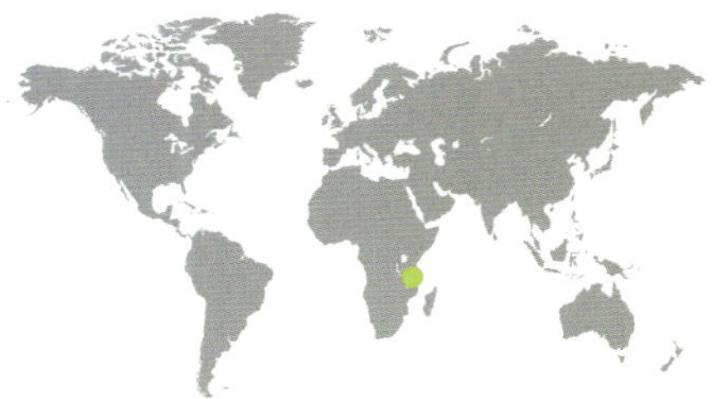

FAMILY	Microhylidae: Hoplophryninae
OTHER NAMES	Roger's Three-fingered Frog, Tanzania Banana Frog, Usambara Blue-bellied Frog
DISTRIBUTION	Eastern Arc Mountains, northeastern Tanzania
ADULT HABITAT	Forest
LARVAL HABITAT	Tiny pools in tree-holes and broken bamboo stems
CONSERVATION STATUS	IUCN Endangered. Has a very restricted distribution and is threatened by deforestation

ADULT LENGTH
⅞–1⅛ in (23–28 mm)

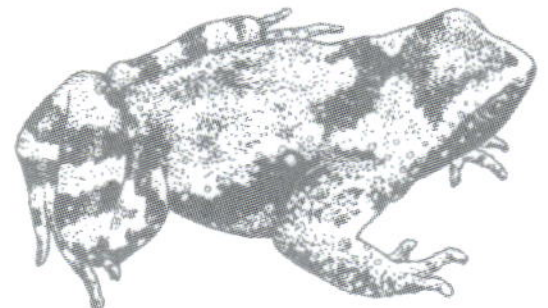

HOPLOPHRYNE ROGERSI

USAMBARA BANANA FROG

BARBOUR & LOVERIDGE, 1928

Actual size

This tiny frog is unusual in that the female has only three fingers. The male's thumb has been replaced by a spine that he presumably uses to hold onto the female during mating. Breeding males also have large glands on their chests and arms. Known only from the East Usambara and South Nguru mountain ranges in Tanzania, the species is found in leaf litter and in the leaf axils of banana trees. It breeds in water-filled tree-holes and in the pools formed where bamboo stems have died and broken off. The eggs are laid on a vertical surface, just above the water.

SIMILAR SPECIES

There is one other species in the genus *Hoplophryne*, the Uluguru Three-fingered Frog (*H. uluguruensis*), also classed as Endangered. It is very similar in appearance and behavior to *H. rogersi* and, though it is found in a number of places in the Eastern Arc Mountains, the two species do not occur together. It has been reported that females of *H. uluguruensis* guard their eggs.

The Usambara Banana Frog has a short, pointed snout. The upper surfaces are bluish gray with darker blotches, taking the form of cross-stripes on the legs. The underside is pale blue with a complex, marbled pattern in black.

FAMILY	Microhylidae: Kalophryninae
OTHER NAMES	Rufous-sided Sticky Frog, Sarawak Grainy Frog
DISTRIBUTION	Sarawak, Malaysia
ADULT HABITAT	Lowland forest
LARVAL HABITAT	Thought to be small pools
CONSERVATION STATUS	IUCN Least Concern. Has a small range that is subject to deforestation

KALOPHRYNUS INTERMEDIUS

INTERMEDIATE STICKY FROG

INGER, 1966

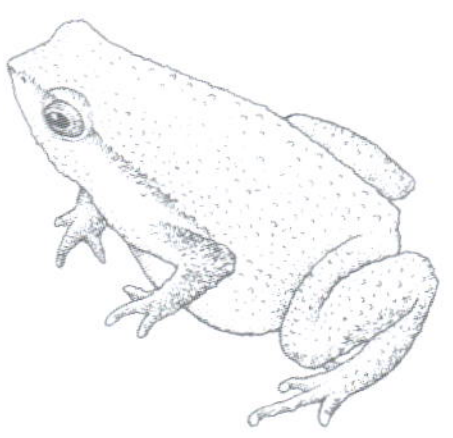

ADULT LENGTH
Male
average 1$\frac{1}{16}$ in (27 mm)

Female
1½–1⅝ in (38–41 mm)

This small, scarcely known frog lives in leaf litter on the forest floor. It feeds on ants and termites. It gets its name from the sticky mucus that it produces from glands on its back when it is handled. This makes it an unattractive prey for predators such as snakes. Its breeding behavior has not been reported, but other members of this genus lay their eggs in small pools, where they float on the surface. Each egg contains enough yolk for the tadpoles to complete their development without having to feed.

Actual size

The Intermediate Sticky Frog has a narrow head, a pointed snout, and short limbs. The skin has a coarse, granular texture and there is a conspicuous fold running from the tip of the snout to the groin. The back is brown with black markings, and the flanks are red, yellow, or cream. The chest and belly are cream.

SIMILAR SPECIES

To date, 27 species have been described in the Asian genus *Kalophrynus*, many of them known as sticky frogs. The Black-spotted Narrow-mouthed Frog (*K. pleurostigma*) is common and widespread throughout Southeast Asia. The Small Sticky Frog (*K. minusculus*) has a very restricted range in southern Sumatra and northern Java in Indonesia. Subject to habitat destruction, it is listed as Vulnerable. *Kalophrynus barioensis* is a very small species from Sarawak, and was described as recently as 2011.

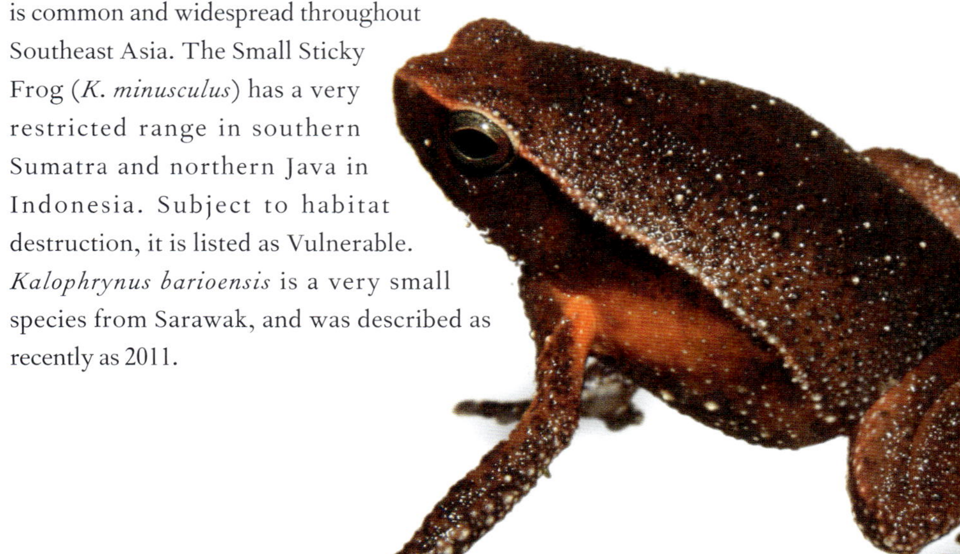

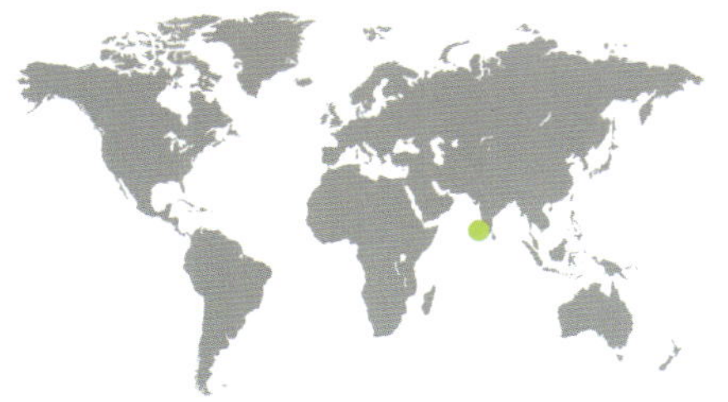

FAMILY	Microhylidae: Melanobatrachinae
OTHER NAMES	Black Microhylid Frog, Black Narrow-mouthed Frog
DISTRIBUTION	Kerala state, India
ADULT HABITAT	Forest at altitudes of 3,000–4,000 ft (900–1,200 m)
LARVAL HABITAT	Thought to be pools in streams
CONSERVATION STATUS	IUCN Vulnerable. A very rare frog, thought to occur in some protected areas

ADULT LENGTH
average 1⁵⁄₁₆ in (34 mm)

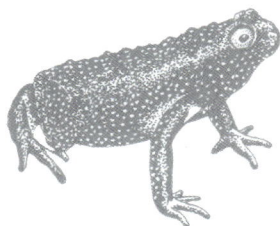

MELANOBATRACHUS INDICUS

KERALA HILLS FROG

BEDDOME, 1878

Actual size

The Kerala Hills Frog has a small head with a short snout. Its back is black and it has a broad red patch across its thighs near the groin. It also has red blotches on its chest. The skin on the back is warty, while the sides and stomach are smooth. The hind legs are short, and there is no webbing between the short toes and fingers.

An inhabitant of India's Western Ghats, the mountain range running down the western coast of India, this small frog was discovered in 1878 and has since been seen just three times, in 1928, 1992, and 1996. Like many other frog species in the Western Ghats, it does not occur anywhere else in the world. Most individuals that have been found were hiding under logs, and it is believed that the species lives in leaf litter on the forest floor. Males lack vocal sacs and do not appear to call to attract females.

SIMILAR SPECIES

The only species in the genus *Melanobatrachus*, *M. indicus* is not closely related to other frogs living in the Western Ghats. Its closest relatives are probably the banana frogs of Tanzania (page 502), from which it separated around 140 million years ago.

FAMILY	Microhylidae: Microhylinae
OTHER NAMES	Brown Thorny Frog, Saffron-bellied Frog, Spiny-heeled Froglet
DISTRIBUTION	Borneo, Philippines, Peninsular Malaysia
ADULT HABITAT	Rainforest
LARVAL HABITAT	Small ephemeral pools
CONSERVATION STATUS	IUCN Least Concern. Threatened in many places by deforestation, but occurs within some protected areas

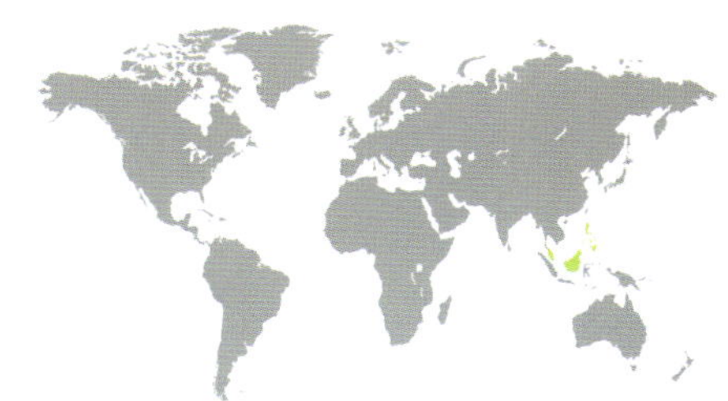

ADULT LENGTH
Male
11/16–7/8 in (18–21 mm)
Female
3/4–15/16 in (20–24 mm)

CHAPERINA FUSCA

YELLOW-SPOTTED NARROW-MOUTHED FROG

MOCQUARD, 1892

This very small frog has a soft, flexible spine on each elbow and heel; the function of these is not known. It is very unusual in that, when picked up, the yellow color on its belly comes off on the handler's fingers. It lives on the forest floor and also climbs up into vegetation. It breeds in small pools, including those that form in logs and dead bamboo. Males gather around such pools, producing a rather faint, insect-like buzz. The tadpoles are filter-feeders, swimming suspended in stagnant water and feeding on tiny aquatic organisms.

SIMILAR SPECIES

Chaperina fusca is the only species in its genus. It is distinguished from other frogs in its family by the spines on its heels and elbows.

Actual size

The Yellow-spotted Narrow-mouthed Frog has an almond-shaped body and widened tips to its fingers and toes. Its back is black, often with a marbled green or blue pattern, though some animals have many small white spots. The flanks and belly are black with large yellow spots. The arms and legs are brown with pale stripes and blotches.

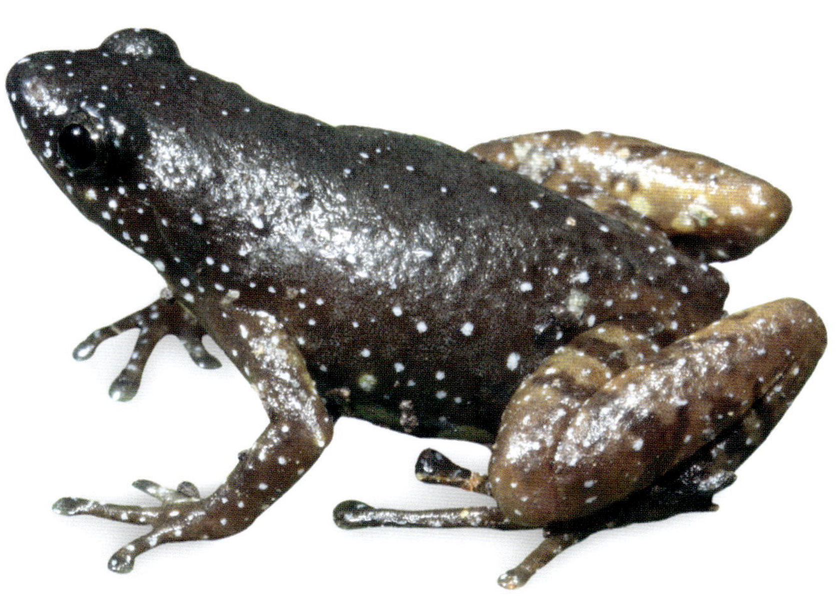

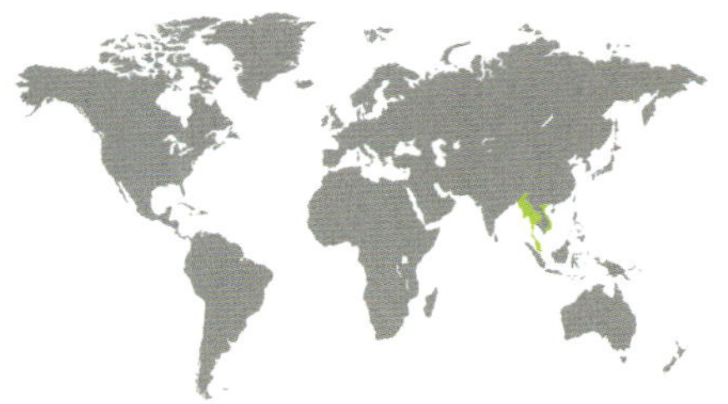

FAMILY	Microhylidae: Microhylinae
OTHER NAMES	Previously *Calluella guttulata*, Orange Burrowing Frog, Striped Spadefoot Froglet
DISTRIBUTION	Myanmar; Thailand, Vietnam, Peninsular Malaysia
ADULT HABITAT	Forest, disturbed habitats, gardens
LARVAL HABITAT	Ephemeral pools
CONSERVATION STATUS	IUCN Least Concern. Much of its habitat has been lost through deforestation, but its range includes some protected areas

ADULT LENGTH
Male
average $1^{11}/_{16}$ in (43 mm)
Female
average $1^{15}/_{16}$ in (50 mm)

GLYPHOGLOSSUS GUTTULATUS

BURMESE SQUAT FROG

(BLYTH, 1856)

Rarely seen because it lives mostly underground, this very plump frog is collected for human consumption in parts of its range. Such collection occurs when it emerges to breed, sometimes in large numbers, in temporary pools formed after heavy rain. The male has a single vocal sac that inflates under his chin and, during mating, he clasps the female around the pelvis. Otherwise, nothing is known of the species' reproductive behavior. When molested, the frog inflates its body and presents its rear end to its tormentor.

SIMILAR SPECIES

To date, ten species have been described in the Asian genus *Glyphoglossus* known as squat frogs, although until recently most species were in the genus *Calluella*. The Bornean Chili Frog (*G. capsus*) gets its name from its bright red and olive-green coloration. It was discovered in Sarawak and described in 2014. The Pahang Squat Frog (*G. minutus*), from Peninsular Malaysia, is a very small species. The most recently described species is the Huadianba Squat Frog (*G. huadianensis*) from China, described in 2021.

Actual size

The Burmese Squat Frog has a wide, slightly flattened body, a short, blunt head, and very short legs. A large fold of skin on the side of the head overhangs and hides the tympanum. The skin is variable in color, ranging from reddish brown, through pink, to lavender, with darker, wavy markings on the back. The head is darker than the body.

FAMILY	Microhylidae: Microhylinae
OTHER NAMES	Blunt-headed Burrowing Frog, Broad-lipped Frog
DISTRIBUTION	Cambodia, Myanmar, Thailand, Vietnam
ADULT HABITAT	Forest
LARVAL HABITAT	Ephemeral pools
CONSERVATION STATUS	IUCN Near Threatened. Has declined largely because of consumption by people, but its range does include some protected areas

GLYPHOGLOSSUS MOLOSSUS

BALLOON FROG

GÜNTHER, 1869

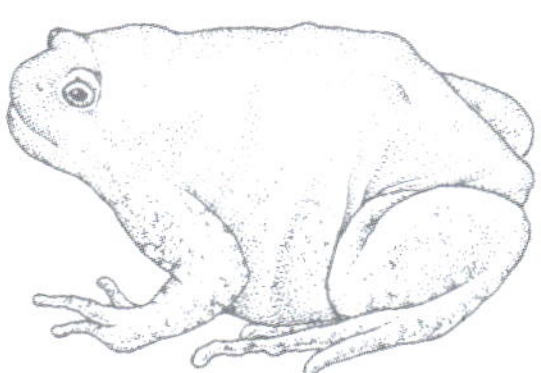

ADULT LENGTH
Male
2½–3 in (63–77 mm)
Female
up to 3⁵⁄₁₆ in (86 mm)

Considered to be a delicacy, this burrow-living frog has declined as a result of overharvesting by local people. As its numbers have fallen, individuals sold at markets have become gradually smaller. An "explosive breeder," it emerges after heavy rain to mate in flooded areas. Males call while floating in open water, their whole body inflated like a balloon—although they are very wary and sink beneath the surface at the slightest disturbance. The male glues himself to the female's back and the pair repeatedly go through an egg-laying procedure, their heads dipped downward, that releases 200–300 eggs at a time onto the water's surface.

The Balloon Frog has a globular body, a short, blunt snout, and large, protruding eyes that point slightly upward. Its characteristic very thick lower lip makes it quite unlike any other frog. The upper surfaces are dark brown and may be decorated with pale brown blotches. The legs are dappled with yellowish-brown blotches and spots.

SIMILAR SPECIES

Glyphoglossus molossus was the only species in *Glyphoglossus* until the squat frog genus *Calluella* was synonymized with *Glyphoglossus*. See the Burmese Squat Frog (*G. guttulatus*; page 506).

Actual size

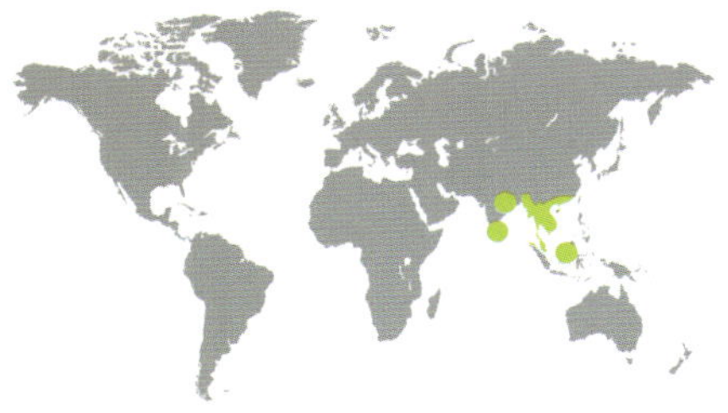

FAMILY	Microhylidae: Microhylinae
OTHER NAMES	Banded Bullfrog, Beautiful Kaloula, Malayan Bullfrog, Malaysian Narrowmouth Toad, Ox Frog
DISTRIBUTION	India, Sri Lanka, Cambodia, Indonesia, Myanmar, Thailand, Peninsular Malaysia. Introduced to Borneo
ADULT HABITAT	Woodland, wetlands, agricultural and urban areas
LARVAL HABITAT	Ephemeral ponds
CONSERVATION STATUS	IUCN Near Threatened. Has a huge range and in places its numbers are increasing

ADULT LENGTH
Male
$2\frac{1}{8}$–$2\frac{13}{16}$ in (54–70 mm)
Female
$2\frac{3}{16}$–$2\frac{15}{16}$ in (57–75 mm)

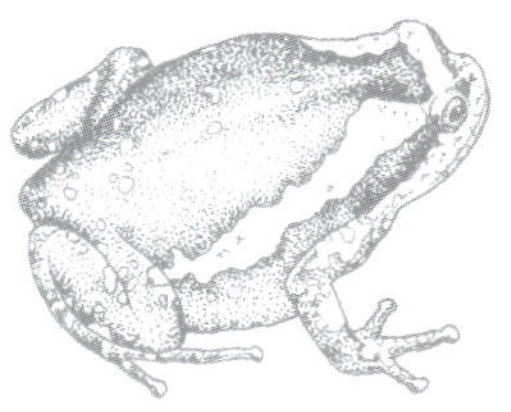

KALOULA PULCHRA

ASIAN PAINTED FROG

GRAY, 1831

This strikingly colored frog can inflate its already plump body with air when it is threatened or molested. It also produces a secretion from its skin, but this is noxious rather than toxic. It hides by day, emerging at night to feed primarily on ants. After heavy rain, males gather around temporary pools, producing a loud, low-pitched call to attract females. They sometimes call synchronously, and at other times alternately. During mating, the male glues himself to the female's back. The eggs float on the water's surface. This is an adaptable species that thrives around human settlements.

SIMILAR SPECIES

There are 18 species in the genus *Kaloula*, the Asian narrowmouth toads. The Smooth-fingered Narrow-mouthed Frog (*K. baleata*) is native to Borneo and other parts of Southeast Asia. The Boreal Digging Frog (*K. borealis*) occurs in China and Korea, and is common in rice fields. See also the Sri Lankan Bullfrog (*Uperodon taprobanica*; page 515).

The Asian Painted Frog has a plump body, a small head, large eyes, and a short, rounded snout. The back is dark brown with irregular yellow patches, and there are bold yellow or pink stripes running along each flank from the eye to the groin. The underside is yellowish brown.

Actual size

FAMILY	Microhylidae: Microhylinae
OTHER NAMES	Borneo Tree Frog
DISTRIBUTION	Western and northern Borneo, Sumatra, Indonesia
ADULT HABITAT	Lowland forest up to 650 ft (200 m) altitude
LARVAL HABITAT	Small pools in tree-holes
CONSERVATION STATUS	IUCN Least Concern. However, its lowland forest habitat is subject to widespread deforestation

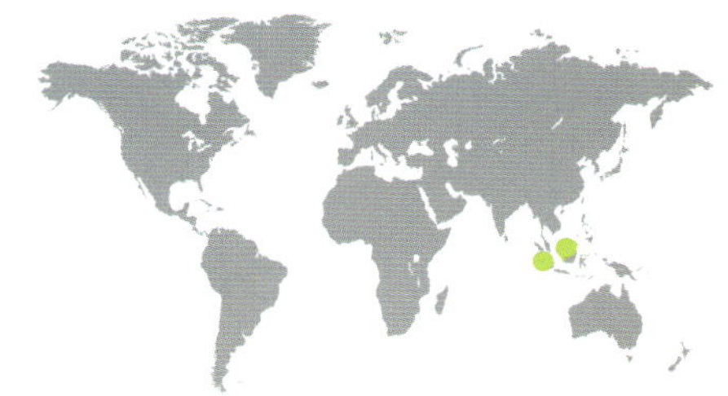

ADULT LENGTH
Male
¾–1 in (19–26 mm)
Female
⅞–1⅛ in (23–29 mm)

METAPHRYNELLA SUNDANA

BORNEO TREE-HOLE FROG

(PETERS, 1867)

The simple, piping call of this small frog is a characteristic sound of Borneo's lowland forest. Unusually for a frog, the call is very variable in pitch. Males defend holes in trees that contain water, and they vary the frequency of their call according to the acoustic properties of their tree-hole, which acts as a resonance chamber. In this way, each male maximizes the distance from which his call can be heard: This can be as far as 160 ft (50 m). Females prefer lower-pitched calls. They lay their eggs in their chosen male's tree-hole, where the tadpoles develop.

Actual size

SIMILAR SPECIES

There is one other species in the genus *Metaphrynella*, the Malaysian Tree-hole Frog (*M. pollicaris*), which occurs in Peninsular Malaysia and Thailand. It breeds in water-filled hollows in trees and bamboo plants, and it is thought that the limited availability of suitable breeding sites restricts the size of its population.

The Borneo Tree-hole Frog is a stocky frog with a pointed, slightly upturned snout and large eyes. There are many warts on the body and limbs, adhesive pads on the fingers and toes, and the feet are webbed. The upper surfaces are light to dark brown with darker blotches, taking the form of cross-stripes on the legs.

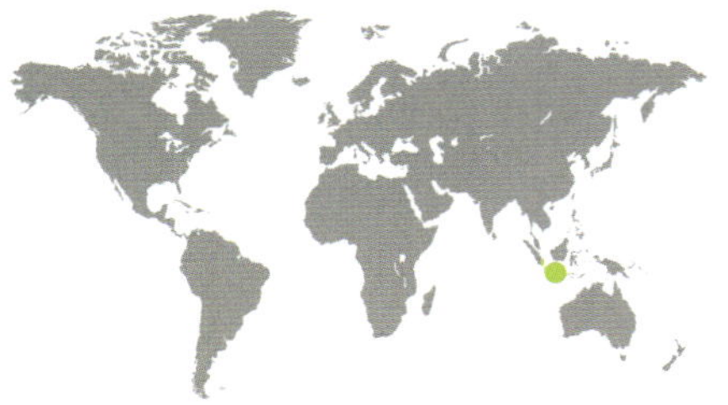

FAMILY	Microhylidae: Microhylinae
OTHER NAMES	Javan Chorus Frog, Javan Rice Frog
DISTRIBUTION	Java and southern Sumatra, Indonesia
ADULT HABITAT	Forest up to 5,250 ft (1,600 m) altitude
LARVAL HABITAT	Pools, ponds, marshes
CONSERVATION STATUS	IUCN Least Concern. Its forest habitat is subject to deforestation but it has adapted to altered habitats

ADULT LENGTH
Male
average ¾ in (20 mm)

Female
average 1 in (25 mm)

MICROHYLA ACHATINA

JAVAN NARROW-MOUTHED FROG

TSCHUDI, 1838

Actual size

The Javan Narrow-mouthed Frog has a plump body, a narrow head, and small eyes. The back is yellowish brown with darker patterns and sometimes there is a narrow, pale stripe running down the middle of the back. The flanks are usually darker than the back.

While its natural habitat is forest, this small frog is often found in rice fields and domestic gardens, suggesting that it has been able to adapt to the widespread deforestation that is occurring in Southeast Asia. It lives in the leaf litter on the forest floor and feeds on ants, termites, and other small insects. When breeding, it sometimes gathers in very large numbers around pools and ponds. Males call at night, forming a chorus of cricket-like sounds. Females produce clutches of around 20 eggs.

SIMILAR SPECIES

There are 54 species currently recognized in the genus *Microhyla*, distributed across Asia. Karunaratne's Narrow-mouthed Frog (*M. karunaratnei*) has declined due to the destruction of its forest habitat, and is known from only two locations. It is listed as Endangered. Some species have been transferred to the closely related genus *Nanohyla* which contains ten species including the Least Narrow-mouthed Frog (*Nanohyla perparva*) in low-elevation forest in Borneo. See also the Borneo Narrow-mouthed Frog (*M. borneensis*; page 511).

FAMILY	Microhylidae: Microhylinae
OTHER NAMES	*Microhyla nepenthicola*, Bornean Chorus Frog, Borneo Rice Frog
DISTRIBUTION	Borneo
ADULT HABITAT	Forest
LARVAL HABITAT	Pools inside pitcher plants
CONSERVATION STATUS	IUCN Least Concern. Its forest habitat is subject to deforestation but its range includes some protected areas

MICROHYLA BORNEENSIS

BORNEO NARROW-MOUTHED FROG

PARKER, 1928

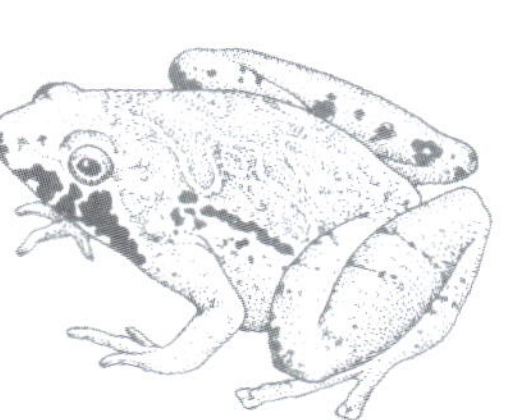

ADULT LENGTH
Male
3/8–1/2 in (10–13 mm)
Female
11/16–3/4 in (17–19 mm)

Once a candidate for the title of world's smallest frog, this tiny amphibian has an intimate relationship with the insectivorous pitcher plant *Nepenthes ampullaria*. The latter feeds on plant detritus and insects that fall into its pitchers. The frog's small size enables it to move around inside the pitcher, clinging to its slippery sides and avoiding the waxy area where insects get trapped. Males call year-round, except on dry nights, attracting females into pitchers, where they lay their eggs on the pitcher walls. The tiny tadpoles feed only on yolk from the egg and metamorphose into frogs just 3/16 in (3.5 mm) long.

Actual size

SIMILAR SPECIES

Microhyla borneensis is the smallest of the 54 members of the genus described to date. The Pothole Narrow-mouthed Frog (*M. petrigena*) is another very small frog also found in Borneo. It breeds in potholes beside small, clear mountain streams. The Narrow-mouthed Frog (*M. rubra*) is a common species found in India, Sri Lanka, Myanmar, and Bangladesh, and is often found close to human settlements. See also the Javan Narrow-mouthed Frog (*M. achatina*; page 510).

The Borneo Narrow-mouthed Frog has a plump, slightly flattened body, a narrow head, and a pointed snout. The skin on the back is warty. The upper surfaces are reddish brown with paler markings, and there are dark cross-stripes on the legs. There are black markings on the lower lip and along the flanks.

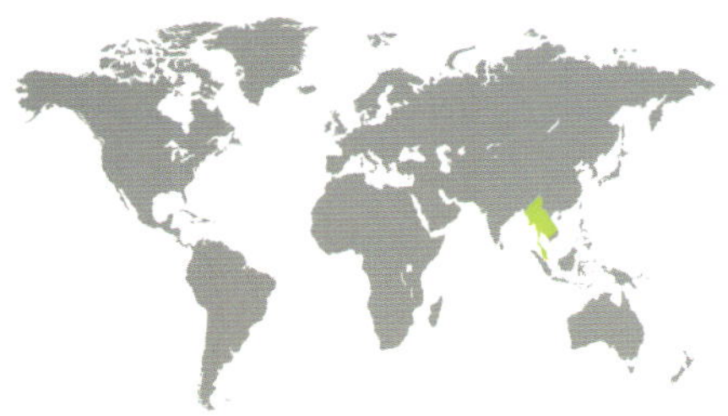

FAMILY	Microhylidae: Microhylinae
OTHER NAMES	Boulenger's Ornate Narrow-mouthed Frog, Deli Little Pygmy Frog, False Ornate Narrow-mouthed Frog
DISTRIBUTION	Myanmar, Thailand, Cambodia, Laos, Vietnam, Peninsular Malaysia, Andaman and Nicobar Islands
ADULT HABITAT	Lowland forest
LARVAL HABITAT	Ephemeral rain pools
CONSERVATION STATUS	IUCN Least Concern. Declining in places but occurs within some protected areas

ADULT LENGTH
Male
average ⅞ in (23 mm)

Female
average 1³⁄₁₆ in (30 mm)

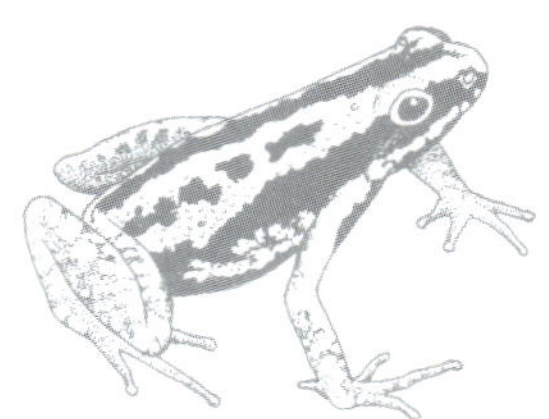

MICRYLETTA INORNATA

DELI PADDY FROG

(BOULENGER, 1890)

Actual size

The Deli Paddy Frog has a flattened body, smooth skin, a rounded snout, and long, thin limbs, fingers, and toes. The color of the upper surfaces is variable, ranging from dark reddish brown to violet, with or without darker markings or spots. There is often a broad black stripe running through the eye and along the flank.

Little is known about this inconspicuous little frog. It is typically found at the edges of lowland forest and in areas where the habitat has been disturbed. It lives on the ground and in low vegetation, and probably feeds on a variety of small insects. The species is an "explosive breeder," meaning that it breeds over a very short period, often in large numbers, after heavy rain. Its eggs float on the surface of the temporary rain pools in which it breeds.

SIMILAR SPECIES

Thirteen other species have been described in the genus *Micryletta*. Stejneger's Paddy Frog (*M. steinegeri*) occurs in Taiwan. As a result of the clearing of its forest habitat to make way for human settlements and agricultural land, it has become very rare and is listed as Vulnerable. The Black-spotted Paddy Frog (*M. nigromaculata*), from Vietnam, is Endangered.

FAMILY	Microhylidae: Microhylinae
OTHER NAMES	Previously *Ramanella variegata*, Marbled Narrow-mouthed Frog, Termite Nest Frog, Variegated Ramanella
DISTRIBUTION	India, Sri Lanka
ADULT HABITAT	Lowland forest up to 3,300 m (1,000 m) altitude (in India)
LARVAL HABITAT	Small ephemeral pools
CONSERVATION STATUS	IUCN Least Concern. Locally abundant, with a stable population

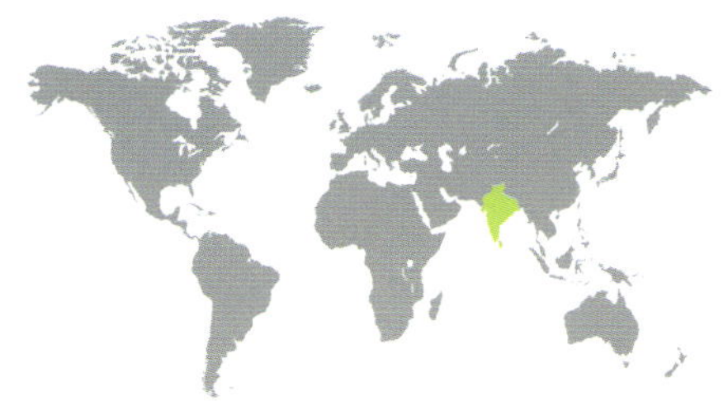

ADULT LENGTH
average 1³⁄₈ in (35 mm)

UPERODON VARIEGATUS

ELURU DOT FROG

(STOLICZKA, 1872)

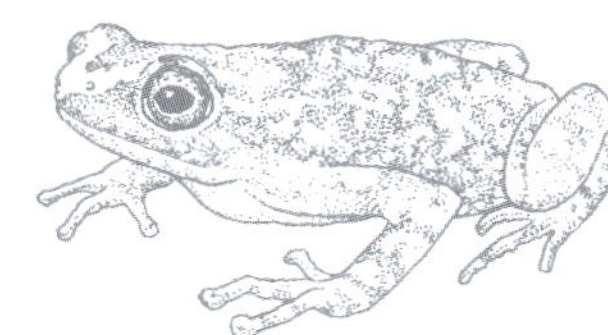

This small frog has been reported from a variety of habitats, including leaf litter, the trunks of trees, holes in trees, termite mounds, and rice fields. It has even been reported swimming in a toilet bowl. It is generally found in areas with dry soil and ground cover, such as rocks and logs. It breeds immediately after heavy rain has created small pools. Males have been observed calling while floating in water, and also from grass stems and low vegetation close to water.

SIMILAR SPECIES

Thirteen species have been described in the genus *Uperodon*, distributed across India and Sri Lanka. The Half-webbed Pug-snouted Frog (*U. palmatus*) and Nagao's Pug-snouted Frog (*U. nagaoi*), both from Sri Lanka, are listed as Endangered.

Actual size

The Eluru Dot Frog has a short head with a blunt snout. It has well-developed triangular pads on the end of its fingers, but not on its toes. The upper surfaces are dark brown, usually with an irregular pattern of light brown patches. There are pale spots on the hands and feet, and the underside is white and is not spotted.

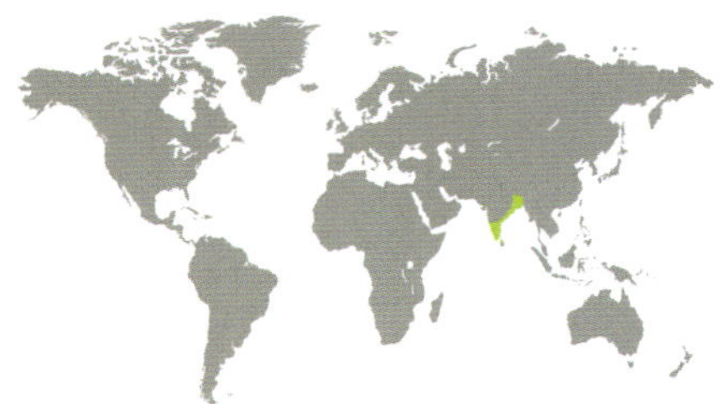

FAMILY	Microhylidae: Microhylinae
OTHER NAMES	Globular Frog, Lesser Balloon Frog
DISTRIBUTION	Southern and eastern India, northern Sri Lanka
ADULT HABITAT	In the soil
LARVAL HABITAT	Ponds, rice fields
CONSERVATION STATUS	IUCN Least Concern. Its population is believed to be stable, but it is threatened in places by urbanization

ADULT LENGTH
Male
1¾–2$^{1}/_{16}$ in (45–53 mm)
Female
1$^{15}/_{16}$–2$^{11}/_{16}$ in (50–66 mm)

UPERODON SYSTOMA

MARBLED BALLOON FROG

(SCHNEIDER, 1799)

Actual size

Adapted for burrowing into soil, this extremely plump frog moves clumsily on land and swims weakly in water. It feeds largely on termites and sometimes takes up residence in termite nests. It emerges on the surface only in the monsoon period, breeding between May and July. Males call from the edge of a pond or rice field, producing a sound likened to the bleating of a goat. They have a particularly large vocal sac. The species is able to survive underground for very long periods. One individual, found more than 3 ft (1 m) down in the ground, is believed to have survived without food for 13 months.

SIMILAR SPECIES

A related species in the genus *Uperodon* is the Greater Balloon Frog (*U. globulosus*), found in India, Nepal, and Bangladesh. Larger than *U. systoma*, it is dull gray in color. It is able to blow itself up like a balloon, its lungs inflating to a level above its spine. It is rarely seen because of its burrowing habits, but is probably more common than previously thought.

The Marbled Balloon Frog has a rotund body, small eyes, and a short, blunt snout. It has two horny tubercles on each of its hind feet that it uses to dig backward into the soil. The upper surfaces are pinkish brown with a complex marbled pattern in dark brown. The underside is white and is not spotted.

FAMILY	Microhylidae: Microhylinae
OTHER NAMES	Previously *Kaloula taprobanica*
DISTRIBUTION	Sri Lanka, India, Bangladesh, Nepal
ADULT HABITAT	Woodland, wetlands, agricultural and urban areas
LARVAL HABITAT	Ephemeral ponds
CONSERVATION STATUS	IUCN Least Concern. Generally common and its range includes several protected areas

ADULT LENGTH
Male
$2\frac{1}{8}$–$2\frac{13}{16}$ in (54–70 mm)
Female
$2\frac{3}{16}$–$2\frac{15}{16}$ in (57–75 mm)

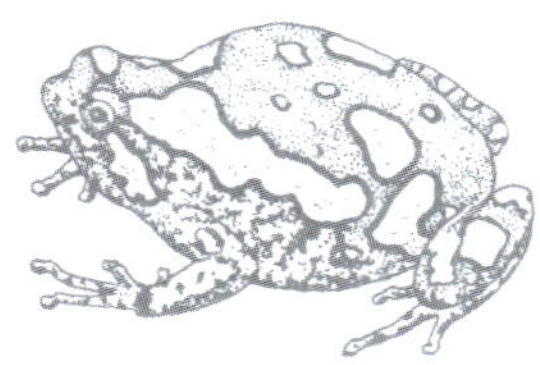

UPERODON TAPROBANICUS

SRI LANKAN BULLFROG

(PARKER, 1934)

This colorful frog lives mostly underground, emerging at night to feed. Despite its somewhat ungainly build, it is adept at climbing. It breeds immediately after heavy rain, often in rice fields and roadside ponds. The males call while swimming in open water. Females produces large numbers of small eggs that float on the water's surface. When attacked or molested, the frog blows itself up with air and tucks its hands and feet under its body. It also produces a skin secretion that causes a burning sensation on human skin.

SIMILAR SPECIES

This species was previously included in the genus *Kaloula* as a subspecies of the Asian Painted Frog (*Kaloula pulchra*; page 508), before being transferred to genus *Uperodon*. The Assam Painted Frog (*U. assamensis*) is another species previously included in *Kaloula*.

The Sri Lankan Bullfrog has a plump body, a small head, large eyes, and a short, rounded snout. The back is gray to black, and there is a striking pattern of red or orange blotches, sometimes forming a stripe, on its flanks. The underside is yellowish brown with black spots.

Actual size

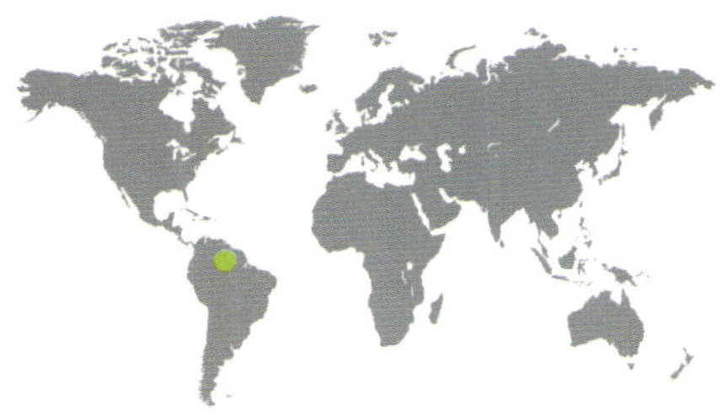

FAMILY	Microhylidae: Otophryninae
OTHER NAMES	None
DISTRIBUTION	Colombia, Venezuela, French Guiana, northern Brazil
ADULT HABITAT	Rainforest with sandy soil, up to 3,600 ft (1,100 m) altitude
LARVAL HABITAT	Sandy streams
CONSERVATION STATUS	IUCN Least Concern. Declining in places due to habitat destruction, but occurs within some protected areas

ADULT LENGTH
Male
up to 2⅛ in (55 mm)
Female
up to 2⅜ in (61 mm)

OTOPHRYNE PYBURNI

PYBURN'S PANCAKE FROG

CAMPBELL & CLARKE, 1998

In shape and coloration, this frog resembles a leaf, making it well camouflaged on the forest floor. Males call during the day from concealed shelters close to a stream. The female lays large eggs, which are thought to be deposited in cavities near streams. They hatch into tadpoles that are highly unusual, being specialized for burrowing into sand. They have impressive arrays of teeth on both jaws that are not for feeding but serve to prevent sand from entering the gut. They also have a long siphon that extends upward above the sand, enabling them to breathe.

SIMILAR SPECIES

There are two other species of pancake frogs. Like *Otophryne pyburni*, they feed primarily on ants and breed in streams. The Robust Pancake Frog (*O. robusta*) occurs in the forests of Venezuela, French Guiana, and Guyana. Steyermark's Pancake Frog (*O. steyermarki*) is a high-altitude species, occurring on several of the tepuis, or flat-topped mountains, that are dotted across Venezuela and Guyana. It has been reported that the Wayampi people of French Guiana eat pancake frogs.

Pyburn's Pancake Frog has a broad body, short legs, and a very pointed snout. A pale yellowish stripe runs from the tip of the snout to the groin. This separates the dark brown flanks from the back, which is reddish brown to yellow, sometimes with darker markings. The underside of the hind legs and feet is orange.

Actual size

FAMILY	Microhylidae: Otophryninae
OTHER NAMES	None
DISTRIBUTION	Colombia, Venezuela, Brazil
ADULT HABITAT	Rainforest
LARVAL HABITAT	Within the egg, then in a burrow away from water
CONSERVATION STATUS	IUCN Least Concern. Though rarely seen, it is believed to be common and not under threat

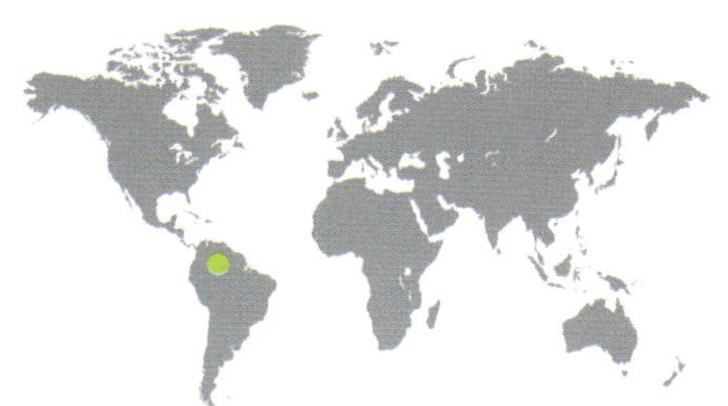

ADULT LENGTH
average 1⅛ in (28 mm)

SYNAPTURANUS SALSERI

TIMBO DISC FROG

PYBURN, 1975

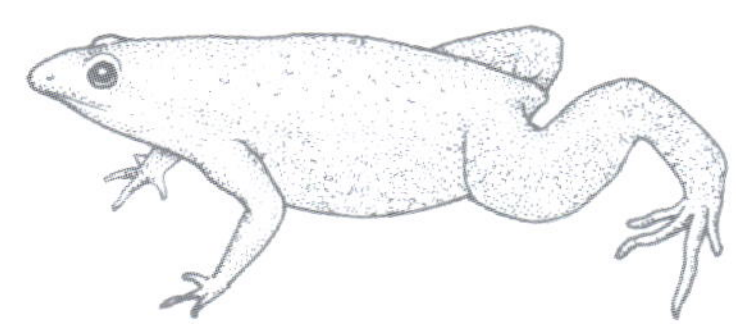

The secretive habits of this small frog mean that it is rarely seen. It lives in burrows and hollows just below the forest floor, feeding on ants and spiders. Breeding occurs after sustained, heavy rain, males calling from burrows beneath matted tree roots. Their call is a long, pure whistle. The female lays a clutch of around eight large eggs in the male's burrows and he remains close to them. The eggs contain a large amount of yolk that sustains the developing tadpoles throughout their development. They hatch from the egg at a late stage and metamorphosis occurs in the burrow, away from water.

Actual size

The Timbo Disc Frog has a stout body, long, muscular hind legs, a narrow head, and a pointed snout. The fingers and toes are long and thin, and are not webbed. The upper surfaces are reddish brown or gray, speckled with small cream spots. The tip of the snout is often very pale.

SIMILAR SPECIES

There are nine other species of disc frog, genus *Synapturanus*, found in northern South America, seven of which have been described since 2021. Their habits are very similar to those of *Synapturanus salseri*. Miranda's Disc Frog (*S. mirandaribeiroi*) is found in Colombia, Brazil, Guyana, Suriname, and French Guiana. It can be distinguished from *S. salseri* by its call, which is a shorter whistle. The Vaupes Disc Frog (*S. rabus*) occurs in Colombia, Ecuador, and Peru.

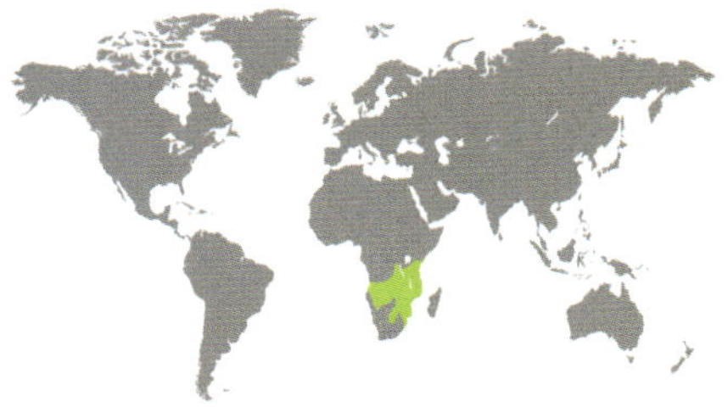

FAMILY	Microhylidae: Phrynomerinae
OTHER NAMES	Banded Rubber Frog
DISTRIBUTION	Central and East Africa, from Somalia in the north to South Africa in the south, and west to Angola and Namibia
ADULT HABITAT	Grassland and savanna up to 4,800 ft (1,450 m) altitude. Also occurs in agricultural land
LARVAL HABITAT	Ephemeral pools and ponds
CONSERVATION STATUS	IUCN Least Concern. Its population appears to be stable and its range includes several protected areas

ADULT LENGTH
Male
up to $2\frac{1}{16}$ in (53 mm)
Female
up to $2\frac{5}{8}$ in (65 mm)

PHRYNOMANTIS BIFASCIATUS

RED-BANDED RUBBER FROG

(SMITH, 1847)

Adapted for walking and running, rather than hopping or jumping, this colorful frog has short, thin legs. Usually found under logs or in termite mounds, it can also climb. It breeds after heavy summer rain in any available pool, including elephant footprints. The male's call is a melodious trill lasting up to three seconds and audible from half a mile (1 km) away. The female lays up to 1,500 eggs attached to floating vegetation. When molested, the frog adopts a defensive posture in which it extends its legs and presents its rear end to its enemy. It also produces a toxic skin secretion when attacked.

SIMILAR SPECIES

The Spotted Rubber Frog (*Phrynomantis affinis*) is a rarely seen species found in the Democratic Republic of the Congo, Namibia, and Zambia. The Somali Rubber Frog (*P. somalicus*) occurs in Ethiopia and Somalia. At least five of the six *Phrynomantis* species have tadpoles that are filter-feeders and are usually seen suspended in open water. They cluster in tight schools when threatened by predators. Newton's Rubber Frog (*P. newtoni*) was only described in 2021. See also the Red Rubber Frog (*P. microps*; page 519).

The Red-banded Rubber Frog has an elongated, slightly flattened body, a pointed head, small limbs, and expanded tips to its fingers and toes. The back is black or brown with two pink, orange, or red stripes. There is a large spot above the anus, and smaller spots on the limbs. The skin is smooth and glossy.

Actual size

FAMILY	Microhylidae: Phrynomerinae
OTHER NAMES	Accra Snake-necked Frog, West African Rubber Frog
DISTRIBUTION	West and Central Africa, from Senegal in the west to Democratic Republic of the Congo and South Sudan in the east
ADULT HABITAT	Grassland, savanna
LARVAL HABITAT	Ephemeral pools and ponds
CONSERVATION STATUS	IUCN Least Concern. Its population appears to be stable and its range includes several protected areas

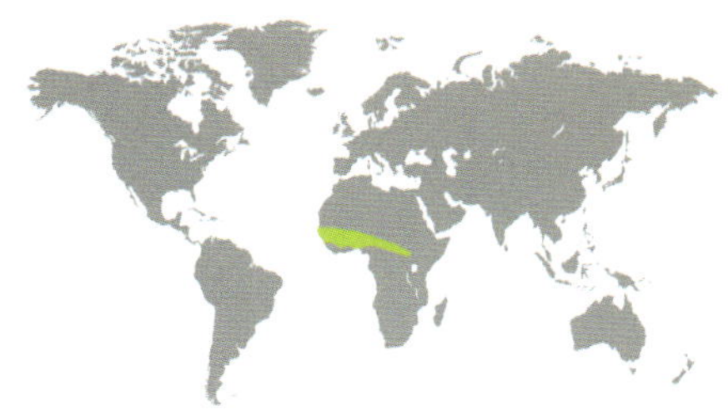

PHRYNOMANTIS MICROPS

RED RUBBER FROG

PETERS, 1875

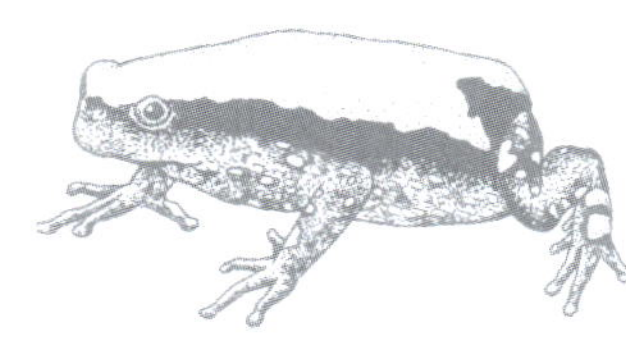

ADULT LENGTH
Male
1 7/16–1 7/8 in (37–47 mm)
Female
1 5/8–2 1/2 in (41–63 mm)

The savanna of West Africa is dry for much of the year, and consequently damp hiding places suitable for frogs are few and far between, and are often occupied by ants. This colorful frog is able to coexist with ants during the dry season. On entering an ant nest, it adopts a hunched posture and allows the insects to inspect it closely. Any ant that bites it soon dies because of its toxic skin. In general ants do not bite it, though they will bite and kill frogs of other species. The tadpoles of *Phrynomantis* species are noted for the way they form dense schools when at risk from predators.

SIMILAR SPECIES

The six species in the African genus *Phrynomantis*, known as rubber frogs or snake-necked frogs, are unusual in having a flexible neck that enables them to move their head from side to side. The Marbled Rubber Frog (*P. annectens*) lives in desert areas in Namibia, Angola, and South Africa. A "crevice-creeper," it has a flattened body that enables it to crawl into deep cracks in rocks where there is moisture. See also the Red-banded Rubber Frog (*P. bifasciatus*; page 518).

The Red Rubber Frog has an elongated body, a long neck, a blunt snout, and short legs. The tips of its fingers are enlarged and triangular, and there is no webbing on its toes. The skin is smooth and shiny. The back is red or orange, and the flanks and limbs are black, with a variable number of orange or red spots.

Actual size

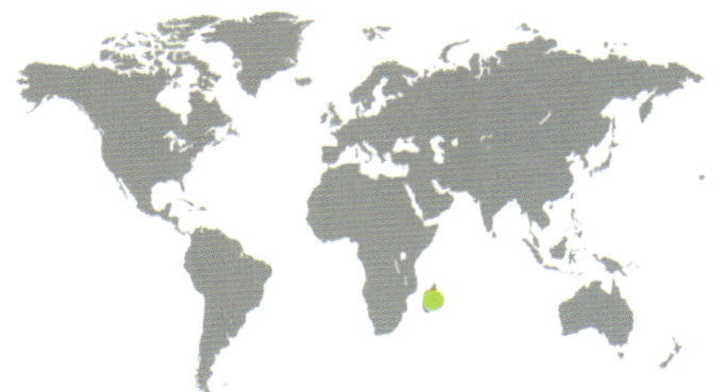

FAMILY	Microhylidae: Scaphiophryninae
OTHER NAMES	None
DISTRIBUTION	Eastern Madagascar up to 3,100 m (950 m) altitude
ADULT HABITAT	Rainforest
LARVAL HABITAT	Ponds
CONSERVATION STATUS	IUCN Least Concern. Has lost much of its habitat through deforestation but its range includes some protected areas

ADULT LENGTH
Male
¾–⅞ in (19–22 mm)
Female
⅞–1 in (22–26 mm)

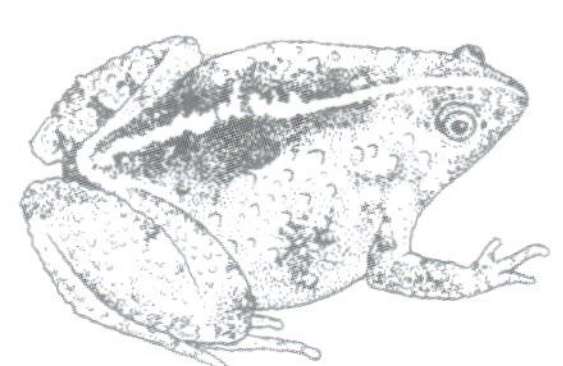

PARADOXOPHYLA PALMATA

WEB-FOOT FROG

(GUIBÉ, 1974)

Actual size

Although fully adapted for an aquatic life, this small frog visits water for only brief periods, to breed. For most of the year it lives on the forest floor, digging beneath the surface during dry periods. A strong swimmer, it has a streamlined body shape, long, powerful legs, and fully webbed feet. Breeding is triggered by heavy rain. The male produces a loud, melodious trill likened to the song of a cricket. The female lays 100–400 eggs that float on the surface of the water.

SIMILAR SPECIES

The only other species in this genus is the Masoala Aquatic Frog (*Paradoxophyla tiarano*), found in northeastern Madagascar and described in 2006. It is very similar to *P. palmata* but has less webbing on its hind feet. Both species are found only in the few remaining patches of Madagascar's once extensive rainforest.

The Web-foot Frog has a wide, slightly flattened body, a narrow head with small eyes, and long, muscular hind legs. The feet are fully webbed. Its coloration is very variable, the upper surfaces usually gray or brown with black spots. There are dark cross-stripes on the legs. The underside is white with gray patches.

FAMILY	Microhylidae: Scaphiophryninae
OTHER NAMES	Malagasy Rainbow Frog, Red Rain Frog
DISTRIBUTION	Central southern Madagascar
ADULT HABITAT	Rocky canyons at altitudes of 2,300–3,300 ft (700–1,000 m)
LARVAL HABITAT	Pools
CONSERVATION STATUS	IUCN Endangered. Has a very restricted range and was subject to intensive collecting for the pet trade. Listed as Critically Endangered until 2008. Collecting is now restricted

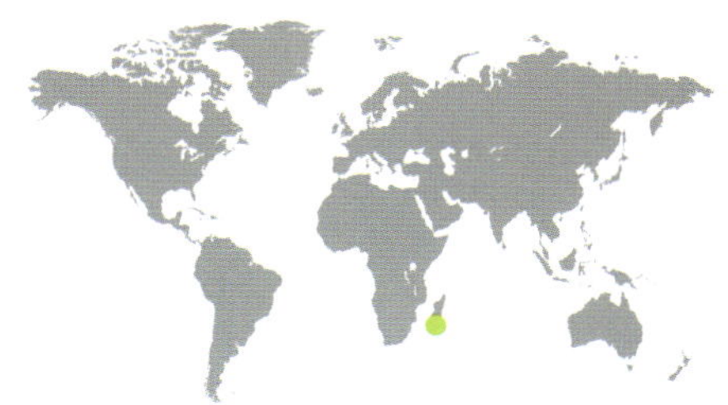

ADULT LENGTH
Male
¾–1³⁄₁₆ in (20–30 mm)
Female
1³⁄₁₆–1⁹⁄₁₆ in (30–40 mm)

SCAPHIOPHRYNE GOTTLEBEI

GOTTLEBE'S NARROW-MOUTHED FROG

BUSSE & BÖHME, 1992

The distinctive coloration of this very rare frog has made it popular in the international pet trade, despite the fact that captive animals typically die after a short time. It lives in Madagascar's Isalo massif, in narrow canyons that provide cool, moist conditions in contrast to the surrounding hot, dry landscape. Broadened fingertips equipped with claw-like, horny skin flaps enable it to climb rock faces at night. By day, it burrows into the ground with the spade-like tubercles on its hind feet. When the flow in streams running through the canyons has dwindled to a trickle, it breeds in the remaining pools.

SIMILAR SPECIES

To date, ten species have been described in the genus *Scaphiophryne*, commonly known as the rain frogs of Madagascar. The Marbled Burrowing Frog (*S. marmorata*) has an attractive green and brown skin pattern. A nocturnal frog, it lives in the rainforest of eastern Madagascar and is listed as Vulnerable. The Madagascar Rain Frog (*S. madagascariensis*) is found at higher altitudes and is listed as Near Threatened.

Actual size

Gottlebe's Narrow-mouthed Frog has a rounded body, a short snout, and large, prominent eyes. The toes are webbed and the fingers have expanded tips. There is a horny tubercle on each hind foot. The upper surfaces are marked in bold patches of red, green, and white, bordered by broad black stripes. The belly is gray.

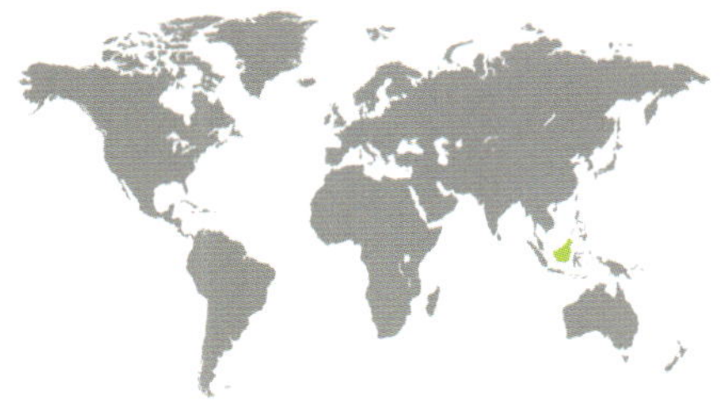

FAMILY	Ceratobatrachidae: Alcalinae
OTHER NAMES	Balu Eastern Frog, Dwarf Mountain Frog
DISTRIBUTION	Sarawak, Sabah, Brunei, and Kalimantan, Borneo
ADULT AND LARVAL HABITAT	Lowland and mid-montane rainforest streamside leaf litter at altitudes of 330–3,000 ft (100–900 m)
CONSERVATION STATUS	IUCN Least Concern. Possibly threatened by habitat loss

ADULT LENGTH
Male
⅞–1 in (21–25 mm)
Female
1–1¼ in (26–31 mm)

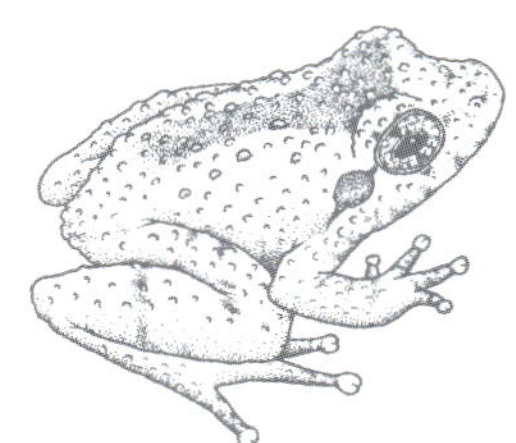

ALCALUS BALUENSIS

KINABALU MOUNTAIN FROG

(BOULENGER, 1896)

Actual size

The Kinabalu Mountain Frog is a small, wrinkled, "toad-like" species with a short, stout body, a moderately pointed snout, eyes with horizontal pupils, and small tympana. It is generally pale brown with dark brown dorsal markings that may break up its outline in the leaf litter.

The Kinabalu Mountain Frog is a small, semiaquatic, leaf-litter frog that was first described from Mount Kinabalu in Sabah. At 13,400 ft (4,095 m) above sea level, Kinabalu is the highest mountain between the Himalayas and New Guinea and a center of montane herpetological diversity. Since its discovery, *Alcalus baluensis* has been found in numerous scattered locations in northern and western Borneo. It is believed to have direct development like its congenerics, females laying large, nutritious eggs on land, which then hatch into small froglets, but this has yet to be confirmed. Males do not call. Juvenile *A. baluensis* have been documented being predated by huntsman spiders (*Heteropoda*).

SIMILAR SPECIES

Alcalus is the sole genus in the ceratobatrachid subfamily Alcalinae, and both are named for the Filipino herpetologist Angel Alcala (1929–2023). There are five other species in *Alcalus*, distributed from India to Palawan in the Philippines, with two other Bornean species, the King Mountain Frog (*A. rajae*), from east Kalimantan, and the Saribau Mountain Frog (*A. sariba*), from Sarawak. All are small, leaf-litter species.

FAMILY	Ceratobatrachidae: Ceratobatrachinae
OTHER NAMES	Previously *Discodeles guppyi*, Shortlands Island Webbed Frog
DISTRIBUTION	Solomon Islands, Bougainville and New Britain islands, Papua New Guinea
ADULT HABITAT	Lowland rainforest
LARVAL HABITAT	Within the egg
CONSERVATION STATUS	IUCN Least Concern. Declining in places but thrives in altered habitats, such as degraded forest and gardens

ADULT LENGTH
up to 6⅝ in (165 mm)

CORNUFER GUPPYI

SOLOMONS GIANT WEBBED FROG

(BOULENGER, 1884)

This large frog has long been collected, both for food and for sale in the international pet trade. In the 1990s there was much concern about the export of amphibians, reptiles, and butterflies from the Solomon Islands; since then, international agreements have brought such trade under greater control. Despite being collected, and despite the destruction of much of its forest habitat, this frog has remained reasonably common and has adapted to altered habitats. A species with direct development, it lays its eggs on the ground and they hatch directly into small frogs.

SIMILAR SPECIES

There were five species in the genus *Discodeles* when it was synonymized with *Cornufer*, all found in Papua New Guinea and the Solomon Islands. The Warty Webbed Frog (*C. bufoniformis*) occurs in the Solomon Islands and on the island of Bougainville. It is very similar to *C. guppyi* and, like this species, is eaten by people. The Malukuna Webbed Frog (*C. malukuna*) is little known and is confined to the island of Guadalcanal.

Actual size

The Solomons Giant Webbed Frog is a large frog with a large head, very large eyes, and powerful hind legs. The hind feet are webbed and there are many warts on the back. The upper surfaces are olive-green with darker patches on the back and there are usually stripes on the legs. The belly is dirty white in color.

FAMILY	Ceratobatrachidae: Ceratobatrachinae
OTHER NAMES	Previously *Cornufer solomonis*
DISTRIBUTION	Solomon Islands, Papua New Guinea (Bougainville). 0–250 m asl.
ADULT HABITAT	Rainforest
LARVAL HABITAT	Within the egg
CONSERVATION STATUS	IUCN Least Concern. Threatened locally by deforestation.

ADULT LENGTH
Male
1 11/16–2 1/8 in (43–54 mm)
Female
2 15/16–3 1/4 in (75–84 mm)
all measurements based on very few specimens

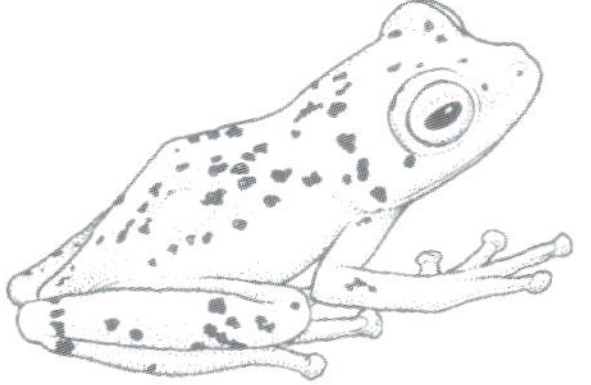

CORNUFER HEFFERNANI

SOLOMON ISLANDS PALM FROG

(KINGHORN, 1928)

This extremely rare frog is more often heard than seen. The male's call, a "whoo-ee" rising from low to high pitch, was recorded during a survey of some of the islands in the Solomon group in 2009. Older accounts report that males call from fallen logs, rocks, and other prominent positions. The species has been reported from several islands in the Solomons, as well as the nearby islands of Bougainville and Buka, which are part of Papua New Guinea. It does not have free-living tadpoles, but has direct development, the eggs hatching to produce tiny frogs.

The Solomon Islands Palm Frog has a broad head, large eyes with horizontal pupils, and well-developed disks on its fingers and toes. The back, arms, and legs are green with scattered black spots, and the flanks and underside are yellow. The pupils of the eyes are pale blue, gray or brown.

SIMILAR SPECIES

Palmatorappia solomonis was the only species in its genus. In 2015, both *P. solomonis* and the Solomons Wrinkled Ground Frog (*Platymantis solomonis*) were synonymized into the genus *Cornufer*. This resulted in potentially two species called *C. solomonis*, and being the older name the former *Platymantis*, described by Boulenger in 1884, had priority and became *C. solomonis*. *Palmatorappia solomonis* was described by Sternfeld in 1920, but it contains a junior synonym, described by Kinghorn a few years later, *Hypsirana heffernani*, which now takes precedence as *Cornufer heffernani*, a substitute name.

Actual size

FAMILY	Ceratobatrachidae: Ceratobatrachinae
OTHER NAMES	Previously Island Forest Frog
DISTRIBUTION	Gigante Sur Island, Philippines
ADULT HABITAT	Rock crevices and caves in forest
LARVAL HABITAT	Within the egg
CONSERVATION STATUS	IUCN Critically Endangered. Its range covers less than 4 sq miles (10 sq km), and its forest habitat has been totally destroyed

ADULT LENGTH
Male
1 7/16–1 5/8 in (37–42 mm)
Female
1 9/16–1 13/16 in (40–46 mm)

PLATYMANTIS INSULATUS

GIGANTE ISLAND GROUND FROG

BROWN & ALCALA, 1970

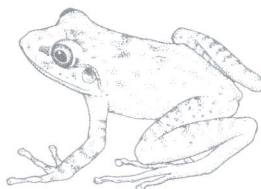

Gigante Sur Island in the Philippines is a volcanic island with an area of only 740 acres (300 ha). Composed of limestone, it contains many caves and was formerly covered in forest. The caves provide roosting sites for millions of bats that fill the caves with their droppings. The forest has been largely destroyed and the island is subject to quarrying for limestone and bat guano. In this unpromising environment, this extremely rare frog somehow maintains its existence, living in caves and rock crevices where moisture persists. Like other frogs in its family, there are no free-living tadpoles. The eggs, laid on land, hatch directly into frogs.

SIMILAR SPECIES

Previously including species from New Guinea to Fiji, the genus *Platymantis* was recently confined to the Philippines. A diluted *Platymantis* now contains 32 species including the Rough-backed Forest Ground Frog (*P. corrugatus*), a widespread species, and the Luzon Forest Frog (*P. luzonensis*), an arboeal and Near Threatened species found only on Luzon. Former New Guinea and Pacific species, such as the Papuan Wrinkled Frog (formerly *P. papuensis*), are now placed in genus *Cornufer*.

The Gigante Island Ground Frog has long hind legs, a narrow head with a rounded snout and large eyes. The fingers and toes are long and thin and have enlarged tips. The upper surfaces are variable in color, most commonly green or brown, with irregular darker blotches and there are stripes on the legs.

Actual size

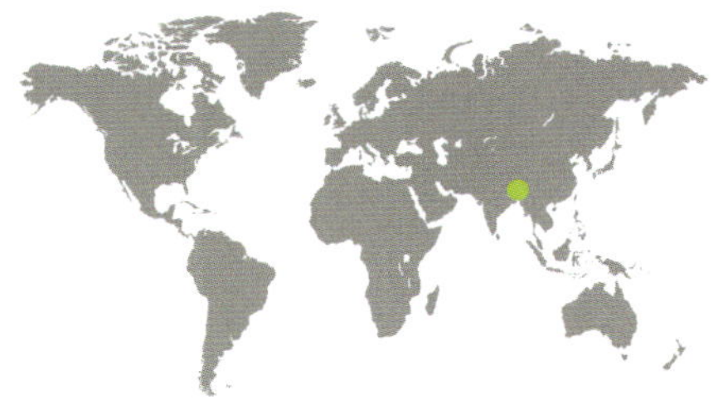

FAMILY	Ceratobatrachidae: Liuraninae
OTHER NAMES	None
DISTRIBUTION	Arunachal Pradesh, India
ADULT AND LARVAL HABITAT	High-elevation evergreen and temperate forest leaf litter at altitudes of 5,900–7,900 ft (1,800–2,400 m)
CONSERVATION STATUS	Data Deficient. The status of this species is unknown, but two Chinese *Liurana* are Near Threatened and Vulnerable, respectively

ADULT LENGTH
Male
9/16–3/4 in (14–20 mm)
Female
unknown

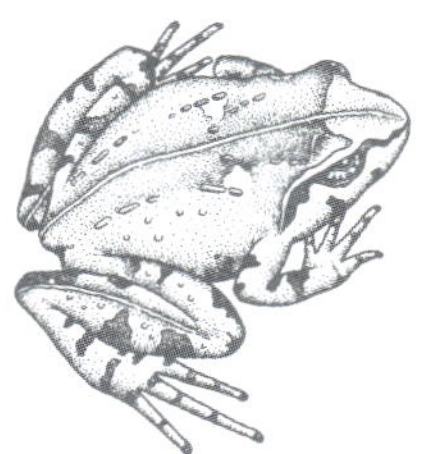

LIURANA HIMALAYANA

HIMALAYAN PAPILLA-TONGUED FROG

SAIKIA & SINHA, 2019

The Himalayan Papilla-tongued Frog was described as recently as 2019, from the Talle Valley Wildlife Sanctuary in northeast India. The habitat comprises high-elevation evergreen forest with scattered pines and ferns or open pine woodland with grassland. *Liurana himalayana* are active nocturnally, although the region where they occur may easily drop below freezing at night and retain heavy frosts during the morning. The common name comes from the presence of numerous papillae in the frog's tongue, the purpose of which is unknown. *Liurana himalayana* has direct development, the female laying eggs in damp leaf litter, which subsequently hatch into froglets, without a tadpole stage.

The Himalayan Papilla-tongued Frog resembles a small ranid frog, being relatively smooth-skinned, with a sharp ridge extending from the eye to the nostril, a broad, black stripe running from the tip of the nose to behind the eye, and a fine stripe running down the middle of the back. Its patterning is a cryptic mix of russet reds and browns, with black spots. It has small, hidden tympana, while its two Indian congenerics have large, distinct tympana. It has powerful limbs and long, unwebbed fingers and toes.

SIMILAR SPECIES

The ceratobatrachid subfamily Liuraninae contains a single genus (*Liurana*) and eight species, from Tibet (Xizang), in China, and northeast India. Two other species were described at the same time and from the same locality as *L. himalayana*, i.e. the Indian Papilla-tongued Frog (*L. indica*), from the higher elevation of 8,061 ft (2,457 m), and the Minute Papilla-tongued Frog (*L. minuta*), from 6,220 ft (1,896 m).

Actual size

FAMILY	Conrauidae
OTHER NAMES	Giant Slippery Frog
DISTRIBUTION	Southwestern Cameroon, Equatorial Guinea
ADULT AND LARVAL HABITAT	Rivers in rainforest
CONSERVATION STATUS	IUCN Endangered. Subject to collection for food and destruction of its forest habitat

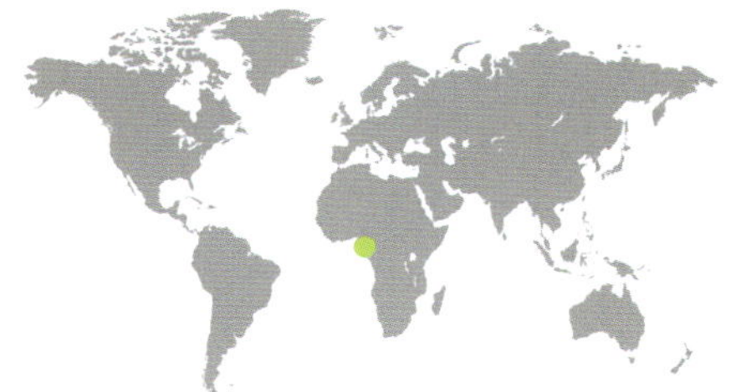

ADULT LENGTH
6¾–13 in (170–330 mm)

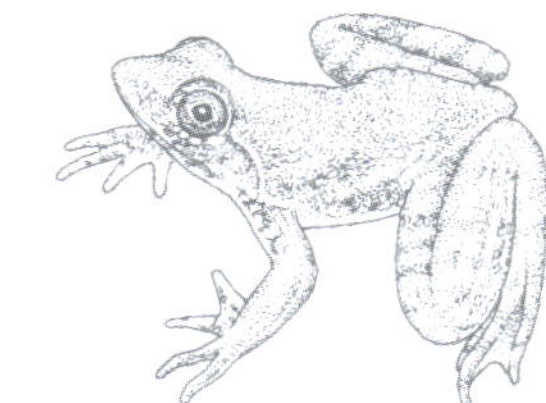

CONRAUA GOLIATH

GOLIATH FROG

(BOULENGER, 1906)

This is the world's largest frog, some individuals reaching 6½ lb (3 kg) in weight. It lives in rapids and cascades in sandy-bottomed rivers that flow through rainforest. At night, it emerges onto land in search of food, and it is then caught by local people, for whom it is a significant source of food. Its mating behavior has not been observed. It lays its eggs—several hundred of them—on submerged vegetation. The tadpoles feed only on the plant *Dicraea warmingii*, which grows on rocks. There appear to be no obvious differences between the sexes and neither sex calls.

The Goliath Frog is a very large, powerfully built frog with very long hind legs. The eyes are large and the toes are fully webbed. The upper surfaces are dark green to black. The belly and the undersides of the limbs are yellow or orange.

SIMILAR SPECIES

There are eight species in the genus *Conraua*, all found in Africa. The Abo Slippery Frog (*C. crassipes*) lives in fast-flowing rivers from Nigeria, through Cameroon, to the Democratic Republic of the Congo. The Togo Slippery Frog (*C. derooi*), which is about half the size of *C. goliath*, lives in forest streams in Ghana and Togo, and the recently described Atewa Slippery Frog (*C. sagyimase*) from Ghana, are listed as Critically Endangered.

Actual size

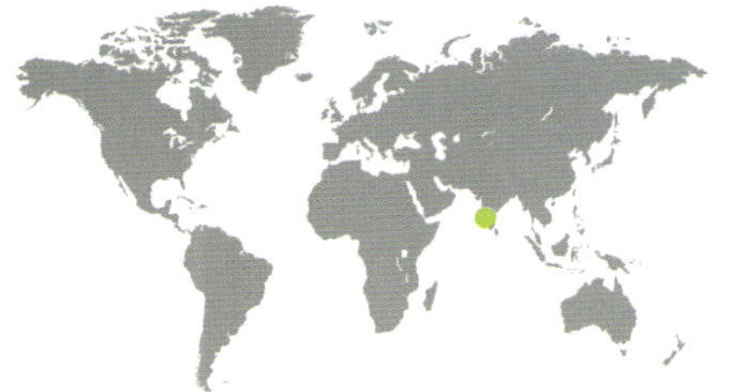

FAMILY	Micrixalidae
OTHER NAMES	Brown Tropical Frog, Kalakkad Dancing Frog
DISTRIBUTION	Western Ghats, western India
ADULT HABITAT	Around fast-flowing streams in dense forest
LARVAL HABITAT	In fast-flowing streams in dense forest
CONSERVATION STATUS	IUCN Endangered. Its forest habitat is subject to deforestation

ADULT LENGTH
average 1¼ in (32 mm)

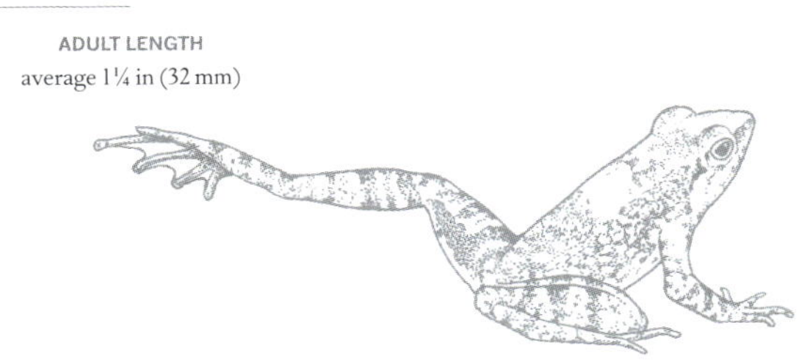

MICRIXALUS FUSCUS

KALAKKAD TORRENT FROG

(BOULENGER, 1882)

Actual size

This small frog is active by day, in the breeding season calling from 6 a.m. to 6 p.m. Males vigorously defend rocks from any rivals that try to displace them. They do so by calling and also by foot-flagging, extending each hind leg alternately and spreading the webbed foot. This bimodal pattern of communication, using both auditory and visual signals, is associated with the species' habitat: The noise created by fast-flowing streams and waterfalls means that calls on their own are a rather ineffective way of deterring rivals, especially at a distance.

SIMILAR SPECIES

To date, 26 species of *Micrixalus*, the Indian dancing frogs, have been described, many of them very recently—no fewer than 14 new species were described in 2014 alone. The Spotted Dancing Frog (*M. specca*) and the Cave Dancing Frog (*M. spelunca*)both occur only in the Western Ghats region of southern India, where they are threatened by deforestation; they are both listed as Critically Endangered.

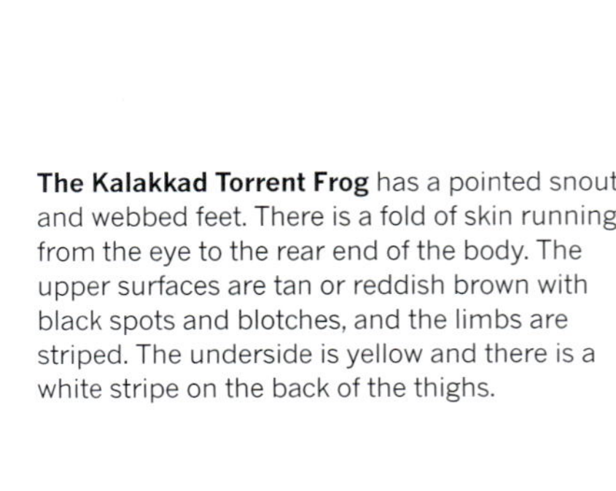

The Kalakkad Torrent Frog has a pointed snout and webbed feet. There is a fold of skin running from the eye to the rear end of the body. The upper surfaces are tan or reddish brown with black spots and blotches, and the limbs are striped. The underside is yellow and there is a white stripe on the back of the thighs.

FAMILY	Nyctibatrachidae: Astrobatrachinae
OTHER NAMES	Starry Dwarf Triangle Disk Frog
DISTRIBUTION	Wyanad Plateau, Kerala, India
ADULT HABITAT	Monsoon forest leaf litter close to water at altitudes of 4,300–4,600 ft (1,300–1,400 m)
LARVAL HABITAT	Unknown
CONSERVATION STATUS	Data Deficient. Described too recently to have been assessed

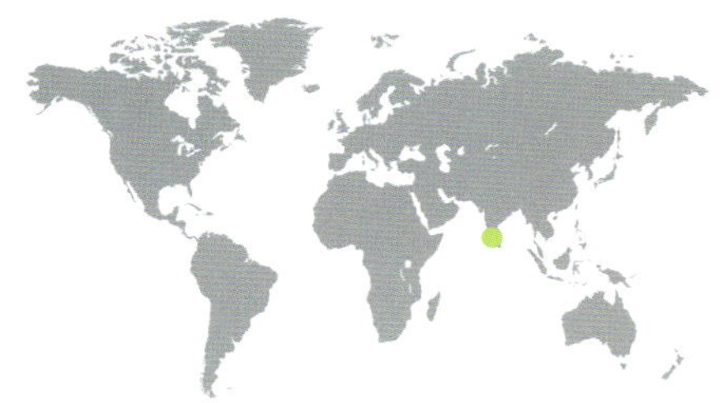

ADULT LENGTH
Male
¾–1¹⁄₁₆ in (20–27 mm)

Female
¹⁵⁄₁₆–1⅛ in (24–29 mm)

ASTROBATRACHUS KURICHIYANA

STARRY DWARF FROG

VIJAYAKUMAR, PYRON, DINESH, TORSEKAR, SRIKANTHAN, SWAMY, STANLEY, BLACKBURN, & SHANKER, 2019

The Starry Dwarf Frog was described in 2019, from the Kurichiyarmala district of Kerala province, in the Western Ghats of southern India. It is the sole representative of the nyctibatrachid subfamily Astrobatrachinae. The Western Ghats are well known as ancient refugia and centers of endemism and biodiversity, and this small frog represents a deeply divergent lineage. Nocturnal and secretive, it inhabits monsoon forest leaf litter, near water. Its reproductive habits are unstudied. The entire range of this species, and therefore of the entire subfamily, is small, and being listed as Data Deficient by the IUCN means that it is in urgent need of study because it may quickly become endangered due to climate or habitat changes.

Actual size

SIMILAR SPECIES

The closest relatives of *Astrobatrachus kurichiyana*, the sole species in the Astrobatrachinae, are the genera *Lankanectes* (page 530) and *Nyctibatrachus* (page 531), which belong to two different subfamilies.

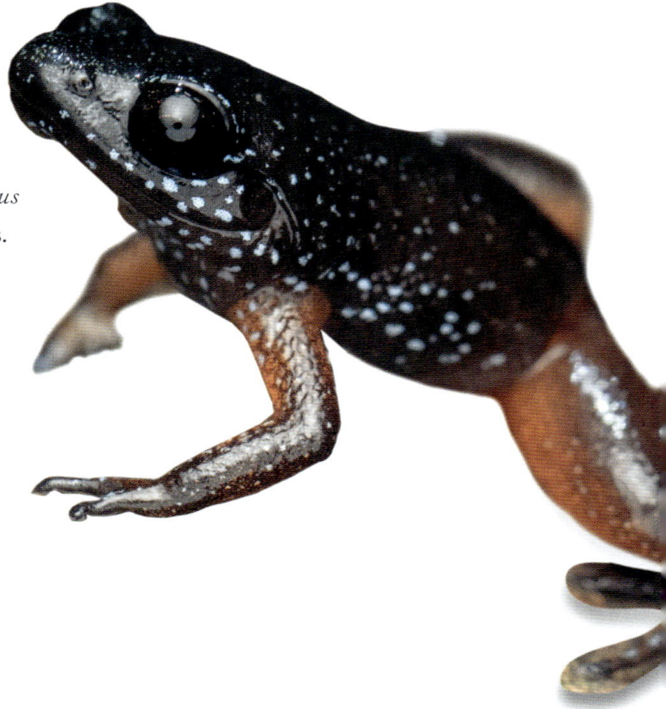

The Starry Dwarf Frog is a small species with smooth skin, a pointed snout, large, black eyes, small but distinct tympana, and well-developed limbs, with minimally webbed fingers and toes that terminate in small, pointed disks. The body is black or brown above, and around the eyes and on the flanks and throat are pale blue or white spots that resemble a starry night sky, while the limbs and belly are orange. The eyes are black with a horizontal pupil.

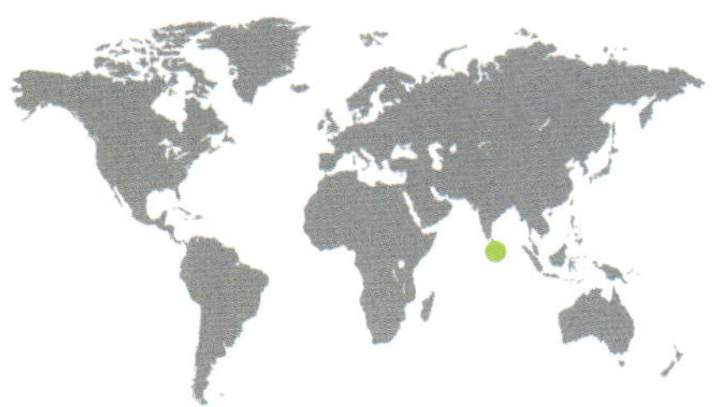

FAMILY	Nyctibatrachidae: Lankanectinae
OTHER NAMES	Sri Lanka Wart Frog
DISTRIBUTION	Southern Sri Lanka
ADULT HABITAT	Slow-moving rivers and marshy areas
LARVAL HABITAT	Pools and puddles
CONSERVATION STATUS	IUCN Near Threatened. Declining in places as a result of drainage of wetlands and pollution by agrochemicals

ADULT LENGTH
Male
average 1⅜ in (35 mm)
Female
1 7/16–2 13/16 in (37–71 mm)

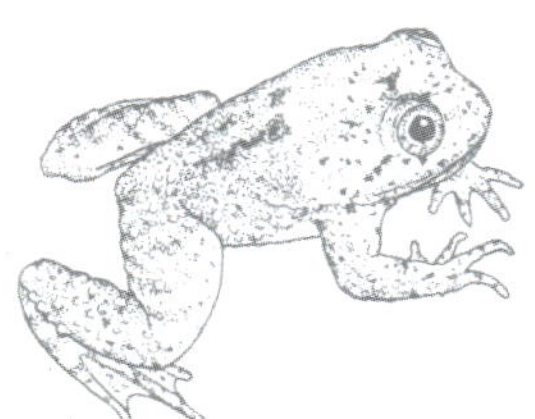

LANKANECTES CORRUGATUS

CORRUGATED WATER FROG

(PETERS, 1863)

This frog gets its name from the transverse folds of skin that run across its back. It is a wholly aquatic species and is well adapted for life in water. Its nostrils and eyes are positioned on its head in such a way that they protrude above the water's surface only when it is at rest. Its skin contains lateral line organs, tiny receptors found in amphibian larvae that detect vibrations in the water. The frog's coloration and skin patterns make it very well camouflaged when it is sitting in muddy water. Its tadpoles have been found in puddles.

The Corrugated Water Frog has very muscular legs, a short snout, and upwardly directed eyes. Its feet are fully webbed. Its upper surfaces are brown or orange-brown in color, with black blotches and spots, and there is often a broad yellow or orange stripe down the middle of the back. The underside is pale brown or white.

SIMILAR SPECIES

A second species, the Knuckles Range Water Frog (*L. pera*), was described in 2018, also from Sri Lanka. It is listed as Critically Endangered.

Actual size

FAMILY	Nyctibatrachidae: Nyctibatrachinae
OTHER NAMES	Wrinkled Frog
DISTRIBUTION	Western Ghats, western India
ADULT HABITAT	Close to fast-flowing streams in tropical forest
LARVAL HABITAT	Streams
CONSERVATION STATUS	IUCN Endangered. Has a very restricted range that is subject to habitat deterioration through deforestation, mining, and tourism

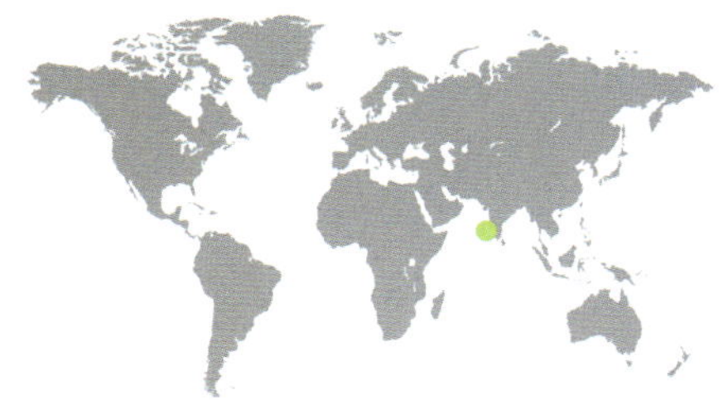

NYCTIBATRACHUS KARNATAKAENSIS

GIANT WRINKLED FROG

DINESH, RADHAKRISHNAN, MANJUNATHA REDDY & GURURAJA, 2007

ADULT LENGTH
Male
average 2¾ in (68 mm)

Female
average 3$^{5}/_{16}$ in (85 mm)

This large and very rare frog is most commonly found in cavities under boulders. When disturbed, it dives into water, where it can remain submerged for a long time. It is most active at night. During mating, the male makes a "wok... wok... wok" call and the female lays her eggs on leaves overhanging streams, where the larvae develop. This species occurs only within a small area of Karnataka state, in the Western Ghats of western India. This mountainous, forested region is home to many frogs that occur nowhere else in the world, many of them endangered.

SIMILAR SPECIES

Nyctibatrachus karnatakaensis is the largest of the 34 members of the genus so far described. The Miniature Night Frog (*N. minimus*), at ½ in (12 mm) long, is the smallest known frog in India. It is mostly terrestrial in its habitats. The Kumbara Night Frog (*N. kumbara*), a stream-living species, is thought to be unique in that the male covers the eggs with mud. Both species are listed as Vulnerable.

The Giant Wrinkled Frog is a plump frog with wrinkled skin on its back and throat, a rounded snout, large, forward-pointing eyes, and webbed feet. There are adhesive disks on the fingers and toes. The back is reddish brown and the flanks and legs are brown, flecked with yellow. The underside is yellow.

Actual size

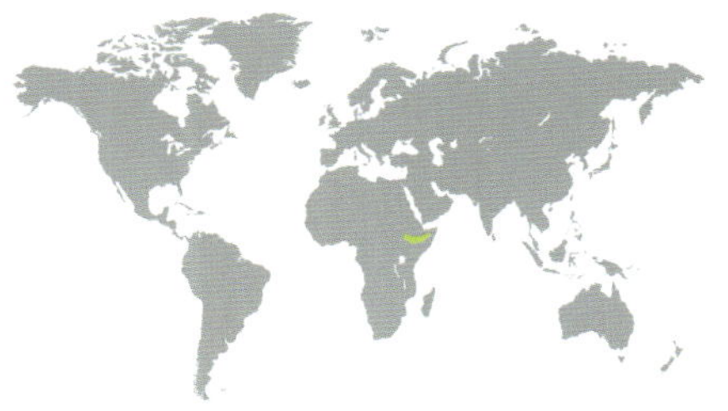

FAMILY	Ericabatrachidae
OTHER NAMES	None
DISTRIBUTION	Southern Ethiopia
ADULT HABITAT	Woodland and forest at altitudes of 7,900–10,500 ft (2,400–3,200 m)
LARVAL HABITAT	Unknown
CONSERVATION STATUS	IUCN Critically Endangered. Occurs in a very restricted area, and has been seen only rarely since 1986

ADULT LENGTH
Male
¾–⅞ in (19–22 mm)
Female
⅞–1¹⁄₁₆ in (23–27 mm)

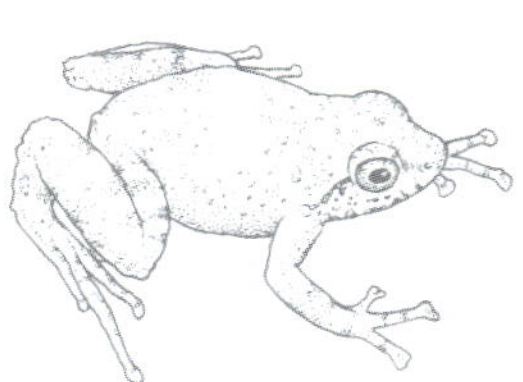

ERICABATRACHUS BALEENSIS

BALE MOUNTAINS FROG

LARGEN, 1991

Actual size

The Bale Mountains Frog has a small head with a rounded snout, a plump body, and numerous tiny warts on its back. The fingers and toes are long and thin, and they end in adhesive disks. The upper surfaces are green, brown, or reddish in color, dappled with tiny white spots. There is a dark stripe running from the eye to the armpit.

The Bale Mountains in southern Ethiopia are notable for a variety of animal species that occur nowhere else on the planet, including the Ethiopian Wolf (*Canis simensis*) and the recently described Bale Mountains Viper (*Bitis harenna*). The Bale Mountains Frog is found on the banks of small, fast-flowing streams. It was quite numerous in 1986, but has virtually disappeared since then. The streams in which it lives have been severely degraded by grazing cattle, and the disease chytridiomycosis was detected in several local frog species in 2008. Its reproductive biology is unknown, but it is suspected that females deposit their eggs on land.

SIMILAR SPECIES

Ericabatrachus baleensis is the only species in its genus, which is closely related to the African water frogs—genus *Petropedetes*—see Fuchs' Water Frog (*P. vulpiae*; page 533).

FAMILY	Petropedetidae
OTHER NAMES	None
DISTRIBUTION	West Africa, from eastern Nigeria to southern Gabon
ADULT HABITAT	Lowland forest
LARVAL HABITAT	Streams
CONSERVATION STATUS	IUCN Least Concern

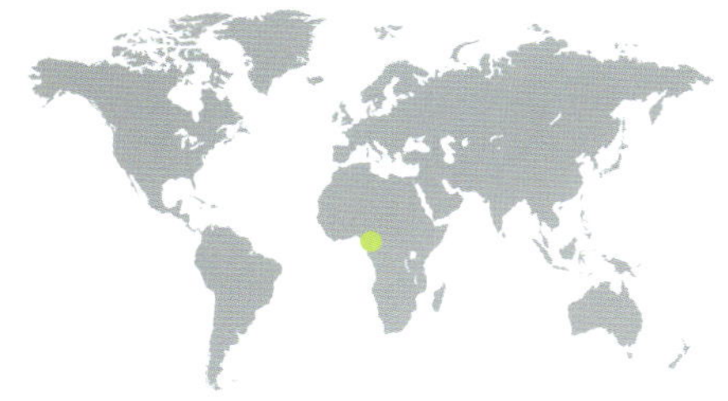

PETROPEDETES VULPIAE

FUCHS' WATER FROG

BAREJ, RÖDEL, GONWOUO, PAUWELS, BÖHME & SCHMITZ, 2010

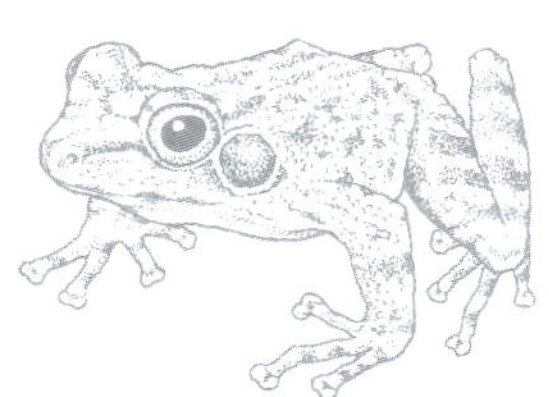

ADULT LENGTH
Male
1 5/16–1 3/4 in (33–44 mm)
Female
7/8–1 13/16 in (21–46 mm)

The male of this little-known frog undergoes several marked physical changes in the breeding season, which coincides with the rainy season. His forearms become much larger and more muscular than the female's, and a sharp spine develops on each thumb. In addition, a fleshy papilla develops near the center of his very large tympanum; the function of this is unknown. He calls from within rock cavities, producing a soft "douc... douc... douc" sound. The eggs are laid on moist rocks and the male sits beside them during the night.

Actual size

SIMILAR SPECIES

Nine species have been described in the genus *Petropedetes*, all found in West and Central Africa. The Efulen Water Frog (*P. palmipes*) occurs in forest streams in Cameroon, Equatorial Guinea, and Gabon, and is listed as Vulnerable. Perret's Water Frog (*P. perreti*) occurs in Cameroon. Unlike *P. vulpiae*, both have webbed feet, suggesting that they are aquatic in their habits. Perret's Water Frog is listed as Critically Endangered.

Fuchs' Water Frog has a large, wide head and prominent warts on its back and sides. The eyes are large and the tympanum is very well defined. The fingers and toes are long and thin with widened tips. The upper surfaces of the frog are brown with dark brown blotches, and the underside is white.

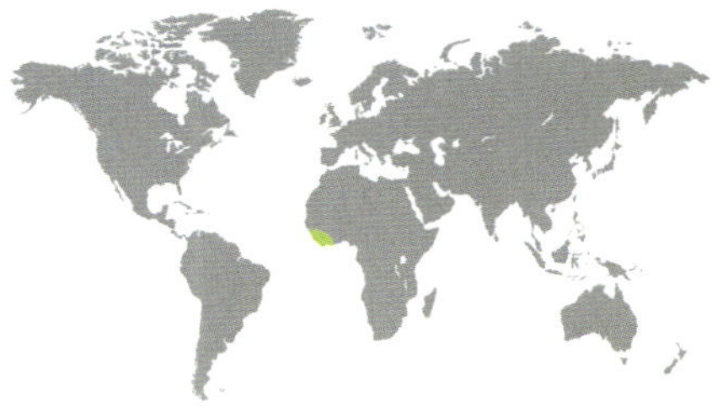

FAMILY	Odontobatrachidae
OTHER NAMES	Sierra Leone Water Frog, Swimmer, Saber-toothed Frog
DISTRIBUTION	Sierra Leone, Liberia, Guinea
ADULT HABITAT	Fast-flowing mountain streams and cascades up to 4,600 ft (1,400 m) altitude
LARVAL HABITAT	Cascades
CONSERVATION STATUS	IUCN Least Concern. The four congenerics are Near Threatened, Vulnerable, or Endangered

ADULT LENGTH
Male
1¹¹⁄₁₆–2⅟₁₆ in (43–53 mm)
Female
1¾–2⅜ in (45–61 mm)

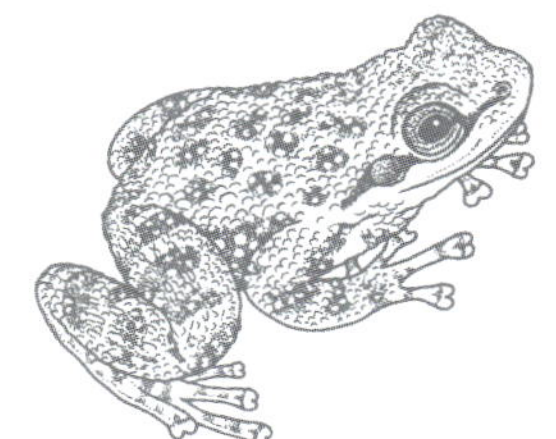

ODONTOBATRACHUS NATATOR

COMMON TOOTHED FROG

(BOULENGER, 1905)

Actual size

The Common Toothed Frog is fairly large, with extremely rough or granular skin, large eyes, small tympana, and powerful limbs with unwebbed digits that terminate in a bilobed disk. It is cryptically patterned with green and browns.

The Common Toothed Frog is a relatively large species. Females are larger than males, with a more robust body, longer limbs, broader heads, and more elongate snouts. Males, and occasionally females, possess oval glands on the undersides of their femurs, though their purpose in unknown. Inside the mouth both sexes exhibit backward-facing teeth on their maxillary and premaxillary bones and a pair of tusklike fangs on the mandibles. Using this effective armory, toothed frogs can overpower and swallow other frogs, and it is also used by males in territorial disputes. It is surprising that such formidable frogs also include vegetation in their diets. Toothed frogs are nocturnal, but they may adopt a terrestrial, aquatic, or even arboreal lifestyle. These frogs live close to cascades, laying their eggs in the splash zone from where the streamlined tadpoles can escape into the water and use the large suckers on their mouths to anchor themselves to rocks in the fast current.

SIMILAR SPECIES

The genus *Odontobatrachus* contains four additional species, all described in a single paper in 2015, before which they were included in *O. natator*. The other four species occur in Guinea, Liberia, and Ivory Coast, and while *O. natator* is considered as of Least Concern, the other species range from Near Threatened to Endangered.

FAMILY	Phrynobatrachidae
OTHER NAMES	Eastern Puddle Frog, Small Puddle Frog, Zanzibar Puddle Frog
DISTRIBUTION	East Africa, from Kenya and Somalia in the north to Mozambique and South Africa in the south
ADULT HABITAT	Near permanent water in savanna and grassland
LARVAL HABITAT	Temporary or permanent pools and ponds
CONSERVATION STATUS	IUCN Least Concern. Common across its large range; has declined only in South Africa

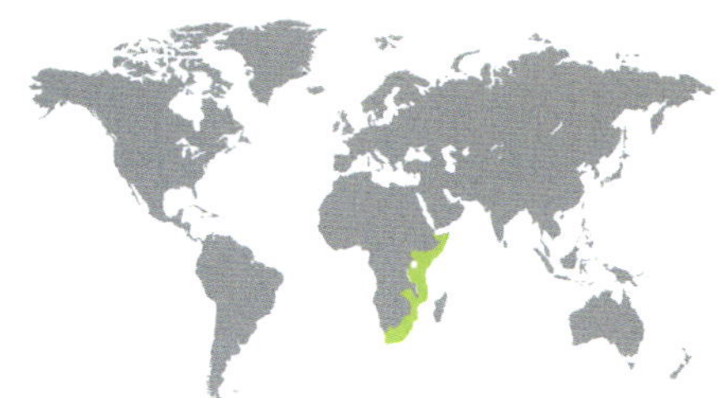

ADULT LENGTH
Male
$\frac{11}{16}$–$1\frac{1}{8}$ in (18–28 mm)
Female
$\frac{7}{8}$–$1\frac{3}{16}$ in (21–30 mm)

PHRYNOBATRACHUS ACRIDOIDES

EAST AFRICAN PUDDLE FROG

(COPE, 1867)

In common with the other puddle frogs of Africa, this small species is somewhat toad-like in appearance, with warty skin. It occurs in lowland areas, wherever there is permanent water. Its coloration makes it well camouflaged in muddy pools. Puddle frogs can breed at almost any time of year and can reach sexual maturity at around five months of age, so that, under favorable conditions, they can be very common. This species has a coarse, rasping call similar to that of crickets. It lays its eggs in clumps attached to aquatic vegetation.

Actual size

SIMILAR SPECIES

Currently there are 96 species in the genus *Phrynobatrachus* occur across Africa, south of the Sahara. The Guinea River Frog (*P. guineensis*) is found in Côte d'Ivoire and Sierra Leone. It lives in dry forest and is emancipated from ponds and streams, laying its eggs in water-filled tree-holes, fruit capsules, and snail shells. See also the Snoring Puddle Frog (*Phrynobatrachus natalensis;* page 536).

The East African Puddle Frog has large eyes and warty skin on its back. There are small disks on the tips of its fingers and toes. The upper surfaces are grayish green or grayish brown and there may be a broad or narrow yellowish stripe down the middle of the back. The underside is white.

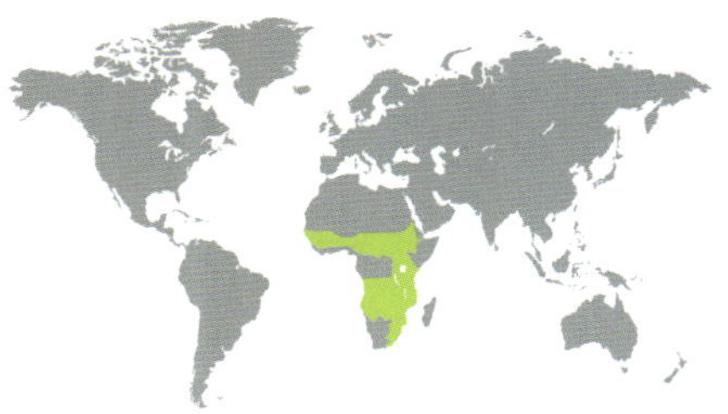

FAMILY	Phrynobatrachidae
OTHER NAMES	Natal Frog, Smith's Frog
DISTRIBUTION	Much of sub-Saharan Africa, excluding tropical forest and desert areas
ADULT HABITAT	Near permanent water in savanna and grassland
LARVAL HABITAT	Temporary or permanent pools and ponds
CONSERVATION STATUS	IUCN Least Concern. Common across its large range and occurs in many protected areas

ADULT LENGTH
Male
1–1$\frac{5}{16}$ in (25–34 mm)
Female
1–1$\frac{9}{16}$ in (26–40 mm)

PHRYNOBATRACHUS NATALENSIS

SNORING PUDDLE FROG

(SMITH, 1849)

Actual size

The Snoring Puddle Frog is a chubby frog with a small head, a pointed snout, and close-set eyes. It usually has a number of warts on its back. The upper surfaces are gray, green, or brown, and there is often a yellow or pale green stripe running down the middle of the back. The underside is cream.

This small frog gets its name from its mating call, a slow, quiet snore. Males call both by day and night during wet weather. They also engage in frequent aggressive interactions, in defense of their call sites around ponds. The female lays around 200 eggs, which form a thin film on the water's surface. They reach metamorphosis after 27–40 days. There is considerable variation in this species, particularly in terms of coloration, body size, and time of breeding, and it is likely that it actually comprises several species.

SIMILAR SPECIES

Some species of *Phrynobatrachus* are very small. The Dwarf Puddle Frog (*P. parvulus*) never exceeds 1 in (25 mm) in length. Its call is a cricket-like trill and it occurs across much of southern Africa. The Near Threatened *P. phyllophilus* is found in Côte d'Ivoire, Liberia, and Sierra Leone, and has a maximum length of $\frac{7}{8}$ in (23 mm). It lays small clutches of yolk-rich eggs on leaves close to small puddles. See also the East African Puddle Frog (*P. acridoides*; page 535).

FAMILY	Ptychadenidae
OTHER NAMES	Hildebrandt's Burrowing Frog
DISTRIBUTION	Across much of sub-Saharan Africa
ADULT HABITAT	Savanna, grassland
LARVAL HABITAT	Ephemeral pools
CONSERVATION STATUS	IUCN Least Concern. There is no evidence that it is declining

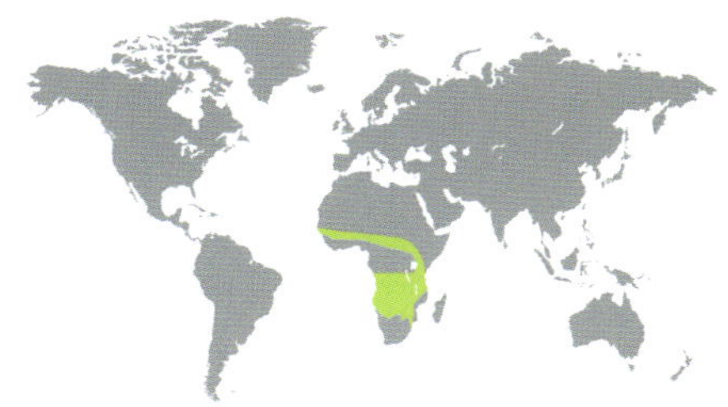

ADULT LENGTH
Male
2–2⅝ in (52–65 mm)

Female
1¹³⁄₁₆–2¹³⁄₁₆ in (46–70 mm)

HILDEBRANDTIA ORNATA

ORNATE FROG

(PETERS, 1878)

This colorful frog is rarely seen, as it spends most of its life underground. It uses large, spade-like tubercles on its hind feet to burrow backward into soft soil. It emerges after rain in early summer to gather, sometimes in large numbers, around shallow temporary pools. The male has a raucous call, likened to a hoot or a quack, that can be heard over a large distance. The eggs are laid singly, scattered in shallow water. The tadpoles are carnivorous in captivity but it is not certain that they are in the wild. In Burkina Faso, this species is eaten as food and is used in traditional medicine.

SIMILAR SPECIES

There are two other species in the genus *Hildebrandtia*. They are barely known but are thought to live in a similar habitat, and to have similar breeding habits, to *H. ornata*. The Somali Ornate Frog (*H. macrotympanum*) is found in Ethiopia, Somalia, and Kenya. The Angola Ornate Frog (*H. ornatissima*) is a particularly colorful frog, with a complex pattern of green, pink, and yellow patches and black spots.

The Ornate Frog has a very plump body, a blunt snout, and large eyes. The fingers are short. Its coloration is very variable. Most commonly it has a broad green or golden-brown stripe down the middle of its back, and other stripes of brown, white, and reddish brown. Some individuals are largely green, with pale yellow mottling. The underside is white.

Actual size

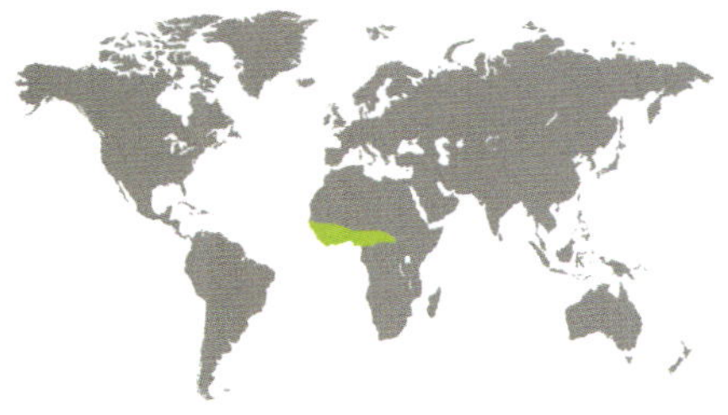

FAMILY	Ptychadenidae
OTHER NAMES	None
DISTRIBUTION	West and Central Africa, from Senegal in the west to South Sudan in the east
ADULT HABITAT	Wooded savanna
LARVAL HABITAT	Small, stagnant, ephemeral pools
CONSERVATION STATUS	IUCN Least Concern. Shows no evidence of a general decline

ADULT LENGTH
Male
1 5/16–2 in (34–52 mm)

Female
1 5/16–2 9/16 in (34–64 mm)

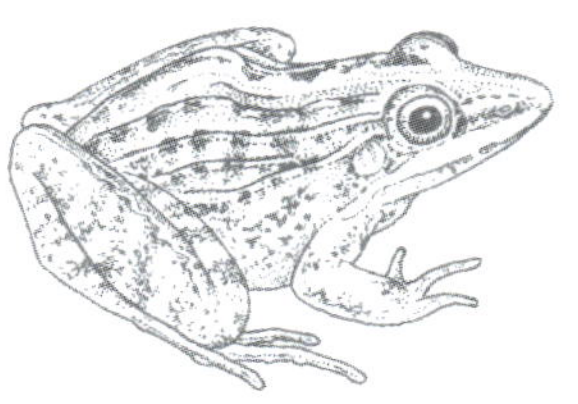

PTYCHADENA BIBRONI

BROAD-BANDED GRASS FROG

(HALLOWELL, 1845)

Like other grass frogs, this species is capable of prodigious leaps. Normally relying on its cryptic coloration to hide in the grass, it jumps away, often toward water, when disturbed. Mating occurs at night after heavy rain, the male producing a squeaky call. Females lay 800–1,500 eggs that float on the water's surface. In parts of its range, for example, in Burkino Faso, this species is an important source of food for local people and is also used in various aspects of traditional medicine.

SIMILAR SPECIES

Sub-Saharan Africa and the Indian Ocean islands are home to 59 species of Ptychadena. Among several species of grass frogs found in West Africa is the forest-dwelling Snouted Grassland Frog (*Ptychadena longirostris*), which has a very pointed head and extremely long legs, and Tournier's Rocket Frog (*P. tournieri*), a small, slender frog found in savanna. The latter also occurs in flooded rice fields and gets its common name from its leaping ability.

The Broad-banded Grass Frog has powerful, muscular hind legs, a triangular head with a pointed snout, and four ridges running along its back. The upper surfaces are gray or pale brown with brown or black patches, and there is often a pale stripe running from the tip of the snout along the whole length of the body.

Actual size

FAMILY	Ptychadenidae
OTHER NAMES	Mascarene Ridged Frog
DISTRIBUTION	Madagascar, introduced to Seychelles and Mascarenes
ADULT HABITAT	Any habitat where there is water
LARVAL HABITAT	Puddles, ditches, water-filled ruts
CONSERVATION STATUS	IUCN Least Concern. Has a huge range, shows no evidence of a general decline, and is common in disturbed habitats

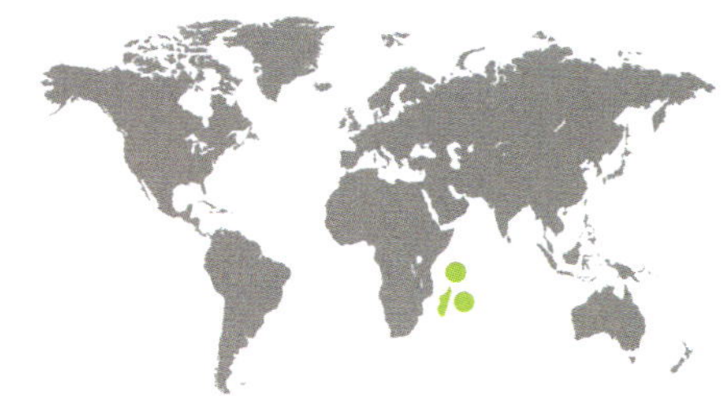

PTYCHADENA MASCARENIENSIS

MASCARENE GRASS FROG

(DUMÉRIL & BIBRON, 1841)

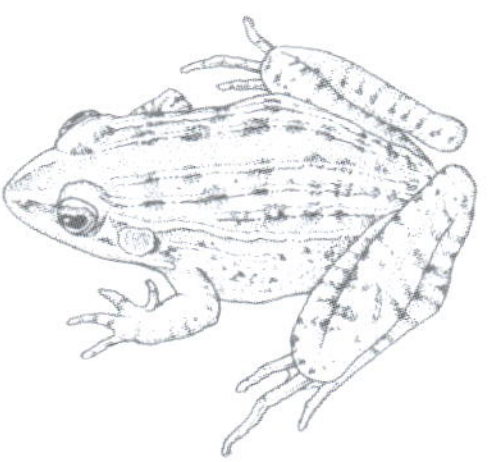

ADULT LENGTH
Male
$1\frac{11}{16}$–$2\frac{3}{16}$ in (43–57 mm)
Female
$1\frac{11}{16}$–$2\frac{3}{4}$ in (43–68 mm)

First described on the Mascarene Islands (Mauritius, Réunion, and Rodrigues), this frog has two distinct defensive strategies. One is to empty its bladder at the same time as it jumps, confusing its attacker. The other, called "foam and moan," involves secreting a foamy secretion, making a moaning sound, and going rigid. The species breeds in the rainy season. The male's call varies across its large range: In South Africa it is a nasal bray followed by a series of clucks. The female lays around 1,100 eggs attached to aquatic plants in shallow water.

SIMILAR SPECIES

Mainland African populations of *Ptychadena mascareniensis* are now thought to represent as yet undescribed species. Although named for the Mascarene Islands this frog is thought to have been introduced there from Madagascar. Many grass frogs thrive in human-altered habitats, provided they have access to water. Just one species, Newton's Grassland Frog (*P. ewtoni*), occurs only on the small island nation of São Tomé and Príncipe, in the Gulf of Guinea; it is listed as Endangered, because of the draining of wetlands.

The Mascarene Grass Frog has a pointed snout, long legs, and two pairs of ridges running the length of its back. The tympanum is large and the toes are webbed. The back is brown with a broad stripe down the midline, which may be white, beige, yellow, orange, or green. There are dark spots on the back and dark stripes on the legs.

Actual size

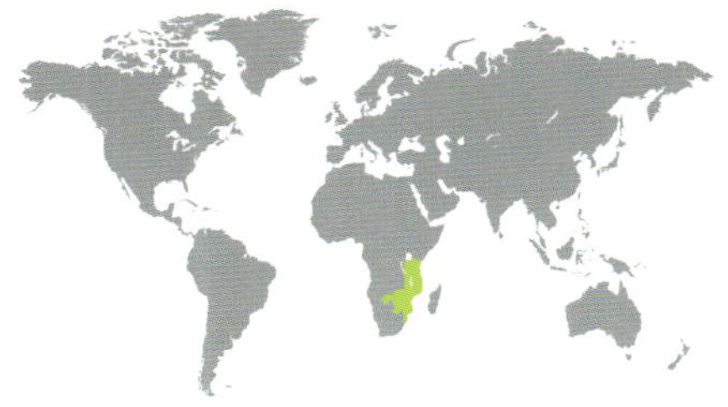

FAMILY	Ptychadenidae
OTHER NAMES	Broad-banded Grass Frog, Mozambique Grass Frog, Single-striped Grass Frog
DISTRIBUTION	Eastern Africa, from Somalia in the north to South Africa and Namibia in the south
ADULT HABITAT	Grassland, savanna
LARVAL HABITAT	Shallow pans, flooded grassland
CONSERVATION STATUS	IUCN Least Concern. Has a large range, shows no evidence of general decline, and is tolerant of habitat disturbance

ADULT LENGTH
Male
1⅛–1¾ in (29–44 mm)

Female
1$\frac{5}{16}$–2$\frac{1}{16}$ in (33–53 mm)

PTYCHADENA MOSSAMBICA

MOZAMBIQUE RIDGED FROG

(PETERS, 1854)

Actual size

The Mozambique Ridged Frog has a very pointed snout and prominent ridges along each side of its back. The back is gray, brown, or green, with a broad cream-colored stripe down its middle that is bordered by black spots. There is a dark stripe running from the tip of the snout to the tympanum and a white stripe along the upper lip.

This frog has been reported to achieve a height of 10 ft (3 m) when jumping. When disturbed, it tends to jump toward dense vegetation rather than into water. After rain, it can form very large breeding groups around shallow water, males producing an incessant quacking sound from well-concealed calling sites. Calling activity reaches a peak around midnight. Like many frogs, this species varies considerably in size across its range: It is smaller in Zambia and Tanzania than it is in South Africa. It is tolerant of habitat change and thrives in agricultural land.

SIMILAR SPECIES

The Dwarf Grass Frog (*Ptychadena taenioscalis*), as its common English name implies, is smaller than other grass frogs. It is found in savanna and grassland from Angola in the west to Tanzania in the east. It also occurs in savanna woodland in KwaZulu-Natal, South Africa. The Striped Grass Frog (*P. porosissima*) is found from South Africa's Eastern Cape in the south to Kenya in the north. There is also an old isolated population from Lake Tana, Ethiopia.

FAMILY	Ranidae
OTHER NAMES	Previously *Rana aurora*, Red-legged Frog
DISTRIBUTION	Western North America, from British Colombia south to northern California
ADULT HABITAT	Near ponds, streams, and wetlands in lowlands and hilly country
LARVAL HABITAT	Pools and ponds
CONSERVATION STATUS	IUCN Least Concern. Has declined in many parts of its range, as a result of urbanization and draining of wetlands

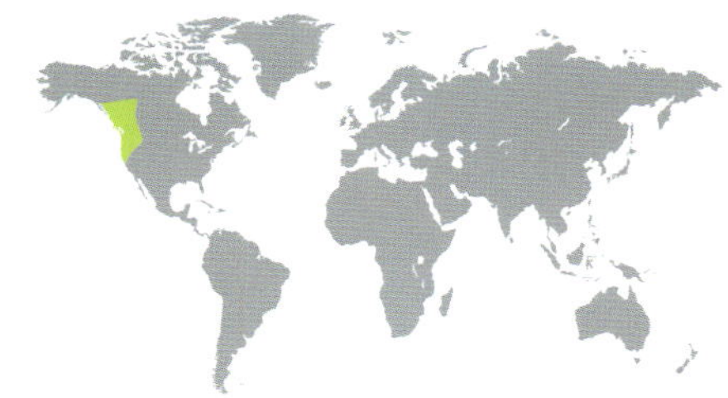

ADULT LENGTH

Male

1¾–2$\frac{13}{16}$ in (44–70 mm)

Female

1⅞–3⅞ in (48–100 mm)

AMERANA AURORA

NORTHERN RED-LEGGED FROG

(BAIRD & GIRARD, 1852)

Northern Red-legged Frogs have a brief breeding season in early spring, when they gather in small ponds in areas with good cover. Calling males make less noise than many frogs—they lack vocal sacs and call from under the water's surface, producing a mixture of chuckles and groans. Once a very abundant species, the frog has declined over much of its range, primarily through loss of its habitat; for example, it has nearly vanished from Oregon's Willamette Valley due to the draining of wetlands. Another threat, affecting it at the tadpole stage, is the Western Mosquitofish (*Gambusia affinis*), introduced to freshwater ecosystems to control malaria.

The Northern Red-legged Frog gets its name from a red flush on its abdomen and the underside of its hind legs. Otherwise, it is brown, gray, olive-green, or reddish in color, with many dark spots. The eyes are large and protruding, and there are well-developed ridges along each side of the back.

SIMILAR SPECIES

The California Red-legged Frog (*Amerana draytonii*) was formerly regarded as a subspecies of *A. aurora*. A larger frog, it once ranged from northern California to Mexico and was so abundant that, in 1895, an estimated 120,000 were eaten by locals in California. Now listed as Vulnerable, it has vanished from much of its range, largely as a result of pollution from insecticides blown on the wind from California's Great Central Valley, an area of intensive agriculture. Genus *Amerana* contains eight species.

Actual size

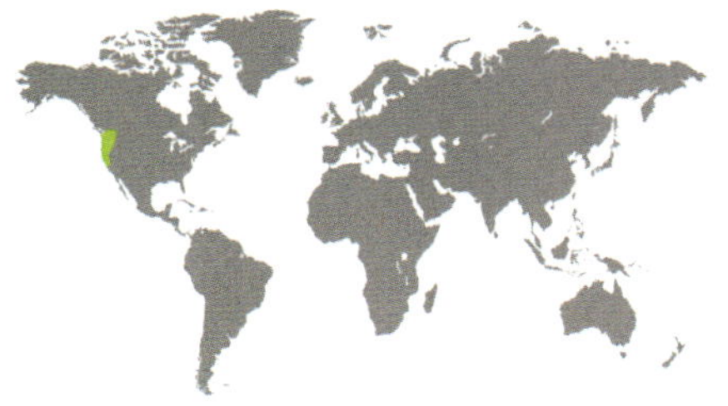

FAMILY	Ranidae
OTHER NAMES	Previously *Rana cascadae*
DISTRIBUTION	California, Oregon, and Washington, USA
ADULT HABITAT	Montane forest and meadows, up to the timberline, near ponds and streams
LARVAL HABITAT	Shallow pools formed by melting snow
CONSERVATION STATUS	Not listed. Has declined dramatically in many parts of its range due to introduced trout and the North American Bullfrog (*Aquarana catesbeiana*), and elevated UV radiation

ADULT LENGTH
Male
1¾–2¼ in (44–58 mm)

Female
1¹³⁄₁₆–2¹⁵⁄₁₆ in (46–75 mm)

AMERANA CASCADAE

CASCADES FROG

(SLATER, 1939)

Although its habitat is relatively unaffected by human activities, the Cascades Frog has declined dramatically over most of its range. A detailed survey carried out in Oregon in 1990 revealed that it had disappeared from 80 percent of 30 localities in which it had been recorded previously. It is very likely that a major factor in this decline has been increased levels of UV-B radiation, resulting from the thinning of the ozone layer. UV-B damages the genetic material in amphibian eggs, especially in species that, like the Cascades Frog, lay their eggs close to the surface of open, sunlit water.

SIMILAR SPECIES

The Oregon Spotted Frog (*Amerana pretiosa*) occurs in the same US states as *A. cascadae*, but at lower altitudes. It has declined over much of its range, especially in California, as a result of drainage of its wetland habitat and is listed as Vulnerable. Less threatened is the Colombia Spotted Frog (*A. luteiventris*), whose range extends into British Columbia.

The Cascades Frog has prominent ridges along each side of its back, which is pale brown or green in color, with several well-defined black spots. The throat and belly are yellow or white, and the legs have dark transverse stripes. Most individuals have a dark "mask" running from the snout, through the eye, to the tympanum.

Actual size

FAMILY	Ranidae
OTHER NAMES	Previously *Rana muscosa*, Sierra Madre Yellow-legged Frog, Southern Mountain Yellow-legged Frog
DISTRIBUTION	In two distinct areas of central and southern California, USA
ADULT HABITAT	Wet meadows, streams, and lakes at high altitudes up to 7,550 ft (2,300 m)
LARVAL HABITAT	Streams and lakes
CONSERVATION STATUS	IUCN Near Threatened (as a subspecies of *Amerana boylii*). Nearly extinct in the southern part of its range; its decline is caused by introduced fish, chytridiomycosis, and pesticides

AMERANA MUSCOSA

MOUNTAIN YELLOW-LEGGED FROG

(CAMP, 1917)

ADULT LENGTH

Male
$1\frac{9}{16}$–$2\frac{15}{16}$ in (40–75 mm)

Female
$1\frac{3}{4}$–$3\frac{1}{2}$ in (45–89 mm)

This endangered frog is very aquatic in its habits and starts to breed as soon as winter snow has melted from its mountain habitat. Beginning in the late 1800s, trout were introduced to many permanent waterbodies in the Sierra Nevada with catastrophic effect, as the trout ate tadpoles. More recently, a concerted effort to remove trout from a number of small lakes has resulted in some frogs returning. However, this species is also very susceptible to the disease chytridiomycosis and, like many other Californian frogs, is probably adversely affected by pesticides drifting on the wind from agricultural areas.

SIMILAR SPECIES

Found in northern California and Nevada, the Sierra Nevada Yellow-legged Frog (*Amerana sierrae*) is threatened by the same factors as *A. muscosa* and is also listed as Endangered. It has shorter legs and a distinct mating call. At lower altitudes, the Foothill Yellow-legged Frog (*A. boylii*), of which *A. muscosa* was once a subspecies, has declined over much of its range because of deterioration in its stream habitat; it is listed as Near Threatened.

The Mountain Yellow-legged Frog emits a garlic-like odor when handled. It has a netlike pattern of brown or black markings on its back, which is otherwise yellowish, reddish brown, or olive-green. The long hind legs are striped. The undersides of the legs and belly are flushed with yellow or orange.

Actual size

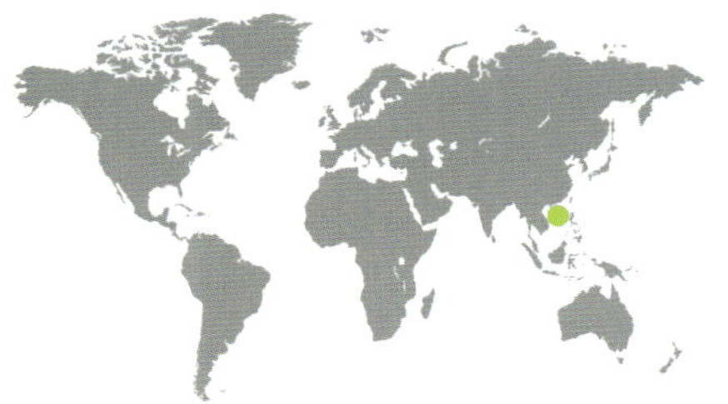

FAMILY	Ranidae
OTHER NAMES	Hainan Sucker Frog
DISTRIBUTION	Hainan Island, off southern China
ADULT AND LARVAL HABITAT	Fast-flowing forest streams
CONSERVATION STATUS	IUCN Endangered. Its habitat has been reduced and fragmented by deforestation and hydroelectric schemes

ADULT LENGTH
Male
average 2⅞ in (73 mm)
Female
average 3⅛ in (80 mm)

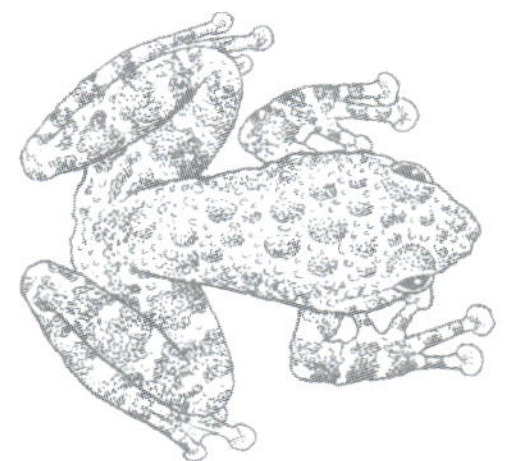

AMOLOPS HAINANENSIS

HAINAN TORRENT FROG

(BOULENGER, 1900)

This frog gets its common name from its habitat: It lives in fast-flowing streams and the adjacent splash zone. It breeds between April and August, and the eggs are attached to wet rocks close to waterfalls. The tadpoles have very large abdominal suckers that enable them to cling to rocks. So effective are these that, when picked up by its tail, a tadpole retains its grip on a stone up to 60 times its weight. The tadpoles are also unusual in having poison glands. In addition to threats to its habitat, this frog is eaten by local people.

SIMILAR SPECIES

There are 85 species in the genus *Amolops*, found in Nepal, India, China, and the Malay Peninsula. Called torrent frogs, they live in and around mountain streams. The Rufous-spotted Torrent Frog (*A. loloensis*) is found at high altitudes in Sichuan province, China, and is listed as Vulnerable. Its habitat is being destroyed and it is also adversely affected by chemical pollution from mining.

Actual size

The Hainan Torrent Frog has a flattened body, muscular legs, and a warty skin. Both fingers and toes end in very large adhesive disks, and the toes are webbed. The back is olive-green or dark brown with large, dark spots and small, pale flecks. The underside is reddish.

FAMILY	Ranidae
OTHER NAMES	Previously *Rana catesbiana*, *Lithobates catesbeianus*
DISTRIBUTION	Central and eastern USA, Canada. Has spread into many western areas of the USA and has been introduced to other parts of the world
ADULT AND LARVAL HABITAT	Larger ponds and lakes
CONSERVATION STATUS	IUCN Least Concern. One of very few frogs whose population is increasing. Regarded as a pest in much of the world

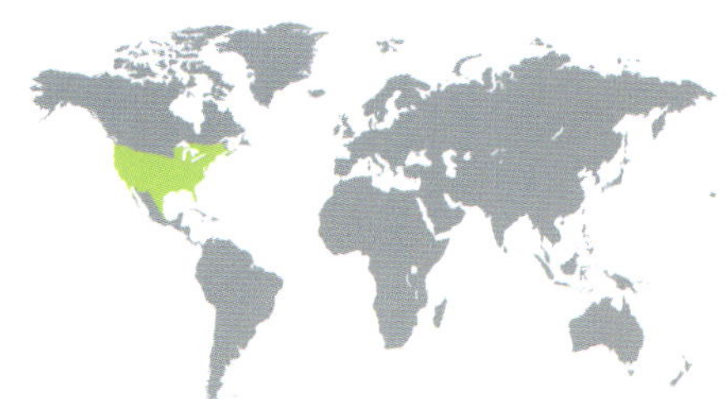

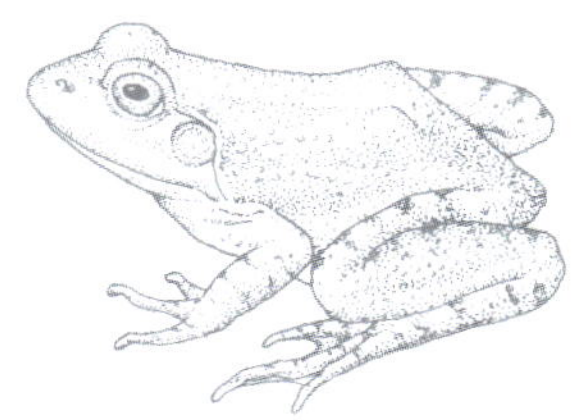

ADULT LENGTH
Male
3½–7³⁄₁₆ (90–180 mm)
Female
4¾–7⅞ in (120–200 mm)

AQUARANA CATESBEIANA

NORTH AMERICAN BULLFROG

(SHAW, 1802)

Though declining in parts of its natural range, this very large frog is a destructively invasive species in many other parts of the world, where it was initially introduced to be farmed as a source of food. It thrives in new areas, because of an absence of natural predators, and displaces native frog species. Its fast-growing tadpoles compete with native tadpoles for food, and adults eat smaller native frogs. Most importantly, North American Bullfrogs carry, but are immune to, the fungus that causes the disease chytridiomycosis. The introduction of this species to other areas of the world has thus been a major cause of the spread of this lethal disease.

The North American Bullfrog is the largest frog in North America. It has long, powerful hind limbs and prominent tympana, which are larger in the males than the females. It is usually green above with brown markings, and it always has a green head. It is white below, often with a yellow tinge. The male has thicker forelimbs than the female.

SIMILAR SPECIES

Though not reaching the same very large size as *Aquarana catesbeiana*, the Pig Frog (*A. grylio*) is in the same genus and often confused with it. It occurs in lowland swamps and lakes in southern USA, from Texas to South Carolina. It gets its name from the low-pitched grunts that males make during breeding. It can be heard at any time of year but most breeding occurs between March and September. Genus *Aquarana* now contains seven species. See also the Green Frog (*A. clamitans*; page 546).

Actual size

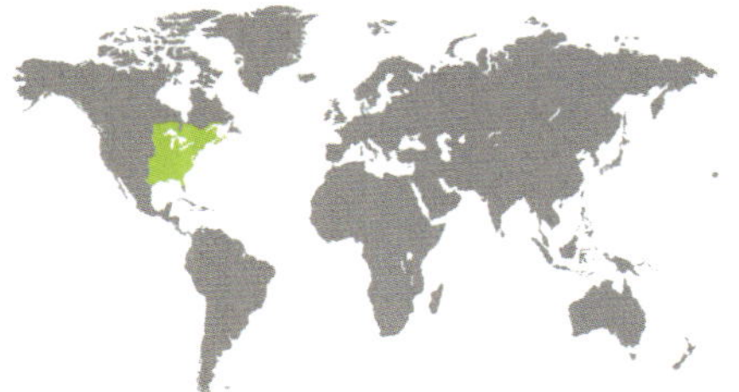

FAMILY	Ranidae
OTHER NAMES	Previously *Rana clamitans*, *Lithobates clamitans*, Bawling Frog, Belly Bumper
DISTRIBUTION	Eastern USA, eastern Canada
ADULT HABITAT	In or near shallow, permanent waterbodies
LARVAL HABITAT	Shallow, permanent waterbodies
CONSERVATION STATUS	IUCN Least Concern. Shows no evidence of a general decline across its large range

ADULT LENGTH
Male
2³⁄₁₆–4 in (57–103 mm)

Female
2⅜–4⅛ in (60–105 mm)

AQUARANA CLAMITANS

GREEN FROG

(LATREILLE, 1801)

The Green Frog is variable in color: Some individuals are more brown, dark gray, or bronze than green. The upper lip is green and the underside is white. There are well-developed ridges along each side of the back that do not reach the groin. The tympana are conspicuous and are particularly large in males.

This large frog breeds in ponds and shallow lakes, males defending territories in which females will lay their eggs. The male's call is likened to the sound of a loose banjo string being plucked. Larger males have deeper voices than small males, and the latter generally retreat when they hear a low-pitched call. The longer a male successfully defends a territory, the more females he mates with. There are two subspecies, the Northern Green Frog (*Rana clamitans melanota*) and the Southern Green Frog or Bronze Frog (*R. c. clamitans*). The latter is smaller and is often bronze in color.

SIMILAR SPECIES

The Carpenter Frog (*Aquarana virgatipes*) gets its common name from its call, likened to the sound of a nail being hit repeatedly. A very aquatic species, it is found in wetter parts of the coastal states of eastern USA, and consequently is sometimes known as the Sphagnum Frog. The Pickerel Frog will be mentioned under *Lithobates* where it is now placed.

Actual size

FAMILY	Ranidae
OTHER NAMES	Amami-Oshima Frog
DISTRIBUTION	Amami-Oshima and Kakeromajima islands, southern Japan
ADULT HABITAT	Mountain forest
LARVAL HABITAT	Small pools
CONSERVATION STATUS	IUCN Endangered. Has a very small range and its habitat is threatened by deforestation; also preyed upon by introduced mongoose

BABINA SUBASPERA

OTTON FROG

(BARBOUR, 1908)

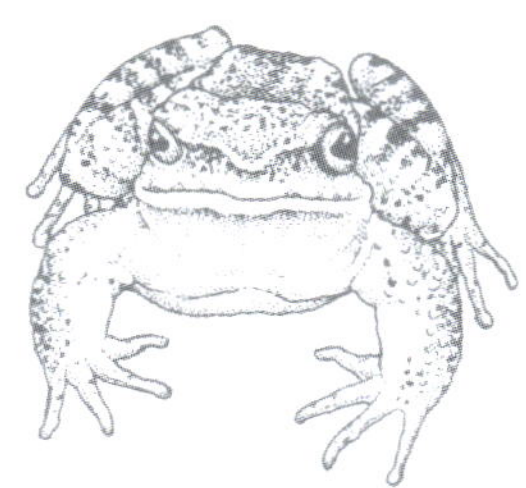

ADULT LENGTH
Male
3⅝–5 in (93–126 mm)
Female
4⅜–5½ in (111–140 mm)

This large frog is unusual in appearing to have five digits on its hands, frogs typically only having four. The extra digit is a "pseudothumb" containing a sheathed spine that can be pushed out. It is longer and thicker in males than it is in females and appears to be used by males in two contexts: while clasping the female in amplexus and when fighting other males. It appears to serve no function in females. The species breeds from April to August, females laying around 1,300 eggs in a specially dug pit about 12 in (30 cm) across.

SIMILAR SPECIES

The Asian genus *Babina* contains just one other species, Holst's Frog (*B. holsti*), which is a large, robust frog found in the mountains in some of the southern islands of Japan. Also called the Dagger Frog, it has a pseudothumb like that of *B. subaspera*. It is also listed as Endangered. Eight other species have been transferred to the genus *Nidirana*, including the East China Music Frog (*Nidirana adenopleura*; page 564), from China, Vietnam, Laos, and Thailand.

The Otton Frog has a robust body, a large head, and powerful limbs. Males have thicker, more muscular arms than females and there is partial webbing between the toes. The back is covered in warts and is greenish brown in color with darker markings. There are dark stripes on the legs.

Actual size

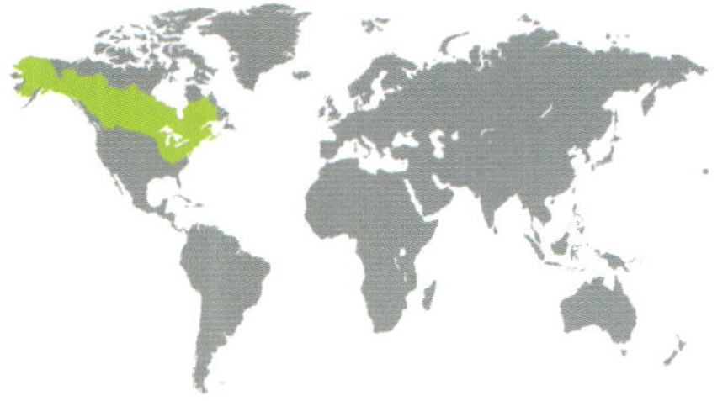

FAMILY	Ranidae
OTHER NAMES	Previously *Rana sylvatica*, *Lithobates sylvaticus*
DISTRIBUTION	Alaska and northeastern states, USA; Canada
ADULT HABITAT	Woodland
LARVAL HABITAT	Shallow ephemeral ponds
CONSERVATION STATUS	IUCN Least Concern. Shows no evidence of a general decline across its large range

ADULT LENGTH
1⅜–3¼₆ in (35–82 mm)
males are generally smaller

BOREORANA SYLVATICA

WOOD FROG

(LECONTE, 1825)

This is the only North American frog that occurs north of the Arctic Circle. It is found right up to the northern timberline and is able live in such a cold environment because its body contains large amounts of glucose and urea, which act as antifreeze, allowing it to survive in the frozen ground during winter. It is the earliest frog to breed in spring, emerging and gathering in ponds as the snow melts. The mating period is brief and involves much competition among males, which outnumber females. Some females get trapped inside "mating balls" of males and are killed as a result.

SIMILAR SPECIES

The Mink Frog (*Aquarana septentrionalis*) is also called the North Frog because it occurs in eastern Canada and northeastern USA, but it lacks the Wood Frog's tolerance of frozen conditions. When handled, it emits a smell likened to that of a mink or of rotting onions. It is a species in which major deformities, such as missing or extra limbs, have been found, the causes of which are little understood.

The Wood Frog has a characteristic "bandit's mask," a dark stripe running from the tip of its nose to just behind the tympanum. There are prominent folds of skin along each side of the back, which is brown, tan, pinkish, or bronze in color, sometimes with black spots. There is a pale stripe along the upper jaw and the legs are striped.

Actual size

FAMILY	Ranidae
OTHER NAMES	Malabar Frog
DISTRIBUTION	Western Ghats, India
ADULT HABITAT	Tropical forest at altitudes of 1,600–6,600 ft (500–2,000 m)
LARVAL HABITAT	Natural and artificial ponds and lakes
CONSERVATION STATUS	IUCN Least Concern. Much of its habitat has been lost through deforestation; suffers high mortality on roads

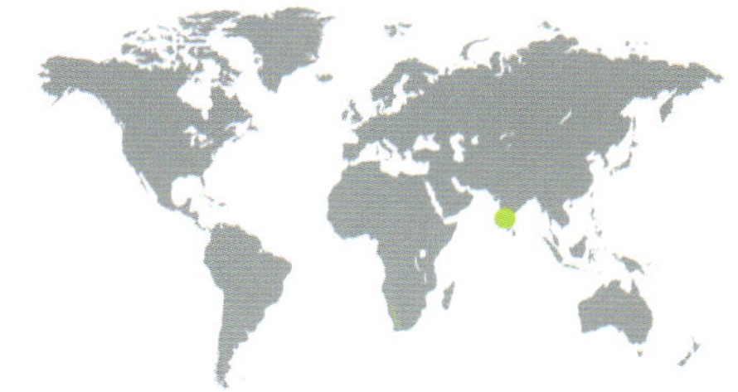

ADULT LENGTH
average 3⅛ in (80 mm)

CLINOTARSUS CURTIPES

BICOLORED FROG

(JERDON, 1853)

This large, ground-living frog is found in the leaf litter in tropical forest. It migrates to ponds and lakes to breed, when many individuals are killed on roads. Little is known about its reproductive behavior, other than that the male's call is a series of single notes and that males wrestle with one another. The species' tadpoles form large, dense schools in streams and are eaten by local people. Like many other frogs, it has antibacterial peptides in its skin, making it of interest to the pharmaceutical industry.

SIMILAR SPECIES

There are two other species in the genus *Clinotarsus*. The Assam Hills Frog (*Clinotarsus alticola*) is associated with forest streams across northern India, northern Bangladesh, Myanmar, and Vietnam. The recently described Phang Nga Stream Frog (*C. penelope*) is found further south in peninsular Thailand and Myanmar. Both species have large tadpoles, with large eyespots at the base of their tails.

The Bicolored Frog has large eyes and even larger tympana. The legs are slender and the toes are webbed. There is a fold of skin running from the eye to the groin that separates the pale back from the dark flanks. The back may be yellow, orange, pink, or pale brown, and the flanks are brown or black.

Actual size

FAMILY	Ranidae
OTHER NAMES	Previously *Rugosa rugosa*, Wrinkled Frog
DISTRIBUTION	Japan. Introduced to Hawaii
ADULT HABITAT	Near water in lowlands and on lower mountain slopes
LARVAL HABITAT	Ponds, ditches, and rice fields
CONSERVATION STATUS	IUCN Least Concern. Shows no evidence of a general decline

ADULT LENGTH
Male
1³⁄₁₆–1⅞ in (30–47 mm)
Female
1¾–2⅜ in (44–60 mm)

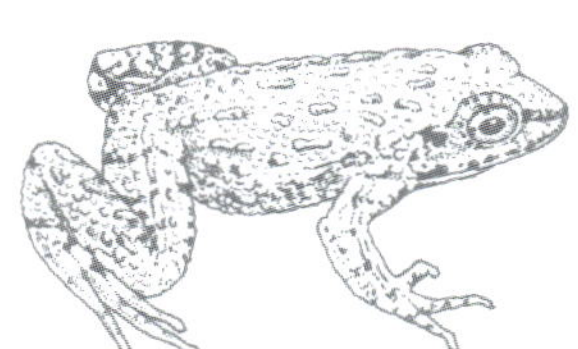

GLANDIRANA RUGOSA

JAPANESE WRINKLED FROG

(TEMMINCK & SCHLEGEL, 1838)

A native of Japan, this frog was introduced to Hawaii in the late nineteenth century and has become well established there. An adaptable species, it lives and breeds in rice fields and artificial ponds. The female lays her eggs in small clumps. In Japan, where the species has long been eaten by local people, it spends the winter in water. In some places, its tadpoles also overwinter in water, metamorphosing the year after they hatch. In the warmer climate of Hawaii, however, this does not seem to occur. The frog is reported to have a distasteful odor.

SIMILAR SPECIES

There are six species in the genus *Glandirana*, distributed across China, Korea, Russia, and Japan. The Northeast China Rough-skinned Frog (*G. emelianjovi*) is a common frog in Korea and northeastern China, where it occurs in a variety of freshwater habitats, including slow-moving rivers and rice fields. Both the Little Rough-skinned Frog (*G. minima*) from China, and the Sado Island Wrinkled Frog (*G. susurra*), from Japan, are Endangered.

The Japanese Wrinkled Frog gets its name from a number of very distinct longitudinal wrinkles on its back and sides. It has a slender body, a relatively large head, a pointed snout, and webbed feet. It is variable in color, being gray, green, or brown, with transverse stripes on its legs.

Actual size

FAMILY	Ranidae
OTHER NAMES	None
DISTRIBUTION	Northeastern Borneo
ADULT HABITAT	Forest in hilly country at altitudes of 800–2,500 ft (250–750 m)
LARVAL HABITAT	Fast-flowing streams
CONSERVATION STATUS	IUCN Least Concern. Shows no evidence of a general decline

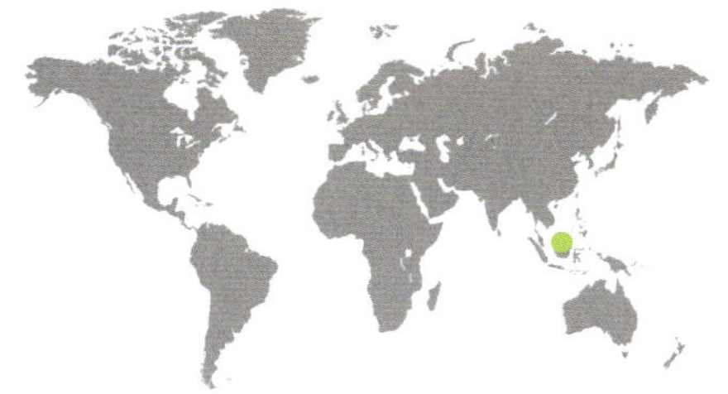

HUIA CAVITYMPANUM

HOLE-IN-THE-HEAD FROG

(BOULENGER, 1896)

ADULT LENGTH

Male
1⅝–2 in (42–52 mm)

Female
2 15/16–3⅛ in (75–80 mm)

This highly unusual frog breeds in very fast-flowing streams, a habitat that presents two major challenges. First, the sound frequencies that male frogs typically use to attract mates would be inaudible. The frog therefore produces ultrasound, sound frequencies that are too high for the human ear to hear, but that are audible despite the sound of rushing water. To hear such sounds it has highly unusual ears, with the tympanum deep inside the head. There is a hole where the tympanum would normally be, hence the species' common name. Second, to survive without being washed away, the tadpoles have an abdominal sucker that enables them to cling to rocks.

SIMILAR SPECIES

The genus *Huia* used to contain at least seven additional species, all associated with torrents in Southeast Asia, but they have all been transferred to the genera *Wijayarana* or *Odorrana*, so that *H. cavitympanum* remains the only species in its genus. The only other frog known to use ultrasound in communication is the Concave-eared Torrent Frog (*Odorrana tormota*; page 565), from China.

The Hole-in-the-head Frog gets its name from the deep hole behind the eye. It has a rounded snout, long legs, and slightly pointed pads at the tips of its fingers and toes. The toes are fully webbed. The back and the top of the head are chocolate-brown, the flanks are tan or pinkish with dark brown spots, and the throat, chest, and belly are white or yellowish.

Actual size

FAMILY	Ranidae
OTHER NAMES	*Papurana daemeli*, Australian Bullfrog, Water Frog, Wood Frog
DISTRIBUTION	New Guinea; New Britain; Northern Queensland and Northern Territory, Australia; Babar Is., Indonesia
ADULT HABITAT	Lowland rainforest
LARVAL HABITAT	Temporary and permanent streams
CONSERVATION STATUS	IUCN Least Concern. Its population is stable

ADULT LENGTH
Male
1¹¹⁄₁₆–2¼ in (43–58 mm)
Female
2¼–3⅛ in (58–81 mm)

HYLARANA DAEMELI

AUSTRALIAN WOOD FROG

(STEINDACHNER, 1868)

This elegant frog is the only member of the very large family Ranidae, known as true frogs, that occurs in Australia, where it is most common in the Cape York Peninsula. It feeds on arthropods and smaller frogs, and is active only at night. It breeds in spring and summer. Males have paired vocal sacs and their mating call, given from a perch over water, is a series of low-pitched "quacks." Females lay several thousand eggs in a shapeless mass. The species is also common in New Guinea, where it is eaten by local people.

SIMILAR SPECIES

There are 107 species in the genus *Hylarana*, many of them occurring in Southeast Asia. The Baram River Frog (*H. baramica*) occurs in Indonesia, Malaysia, and Singapore; the Sulawesi Frog (*H. celebensis*) occurs in Indonesia; and the San Cristobal Frog (*H. kreffti*) is found in the Solomon Islands, and New Ireland (Papua New Guinea).

Actual size

The Australian Wood Frog has an elongated body, a pointed snout, and long, muscular legs. The eyes are large and the tympana are conspicuous. There is a fold of skin running from the eye to the groin. The upper surfaces are olive-green or brown, with irregular dark blotches on the back and dark stripes on the legs. The underside is white.

FAMILY	Ranidae
OTHER NAMES	Golden-lined Frog, Green Lotus Frog
DISTRIBUTION	Southeast Asia, from Myanmar in the north to Indonesia and the Philippines in the south
ADULT HABITAT	Floodplains up to 4,000 ft (1,200 m altitude)
LARVAL HABITAT	Ponds
CONSERVATION STATUS	IUCN Least Concern. Its population is stable

HYLARANA ERYTHRAEA

GREEN PADDY FROG

(SCHLEGEL, 1837)

ADULT LENGTH
Male
1¼–1⅞ in (32–48 mm)
Female
1⅞–3¹⁄₁₆ in (48–78 mm)

This streamlined frog is very wary and difficult to approach. It thrives in such human-altered habitats as irrigation ditches and flooded rice fields, where it is often seen in floating vegetation or sitting at the water's edge. In Sarawak and the Philippines it can breed at any time of year. Males call from open water late at night, producing what has been described as a squeaky warble. In common with some other ranid frogs, blue individuals of this species are quite common. This is caused by a lack of xanthophores, cells in the skin that contain yellow pigment.

SIMILAR SPECIES

Several *Hylarana* species occur in Central Asia. The Malabar Hills Frog (*H. malabarica*) is a particularly colorful frog that occurs in India. The Thai Stream Frog (*H. cubitalis*), found in Myanmar and Thailand, is declining due to the widespread destruction of its forest habitat.

The Green Paddy Frog has a slender body, a pointed snout, and long, muscular legs. The eyes are large and the tympana are conspicuous. There is a white fold of skin running from the eye to the groin. The back and upper flanks are bright green, while the lower flank and belly are white with black markings. The arms and legs are brown.

Actual size

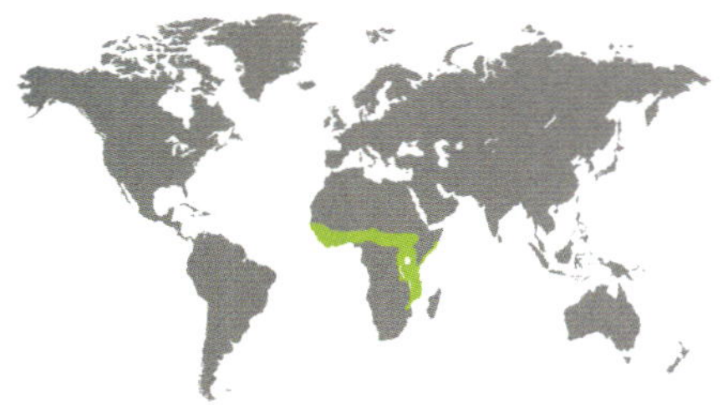

FAMILY	Ranidae
OTHER NAMES	Golden-backed Frog, Lake Galam Frog, Marble-legged Frog
DISTRIBUTION	Africa, from Senegal in the west to Eritrea in the northeast, and south to Malawi and Mozambique
ADULT HABITAT	Savanna
LARVAL HABITAT	Ponds
CONSERVATION STATUS	IUCN Least Concern. Its population is stable

ADULT LENGTH
Male
average 3¹⁄₁₆ in (78 mm)
Female
average 3⁵⁄₁₆ in (86 mm)

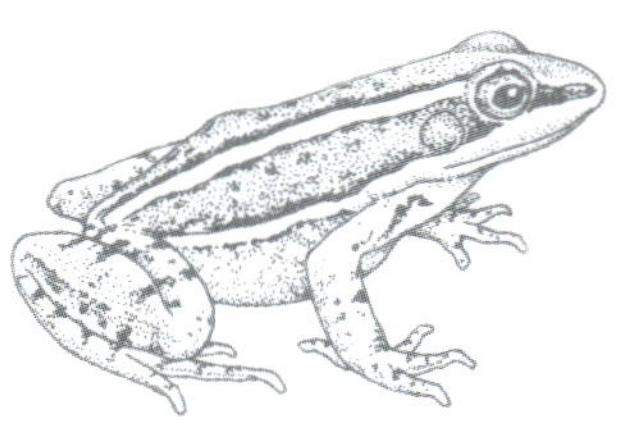

HYLARANA GALAMENSIS

GALAM WHITE-LIPPED FROG

(DUMÉRIL & BIBRON, 1841)

This large frog, named after a lake in Senegal, is an important source of food and traditional medicines in parts of West Africa. An aquatic species, it is commonly seen on the banks of rivers and ponds. Males begin to call with the first rains, but mating and spawning often do not occur until several weeks or months later. The call, described as a "nasal bleat," is produced at night, from hidden places near water. The female lays between 1,500 and 4,000 eggs that float on the water's surface.

SIMILAR SPECIES

The Nkongsamba Frog (*Hylarana asperrima*) is listed as Vulnerable. It breeds in fast-flowing streams in the forests of Cameroon and Nigeria, and its habitat is being destroyed by deforestation. Darling's White-lipped Frog (*H. darlingi*) is a common frog, living in a wide variety of habitats in Angola, Zambia, Zimbabwe, and Mozambique.

The Galam White-lipped Frog has a plump body, a pointed snout, and relatively short hind legs. The eyes are large and the tympana are conspicuous. There is a flat yellow ridge running from the eye to the rear of the body. The upper surfaces are pale or dark brown, with irregular dark blotches on the back and legs. The upper lip is white.

Actual size

FAMILY	Ranidae
OTHER NAMES	*Sylvirana guentheri*, Günther's Amoy Frog
DISTRIBUTION	Southern China, including Macau and Hong Kong; Vietnam. Introduced to Guam
ADULT HABITAT	Pools, marshes, ditches, and rice fields, up to 3,600 ft (1,100 m) altitude
LARVAL HABITAT	Ponds, rice fields, and slow-moving streams
CONSERVATION STATUS	IUCN Least Concern. Its population is generally stable, but is declining in China

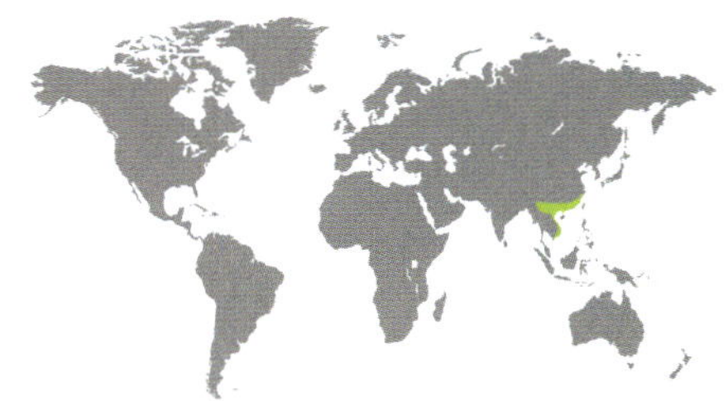

HYLARANA GUENTHERI

GÜNTHER'S FROG

(BOULENGER, 1882)

ADULT LENGTH
Male
2½–2¾ in (63–68 mm)

Female
2$^{15}/_{16}$–3 in (75–76 mm)

This athletic frog breeds in May and June, males producing a low-pitched croak and females laying 2,000–3,000 eggs. While males live for 1–4 years, the larger females live for 2–6 years. The species is widely collected for food in China and this has contributed to its decline in that country. Eating it entails some risk, because it is one of many intermediate hosts for a nematode worm that causes Angiostrongyliasis, a disease that affects the central nervous system of humans. The frog is of interest to the pharmaceutical industry because its skin contains antimicrobial compounds.

SIMILAR SPECIES

The Similar Frog (*Hylarana attigua*) is a stream-breeding species, known from only four localities in Laos and Vietnam. Its forest habitat is being destroyed but it is listed as Least Concern. The Guangdong Frog (*H. macrodactyla*) is a common species found in a variety of wet habitats in Myanmar, Thailand, Cambodia, Laos, Vietnam, and southern China.

Günther's Frog has a long, pointed snout, a flattened head, and long hind legs. The eyes are very large and the tympana are conspicuous. There is a prominent fold of skin running from the eye along the length of the body. The upper surfaces are brown or yellow-brown, with dark blotches on the back and dark stripes on the legs. There is a pale stripe along the upper lip.

Actual size

FAMILY	Ranidae
OTHER NAMES	*Pulchrana picturata*
DISTRIBUTION	Borneo, Sumatra, Indonesia, possibly also Peninsular Malaysia
ADULT HABITAT	Rainforest up to 3,300 ft (1,000 m) altitude
LARVAL HABITAT	In quiet pools beside streams
CONSERVATION STATUS	IUCN Least Concern. Its population is declining due to deforestation, but it occurs in some protected areas

ADULT LENGTH
Male
$1\frac{5}{16}$–$1\frac{7}{8}$ in (33–47 mm)
Female
$1\frac{15}{16}$–$2\frac{3}{4}$ in (49–68 mm)

HYLARANA PICTURATA

SPOTTED STREAM FROG

(BOULENGER, 1920)

This colorful frog is found among leaf litter along the banks of the small lowland streams where it breeds. Males call from elevated positions beside the water. Solitary males can be heard on most nights but, occasionally, large choruses gather to call together. The tadpoles develop in quiet side-pools beside streams that contain leaf litter. They hide by day beneath the leaves. The skin of this species has been found to have antimicrobial properties. The frog has tested positive for the fungus that causes the disease chytridiomycosis but it is not known if it is susceptible to infection.

SIMILAR SPECIES

Hylarana picturata is very similar to, and has been confused with, the Striped Stream Frog (*H. signata*). The latter has very similar habits but differs in appearance, having a continuous narrow stripe along each side of the body from the tip of the snout to the groin. It is found on the Malay Peninsula and in Sumatra and Borneo.

Actual size

The Spotted Stream Frog has a triangular head, large eyes, and long, slender limbs. The fingers and toes are long and have swollen tips. The upper surfaces are brown or black with a complex pattern of green, yellow-orange, or red spots that may coalesce into a mottled pattern. The underside is pale gray with white markings.

FAMILY	Ranidae
OTHER NAMES	Previously *Rana areolata*, Hoosier Frog
DISTRIBUTION	Central USA
ADULT HABITAT	Prairie, grassland, pine forests, woodland, and river floodplains
LARVAL HABITAT	Ephemeral ponds and flooded areas
CONSERVATION STATUS	IUCN Least Concern. Has declined as a result of draining of wetlands and the introduction of predatory fish

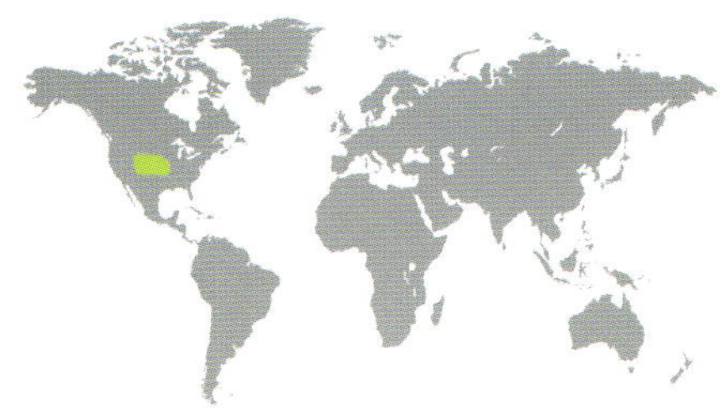

LITHOBATES AREOLATUS

CRAWFISH FROG

(BAIRD & GIRARD, 1852)

ADULT LENGTH
Male
2⅜–4⅜ in (61–112 mm)
Female
$2^{13}/_{16}$–4¾ in (70–121 mm)

This large frog gets its common name from the fact that it lives, for more than ten months of the year, in crayfish burrows and other holes in the ground. Individuals are strongly attached to their own burrows, situated in upland sites, returning to them after making a hazardous migration, of up to 0.75 mile (1.2 km), to a breeding pond and back again. This strong site fidelity has made it possible for biologists to observe their behavior, by photography and radiotelemetry, to an extent impossible with other frogs. This has revealed a remarkable ability among Crawfish Frogs to find their way to and from their home each year.

SIMILAR SPECIES

Lithobates areolatus is closely related to the Carolina Gopher Frog (*L. capito*; page 558) and the Dusky Gopher Frog (*L. sevosus*). All are "explosive breeders," gathering in large choruses after heavy rain. While a chorus of Crawfish Frogs has been likened to "a sty full of hogs at feeding time"(Roger Conant), that of gopher frogs is like a deep, rattling snore. The Dusky Gopher Frog is listed as Critically Endangered.

Actual size

The Crawfish Frog has a chunky, toad-like body shape, but, unlike a toad, its skin is smooth and is not warty. The eyes are large and the tympana are clearly visible. The upper surfaces are gray or greenish, decorated with large, pale-bordered dark spots. The belly is white and is not spotted.

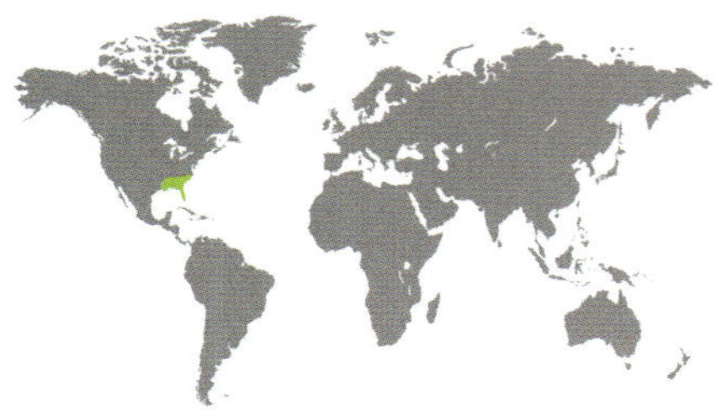

FAMILY	Ranidae
OTHER NAMES	Florida Gopher Frog
DISTRIBUTION	Southeastern USA
ADULT HABITAT	Pinewoods and old fields on sandy soils, close to ephemeral and permanent ponds
LARVAL HABITAT	Ephemeral or permanent ponds without fish
CONSERVATION STATUS	IUCN Vulnerable. Has declined over much of its range due to destruction of its habitat

ADULT LENGTH
$2\frac{1}{2}$–$3\frac{15}{16}$ in (63–102 mm)

LITHOBATES CAPITO

CAROLINA GOPHER FROG

(LECONTE, 1855)

This toad-like frog gets its common name from living in the burrows of other animals, particularly the Gopher Tortoise (*Gopherus polyphemus*). It hides underground by day, emerging at night to feed close to its burrow. In late winter or early spring, heavy rain triggers a brief breeding period, individuals migrating to ponds that do not contain predatory fish. The male's mating call is a deep, rattling snore. Females lay clumps of 1,000–2,000 eggs attached to the stems of emergent plants. The Florida Gopher Frog is regarded as a distinct subspecies, *Lithobates capito aesopus*.

The Carolina Gopher Frog has a short, plump body, a rounded snout, and short legs. Its skin is warty, like a toad's. There are prominent folds of skin along each side of the back, which may be brown or bronze in color. The back is cream, gray, or dark gray with many round black spots.

SIMILAR SPECIES

The Dusky Gopher Frog (*Lithobates sevosus*) was formerly regarded as a subspecies of *L. capito*, but genetic studies have shown that it is a distinct species. It is darker in color than *L. capito*, and is sometimes almost black. Destruction of its habitat has reduced its numbers to a single population living in a single pond; there may be as few as 100 individuals still alive.

Actual size

FAMILY	Ranidae
OTHER NAMES	Previously *Rana kauffeldi*
DISTRIBUTION	New York City and the surrounding metropolitan area, USA
ADULT HABITAT	Freshwater habitats of all kinds
LARVAL HABITAT	Shallow pools
CONSERVATION STATUS	IUCN Least Concern. Has a very restricted range within an urban area

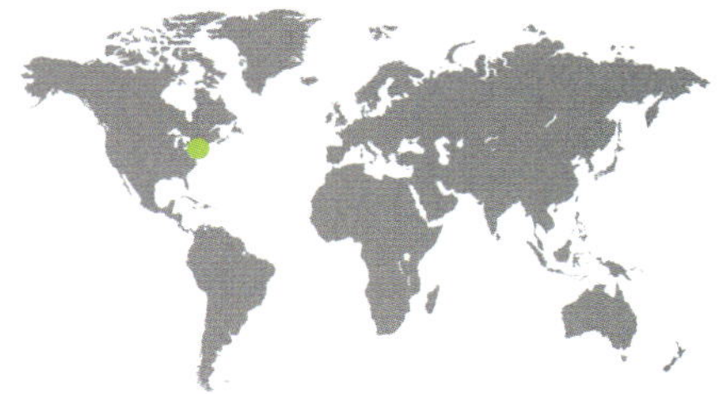

ADULT LENGTH
¾–3$\frac{5}{16}$ in (20–85 mm); males are generally smaller

LITHOBATES KAUFFELDI

ATLANTIC COAST LEOPARD FROG

(FEINBERG, NEWMAN, WATKINS-COLWELL, SCHLESINGER, ZARATE, CURRY, SHAFFER & BURGER, 2014)

The announcement of this new frog species in 2014 aroused considerable media attention because it was found in one of the world's most heavily populated areas. It is an example of a "cryptic species," meaning that it had previously been incorrectly included within one or more other species. In this instance, this new frog was included with the Southern Leopard Frog (*Lithobates sphenocephala*; page 562). It migrates to breeding sites in February and calling is most intense in March. Males gather, in groups of five or more, in shallow pools and call at night. Females are attracted to these choruses and lay their eggs in clumps, close to those of other females.

SIMILAR SPECIES

Lithobates kauffeldi is very similar in appearance to the Northern Leopard Frog (*L. pipiens*; page 561) and the Southern Leopard Frog (*L. sphenocephalus*). The three species can be differentiated by their calls: *L. kauffeldi* produces a single-note "chuck;" *L. sphenocephalus* utters a multi-note "ak… ak… ak;" and the call of *L. pipiens* is a prolonged snore. The three species are also quite distinct genetically. The genus *Lithobates* contains 47 species.

The Atlantic Coast Leopard Frog is a slim, athletic frog with well-defined folds of skin along each side of the back, and long, muscular legs. The back is brown, gray, or green in color, with large, dark spots. The underside is white or cream, and there is a faint, pale spot in the center of the tympanum.

Actual size

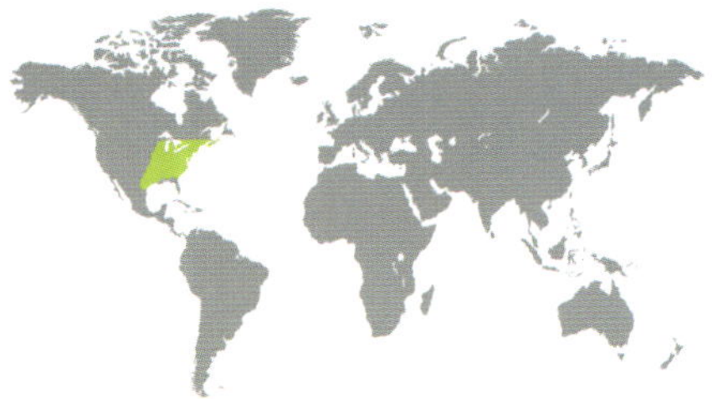

FAMILY	Ranidae
OTHER NAMES	Previously *Rana palustris*, Yellow-legged Frog
DISTRIBUTION	Eastern USA, parts of eastern Canada
ADULT HABITAT	In or near permanent shallow waterbodies
LARVAL HABITAT	Woodland ponds and pools
CONSERVATION STATUS	IUCN Least Concern. Shows no evidence of general decline across its large range

ADULT LENGTH
Male
1¹¹⁄₁₆–2³⁄₁₆ in (43–57 mm)
Female
2⅛–3¹⁄₁₆ in (54–78 mm)

LITHOBATES PALUSTRIS

PICKEREL FROG

(LECONTE, 1825)

Actual size

Unusual among *Rana* species in having toxic skin secretions, the Pickerel Frog has warning coloration in the form of a yellow or orange patch on the inside of its legs and belly. Migration to breeding sites occurs following a rise in temperature in early spring and thus takes place much earlier in the south than in the north of its range. Males call from beneath the water's surface, producing a soft, grating growl. Females lay 2,000–3,000 eggs in a submerged globular mass. Outside the breeding season, the species is often found in caves.

SIMILAR SPECIES

The Pickerel Frog is easily confused with the Atlantic Coast Leopard Frog (*Lithobates kauffeldi*; page 559), the Northern Leopard Frog (*L. pipiens*; page 561), and the Southern Leopard Frog (*L. sphenocephalus*; page 562), but the large blotches on its back tend to be square in shape, rather than round or oval. It also differs from those species in having toxic skin secretions.

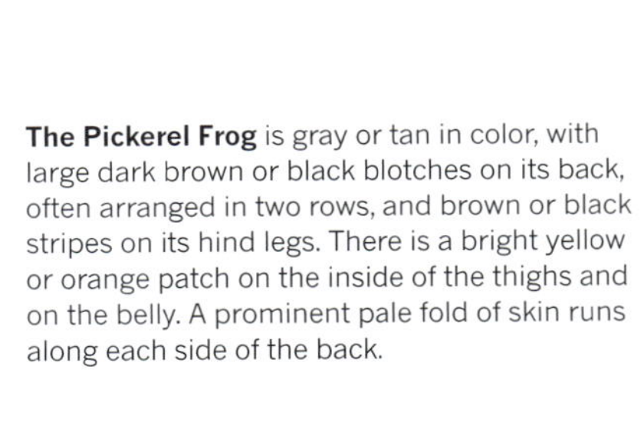

The Pickerel Frog is gray or tan in color, with large dark brown or black blotches on its back, often arranged in two rows, and brown or black stripes on its hind legs. There is a bright yellow or orange patch on the inside of the thighs and on the belly. A prominent pale fold of skin runs along each side of the back.

FAMILY	Ranidae
OTHER NAMES	Previously *Rana pipiens*
DISTRIBUTION	Canada, USA
ADULT HABITAT	Grassland, brushland, forest
LARVAL HABITAT	Permanent ponds and slow-moving streams
CONSERVATION STATUS	IUCN Least Concern. Has a very large range but has declined dramatically in many areas, especially in western USA

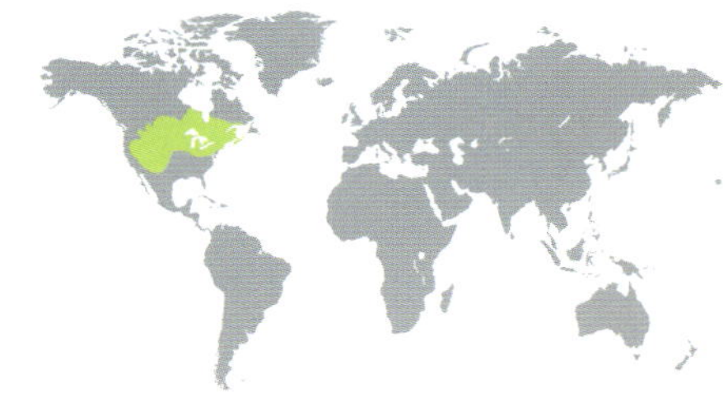

ADULT LENGTH
1 15/16–4 5/16 in (50–110 mm)
males are generally smaller

LITHOBATES PIPIENS

NORTHERN LEOPARD FROG

(SCHREBER, 1782)

Although it has a huge range and is common in much of eastern USA, there is cause for concern over the decline of this species in western North America. This is partly due to habitat loss and the introduction of North American Bullfrogs (*A. catesbeiana*; page 545) to the region, but more subtle factors are also involved. Atrazine is the most widely used herbicide in North America and it disrupts the reproductive development of leopard frogs in agricultural areas. Many males have become hermaphrodites, containing ovaries as well as testes. Nitrate fertilizers impair the growth and development of tadpoles. The cause of severe deformities that are quite common among young leopard frogs is not known.

The Northern Leopard Frog is a slim, athletic frog with well-defined folds of skin along each side of the back, and long, muscular legs. The back is green or brown in color, with large, pale-bordered dark spots. The underside is white or cream, and there is a white stripe along the upper jaw.

SIMILAR SPECIES

The Southern Leopard Frog (*Lithobates sphenocephalus*; page 562) occurs across much of southern USA. Less widely distributed is the Plains Leopard Frog (*L. blairi*), an inhabitant of the prairies and grasslands of central USA. The Relict Leopard Frog (*L. onca*) was thought to be extinct but persists in a few locations in Nevada, Arizona, and Utah. It is listed as Endangered.

Actual size

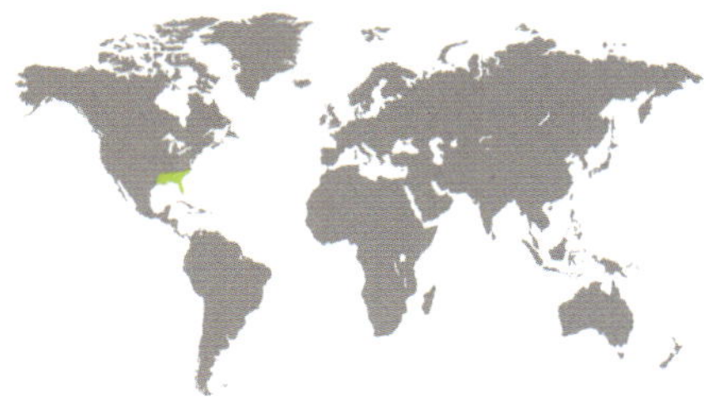

FAMILY	Ranidae
OTHER NAMES	Previously *Rana sphenocephala*
DISTRIBUTION	Southeastern USA
ADULT HABITAT	Freshwater habitats of all kinds
LARVAL HABITAT	Shallow pools
CONSERVATION STATUS	IUCN Least Concern. Shows no evidence of a general decline across its large range

ADULT LENGTH
1 15/16–3½ in (50–90 mm)
males are generally smaller

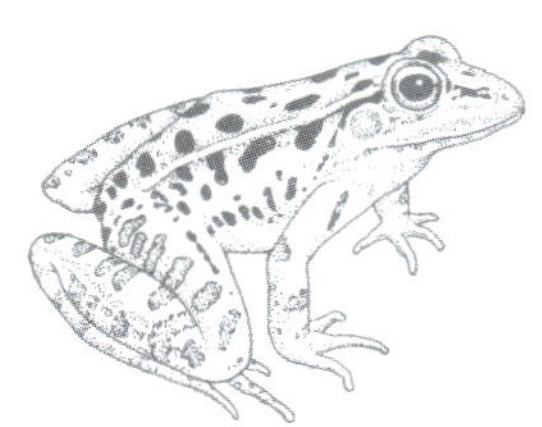

LITHOBATES SPHENOCEPHALUS

SOUTHERN LEOPARD FROG

(COPE, 1889)

Often found in grassland some distance from water, this leopard frog breeds in spring in the north of its range, and at any time of year after heavy rain in the south. Males have two vocal sacs and call while floating in water or while perched on floating vegetation. The call is a series of harsh, guttural grunts that are produced at a faster rate at higher temperatures. The eggs are eaten by crayfish, but the developing embryos have the ability to detect the predators and hatch earlier when they are present. When disturbed, leopard frogs escape by means of a series of zigzag leaps.

The Southern Leopard Frog is a slim, athletic frog with well-defined folds of skin along each side of its back, and long, muscular legs. The back is brown, gray, or green in color, with large, dark spots. The underside is white or cream, and there is a pale spot in the center of the tympanum.

SIMILAR SPECIES

Lithobates sphenocephalus is typically smaller than its northern counterpart, the Northern Leopard Frog (*L. pipiens*; page 561). It shares part of its range with the Rio Grande Leopard Frog (*L. berlandieri*), a paler frog that breeds earlier in the year and is found in southern USA, Mexico, and Central America. The Lowland Leopard Frog (*L. yavapaiensis*) is found in near-desert areas of Arizona and Mexico.

Actual size

FAMILY	Ranidae
OTHER NAMES	Kiau Borneo Frog, Kinabalu Torrent Frog
DISTRIBUTION	Northeastern Borneo
ADULT HABITAT	Montane forest
LARVAL HABITAT	Small, clear, rocky streams
CONSERVATION STATUS	IUCN Least Concern. Has a restricted range and is threatened by deforestation, but occurs within some protected areas

ADULT LENGTH
Male
2¼–2¾ in (58–68 mm)
Female
2¹⁵⁄₁₆–3⅝ in (75–93 mm)

MERISTOGENYS KINABALUENSIS

MONTANE TORRENT FROG

(INGER, 1966)

The tadpoles of this green-eyed frog are adapted for life in fast-flowing mountain streams. They have a larger sucker on their abdomen for sticking themselves to rocks and very muscular tails, enabling them to swim against the current. The horny beaks around their mouths are used to scrape algae off rocks. The tadpoles can grow up to 2⅜ in (60 mm) long. Adults and juvenile frogs are often found wandering in the forest some way from streams. They feed on insects, centipedes, and scorpions. Males call from rocks or vegetation close to streams, and the eggs are attached to submerged rocks.

SIMILAR SPECIES

To date, 13 species have been described in the genus *Meristogenys*, several of them in the last few years. Known as torrent frogs, all are found only in Borneo, breed in mountain streams, and have similar tadpoles. The Western Torrent Frog (*M. jerboa*) is found at only one location, within a protected area, and is listed as Vulnerable. The Northern Torrent Frog (*M. orphnocnemis*) is a small species and is listed as of Least Concern.

The Montane Torrent Frog has a blunt snout, large eyes, long hind legs, and fully webbed feet. The long fingers and toes end in triangular pads. The upper surfaces are olive-green with reddish-brown patches, the flanks are green, and the belly is yellow. The limbs are brown with dark cross-stripes. The irises of the eyes are bright green.

Actual size

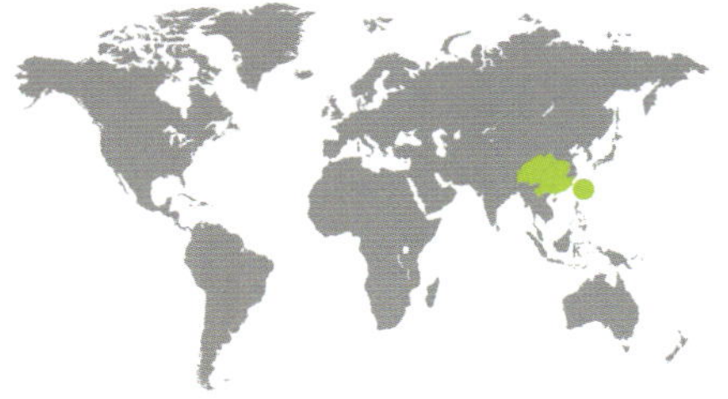

FAMILY	Ranidae
OTHER NAMES	Olive Frog, Fuhacho Frog
DISTRIBUTION	Central and southern China, Taiwan
ADULT HABITAT	Marshes, rice fields, ditches, and ponds, up to 5,900 ft (1,800 m) altitude
LARVAL HABITAT	Ponds and pools
CONSERVATION STATUS	IUCN Least Concern. Has a large range, its population is stable, and it occurs in some protected areas

ADULT LENGTH
Male
$1\frac{5}{8}$–$2\frac{1}{4}$ in (42–59 mm)
Female
$1\frac{7}{8}$–$2\frac{5}{8}$ in (47–65 mm)

NIDIRANA ADENOPLEURA

EAST CHINA MUSIC FROG

(BOULENGER, 1909)

In the great majority of frogs the act of mating is preceded by a period of amplexus, in which the male mounts the female's back and clasps her firmly. The duration of amplexus varies enormously among species: In the European Common Toads (*Bufo bufo*; page 157) it can last for two or three days; in the Harlequin Frog (*Atelopus varius*; page 152) it can last for a month. In this Chinese frog, amplexus lasts only 11 minutes on average, of which three minutes are devoted to egg-laying. Intrusions by rival males, attempting to displace amplectant males, are common in this species and abbreviated amplexus may have evolved to reduce the risk of such displacement.

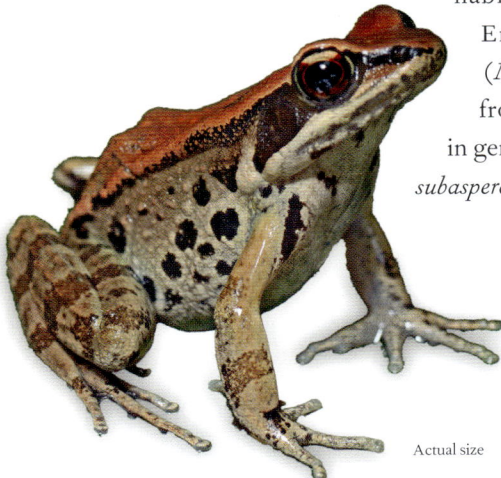

The East China Music Frog has a pointed snout, large eyes, a conspicuous tympanum, and powerful hind limbs. There is a distinct fold of skin running along each side of the back. The upper surfaces are green or brown, with brown or black spots on the back, and there is a broad, dark stripe through the eye. There are dark stripes on the legs.

SIMILAR SPECIES

The Ryukyu Brown Frog (*Nidirana okinavana*) is a small, stocky toad, found in the smaller southern islands of Japan. It lays its eggs in holes constructed in wet mud next to ponds. The tadpoles require rain to wash them into open water. Its habitat is being destroyed and it is listed as Endangered. The Yunkwei Plateau Frog (*N. pleuraden*) is a common but declining toad from China. Many *Nidirana* species used to be in genus *Babina*. See also the Otton Frog (*Babina subaspera*; page 547).

Actual size

FAMILY	Ranidae
OTHER NAMES	Anhui Sucker Frog
DISTRIBUTION	Zhejiang province, eastern China
ADULT HABITAT	Wooded mountains and hills
LARVAL HABITAT	Fast-flowing streams
CONSERVATION STATUS	IUCN Least Concern. Its range has been fragmented by deforestation, and it is now known in only five localities

ADULT LENGTH
Male
average 1¼ in (32 mm)
Female
average 2³⁄₁₆ in (56 mm)

ODORRANA TORMOTA

CONCAVE-EARED TORRENT FROG

(WU, 1977)

Frogs living and breeding in noisy, fast-flowing streams have a problem: They cannot be heard. Some males have evolved visual signals to attract females, as in the Kalakkad Torrent Frog (*Micrixalus fuscus*; page 528), while others, like this species and the Hole-in-the-head Frog (*Huia cavitympanum*; page 551) use very high-frequency ultrasound that can be heard above the sound of rushing water. The male produces a complex birdlike call that contains high-frequency components and it is likely that he can differentiate between familiar neighbors and unfamiliar intruders. Unusually among frogs, females also call, and the high-frequency elements in their calls enable males to locate them very accurately.

Actual size

The Concave-eared Torrent Frog has a prominent, largely black fold of skin running along each side of the back from eye to groin. The ear of the male is a deep hollow, the tympanum being recessed inside the head. The back is tan in color, often with black dots, and there is a white stripe along the upper lip.

SIMILAR SPECIES

There are 66 species in the genus *Odorrana*, known as torrent frogs or odorous frogs, all found in eastern Asia. The Large Odorous Frog (*O. graminea*), from China, has high frequencies in its call, but it does not have recessed ears like those of *O. tormota*. The Amami Tip-nosed Frog (*O. amamiensis*), from Japan, is listed as Endangered, much of its natural habitat having been destroyed.

FAMILY	Ranidae
OTHER NAMES	Iberian Green Frog, Perez's Frog
DISTRIBUTION	Spain, Portugal, southern France. Introduced to the Canary Islands, Madeira, and the Balearic Islands
ADULT HABITAT	In and close to ponds and slow-moving rivers
LARVAL HABITAT	Ponds and slow-moving rivers
CONSERVATION STATUS	IUCN Least Concern. Declining in some parts of its range where agrochemicals are used

ADULT LENGTH
Male
1⅜–2¹³⁄₁₆ in (35–70 mm)
Female
1¾–3⁵⁄₁₆ in (45–85 mm)

PELOPHYLAX PEREZI

IBERIAN WATER FROG

(LÓPEZ-SEOANE, 1885)

As its common name implies, this frog spends most of its life in water, remaining there through the winter. Breeding occurs in spring, when males develop horny nuptial pads on their thumbs, enabling them to clasp females securely. Their call has been likened to a deep growl or a laugh. Depending on their size, females lay 800–10,000 eggs in a large clump. Europe's water frogs are remarkable in that some species hybridize to produce quite distinct and recognizable forms, called hybridogenetic species. In southern France, *Pelophylax perezi* breeds with the Marsh Frog (*P. ridibundus*; page 567) to produce Graf's Frog (*P. grafi*).

The Iberian Water Frog has large, prominent eyes, placed close together on the top of the head, enabling it to scan its environment while floating in water. There are conspicuous skin folds running along each side of the back. Its color is very variable, being green, brown, gray, or yellow with dark blotches.

SIMILAR SPECIES

To date, 13 species of *Pelophylax*, known as water frogs or green frogs, have been described, from Europe, Asia, North Africa, and the Middle East. Some Mediterranean islands have their own endemic water frogs. For example, the Cretan Frog (*P. cretensis*) is an Endangered species found only on Crete, but the Endangered Karpathos Water Frog (*P. cerigensis*), from the Greek island of Karpathos, is now a subspecies of the widespread Marsh Frog (*P. ridibundus*) and the Sahara Frog (*P. saharicus*; page 568).

Actual size

FAMILY	Ranidae
OTHER NAMES	Lake Frog, Laughing Frog
DISTRIBUTION	From western Europe to Central Asia
ADULT AND LARVAL HABITAT	Still and flowing water
CONSERVATION STATUS	IUCN Least Concern. Tolerates high levels of pollution, thrives in artificial habitats, and is expanding its range in many places

PELOPHYLAX RIDIBUNDUS

MARSH FROG

(PALLAS, 1771)

ADULT LENGTH
Male
up to 4¾ in (120 mm)

Female
up to 6¹¹⁄₁₆ in (170 mm)

Europe's largest native frog, the Marsh Frog lives in water but feeds mostly on land, eating insects and other invertebrates. It basks in the sun on the banks of ponds, lakes, streams, and rivers, leaping into the water with a resounding "plop" when disturbed. Breeding occurs in spring, males defending large territories by calling while floating in the water. They have a vocal sac on each side of the mouth and have a large repertoire of calls. Receptive females have a call of their own, which they produce when approaching a territorial male. Depending on their size, females lay up to 16,000 eggs.

The Marsh Frog has very long legs and granular, often warty skin. There is a prominent ridge along each side of the back and the hind feet are webbed. It is variable in color, being brown or gray, usually with patches of green or yellow and with many large, dark spots. There is often a pale green stripe down the middle of the back.

SIMILAR SPECIES

Pelophylax ridibundus shares much of its range with the Pool Frog (*P. lessonae*), a species of similar habits. In many places the two species hybridize to produce the hybridogenetic species, the Edible Frog (*P. esculentus*). In most instances, Edible Frogs cannot breed successfully with one another and are thus dependent on the presence of their parent species if they are to persist. See also the Iberian Water Frog (*P. perezi*; page 566) and the Sahara Frog (*P. saharicus*; page 568).

Actual size

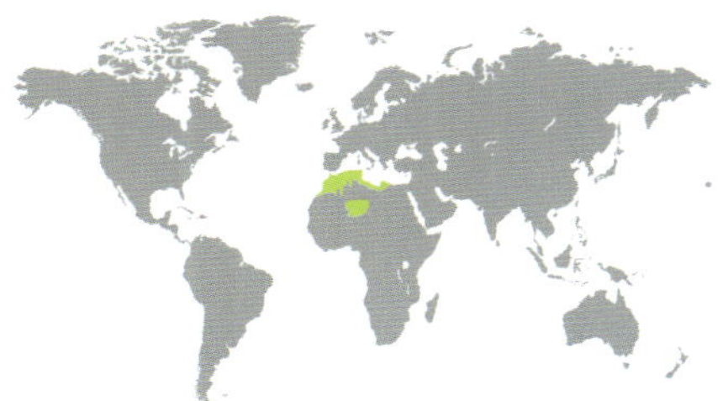

FAMILY	Ranidae
OTHER NAMES	Moroccan Green Frog, North African Green Frog
DISTRIBUTION	North Africa, from Morocco in the west to western Egypt in the east; also southern Algeria
ADULT HABITAT	In and around ponds, streams, irrigation ditches, and reservoirs, up to 8,760 ft (2,670 m) altitude
LARVAL HABITAT	Ponds, streams, irrigation ditches, reservoirs
CONSERVATION STATUS	IUCN Least Concern. Has an extensive range and there is no evidence of a general population decline

ADULT LENGTH
Male
1$\frac{9}{16}$–2$\frac{15}{16}$ in (40–75 mm)
Female
1$\frac{9}{16}$–4$\frac{1}{8}$ in (40–105 mm)

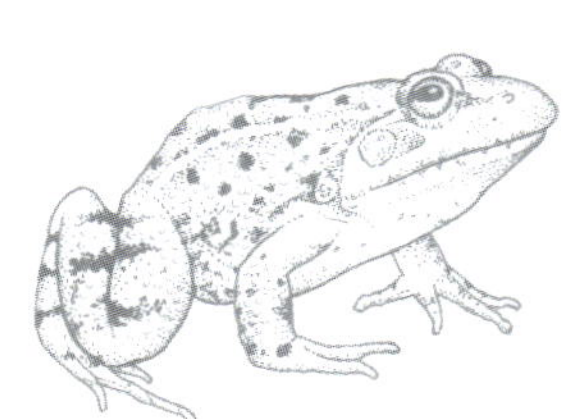

PELOPHYLAX SAHARICUS

SAHARA FROG

(BOULENGER, 1913)

The Sahara Desert is an unlikely place to find a frog, but this close relative of Europe's water frogs thrives in the Maghreb region of North Africa and also occurs in some isolated oases in the Sahara itself. It is unusual in that it is active year-round, appearing not to have a "resting period" as do similar frogs. In many places it is dependent on waterbodies, such as ditches and reservoirs, that humans have created to collect and store scarce water supplies. It lives mostly in the water, emerging onto land to feed.

The Sahara Frog has a pointed snout and eyes positioned close together on the top of its head. There are prominent ridges of skin along each side of the back, and the toes are webbed. It is most commonly green with large brown blotches on its body and brown stripes on its legs.

SIMILAR SPECIES

Pelophylax saharicus is very similar to the Iberian Water Frog (*P. perezi*; page 566) but is genetically distinct. Exportation for food of the Epirus Water Frog (*P. epeiroticus*) from Albania and Greece has contributed to its being listed as Near Threatened. See also the Marsh Frog (*P. ridibundus*; page 567).

Actual size

FAMILY	Ranidae
OTHER NAMES	Swedish Swamp Frog
DISTRIBUTION	Central and eastern Europe, Sweden, Finland, much of Russia
ADULT HABITAT	Tundra, woodland, steppe
LARVAL HABITAT	Still waterbodies, from small ponds to large lakes
CONSERVATION STATUS	IUCN Least Concern. Has a very large range, is very abundant in many places, and shows no evidence of general decline

ADULT LENGTH
$2\frac{1}{8}$–$2\frac{13}{16}$ in (55–70 mm)

RANA ARVALIS

MOOR FROG

NILSSON, 1842

This small brown frog is remarkable for the fact that, for just a few days in the breeding season, males turn bright, pale blue. It has been suggested that females prefer the most brightly colored males but there is little evidence for this. It is more likely that the blue color simply enables males to distinguish females from other males during their hectic competition for mates. Mating in this species is "explosive," occurring over just a few days, and often involves large numbers of individuals scrambling to find partners. It takes place from March to June, depending on latitude, and females lay 500–3,000 eggs.

SIMILAR SPECIES

Rana arvalis is one of many species of so-called brown frogs, found across Europe and Asia, all rather similar in appearance and natural history. This frog is very similar to the European Common Frog (*R. temporaria*; page 571) and also the Agile Frog (*R. dalmatina*; page 570), but it has shorter legs and a more pointed snout. Even after transfer of many species to the genera *Lithobates*, *Pelophylax*, *Amerana*, etc. the genus *Rana* still contains 52 species.

The Moor Frog has a pointed snout, large eyes, smooth skin, and small ridges along each side of its back. Generally reddish brown in color, it has a conspicuous dark "bandit" stripe from nose tip to tympanum. The underside is white and is not spotted. The hind legs are rather short and have dark brown transverse stripes.

Actual size

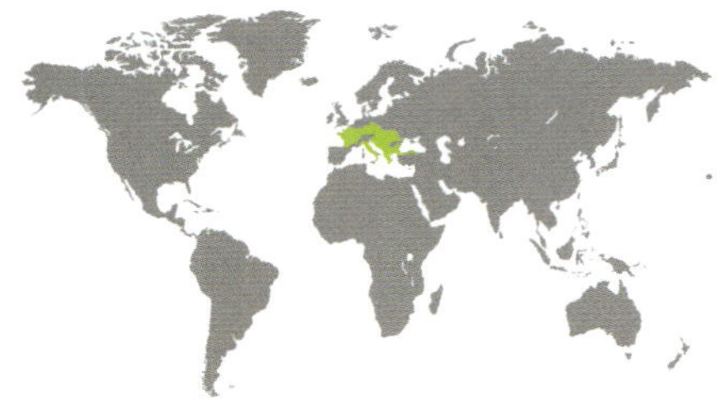

FAMILY	Ranidae
OTHER NAMES	None
DISTRIBUTION	Central and southern Europe, with a few isolated populations in northern Europe
ADULT HABITAT	Open deciduous woodland, swampy meadows
LARVAL HABITAT	Ponds and swamps in woodland and at forest edges
CONSERVATION STATUS	IUCN Least Concern. Has declined in parts of its range due to destruction of its habitat

ADULT LENGTH

Male
$1\frac{9}{16}$–$2\frac{5}{8}$ in (40–65 mm)

Female
$1\frac{5}{8}$–$3\frac{1}{2}$ in (42–90 mm)

RANA DALMATINA

AGILE FROG

FITZINGER, 1839

This member of a group of European frogs known as brown frogs gets its common name from its ability to leap long distances when escaping a threat. It breeds very early in the spring, the male calling "quar… quar… quar" to attract females. Females lay clutches of 450–1,800 eggs. As in many other frogs, it appears that females are usually monogamous, with all of their eggs being fertilized by one male. However, genetic studies have revealed that around 18 percent of clutches are fathered by two males.

SIMILAR SPECIES

Rana dalmatina shares much of its range with the European Common Frog (*R. temporaria*; page 571), but typically breeds earlier in the spring. The Italian Agile Frog (*R. latastei*) lives in lowland areas of northern Italy and adjoining countries. Its call is like the mewing of a cat, and it is listed as Vulnerable.

Actual size

The Agile Frog has a pointed snout and tympana that are very close to its eyes. It has exceptionally long hind limbs with webbed toes. It is pale brown in color with yellow flanks; there are dark brown spots on the back and dark brown stripes on the legs. It has a dark "bandit mask" on the side of the head, running through the eye.

FAMILY	Ranidae
OTHER NAMES	Brown Frog, Grass Frog
DISTRIBUTION	Much of Europe, including Scandinavia and western Russia. Introduced to Ireland
ADULT HABITAT	Damp places in habitats of all kinds
LARVAL HABITAT	Standing water, from lakes to ponds, ditches, and wheel ruts
CONSERVATION STATUS	IUCN Least Concern. Shows no evidence of a general decline across its very large range

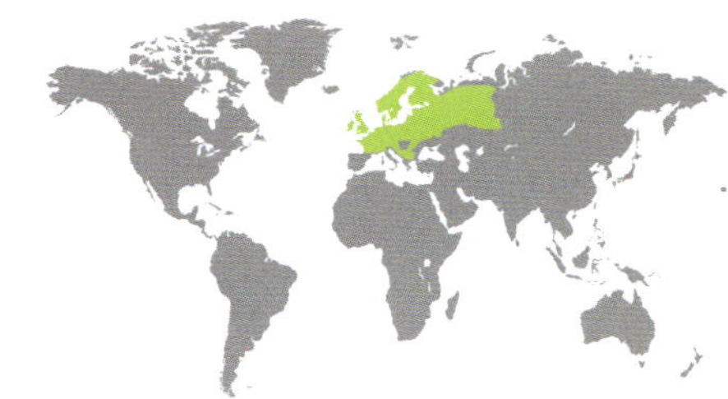

ADULT LENGTH
$1\frac{15}{16}$–$4\frac{5}{16}$ in (50–110 mm); males are generally smaller

RANA TEMPORARIA

EUROPEAN COMMON FROG

LINNAEUS, 1758

Common frogs migrate, often in large numbers, to their breeding ponds in early spring. Males arrive first and call from the water, sometimes while submerged, producing a deep growl. They have thicker, more muscular arms than females and large black nuptial pads on their thumbs. These enable them to clasp females in an amplexus so firm that they are hard to separate, and the female is left with two open wounds on her chest. Females produce 1,000–4,000 eggs, very few of which survive to adulthood. As tadpoles, the young are eaten by newts, fish, and insect larvae, and as adults, the frogs are a favorite food of herons.

The European Common Frog has long, powerful hind legs. It is very variable in color, and while usually green or brown, it may be red, yellow, or blue. There are large black spots on its body and cross-bands on its legs. The longitudinal folds of skin on its back are closer together than in most *Rana* frogs.

SIMILAR SPECIES

Rana temporaria is one of several so-called brown frogs found across Europe. These include the Moor Frog (*R. arvalis*; page 569) and the Agile Frog (*R. dalmatina*; page 570). The Pyrenean Frog (*R. pyrenaica*), which is confined to the mountains of southern France and northern Spain, is about half the size of *R. temporaria*. It has declined due to habitat loss, pollution, and the introduction of trout, and is consequently listed as Endangered.

Actual size

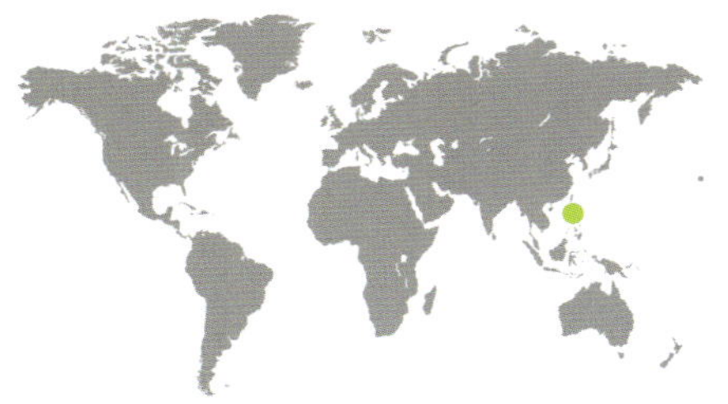

FAMILY	Ranidae
OTHER NAMES	None
DISTRIBUTION	Luzon Island, Philippines
ADULT HABITAT	In and around streams in montane forest
LARVAL HABITAT	Presumed to be mountain streams
CONSERVATION STATUS	IUCN Vulnerable

ADULT LENGTH

Male
$1\frac{7}{8}$–$2\frac{1}{16}$ in (47–53 mm)

Female
$2\frac{1}{4}$–$2\frac{13}{16}$ in (59–71 mm)

SANGUIRANA AURANTIPUNCTATA

ORANGE-SPOTTED STREAM FROG

FUITEN, WELTON, DIESMOS, BARLEY, OBERHEIDE, DUYA, RICO & BROWN, 2011

In the last 20 years, our knowledge of the world's amphibians has undergone a revolution. By 2002, some 5,400 species had been described and named. By mid-2014, however, this figure was approaching 7,300 and is being added to every week. In part this is due to the careful exploration of parts of the world whose biodiversity was previously little known. This remarkably pretty frog is an example of this, and was unknown to science until as recently as 2011. A nocturnal animal, it is found on rocks and vegetation close to fast-flowing streams, where it is presumed to lay its eggs.

SIMILAR SPECIES

There are eight species in the genus *Sanguirana*, most of which are endemic to the Philippines, although the Calamianes Frog (*Sanguirana sanguinea*) also occurs in Eastern Indonesia. Alcala's Sierra Madre Frog (*S. tipanan*) and the Balbalan Frog (*S. igorata*) are both declining due to the destruction of their forest habitat to make way for agriculture and human dwellings. Both are listed as Vulnerable.

The Orange-spotted Stream Frog has a slender body, large, protuberant eyes, and slender hind legs with webbed feet. Its hands are particularly large, as are the adhesive disks at the end of its fingers. The upper surfaces are bright yellowish green, and while the male is decorated with small gray, purple, or orange spots, females have large flower-shaped orange spots as shown here.

Actual size

FAMILY	Ranidae
OTHER NAMES	Black-spotted Rock Frog
DISTRIBUTION	Borneo, Philippines
ADULT HABITAT	Forest, up to 4,600 ft (1,400 m) altitude
LARVAL HABITAT	Pools in fast-flowing streams
CONSERVATION STATUS	IUCN Least Concern

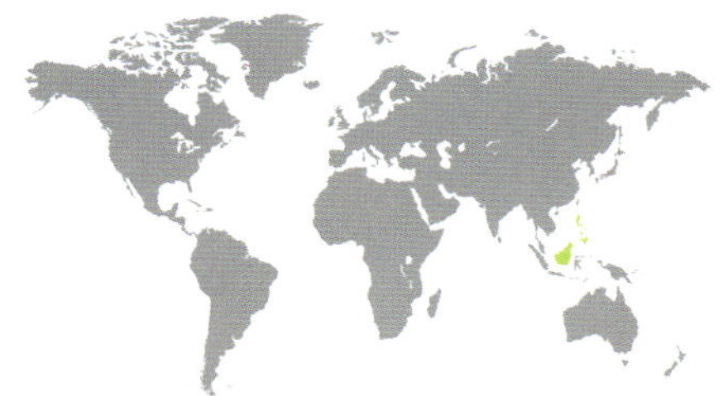

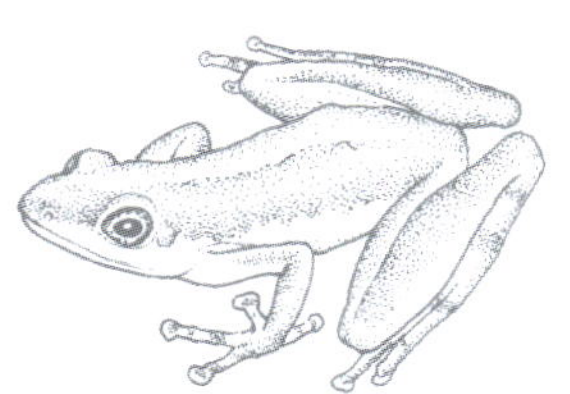

ADULT LENGTH
Male
1⅛–1⁷⁄₁₆ in (29–37 mm)
Female
1¾–2⅛ in (44–55 mm)

STAUROIS GUTTATUS

BORNEAN FOOT-FLAGGING FROG

(GÜNTHER, 1858)

This small frog is often seen sitting on rocks in the splash zone of fast-flowing streams. Active by day, males and females communicate by means of both visual and auditory signals. The male produces sharp chirps, alerting females to his presence, and he then signals to any female that approaches by adopting a variety of postures, including waving one of his feet in the air, spreading the pale turquoise-blue webbing between his toes. The female also signals to the male by means of postures and calls. The tadpoles of this species live in thick deposits of leaf litter in streams, and are red and iridescent blue in color.

Actual size

SIMILAR SPECIES

There are six species in the genus *Staurois*, all found in the forests of Borneo and the Philippines. Because of their association with streams, they are known as splash frogs. The Green-spotted Rock Frog (*S. tuberilinguis*) and the Rock Skipper (*S. latopalmatus*) both occur in Borneo. All species in this genus are threatened by destruction of their forest habitat.

The Bornean Foot-flagging Frog has a narrow body, a pointed snout, and long, slender hind limbs. There are well-developed adhesive disks on the fingers and toes. It is bright green in color, with large black spots on its back. The webbing between the toes is turquoise-blue.

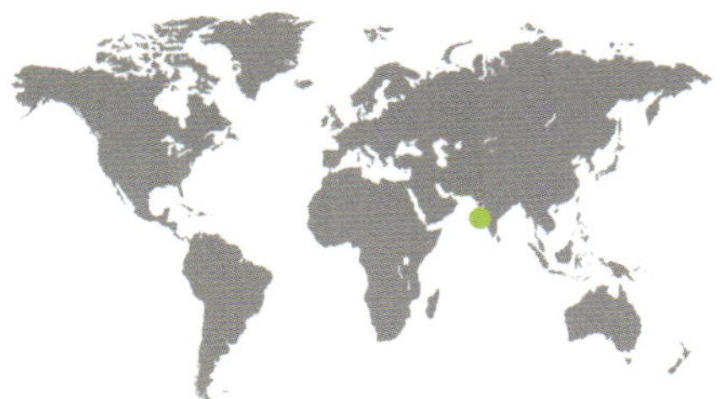

FAMILY	Ranixalidae
OTHER NAMES	Boulenger's Brown Frog
DISTRIBUTION	Western Ghats, western India
ADULT HABITAT	Forest at altitudes of 1,300–4,000 ft (400–1,200 m)
LARVAL HABITAT	Presumed to be streams
CONSERVATION STATUS	IUCN Least Concern. Its small range is being reduced and fragmented by deforestation

ADULT LENGTH
1¼–1½ in (32–38 mm)

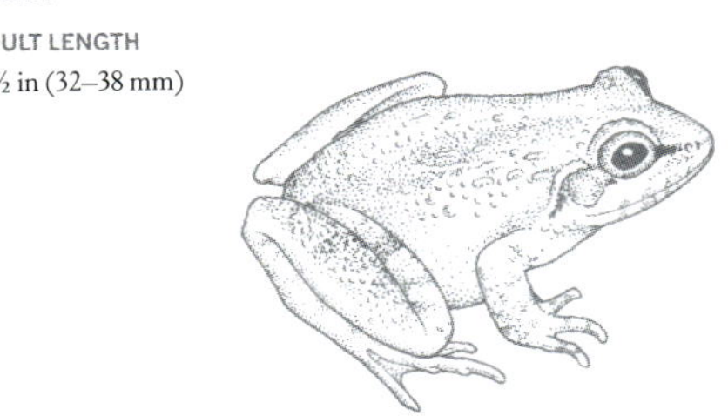

INDIRANA LEITHII

LEITH'S LEAPING FROG

(BOULENGER, 1888)

Actual size

Leith's Leaping Frog has several elongated warts on its back and a glandular fold of skin running from its eye to its shoulder. The toes are partially webbed, and both toes and fingers are tipped by adhesive disks. The upper surfaces are brown with small, dark spots, and the legs have dark transverse stripes.

The tadpoles of this rare and little-known frog behave in a very unusual way. They are found on the faces of rocks kept wet by spray from nearby hill streams. They can use their long tails to jump into pools as much as 6 ft (2 m) away, then after a few minutes they climb back up to a rock face. The adults are ground-dwelling, living among leaf litter and grass, and in ditches on hillsides. Mating has not been described, but the eggs are laid in pools in streams. Individuals of this species have been found to be infected with the fungal disease chytridiomycosis, but they do not show external symptoms.

SIMILAR SPECIES

Fourteen species in the genus *Indirana* have been described to date, but it is likely that more will be discovered. They are found in central and southern India, and most are threatened to varying degrees by deforestation. The Small-handed Leaping Frog (*I. semipalmata*) is found in the Western Ghats and is listed as of Least Concern. By contrast, Beddome's Leaping Frog (*I. beddomii*) and the Rocky-terrain Leaping Frog (*I. paramakri*), both from southern India, are Endangered.

FAMILY	Dicroglossidae: Dicroglossinae
OTHER NAMES	Baluch Mountain Frog, Murray's Frog
DISTRIBUTION	Afghanistan, Pakistan, India
ADULT HABITAT	Mountain streams
LARVAL HABITAT	Pools in mountain streams
CONSERVATION STATUS	IUCN Least Concern. May be adversely affected in Pakistan by agrochemicals and domestic detergents. Declining in India

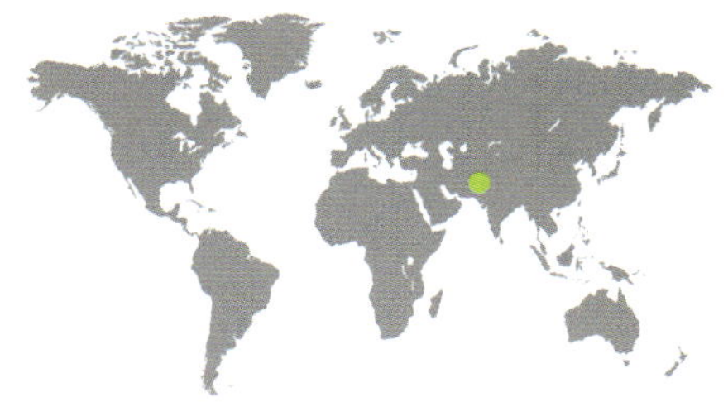

ADULT LENGTH
Male
2¾–3⅛ in (68–81 mm)
Female
3¹⁄₁₆–3⁹⁄₁₆ in (79–91 mm)

CHRYSOPAA STERNOSIGNATA

KAREZ FROG

(MURRAY, 1885)

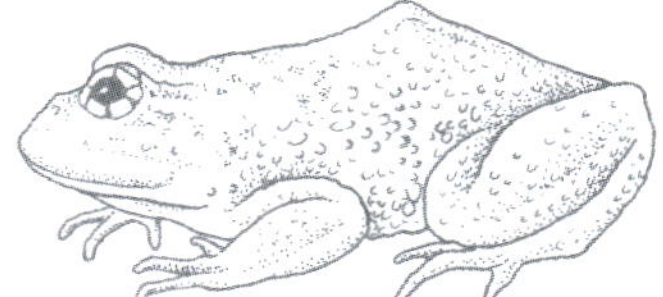

This frog is named after the karez, an ancient irrigation system in arid areas of Pakistan that includes extensive underground tunnels. It is wholly aquatic in its habits and remains in streams even in winter, when it moves around slowly beneath ice. It breeds between April and June, with males calling at sunset and producing a low-pitched, melodious sound. The eggs are large and are attached to submerged vegetation. This species has dramatically declined in India, partly because its tadpoles are killed by toxic chemicals that are used to catch fish.

SIMILAR SPECIES

The very distinctive *Chrysopaa sternosignata* is the only species in its genus. It is closely related to the genus *Nanorana*, most of which are high-altitude, stream-living frogs found across the Asian mainland. See the Yunnan Spiny Frog (*Nanorana yunnanensis*; page 582).

The Karez Frog has a broad, slightly flattened head, upwardly directed eyes, and fully webbed feet. Its skin appears rather loose and forms folds out of water. There are several warts on its back, tipped with sharp spines. The upper surfaces are olive-brown or dark green, with several small yellow, orange, or red spots.

Actual size

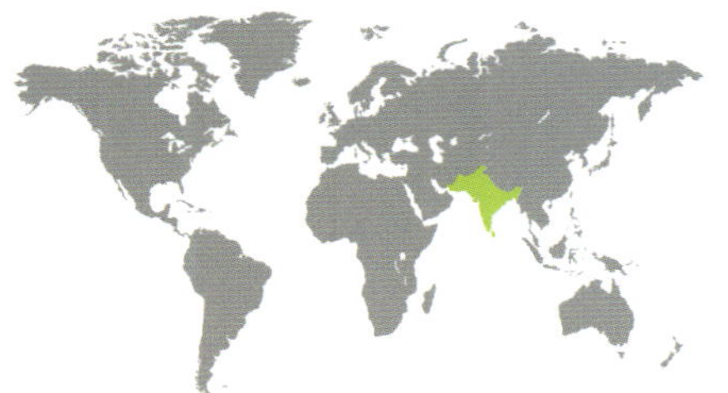

FAMILY	Dicroglossidae: Dicroglossinae
OTHER NAMES	Green Wart Frog (Vietnam), Water Skipper (India)
DISTRIBUTION	Iran, Afghanistan, Pakistan, India, Nepal, Bangladesh, Sri Lanka
ADULT HABITAT	Small and large waterbodies up to 8,200 ft (2,500 m) altitude
LARVAL HABITAT	Small and large waterbodies
CONSERVATION STATUS	IUCN Least Concern. Very common over much of its range, but threatened in places by draining of wetlands and pollution

ADULT LENGTH
Male
1$\frac{3}{8}$–1$\frac{3}{4}$ in (35–45 mm)

Female
1$\frac{15}{16}$–2$\frac{5}{8}$ in (50–65 mm)

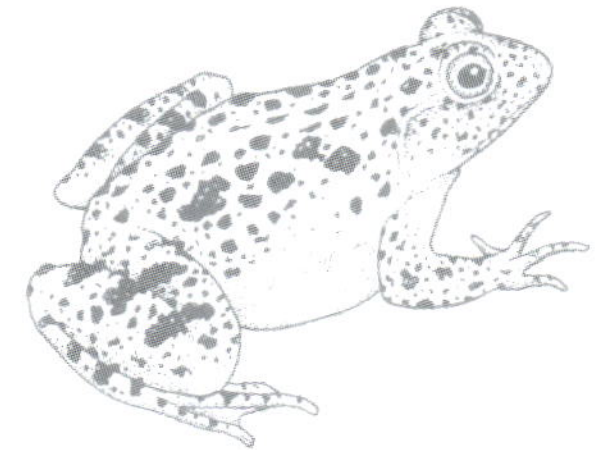

EUPHLYCTIS CYANOPHLYCTIS

COMMON SKITTERING FROG

(SCHNEIDER, 1799)

This very aquatic frog gets its name from its means of escaping danger, skittering at high speed across the water's surface, its flattened body inflated with air and propelled by its hind feet. Its call can be heard at any time of year, but breeding occurs as temperatures rise in early summer. Males gather together, jumping over one another and calling repeatedly. It is reported that females call in response to males and that they may mate with more than one male, but there are no detailed descriptions of mating behavior. At night, this frog leaves the water to forage for insects.

SIMILAR SPECIES

There are eight species in the genus *Euphlyctis*, also known as five-fingered frogs, mostly found across Asia. The Arabian Skittering Frog (*E. ehrenbergii*) occurs in Saudi Arabia and Yemen. The Indian Bullfrog (*E. hexadactyla*) is a large frog that is very unusual in that it feeds largely on plants—up to 80 percent of its gut contents may be plant material.

Actual size

The Common Skittering Frog has a slightly flattened body and head, with eyes on top of its head. Its feet are fully webbed. The upper surfaces are gray, olive-green, or brown, with numerous irregular black spots. The underside is white and there is often a pale band along the flanks.

FAMILY	Dicroglossidae: Dicroglossinae
OTHER NAMES	Asian Brackish Frog, Java Wart Frog, Mangrove Frog, Rice Field Frog
DISTRIBUTION	Thailand, Malaysia, Borneo, Philippines, Indonesia
ADULT HABITAT	Mangrove swamps, lowland wetlands, rice fields
LARVAL HABITAT	Salty, brackish, or freshwater pools
CONSERVATION STATUS	IUCN Least Concern. Has a very large range and its population is increasing in places

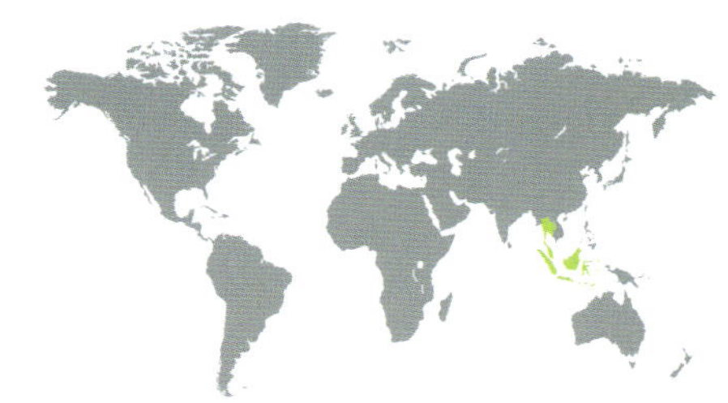

ADULT LENGTH
Male
2–2$^{13}/_{16}$ in (51–70 mm)

Female
2$^{1}/_{16}$–3$^{3}/_{16}$ in (53–82 mm)

FEJERVARYA CANCRIVORA

CRAB-EATING FROG

(GRAVENHORST, 1829)

This is thought to be the only frog that can live permanently in and around salt water. It occurs in coastal mangrove swamps, where it feeds on crabs. It also occurs in a variety of freshwater habitats, including rice fields, where it eats insects and small frogs. It may breed at any time of year but is most likely to do so in the rainy season. Males do not form choruses, but space themselves around waterbodies, producing a gargle-like call. The species is eaten across much of its range and is collected in large numbers in Java for export as frog legs.

SIMILAR SPECIES

There are fourteen species in the genus *Fejervarya*, following the transfer of a number of species to related genera, such as the Southern Cricket Frog (*Minervarya syhadrensis*; page 580). The Indian Cricket Frog (*F. limnocharis*), also known as the Grass Frog, has a huge range from Pakistan to China, and south to Indonesia. A small frog, it is common around human settlements and is one of the first frogs to call at the start of the monsoon.

The Crab-eating Frog has many small warts and folds in the skin on its back and legs. It has a large head, a long snout, large eyes, and a conspicuous tympanum. It has powerful, muscular legs, ending in webbed feet. In color, it is gray or brown with irregular dark patches, and there are transverse stripes on the legs.

Actual size

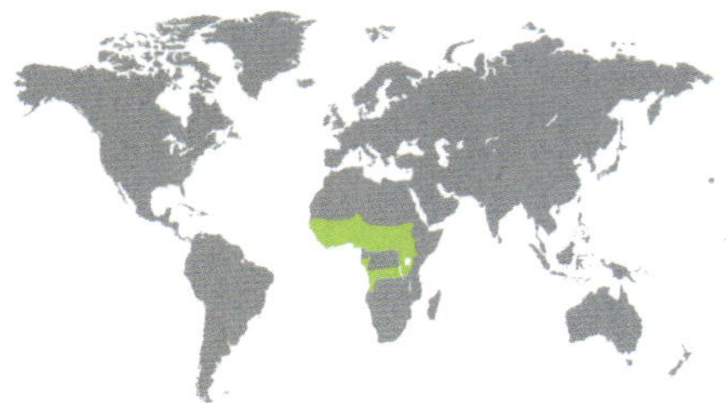

FAMILY	Dicroglossidae: Dicroglossinae
OTHER NAMES	African Tiger Frog, Giant Swamp Frog, Groove-crowned Bullfrog
DISTRIBUTION	Central Africa south of the Sahara, from Senegal in the west to Kenya in the east
ADULT HABITAT	Savanna; sometimes in forest
LARVAL HABITAT	Ephemeral fish-free pools and ponds
CONSERVATION STATUS	IUCN Least Concern. Has declined in some parts of its large range due to consumption by people

ADULT LENGTH
Male
2¾–4⁵⁄₁₆ in (68–110 mm)

Female
4⁵⁄₁₆–5⁵⁄₁₆ in (110–135 mm)

HOPLOBATRACHUS OCCIPITALIS

CROWNED BULLFROG

(GÜNTHER, 1858)

The Crowned Bullfrog gets its name from the pale green or yellow groove that runs across its head between its eyes. A large, rather flattened frog, it has protruding eyes high on its head, and upward-pointing nostrils. Its skin is very warty, and both hands and feet are fully webbed. Its upper surface is yellowish green, olive, or brown with dark markings, and it is white below.

This very large frog is valued by people in parts of its range for two reasons. First, its large, muscular legs make the adults a valued source of food, especially in West Africa, and second, its tadpoles eat mosquito larvae and may therefore help to control malaria. The larger tadpoles also eat smaller ones and, when laying their eggs, adults choose pools carefully, avoiding those that already contain tadpoles and those likely to dry up quickly. Mucus glands in the frog's skin make it very slippery and, when disturbed, it escapes by leaping across the water, uttering a characteristic whistle as it does so.

SIMILAR SPECIES

There are six species in the genus *Hoplobatrachus*, known as crowned bullfrogs and distributed across Africa and Asia. Jerdon's Bullfrog (*H. crassus*) is a large aquatic frog found in India, Bangladesh, Nepal, and Sri Lanka, and is noted for the aggressiveness of its tadpoles toward those of other species. The Indus Valley Bullfrog (*H. tigerinus*) has a reputation for attempting to eat anything that moves. It occurs across much of Asia and has been introduced to Madagascar and the Maldives.

Actual size

FAMILY	Dicroglossidae: Dicroglossinae
OTHER NAMES	Blyth's River Frog, Blyth's Wart Frog
DISTRIBUTION	Myanmar, Thailand, Peninsular Malaysia, Sumatra, Indonesia, Singapore
ADULT AND LARVAL HABITAT	Forest streams
CONSERVATION STATUS	IUCN Least Concern. Has declined in many parts of its range due to human consumption and deforestation of its habitat

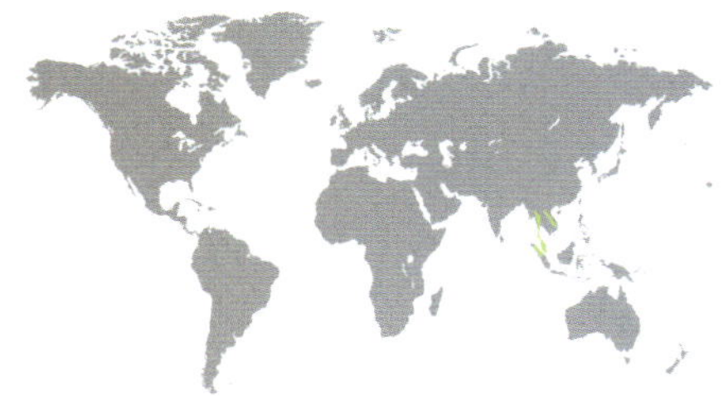

LIMNONECTES BLYTHII

GIANT ASIAN RIVER FROG

(BOULENGER, 1920)

ADULT LENGTH
Male
3$\frac{5}{16}$–4$\frac{15}{16}$ in (85–125 mm)
Female
3½–10¼ in (90–260 mm)

Large specimens of this aquatic frog can weigh more than 2⅕ lb (1 kg), making them an attractive source of food for humans. They have declined in numbers in areas where people harvest them, but have otherwise remained common if their habitat is undisturbed. Where harvesting has ceased, populations have recovered within 5–10 years. This species is unusual in that males do not call but females do. It is thought that the female's call alerts males to her presence. The male creates a hollow in a sandy stream bed, into which the eggs are laid.

SIMILAR SPECIES

To date, 90 species of the Asian genus *Limnonectes* have been described. There is variation among species in whether the male or the female is the larger sex, and in the size of a pair of fangs in the lower jaw. In the Khorat Big-mouthed Frog (*L. megastomias*), found in Thailand, the male is larger than the female and has a huge head and large fangs, which it uses to fight other males.

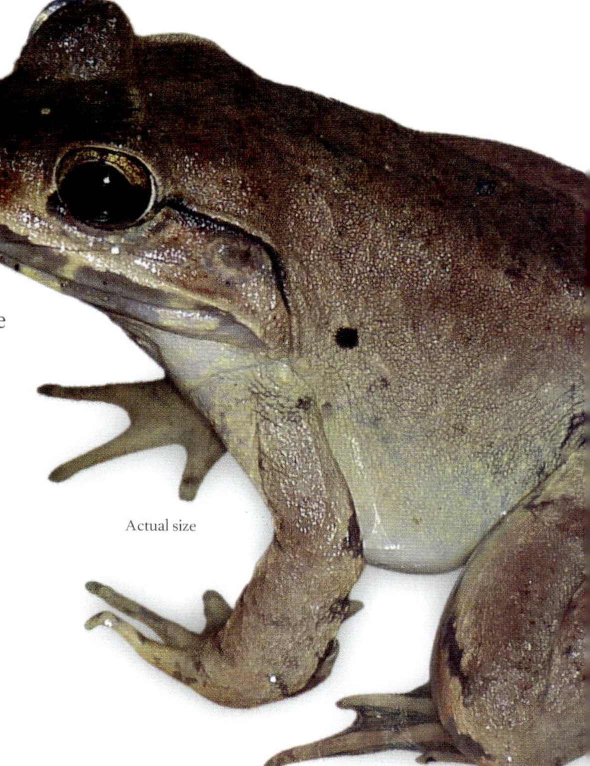

Actual size

The Giant Asian River Frog has long, muscular legs with fully webbed feet. The snout is pointed and the eyes are large. It is variable in color, being green, brown, or reddish. The legs are often striped, and some individuals have a broad yellow band running down the middle of the head and body.

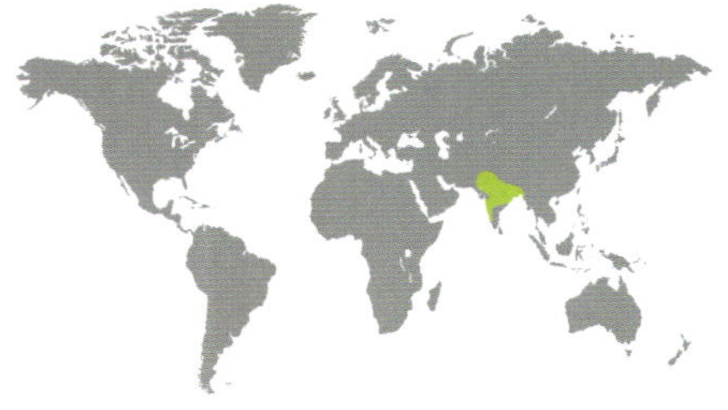

FAMILY	Dicroglossidae: Dicroglossinae
OTHER NAMES	Previously *Fejervarya syhadrensis*, Bombay Wart Frog, Common Paddy-field Frog, Hill Cricket Frog, Long-legged Cricket Frog, Syhadra Frog
DISTRIBUTION	Pakistan, India, Bangladesh, Nepal
ADULT HABITAT	Wetlands, rice fields
LARVAL HABITAT	Ponds
CONSERVATION STATUS	IUCN Least Concern. Has a very large range; its main threat is from agrochemicals

ADULT LENGTH
Male
11/16–3/4 in (17–19 mm)

Female
3/4–7/8 in (20–23 mm)

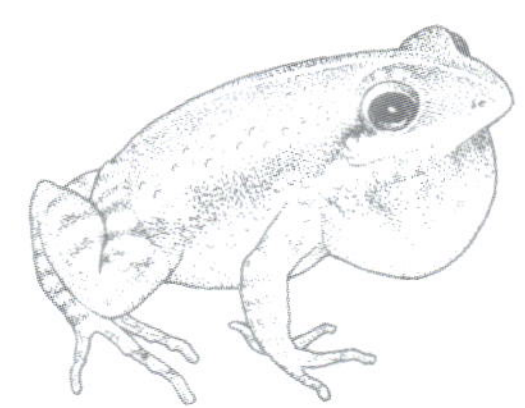

MINERVARYA SYHADRENSIS

SOUTHERN CRICKET FROG

(ANNANDALE, 1919)

Actual size

The Southern Cricket Frog has longitudinal folds in the skin on its back and small warts on its back and legs. It has a pointed snout, large eyes, and a conspicuous tympanum, and long, muscular legs and webbed feet. In color, it is gray with dark spots, and there may be a broad, pale stripe down the middle of the back.

This small frog is common in agricultural areas and is valued by farmers in Asia because it consumes insect pests and their larvae. Breeding is triggered by the first monsoon rains, between April and June, and continues until September or October. Breeding males space themselves about 3 ft (1 m) apart around waterbodies, hidden in marginal vegetation, and call antiphonally with their nearest neighbor. Their call has been likened to the clatter of an old-fashioned typewriter. The eggs are laid in small batches and are attached to waterplants.

SIMILAR SPECIES

Minervarya syhadrensis is widespread and common; other members of the genus are very rare. Sengupta's Cricket Frog (*M. sengupti*) is only known from one location in Meghalaya, India, and it is listed as Endangered. Also Endangered is the Sri Lanka Paddy Field Frog (*M. greenii*), which has declined as a result of deforestation, introduced trout, and the use of agrochemicals. See also the Crab-eating Frog (*Fejervarya cancrivora*; page 577).

FAMILY	Dicroglossidae: Dicroglossinae
OTHER NAMES	Ceylon Streamlined Frog
DISTRIBUTION	Southwestern Sri Lanka
ADULT HABITAT	Wet tropical forest up to 4,000 ft (1,200 m) altitude
LARVAL HABITAT	Rocky mountain streams
CONSERVATION STATUS	IUCN Vulnerable. Has lost much of its habitat through deforestation and is adversely affected by drought

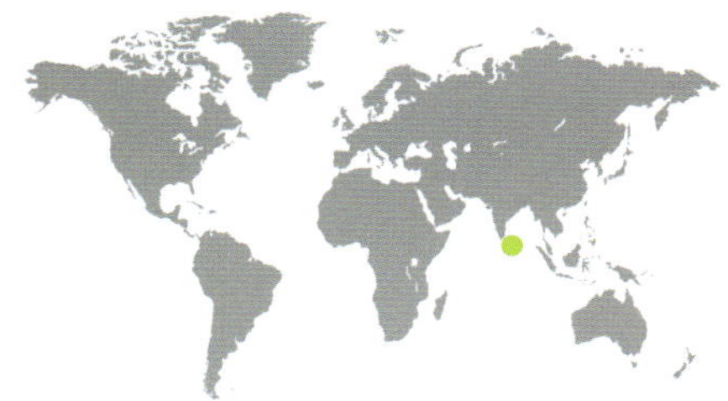

NANNOPHRYS CEYLONENSIS

SRI LANKAN ROCK FROG

GÜNTHER, 1869

ADULT LENGTH
Male
1$\frac{5}{16}$–1$\frac{11}{16}$ in (33–43 mm)
Female
1$\frac{3}{4}$–2$\frac{1}{16}$ in (45–53 mm)

The tadpoles and adults of this rare frog are both adapted for life in rocky mountain streams. The adults, most commonly seen sitting on rocks in cascades, are well camouflaged against their algae- and moss-covered background. They lay their eggs in rock crevices in the splash zone, and the male defends them against predators until they hatch. The tadpoles are semiterrestrial, spending much of their time out of water. They have a large sucker-like disk around the mouth that enables them to cling to rocks. Initially herbivorous, they become carnivorous as they grow larger.

SIMILAR SPECIES

Four species have been described in the Sri Lankan genus *Nannophrys*, known as streamlined frogs. Kirthisinghe's Rock Frog (*N. marmorata*), also known as the Marbled Streamlined Frog, occurs only in the Knuckles Mountain Range and is listed as Endangered. The Sri Lankan Tribal Rock Frog (*N. naeyakai*), described in 2007 and found at just two locations, is Endangered. Günther's Streamlined Frog (*N. guentheri*) has not been seen for more than 100 years and is listed as Extinct.

The Sri Lankan Rock Frog has a wide, flattened body, a large head, large eyes, a blunt snout, and warty skin. The upper surfaces are yellow or olive-green with scattered brown markings, and there are brown cross-stripes on the legs.

Actual size

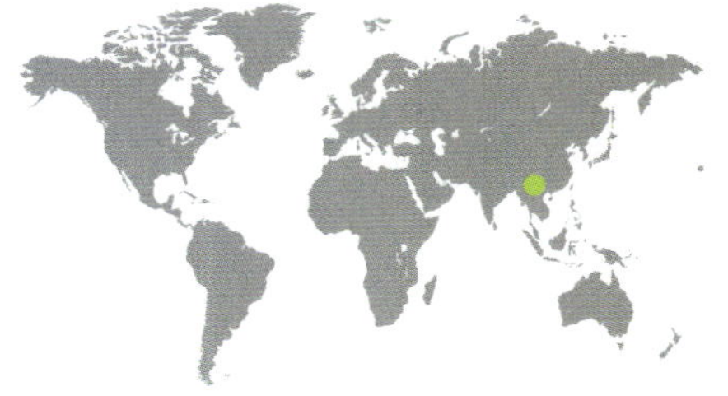

FAMILY	Dicroglossidae: Dicroglossinae
OTHER NAMES	Yunnan Paa Frog
DISTRIBUTION	Southwestern and central China, Myanmar, Vietnam
ADULT HABITAT	Streams in forest and grassland up to 9,800 ft (3,000 m) altitude
LARVAL HABITAT	Rocky streams
CONSERVATION STATUS	IUCN Vulnerable. Has declined as a result of overconsumption by humans and degradation of its habitat

ADULT LENGTH
Male
up to $3\frac{13}{16}$ in (98 mm)
Female
up to $3\frac{7}{8}$ in (99 mm)

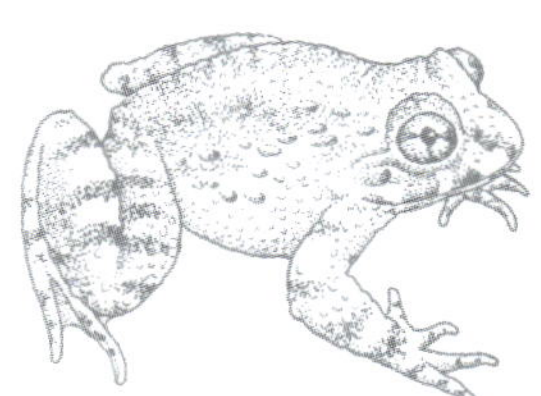

NANORANA YUNNANENSIS

YUNNAN SPINY FROG

(ANDERSON, 1879)

The Yunnan Spiny Frog has a sturdy body, a wide, flat head, muscular hind legs, and fully webbed feet. Breeding males have thicker, more muscular arms than females and spines on their chests. The upper surfaces are covered in warts and are gray or yellowish brown in color. The underside is gray or yellow.

This large frog gets its name from an array of sharp spines on the chest of breeding males. These, together with their very muscular arms, enable males to keep a firm grip on females during mating. The frog is often seen sitting on moss-covered rocks near the fast-flowing streams where it breeds. The breeding season is prolonged, with eggs being found—hidden under stones—from April to June. Collection of the frogs for food by people is a major cause of the species' population decline. Detailed analysis of its tissues suggests that frog meat is more nutritious and tastier than either pork or beef.

SIMILAR SPECIES

The genus *Nanorana* contains 33 species, all found across the Asian mainland. Most of them are high-altitude, stream-living frogs. The Tibetan Frog (*N. pleskei*) occurs in central China, and has declined as a result of overharvesting, but despite this threat it is still listed as Least Concern. The Sikkim Paa Frog (*N. liebigii*) occurs at high altitudes in India, China, and Nepal. The Medog Spiny Frog (*N. medogensis*), from Tibet, and the Kangzian Swelled-vent Frog (*N. kangxianensis*), from China, are Endangered.

Actual size

FAMILY	Dicroglossidae: Dicroglossinae
OTHER NAMES	Common Spiny Frog, Hong Kong Paa Frog, Lesser Spiny Frog
DISTRIBUTION	Central and southern China, including Hong Kong
ADULT HABITAT	Streams in forest and shrubland up to 4,600 ft (1,400 m) altitude
LARVAL HABITAT	Fast-moving streams
CONSERVATION STATUS	IUCN Least Concern. Much of its habitat is being degraded and it is widely collected for food

QUASIPAA EXILISPINOSA

HONG KONG SPINY FROG

(LIU & HU, 1975)

ADULT LENGTH
Male
up to $2\frac{3}{8}$ in (61 mm)

Female
up to $2\frac{3}{16}$ in (57 mm)

Although this frog is commonly collected for food, little is known about its natural history. Males are larger than females, and have greatly enlarged arms, suggesting that combat among them is common. Breeding males also have many sharp spines on their chest and fingers, but the biological function of these is not known; they may simply help the male to maintain a firm grip on his mate. The species is active at night, hiding by day under rocks and leaf litter. It feeds on insects, crayfish, and smaller frogs. The eggs are laid in clusters of 5–10, glued to rocks in shallow pools.

SIMILAR SPECIES

There are 13 species in the genus *Quasipaa*, all found in Asia. Larger than *Q. exilispinosa*, the Giant Spiny Frog (*Q. spinosa*) is a traditional food source in Vietnam and is considered a delicacy in China. It is listed as Vulnerable. Boulenger's Spiny Frog (*Q. boulengeri*) also occurs in China and has tested positive for the fungus that causes chytridiomycosis. It is listed as Vulnerable.

The Hong Kong Spiny Frog is a plump, rather squat frog with a short snout, wrinkled skin, and partially webbed feet. The upper surfaces are grayish or reddish brown. There is a dark, mottled pattern on the back and dark stripes on the hind legs. The underside is yellow.

Actual size

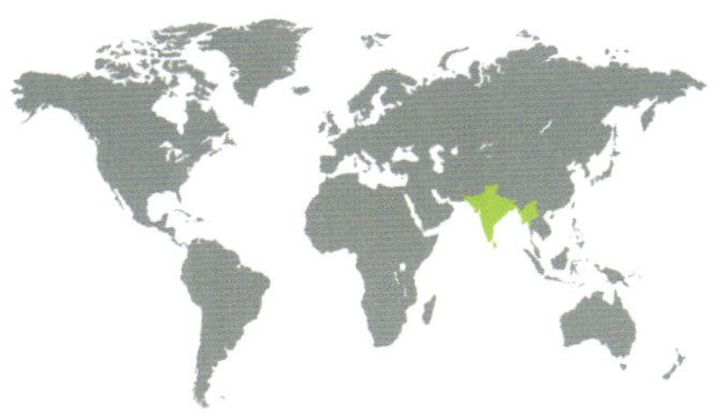

FAMILY	Dicroglossidae: Dicroglossinae
OTHER NAMES	Indian Burrowing Frog, Punjab Bullfrog
DISTRIBUTION	Pakistan, India, Sri Lanka, Nepal, Bangladesh, Myanmar
ADULT HABITAT	Arid areas in a wide variety of habitats
LARVAL HABITAT	Ephemeral pools
CONSERVATION STATUS	IUCN Least Concern. A wide-ranging species with a stable population that occurs within some protected areas

ADULT LENGTH
Male
1⅝–2⅛ in (41–55 mm)
Female
1$^{11}/_{16}$–2$^{3}/_{16}$ in (43–57 mm)

SPHAEROTHECA BREVICEPS

BURROWING FROG

(SCHNEIDER, 1799)

This rotund frog occurs in areas where there is soft, sandy soil, into which it can burrow backward, using shovel-like horny tubercles on its hind feet. It emerges at night to feed and is often observed eating centipedes and millipedes. It breeds after the first showers of the summer monsoon, males gathering around large ponds and spacing themselves out. Their call is a repeated "awang… awing… awang." The large eggs are laid in batches that float and stick to blades of grass.

SIMILAR SPECIES

There are nine species in the genus *Sphaerotheca*, all found in Asia. The Marbled Sand Frog (*S. rolandae*) and Dobson's Burrowing Frog (*S. dobsonii*) are very similar in appearance and habits. *Sphaerotheca dobsonii* is from India and *S. rolandae* occurs in Sri Lanka. In appearance, they resemble frogs in the unrelated genus *Uperodon*—see the Marbled Balloon Frog (*U. systoma*; page 514).

Actual size

The Burrowing Frog is a short-bodied, stocky frog with short hind legs, large eyes, and a short, blunt snout. The upper surfaces are yellowish or pale olive-green, with a darker, marbled pattern on the back and transverse stripes on the legs. There is often a pale stripe running down the middle of the back.

FAMILY	Dicroglossidae: Occidozyginae
OTHER NAMES	Philippine Oriental Frog, Spotted Puddle Frog
DISTRIBUTION	Philippines, Borneo, Peninsular Malaysia, Vietnam
ADULT HABITAT	Lowland forest up to 4,000 ft (1,200 m) altitude
LARVAL HABITAT	Puddles
CONSERVATION STATUS	IUCN Least Concern. Has declined in places as a result of deforestation

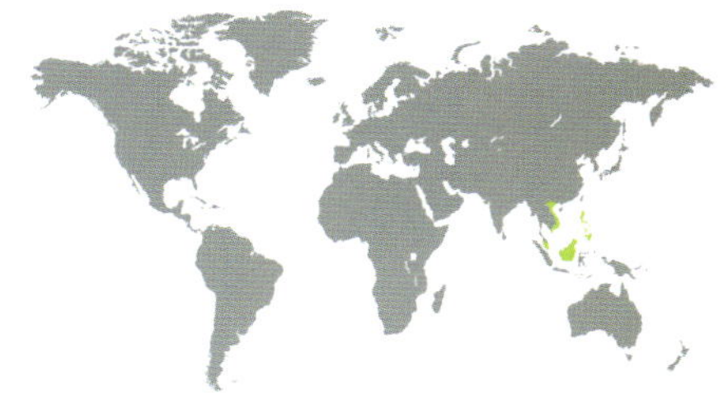

ADULT LENGTH
Male
⅞–1¼ in (21–31 mm)
Female
1⅜–1⅞ in (35–48 mm)

OCCIDOZYGA LAEVIS

YELLOW-BELLIED PUDDLE FROG

(GÜNTHER, 1858)

As its common name implies, this small frog lives in puddles—including those created by wallowing rhinos and pigs, and the footprints of Asian elephants. Its coloration matches that of muddy water and it floats with only its snout and eyes above the water's surface. When it moves around the forest floor, it does so in short hops. It is unusual in that, rather than lunging at its prey with its mouth open, it reaches out with its forelimbs, fingers spread, to scoop prey into its mouth. The tadpoles are long and slender, and are unusual in feeding on aquatic insects and their larvae.

SIMILAR SPECIES

Given the large and fragmented distribution of *Occidozyga laevis*, it is likely that it includes more than one species. Currently, 18 species are recognized in the genus. The Balu Oriental Frog (*O. baluensis*) is found in Borneo and lives among the mud and pebbles that are present where water seeps out of the ground. It is listed as Near Threatened. The Celebes Oriental Frog (*O. celebensis*) lives in Sulawesi, where it thrives and breeds in rice fields. The Mount Tompotika Rice Frog (*O. tompotika*), from Sulawesi, is Critically Endangered.

The Yellow-bellied Puddle Frog has upwardly protruding eyes, a squat, stocky body, and short, fat hind limbs. The toes are fully webbed. The skin has a corrugated texture and there are several oval-shaped bumps on its back. It is dark gray-brown in color, except for the belly and underside of the thighs, which are lemon-yellow.

Actual size

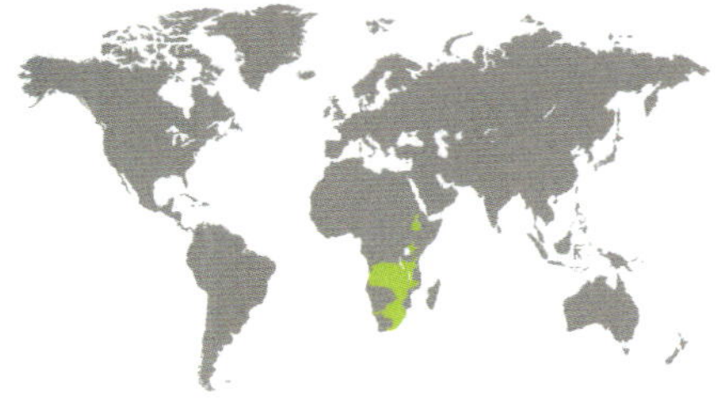

FAMILY	Pyxicephalidae: Cacosterninae
OTHER NAMES	Angola River Frog
DISTRIBUTION	Southern and eastern Africa, from Ethiopia in the north to South Africa in the south and Angola in the west
ADULT HABITAT	Rivers in grassland and savanna
LARVAL HABITAT	Ponds and stream edges
CONSERVATION STATUS	IUCN Least Concern. Has a large range and is not subject to serious threats

ADULT LENGTH
Male
average 2½ in (62 mm)
Female
average 2⅞ in (73 mm);
up to 3½ in (90 mm)

AMIETIA ANGOLENSIS

COMMON RIVER FROG

(BOCAGE, 1866)

This large frog is a strong swimmer and, with its long, muscular legs, is also a very good jumper. It may breed at any time of year, males calling day and night from floating vegetation or from the water's edge. If disturbed, they dive into deep water and bury themselves in the mud at the bottom of the pond or stream. The call consists of a rattling series of clicks followed by a croak. Females lay 400–500 eggs. The tadpoles take two years to reach metamorphosis and can reach up to 3⅛ in (80 mm) in length.

SIMILAR SPECIES

With its enormous range, it is likely that this frog is actually more than one species. The Amani River Frog (*Amietia tenuoplicata*), from Tanzania and Malawi, was identified as a species distinct from *A. angolensis* in 2007. The Large-mouthed Frog (*A. vertebralis*) is a very large frog with a huge head and wide mouth that occurs in Lesotho. The Mulanje River Frog (*A. johnstoni*), also from Malawi, is Endangered. See also the Cape River Frog (*A. fuscigula*; page 587).

Actual size

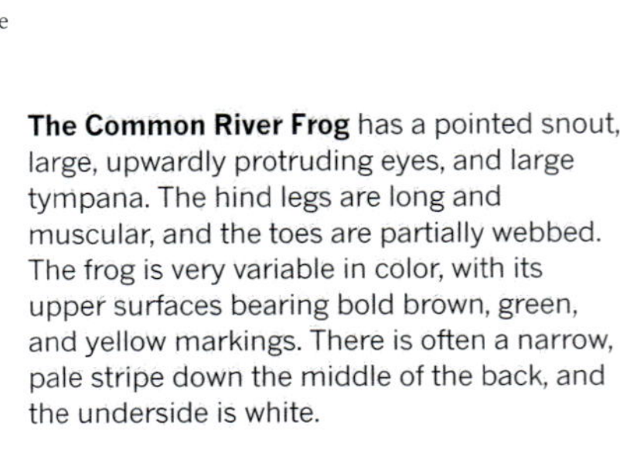

The Common River Frog has a pointed snout, large, upwardly protruding eyes, and large tympana. The hind legs are long and muscular, and the toes are partially webbed. The frog is very variable in color, with its upper surfaces bearing bold brown, green, and yellow markings. There is often a narrow, pale stripe down the middle of the back, and the underside is white.

FAMILY	Pyxicephalidae: Cacosterninae
OTHER NAMES	Dark-throated River Frog
DISTRIBUTION	Eastern Cape and Western Cape provinces, South Africa
ADULT HABITAT	Rivers in grassland and fynbos
LARVAL HABITAT	Ponds and dams
CONSERVATION STATUS	IUCN Least Concern. Has a large range, is not subject to serious threats and has adapted to human-altered habitats

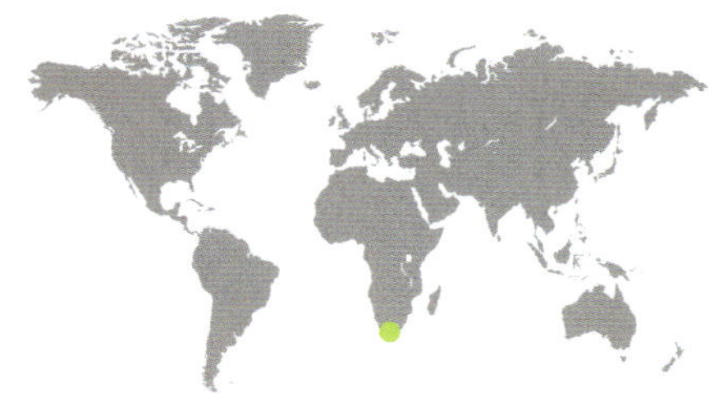

AMIETIA FUSCIGULA

CAPE RIVER FROG

(DUMÉRIL & BIBRON, 1841)

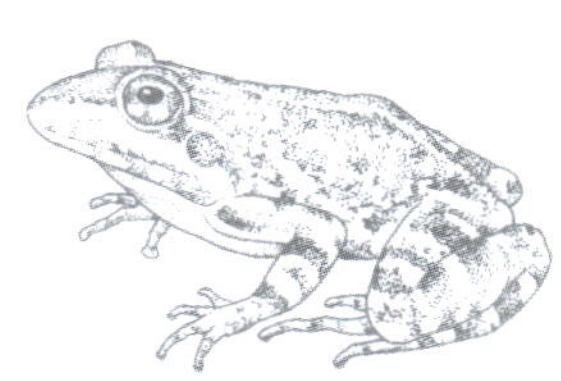

ADULT LENGTH
Male
average $2\frac{15}{16}$ in (75 mm)

Female
average 3½ in (90 mm);
up to $4\frac{15}{16}$ in (125 mm)

This very large frog has a reputation for eating anything that moves; feeding primarily on insects, it will also devour mice, other frogs, and crabs. In dry areas, it is found in permanent springs, ponds, and farm dams. Elsewhere, it occurs along most well-vegetated rivers and streams. The species has adapted quite well to human encroachment and is found in agricultural areas. It can breed at any time of year but most breeding takes place in the rainy season. The frog's large size makes it a preferred prey of herons and other waterbirds.

SIMILAR SPECIES

To date, 16 species of *Amietia* have been identified and it is likely that there will be more. Poynton's Water Frog (*A. poyntoni*) was described in 2013. It occurs in Namibia and northern South Africa. The Molo Frog (*A. wittei*) occurs in the mountains of Kenya, Tanzania, and the Democratic Republic of the Congo. See also the Common River Frog (*A. angolensis*; page 586).

Actual size

The Cape River Frog has a rounded snout, large, upwardly protruding eyes, and large tympana. The hind legs are long and muscular, and the toes are fully webbed. Its skin is smooth, with longitudinal folds on the back. Generally brown or olive-green in color, it has a yellow or white stripe down the middle of its back. The underside is white.

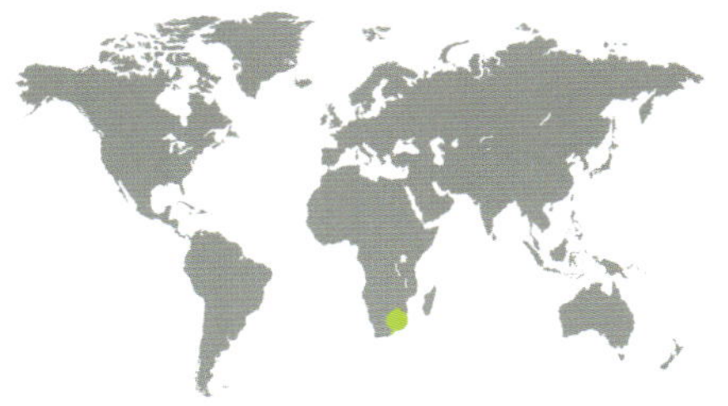

FAMILY	Pyxicephalidae: Cacosterninae
OTHER NAMES	Hewitt's Moss Frog, Natal Moss Frog
DISTRIBUTION	Eastern South Africa
ADULT HABITAT	Wet areas in forest and dense bush up to 8,900 ft (2,700 m) altitude
LARVAL HABITAT	Within the egg
CONSERVATION STATUS	IUCN Least Concern. Occurs at many well-protected, high-altitude areas

ADULT LENGTH
Male
average ⅞ in (22 mm)
Female
average 1⅛ in (29 mm)

ANHYDROPHRYNE HEWITTI

NATAL CHIRPING FROG

(FITZSIMONS, 1947)

Actual size

The Natal Chirping Frog is a tiny frog with long, thin fingers and toes. Its upper surfaces are orange-brown, dark brown, gray, or black, sometimes with a pale stripe down the middle of the back. A dark band runs from snout to armpit, and there is a white stripe on the upper lip. The underside is off-white with a dark, mottled pattern.

This tiny frog is very hard to find, even when it is calling. It is most common in wet areas near waterfalls in the Drakensberg Mountains. It breeds from October to January, males uttering an insect-like "tik… tik… tik" call from concealed positions under vegetation. The female lays 14–40 eggs in moss or leaf litter beside a stream. There are no free-swimming tadpoles; instead, the species shows direct development, its eggs hatching after about 20 days to produce tiny froglets.

SIMILAR SPECIES

There are two other species of *Anhydrophryne*, both from South Africa. The Ngoni Moss Frog (*A. ngongoniensis*) is so secretive, and its call so quiet, that it was not discovered until 1993. It is restricted to the mist belt in South Africa's eastern escarpment, where much of its forest habitat has been cleared for agriculture and plantations; it is listed as Endangered. The Hogsback Frog (*A. rattrayi*), found in Eastern Cape province, is less than ⅞ in (22 mm) long and is categorized as Vulnerable.

FAMILY	Pyxicephalidae: Cacosterninae
OTHER NAMES	Drewes' Chirping Frog
DISTRIBUTION	Western Cape province, South Africa
ADULT HABITAT	Seepages in fynbos
LARVAL HABITAT	Within the egg
CONSERVATION STATUS	IUCN Near Threatened. Known from only two locations

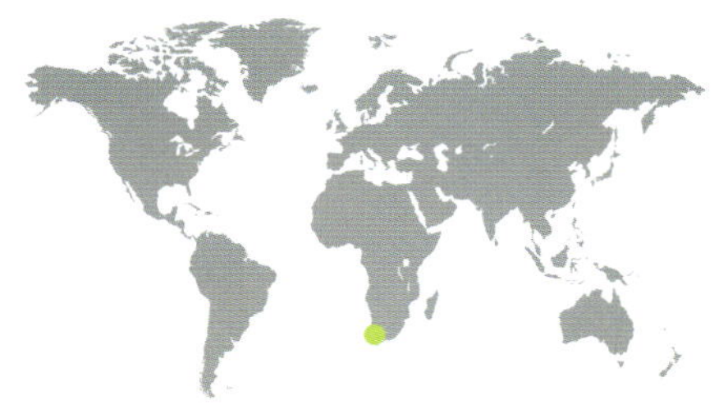

ADULT LENGTH
Male
average 9/16 in (15 mm)

Female
average 11/16 in (17 mm)

ARTHROLEPTELLA DREWESII

DREWES' MOSS FROG

CHANNING, HENDRICKS & DAWOOD, 1994

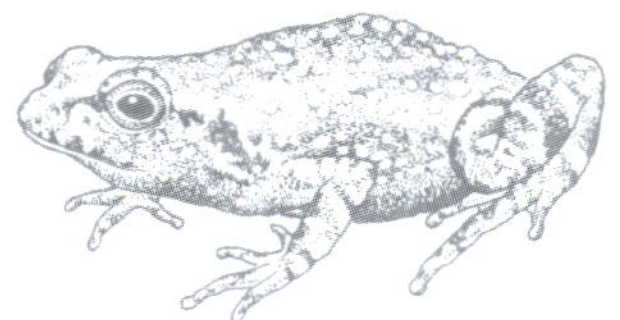

This tiny frog is typically heard rather than seen. Its small size and cryptic coloration make it very hard to find at the base of dense vegetation, where it lives. It can be heard calling in the rainy season, from June to September, its call consisting of five to seven short peeps. It lays its eggs, in clutches of about ten, under moss. They develop directly, hatching into tiny frogs; there is no free-living tadpole stage. The main threat to the species' continued existence is the occasional bushfires that sweep through its fynbos habitat.

Actual size

SIMILAR SPECIES

The moss frogs (genus *Arthroleptella*), of which ten species have been described, are found in the mountains of South Africa's Western Cape province. The Cape Peninsula Moss Frog (*A. lightfooti*) occurs up to the very summit of Table Mountain and often lays its eggs near waterfalls. It is listed as Near Threatened. The Rough Moss Frog (*A. rugosa*) occupies an area of less than 0.4 sq mile (1 sq km) and is Critically Endangered.

Drewes' Moss Frog has a rounded snout, large eyes, and rows of black warts running along its back. The tips of the fingers and toes are slightly expanded. The overall color is brown, the legs paler than the body. There is a black mask-like stripe running through the eye.

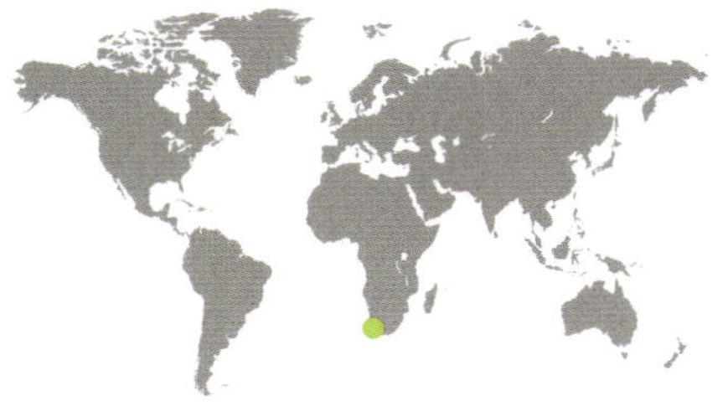

FAMILY	Pyxicephalidae: Cacosterninae
OTHER NAMES	Cape Caco, Cape Metal Frog
DISTRIBUTION	Western Cape province, South Africa
ADULT HABITAT	Low-lying sandy areas
LARVAL HABITAT	Shallow, ephemeral pools
CONSERVATION STATUS	IUCN Near Threatened. More than 90 percent of its former habitat has been lost to urbanization and agriculture

ADULT LENGTH
Male
average 1 in (25 mm)

Female
average 1¼ in (32 mm);
up to 1$^{9}/_{16}$ in (39 mm)

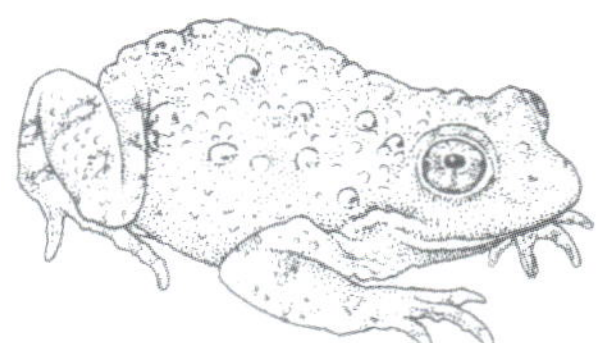

CACOSTERNUM CAPENSE

CAPE DAINTY FROG

HEWITT, 1925

Actual size

The Cape Dainty Frog is distinguished from other cacos by a number of large, blister-like glands on the lower part of its back. It has a slender, elongated body and a relatively small head. The upper surfaces are gray, cream, or brown with dark speckles. The underside is white, with large, irregular olive-green or black patches.

This small frog is actually the largest of the dainty frogs or cacos, found across southern Africa. For much of the year it lives underground, emerging after heavy winter rain between June and August to breed in shallow pools. Males call at night, producing short "creak" sounds at the rate of two per second. Only rarely do they gather together to form choruses. The eggs are laid in jelly clusters attached to submerged vegetation. The fact that other frog species die when kept in the same vivarium as this species suggests that it is toxic.

SIMILAR SPECIES

To date, 17 species in the genus *Cacosternum* have been described. Boettger's Dainty Frog (*C. boettgeri*) is a wide-ranging species, occurring in grassland in much of southern Africa. A very small frog, it never exceeds ⅞ in (23 mm) in length. The Namaqua Dainty Frog (*C. namaquense*) lives in rocky areas and is confined to the Namaqualand region of South Africa and southern Namibia. See also the Karoo Dainty Frog (*C. karooicum*; page 591).

FAMILY	Pyxicephalidae: Cacosterninae
OTHER NAMES	Karoo Caco
DISTRIBUTION	Western Cape and Northern Cape provinces, South Africa
ADULT HABITAT	Dry shrubland, semidesert, rocky areas
LARVAL HABITAT	Ephemeral streams
CONSERVATION STATUS	IUCN Least Concern. Has a wide distribution and is not subject to any major threats

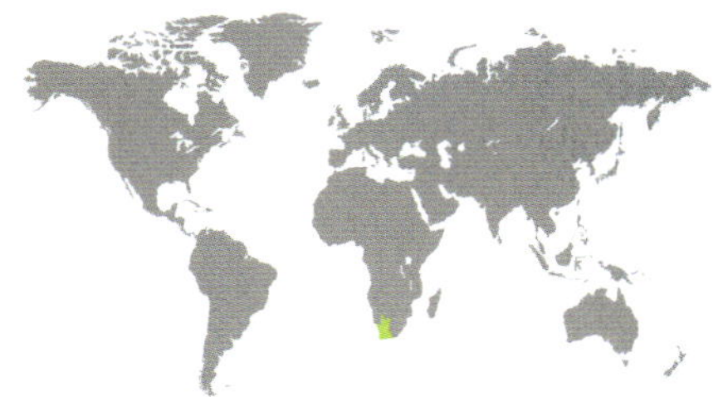

CACOSTERNUM KAROOICUM

KAROO DAINTY FROG

BOYCOTT, DE VILLIERS & SCOTT, 2002

ADULT LENGTH
Male
⅞–1 in (23–26 mm)
Female
⅞–1¼ in (23–31 mm)

Home to this tiny frog, after which it is named, the Karoo is a large area of semidesert in South Africa. The frog's flattened shape enables it to push its way deep into rock crevices, where it survives the dry season. It is an opportunistic breeder, breeding at any time of year when rain is sufficient to cause ephemeral streams to flow. Males call while half-submerged, producing a long, coarse rattle. They also have a shorter territorial call that they use to space themselves out. The eggs are laid attached to submerged vegetation in larger pools that have a chance of persisting long enough for the tadpoles to reach metamorphosis.

Actual size

The Karoo Dainty Frog has a flattened body and head, a rounded snout, and small, round warts on its back. It is olive-green or khaki in color, sometimes with a tinge of bronze or red, and with darker patches on the body and legs. The underside is white with dense black spots concentrated under the chin.

SIMILAR SPECIES

The 17 species of caco that have been described to date are all very small. The Bronze Dainty Frog (*Cacosternum nanum*) has a maximum size of ⅝ in (16 mm). This frog is common in areas of relatively high rainfall, breeding in tiny bodies of water, and has been reported to reach metamorphosis in just 17 days. The equally small Mountain Dainty Frog (*C. parvum*) is a high-altitude species, found in the Drakensberg and in Swaziland. See also the Cape Dainty Frog (*C. capense*; page 590).

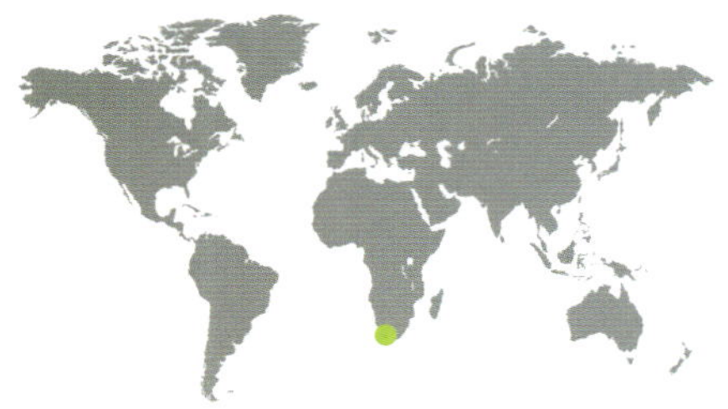

FAMILY	Pyxicephalidae: Cacosterninae
OTHER NAMES	Cape Flats Frog
DISTRIBUTION	Coastal lowlands, from Cape Peninsula to Cape Agulhas, Western Cape province, South Africa
ADULT AND LARVAL HABITAT	Acidic black-water pools associated with fynbos vegetation
CONSERVATION STATUS	IUCN Critically Endangered. Threatened by habitat loss

ADULT LENGTH
Male
average ½ in (12 mm)

Female
average 9/16 in (15 mm);
up to 11/16 in (18 mm)

MICROBATRACHELLA CAPENSIS

MICRO FROG

(BOULENGER, 1910)

Actual size

The Micro Frog has a rather plump body, short, thin legs, and a rounded snout. The skin is smooth and the toes are partially webbed. Its upper surfaces are brown, green, or gray with lighter speckles, and there is often a thin, pale line down the middle of the back. The belly is brown with white spots.

This tiny frog has lost more than 80 percent of its very specific habitat—acidic black-water pools associated with Western Cape province's unique fynbos vegetation—as a result of extensive housing developments. With the arrival of winter rain in May, it emerges from its dry-season hiding places and males start calling from emergent aquatic vegetation. Calling takes place mostly at night, sometimes in large choruses, the call consisting of a series of low-pitched, scratchy notes. The eggs are deposited in clusters of up to 20, attached to submerged vegetation just below the water's surface.

SIMILAR SPECIES

Microbatrachella capensis is the only species in its genus, which is closely related to the genus *Cacosternum* (see pages 590 and 591). It shares its very unusual, and very threatened, habitat with the totally unrelated Cape Platanna (*Xenopus gilli*; page 53).

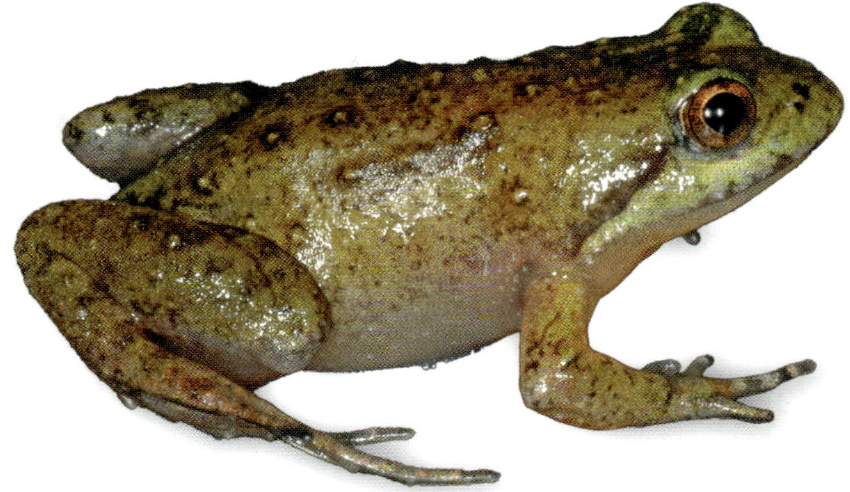

FAMILY	Pyxicephalidae: Cacosterninae
OTHER NAMES	Boneberg's Frog, Natal Diving Frog, Natal Frog
DISTRIBUTION	Coastal KwaZulu-Natal and Eastern Cape provinces, South Africa
ADULT HABITAT	Coastal forest up to 3,000 ft (900 m) altitude
LARVAL HABITAT	Forest streams
CONSERVATION STATUS	IUCN Endangered. Much of its habitat has been lost to agriculture and urbanization

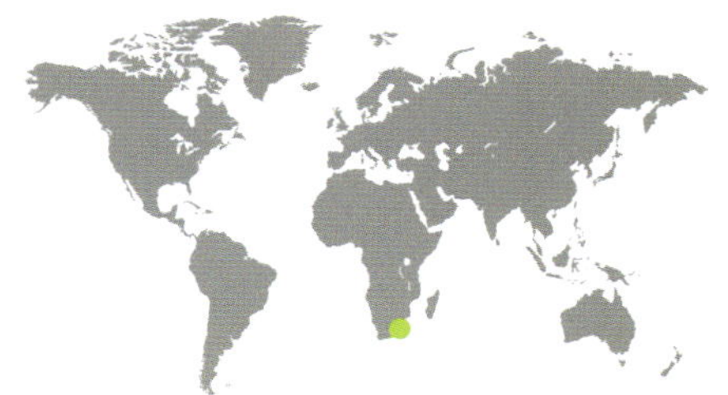

NATALOBATRACHUS BONEBERGI

KLOOF FROG

HEWITT & METHUEN, 1912

ADULT LENGTH
Male
average 1 in (25 mm)

Female
average 1⁵⁄₁₆ in (34 mm);
up to 1⁷⁄₁₆ in (37 mm)

This small, athletic frog lives in forested canyons close to the shallow streams in which it breeds. It is very difficult to catch, being a good jumper on land and a strong swimmer in the water. Between October and May, males call from perches 3–6 ft (1–2 m) above water, producing a rather faint "click." The female lays a clutch of 75–95 eggs, attached to a leaf, twig, or rock overhanging a pool in a stream. They hatch after six days, the tadpoles falling into the water below. Before they hatch, the female visits the eggs and urinates on them to keep them moist.

SIMILAR SPECIES

Natalobatrachus bonebergi is the only species in its genus. In its morphology and habits it is most similar, within its family, to the African stream frogs, genus *Strongylopus* (see page 596).

Actual size

The Kloof Frog has a very pointed snout, a slender body, and long, muscular legs. The long fingers and toes end in large T-shaped disks. There are longitudinal ridges in the skin on the back, which is brown or green in color with a pale stripe running down the middle. A dark stripe runs from the snout to the armpit, and the legs are striped.

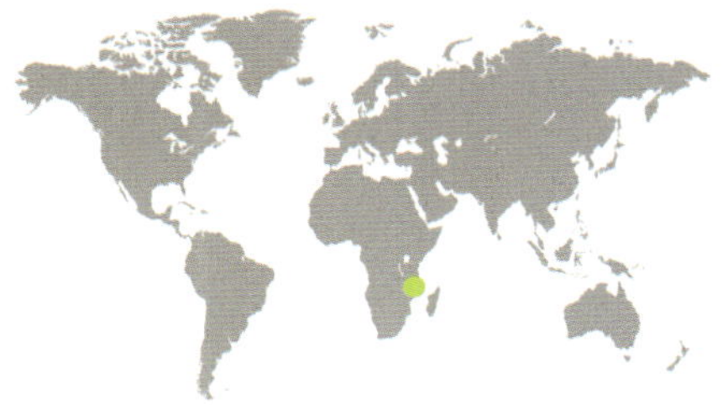

FAMILY	Pyxicephalidae: Cacosterninae
OTHER NAMES	Broadley's Mountain Frog
DISTRIBUTION	Southern Malawi, northern Mozambique
ADULT HABITAT	Forest and grassland at altitudes of 4,000–10,000 ft (1,200–3,000 m)
LARVAL HABITAT	Mountain streams
CONSERVATION STATUS	IUCN Endangered. Has a restricted range that is subject to deforestation and bush fires

ADULT LENGTH
up to 1⅛ in (28 mm)

NOTHOPHRYNE BROADLEYI

MONGREL FROG

POYNTON, 1963

Actual size

The Mongrel Frog has a flattened body, large, protruding eyes, and many large warts on its back. There is no webbing on hands or feet, and the disks on the fingers and toes are small. The upper surfaces are brown or green with a pale stripe down the middle of the back. There is a pale stripe on the head between the eyes, bounded behind by a dark stripe, and there are dark lateral stripes on the legs.

This small frog is known from only five locations on Mt. Mulanje in Malawai and Mt. Ribaue in Mozambique, though it possibly occurs elsewhere. It lives in rocky areas in high-altitude grassland and forest and, at least on Mt. Mulanje, gathers to breed in large numbers. The male's call is a weak chirp. The eggs are laid in clutches of around 30 in wet moss at the edge of streams running over rocks. When the eggs hatch, the tadpoles disperse by wriggling over wet rocks. The fungus that causes the disease chytridiomycosis has been detected in this frog in Malawi.

SIMILAR SPECIES

The genus *Nothophryne* contains five species, four of them described in 2018. They are very specialized in terms of their habitat and breeding biology, and their affinities to other members of their family have not yet been determined.

FAMILY	Pyxicephalidae: Cacosterninae
OTHER NAMES	Kogelberg Reserve Frog
DISTRIBUTION	Southwestern Western Cape province, South Africa
ADULT HABITAT	Montane fynbos at altitudes of 650–5,900 ft (200–1,800 m)
LARVAL HABITAT	Shallow pools or slow-moving streams
CONSERVATION STATUS	IUCN Near Threatened. Has a very restricted range, but much of this lies within protected areas

POYNTONIA PALUDICOLA

MONTANE MARSH FROG

CHANNING & BOYCOTT, 1989

ADULT LENGTH
Male
average $\frac{7}{8}$ in (23 mm)

Female
average 1 in (26 mm);
up to 1$\frac{3}{16}$ in (30 mm)

This small, secretive frog occurs only in an area of very high rainfall at the southern tip of Africa. It is very well camouflaged against the wet moss in which it lives. It may breed at any time of year, during periods of heavy rain. Males produce a distinctive, coarse "kruck... kruck... kruck" mating call, but mating itself has not been observed. The tadpoles have exceptionally long, narrow tails and are typically found hiding in mud. This species requires seepage areas at the bottom of valleys in order to reproduce, and is vulnerable to the damming of streams and consequent flooding of its habitat.

Actual size

The Montane Marsh Frog has a short, rounded snout, a narrow head, upwardly protruding eyes, and short, robust hind legs. The skin on its back, sides, and limbs is very rough and warty. The upper surfaces are gray or brown, and some individuals have a reddish stripe down the middle of the back. There are white or orange stripes between the eye and the upper jaw.

SIMILAR SPECIES

Poyntonia paludicola is the only species in its genus. Its shape and very warty skin make it toad-like in appearance and distinct from other frogs in its family. In terms of its habitat and small size, it is similar to the moss frogs of the genus *Arthroleptella* (see page 589), but can be distinguished from them by its low-pitched call.

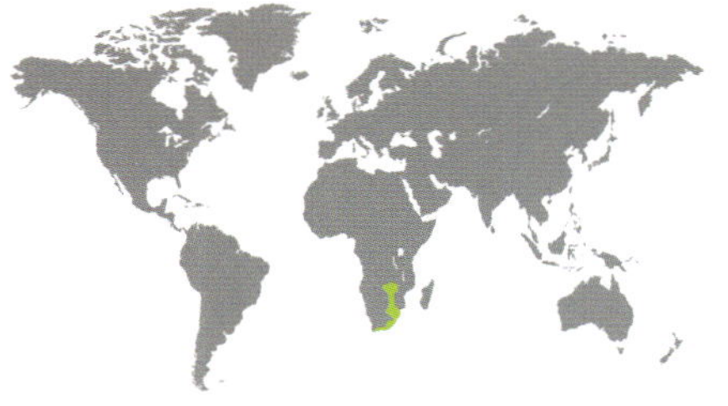

FAMILY	Pyxicephalidae: Cacosterninae
OTHER NAMES	Long-toed Frog, Striped Grass Frog
DISTRIBUTION	South Africa, Zimbabwe, Mozambique, Zambia, Swaziland
ADULT HABITAT	Grassland, savanna, woodland, heath, fynbos
LARVAL HABITAT	Permanent ponds, lakes, and reservoirs
CONSERVATION STATUS	IUCN Least Concern. Has a large range that includes several protected areas

ADULT LENGTH
Male
$\frac{7}{8}$–$1\frac{7}{16}$ in (23–37 mm)
Female
$1\frac{3}{16}$–$1\frac{15}{16}$ in (30–50 mm)

STRONGYLOPUS FASCIATUS

STRIPED STREAM FROG

(SMITH, 1849)

This athletic-looking frog has very long legs and a streamlined body, and is capable of prodigious leaps when disturbed. Found close to permanent waterbodies in a wide variety of habitats, it also thrives in gardens and plantations. It breeds in fall and early winter, in response to a drop in the temperature. The male has a single vocal sac and his call is a series of high-pitched chirps. The eggs are laid singly in shallow grassy pools or streams.

SIMILAR SPECIES

There were 11 species in the African genus *Strongylopus*, but in recent years several have been sunk into synonymy, such as the Namaqua Stream Frog (*S. springbokensis*) into Grey's Stream Frog (*S. grayii*) and the Mount Kilimanjaro Stream Frog (*S. kilimanjaro*) into the Mount Meru Stream Frog (*S. merumontanus*). Only six species remain including the Banded Stream Frog (*S. bonaespei*), which is found in the mountains of the Western Cape province.

The Striped Stream Frog has a very pointed snout and large eyes. The hind legs are very long, and the fingers and toes are extremely long and thin. The upper surfaces are marked with longitudinal stripes of dark brown, contrasting with alternating silvery, yellow, or golden-brown stripes. The legs are brown with darker spots. There is dark stripe through the eye with a white stripe beneath it.

Actual size

FAMILY	Pyxicephalidae: Cacosterninae
OTHER NAMES	Common Sand Frog, Striped Pyxie, Tremolo Sand Frog
DISTRIBUTION	Sub-Saharan Africa, excluding areas of tropical forest
ADULT HABITAT	Savanna, including arid areas
LARVAL HABITAT	Ephemeral pools
CONSERVATION STATUS	IUCN Least Concern. Has a huge range and a stable population

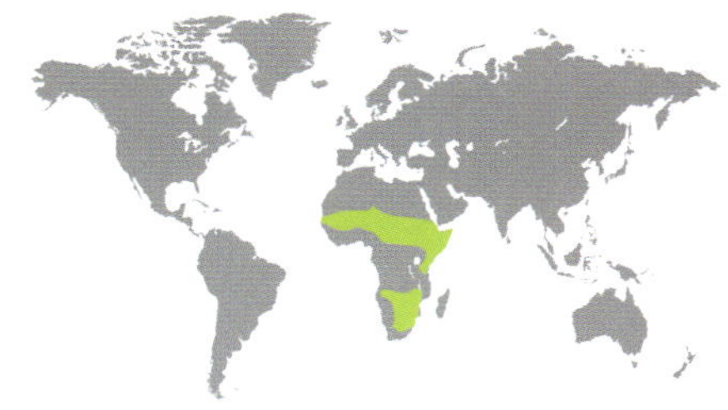

TOMOPTERNA CRYPTOTIS

CRYPTIC SAND FROG

(BOULENGER, 1907)

ADULT LENGTH

Male
1½–1¾ in (38–45 mm)

Female
1 9/16–2¼ in (40–58 mm)

The name of this very common frog refers not to its appearance, but to the fact that its tympanum is hidden beneath the skin. It has well-developed shovel-like tubercles on its hind feet, with which it burrows backward into sand, and it spends most of the year buried deep underground. The first rain in spring triggers breeding in small, temporary pools. Males produce a call consisting of a series of metallic chirps, and females lay 2,000–3,000 eggs in shallow water. In Burkina Faso, the frog is collected for food and as a source of traditional medicines.

Actual size

The Cryptic Sand Frog has a compact, almost spherical body shape, short legs, and very warty skin. There is a large shovel-like tubercle on the bottom of each foot. The upper surfaces are brown or beige with darker blotches, and there is often a pale stripe down the middle of the back. The underside of the body and legs is white.

SIMILAR SPECIES

To date, 15 species have been described in the genus *Tomopterna*. This number is likely to increase; given its very large range, *T. cryptotis* is probably more than one species. In appearance, it is virtually identical to Tandy's Sand Frog (*T. tandyi*) and Delalande's Sand Frog (*T. delalandii*). Both occur in southern Africa and can be identified reliably only by their distinctive calls.

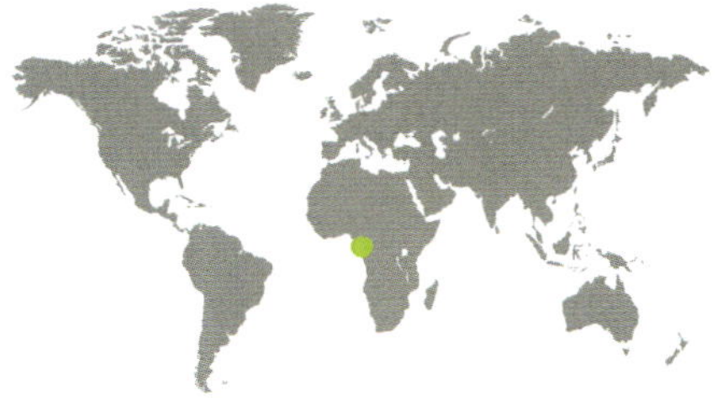

FAMILY	Pyxicephalidae: Pyxicephalinae
OTHER NAMES	West African Brown Frog
DISTRIBUTION	Southern Cameroon, Equatorial Guinea including Bioko Island, western Gabon
ADULT HABITAT	Swamps in forest and bush in farmland
LARVAL HABITAT	Pools and marshes
CONSERVATION STATUS	IUCN Least Concern. Its population is stable and it tolerates a degree of habitat alteration

ADULT LENGTH
Male
2⅝–3⁷⁄₁₆ in (65–88 mm)
Female
3–3¹¹⁄₁₆ in (76–95 mm)

AUBRIA SUBSIGILLATA

BROWN BALL FROG

(DUMÉRIL, 1856)

This highly aquatic species is unusual among frogs in that its diet consists primarily of fish; it also eats smaller frogs and arthropods. It is active only at night, spending the day buried deep in mud. It breeds in pools in swampy and marshy areas, males calling while floating in open water; their call has been likened to the sound of kettledrums. The eggs are laid in long strings. The conservation status of this frog is unclear, but the fact that it is found in areas where the bush is used extensively for farming suggests that it is tolerant of habitat change.

The Brown Ball Frog is a large, stockily built frog with muscular hind legs, a pointed snout, and large, protruding eyes. There is some webbing on the hind feet. The upper surfaces are dark brown, marked with black blotches, and the underside of the body and chin is brown, speckled with large white spots.

SIMILAR SPECIES

There is one other species in the genus *Aubria*. It lives in the same kind of habitat as *A. subsigillata* and shares its habit of eating fish. The Masako Ball Frog (*A. masako*) is a large frog found in the Congo Basin, Cameroon, and Gabon. A former species, *A. occidentalis*, described in 1995, from Liberia to Cameroon, was synonymized with *A. subsigillata*.

Actual size

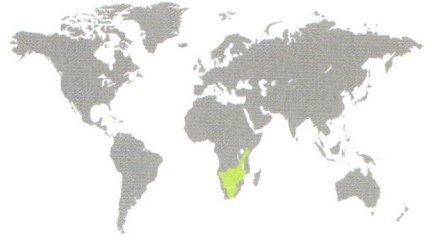

FAMILY	Pyxicephalidae: Pyxicephalinae
OTHER NAMES	Giant Bullfrog, Giant Pyxie Frog
DISTRIBUTION	Eastern and southern Africa, from Kenya in the north to South Africa in the south and Namibia in the west
ADULT HABITAT	Savanna grassland
LARVAL HABITAT	Ephemeral pools
CONSERVATION STATUS	IUCN Least Concern. Declining in parts of its range due to urbanization, and in some places is a source of food for local people

PYXICEPHALUS ADSPERSUS

GIANT AFRICAN BULLFROG

TSCHUDI, 1838

ADULT LENGTH
Male
average 6⅝ in (165 mm);
up to 9⅝ in (245 mm)

Female
average 4½ in (114 mm)

Males of this huge frog, weighing up to 3 lb (1.4 kg), are noted for their aggressiveness to one another and for their diligence as fathers. The male is much larger than the female and, in the breeding season, has two large tusks in his lower jaw. He defends his calling site, eggs, and tadpoles. As the tadpoles grow and develop, he digs a channel, enabling them to reach larger expanses of water. Breeding occurs after heavy rain and may not take place in drier years. This frog can remain underground, encased in a cocoon, for several years if there is no rain.

SIMILAR SPECIES

There are four species in the African genus *Pyxicephalus*, the latest described being Beytell's Bullfrog (*P. beytelli*) in 2024. In 2013, Beira Bullfrog (*P. angusticeps*) was validated as a distinct species. It has a less massive head than the other species and occurs in the coastal lowlands of Kenya, Tanzania, and Mozambique. In the pet trade they are called Pixie Frogs. See also the Edible Bullfrog (*P. edulis*; page 600).

The Giant African Bullfrog is a very large frog with a massive head and a very wide mouth. There are longitudinal ridges on its back and large horny tubercles on its hind feet. The upper surfaces are dark green, gray, or brown, and the underside is white. The male has a bright yellow or orange patch at the base of his forelimbs.

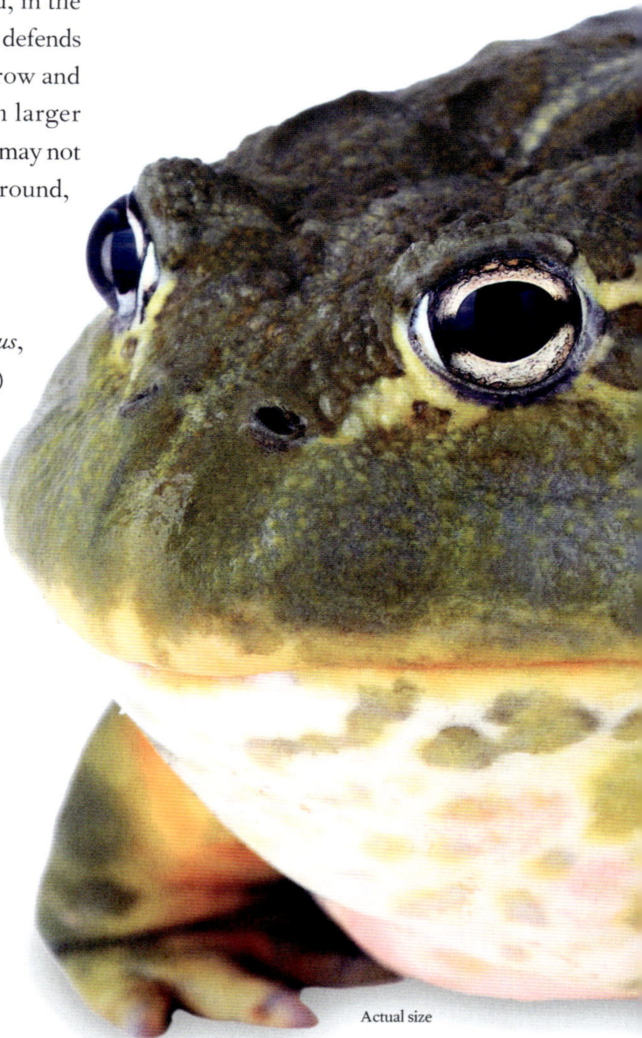

Actual size

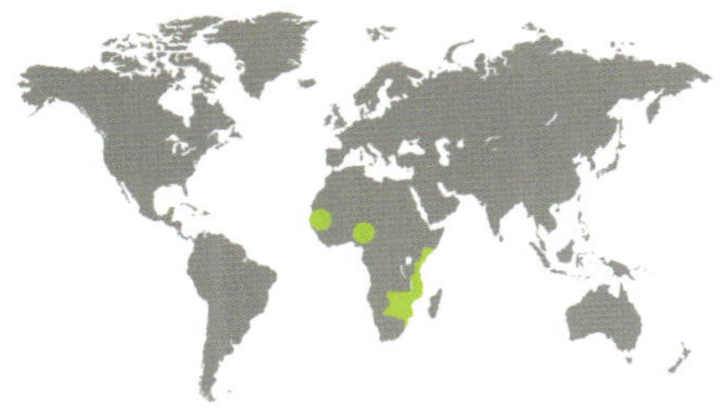

FAMILY	Pyxicephalidae: Pyxicephalinae
OTHER NAMES	African Bullfrog, Lesser Bullfrog, Peter's Bullfrog
DISTRIBUTION	East Africa, from Somalia in the north to northern South Africa; also isolated populations in West Africa, including Senegal, Mauritania, and Nigeria
ADULT HABITAT	Arid savanna grassland
LARVAL HABITAT	Ephemeral pools
CONSERVATION STATUS	IUCN Least Concern. Declining in parts of its range, especially in West Africa, where it is collected for food

ADULT LENGTH
Male
3¼–4¾ in (83–120 mm)
Female
3⅗₁₆–4⁵⁄₁₆ in (85–110 mm)

PYXICEPHALUS EDULIS

EDIBLE BULLFROG

PETERS, 1854

Though not the largest of the African bullfrogs, this is a very big frog and is widely eaten by people. Males can be twice the weight of females. The species does not require as much rain to trigger breeding as the Giant African Bullfrog (*Pyxicephalus adspersus*), and thus breeding tends to take place in response to the first rain, earlier than the larger species. The male's call is like the barking of a small dog and, by moonlight, its white vocal sacs are visible when inflated. In Mozambique, this species breeds in rice paddies. For around ten months of the year it lives underground, encased in a waterproof cocoon consisting of layers of shed skin.

The Edible Bullfrog is a rotund frog with short legs, protruding eyes, and short, longitudinal ridges and round warts on its back. It has large horny tubercles on its hind feet, which are also webbed. The upper surfaces are yellowish or olive-green, and the underside is white. Males tend to be greener and females browner in color, and the male has a yellow throat.

SIMILAR SPECIES

The least well known of the five *Pyxicephalus* species that have been described to date is Calabresi's Bullfrog (*P. obbianus*). It is found only in northern coastal regions of Somalia. An uncommon species, it is nonetheless classed as of Least Concern, partly owing to its tolerance of a wide range of habitats. See also the Giant African Bullfrog (*P. adspersus*; page 599).

Actual size

FAMILY	Rhacophoridae: Buergeriinae
OTHER NAMES	Buerger's Frog
DISTRIBUTION	Honshu, Kyushu, and Shikoku islands, Japan
ADULT HABITAT	Mountains
LARVAL HABITAT	Mountain streams
CONSERVATION STATUS	IUCN Least Concern. A common species with a stable population

ADULT LENGTH
Male
$1\frac{7}{16}$–$1\frac{3}{4}$ in (37–44 mm)
Female
$1\frac{13}{16}$–$2\frac{3}{4}$ in (49–69 mm)

BUERGERIA BUERGERI

KAJIKA FROG

(TEMMINCK & SCHLEGEL, 1838)

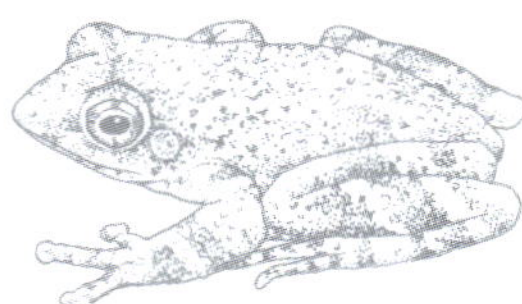

This stream-living frog breeds between April and August. Males develop nuptial pads on their hands and call throughout the day from territories along streams that they defend against other males. After going into amplexus, pairs enter the stream and 200–600 eggs are deposited in several globular masses underneath rocks and stones. While males spend around 20 days at a breeding site, females stay for only one or two nights. As a result, the sex ratio at the breeding site is strongly male-biased. The tadpoles live among pebbles in streams, browsing on algae.

SIMILAR SPECIES

Buergeria is a small genus, containing just six species, the most recently described species being the Yaeyama Kajika Frog (*B. choui*), from Japan and Taiwan. The Strong Stream Frog (*B. robusta*) occurs only on the island of Taiwan. The Ryukyu Kajika Frog (*B. japonica*) is found on Taiwan and a number of small islands in southern Japan. The Red-headed Flying Frog (*B. oxycephala*), from Hainan Island, China, was once common but is now listed as Vulnerable, much of its habitat having been converted to farmland.

The Kajika Frog is a rather slender frog with webbed hind feet and rough skin on its back. The upper surfaces are green or gray with darker green patches, and with dark transverse stripes on the legs. Some individuals have reddish-brown patches on the back and head. The underside is white.

Actual size

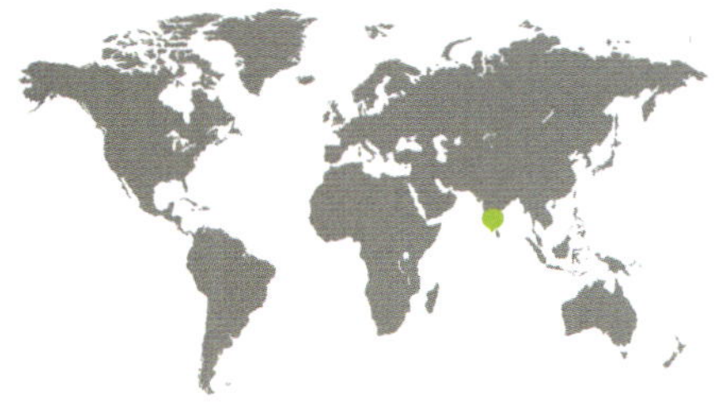

FAMILY	Rhacophoridae: Rhacophorinae
OTHER NAMES	None
DISTRIBUTION	Kerala, southern India
ADULT HABITAT	Forest at altitudes of 3,600–5,250 ft (1,100–1,600 m)
LARVAL HABITAT	Swamps and marshes
CONSERVATION STATUS	IUCN Endangered. Much of its forest habitat has been lost to agriculture

ADULT LENGTH
Male
average 1⁹⁄₁₆ in (40 mm)

Female
2⅜ in (61 mm);
based on a
single specimen

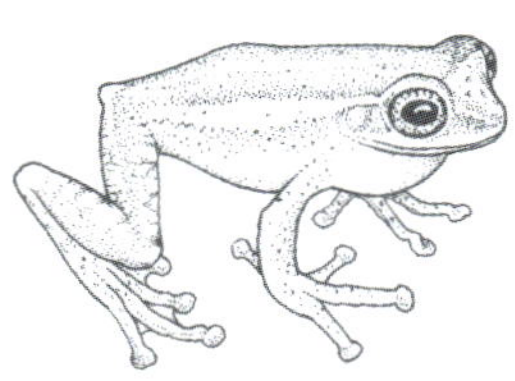

BEDDOMIXALUS BIJUI

BIJU'S TREE FROG

(ZACHARIAH, DINESH, RADHAKRISHNAN, KUNHIKRISHNAN, PALOT & VISHNUDAS, 2011)

This arboreal frog is known from only one location, a tea estate in Kerala, southern India. It breeds between April and June, gathering around swamps and marshes. As well as calling, males produce an odor from glands on their back and flanks that has been described as smelling like burnt rubber. It is not clear what function this serves. Males initially call from high up in trees but, as the night goes on, they descend to the ground. The eggs are laid in clumps attached to submerged grass.

SIMILAR SPECIES

Beddomixalus bijui is the only species in this new genus, described in 2011. It is one of many rare frogs that are restricted to the Western Ghats region of western India, and was formerly included within the genus *Polypedates*. See also the Common Hourglass Tree Frog (*P. cruciger*; page 614), the Chunam Tree Frog (*P. maculatus*; page 615); and Myristica Lowland Tree Frog (*Mercurana myristicapalustris*; page 609).

Biju's Tree Frog has a slender, elongated body, a pointed snout, webbed feet, and well-developed disks on its fingers and toes. The upper surfaces are reddish brown, with a paler stripe along each flank. The underside is white or yellowish. There is a narrow, dark stripe on the top of the head, connecting the eyelids.

Actual size

FAMILY	Rhacophoridae: Rhacophorinae
OTHER NAMES	Western Foam-nest Tree Frog
DISTRIBUTION	West and Central Africa, from Sierra Leone in the west to Uganda in the east
ADULT HABITAT	Rainforest
LARVAL HABITAT	Small pools
CONSERVATION STATUS	IUCN Least Concern. Declining in places due to deforestation, but occurs within some protected areas

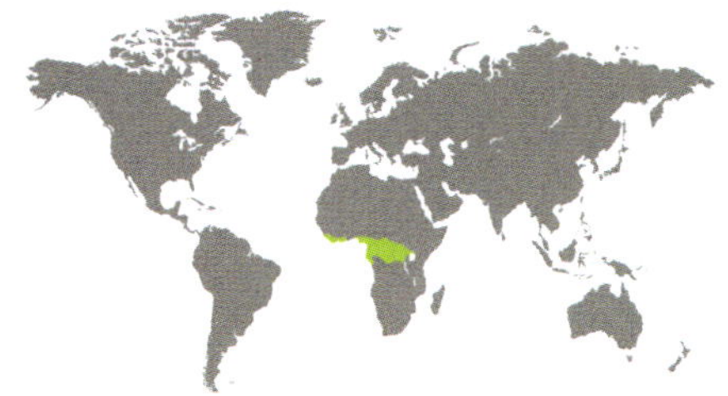

ADULT LENGTH
Male
1 15/16–1 in (44–49 mm)
Female
2–2 3/8 in (51–60 mm)

CHIROMANTIS RUFESCENS

WEST AFRICAN FOAM-NEST TREE FROG

GÜNTHER, 1869

The female of this forest-dwelling frog produces a foam nest, whipped up by rapid leg movements by both her and up to three attendant males. The nest hangs from a branch over a pool, which can be as small as a wheel rut or an elephant's footprint. The female deposits her eggs in the foam and all attending males participate in fertilizing them. After 5–8 days, the eggs hatch and the tadpoles fall into the water below. Foam nests are sometimes made inside the old nests of weaverbirds, possibly to reduce the risk of them being raided by monkeys.

SIMILAR SPECIES

There are four species in the genus *Chiromantis*, occurring across Africa and all producing hanging foam nests. Keller's Foam-nest Frog (*C. kelleri*) and Peter's Foam-nest Frog (*C. petersi*) are species found in dry savanna in eastern and northeastern Africa. The closest genus to *Chiromantis* is the Asian foam-nest frog genus *Feihyla* (page 605). See also the African Grey Tree Frog (*C. xerampelina*; page 604).

The West African Foam-nest Tree Frog has a slim body, large eyes, and long legs. The hands and feet are fully webbed, and there are large adhesive disks on the fingers and toes. The upper surfaces are gray or gray-green, often with an indistinct, darker marbled pattern. The throat and belly are white, and the undersides of the limbs are bright green.

Actual size

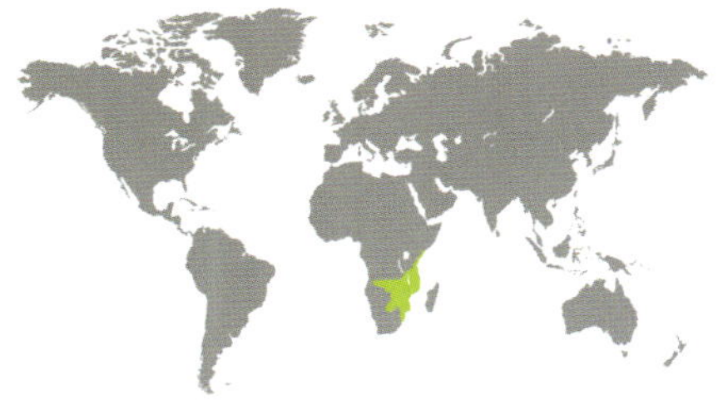

FAMILY	Rhacophoridae: Rhacophorinae
OTHER NAMES	Grey Foam-nest Tree Frog, Southern Foam-nest Tree Frog
DISTRIBUTION	East and southern Africa, from Kenya in the north to South Africa in the south and west to Angola
ADULT HABITAT	Wooded savanna
LARVAL HABITAT	Small pools and ponds
CONSERVATION STATUS	IUCN Least Concern. Has a large range and shows no evidence of decline

ADULT LENGTH
Male
$1\frac{11}{16}$–$2\frac{15}{16}$ in (43–75 mm)
Female
$2\frac{3}{8}$–$3\frac{1}{2}$ in (60–90 mm)

CHIROMANTIS XERAMPELINA

AFRICAN GREY TREE FROG

PETERS, 1854

This large frog has skin adapted to reduce water loss, enabling it to survive perched in trees in very arid conditions. It is able to change color to maintain camouflage and becomes very pale—almost white—during the heat of the day. Males call from a branch over a pool, and may be joined by up to 12 other males. When a female arrives, she is clasped by one of the males. She then produces a secretion that the male or males whip into a foam with their legs. The female ejects up to 1,200 eggs into the foam and one or more of the males shed sperm onto them.

SIMILAR SPECIES

The four species in genus *Chiromantis* are the only foam-nest frogs in Africa, but many more species are found in Asia. For example, Doria's Asian Tree Frog (*Chirixalus doriae*) is a common species found from India in the west to China and Vietnam in the east. The Samkos Bush Frog (*Feihyla samkosensis*) is found only in the Cardamom Mountains of Cambodia at altitudes up to 16,400 ft (5,000 m). See also the West African Foam-nest Tree Frog (*C. rufescens*; page 603) and Dring's Flying Frog (*F. kajau*; page 605).

The African Grey Tree Frog is a compact frog with long, slender limbs, webbed hands and feet, and adhesive disks on all its toes. The protruding eyes have horizontal pupils. It is gray or pale brown with darker markings during the night, and may be almost white during the day.

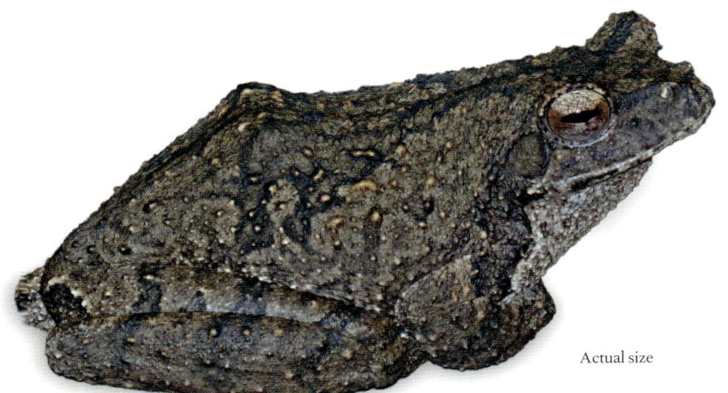

Actual size

FAMILY	Rhacophoridae: Rhacophorinae
OTHER NAMES	White-eared Tree Frog
DISTRIBUTION	Northeastern Borneo
ADULT HABITAT	Forest below 2,500 ft (750 m) altitude
LARVAL HABITAT	Pools or slow-moving streams
CONSERVATION STATUS	IUCN Least Concern. Declining across much of its range due to deforestation, but occurs within some protected areas

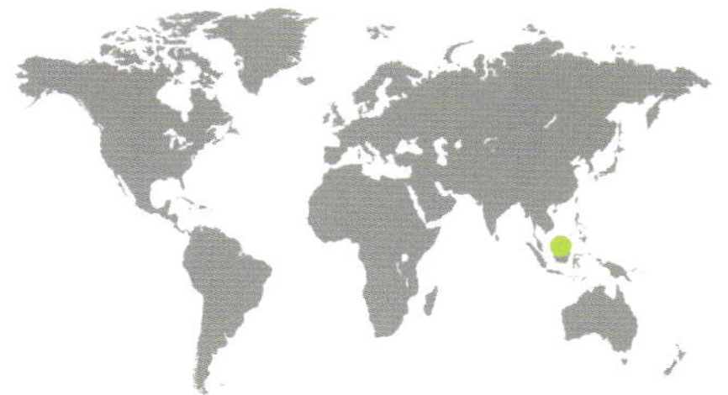

ADULT LENGTH
Male
$^{11}/_{16}$–$^{3}/_{4}$ in (18–20 mm)

Female
unknown

FEIHYLA KAJAU

DRING'S FLYING FROG

(DRING, 1983)

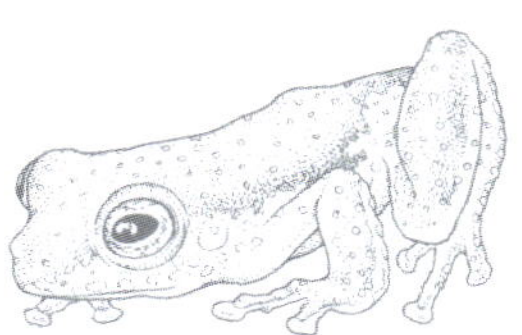

Actual size

This tiny, very rare frog, like others that are called "flying" frogs, cannot fly in the true sense. Rather, it uses its outspread limbs, which are edged with flaps of skin, and its webbed feet to glide downward across gaps between trees. The species breed at night after heavy rain, males gathering to form choruses along small streams. They call from the upper surfaces of leaves, producing a series of clicks. The eggs are attached to the underside of leaves and, when they hatch, the tadpoles fall into the water below.

Dring's Flying Frog has webbed hind feet and wavy flaps of skin along the outside edges of its arms and legs. There are well-developed adhesive disks on its fingers and toes. Its back is green, dappled with tiny white spots, and there is a white patch around the tympanums. The insides of the thighs are orange, and the flanks and belly are unpigmented.

SIMILAR SPECIES

Thirteen species have been described in the genus *Feihyla*, recently differentiated from the genus *Rhacophorus*. The Vietnamese Bubble-nest Frog (*F. palpebralis*) lives in swampy, forested areas in China and Vietnam. It lays its eggs attached to plant stems above water. It is listed as Least Concern, even though its habitat is being threatened by deforestation and pollution. The Samkos Bush Frog (*F. samkosensis*), from Cambodia, is Vulnerable.

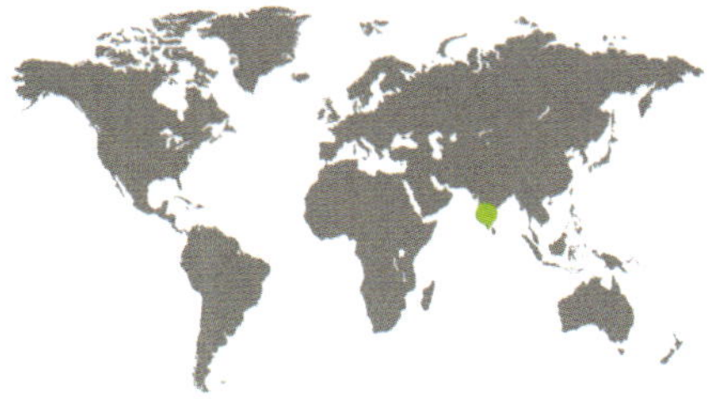

FAMILY	Rhacophoridae: Rhacophorinae
OTHER NAMES	Star-eyed Ghat Frog
DISTRIBUTION	Southern Western Ghats, Tamil Nadu, India
ADULT HABITAT	Montane forest at altitudes of 5,600–6,600 ft (1,700–2,000 m)
LARVAL HABITAT	Mountain streams
CONSERVATION STATUS	IUCN Near Threatened. Likely to be listed as threatened once further data have been collected, as it appears to be declining

ADULT LENGTH
Male
1 9/16–2 in (39–52 mm)
Female
2 1/4–2 11/16 in (58–67 mm)

GHATIXALUS ASTEROPS

GHAT TREE FROG

BIJU, ROELANTS & BOSSUYT, 2008

This frog is found in sholas, isolated patches of forest that occur in the higher mountainous areas of southern India. It lives in vegetation and leaf litter close to streams, and jumps into the water when disturbed. The male's call consists of five to seven birdlike whistles. The eggs are laid in a foam nest attached to rocks on vertical, moss-covered banks up to 10 ft (3 m) above a stream. Four days after being laid, the eggs hatch into tiny tadpoles that fall into the water. The Latin species name *asterops* means "star eyes."

SIMILAR SPECIES

There are two other species in this genus, the Green Tree Frog or Variable Ghat Frog (*Ghatixalus variabilis*), and the more recently described Large Ghat Tree frog (*G. magnus*). Both occur in similar habitats to *G. asterops*, but *G. variablis* occurs to the north and *G. magnus* to the south of the range of *G. asterops*. Neither species has the radiating gold markings in the eye that distinguish *G. asterops*.

Actual size

The Ghat Tree Frog has a rounded snout, large eyes, and a fold of skin running from the eye to the shoulder. The back is gray, brown, or cream with irregular brown patches, the flanks are yellow with brown patches, and the legs are striped. The irises of the eyes are black, adorned with delicate, radiating gold lines.

FAMILY	Rhacophoridae: Rhacophorinae
OTHER NAMES	None
DISTRIBUTION	Vietnam
ADULT HABITAT	Montane forest at altitudes of 2,000–4,300 ft (600–1,300 m)
LARVAL HABITAT	Pools and puddles
CONSERVATION STATUS	IUCN Least Concern

ADULT LENGTH
Male
⅞–1 in (21–25 mm)
Female
1–1 1/16 in (26–27 mm)

GRACIXALUS QUANGI

QUANG'S TREE FROG

ROWLEY, DAU, NGUYEN, CAO & NGUYEN, 2011

This small tree frog is known from only a single location in the mountains of Vietnam but, this being a little-explored part of the world, it may well occur elsewhere. It is an arboreal species that deposits clutches of 7–18 large eggs, encased in jelly, on leaves hanging over pools and puddles. On hatching, the tadpoles fall into the water below. The males' advertisement call is very unusual in being highly variable: They produce such a varied sequence of whistles, clicks, and chirps that no two calls sound the same.

Actual size

Quang's Tree Frog has a flattened body, a pointed, triangular snout, and many small spines on its back, legs, and eyelids. There are large adhesive disks on the fingers and toes, and the feet are partially webbed. The back is olive-green, the flanks are turquoise-blue with green, black, or brown spots, and there are yellow patches in the armpits and groin.

SIMILAR SPECIES

To date, 16 species in the Southeast Asian genus *Gracixalus* have been described, but it is likely that further exploration of the region will reveal more. The Thorny Tree Frog (*G. lumarius*), so called because of the numerous spines on its back, breeds in phytotelmata, tiny water-filled cavities in the trunks and branches of trees. It was first described as recently as 2014 but it is already listed as Endangered, as is the Sapa Bush Frog (*G. sapaensis*). Both species are from Vietnam.

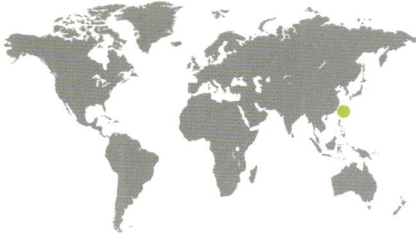

FAMILY	Rhacophoridae: Rhacophorinae
OTHER NAMES	Mientien Tree Frog
DISTRIBUTION	Taiwan
ADULT HABITAT	Shrubland, grassland, and rice fields, up to 2,500 ft (750 m) altitude
LARVAL HABITAT	Shallow pools
CONSERVATION STATUS	IUCN Least Concern. Declining in places but occurs within some protected areas

ADULT LENGTH
$\frac{15}{16}$–$1\frac{11}{16}$ in (24–43 mm)

females are larger than males

KURIXALUS IDIOOTOCUS

TEMPLE TREE FROG

(KURAMOTO & WANG, 1987)

Actual size

The Temple Tree Frog has a slender body, a pointed snout, and large, protruding eyes. There are large adhesive disks at the end of the long fingers and toes. The upper surfaces are gray, light brown, dark brown, or yellow, and there is a darker, hourglass-shaped pattern on the back.

The specific name of this frog is based on the Greek words *idios* and *ooticus*, and means "peculiar egg-laying." The eggs are laid, not in the water, but on the ground up to 1.6 ft (50 cm) from the edge of a pool, usually hidden under leaves or in a crevice. The average clutch size is 180 eggs. The eggs develop but do not hatch until heavy rain triggers them to do so and washes the emerging tadpoles into the pool. The male's call has been likened to the warbling of birds.

SIMILAR SPECIES

Twenty-three species have been described in the genus *Kurixalus*. While *K. idiootocus* lives at low altitudes, Eiffinger's Tree Frog (*K. eiffingeri*) lives at higher altitudes, in Japan as well as Taiwan, and lays its eggs in water-filled tree-holes in montane forests. It feeds its tadpoles on unfertilized eggs, the tadpoles soliciting meals by nibbling at their mother's skin when she comes to visit them.

FAMILY	Rhacophoridae: Rhacophorinae
OTHER NAMES	None
DISTRIBUTION	Southern India
ADULT HABITAT	Lowland swamp forest
LARVAL HABITAT	Pools or small streams
CONSERVATION STATUS	IUCN Endangered. Much of its swamp habitat has already been destroyed

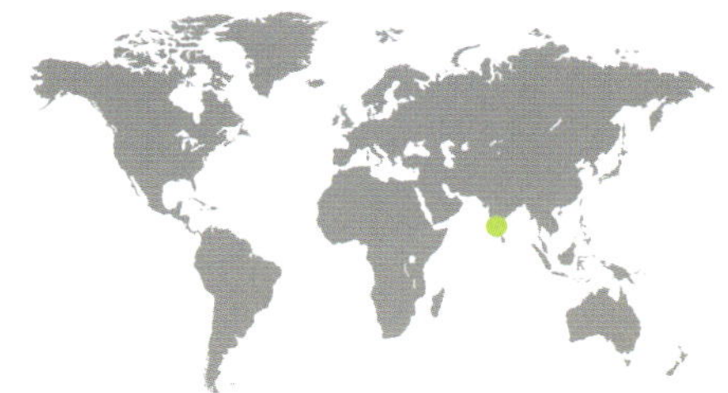

MERCURANA MYRISTICAPALUSTRIS

MYRISTICA LOWLAND TREE FROG

ABRAHAM, PYRON, ANSIL, ZACHARIAH & ZACHARIAH, 2013

ADULT LENGTH
Male
average 1⁷⁄₁₆ in (36 mm)

Female
2⅝ in (65 mm);
based on a
single specimen

This recently discovered tree frog breeds in the pre-monsoon season, males initially calling from perches above small streams and pools. As the night progresses, they descend and call from lower branches and are increasingly likely to fight other males for the possession of calling perches. When pairs form, they descend to the forest floor, gradually becoming less brightly colored until they are hard to see in the mud. The female uses her hind legs to push the fertilized eggs into the wet mud, where they develop.

Actual size

The Myristica Lowland Tree Frog has large, protruding eyes, long, slender legs, webbed feet, and adhesive disks on its fingers and toes. The upper surfaces are rusty brown and there are yellow patches on the flanks and upper arms. The underside is white.

SIMILAR SPECIES

There is only one species in this new genus, described in 2013 and named in honor of the late singer Freddie Mercury. It exemplifies the many formerly unknown species of frog that are being found in the biodiversity hotspot that encompasses the Western Ghats of India and Sri Lanka. See also Biju's Tree Frog (*Beddomixalus bijui*; page 602).

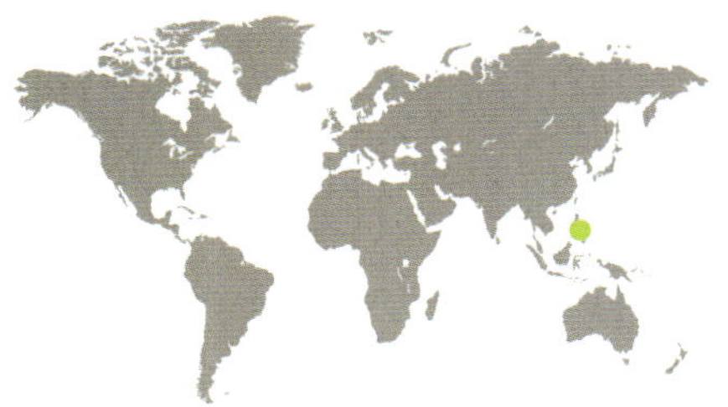

FAMILY	Rhacophoridae: Rhacophorinae
OTHER NAMES	Spiny Indonesian Tree Frog
DISTRIBUTION	Mindanao, Leyte, Bohol, and Basilan islands, Philippines
ADULT HABITAT	Forest at altitudes of 1,600–3,300 ft (500–1,000 m)
LARVAL HABITAT	Water-filled hollows in trees
CONSERVATION STATUS	IUCN Least Concern. Deforestation has caused the destruction and fragmentation of much of its habitat

ADULT LENGTH
$1\frac{3}{16}$–$1\frac{7}{16}$ in (30–37 mm)

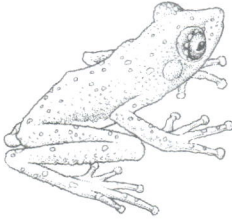

NYCTIXALUS SPINOSUS

PHILIPPINE SPINY TREE FROG

(TAYLOR, 1920)

Actual size

The Philippine Spiny Tree Frog has a pointed snout, long, slender limbs, and bony crests on its head. There are adhesive disks on the fingers and toes, and the feet are partially webbed. The upper surfaces are reddish brown in color, decorated with yellow or orange spots and numerous white spines. The underside is yellow or orange.

This tree frog gets its name from the numerous small white spines scattered over its body, head, and limbs. It is an arboreal species that is active at night, and is found in leaf litter on the forest floor during the day. It is a quick and agile jumper, and feeds mostly on ants. Its mating behavior has not been reported but it is known that females lay 30–40 eggs in water-filled tree-holes, where the tadpoles develop.

SIMILAR SPECIES

There are two other species in the genus *Nyctixalus*, known as Indonesian tree frogs. The Cinnamon Frog (*N. pictus*) is found in low-altitude forest on Palawan in the Philippines, Borneo, Sumatra in Indonesia, and in Peninsular Malaysia. Deforestation is a threat to this species. The Javan Spiny Tree Frog (*N. margaritifer*) is the third species in the genus. All three species are listed as Least Concern.

FAMILY	Rhacophoridae: Rhacophorinae
OTHER NAMES	None
DISTRIBUTION	Sabah, Malaysia
ADULT HABITAT	Forest at altitudes of 2,100–5,900 ft (640–1,800 m)
LARVAL HABITAT	Presumed to be within the egg
CONSERVATION STATUS	IUCN Least Concern. Has a small, fragmented range that includes one protected area

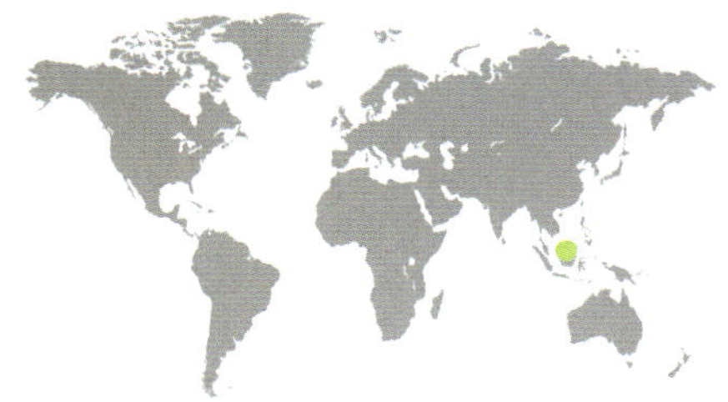

PHILAUTUS BUNITUS

GREEN BUSH FROG

INGER, STUEBING & TAN, 1995

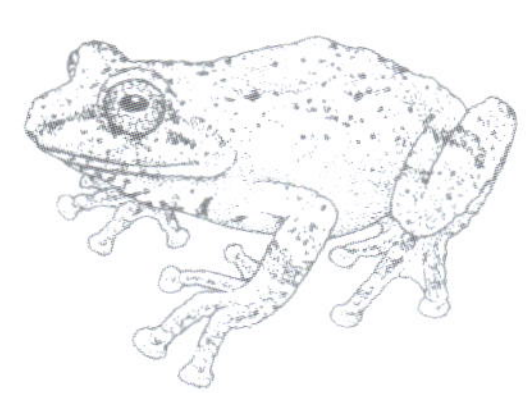

ADULT LENGTH
Male
1 3/8–1 5/8 in (35–41 mm)
Female
1 3/4–1 13/16 in (44–46 mm)

Often heard but seldom seen, this well-camouflaged tree frog lives several feet above the ground. The male's call is a series of loud, harsh notes. Mating has not been observed, but it is believed that, like other species in its genus, it lays large eggs that hatch directly into tiny frogs, the tadpole stage being completed within the egg. It is likely that this frog lays its eggs among moist epiphytic plants growing on trees. While it is abundant on Mt. Kinabalu, it is declining in other parts of its restricted northern Borneo range due to deforestation.

Actual size

The Green Bush Frog is a stocky frog with a broad head and large eyes. It has large adhesive disks on its fingers and toes, and the hind feet are webbed. The upper surfaces are pale green, decorated with brown or black speckles. The underside is green or orange, and the irises of the eyes are orange.

SIMILAR SPECIES

Hose's Bush Frog (*Philautus hosii*), also found in northern Borneo, has distinctive green eyes. It lays its eggs on leaves overhanging streams. The Vermiculated Bush Frog (*P. vermiculatus*) is a common species found in Thailand and Peninsular Malaysia. See also the Palawan Tree Frog (*P. everetti*; page 612) and the Kerangas Bush Frog (*P. kerangae*; page 613).

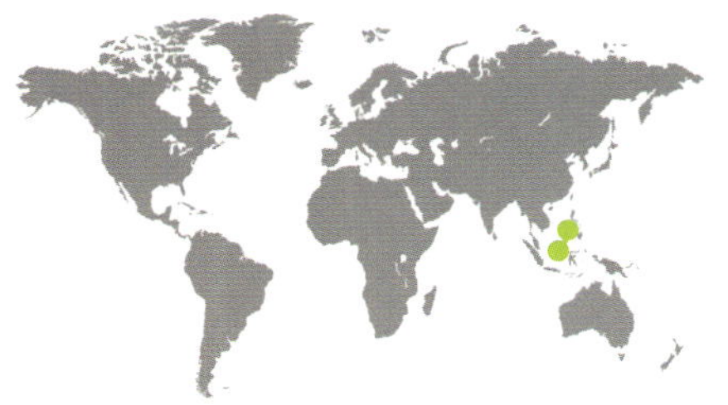

FAMILY	Rhacophoridae: Rhacophorinae
OTHER NAMES	Everett's Flying Frog, Mossy Tree Frog
DISTRIBUTION	Palawan island, Philippines
ADULT HABITAT	Forest
LARVAL HABITAT	Unknown
CONSERVATION STATUS	IUCN Endangered. Subject to habitat loss through deforestation in many parts of its range

ADULT LENGTH
Male
1³⁄₁₆–1¼ in (30–32 mm)
Female
1¾–1¹⁵⁄₁₆ in (45–49 mm)

PHILAUTUS EVERETTI

PALAWAN TREE FROG

(BOULENGER, 1894)

This small tree frog is typically found sitting on mossy logs, where its dappled skin pattern makes it very well camouflaged. The male's call is a series of short, harsh notes. It is known that it has free-swimming tadpoles. It is thought that this frog is hard to find because of its cryptic coloration, and not because it is particularly rare. It has been found in several areas of undisturbed forest, but its habitat is under threat throughout the region due to deforestation, and its future depends on the maintenance of protected areas of forest.

SIMILAR SPECIES

To date, 54 species have been described in the Southeast Asian genus *Philautus*, known as oriental bush or shrub frogs or bubble-nest frogs. The Leyte Bush Frog (*P. leitensis*), also from the Philippines, is a wholly terrestrial species; its eggs develop on land, hatching to produce tiny frogs. The Tura Scrub Frog (*P. kempiae*), from northweastern India, is listed as Critically Endangered. See also the Green Bush Frog (*P. bunitus*; page 611) and the Kerangas Bush Frog (*P. kerangae*; page 613).

Palawan Tree Frog is a stocky frog with long, slender legs and a rounded snout. There are numerous pointed skin projections on its back and head, and a line of spines across the shoulders. The upper surfaces are mottled green and black, the flanks and underside are yellow, and there is a pale stripe on the head between the eyes.

Actual size

FAMILY	Rhacophoridae: Rhacophorinae
OTHER NAMES	Kerangas Bubble-nest Frog
DISTRIBUTION	Sarawak, Malaysia
ADULT HABITAT	Forest at around 650 ft (200 m) altitude
LARVAL HABITAT	Within the egg
CONSERVATION STATUS	IUCN Vulnerable. Has a very restricted range and lives in a habitat that is subject to widespread destruction

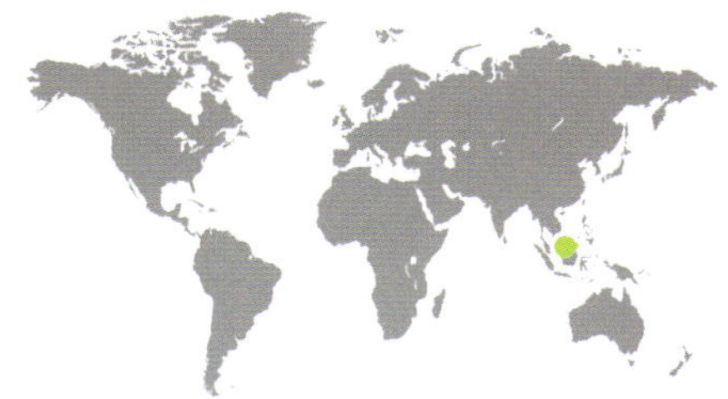

ADULT LENGTH
Male
average 1⁵⁄₁₆ in (33 mm)
Female
average 1¹¹⁄₁₆ in (43 mm)

PHILAUTUS KERANGAE

KERANGAS BUSH FROG

DRING, 1987

This frog lays its eggs in the pitchers of the Fanged Pitcher Plant (*Nepenthes bicalcarata*). The eggs appear to be laid in clutches of six to eight and, on the basis of a single observation, may be guarded by the male. The male's call consists of a series of harsh, loud notes, usually produced 6–16 ft (2–5 m) above the ground. There is no free-living tadpole, the eggs hatching directly into frogs, ⁵⁄₁₆ in (8 mm) long. Pitcher plants grow in peat swamp forest, formed where waterlogged soil prevents fallen leaves and wood from decomposing. This habitat is characteristic of Borneo but is being rapidly destroyed.

Actual size

The Kerangas Bush Frog has warty skin, slender legs, large, protruding eyes, and well-developed disks on its fingers and toes. The upper surfaces are mottled in green and brown, and the underside is white with a mottled brown pattern.

SIMILAR SPECIES

Borneo is home to at least 19 species in the genus *Philautus*. Inger's Bush Frog (*P. ingeri*) is one of the largest of them, reaching 2 in (50 mm) in length. An unrelated frog, the Borneo Narrow-mouthed Frog (*Microhyla borneensis*; page 511) deposits its eggs in *Nepenthes ampullaria*, a pitcher plant related to the Fanged Pitcher Plant used by *Philautus kerangae*. This is not as hazardous as it might seem, as neither plant species is carnivorous. See also the Green Bush Frog (*P. bunitus*; page 611) and Palawan Tree Frog (*P. everetti*; page 612).

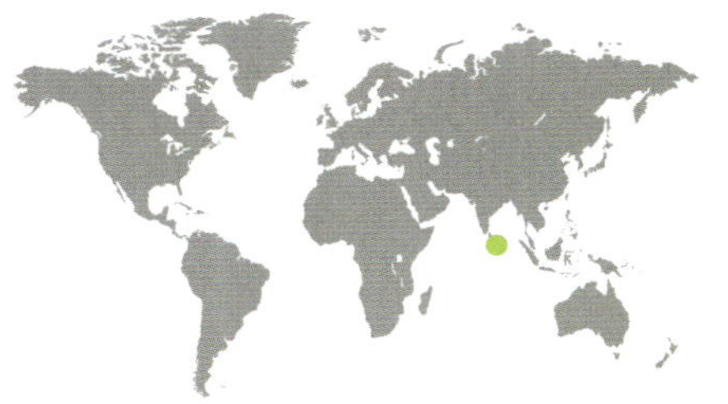

FAMILY	Rhacophoridae: Rhacophorinae
OTHER NAMES	Sri Lanka Whipping Frog
DISTRIBUTION	Southern Sri Lanka
ADULT HABITAT	Woodland and forest up to 5,000 ft (1,525 m) altitude
LARVAL HABITAT	Pools, ponds, and other bodies of still water
CONSERVATION STATUS	IUCN Least Concern. A common species that has adapted well to human-altered habitats

ADULT LENGTH
Male
1$^{15}/_{16}$–2$^{3}/_{8}$ in (50–60 mm)
Female
2$^{7}/_{8}$–3½ in (72–90 mm)

POLYPEDATES CRUCIGER

COMMON HOURGLASS TREE FROG

BLYTH, 1852

This large tree frog gets its name from the distinctive pattern on its head and back. A common species, it thrives around humans, being found in banana plantations, domestic gardens, and even houses. It is arboreal in its habits, and it lays its eggs in a foam nest hanging from vegetation over a pool so that, on hatching, the tadpoles fall into the water below. The only threat it faces seems to be from agrochemicals. As in many frogs, even very low concentrations of insecticides, herbicides, and fertilizers reduce the growth rate of tadpoles and cause young frogs to develop serious deformities, such as missing limbs.

The Common Hourglass Tree Frog has a large head, large, prominent eyes, and long, slender legs. The feet are webbed, and there are adhesive disks on the fingers and toes. The upper surfaces are green-brown, brown, or yellow, with a darker hourglass-shaped pattern on the head and back. There is a white stripe on the upper lip.

SIMILAR SPECIES

There are 25 species in the genus *Polypedates*, sometimes known as whipping frogs, found throughout Asia. The Asian Brown Tree Frog (*P. leucomystax*) is a common species across Southeast Asia, where it is known by different names in different places, such as the Javan Whipping Frog. It is able to change color, being uniformly pale by day, and dark and striped by night. See also the Chunam Tree Frog (*P. maculatus*; page 615).

Actual size

FAMILY	Rhacophoridae: Rhacophorinae
OTHER NAMES	Common Indian Tree Frog, Himalayan Tree Frog, Spotted Tree Frog
DISTRIBUTION	India, Nepal, Bhutan, Bangladesh, Sri Lanka
ADULT HABITAT	Moist forest up to 9,850 ft (3,000 m) altitude
LARVAL HABITAT	Ephemeral pools and rice fields
CONSERVATION STATUS	IUCN Least Concern. A common species that has adapted well to human-altered habitats

POLYPEDATES MACULATUS

CHUNAM TREE FROG

(GRAY, 1830)

ADULT LENGTH
Male
1⁵⁄₁₆–2³⁄₁₆ in (34–57 mm)
Female
1¾–3½ in (44–89 mm)

This tree frog has adapted well to the presence of humans, and has become semi-urban in cities with large gardens. In dry weather it often seeks refuge inside houses. It breeds in the monsoon season, and males can be heard producing a sharp "rat-a-tat" call after sunset between April and October. Up to 850 eggs are laid in a foam nest that is attached to a tree over water, into which the tadpoles fall when the eggs hatch. In dry weather the frog rubs a mixture of mucus and lipids over its skin, reducing water loss to a limited extent.

SIMILAR SPECIES

Of the 25 species of *Polypedates*, found across Asia, one of the most spectacular is the File-eared Tree Frog (*P. otilophus*) of Borneo, the females of which can be as much as 3⅞ in (100 mm) long. The species gets its name from sharp, bony ridges behind the eye and above the tympanum. It emits an unpleasant, musty smell. See also the Common Hourglass Tree Frog (*P. cruciger*; page 614).

The Chunam Tree Frog has a pointed snout, large eyes, and long, slender legs. The feet are webbed, and there are adhesive disks on the fingers and toes. The upper surfaces are olive-green, chestnut-brown, gray, or yellow, with darker spots and patches, and the legs are striped. There is a white stripe on the upper lip.

Actual size

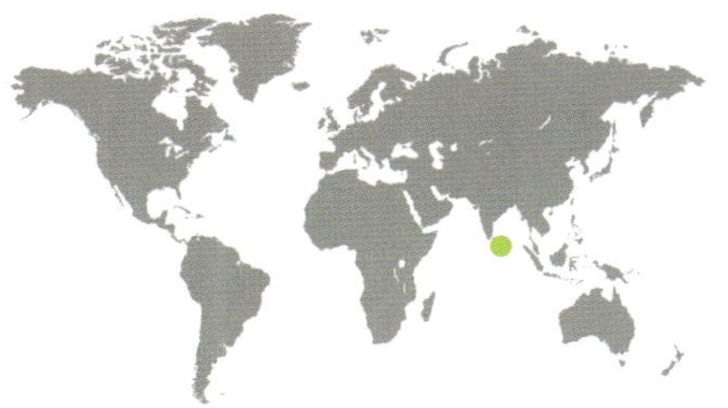

FAMILY	Rhacophoridae: Rhacophorinae
OTHER NAMES	None
DISTRIBUTION	Central Sri Lanka
ADULT HABITAT	Wet forest
LARVAL HABITAT	Within the egg
CONSERVATION STATUS	Not yet assessed. Has a very restricted range and is sensitive to habitat change

ADULT LENGTH
Male
average 1⁵⁄₁₆ in (33 mm)
Female
1⁷⁄₁₆–1¹¹⁄₁₆ in (36–43 mm)

PSEUDOPHILAUTUS HALLIDAYI

HALLIDAY'S SHRUB FROG

(MEEGASKUMBURA & MANAMENDRA-ARACHCHI, 2005)

Actual size

Halliday's Shrub Frog is a small, stout frog with a relatively large head, short body, and large eyes. There are well-developed adhesive disks on the fingers and toes, and some webbing between the toes. The upper surfaces are greenish brown with irregular darker markings. The underside is pale.

This small, secretive frog was named in 2005, for the author of the first edition of this book, the late Tim Halliday. It breeds with the onset of sustained rain. Though often found on boulders near streams, it does not breed in water. The male calls from a perch up to 16 ft (5 m) from the ground. The female finds him there, the pair go into amplexus, and they then descend to the ground. With the male on her back, the female excavates a deep hole in the ground with her hind legs and snout. The eggs are laid and fertilized, and the male then leaves. The female mixes the eggs with soil, covering them up, and leaves them to develop into tiny frogs.

SIMILAR SPECIES

To date, 78 species have been described in the genus *Pseudophilautus*, known as shrub frogs and found in Sri Lanka and India. Most have direct development, with no free-living tadpoles, and lay their eggs in the ground. The Round-snout Pygmy Frog (*P. femoralis*) is unusual in laying its eggs on the underside of leaves. Sixteen species of *Pseudophilautus* are officially Extinct, but the Starry Shrub Frog (*P. stellatus*), from Sri Lanka, thought to be extinct for more than 150 years, was rediscovered in 2009. Such species are referred to as "Lazarus species," returned from the dead. See also the Bigfoot Shrub Frog (*P. macropus*; page 617).

FAMILY	Rhacophoridae: Rhacophorinae
OTHER NAMES	Webtoe Tree Frog
DISTRIBUTION	Central Sri Lanka
ADULT HABITAT	Wet forest at altitudes of 1,980–2,500 ft (603–760 m)
LARVAL HABITAT	Within the egg
CONSERVATION STATUS	IUCN Endangered. Has a very restricted range, and is threatened by habitat destruction and chemical pollution

PSEUDOPHILAUTUS MACROPUS

BIGFOOT SHRUB FROG

(GÜNTHER, 1869)

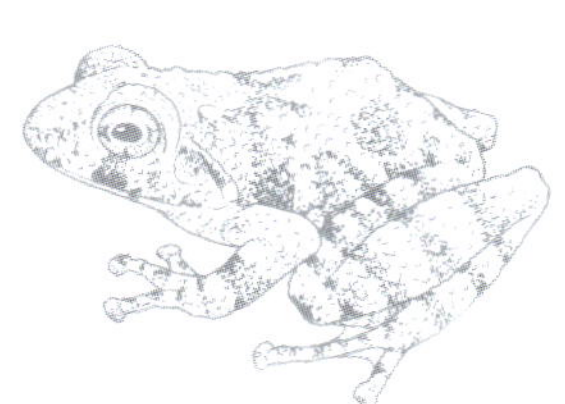

ADULT LENGTH
Male
$^{15}/_{16}$–1$^{3}/_{16}$ in (24–30 mm)
Female
1$^{3}/_{16}$–1$^{5}/_{8}$ in (30–42 mm)

This very rare frog occurs in an area of less than 4 sq miles (10 sq km) in the Knuckles Range of hills in Sri Lanka. It is typically found under leaf litter or sitting on wet boulders. Like its close relatives, it shows direct development, the tadpole stage being completed inside the egg, which hatches to release a tiny froglet. It lays its eggs in an earth-filled hole in the ground close to a stream. Much of its natural forest habitat has been cleared and replaced by cardamom plantations. It appears to be able to live in these, provided they are not sprayed with insecticides.

SIMILAR SPECIES

While *Pseudophilautus macropus* is a smooth-skinned frog, other members of this genus from Sri Lanka are notable for the roughness of their skins. Schmarda's Shrub Frog (*P. schmarda*), found in the central hill country, is covered in pointed warts. It is listed as Endangered. The Tubercle Shrub Frog (*P. cavirostris*), from southwestern Sri Lanka, has many warts and spines on its back and legs, and is listed as Vulnerable. See also Halliday's Shrub Frog (*P. hallidayi*; page 616).

The Bigfoot Shrub Frog has a plump body, a rounded snout, and large eyes. There are well-developed adhesive disks on its fingers and toes, and the hind feet are fully webbed. It is generally dark brown in color, with darker brown patches and often a W-shaped patch behind the head. The legs are striped.

Actual size

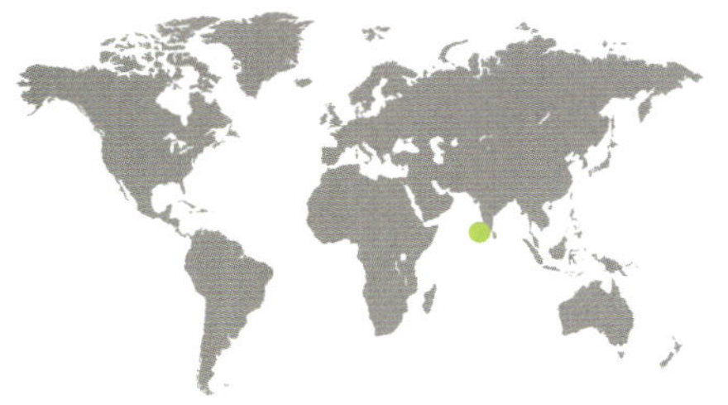

FAMILY	Rhacophoridae: Rhacophorinae
OTHER NAMES	Günther's Bush Frog, White-spotted Bush Frog
DISTRIBUTION	Southern Kerala, India
ADULT HABITAT	Wet forest
LARVAL HABITAT	Presumed to be within the egg
CONSERVATION STATUS	IUCN Vulnerable. Lives in a very restricted area that is threatened by habitat destruction

ADULT LENGTH
Male
average 1 in (25 mm)
Female
average 1⅛ in (28 mm)

RAORCHESTES CHALAZODES

PALE-SPOTTED BUSH FROG

(GÜNTHER, 1876)

Actual size

The Pale-spotted Bush Frog has a rounded snout, large eyes, and well-developed adhesive disks on its fingers and toes. The upper surfaces are bright, fluorescent green, with pale blue patches on the thighs and lower flanks. The irises of the eyes are black, with bright yellow spots, sometimes forming a cross centered on the pupil.

This tiny, colorful frog was discovered in 1874 and was not seen again for more than 130 years. An initiative launched in 2010 to find the world's "lost amphibians" resulted in its rediscovery, along with another four out of 50 targeted species in India. Active at night, it lives among reeds in the forests of the Cardamom Hills of southern India. Nothing is known about its reproductive behavior, but it is presumed that, like its close relatives, it lays its eggs on land and that they hatch directly into frogs.

SIMILAR SPECIES

To date, 76 species have been described in the genus *Raorchestes*. They are small frogs, in which males have a single large vocal sac, and are found in South and Southeast Asia. They are particularly diverse in India's Western Ghats, where new species are still being found. Thirty species of *Raorchestes* have been described since 2012, with the Nilgiri Grassland Bush Frog (*R. primarrumpfi*), from 2014, listed as Critically Endangered.

FAMILY	Rhacophoridae: Rhacophorinae
OTHER NAMES	Black-webbed Tree Frog, Reinwardt's Flying Frog
DISTRIBUTION	Java and Sumatra, Indonesia; Peninsula Malaysia, Sabah, and Sarawak, Malaysia
ADULT HABITAT	Low-altitude forest
LARVAL HABITAT	Semipermanent forest pools
CONSERVATION STATUS	IUCN Least Concern. IUCN Vulnerable to deforestation and does not occur in any protected areas

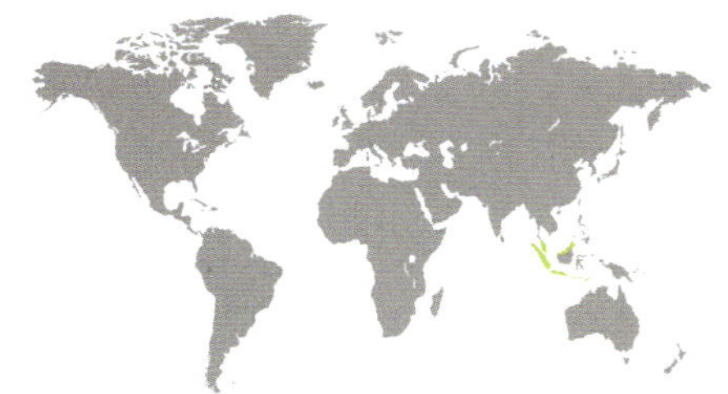

RHACOPHORUS REINWARDTII

GREEN FLYING FROG

(SCHLEGEL, 1840)

ADULT LENGTH
Male
1⅝–2 in (42–52 mm)
Female
2⅛–3⅛ in (55–80 mm)

Flying frogs do not really fly, but are able to traverse quite large distances between trees by making a slow, gliding descent. They spread their fingers and toes, using the webbing between them to slow their descent. For most of their time, they live high up in the canopy and are rarely seen, but they gather in quite large numbers to breed around pools. Male Green Flying Frogs call from the trees, producing a low, crackling chuckle. The eggs are laid in a large foam nest suspended from a branch above the breeding pool so that, when the eggs hatch, the tadpoles fall into the water below.

SIMILAR SPECIES

The Black-webbed Flying Frog (*Rhacophorus kio*), found in India, China, Thailand, Vietnam, Cambodia, and Laos, has only recently been identified as a separate species from *R. reinwardtii* and is very similar in appearance. The original flying frog of Borneo is Wallace's Flying Frog (*R. nigropalmatus*), a large species from Indonesia, Thailand, and Malaysia that has black webbing between its fingers and toes. See also the Vampire Flying Frog (*R. vampyrus*; page 620).

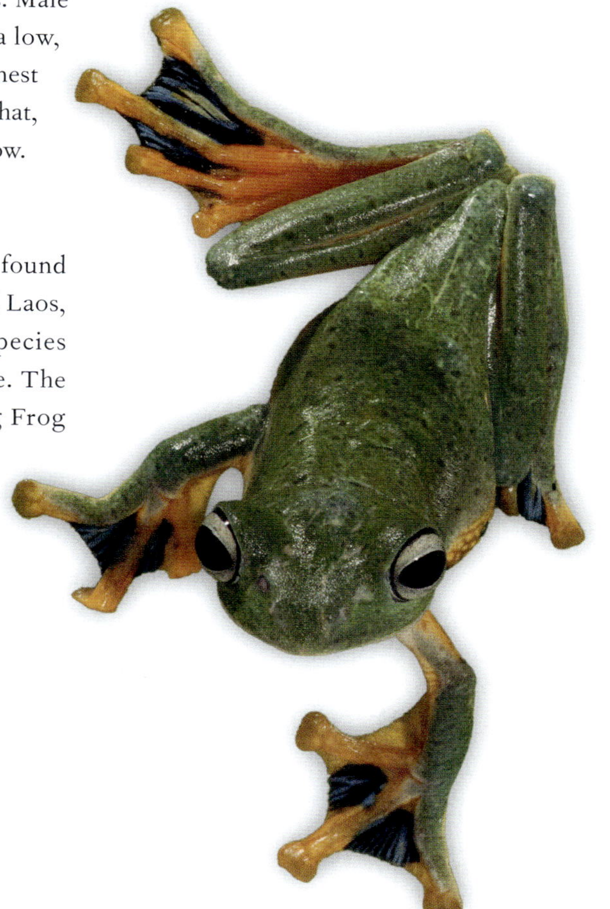

The Green Flying Frog has large, fully webbed hands and feet with large pads at the ends of the fingers and toes. There are flaps of skin on the outer edges of the arms and legs. The back is dark green with scattered dark spots, the flanks are yellow with a black patch decorated with blue spots, and the groin, underside of the legs, and feet are yellow.

Actual size

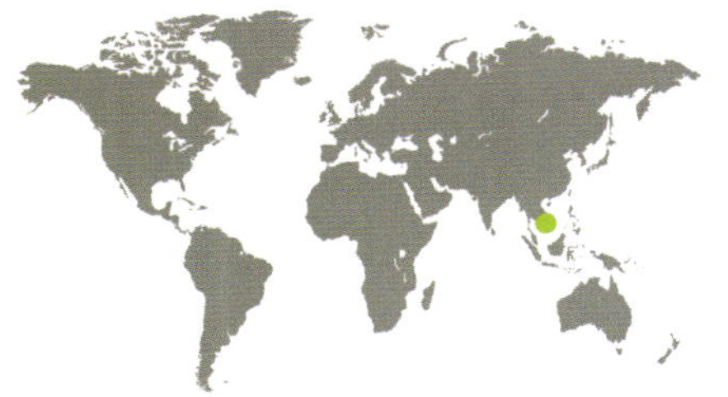

FAMILY	Rhacophoridae: Rhacophorinae
OTHER NAMES	None
DISTRIBUTION	Southern Vietnam
ADULT HABITAT	Evergreen forest at altitudes of 4,820–6,575 ft (1,470–2,004 m)
LARVAL HABITAT	Water-filled tree-holes
CONSERVATION STATUS	IUCN Endangered. Has a small range that is under threat from habitat degradation

ADULT LENGTH
Male
1⅝–1¾ in (42–45 mm)
Female
1½–2⅛ in (38–54 mm)

RHACOPHORUS VAMPYRUS

VAMPIRE FLYING FROG

ROWLEY, LE, TRAN, STUART & HOANG, 2010

This recently discovered frog gets its name from the unusual mouthparts of its tadpoles, not of the adults. The eggs are laid in masses of foam glued to the wall of a water-filled tree-hole. When the eggs hatch, the tadpoles fall into the water and it is thought that they are then fed on unfertilized eggs provided by their mother. The tadpoles have unique mouthparts, comprising a row of backward-pointing teeth in the upper jaw and a pair of forward-pointing "fangs" in the lower jaw. It seems that these are an adaptation for feeding on eggs.

SIMILAR SPECIES

Of the 46 species currently classified in the Asian genus *Rhacophorus*, around a quarter have been described in the last decade. This is largely because remote areas of countries like Vietnam have only recently been explored. The Vietnam Flying Frog (*R. calcaneus*) also occurs in Laos. It is a little-known species, listed as Vulnerable due to deforestation. See also the Green Flying Frog (*R. reinwardtii*; page 619).

The Vampire Flying Frog has a blunt snout, large eyes, well-developed pads on its fingers and toes, and spines on its heels. The upper surfaces are pale tan by day, turning to brick-red at night. The extensive webbing between the fingers and toes is gray, and the irises of the eyes are bright yellow.

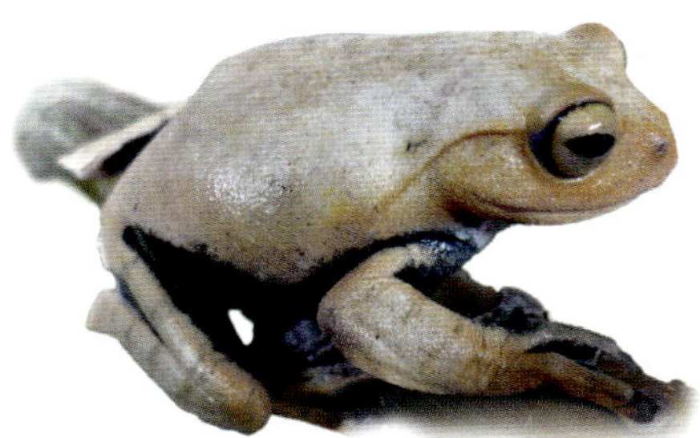

Actual size

FAMILY	Rhacophoridae: Rhacophorinae
OTHER NAMES	Previously *Polypedates eques*, Günther's Whipping Frog, Saddled Tree Frog, Spurred Tree Frog
DISTRIBUTION	Sri Lanka
ADULT HABITAT	Forest at altitudes of 4,000–7,000 ft (1,200–2,135 m)
LARVAL HABITAT	Permanent and ephemeral pools
CONSERVATION STATUS	IUCN Endangered. Has a very restricted range, subject to habitat loss through deforestation

ADULT LENGTH
Male
1¼–1$^{9}/_{16}$ in (32–40 mm)
Female
2½–2$^{13}/_{16}$ in (62–71 mm)

TARUGA EQUES

MONTANE HOURGLASS TREE FROG

(GÜNTHER, 1858)

Found only in the central hills of Sri Lanka, this arboreal frog shows marked sexual dimorphism, females being much larger than males. It lives in the canopy of the forest, descending to breed in foliage just above water. The female makes a hanging foam nest and several males may participate in fertilizing the eggs. The eggs hatch within the nest and the tadpoles fall into the water below. As in many parts of Sri Lanka, native forest within this species' range is being destroyed for its timber and to create land for growing tea and vegetables. However, it does occur within the protected Horton Plains National Park.

SIMILAR SPECIES

There are three species in the recently recognized genus *Taruga*, all of which are found in Sri Lanka. The Sharp-snout Saddled Tree Frog (*T. longinasus*) has a very long, pointed snout and is similar in its natural history to *T. eques*. The Rakwana Sharp-nosed Tree Frog (*T. fastigo*) is known only from a single, small location in southwestern Sri Lanka. Both species are Endangered.

The Montane Hourglass Tree Frog gets its name from a dark, hourglass-shaped marking on its back. It has a very pointed snout, spurs on its heels, and several warts around its cloaca. There is a marked fold of skin running from the eye to the middle of the flank. It is variable in color, being gray, brown, yellowish, or red on its back, and yellow on its underside.

Actual size

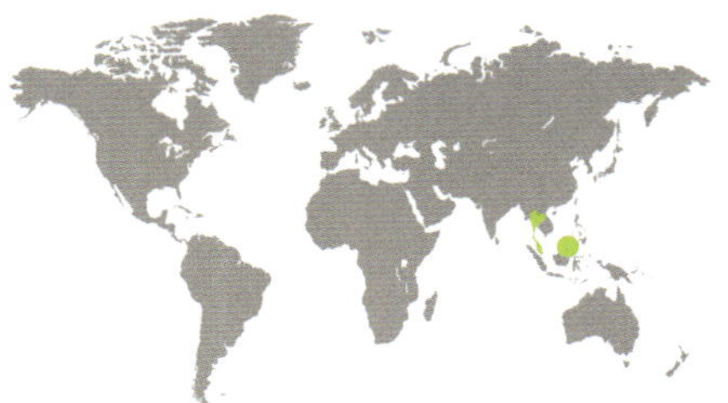

FAMILY	Rhacophoridae: Rhacophorinae
OTHER NAMES	Thorny Bush Frog
DISTRIBUTION	Thailand; Peninsular Malaysia and eastern Sabah, Malaysia; Singapore
ADULT HABITAT	Lowland rainforest up to 2,600 ft (800 m) altitude
LARVAL HABITAT	Water-filled tree-holes
CONSERVATION STATUS	IUCN Least Concern; Critically Endangered in Singapore. Is at risk from deforestation over much of its range

ADULT LENGTH
Male
unknown

Female
1⁹⁄₁₆–1¹⁵⁄₁₆ in (40–49 mm)

THELODERMA HORRIDUM

ROUGH TREE FROG

(BOULENGER, 1903)

Actual size

This frog is very little known. Whether this is because it is very rare or because it is hard to find is not clear. In support of the latter explanation is the fact that it is extremely well camouflaged against the bark of trees. It is reported that the male's call is a series of low grunts and croaks, and that the eggs are deposited in a foam nest positioned above a water-filled hole in a tree. The tadpoles grow to a length of 1¼ in (32 mm).

SIMILAR SPECIES

To date, 30 species have been described in the genus *Theloderma*. Because several of these have been found only very recently, it is likely that there are others yet to be discovered. Nine species have been described since 2016, including the False Warted Tree Frog (*T. pseudohorridum*) from Sumatra, Indonesia, in 2023, and the Cao Bang Tree Frog (*T. woltersi*) from Vietnam, in 2024. Most species are threatened by the extensive logging that occurs in much of Asia's lowland forest.

The Rough Tree Frog gets its name from the many small whitish spines, clustered on warts, on its back, head, and legs. It has very wide, bright orange pads at the end of its fingers and toes. The back is dark brown, there are dark patches on the flanks, and the underside is dark gray, marbled with white.

FAMILY	Rhacophoridae: Rhacophorinae
OTHER NAMES	Kinugasa Flying Frog
DISTRIBUTION	Honshu and Sado islands, Japan
ADULT HABITAT	Forest up to 6,600 ft (2,000 m) altitude
LARVAL HABITAT	In ponds and pools
CONSERVATION STATUS	IUCN Least Concern. Its population is stable and its range includes many protected areas

ADULT LENGTH
Male
1⅝–2⅜ in (42–60 mm)
Female
2¼–3¼₆ in (59–82 mm)

ZHANGIXALUS ARBOREUS

FOREST GREEN TREE FROG

(OKADA & KAWANO, 1924)

Some breeding populations of this frog have become tourist attractions, such is the number of individuals gathering to breed in one place. Males are considerably smaller than females, and produce a call consisting of two to six clicks. Individual females are usually clasped by several males at once. They lay 300–800 eggs, contained in a foamy mass hanging from a branch over a pool. This is formed from an albumin-based substance that the female secretes from her cloaca and whips up with her hind legs. It hardens on the outside until the eggs hatch, when the tadpoles wriggle out to fall into the water below.

SIMILAR SPECIES

The genus *Zhangixalus* contains 44 species. Schlegel's Green Tree Frog (*Z. schlegeli*) is found on the Japanese islands of Honshu, Shikoku, and Kyushu. It lays its eggs in a foam nest that is contained within a hole in the ground. Its call is different from that of *Z. arboreus* and, during mating, the female is usually clasped by only one male.

The Forest Green Tree Frog has a large head, large adhesive disks on its fingers and toes, and more webbing on its hands than on its feet. The upper surfaces are either uniformly green, or green with many brown spots edged in black. The belly is white, and the undersides of the legs are flushed with red.

Actual size

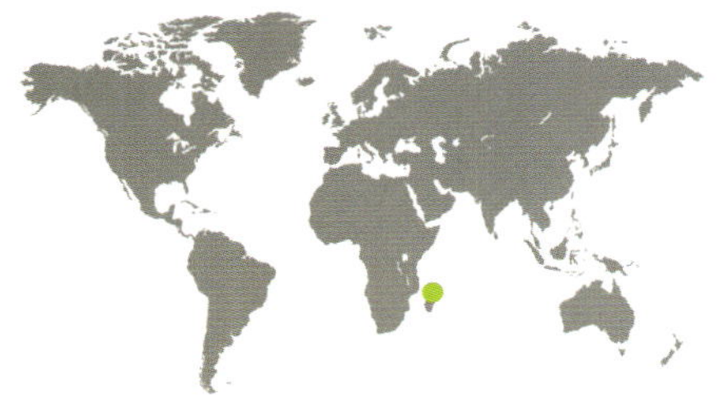

FAMILY	Mantellidae: Boophinae
OTHER NAMES	None
DISTRIBUTION	Northeastern and central Madagascar
ADULT HABITAT	Forest up to 5,600 ft (1,700 m) altitude
LARVAL HABITAT	Pools or slow-moving streams
CONSERVATION STATUS	IUCN Least Concern. Declining in places due to deforestation, but occurs within some protected areas

ADULT LENGTH
Male
2⅜–3⅛ in (60–80 mm)

Female
2⅜–4 in (60–103 mm)

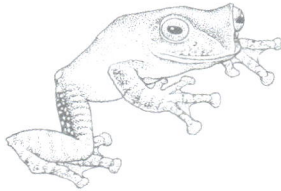

BOOPHIS MADAGASCARIENSIS

MADAGASCAR BRIGHT-EYED FROG

(PETERS, 1874)

One of the largest frogs found in Madagascar, this athletic tree frog can be found up to 30 ft (10 m) above the ground and is capable of leaping large distances. Males gather together in, or close to, water to form a chorus, and females lay around 400 eggs. The call has been described as a groaning sound and recent research has shown that it is unusually complex: No fewer than 28 distinct call types have been identified, an unprecedented level of variation for a frog. The function of the different call types is, however, not known.

The Madagascar Bright-eyed Frog has large eyes, a pointed snout, and long legs. The large hands and feet are partially webbed, and there are large adhesive disks on the fingers and toes. There are pointed skin extensions on the elbows and heels. The upper surfaces are beige, brown, or reddish, and there are dark cross-bands on the legs. The underside is cream.

SIMILAR SPECIES

To date, 87 species have been described in the genus *Boophis*, found in Madagascar and the neighboring island of Mayotte. A smaller, Vulnerable, tree frog, the Nosy Be Bright-eyed Frog (*B. brachychir*), also has the curious elbow and heel extensions seen in *B. madagascariensis*. The White-lipped Bright-eyed Frog (*B. albilabris*) is a large, bright green tree frog with fully webbed feet. See also Williams' Bright-eyed Frog (*B. williamsi*; page 625).

Actual size

FAMILY	Mantellidae: Boophinae
OTHER NAMES	None
DISTRIBUTION	Central Madagascar
ADULT HABITAT	Forest up to 6,900 ft (2,100 m) altitude
LARVAL HABITAT	Mountain streams
CONSERVATION STATUS	IUCN Critically Endangered. Known from only one small, unprotected location

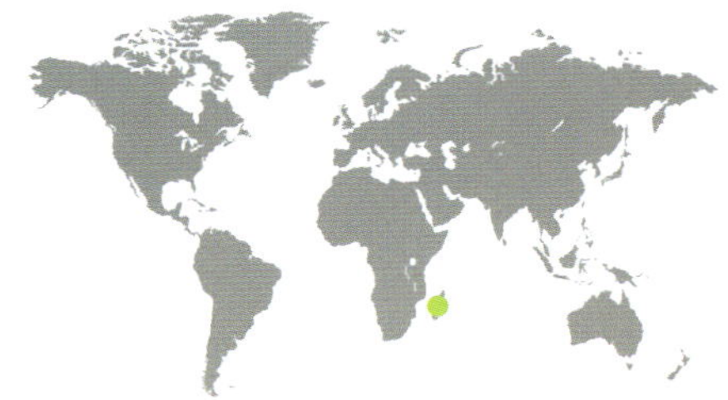

BOOPHIS WILLIAMSI

WILLIAMS' BRIGHT-EYED FROG

(GUIBÉ, 1974)

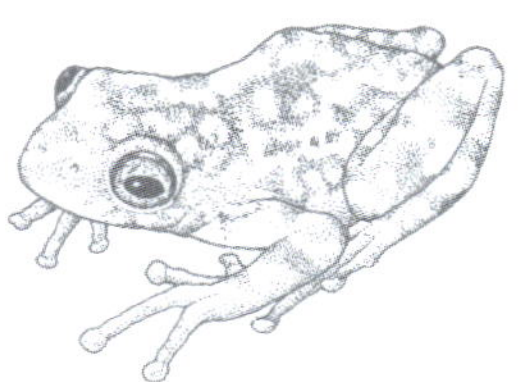

ADULT LENGTH
Male
average $1\frac{7}{16}$ in (37 mm)

Female
$1\frac{9}{16}$–$1\frac{3}{4}$ in (40–44 mm)

One of the rarest of Madagascar's many rare frogs, this is a mountain-top species, found in tiny, scattered remnants of the forest that once covered much of the Ankaratra Massif in the center of the island. It exists in an unprotected area of less than 4 sq miles (10 sq km). Its tadpoles, which develop in cold, fast-flowing mountain streams, grow to a large size. They have large sucker-like mouths that enable them to cling onto rocks while browsing on algae.

Actual size

SIMILAR SPECIES

To date, 87 species have been described in the genus *Boophis*, found only in Madagascar and Mayotte. Goudot's Bright-eyed Frog (*B. goudotii*) is a large frog, reaching up to $3\frac{7}{8}$ in (100 mm) in length, and has been known to eat *B. williamsi*. A poor climber, its natural habitat is forest, but it thrives in human-altered habitats such as rice fields, and is listed as of Least Concern. See also the Madagascar Bright-eyed Frog (*B. madagascarensis*; page 624).

Williams' Bright-eyed Frog has a slender body, long arms and legs, and large eyes. The skin is smooth except for an area of warts on the lower back. There are well-developed disks on the very long fingers and toes. The upper surfaces of the body and legs are brown with scattered orange spots, and the underside is dirty white.

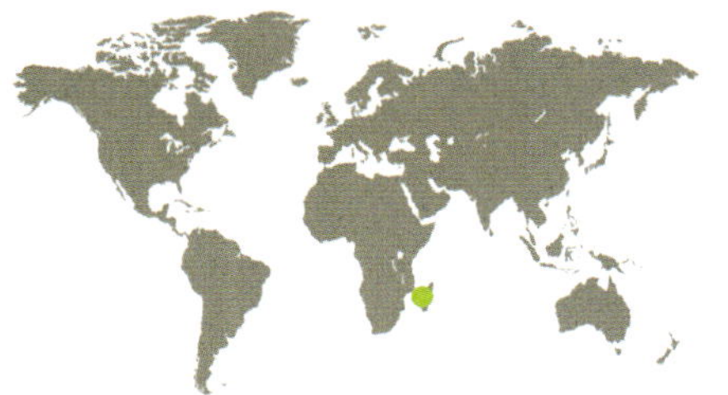

FAMILY	Mantellidae: Laliostominae
OTHER NAMES	None
DISTRIBUTION	Western Madagascar
ADULT HABITAT	Dry forest
LARVAL HABITAT	Ephemeral pools and puddles
CONSERVATION STATUS	IUCN Vulnerable. Found in a single location and under threat from habitat destruction

ADULT LENGTH
Male
1$^{9}/_{16}$–1$^{3}/_{4}$ in (39–45 mm)

Female
average 2$^{3}/_{8}$ in (60 mm)

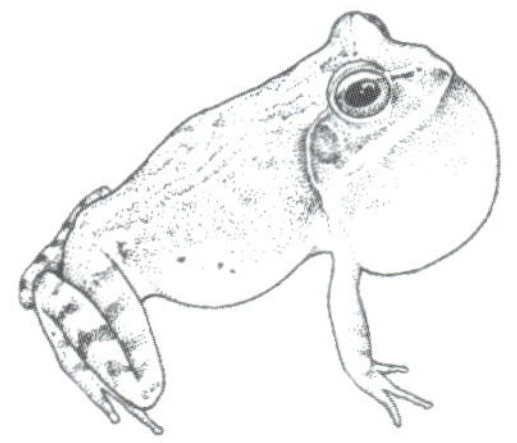

AGLYPTODACTYLUS LATICEPS

BROAD-HEADED MALAGASY JUMPING FROG

GLAW, VENCES & BÖHME, 1998

This ground-living frog, known for its jumping prowess, lives only in the Kirindy Forest near Madagascar's west coast, and is most commonly found among leaf litter on the forest floor. It is an "explosive breeder," gathering in large numbers around small temporary pools and puddles after heavy rain. Females lay around 4,000 small black eggs that soon hatch to produce tadpoles. These take only around 12 days to reach metamorphosis, ensuring that they complete their development before their breeding pool dries up.

SIMILAR SPECIES

There are six species in the genus *Aglyptodactylus*, known as Malagasy jumping frogs. The Madagascar Jumping Frog (*A. madagascariensis*) is a relatively common species with an extensive range in the forests of eastern Madagascar. The Axe-Toed Jumping Frog (*A. securifer*) is also found in the Kirindy Forest and is listed as of Least Concern. When in amplexus, males turn from brown to bright yellow, a change that is soon reversed if a pair is separated.

Actual size

The Broad-headed Malagasy Jumping Frog has a stout body, a large, broad head, very large eyes, and long, muscular hind legs with webbed feet. There are no disks on the long fingers or toes. It has smooth skin and is gray or pale brown in color, with black markings on the flanks and brown stripes on the legs.

FAMILY	Mantellidae: Laliostominae
OTHER NAMES	None
DISTRIBUTION	Western Madagascar
ADULT HABITAT	Dry forest and bushy savanna
LARVAL HABITAT	Ephemeral pools
CONSERVATION STATUS	IUCN Least Concern. Has a large range and occurs in a variety of degraded and altered habitats

LALIOSTOMA LABROSUM

MADAGASCAR BULLFROG

(COPE, 1868)

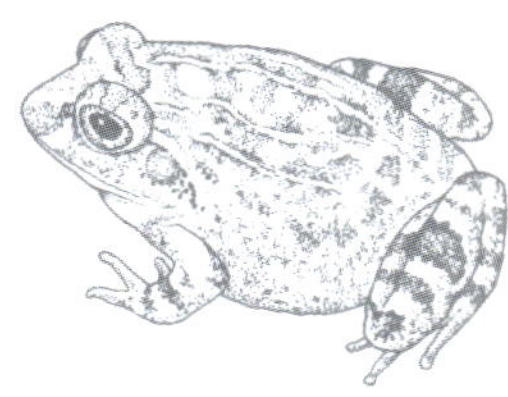

ADULT LENGTH
Male
1¾–2³⁄₁₆ in (44–56 mm)
Female
2³⁄₁₆–3⅛ in (56–80 mm)

This large frog is never seen during the dry season, when it leads a secretive subterranean existence. It emerges onto the surface after heavy rain and gathers, sometimes in large numbers, around temporary pools. Males call at night, sitting on the ground beside their chosen pool. The call has been described as a series of unharmonious notes. The female produces a large number of eggs, and the tadpoles, as they develop and grow, sometimes become carnivorous, eating smaller tadpoles of their own and other species.

SIMILAR SPECIES

The only species in its genus, *Laliostoma labrosum* is unlike any other Madagascan frog. With its robust build and burrowing habits, it resembles, but is not related to, bullfrogs living on other continents.

The Madagascar Bullfrog is a stout frog with a large head, a wide mouth, and large, protruding eyes. There are longitudinal folds of skin on the back, and the hind feet are webbed. The upper surfaces are gray or pale brown with irregular lighter and darker markings. The legs are striped and the underside is white.

Actual size

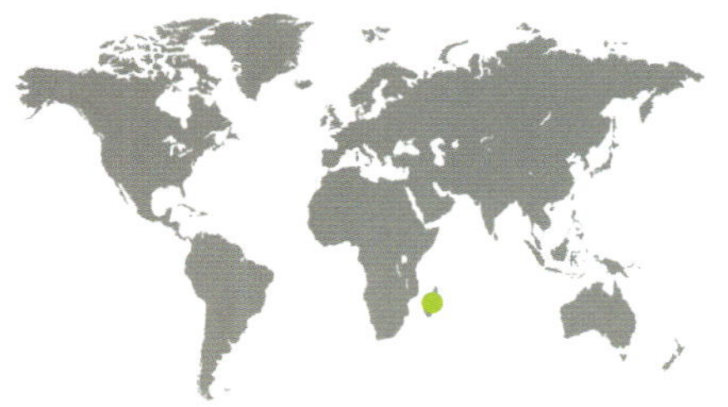

FAMILY	Mantellidae: Mantellinae
OTHER NAMES	None
DISTRIBUTION	Eastern Madagascar
ADULT HABITAT	Forest
LARVAL HABITAT	Pools
CONSERVATION STATUS	IUCN Least Concern. Has disappeared where forest has been cleared but survives in some degraded habitats and in at least one protected area

ADULT LENGTH
¾ in (19–20 mm)

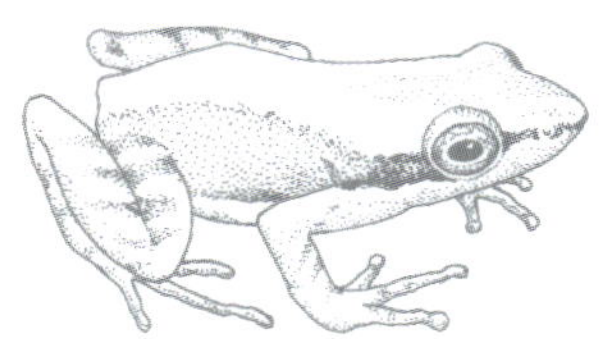

BLOMMERSIA BLOMMERSAE

MORAMANGA MADAGASCAR FROG

(GUIBÉ, 1975)

Actual size

The Moramanga Madagascar Frog has a pointed snout, webbed hind feet, and enlarged tips to its fingers and toes. Its smooth skin is very variable in color, being differing shades of brown. There is a dark stripe that runs from the tip of the snout, through the eye, to the tympanum.

This small frog is found in swampy areas and around pools within and on the edge of the remaining areas of Madagascar's forest. Males call, both by day and night, from low vegetation and, after heavy rain, can gather in very large numbers to form a chorus. Their call consists of two or three very loud chirps. The eggs are laid in masses of jelly, stuck to leaves hanging over a pool, so that, when they hatch, the tadpoles fall into the water.

SIMILAR SPECIES

To date, 12 species have been described in the genus *Blommersia*. Of these, the Angolafo Palm Frog (*B. angolafa*), described in 2010, is unusual in two respects. First, males and females differ in color, males being yellow-orange and females brown. And second, it is found in and around the fallen leaves and bracts of palm trees that have filled with water; in these tiny pools the eggs are laid and the tadpoles develop.

FAMILY	Mantellidae: Mantellinae
OTHER NAMES	Angel's Madagascar Frog
DISTRIBUTION	Southeastern Madagascar
ADULT HABITAT	Forest up to 3,300 ft (1,000 m) altitude
LARVAL HABITAT	Unknown
CONSERVATION STATUS	IUCN Vulnerable. Its range has been reduced to a small area by deforestation but it occurs in two national parks

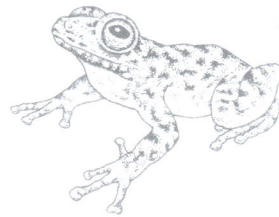

ADULT LENGTH
$2^{3}/_{8}$–$3^{1}/_{8}$ in (60–80 mm)

BOEHMANTIS MICROTYMPANUM

MADAGASCAR SMALL-EARED FROG

(ANGEL, 1935)

This large frog gets its Latin species name from its small, indistinct tympanums. This, and the fact that it lives close to noisy, fast-flowing mountain streams, suggests that its hearing is poor, and it is thought that males do not call. Active at night, it is usually seen on large rocks in torrents. It has been collected for food, but this is not thought to be a major factor in its decline, which is primarily due to destruction of its forest habitat to make way for agriculture.

SIMILAR SPECIES

The only species in its genus, this frog was formerly included in the genus *Mantidactylus* (see Ankaratra Ground Frog *M. pauliani*; page 636). It is rather similar in appearance to Grandidier's Ground Frog (*M. grandidieri*), which also lives in fast-flowing streams in pristine forest, but is distinguished from that species by the much larger disks on its fingers and toes.

The Madagascar Small-eared Frog has a stout body, a wide head, and large eyes. The hind feet are webbed, and there are well-developed disks on the fingers and toes. The skin is smooth, and the upper surfaces are olive-green in color with indistinct paler and darker markings. The underside is white.

Actual size

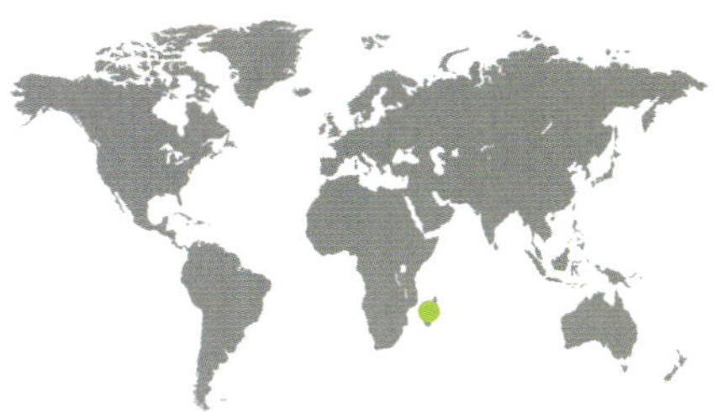

FAMILY	Mantellidae: Mantellinae
OTHER NAMES	None
DISTRIBUTION	Eastern Madagascar
ADULT HABITAT	Rainforest up to 2,300 ft (700 m) altitude
LARVAL HABITAT	Presumed to be within the egg
CONSERVATION STATUS	IUCN Least Concern. Declining in places due to deforestation, but occurs within some protected areas

ADULT LENGTH
Male
1 7/16–1 11/16 in (36–43 mm)
Female
1 5/8–1 7/8 in (41–47 mm)

GEPHYROMANTIS LUTEUS

WHITE MADAGASCAR FROG

(METHUEN & HEWITT, 1913)

Actual size

The White Madagascar Frog has a pointed snout, long, muscular legs, and large, prominent eyes. The male has paired vocal sacs, one on each side of the mouth. The back and legs are light or reddish brown with dark markings, especially on the legs. There are adhesive disks on the fingers and toes, and the feet are partially webbed.

This frog lives in pristine forest along the eastern coast of Madagascar. It is found by day on the ground, among leaf litter, and is able to leap considerable distances. At night, it moves up into the trees, males calling from perches 3–6 ft (1–2 m) above the ground. Their call is a series of melodious notes. They do not form choruses and do not call close to water, suggesting that, as in closely related species, the eggs are laid on land and hatch directly into small frogs.

SIMILAR SPECIES

To date, 61 species have been described in the genus *Gephyromantis*, known as grainy frogs and exemplified by the Grainy Madagascar Frog (*G. granulatus*), which has warty, wrinkled skin. The Resilient Grainy Frog (*G. mafy*), from eastern Madagascar, is Critically Endangered because its entire range lies within degraded habitat.

FAMILY	Mantellidae: Mantellinae
OTHER NAMES	None
DISTRIBUTION	Eastern Madagascar
ADULT HABITAT	Rainforest up to 4,600 ft (1,400 m) altitude
LARVAL HABITAT	Water-filled leaf axils of plants
CONSERVATION STATUS	IUCN Least Concern. Declining in places but occurs within several protected areas

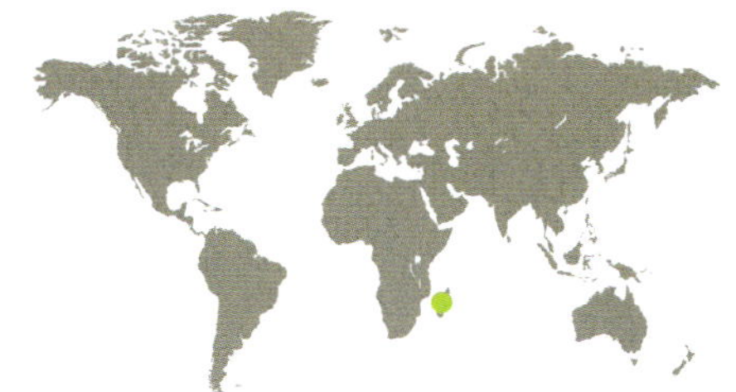

GUIBEMANTIS PULCHER

TSARAFIDY MADAGASCAR FROG

(BOULENGER, 1882)

ADULT LENGTH
Male
average 1 in (25 mm)

Female
7/8–1 1/8 in (22–28 mm)

This small, colorful frog deposits its eggs in the water-filled leaf axils of screw pines (*Pandanus* spp.). Also known as pandans, these palm-like trees occur in the tropics and, in Madagascar, their leaves are widely used to roof huts. It is not known whether the tadpoles feed themselves in their tiny pools or whether they are fed by their parents, as they are in many frog species that breed in plants. Adults in this and related species have conspicuous femoral glands on the underside of their thighs; it is not known what function these serve.

Actual size

The Tsarafidy Madagascar Frog has a pointed snout, large, protruding eyes, and smooth skin. There are well-developed adhesive disks on the fingers and toes. The head, body, and legs are green with brown or purple spots, and there is a purple band running from the tip of the snout, through the eye and along the flank.

SIMILAR SPECIES

It is likely that further species will be added to the 25 so far described in the Madagascan genus *Guibemantis*. The Free Madagascar Frog (*G. liber*) is noted for its very loud call. It is unusual within the genus in that it lays its eggs in ponds rather than in the axils of leaves. The Dotted Madagascar Frog (*G. punctatus*) gets its name from the many spots that decorate its body. It is Critically Endangered.

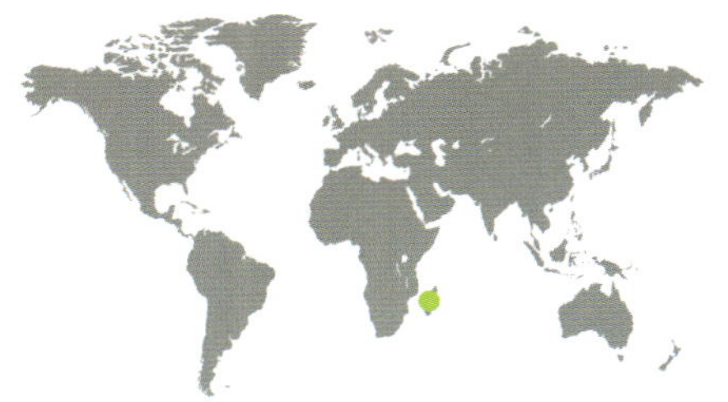

FAMILY	Mantellidae: Mantellinae
OTHER NAMES	Red Mantella
DISTRIBUTION	Eastern Madagascar
ADULT HABITAT	Wet forest
LARVAL HABITAT	Small pools
CONSERVATION STATUS	IUCN Endangered. Its habitat has been greatly reduced and fragmented by deforestation and it is very popular in the pet trade

ADULT LENGTH
Male
¾–¹⁵⁄₁₆ in (19–24 mm)
Female
¾–1¼ in (19–31 mm)

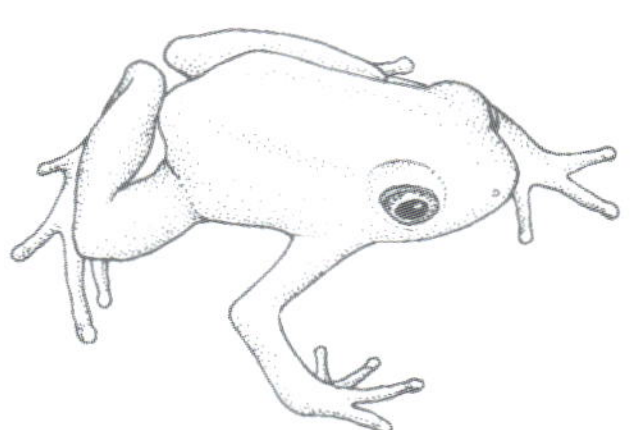

MANTELLA AURANTIACA

GOLDEN MANTELLA

MOCQUARD, 1900

Actual size

The Golden Mantella is a small, plump frog with skin that varies in color from yellow, through orange, to red. The iris of the eye is black. The hands and feet are not webbed, and the slender fingers and toes are tipped by small disks. Females are slighter broader than males, and juveniles are green and black.

This small, brightly colored frog is most active early in the morning and late in the afternoon. It is wholly terrestrial and is most commonly seen in sunny patches in swampy areas within the rainforest. The male's call consists of short, cricket-like chirps. A clutch of 20–60 eggs is laid in moist leaf litter. When these hatch, the tadpoles are washed into small pools by heavy rain. Young frogs reach sexual maturity within a year. The vivid coloration of this frog warns potential predators that its skin contains a variety of toxic compounds, derived from its insect prey.

SIMILAR SPECIES

The color of *Mantella aurantiaca* distinguishes it from the 15 other species in the genus. Most are threatened by extinction to varying degrees as a result of habitat destruction and collection for the international pet trade. Such collection is now subject to more effective control by international regulations than it was previously. See also Cowan's Mantella (*M. cowanii*; page 633), the Madagascan Mantella (*M. madagascariensis*; page 634), and the Green Mantella (*M. viridis*; page 635).

FAMILY	Mantellidae: Mantellinae
OTHER NAMES	Black Golden Frog, Cowan's Golden Frog, Harlequin Mantella
DISTRIBUTION	Central Madagascar
ADULT HABITAT	Forest along streams at altitudes of 3,300–6,600 ft (1,000–2,000 m)
LARVAL HABITAT	Streams
CONSERVATION STATUS	IUCN Endangered. Threatened by habitat destruction and overcollection for the pet trade

ADULT LENGTH
⅞–1⅛ in (22–29 mm)

MANTELLA COWANII

COWAN'S MANTELLA

BOULENGER, 1822

This extremely rare frog inhabits the high plateau in east-central Madagascar, living in strips of forest along mountain streams. Its population is estimated to have declined by 80 percent over a period of 15 years and its range has been reduced to less than 4 sq miles (10 sq km). Although some new locations were reported in 2009, the species remains very rare and does not occur in any protected areas. Much of its forest habitat has been destroyed and it has been subject to intensive collecting for the global pet trade. A moratorium on its export from Madagascar was declared in 2003.

Actual size

Cowan's Mantella is a small, rather stout frog with a pointed snout and glossy skin. The head, body and legs are black and there are red, sometimes orange or yellow, patches on the legs and flanks where the legs join the body. The irises of the eyes are black and there are blue spots on the underside.

SIMILAR SPECIES

M. cowani is similar to, and hybridizes with, Baron's Mantella (*M. baroni*), a common species that is more tolerant of habitat change and is listed as of Least Concern. The Yellow Mantella (*M. crocea*) is predominantly yellow or green, breeds in swamps, and is listed as Endangered. See also the Golden Mantella (*M. aurantiaca*; page 632), the Madagascan Mantella (*M. madagascariensis*; page 634), and the Green Mantella (*M. viridis*; page 635).

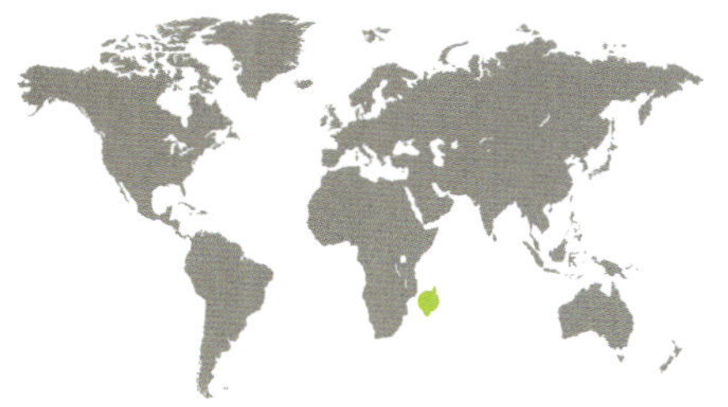

FAMILY	Mantellidae: Mantellinae
OTHER NAMES	Malagasy Painted Mantella
DISTRIBUTION	Eastern Madagascar
ADULT HABITAT	Near streams in primary rainforest
LARVAL HABITAT	Streams
CONSERVATION STATUS	IUCN Vulnerable. Threatened by habitat destruction and overcollection for the pet trade

ADULT LENGTH
Male
up to ⅞ in (22 mm)
Female
up to 1 in (25 mm)

MANTELLA MADAGASCARIENSIS

MADAGASCAN MANTELLA

(GRANDIDIER, 1872)

Actual size

The Madagascan Mantella is a small, rather stout frog with a pointed snout. It is extremely variable in color, but most individuals have a black head, back, and broad band around the body. A narrow yellow or green stripe runs from behind the eye to the tip of the snout. The arms, and a large patch on each flank, are yellow or green, with black markings. The legs, and a large patch around each groin, are green, yellow, or orange, with black cross-stripes. The undersides of the legs may be red.

This tiny, brightly colored frog is active by day, hopping about on the ground in the few patches of natural Madagascan forest that have survived extensive deforestation on the island. It is very variable in color, some individuals being green or yellow and black. Males are territorial and produce a chirp-like call. During mating, eggs are deposited on land but the tadpoles develop in streams. Mantellas represent a remarkable example of convergent evolution with the unrelated poison-dart frogs (pages 363–388) of South America. Both are active by day, have toxic skin secretions, and are brightly colored. Their vivid hues warn potential predators that they are poisonous.

SIMILAR SPECIES

Mantella madagascariensis is similar in color pattern to, and can be confused with, Baron's Mantella (*M. baroni*), a relatively common species, and Parker's Mantella (*M. pulchra*), which is classified as Near Threatened. In all, there are 16 species of mantella, most of which are threatened to some degree by the habitat destruction and degradation that affects much of Madagascar. See also the Golden Mantella (*M. aurantiaca*; page 632), Cowan's Mantella (*M. cowanii*; page 633), and the Green Mantella (*M. viridis*; page 635).

FAMILY	Mantellidae: Mantellinae
OTHER NAMES	Lime Mantella
DISTRIBUTION	Northern Madagascar
ADULT HABITAT	Dry lowland forest up to 1,000 ft (300 m) altitude
LARVAL HABITAT	Temporary streams
CONSERVATION STATUS	IUCN Endangered. Threatened by habitat destruction and overcollection for the pet trade

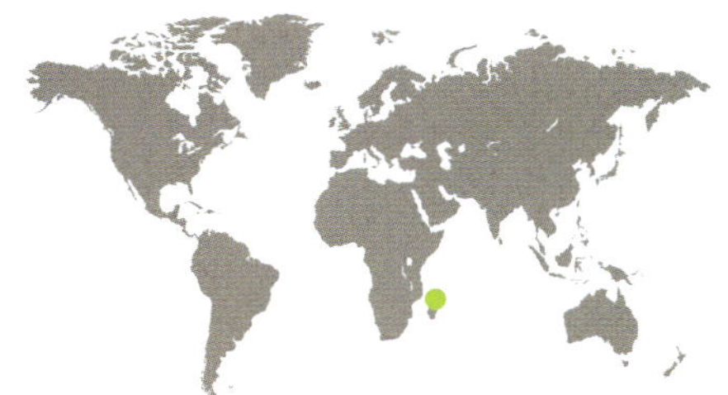

MANTELLA VIRIDIS

GREEN MANTELLA

PINTAK & BÖHME, 1988

ADULT LENGTH
Male
7/8–1 in (22–25 mm)
Female
1 1/16–1 3/16 in (27–30 mm)

Confined to a very small area at the northern tip of Madagascar, this small terrestrial frog lives near streams. It feeds on insects and other invertebrates, some of which provide the toxic compounds that make it distasteful to predators. The male attracts females by means of a call consisting of a series of clicks. Mating occurs on land and the eggs are laid on the ground close to streams. They hatch during heavy rain, which washes the tadpoles into the streams. In the past, intensive collecting for the pet trade depleted the population, but this is now controlled and the species is also being kept in several captive populations.

Actual size

The Green Mantella has a green to yellow back, head, and legs; a black "mask" through the eyes; and a white stripe on the upper lip. The underside is black, patterned with blue spots. There are small adhesive disks on the fingers and toes. The female is broader and plumper than the male.

SIMILAR SPECIES

Mantella viridis is closely related to Ebenau's Mantella (*M. ebenaui*), which also occurs in the extreme north of Madagascar. It is listed as of Least Concern, as is the Bronze Mantella (*M. betsileo*), which is found in a number of areas in eastern and central Madagascar. See also the Golden Mantella (*M. aurantiaca*; page 632), Cowan's Mantella (*M. cowanii*; page 633); and the Madagascan Mantella (*M. madagascariensis*; page 634).

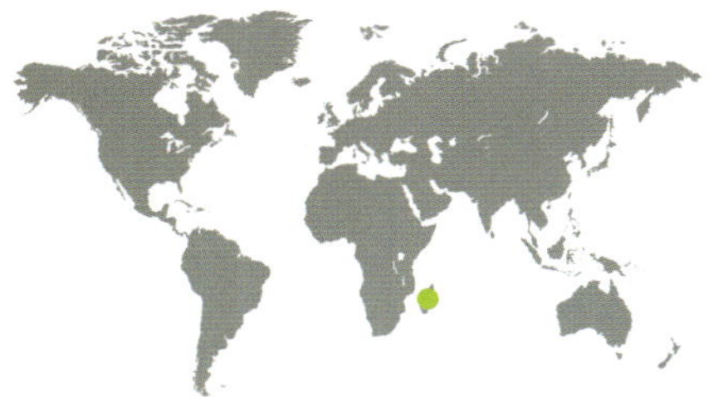

FAMILY	Mantellidae: Mantellinae
OTHER NAMES	None
DISTRIBUTION	Central Madagascar
ADULT HABITAT	Montane forest
LARVAL HABITAT	Slow-moving streams
CONSERVATION STATUS	IUCN Critically Endangered. Its range has been reduced by habitat destruction to a single location

ADULT LENGTH
Male
1–1¼ in (25–32 mm)
Female
15⁄16–15⁄16 in (24–34 mm)

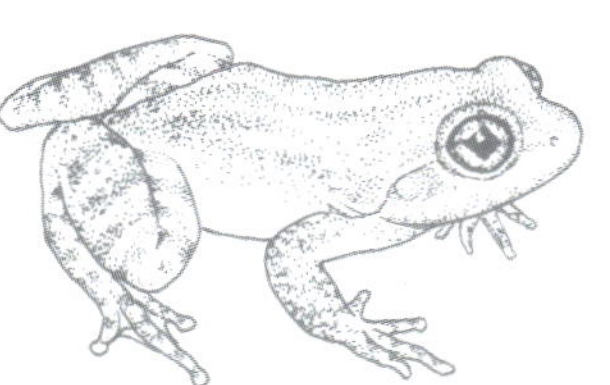

MANTIDACTYLUS PAULIANI

ANKARATRA GROUND FROG

GUIBÉ, 1974

Actual size

The Ankaratra Ground Frog is a small, rather plump frog with a short snout and protruding eyes. The fingers and toes have blunt tips and the feet are webbed. The upper surfaces are brown with large, dark brown spots, and there are dark brown stripes on the legs. The underside is pale.

This tiny frog epitomizes a category of amphibians that are particularly threatened by extinction: those that live at high altitude. Mountain-living frogs tend naturally to occupy small ranges and, if their habitat alters, either through human activities or climate change, they are unable to disperse to new locations. The Ankaratra Massif in the center of Madagascar is home to around 15 frog species, but much of its natural forest habitat has been destroyed by tree-felling and fires. This aquatic frog appears to persist in a single stream bordered by remnants of forest.

SIMILAR SPECIES

To date, 58 species of the Madagascan genus *Mantidactylus* have been described. Most closely related to *M. pauliani* is the Andringitra Madagascar Frog (*M. madecassus*). This is found at altitudes of 4,900–8,200 ft (1,500–2,500 m), typically in mountain streams above the tree-line. It is known from only about ten locations and is listed as Endangered.

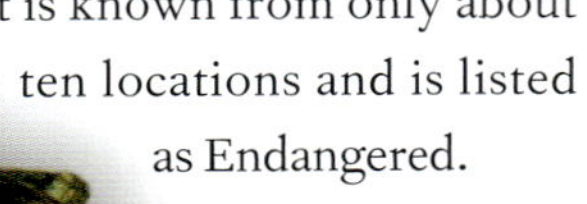

FAMILY	Mantellidae: Mantellinae
OTHER NAMES	None
DISTRIBUTION	Southeastern Madagascar
ADULT HABITAT	Rocky outcrops in forest and above the tree-line at altitudes of 4,430–8,200 ft (1,350–2,500 m)
LARVAL HABITAT	Streams
CONSERVATION STATUS	IUCN Near Threatened. Its habitat has been reduced and fragmented, but it occurs in some protected areas

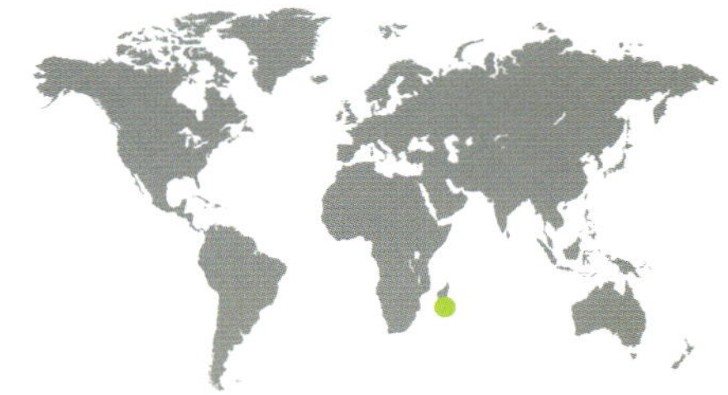

SPINOMANTIS ELEGANS

ELEGANT MADAGASCAR FROG

(GUIBÉ, 1974)

ADULT LENGTH
Male
unknown

Female
1$^{15}/_{16}$–2$^{3}/_{8}$ in (50–60 mm)

So rare is this handsome frog that no adult males have ever been collected. It lives among boulders and caves in the mountains of Madagascar, and males call from within caves and rock crevices. Its tadpoles live in streams and take at least a year to reach metamorphosis. In doing so they grow very large, reaching 4⅛ in (106 mm) in total length. They are also black in color, a feature of many tadpoles living at high altitudes. Black pigment in the skin may be a protection against ultraviolet radiation, a means of maximizing absorption of the sun's heat, or a defense against predators.

SIMILAR SPECIES

To date, 14 species have been described in the genus *Spinomantis*, found only in eastern Madagascar. Bruna's Madagascan Stream Frog (*S. brunae*) occurs at low altitudes in association with rocky streams in pristine forest. It is listed as Endangered. Peracca's Madagascar Frog (*S. peraccae*) has an extensive range, breeds in slow-moving forest streams, and is listed as of Least Concern.

The Elegant Madagascar Frog has a slim, elegant body, long, slender limbs, and large, protruding eyes. The fingers and toes are long and thin, and end in well-developed adhesive disks. The upper surfaces are brown, with large, darker patches edged in yellow, forming a delicate reticulated pattern. The underside is pale brown.

Actual size

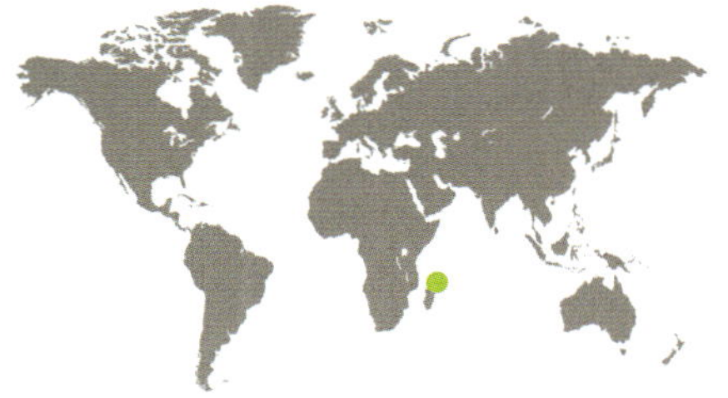

FAMILY	Mantellidae: Mantellinae
OTHER NAMES	None
DISTRIBUTION	Northern Madagascar
ADULT HABITAT	Limestone outcrops
LARVAL HABITAT	Small pools
CONSERVATION STATUS	IUCN Endangered. A rare species with a very restricted range

ADULT LENGTH
Male
2¹⁄₁₆–2³⁄₁₆ in (53–56 mm)

Female
2¹¹⁄₁₆ in (66–67 mm)

based on four specimens of each sex

TSINGYMANTIS ANTITRA

ANKARANA KARST FROG

GLAW, HOEGG & VENCES, 2006

This recently discovered frog gets its name from its habitat, karst limestone rock formations, called *tsingy* in Malagasy. It is known from only one location, the Ankarana Special Reserve in northern Madagascar. Evidence of its breeding behavior has been found in a rock-strewn dry riverbed. The male's call consists of rather quiet, short notes, repeated at regular intervals. Tadpoles have been found in small rockpools in the riverbed, an environment that seems to provide little food. The form of the tadpoles' mouthparts is similar to that of tadpoles that feed on eggs, but there is no direct evidence that this is how they feed.

SIMILAR SPECIES

Tsingymantis antitra is the only species in its genus. Analysis of its skeleton, supported by genetic data, suggests that it separated from other frogs in the family Mantellidae around 40 million years ago, early in the family's evolution. In living in a relatively dry habitat it differs from other mantellids, which are found mostly in wet forest habitats.

The Ankarana Karst Frog has very large disks on its fingers and toes, large eyes, and conspicuous tympanums. The skin on the back is smooth in females and granular in males. The upper surfaces are brown with irregular green or light brown markings. The underside is bluish white on the belly, and brown on the throat and limbs. The irises of the eyes are silvery.

Actual size

FAMILY	Mantellidae: Mantellinae
OTHER NAMES	None
DISTRIBUTION	Northwestern Madagascar
ADULT HABITAT	Forest
LARVAL HABITAT	Unknown
CONSERVATION STATUS	Data Deficient. Known from only one restricted area

WAKEA MADINIKA

WAKE'S SMALL MALAGASY FROG

(VENCES, ANDREONE, GLAW & MATTIOLI, 2002)

ADULT LENGTH
Male
7/16–1/2 in (11–13 mm)
Female
9/16–5/8 in (15–16 mm)

Actual size

This tiny frog is known from only one locality, a cacao plantation in the Sambirano region of northwestern Madagascar. That it was found in a plantation suggests that it is not wholly dependent on forest, which is presumed to be its natural habitat. All specimens have been found close to a pond, which may be its breeding site. The word *madinika* is Malagasy for "small," reflecting the fact that this is the smallest species in the family Mantellidae yet described. It lives among leaf litter and its call, heard only at night, is described as a feeble chirp.

Wake's Small Malagasy Frog has a slender body and slender arms. The back is pale brown with gray markings, in the form of an "X" on the shoulders and an inverted "V" lower down. The flanks are grayish brown with dark spots and there is a dark stripe through the eye. The legs are brown and gray with dark markings, and the belly is silvery or white.

SIMILAR SPECIES

Wakea madinika is the only species in its genus. It is most closely related to frogs in the genera *Mantidactylus* (see page 636) and *Blommersia* (see page 628). It is very similar to the Malagasy Difficult Frog (*B. sarotra*), an inconspicuous, small brown frog found in central Madagascar, which is thought to lay its eggs on the ground near water.

APPENDICES

GLOSSARY

Amphibia The taxonomic class of organisms that includes the orders Anura (frogs and toads), Caudata (salamanders and newts), and Gymnophiona (caecilians). Amphibians are ectothermic vertebrates that typically live partly on land and partly in water, and that require water to breed.

amplexus A behavior in which the male clasps the female with his forelegs during mating, either under her arms or around her abdomen. In most species the eggs are fertilized externally as they are released from the female's body.

anthropogenic Caused by man; used to refer to environmental change caused by human activities.

Anura The taxonomic order of organisms within the class Amphibia that includes the frogs and toads. Anura means "without tail" and refers to the most obvious difference between frogs and toads and other amphibians.

armpit The area below the joint where the forelimb joins the body.

biodiversity The variety of living organisms found on Earth or within a particular habitat, most commonly measured as the number of species.

biological control agent A living organism that is used to control pests. The use of such agents is not always successful, as in the introduction of the Cane Toad (*Rhinella marina*; page 201) to Queensland, Australia, in an effort to control pests on sugarcane.

Bushveld An ecoregion in southern Africa characterized by high elevation, low rainfall, and subtropical woodland vegetation.

Caatinga An ecoregion of northeastern Brazil characterized by low annual rainfall, high temperatures, and desert scrub vegetation.

Cerrado An ecoregion of central Brazil characterized by a semi-humid climate and savanna vegetation.

Chaco An ecoregion of central South America characterized by high temperatures, arid or semiarid conditions, and vegetation ranging from forests to savanna and desert plants. Also called the Gran Chaco, and subdivided into the Chaco Austral, Chaco Central, and Chaco Boreal.

chytridiomycosis An infectious disease affecting amphibians and caused by the fungal pathogen *Batrachochytrium dendrobatidis*. Over the past 30 years it has caused the catastrophic decline and extinction of frogs around the world. Species that breed in streams at higher altitudes seem to be most susceptible.

CITES The Convention on International Trade in Endangered Species of Wild Fauna and Flora. The multilateral agreement was enacted in 1975 and aims to protect animals and plants through monitoring and regulating their trade.

cloaca The single external posterior opening of the digestive, urinary, and genital tracts in some animals, including frogs and toads, through which urine, feces, and sperm and eggs pass out of the body.

cloud forest A tropical, subtropical, or montane forest characterized by low cloud cover, resulting in humid conditions that favor the growth of mosses.

convergent evolution The independent evolution among unrelated species of adaptations such as analogous body forms or colors. An example in frogs is the similar bright coloration displayed by species in the Madagascan genus *Mantella* and the South American poison-dart frogs (family Dendrobatidae).

co-ossification Joining through bone formation. In Bruno's Casque-headed Tree Frog (*Aparasphenodon brunoi*; page 307), the skin on the top of the head is co-ossified with the skull.

crypsis (adj. cryptic) The ability to avoid detection by predators through camouflage and/or behavior.

cryptic species A species that is similar in appearance to another, genetically distinct species, and hence was mistakenly taxonomically included with that species, e.g. the Atlantic Coast Leopard Frog (*Lithobates kauffeldi*; page 559) was previously mistakenly included with the Southern Leopard Frog (*L. sphenocephala*; page 562).

dear enemy recognition The different response of a territorial animal to an intruder it recognizes compared to an unknown intruder, e.g. the North American Bullfrog (*Aquarana catesbeiana*; page 545) recognizes the calls of its neighbors and is less aggressive toward them than to unfamiliar intruders.

dichromatism The occurrence of two different colorations. Sexual dichromatism is when males and females of the same species differ in color, e.g. as in the Argus Reed Frog (*Hyperolius argus*; page 443).

direct development Transition from the egg to the adult form in amphibians without passing through a free-living tadpole stage, as in the Puerto Rican Coqui (*Eleutherodactylus coqui*; page 402).

dorsal Relating to the back or upper surface.

dry forest Forest growing in an area of low rainfall and characterized by species adapted to these conditions.

ecosystem A community of living organisms that interact with one another and the physical environment in which they live.

ectothermic Regulating body temperature by exchanging heat with the surroundings, as is the case with frogs (cf. endothermic).

endemic Found only in a specific geographic location.

endothermic Regulating body temperature largely through the heat produced internally through metabolic functions (cf. ecothermic).

endotrophic Gaining food from within another organism. The tadpoles of Boulenger's Climbing Frog (*Anodonthyla boulengerii*; page 486) have endotrophic development, gaining nourishment solely from the egg yolk provided by the mother rather than feeding externally.

epibatidine A toxic alkaloid produced in the skin of Anthony's Poison Arrow Frog (*Epipedobates anthonyi*; page 367) and the Phantasmal Poison Frog (*E. tricolor*; page 368). Synthetic versions of the toxin are used to make non-lethal painkillers.

estivation A state of dormancy or torpor entered into by animals during hot, dry periods, e.g. in summer. The Shoemaker Frog (*Neobatrachus sutor*; page 98) spends most of its life in a state of estivation, protected underground in the arid conditions of Western Australia and emerging only after rain to breed (cf. hibernation).

explosive breeding A reproductive strategy in which large congregations of frogs gather over a short time period in a fierce competition for mates; males sometimes attempt to mate with each other or even inanimate objects in their frenzy to mate.

filter-feeder An organism that strains suspended food from water, as in the case of tadpoles of the Yellow-spotted Narrow-mouthed Frog (*Chaperina fusca*; page 505).

froglet A young frog, either one that has recently metamorphosed from a tadpole, or that has hatched from an egg during direct development.

fynbos An ecoregion of South Africa's Western Cape, characterized by mountainous terrain and a Mediterranean climate with cool, wet winters and hot, dry summers. Plants here are adapted to variable rainfall and poor soils, and more than 70 percent are endemic. Frog species endemic to the region include the Critically Endangered Table Mountain Ghost Frog (*Heleophryne rosei*; page 85).

granular gland A gland in the skin that secretes poisons or serums, often as a defense mechanism.

groin The area where the hind legs join the abdomen.

herpetology The study of amphibians and reptiles.

hibernation A state of dormancy or torpor entered into by animals to conserve energy. Hibernation often takes place in winter, when temperatures fall and food may be scarce (cf. estivation).

hybridogenesis A hemiclonal method of reproduction in which half the genome is passed on sexually and the other half clonally. The Marsh Frog (*Pelophylax ridibundus*; page 567) breeds with the Pool Frog (*P. lessonae*) to produce the hybridogenetic Edible Frog (*P. esculentus*). Most Edible Frogs cannot breed with one another and instead backcross with either one of their parent species.

incertae sedis This Latin term is used alongside taxa (species, genera, family, etc.) that are of "uncertain taxonomic placement" within the phylogenetic tree containing all related taxa. Many such taxa have incomplete evolutionary histories.

intracapsular resistance A process whereby development in the egg halts until particular environmental conditions are met, e.g. full immersion in water.

IUCN Red List of Threatened Species An inventory of species' conservation status. The classification categories used are Extinct, Extinct in the Wild, Critically Endangered, Endangered, Vulnerable, Near Threatened, Least Concern, Data Deficient, and Not Evaluated. "Threatened" species are those in either the Critically Endangered, Endangered, or Vulnerable categories. For each of the species listed in this guide, the IUCN Red List category is given in the "Conservation Status" section of the fact file.

kl. When used within a scientific name, i.e. *Pelophylax* kl. *esculentus*, the use of kl. indicates that the species is a "klepton," a species that relies on closely related species for is reproductive cycle. *Pelophylax* kl. *esculentus* is the fertile hybrid of hybridization between the Marsh Frog (*P. ridibundus*) and Pool Frog (*P. lessonae*). There are several klepton species in *Pelophylax*.

lateral-line organs The organs of a tadpole sensitive to water movements, such as those made by predators. The organs usually disappear during metamorphosis, as most adult frogs do not feed underwater; examples of adults that retain lateral-line organs include the African Clawed Frog (*Xenopus laevis*; page 54).

lek A group of males that gather together in a confined area to display to (by calling in frogs) and compete for females, e.g. as occurs among Italian Tree Frogs (*Hyla intermedia*; page 300).

mass extinction event A rapid, widespread decrease in the diversity and abundance of life on Earth, as in the event 66 million years ago that resulted in the extinction of 75 percent of all species, including most dinosaurs. Scientists believe that the ongoing extinction of large numbers of animals—including frogs—since the end of the last Ice Age represents a further mass extinction event, called the Holocene, or Sixth, Extinction.

metamorphosis A biological process resulting in the change of form of an animal. Frogs and toads typically undergo metamorphosis at the end of the larval stage, when the tadpole develops lungs, grows legs, and absorbs its tail. This physical change is generally accompanied by changes in behavior and habitat.

miniaturization An evolutionary trend that has arisen independently in many frog and toad species, especially in wet tropical forested regions, allowing them to occupy microhabitats unavailable to most vertebrates. Examples include Amau Miniature Frog (*Paedophryne amauensis*; page 481), believed to be the world's smallest vertebrate at just ¼ in (7 mm) in length. Amphibians that have undergone miniaturization in their evolution often have reduced numbers of bones, particularly in their hands and feet.

mucus gland A gland in the skin that secretes mucus, protecting the skin and allowing oxygen absorption to take place through it.

Müllerian mimicry A form of mimicry in which two or more poisonous species that share a predator, or predators, have evolved to resemble one another to avoid predation, e.g. the Imitating Poison Frog (*Ranitomeya imitator*; page 386) very closely resembles the variable Poison Frog (*R. variabilis*), Crowned Poison Frog (*R. fantastica*), and Reticulated Poison Frog (*R. ventrimaculata*).

native Occurring in a geographic location solely as a result of natural processes, and with no form of human assistance.

nuptial pad A rough patch of skin that develops on the forearm of breeding male frogs to aid grip during amplexus.

nuptial spine An extreme form of nuptial pad in the form of a spiked swelling. These spines may also be used in male combat.

ocelli Tiny spots on species of glass frog; in the Napo Glass Frog (*Nymphargus anomalus*; page 267) these are arranged in rings on the back and legs.

oviparity (adj. oviparous) Laying eggs rather than giving birth to live young. In amphibians the eggs are usually fertilized externally on release from the female's body.

ozone layer The region of the Earth's stratosphere containing high levels of ozone that absorbs ultraviolet radiation.

papilla A nipple-like protrusion.

páramo An ecoregion below the snowline and above the timberline in the northern Andes of South America, characterized by a wide daily fluctuation in temperature, mountainous terrain and poor soils, and grassy and shrubby vegetation.

parotoid gland An external warty gland in the skin near the tympanum in certain frog and toad species that secretes a milky toxin or irritant as a defense mechanism.

phragmosis A strategy in which a burrowing animal uses its own body to block its entrance hole as a defensive barrier, e.g. Bruno's Casque-headed Tree Frog (*Aparasphenodon brunoi*; page 307) hides in a bromeliad or tree-hole by day, sealing the entrance with the top of its head.

phytotelma (pl. phytotelmata) A waterbody inside a terrestrial plant, used by some anurans for breeding and as a larval habitat, e.g. as in the Red-backed Poison Frog (*Ranitomeya reticulata*; page 387).

radiotelemetry A method in which radio transmitters are attached to frogs in order to track their movements.

ranaviruses A range of viruses, thought to have evolved from fish viruses, that infect amphibians and have caused mass mortalities in some species, e.g. in the European Common Frog (*Rana temporaria*; page 571) in the southeastern United Kingdom in the late 1980s.

red-legged disease Also called red-leg, this disease is caused by the bacterium *Aeromonas hydrophila* and has resulted in mass mortality in some anurans, e.g. in the Western Toad (*Anaxyrus boreas*; page 136) in Colorado in the 1970s and 1980s.

release call The call made by individuals when clasped by an inappropriate partner (of the same sex or different species) during mating, causing the partner to release them.

restinga A coastal region of tropical and subtropical forest in northeast Brazil.

Saprolegnia A fungal pathogen that affects some amphibian eggs and embryos.

satellite strategy An energy-saving strategy during breeding whereby a silent male sits close to a calling male in order to intercept any females attracted by the calls.

sexual dimorphism In which the male and female of a species differ in body size or coloration; e.g. the male Argus Reed Frog (*Hyperolius argus*; page 443) is typically green above with a pale midline stripe and white below, while females are light brown or reddish above and orange below, and have red limbs and feet.

shola Isolated patches of forest in higher mountainous regions of southern India. Because of their isolated nature, sholas are home to many endemic and threatened species.

spicule A small needle-like structure, e.g. as found in the ocelli of the Napo Glass Frog (*Nymphargus anomalus*; page 267).

synergy (adv. synergistically) The interaction between individual factors producing a combined effect that is greater than the sum of their individual contributions.

tarsal bones Ankle bones; these are greatly lengthened in anurans.

tepui Flat-topped mountains (3,300–10,000 ft, or 1,000–3,000 m altitude) characteristic of eastern Venezuela and western Guyana. Their isolated nature has led to a high degree of endemism among the species they support.

tetrodotoxin A nerve toxin produced by some fish and amphibians, e.g. Chiriqui Harlequin Toad (*Atelopus chiriquiensis*; page 146).

toadlet A young toad, either one that has recently metamorphosed from a tadpole, or that has hatched from an egg during direct development.

tsingy A region of Madagascar characterized by needle-shaped karst limestone formations.

tubercle A small rounded projection or process. The metatarsal tubercle, a projection from the base of the hind foot, is an important diagnostic indicator in some species.

tympanum An oval-shaped membrane behind the eye that transmits sound to the frog's inner ear.

unken reflex A defensive posture in some frog species in which it contorts its body to expose its brightly colored belly in an effort to warn off predators, e.g. as seen in the European Fire-bellied Toad (*Bombina bombina*; page 44).

UV-B A wavelength of ultraviolet radiation that causes sunburn and skin cancer, and that in frogs damages DNA in eggs, leading to abnormal development and early death, and may also be an environmental stressor, making them more susceptible to diseases such as chytridiomycosis.

ventral Relating to the stomach or lower surface.

viviparity (adj. viviparous) In which the embryo develops inside the mother, which gives birth to live offspring rather than laying eggs, e.g. as seen in Tornier's Forest Toad (*Nectophrynoides tornieri*; page 184).

vlei A shallow seasonal lake in South Africa.

vocal sac A flexible membrane open to the mouth cavity in a frog that it fills with air from its lungs to amplify the sounds made by its larynx. Some frogs have a single vocal sac under the chin; others have paired sacs, one on each side of the mouth.

wet forest Forest growing in an area of high rainfall and characterized by species that are adapted to these conditions.

xanthophore A cell that contains yellow pigment. Individual frogs that lack xanthophores appear blue, e.g. some Green Paddy Frogs (*Hylarana erythraea*; page 553).

xeric Very dry and desert-like. Xeric habitats are characterized by extremely low rainfall, and by plants and animals that have special adaptations allowing them to live in such conditions.

zetekitoxin A lethal nerve poison produced exclusively in nature by the Panamanian Golden Frog (*Atelopus zeteki*; page 153).

RESOURCES

The following is a selection of the many resources available to those with an interest in frogs.

ESSENTIAL WEB SITES

At a time when new species are being described every week, and when amphibian taxonomy is subject to frequent change, the Internet is the best source of the most up-to-date information. In particular, the following three linked web sites are essential:

AmphibiaWeb
Provides information on the appearance, natural history, and distribution of all described amphibian species, with range maps and, for many species, numerous photographs. Also contains well-referenced accounts of the causes of amphibian population declines.
www.amphibiaweb.org

Amphibian Species of the World 6.0, an Online Reference
Provides the latest information on amphibian nomenclature and taxonomy. For each species, genus, and family, etc. of amphibian, it provides a complete, fully referenced account of how its name has changed over time.
https://amphibiansoftheworld.amnh.org

IUCN Red List of Threatened Species
Provides information for each species on: Red List category, geographic range, population trend, habitat and ecology, threats, and actions being taken to conserve it.
www.iucnredlist.org

OTHER WEB SITES

African Amphibians
Provides information on African amphibians.
www.africanamphibians.myspecies.info

Amphibian Ark
Site created by the World Association of Zoos and Aquariums, which focuses on the captive breeding of endangered amphibians.
www.amphibianark.org

Amphibian Survival Alliance
Joint site of the Amphibian Specialist Group and the Amphibian Survival Alliance, groups that work together on amphibian conservation, research, and education.
www.amphibians.org

Frogs of Borneo
Provides information, with many pictures, on the frogs of Borneo.
www.frogsofborneo.org

Save the Frogs!
An organization devoted to the active conservation of frogs in all parts of the world.
www.savethefrogs.com

FURTHER READING: GENERAL INTEREST

Collins, J. P. & Crump, M. L.
Extinction in Our Times. Global Amphibian Declines
New York, USA: Oxford University Press, 2009.

Carroll, R. L.
The Rise of Amphibians: 365 Million Years of Evolution.
Baltimore, USA: John Hopkins University Press, 2009.

Dodd, C. K.
Amphibian Ecology and Conservation. A Handbook of Techniques
Oxford, UK: Oxford University Press, 2010.

Duellman, W. E.
Patterns of Distribution of Amphibians: A Global Perspective.
Baltimore, USA: John Hopkins University Press, 1999.

Duellman, W. E. & Trueb, L.
Biology of Amphibians
New York, USA: McGraw-Hill, 1986.

Halliday, T. & Adler, K.
The New Encyclopedia of Reptiles and Amphibians
Oxford, UK: Oxford University Press, 2002.

Lannoo, M. (Ed.)
Amphibian Declines. The Conservation Status of United States Species
Berkeley, USA: University of California Press, 2005.

Moore, R.
In Search of Lost Frogs: the Quest to Find the World's Rarest Amphibians
London, UK: Bloomsbury, 2014.

O'Shea, M. & Maddock, S.
Frogs of the World: A Guide to Every Family.
Princeton, USA: Princeton Univeristy Press, 2024.

Richardson, M.
Threatened and Recently Extinct Vertebrates of the World: A Bibliogeographic Approach.
Cambridge, UK: Cambridge University Press, 2023.

Semlitsch, R. D. (Ed.)
Amphibian Conservation
Washington, USA: Smithsonian Institution, 2003.

Smith, R. K. & Sutherland, W. J.
Amphibian Conservation. Global Evidence for the Effects of Interventions
Exeter, UK: Pelagic Publishing, 2014.

Stebbins, R. C. & Cohen, N. W.
A Natural History of Amphibians
Princeton, USA: Princeton University Press, 1995

Stuart, S., Hoffmann, M., Chanson, J., Cox, N., Berridge, R., Ramani, P. & Young, B.
Threatened Amphibians of the World
Barcelona, Spain: Lynx Editions, 2008.

Wells, K. D.
The Ecology and Behavior of Amphibians.
Chicago, USA: University of Chicago Press, 2007.

FURTHER READING: REGIONAL FIELD GUIDES

Spreybroeck, J., Beukema, W., Bok, B., & Van de Voort, J.
Field Guide to the Amphibians and Reptiles of Britain and Europe. London, UK: Bloomsbury Publishing, 2016

Tyler, M. J., & Knight, F.
Field Guide to the Frogs of Australia. Clayton, Australia: CSIRO Publishers, 2020.

Channing, A.
Amphibians of Central and Southern Africa
Ithaca, USA: Cornell University Press, 2001.

Channing, A., & Rödel, M-.O.
Field Guide to Frogs and Other Amphibians of Africa.
Cape Town, South Africa: Struik, 2019.

Dufresnes, C.
Amphibians of Europe, North Africa & the Middle East.
London, UK: Bloomsbury Publishing, 2019.

Du Preez, L. & Carruthers, V.
A Complete Guide to the Frogs of Southern Africa
Cape Town, South Africa: Struik Nature, 2009.

Elliott, L., Gerhardt, C. & Davidson, C.
The Frogs and Toads of North America
Boston, USA: Houghton Mifflin Harcourt, 2009.

Heatwole, H. & Das, I.
Conservation Biology of Amphibians of Asia: Status of Conservation and Decline of Amphibians: Eastern Hemisphere. Malaysia: Natural History Books (Borneo), 2014

SCIENTIFIC JOURNAL ARTICLES

Kiesecker, J. M., Blaustein, A. R. & Belden, L. K. (2001) "Complex causes of amphibian population declines. " *Nature*: 410; 681–684.

Leudtke, J. A., et al. (2023) "Ongoing declines for the world's amphibians in the face of emerging threats." *Nature* 622(7982): 308–314.

Wake, D. B. & Vredenburg, V. T. (2008) "Are we in the midst of the sixth mass extinction? A view from the world of amphibians." *Proceedings of the National Academy of Sciences*: 105: 11466–11473.

A NOTE *on* NOMENCLATURE

SCIENTIFIC AND COMMON NAMES

Every species has a scientific two-part name, comprising a genus name (e.g., *Rana*) plus a species epithet (e.g., *temporaria*). The resulting species name (*Rana temporaria*) is internationally recognized in all languages. The names of many frogs and toads have changed over time. For example, Halliday's Shrub Frog was named *Philautus hallidayi* when it was first described in 2005, but renamed *Pseudophilautus hallidayi* in 2009. The names of many species that were described less recently have been changed several times. A name may change only as a result of new, published research and the rules for creating and changing a name are governed by an international code. Some name changes are currently controversial and are not recognized by all authorities. This book follows the scientific nomenclature used by the Amphibian Species of the World (ASW) web site (https://amphibiansoftheworld.amnh.org). This occasionally differs from the nomenclature used by the AmphibiaWeb web site (www.amphibiaweb.org). For example, the North American Bullfrog is called *Aquarana catesbeiana* in this book and by ASW, but *Rana catesbeiana* by AmphibiaWeb. In this and other instances, the alternative scientific name is given under "Other Names" in the information chart at the top of each entry. The same applies where a name has changed between the first and second editions of this book.

Many frogs and toads have at least one common name. There are no conventions as to which common name is correct. In this book, we have listed alternative English names for many species, but have not listed non-English names that may be used locally. Where no common name currently exists, one has been coined based on the physical characteristics of the frog, or its collection locality, or the person for whom it was named, its eponym.

AUTHOR CITATIONS

The name of each species is followed by the name of an author and a date, as in *Rana temporaria* Linnaeus, 1758. This means that this species was first described by Carl Linnaeus in 1758. The fact that the name and date are not bracketed indicates that this species, the Common European Frog, is known by the scientific name that Linnaeus gave it in 1758. For some species, the name and date appear in brackets, as in European Tree Frog, *Hyla arborea* (Linnaeus, 1758). This means that Linnaeus was the first author to describe this species, in 1758, but that he gave it a different scientific name to the one that is currently accepted (he called it *Rana arborea*).

INDEX *of* COMMON NAMES

INDEX *of* SCIENTIFIC NAMES

INDEX *of* FAMILY NAMES

ACKNOWLEDGMENTS

AUTHOR ACKNOWLEDGMENTS

Professor Tim Halliday, the author of the first edition of *The Book of Frogs* (2016), tragically passed away in April 2019. I had known Tim for many years, and we had co-authored the Dorling Kindersley *Guide to Reptiles and Amphibians* in 2000. I was greatly honored to be invited, by the University of Chicago Press, to revise *The Book of Frogs* for this second edition, and I did this whilst retaining Tim's legacy throughout.

A big thank you to all the herpetologists and photographers who provided images for both the first and second editions of this book, and all the Bright Press staff named on p.4 who have contributed to its metamorphosis from egg to tadpole to fully formed frog. Especially thanks to the late Tim Halliday for his massive contributions to batrachology.

Mark O'Shea

PICTURE CREDITS

The publisher would like to thank the following individuals and organizations for their kind permission to reproduce the images in this book. Every effort has been made to acknowledge the images, however, we apologize if there are any unintentional omissions and would be grateful to be notified of any corrections that should be incorporated in future reprints or editions of this book.

Hanyrol H. Ahmad Sah /Wikimedia Commons/CC BY 4.0: 522. **Alamy** /© blickwinkel: 323; /imageBROKER: 18; /Minden Pictures: 43; /© National Geographic Image Collection: 220; /Nic Hamilton Photographic: 31; / piemags/nature: 496; /© Jim Zuckerman: 383. **American Museum of Natural History** /C. W. Myers: 360. **Carlos Henrique Almeida**: 255. **Esteban Alzate Basto**: 269, 287. **Franco Andreone**: 489, 491, 633. Marion Anstis: 95, 103, 338, 475. **Alejandro Arteaga** /www.tropicalherping.com: 413, 418. **Auscape** /ardea.com /Greg Harold: 104; /John Wombey: 117, 119. **Champika Bandara**: 131. **Néstor G. Basso**: 129. **Shannon Behmke**: 259. **Michal Berec**: 361, 436. **Wouter Beukema**: 49, 50, 81, 154, 512, 610. **David P. Bickford**: 473, 476, 482. **Bigal River Biological Reserve, Orellana, Equador** /www.bigalriverbiologicalreserve.org /CC BY 3.0: 263. **David C. Blackburn**: 55, 426, 536. **Wolfgang Böhme**: 215. **© Professor Amaël Borzée**: 76. **Richard Boycott**: 85, 163, 213, 590. **Forrest Brem**: 32. **Thomas Brown** /CC BY 3.0: 74. **Wolfgang Buitrago González**: 371. **Alessandro Catenazzi**: 261, 409, 412. **Zachary A. Cava**: 80. **Alan Channing**: 451, 463, 592. **John Clare / Amphibianphotos.com**: 61, 141, 183, 274, 385–6, 541, 599–600. **Adam G. Clause**: 542. **Nick Clemann**: 108, 118. **Coastal Plains Institute and Land Conservancy** /D. Bruce Means: 393. **Martin Cohen** /Wild About Australia: 120. **Luis A. Coloma (Centro Jambatu)**: 188, 197, 302, 421, 494. **Werner Conradie**: 438, 449, 535, 540, 594. **Camilo Contreras Carrillo**: 256. **M. A. Cowan**: 116. **Paul Crump**: 140. **Yuriy Danilevsky** /CC BY 3.0: 160. **Dr. Indraneil Das**: 161, 601, 611. **Davidvraju** /Wikimedia Commons/CC BY-SA 4.0: 529. **Mrs Percy Dearmer** /Public domain: 22. **Maximilian Dehling**: 551, 613. **Dr. Rudolf Diesel** /ScienceMedia: 403. **K. P. Dinesh**: 156, 531, 549. **Naomi Doak, /**Calumma Images: 87–88. **Maik Dobiey**: 150, 201, 258, 278, 321, 368, 375, 508, 569. **Cliff and Suretha Dorse**: 464, 595. **Dreamstime** / Natallia Vintsik: 354. **Robert Drewes**: 180, 502, 589. **Bernard Dupont**: 5, 192, 347, 350, 376, 462, 553, 579, 586, 604. **Peter Eaton** /CCBY-SA 2.0: 247. **Devin Edmonds**: 159, 286, 384, 486, 627–8. **Scott Eipper** /NATUREFORYOU: 113, 114, 121, 123. **Tobias Eisenberg**: 229, 349, 414. **Endangered Species International** /Pierre Fidenci: 550, 585. **Albert Feng**: 565. **Juan David Fernandez**: 408. **Don Filipiak**: 271, 419. **Aah-Yeah** /CC BY 3.0: 46. **FLPA**: / Chris Mattison: 25 (bottom); /Pete Oxford /Minden Pictures: 36. **Edgardo Patricio Flores** /CC BY-SA 2.0: 90. **Stephen Forder**: 209. **Paul Freed**: 135, 218, 331, 622. **Sarig Gafny, School of Marine Sciences, Ruppin Academic Center, Israel**: 51. **Biswarup Ganguly** /CC BY 3.0: 515. **Philip-Sebastian Gehring**: 634. **Arthur Georges**: 234, 246. **Getty Images** /Joel Sartore: 33; / Paul Starosta: 430. **Prathamesh Ghadekar**: 574, 615. **Ariovaldo A. Giaretta**: 231. **LeGrand Nono Gonwouo**: 425. **© Professor Marcelo Gordo**: 468. **Philip Gould**: 362, 499, 538, 578. **Girish Gowda**: 155. **Lucas Grandinetti** /CCBY-SA 2.0: 221. **Brian Gratwicke** /CC BY 3.0: 146, 211, 226, 281, 366, 372, 453, 596. **Michael Graziano**: 40, 276, 497. **Eli Greenbaum**: 461. **Joyce Gross**: 136. **Rainer Günther**: 469, 472. **Václav Gvoždík**: 56, 165, 427, 428, 533. **Wulf Haacke**: 193. **A. Haas** /CeNak, Hamburg: 65, 68, 71, 145, 175, 189, 509, 563, 573, 583, 612. **Célio F. B. Haddad**: 250, 252, 333, 391. **Amir Hamidy**: 470. **James H. Harding**: 134, 548. **S. Blair Hedges**: 416. **Jean-Marc Hero**: 122, 343. **Hans Werner Herrmann**: 527. **Andreas Hertz**: 422. **Fabio Hidalgo Wildlife Photographer**: 304. **Mareike Hirschfeld**: 444. **Jeremy Holden**: 507, 605. **Greg Hollis**: 100. **Alan Huettt** /CC BY-SA 2.0: 19. **Huw** /CC BY-SA 2.0: 30. **Images of Africa** /Leonard Hoffman: 53. **Instituto Biotrópicos** /Izabela M. Barata: 249. **Instituto de Biodiversidad Tropical** /César Luis Barrio Amorós: 147, 319, 325, 517. **iStock** /Aralip: 25 (top); /Fabio Maffei: 13; /phloxii: 23. **Peter Janzen**: 176, 182, 196, 280, 324, 330, 336, 370, 378–9, 399, 402, 439, 445, 506, 576, 580, 614, 623, 631–2. **Andres Felipe Jaramillo Martinez**: 266, 369. **John B. Jensen**: 558. **Sun-Jiajie**: 544. **Tim Johnson**: 547. **Jordan de Jong**: 93, 97, 110–11, 105, 107. **Greg F. M. Jongsma**: 207, 441. **Karl-Heinz Jungfer**: 224. **F. Kacoliris**: 240. **Muhammad Sharif Khan**: 575. **Andreas Kay**: 418. **Jörn Köhler**: 151, 301, 317. **Yasunori Koide** /CC BY 3.0: 158. **Philippe J. R. Kok, Vrije Universiteit, Brussel**: 187, 358. **Jonathan Kolby**: 3, 14, 296, 397. **Fred Kraus** 478, 483, 484, 523, 552. **Kerry Krieger** / www.savethefrogs.com: 91. **H. Krisp** /CC BY 3.0: 24. **Tim Krynak**: 420. **Axel Kwet**: 126, 217, 220, 493. **Malcolm Largen**: 132, 459. **Ignacio Jose de La Riva**: 265, 446. **Ch'ien C. Lee**: 77. **Twan Leenders**: 45, 78, 143, 153, 219, 235, 270, 273, 288, 293, 311, 329, 332, 374, 380, 395–6, 454, 518, 545–46, 561, 619. **J. P. van Leeuwen**: 184. **Vincent Lin**: 564. **Nathan Litjens**: 94, 101, 340, 344, 505, 511. **Ray Lloyd**: 106, 109, 112. **Victor Fabio Luna-Mora**: 222, 262, 359. **John D. Lynch**: 199. **Raúl Maneyro**: 179. **D. Mahsberg**: 457. **Shawn Mallan**: 498. **Rafael Márquez**: 568. **Chris Mattison**: 47, 148. **Buddhika Mawella** /CC BY 3.0: 581. **Adam McArthur**: 524. **Lindley McKay**: 195. **Michele Menegon**: 465, 532. **Jan Van der Meulen**: 326, 373, 381. **Zeeshan A. Mirza**: 528. **Robin Moore** /iLCP: 405. **Georg Moser**: 177. **Arturo Muñoz Saravia**: 415. **National Biodiversity Institute, Quito-Equador** /Jorge Brito M.: 367. **Nature Picture Library** /MYN: Andrew Snyder: 58; /Visuals Unlimited: 15. **Truong Nguyen**: 67, 555. **Matthew Niemiller**: 142, 173, 406. **David Nixon** /midlandsreptiles.com: 92, 99. **Andreas and Christel Nöllert**: 48, 79, 82, 162, 299, 300, 337, 556, 567, 621. **Mark O'Shea**: 334. **Kristiina Ovaska**: 133. **Ted Papenfuss**: 204. **Michel de Aguiar Passos**: 169. **Pedro L. V. Peloso**: 285, 492. **Paola Peltzer & Lajmanovich R., Laboratorio de Ecotoxicologia. FBCB-UNL-ConICET. Santa Fe, Argentina**: 309. **Weslei Pertel**: 312. **Trung My Phung**: 75. **Martin Pickersgill**: 166, 198, 203, 206, 434, 440, 514. **Todd W. Pierson**: 73, 157, 284, 292, 294, 315, 322. Tyrone Ping: 466, 588. **Jose Pombalen**: 446. **Daniel Portik**: 52, 429, 431, 430, 432–33, 455, 554, 598, 603. Serban Proches 84, 208. **Diogo B. Provete**: 236, 241, 495. **Robert Puschendorf**: 152, 351. **Alonso Quevedo**: 275. **David V. Raju**: 86, 168, 602, 606, 609, 618. **Pooja Rathod**: 504. **© Dr. Renato Recoder**: 423, 424. **Lawrence Reeves**: 525. **Sean B. Reilly**: 479, 510. **Samuel Renner**: 178. **Stephen Richards**: 477, 480. **Rittmeyer E.N., Allison A., Gründler M.C., Thompson D.K., Austin C.C.** /CC BY 3.0: 481. **Mark-Oliver Rödel**: 185, 210, 534. **Ariel Rodríguez**: 191, 400, 404, 407. **Santiago Ron** /FAUNAWEBEQUADOR: 267, 290, 308, 390. **Gonçalo M. Rosa**: 452, 539, 562, 630. **Daniel Rosenberg**: 149. **Sean Rovito**: 170, 303, 394. **Jodi J. L. Rowley**: 64, 607. **Paddy Ryan** /Ryan Photography: 59. **Mario Sacramento**: 200, 238, 254, 277, 283, 353, 500. **Richard Sage**: 60, 127–8, 130, 139, 167, 181, 543, 557. **MSC. Rodrigo DEO. L. Salles**: 410. **Ryan Sawby**: 242, 248. **Ivan Sazima**: 282. **Greg Schechter** /CC BY 3.0: 291. **Rob Schell Photography**: 137, 144, 348. **Andreas Schlüter**: 279. **Ingrid Sellschop**: 447. **Chintan Seth**: 69. **Shi Jingsong**: 70. **Shutterstock** /Ryan M. Bolton: 20, 310, 327; /Mark Bridger: 7 (bottom); /Patrick K. Campbell: 37; /© CreativeNature R.Zwerver: 571; /davemhuntphotography: 11 (bottom right); /© Dirk Ercken: 6, 11 (top right), 17, 382, 388; /Ginger Livingston Standards: 10 (bottom left); /© Neil Hardwick: 566; /© Vitalii Hulai: 44, 339; /ilikestudio: 7 (top); /Mirek Kijewski: 2; /Brian Lasenby: 10 (bottom right); / Fabio Maffei: 11 (bottom left); /Jason Mintzer: 29; /Morphart Creation: 640–1, 655; /Klaus Ulrich Mueller: 11 (top left); /Hein Nouwens: 38–39, 643; /© Dr. Morley Read: 28, 363; /Aleksey Stemmer: 1, 9 (bottom left), 16, 328; / StevenRussellSmithPhotos: 560; /Andrew Snyder 230; /TessarTheTegu: 21; / viktori-art: 8; /Hector Ruiz Villar: 12; /xpixel: 9 (bottom right). **L. Shyamal** / CCBY-SA 2.0: 513. **© Dr. Bikraujit Sinha**: 526. **Joanna McLellan Smith**: 223. **Mirco Sole**: 57, 125, 245, 305, 501. **Angel Solis**: 172, 264, 320. **Ruchira Somaweera**: 66, 115, 345, 503, 530. **Matt Summerville**: 346. **James Lawrence Taylor**: 244. **Frank Teigler**: 54, 212, 316, 448, 537. **Mauro Teixeira Junior**: 164, 216, 232–3, 260, 307, 352, 355, 356, 392, 398. **Rodrigo Tinoco**: 253. **Anna Todd**: 577. **Josiah Townsend**: 295, 313–14. **Dao Tran**: 620. **Robert Tropek**: 458. **Evan Twomey**: 389. **Kanishka Ukuwela**: 617. **U.S. Fish and Wildlife Service** /Charles H. Smith /Public domain: 27, 174; /Jan P. Zegarra: 401. **Jean-Pierre Vacher**: 516. **Eric Vanderduys**: 96, 102, 335, 341–2, 471, 474. **Jhonattan Vanegas**: 268. **Carlos R. Vásquez Almazán**: 298, 318. **Miguel Vences**: 41–2, 89, 98, 202, 205, 435, 437, 442–3, 450, 456, 485, 487–8, 490, 521, 587, 593, 624–6, 629, 635–9. **Luke & Ursula Verburgt (Enviro-Insight)**: 83, 194, 289, 387, 460, 467, 597. **Atherton de Villiers**: 591. **Jan Van Der Voort**: 138, 570. **Chung-Wei Yu**: 608. **Nayana Wijayathilaka**: 584, 616. **E. R. Wild**: 227. **Bert Willaert** /www.bertwillaert.com: 63, 124, 257. **Christopher J. Williams**: 357. **Brad Wilson**: 186, 297, 364–5, 377. **Germano Woehl Jr**: 251, 272. **Sebastian Wolf**: 520. **University of Kansas** / Rafe Brown: 572; University of Kansas Digital Archive: 239; William Duellman: 171, 214, 225, 228, 237, 243, 306. **USCS** / Chris Brown / Public domain: 62. **Huang-Yaohua**: 72. **Liu Ye**: 582. **Brian Zarate**: 559. **Ginny Zeal**: 35.

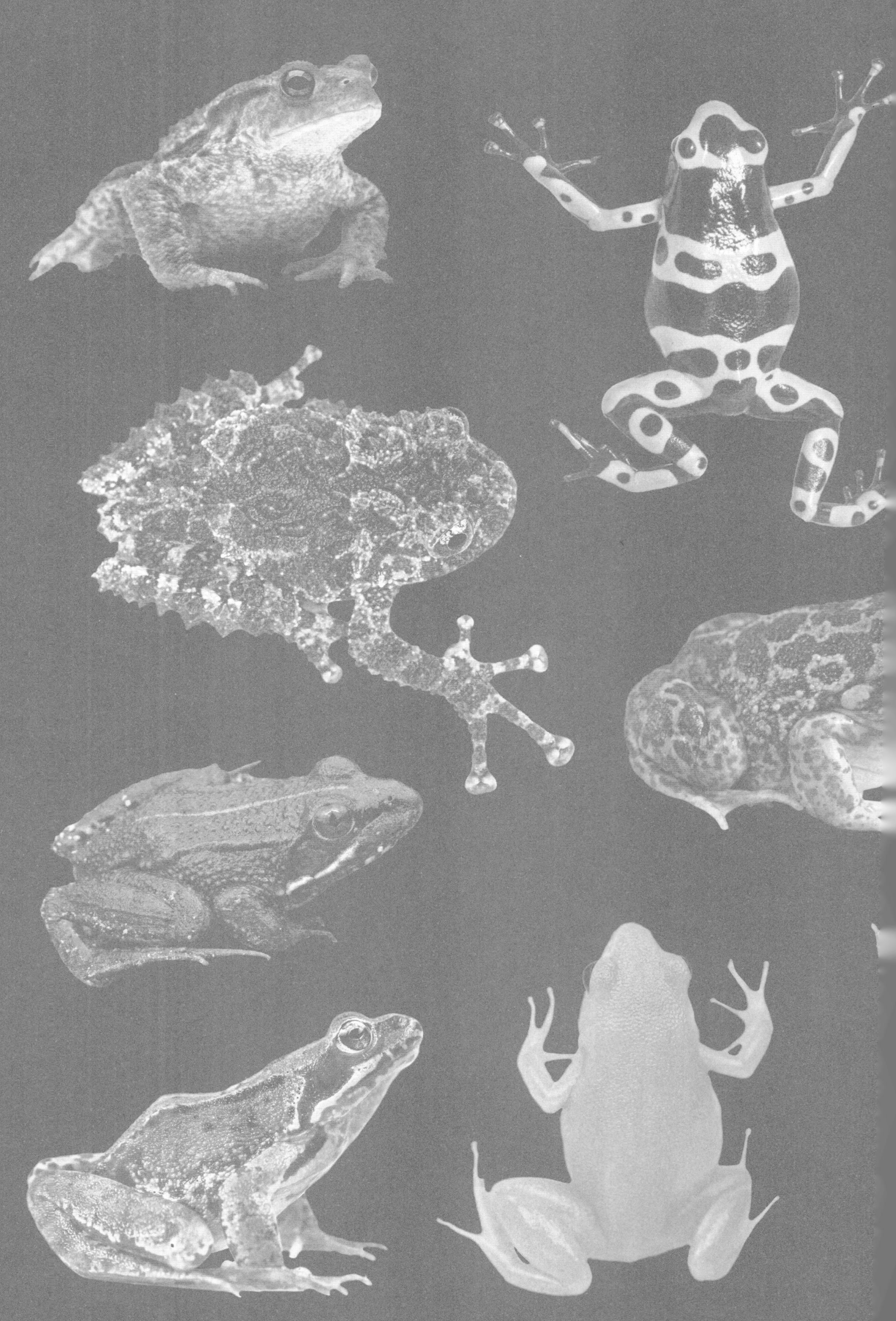